LEE-ENFIELD STORY

Ian Skennerton

First published in 1993 by Ian D. Skennerton, P.O. Box 433, Ashmore City 4214, Australia.

Distributed in Great Britain, Europe and South Africa by Greenhill Books, Park House, 1 Russell Gardens, London NW11 9NN, England.

Distributed in North America by I.D.S.A. Books, P.O. Box 1457, Piqua, Ohio 45356, U.S.A.

Distributed in New Zealand by Bruce Gorton, P.O. Box 917, Invercargill.

National Library of Australia
Cataloguing-in-Publication data:
Skennerton, Ian D.
Includes index.
ISBN 949749 15 X.

British Library
Cataloguing-in-Publication data:
Skennerton, Ian D.
The Lee Enfield Story.
Greenhill Books ISBN 1 85367 138 X

Produced by Ian D. Skennerton, P.O. Box 433, Ashmore City 4214, Australia.
Typesetting processed by John the Typesetter, 3/65 Mary St., Brisbane 4000, Australia.
Layout and design by Ian D. Skennerton.
Printed and bound by Thai Watana Panich Press Pty. Ltd., 891 Rama I Rd., Bangkok, Thailand.

Front Cover:	*Airborne unit soldier in training for invasion of Europe. (Imperial War Museum, London)*
Title Page:	*The No. 4 rifle in active service, Italy, 1944. (Imperial War Museum, London)*
Endpapers:	*Australian fighting column, Western Desert, October 1940. (Australian War Memorial, Canberra)*
Back Cover from top:	*Lee-Enfield Rifle lineage. Magazine Lee-Metford, Short Magazine Lee-Enfield, No. 4 and No. 5 rifles. (MOD Pattern Room, Nottingham)*

PREFACE & AUTHOR'S NOTE

With updated and more recent information on the lineage of this famous rifle and its designers, along with better illustrations and descriptions of certain rifles and parts, "The British Service Lee" has been fully revised, improved and updated over the past ten years, to produce this new compilation of "The Lee-Enfield Story"

The chapters remain in specific family clusters so that any particular development may be followed to its conclusion even after the original pattern may have become obsolete upon the introduction of new models or types. Any study of the Lee-Metford and Lee-Enfield series can easily become confused due to the involved association of various rifles, carbines, conversions, upgraded patterns and small bore trainers, further complicated by changes in nomenclature and the distinct production and development in Australia, Canada, India and the United States.

Separate sections are devoted to the Australian, North American and Indian production and a register of the appropriate stamps and markings has been included to assist with identification of the various rifles in addition to the British examples. An individual chapter has been devoted to the relative technical information with individual illustrations of the service patterns and troop trials models, grouped into families or types; so as to permit ready and accurate identification or the confirmation of correct and original features and parts of any rifle or carbine at hand.

The largest single published reference source for this study has been the British War Office "List of Changes in War Materiel and of Patterns of Military Stores which have been approved and sealed; with instructions relating thereto". The unabridged title name is usually abbreviated to "List of Changes" and a further compression used herein is "LoC". This series of monthly issues was commenced in the year 1860.

The List of Changes is truly a British military bible and the ultimate reference source for students of arms and ammunition and the related accoutrements. In the particular LoC references, the ensuing number is the paragraph reference number, indicated in this text by the original List of Changes indicator for paragraph, "§". The sign used for the plural form, paragraphs, was originally "§§" and that has been applied here also.

The Lee-Enfield Rifle has plausibly earned a place in military history as the principal implement in British and Empire weaponry. There may have been stronger actions and inherently more accurate rifles, but for reliability, speed of operation and the soldier's battlefield tool through the two world wars, the Lee-Enfield was second to none. "The Lee-Enfield Story" is a detailed technical record of its development. Regarding its accomplishments and successes, one has only to look at history.

Ian Skennerton
Albany Creek, 1991

ACKNOWLEDGMENTS

Nearly all of the rifles and carbines described and illustrated here were photographed, studied and researched at the Ministry of Defence Pattern Room, now housed at Nottingham. Without access to this collection and the excellent, specialised library there, such a detailed and comprehensive study would not have been possible; my sincere thanks to the management and prevailing policies of this splendid, working museum and reference collection.

After more than ten years of further investigation, critical review and examination of specimens, I now regard this as the definitive tome. Much of the additional material in this fully revised and updated study is due to correspondence with enthusiasts and researchers in Great Britain, Australia and the United States, as well as subsequent visits to the M.O.D. Pattern Room (now housed at Nottingham). Again, a special "thank you" to Herb Woodend and Carl Woods for their assistance, and to some new staff; the librarian, Howard Mitchell and assistant custodian, Paul Ellis. The following credits include some additional to those listed in the "British Service Lee"; my appreciation to the students and collectors who have recently assisted with some of the text expansion and clarification, especially Peter Laidler and Charles Woodhouse.

A special thanks also to my wife, Nicky, for her help with the computer listings and tabulations, the illustrations, film processing and print enlargements. And for her realistic perspective throughout my ongoing difficulties since the accident on "D-day", 6th June 1987.

AUSTRALIA:-
Barry Temple, Charles Woodhouse and Tom Williams for their initial interest and advice;
Brian Labudda, Ray Jarosz, Peter Burow, Reg Chapman and fellow collectors over many years;
Bob Etherington, Astrid Vallati, Bob Richardson and Peter Richardson for their comments and opinions;
Alan Stein, Alan Thorburn, Fred Gibbs, D.A. Justins, Doug Lynn and other collectors and researchers;
S.A.F. Lithgow museum and personnel, especially Bill Piggott, John Wray, Tom Skeene and Ray Leggatt;
Staff at Australian Archives A.C.T. Regional Office, Canberra, and also at Brighton, Melbourne;
W.O. Ian Kuring, personnel and management at the Royal Infantry Centre Museum, Singleton;
Robert Courtney and Australian War Memorial staff and Trustees for their practical assistance;
Major Steve Larkins of the S.A.R.P. team, formerly from E.D.E. Maribyrnong, Victoria;
and more recently to Brian Labudda and the other keen collectors on the 1991 U.K. Museums trip.

BRITAIN:-
The Management and Staff, M.O.D. Pattern Room, at Royal Ordnance, Nottingham; and also to those at The National Army Museum, Tower of London, Imperial War Museum, (formerly) Q.A.D. Enfield Pattern Room, British Library, Public Record Office, India Office Library, British Patent Office, Birmingham Library, School of Infantry Museum at Warminster and the Royal College of Military Science at Schrivenham, Scottish United Services Institute, Edinburgh Castle, and various regimental museums for their co-operation;
and to the many collectors, museum personnel and members of the trade who made the task much more rewarding;
especially to Herb Woodend, Howard Mitchell, Paul Ellis, Carl Woods, Peter Laidler, Roger Smith, David Penn, Mike Hibberd, Mike Baldwin, Alan Guy, Peter Hayes, De Witt Bailey, Jeremy Tenniswood, John Watts, Peter White, Doug Nie, John Beards, Roger Cleverly, David Booth, J. Anthony Carter, Jim Stonely, Mike Payne, Miss Edna Parker, Fred Davies, Major Maurice Fogwell and Major John Oldfield;
and to B.S.A. Guns, Holland & Holland, Cogswell & Harrison, Sterling Armament, Weller & Dufty, A.J. Parker Ltd., and the Birmingham Proof House, for their considerate attention.

CANADA:-
Derek Complin, Alf Savage, Stephen Alford, Ron Sanderson, Bob Miller, Robert Hanna and Peter Heaton for many letters;
Bob Dynes, John Denner, John Chown, Blake Stevens, John Kirton, Tom Dugelby and John Minnery for their contributions;
John Taylor, Harry Toye, Bruce Toye and Danny Williamson for investigation of many Long Branch rifles and listing of serial number sequences;
and to the Canadian War Museum, Ottawa, for the considerate assistance of the Weapons Department and Library.

FRANCE & GERMANY:-
R. Stevens Arvoy for his prompt communication and also to Hermann Mollenbeck and others who have written.

INDIA:-
Col. HRS Kalkat and the High Commission of India in Australia.
Also to their London counterparts. However, my various requests to visit the Rifle Factory and Service Museums in India itself have been refused.

NEW ZEALAND:-
John Osborne, Ed Glover, Bruce Gorton, Ed Shoebridge and Peter Fox for their correspondence and keen interest.
Also to Bob Branney, now residing in Australia, for his recollections of service in the N.Z. Armed Forces.

SOUTH AFRICA:-
Terence Willson for his association, mutual interest and letters over past years;
Barry Berkovich and members of the Historical Firearms Society of South Africa.

UNITED STATES OF AMERICA:-
James Alley and Bob Faris for years of correspondence and research, also for their support and valued friendship;
Paul Breakey, Robert Rubendunst, Ed Ezell, Pete Dickey, Robert Adams, Dudley Cooke, Stephen Alford, Michael Bolger, Ed Furler, Harmon Remmel, Roe Clark, William Meuse, Mineard Smith, Jack Attias, Dudley Cooke, William Kerr and the many U.S. collectors and students who have corresponded with me over many years.
And again to Paul Breakey and the other Americans on the Collectors Club 1991 U.K. Museums Tour.

THE LEE – ENFIELD STORY

Ian Skennerton

Page

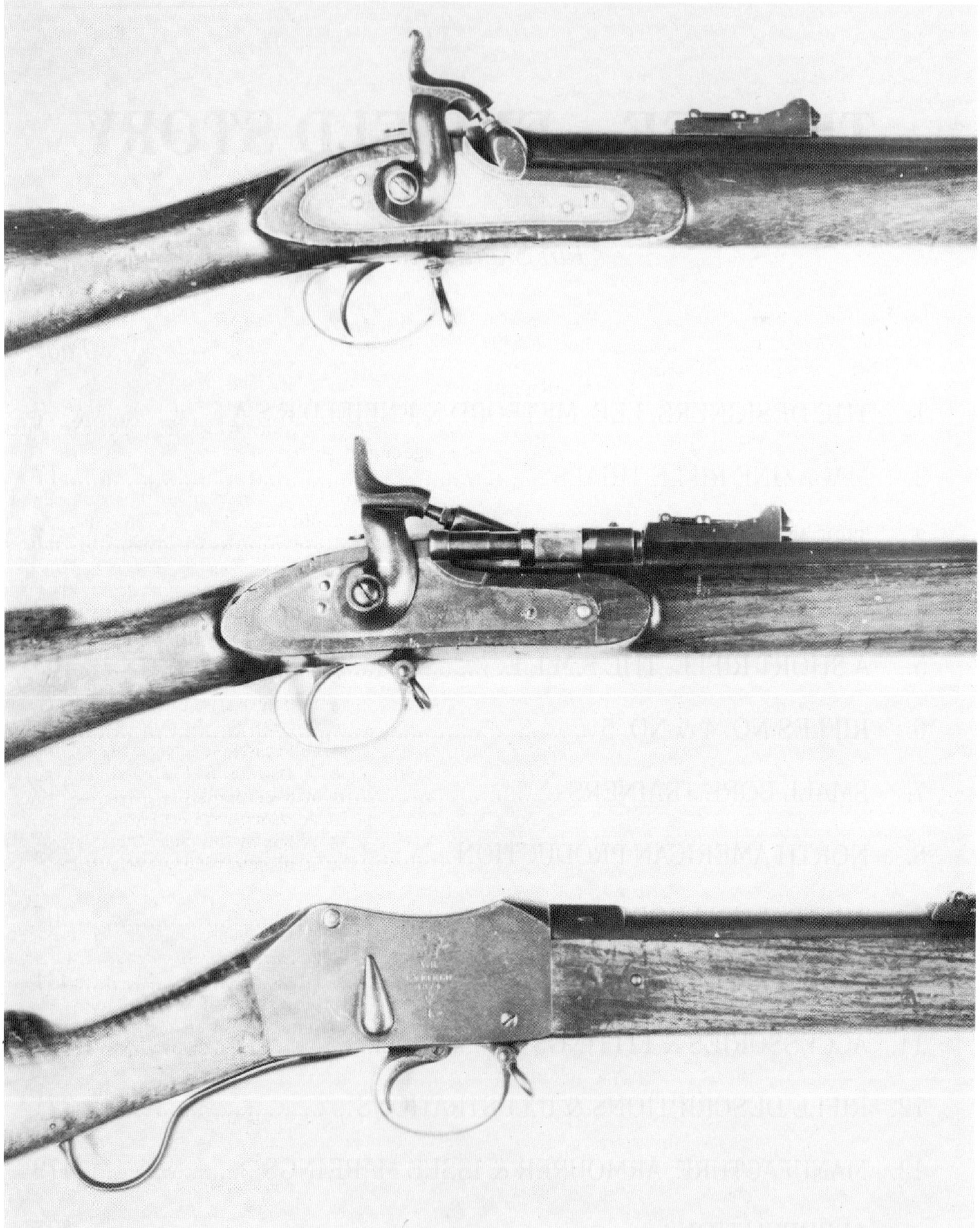

Previous Enfield Production
From top:- ***.577 Enfield Patt. 1853 Rifled Musket***
.577 Snider-Enfield Rifle
.450 Martini-Henry Rifle

MOD Pattern Room

THE DESIGNERS; LEE, METFORD & ENFIELD R.S.A.F.

## CHAPTER 1

The British service Magazine Lee rifle series has seen the most extended commission of any contemporary military weapons family; the 7.62mm NATO conversions of the .303 No. 4 rifle are still in service today in particular specialist roles. The armed forces of Great Britain have always been able to depend upon reliable and well-tried weapons; the coming of the era of the repeating rifle was to prove no exception.

Notable examples of well-made, reliable and accurate British firearms are the rifles and carbines of the Pattern 1853 series which also saw the introduction of component part interchangeability, a policy which has continued ever since. Although some foreign powers may have adopted or used more advanced weapons systems at various times before Britain, the soldiers of the Empire could always rely upon thoroughly tried and tested arms, even under such adverse conditions as the sand and the snow of the North-West Frontier and Afghanistan, the dust of the South African veldt, and the mud of Flanders.

With the adoption of breech-loading rifles by many foreign powers in the 1860's, it was decided to convert existing stocks of muzzle-loading rifles and carbines to a new breech-loading system until time permitted the development and testing of a new breech-loader. So, the .577 Snider became such a stop-gap measure in 1866 and it also saw the introduction of a new type of cartridge designed by Colonel Boxer, which was even to have some applications after the turn of the century.

The .450 Martini-Henry, circa 1871, was a milestone in arms development and production, especially from an engineering viewpoint. Because of its design features, only the most trying exposure to the elements created any difficulties in its operation; this was mainly related to the foil or rolled brass Boxer case upon extraction. These problems were later overcome by the introduction of the drawn brass cartridge case and the combination was successful enough to warrant an updated, smaller bore (.402-in.) Martini. The upgraded model was fitted with a quick-loading magazine and it was being considered as the successor to the Martini-Henry in 1886. This new rifle was designated the Enfield-Martini but its final approval was forestalled by the emergence of the magazine rifle and an even smaller reduced-bore cartridge.

Initially, most other European countries had adopted the bolt systems as their new single-shot breech-loaders but such systems were not particularly favoured by the War Office due to some failures in earlier trials, especially during the 1860's. However, by the late 1880's, the Lee bolt action, coupled with a satisfactory magazine, brought the age of the repeating rifle to the British service.

The most prominent personalities in the origin and development of Britain's new magazine repeating rifle were James Paris Lee, William Ellis Metford, Joseph Speed and John Rigby; Lee was a Scottish-born American, Metford was a British engineer and keen competitive rifleman, and Speed and Rigby were on the management staff at the Royal Small Arms Factory at Enfield Lock.

Up until the 1880's, the combination of lead bullets, black powder and barrel rifling had resulted in certain fouling problems, especially with the smaller (.45-in. and less) bores. As was often the case, the target shooting fraternity paved the way for a solution. W. E. Metford had achieved notable success over the fouling problems by using a hardened lead bullet along with his shallow, five-groove rifling in .45-in. calibre; at this time, the service calibre was .577 and the "small bore" .450 was favoured for match shooting.

Henry's rifling had been adopted with the Martini action in 1871 but Metford's rifling was considered successful enough to warrant its use in the .402 Enfield-Martini trials in 1886; this rifle might well have become the new service arm had it not been for the sudden progress made with the smaller .3-in. calibre and the bolt actions with magazine feed.

William Ellis Metford was born on Monday, 4th October 1824, to a professional family; his father was a keen sportsman and an excellent shot. William was interested in firearms and mechanics from an early age and after his schooling he was apprenticed to an engineer. In this capacity, he was employed on the Wilts, Somerset and Weymouth railway in the south-west of England.

Elected as an Associate of the Institute of Civil Engineers in March 1856, it was during the period 1852-57 that Metford accomplished many successes in the field of rifle experiments. He married and then took up a post in India,

William Ellis Metford: 1824 - 1899

James Paris Lee: 1831 - 1904

arriving in April of 1857 at the height of the Mutiny, where he was besieged in Monghyr on the Ganges. His detailed knowledge of bullet making and pyrotechnics was put to a most practical purpose. Poor health forced his return to England in summer 1858 and it dogged him for the rest of his life.

Metford's suggestion to Mr. Pritchett in 1852 or 1853, of a hollow-based bullet for the Enfield rifle, resulted in the Pritchett bullet for the .577 Enfield; Metford also invented an explosive bullet and shell later adopted by H.M. Government. He was probably best known for his work with target rifles and the National Rifle Association; his special bullets were renowned for their accuracy. It was claimed that Whitworth's patented bullet of 1864 was in fact Metford's design, as Metford had not bothered to take out a patent on it.

He was a close friend of Sir Henry Halford, a prominent figure on various Committees formed to deliberate upon proposals, trials and the ultimate recommendation for adoption of service arms and in September 1891 he was finally awarded the sum of £6,500 by the British Government in royalties for the Service use of his rifling. Metford passed away at Bristol on Saturday, 14th October 1899 after years of failing health.

During the mid-1880's, experiments on the Continent with reduced calibres indicated that a bore of less than .4-in. could be quite practical for service use, using a jacketed bullet. This had certain advantages over larger bores, particularly for military service. In 1887, the Swiss Colonel Rubin's rifle and ammunition were tested and this led to the recommendation of a .3-in. calibre, using a copper jacketed lead bullet. Metford's rifling was adapted to the new small bore and then adopted with the new magazine rifle in 1889.

James Paris Lee was born on Tuesday, 9th August 1831 in the Scottish border town of Hawick on the River Teviot, where his father, George, was a watchmaker and jeweller. The family emigrated to Canada in 1835 or 1836 and settled in Galt (now Cambridge), Ontario where James spent his boyhood; his natural tastes and the character of his new country brought him a good deal into contact with firearms. He made his first gun at the age of twelve years, while still in "linen pinafore and block belt." His father was something of an antiquarian and had a room at his shop filled with old guns, swords, pistols, coins and the like.

An old horse pistol provided a barrel and the stock was fashioned from a piece of walnut which cost him one penny; the bushed fire communication hole was bored out

Above: Hawick, in the Scottish Border country.
Below: Buccleuch St., Hawick; Lee's birthplace.

of the barrel, making it somewhat oversize. A priming pan was fashioned from a half-penny and the "new gun" liberally loaded with powder, a paper wad, and a piece of lead hammered to form a bullet. The target was a large cedar tree at the bottom of the garden but when fire was applied to the pan by his brother Jack, the enlarged fire hole caused James Lee to be wounded in the chin; the cedar tree escaped unscathed.

However, this mishap did not dampen his enthusiasm for, and fascination with, firearms; young James often went hunting. On one foray into the woods he used some gunpowder to hasten kindling on a fire and the resulting explosion injured young Lee to such an extent that he was laid up for some months with large patches of burned and blackened skin.

At the age of 16, out hunting in Dicksons Woods on a cold October day, his numbed hands caused the gun he was carrying to fall onto hard ground, upon which it discharged. This time, the accident-prone James was hospitalised for some 18 months as the ball passed clean through his heel and he suffered considerable shock and loss of blood in making his way home. This injury affected him for many years and he sometimes used a cane because of it; some 50 years after the mishap, a New York surgeon removed some lead shot from his heel, this improved his health and disposition markedly.

James Lee was apprenticed in 1848 and learned his father's clock and watch trade, then moved to Toronto in 1849 to further his apprenticeship and in 1850 he commenced his own business in Chatham, Ontario. Here he married Caroline Chrysler (a forebear of the Chrysler Auto Corporation family) in 1852 or 1853 and obtained a fair measure of success in that district.

St. Mary's Church, Hawick and the commemorative plaque outside. This was unveiled on the 150th anniversary of Lee's birth.

Not long afterwards, the new family moved north to Owen Sound on Georgian Bay for a short stay but Lee then transferred his growing family and business to Janesville, Wisconsin in the United States by 1858. It is reported that he filed for U.S. citizenship, but no record of its finalisation has been found. The first son, William, was born in 1855, followed by George in 1859. The following year, 1860, saw the family move to Stevens Point, Wisconsin.

Experimenting with making his own guns, Lee was obsessed with repeating arms and developed a model with a hollow, sheet-steel stock containing a number of magazines. He claimed that he would not be satisfied until he had a rifle with a 40-round magazine. At Stevens Point in Wisconsin he constructed a successful breech-loading conversion of the Springfield muzzle-loading rifle in 1861 and during the following year, Lee contrived a breech-loading carbine with sideways moving barrel to operate the breech for loading the cartridge. The Sharps Rifle Company produced Lee's early rifle prototypes but not long afterwards he arranged a more satisfactory agreement with the Remington factory.

Backed by a consortium of Milwaukee businessmen, he set up the Lee Firearms Co. with its Armoury at 454 Canal St., Milwaukee and also collaborated with Philo Remington who produced many of the barrels for Lee's prototype and experimental rifles and carbines. Some of the Lee patents were later assigned to Remington at Ilion and Lee also worked there for a number of years as a Remington inventor and designer.

After the outbreak of the American Civil War in 1861, Lee had agreed to travel to Washington, refining and promoting his breech-loading conversion for the Springfield musket. Later in 1862, he finally went East; the conversion impressed the Secretary of War and further tests resulted in the eventual order for 1,000 breech-loading carbines in mid-1864. However, after the conclusion of the Civil War, the U.S. Army cancelled the contract and Lee was forced to sell his converted carbines on the flooded surplus market, to defray expenses. Subsequent legal action against the Department resulted in Lee being awarded less than half of the $15,000 expenses and damages sought.

During an interview in Britain in 1896, Lee indicated that his concept of a quick-reloading rifle using a vertical magazine with the ammunition positioned atop each other in single line, evolved during the early 1860's. In 1875 he presented a modification of the Martini-Henry block to the British Government for consideration; this used a hammer instead of a lever and had a shoulder-slung magazine which permitted a rate of fire of more than 28 rounds per minute.

The main, distinguishing features of Lee's famous bolt action are the rear locking lugs and detachable box magazine; patents for his first successful rifle were taken out in 1877. His first production rifle, the Model 1879, was adopted by China and soon afterwards by the U.S. Navy. This was manufactured by Remington at Ilion, their New York factory, and it was the basis for the British trials rifles; Lee carried out much of his experimental work at the Remington Ilion plant. He also took out patents in England, Belgium and Russia and the Improved Lee was manufactured in 7x57mm, 7.65mm Mauser, 30-40 Krag and 6mm Navy; all these were for military contracts.

Quite a few of the Lee magazine rifles were also produced for the sporting market in a variety of other commercial calibres. His Model 1895 rifle was a straight-pull design which was adopted by the U.S. Navy in 6mm calibre but this model was not as popular with the military or the sporting fraternities. This was the commencement of a rifle lineage which continued for many years and served throughout the two major world conflicts.

Wallaceburg, Ontario. Lee commemorative plaque (above & below). It is claimed that Lee fired the first shot from his prototype Model 1879 rifle into an oak plank on the opposite bank of the Great Lakes waterway, also illustrated. The local museum there also houses some Lee tools and a prototype Lee rifle. *James Alley*

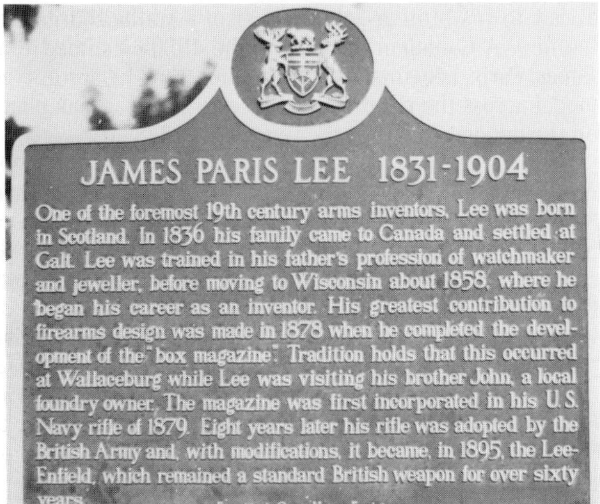

Lee prototype carbine. *Lee family estate*

James Lee's talents were not confined to the rifle field; in northern Canada, he had worked on a process to extract leather tanning solutions from hemlock bark and collected tan bark some 30 miles from his home and sold it at thirty dollars per barrel. He also invented some heating and air-temperature equipment, a jockstrap, and was involved in prototypes of the typewriter made and marketed by Remington, with whom he had close association.

Just across the street from his workplace in the Remington Armoury at Ilion, New York, Lee had lodgings in the Osgood Hotel. In 1878 or 1879, it is claimed that the room above his hotel suite was taken by one of the Mauser brothers who drilled a hole in the floor so as to observe Lee working on his rifle patent drawings and models.

Lee travelled to Britain and the Continent with his wife, Caroline, during the 1880's, pursuing rifle orders and evaluating specific market requirements. Caroline's death in London in 1888, due to heart failure brought on by a severe chill, however, saw him return to New York, never to venture abroad again. He is reported to have been very devoted to his wife and her death was a cruel blow from which he never really recovered.

A few years after his seventieth birthday, Lee died on Wednesday, 24th February, 1904. A commemorative plague has been erected in a park in Wallaceburg, in the Canadian province of Ontario, to record the contributions of James Paris Lee. Some descendants of the Lee family still reside in Wallaceburg where it is claimed that James tested his prototype of the 1879 Model rifle in 1878 during a visit to his brother John, owner of a Wallaceburg steel foundry. A Canadian journal in the 1930's claimed that James Paris Lee fired the first shot from his prototype model across the river near Wallaceburg into an oak plank on the opposite bank, at the time of his 1878 visit.

Another plaque has also been unveiled in Hawick, Scotland, during celebrations of the 150th anniversary of his birth in that city. On 9th August 1981, during this commemoration, the marker was placed outside St, Mary's Church of Scotland, where he had been baptised. It is worthy of mention that this plaque described Lee as "inventor," in the much broader sense and that the church rector refused to have the marker placed inside the church, because of the nature of some of his inventions.

By the end of 1887, after a series of successful trials in England, Lee's magazine rifle was decided upon as Britain's new magazine rifle; the Royal Small Arms Factory at Enfield had also made a number of changes to the layout and certain features of the U.S. Lee. These improvements and preparations for manufacture at Enfield Lock involved other notables in the Lee-Metford and Lee-Enfield rifle story, two Englishmen, Joseph Speed and John Rigby.

James Speed   Manager R.S.A.F.

John Rigby, M.A.   Superintendent R.S.A.F.

Joseph Speed first came to the Enfield small arms factory as an engineer at the age of 25 and he eventually worked his way up to the posts of Assistant Manager and Manager, R.S.A.F. through his ability and application; he was Manager from 1891 until 1909. There was some controversy over his taking out patents on the improvements he had made to the Lee-Enfield rifle whilst a Government employee; "The Times" made a violent attack upon his motives and character. Despite his important contributions to the British service rifle, the War Office did not make any monetary award or grant to Speed, unlike Metford and other contemporary inventors who assisted the Ordnance and War Office. Speed's major contributions relate to the magazine and magazine cut-off, bolt, long range dial sight, bolt dustcover, handguard and safety catch.

John Rigby, M.A. was Superintendent at the R.S.A.F. from 1887 until 1895 and he came from a family of traditional gunmakers rather than a service background like previous Superintendents who were mostly Royal Artillery officers. A partner in John Rigby & Co., London and Dublin, his term of management at the Factory saw the transition through trials, the Lee-Metford and the introduction of the Lee-Enfield. Rigby took out patents on his combination nosecap/sword bayonet bar design which fixed the Pattern 1888 bayonet.

Col. H. Watkin held the post of R.S.A.F. Superintendent from 1899 to 1905 during a critical period of the Lee-Enfield and Short Lee-Enfield development; he also took out some patents on improvements, relating to the "double-pull" trigger, magazine charger guide and special nosecap/sword bar of the Mk I S.M.L.E. rifle. Another Superintendent, Dr. Ashton, was involved in the leaf rearsight of the early Short Lee-Enfield rifle, although it must be pointed out that Joseph Speed remained Factory Manager during this period and he was still involved in the ongoing development.

Most of the British military Lee rifle development was undertaken at Enfield; various members of the management staff, notably the Superintendents, Assistant Superintendents and Managers, figured in many of the subsequent improvements in the small arms field. Due to the significance of some such personnel in the design and improvements of earlier small arms, the following list has been commenced at 1818.

Superintendents

1818	Mr. John Colgate
1823	Mr. W. Nobel
1838	Mr. R. W. Gunner
1854	Mr. J. Gunner
1855	Colonel W. H. M. Dixon, R.A.
1872	Colonel G. H. J. A. Fraser, R.A.
1875	Colonel F. Close, R.A.
1880	Colonel Arbuthnot, R.A.
1887	Colonel King-Harmon
1887	Mr. John Rigby, M.A.
1895	Mr. S. Clarke (Staff Officer-in-Charge)
1897	Mr. G. N. Tapp (Acting)
1898	Mr. (later Sir) F. Donaldson, M.Inst. C.E.

1899	Colonel H. S. S. Watkin, C.B.
1905	Dr. H. T. Ashton, D.Sc., M.Inst. C.E.
1909	Colonel Sir F. Nathan, K.C.B.
1909	Colonel F. T. Fisher
1917	Lt.-Colonel C. J. Newton
1921	Mr. F. Carnegie, C.B.E., M.I. Mech.E.

Assistant Superintendents

1861	Major J. F. L. Baddeley, R.A.
1862	Major Arbuthnot, R.A.
1866	Captain M. P. Eden, R.A.
1871	Colonel Dyer
1875	Major King-Harmon
1880	Captain McClintock
1886	none until 1901
1901	Captain J. B. Pym, R.M.
1906	Colonel A. L. Tisdall, R.A.
1910	Captain C. J. D. Freeth

Managers

1859	Mr. Burton
1876	Mr. Perry
1880	Mr. McGee
1891	Mr. J. Speed
1909	Mr. F. Carnegie

Chief Inspectors

1888	Colonel W. King-Harmon, R.A.
1894	Colonel W. N. Locker, R.A.
1900	Colonel J. D. Holston
1907	Lt.-Colonel W. B. Wallace
1910	Lt.-Colonel A. L. Tisdale, R.A.
1914	Colonel G.H.S. Browne
1919	Lt.-Colonel J. H. Robinson

The names of some of the Superintendents, Managers and Inspectors who held office from 1877 appear in several places in this study, relating to various improvements, but others who are perhaps lesser known are worthy of mention at this juncture also. Major Baddeley was associated with the recessed screw barrel band first used on the Enfield rifle and utilised until the introduction of the Short Magazine Lee-Enfield. Colonels Dixon, Fraser and Arbuthnot figured prominently in the development of the Martini, and the designers of the .276-inch Pattern 1913 Enfield were Messrs. Carnegie and Reavill (Chief Designer, R.S.A.F.), assisted by Lt.-Colonel Newton. An outline of the history of the Royal Small Arms Factory at Enfield will be found in Chapter 3.

Until 1888, the Inspectorate was responsible to the Small Arms Factory management, but from this year, which incidentally corresponded with the commencement of the Lee magazine rifle production, a separate Inspection Department was instituted. These Chief Inspectors were Army officers, many from the Royal Artillery, and they answered to the Services and end users rather than to the factory system. Another "check and balance" which regulated the construction and made the system so effective.

The establishment of Sealed Patterns and the Pattern Room is also associated with the factory regulation and small arms production; it also extended to control external manufacture by other contractors such as B.S.A. and L.S.A. The modern-day Ministry of Defence Pattern Room has also been the major source of examples and records of the British Service Lee rifles and carbines described here.

Sealed Patterns are the original manufacturing samples which were produced to govern manufacture; originally

R.S.A.F. Enfield, Main Machine Shop.
The pond (foreground) was later filled in. *MOD Pattern Room*

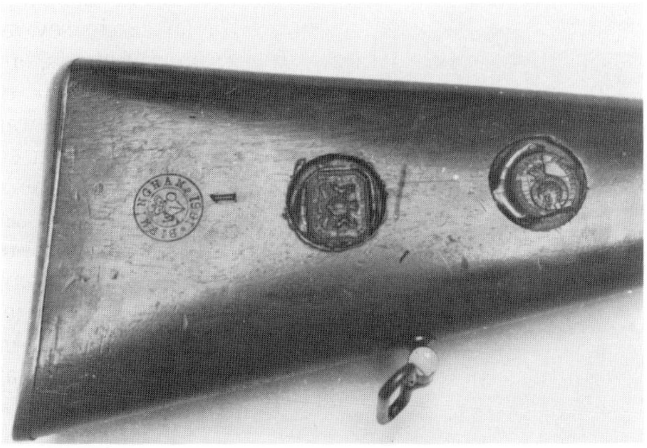

Sealed Pattern markings on some Snider era rifles and carbines. *MOD Pattern Room*

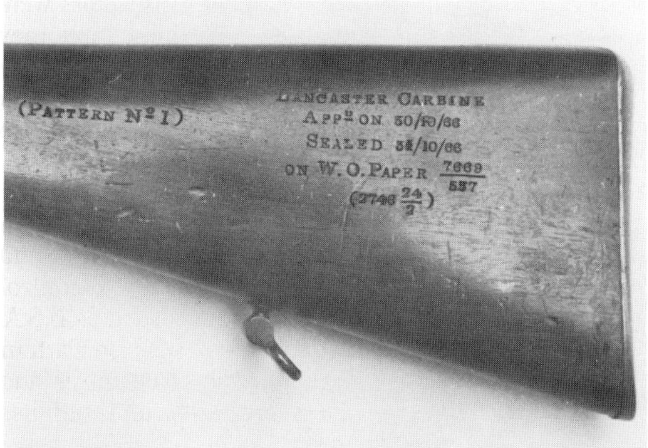

Above: Sealed Pattern tags on Lee-Metford, Lee-Enfield and Short Lee-Enfield rifles. The round brass and paper discs are old museum catalogue tags. *MOD Pattern Room*

they had a red wax seal attached to the arm along with the item's official description, List of Changes reference, date of approval, and the signature of the appropriate Inspector. In some instances in the mid-1800's, rifle and carbine Sealed Patterns had such details stamped on the butts and the red wax seal was placed into a recess in the butt. Later, these details and the seal were attached to calico labels and felt tags.

Origins of the Sealed Patterns and the Pattern Room have been traced back as far as 1631 when King Charles I, disturbed by the variations and absence of uniformity of arms and equipment in military service, decreed that the army should be equipped with standard patterns of arms and armour based upon samples held "in the office of our ordinance."

Such Sealed Patterns of small arms were held at the Tower of London but the tragic 1841 fire which destroyed the Ordnance storehouses resulted in the transfer of responsibility and storage of the Pattern Arms to the Royal Small Arms Factory at Enfield Lock. When the mechanisation of small arms production commenced at Enfield in the 1850's, the role of these Sealed Patterns assumed a greater significance and it was appropriate that they should be kept at the place of manufacture.

In conjunction with the Inspection Department, also located at R.S.A.F. Enfield, this served to govern the manufacture and subsequent changes during Lee-Metford, Lee-Enfield and Short Lee-Enfield production. This influence later extended to Lee-Enfield rifle manufacturing in India, Canada and Australia as well, although the practice of physically sealing Pattern Arms was discontinued in the 1920's.

The success and achievements of the British military Lee rifle was the consequence of the fusion of a number of talents; the genius of James Lee, the engineering skill and practical experience of William Metford, Major Rubin's cartridge, the proficiency of R.S.A.F. Enfield design staff and management, especially Speed, Rigby and Watkins, and the thoroughness and application of the British system. Perhaps the best measure of merit for the Lee rifle is that it remains in British service today in certain 7.62mm sniper and training roles, a basic action and magazine system designed more than a century ago.

Differences in the U.S. production Lee actions.
From top: Models 1879, 1882 & 1885. *R. Faris*

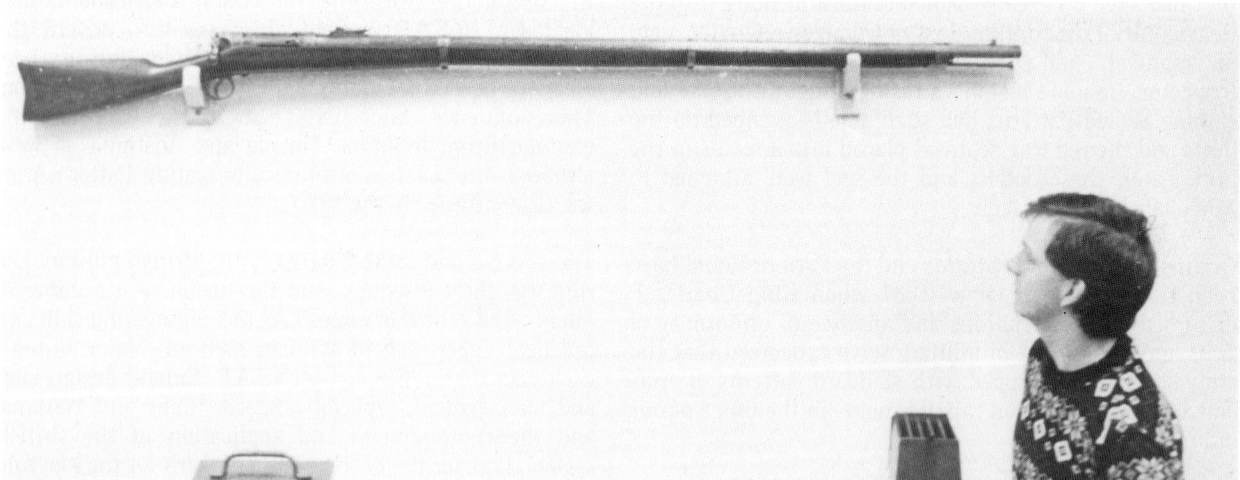

Probably the largest Lee action rifle; a Chinese wallgun or "Jingal" gun.
These were locally produced, large scale replicas of various European rifles.
Some of these guns were captured during the Boxer rebellion. *MOD Pattern Room*

MAGAZINE RIFLE TRIALS

CHAPTER 2

Colonel Frank Hyde of the Sharps Rifle Co. presented a Lee repeating rifle to the War Office for consideration in July 1879 and the following month, the order for six such rifles was confirmed. The first serious consideration by the Small Arms Committee of a general issue repeating rifle was during this period, when six such systems were examined; of these the Winchester Model of 1866 was preferred. However, it was considered to be too complicated and vulnerable to injury during the normal wear and tear of service use and so the question of repeating arms was shelved. In simple terms, they were not considered to be "soldier proof" or advanced enough in design to be sufficiently reliable to undergo the rigours of service.

Some other systems submitted for these trials include the Bacon, the Wilson, the Kerr and the Carter & Edwards; during these trials the Bacon action blew up when the bolt was closed behind a cartridge which had already been inserted in the chamber. The Wilson, too, burst during firing, after exposure tests, and the Kerr and Improved Kerr rifles experienced difficulties with unreliable detonation of the primer. Quite understandably, the Committee's conclusion was that bolt systems were not as safe or as reliable as the block, although high regard was expressed for the excellent qualities of the Kerr and the Carter & Edwards, which were perhaps the best of the British designs.

The Model 1879 Lee (British Pat. 1786) prototypes were manufactured by the Sharps Rifle Company but Lee formed a business arrangement with the Remington Arms Company soon afterwards and this continued for many years, through various commercial and military models. The Lee action employs a rear-locking bolt with long rib and locking lug opposite each other, providing a symmetrical locking system. The early patent date, 4th November, 1879 is marked on the left side of the action body on the following U.S. production rifles and subsequent improved models. Other dates of the various incorporated patents marked on this series of commercial production Lee rifles are: Aug 26th 1884, Sept 9th 1884, March 17th 1885 and Jan 18th 1887.

The United States Remington Arms Company commercial catalogues from this period provide some relevant descriptions and illustrations of the contemporary models, configurations and calibres available, principally of the 1879 model and its immediate subsequent improvements; these have been designated the Models of 1882 and 1885. After the 1879 model, the bolthandle was moved to the rear of the bolt and the extractor system improved, but all War Office reports and references from 1882 are merely to the "Improved Lee" rather than to specific model designations.

The question of magazine arms was referred to the Machine Gun Committee on 23rd October 1879, regarding two points:

(a) Whether they should be introduced into Service;
(b) In the absence of these arms taking the Martini-Henry cartridge, that they be tested with their own cartridges.

It was decided by the Committee to compare the different arms against the Martini-Henry rifle and afterwards consider the general question. This program was approved on 29th March 1880 and the services of a party comprising of a sergeant and three men of the 1/23rd Royal Welsh Fusiliers was placed at the disposal of the Committee for carrying out the trials which involved the shooting and testing of the various rifles.

Those rifles submitted to the Committee for these trials were:

1. Kropatschek
2. Hotchkiss
3. Winchester, Model 1876
4. Lee Rifle
5. Lee Carbine
6. Gardner Rifle
7. Green Rifle
8. Vetterli Rifle

The Krnka Quick-loader and the Mayhew Cartridge Holder were quick-loading apparatus also tested on the Martini-Henry rifle so as to increase its rapidity of fire. In addition, the Godwin and Warry designs were considered, although not recommended for trial.

Pte. Gregory, 1/23rd Royal Welsh Fusiliers, was a member of the testing party and on 19th May 1880 he was injured when a cartridge exploded in the tube magazine of a Winchester at the Woolwich proof putt. The Committee then decided to eliminate those systems in which the cartridges were positioned with the projectiles touching the bases or caps of the cartridges in front, namely the tube magazine systems of the Kropatschek, Hotchkiss and Winchester. For the purpose of comparison with the remaining trials rifles, two Mauser rifles with detachable magazines were ordered from the German government, each with 2,000 rounds of ammunition.

Examples Nos. 4 and 5 of the trials arms were the Lee rifle and carbine fitted with barrels supplied from Enfield and chambered for the .45-in. brass-drawn Gatling cartridge, but extraction difficulties were encountered. No. 6, the Gardner was rendered inoperable early in the trial and as Mr. Gardner was busy with machine gun development, he withdrew the rifle.

The Green rifle, No. 7, was tried at Woolwich on 20th May 1880, but suffered extraction problems which persisted even after minor alterations; the Green was entered by the Henry Rifle Barrel Company. No. 8 was the Vetterli, which had a 10-round magazine and was tested at the Woolwich proof butts on 24th September 1880, but practice was discontinued as it was pointed out that tube magazines were not acceptable, even though the cartridges were rimfire. Of the eight models submitted for trials, three specimens of each type were required.

The Lee rifle, Mauser with Lee magazine and the Green were considered to be the most promising by the end of these 1880 trials, in that order of preference, although the bolt system was not favoured in principle due to some mishaps with earlier trials models. This particular Committee was actually a Committee on Machine Guns and the question of magazine rifles was not of primary importance for them. In their final report dated 21st March 1881, it was stated that no further trial with magazine rifles had been carried out since 9th December 1880.

The Lee rifles and carbine specimens of the 1880 trials utilised Martini-Henry pattern barrels made at Enfield and chambered for the solid-drawn Gatling service ammunition; although some extraction difficulties were encountered, the operation of the magazine and its configuration was generally well-received. The magazine was detachable and had a 5-round capacity; the average rate of fire with an imperfectly drilled man was reported as being 20 rounds in 52 seconds. A cracked stock necessitated one rifle being sent to Enfield for repair.

The formal report on the Lee describes:

The action is on the bolt principle, and is so arranged as to compress the mainspring at the same time as the cartridge is being pushed into the chamber. This is a disadvantage as the force necessary to compress the spring entirely prevents feeling the cartridge into the chamber, and if a cartridge should stick or jam from any cause, it is essential that it should be felt and withdrawn instead of being driven home by the hand, which is liable to cause premature explosion before the bolt is locked. The parts seem strong and easy to manufacture. The action of the extractor is not satisfactory and this is due partly to the chambers not being properly machined out to gauge, but it is chiefly due to the form of the extractor. There does not seem to be sufficient power in the first movement of the bolt to withdraw the empty case. This could be improved by altering the angle at the end of the shoe against which the bolt rubs in turning to produce the first withdrawal of the case. All that is wanted is just to loosen the cartridge, and this could be done by drawing it out 1/16th of an inch instead of 3/16th as at present.

The magazine is placed in an excellent position, as it does not interfere with the balance of the rifle whether full or empty, and is also readily charged. It is strong and serviceable.

LEE RIFLE & CARBINE

Specifications:

	RIFLE	CARBINE
Calibre:	0.45 inch	0.45 inch
Grooves, number:	7	7
form:	7 ribs, 7 grooves	7 ribs, 7 grooves
depth:	0.0075 inch	0.0075 inch
length of 1 turn:	22 inches	22 inches
Mechanism, closing:	By bolt	By bolt
striking:	Spiral spring & firing pin	Spiral spring & firing pin
Length of arm without bayonet:	53 inches	43.5 inches
Weight of arm without bayonet:	9 lb. 5 ozs.	6 lb. 8 ozs.
Bullet form:	Cylindrical hemi-spherical head	Cylindrical hemi-spherical head
length:	2.82 calibres	2.82 calibres
diameter:	0.45 inches	0.45 inches
weight:	480 grains	480 grains
d2/w:	2.953	2.953
Action of grooves:	Compression	Compression
Powder charge:	85 grains	85 grains
Weight of cartridge:	718 grains	718 grains
Muzzle Velocity:	Not taken	Not taken
Sighted up to:	1,400 yards	800 yards
Weight of magazine, empty:	5 oz.	5 oz.
Number of rounds in magazine:	5	5

Note: The barrels are of Enfield Martini-Henry pattern and manufacture, chambered to take the solid drawn Gatling service ammunition.

The 1881 report of the Machine Gun Committee indicated that, for the moment anyway, the subject was closed, but the Navy in particular was anxious to adopt a magazine rifle, even if long range accuracy had to be sacrificed in order to acquire rapidity of fire. Pressure exerted by the Navy resulted in the question of a magazine rifle being raised again, and this time, a special Sub-Committee was constituted for its investigation and consideration.

This occurred on 1st August, 1882, and it was recommended that one or two of the best of the magazine rifles submitted to the Machine Gun Committee be tried again, in competition with other new rifles being acquired in America. New European arms were also submitted and by 11th November 1882, the following rifles had already been submitted for trial:

Mannlicher
Schulhoff (3)
Improved Lee
Spencer-Lee
Chafee-Reece
Gardner
Jarmann

During a meeting at the War Office on 24th January 1883, the question of an advertisement for inventors to submit magazine rifles for consideration was discussed and it was pointed out that all the rifles likely to be submitted were known and so the question was put to the meeting, which resulted in a majority decision that it was not necessary to advertise. Advertisements had been lodged in certain journals in 1864 for submissions which led to the first breechloader, the Snider, and then again in 1866 for the trials that resulted in the adoption of the Martini-Henry.

On 12th March, 1883, after the Sub-Committee was disbanded, the Secretary of State directed that a Small Arms Committee be assembled. The newly formed Committee was to consist of:–

President:	Col. P. Smith, C.B. (Grenadier Guards)
Members:	Col. H.T. Arbuthnot, R.A. (Supt. R.S.A.F.)
	Capt. T.S. Jackson, R.N.
	Lt.-Col. C.G. Slade (2nd Bn. Rifle Brigade)
	Lt.-Col. Sir H. Halford (1st Vol. Bn. Leicesters)
Secretary:	Maj. F. Salmond, Dep-Asst. Adj-Gen. for Musketry, North British District.

This Committee was appointed with the following objectives:

1. To take up the question as to the desirability or otherwise of introducing a magazine rifle for naval or military service, or both, together with the best form of such should it be decided to adopt, one, which question has, up to the present, been dealt with by the Committee presided over by Lieut.-General Sir F. Campbell K.C.B. This should be proceeded with at once, as the Lords Commissioners of the Admiralty have expressed a strong opinion that such an arm is necessary for the Navy, and press for an early decision in the matter. Records of the steps already taken in connection with this subject will be furnished for the information of the Committee.
2. Certain experimental rifles, recently under trial by troops and the Royal Navy, are being altered for re-issue for further trial. At the conclusion of the proposed trial, if satisfactory reports are received as to the general principle of the arm, the elucidation and settlement of minor details connected with it, and its ammunition, will be undertaken by the Committee, under whose direction any experiments connected with the object in view will be conducted.

The various breechloading systems and magazine rifles and carbines, along with the results of the different trials are contained in the Committee's final report. These are reproduced in the following pages.

In Europe, a Colonel Hyde (U.S. Army) was acting as agent for Lee's rifle and the latest pattern rifle was ordered from the United States for further trials, although this rifle was chambered for the .45 Springfield cartridge. On 6th April 1883, this Lee rifle was scheduled for trial at R.S.A.F. Enfield, but Col. Hyde was unable to attend due to illness so the tests were carried out in his absence. On 30th April, the Committee reported that they had found various points of the Lee system that they could improve and requested permission to make these alterations at R.S.A.F. Enfield.

Proper evaluation of the Lee against the other systems was not possible as it used different ammunition, so an "Improved Lee" was fabricated at Enfield and set up with a Henry barrel chambered for the Gatling cartridge, previously tried at Woolwich. Another Lee was also made up at Enfield, fitted with the Bethel-Burton hopper magazine. Both these Lee rifles, with the different magazine systems, showed up well in the trials and were among the top three contestants under consideration.

The Owen Jones was the third system still under consideration, with a block system as opposed to the bolt of the Lee and Lee-Burton. An obvious adaption of the Martini, the Owen Jones was operated by a slide under the butt and it had a 5-round hopper-type magazine. The mechanism was quite complex and therefore not easy or economical to manufacture or repair. Until this time, the trials rifle calibre had been .450-in. Gatling but it was then directed that the trials rifles be made in .402-in. calibre, similar to the Enfield-Martini which was undergoing development at the same time.

Owen Jones was an American gun designer who is probably best known for his revolver designs which resulted in the adoption of the .476-inch Enfield revolver into Her Majesty's service in 1880. This undeniably influenced British service revolver design for many years. When it became apparent that the Small Arms Committee was investigating repeating rifle design, Owen Jones also began to submit such designs for consideration.

The Small Arms Committee report provides a detailed description of these three most successful rifles in the trials:

The **Owen Jones** rifle is composed of the following parts:–

Slide bar, elevator catch, rivet & supporting screw	4
Blocks, front & back, connecting links & pins	6
Body, deflector, axis pin & spring	4
Grip, pins connecting with back block, stop spring and axis pin	4
Trigger, lever, stop, spring, axis pins (2)	6
Cut-off lever, axis, nut and spring	4
Elevator (5 pieces), axis pins (2)	7
Extractors (4 pieces), axis screws (2)	6
Main spring, chamber, nut & plug	4
Cocking piece & axis pin	2
Plunger, stop, stop axis pin, spring	4
Striker, spring return, stop pin	3
Magazine, cover, 2 screws	4
Bottom cover & 1 fixing screw	2
Cartridge stop & axis pin	2
Abutment piece & fixing screw	2
Body	1
– Furniture & Parts common to other rifles –	
Barrel	1
Sight bed, leaf, slide, cap, spring, 3 screws & joint pin	9
Fore-end, nosecap, screw, rod holder screws (2), fore-end hook & screws (2)	9
Butt & swivel	2
Band upper, swivel, pin, screw & nut	5
Hand-guard fittings & fixing pin	16
Butt plate, 2 screws	3
Rod	1
Total	113

From the illustrations of the Owen Jones it is obvious that the "Furniture and Parts common to other rifles" mainly refers to those Enfield-Martini rifle components.

The original description of the Owen Jones continues:

To use this rifle as a single loader the cut-off lever is pushed forward, this shortens the travel of the grip or handle, and prevents the breech-block coming in contact with the elevator. In the backward movement of the grip or handle, the cocking piece in the back of the block engages with the projection on the plunger and moves it backward, at the same time compressing the main spring until the stop-piece, which is pivoted in the plunger, is pulled past the step provided for it in the tube. Just after this point is passed, the cocking-piece, travelling at a different angle from the projection on the plunger, slips by the latter, and the stop then drops into the step in the tube and holds the main spring compressed. A cartridge is now inserted in the breech of the barrel, as in the Martini-Henry rifle, and the grip or handle is pushed forward, and with it, by means of the back block and the links, the breech-block, causing the deflector to rise and pushing the cartridge into the barrel.

As soon as the breech-block has reached the breech of the barrel, its forward movement is arrested and its rear end is pushed up by means of the links until it is firmly seated against the back part of the body or shoe; in doing this the back block passes under the rear end of the breech-block and locks it in position for firing. As soon as the forward movement of the grip or handle is finished, the catch in the grip or handle engages with the guide bar and holds all the action firmly in position. To pull on the trigger communicates motion to the intermediate lever, and the latter releases the stop in the plunger, which is then forced forward by the main spring, and acts on the striking pin in the breech-block, driving it against the cap in the cartridge. The pressure of the hand is now brought back against the catch in the grip or handle, releasing the action and permitting the parts to be moved backwards.

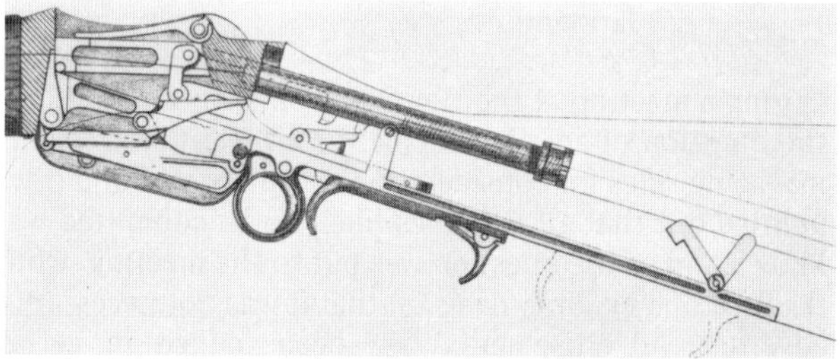

As soon as the grip or handle begins to move backwards, the cocking-piece engages with the projection on the plunger, and begins the operation of cocking, and as soon as the back block is moved from under the breech-block, the rear end of the latter is pulled down and the front end bearing on the starting lever (or extractor) moves the cartridge backwards about 1/32-inch, and the continued backward movement of the block causes the stop in the plunger to drop into the step in the tube, and also permits the deflector to descend until its front end comes in contact with the guide bar. Just after this, the breech-block comes in contact with the extractor lever, and by it communicates a sudden and rapid movement to the extractor, causing the empty cartridge shell to be thrown out. The rifle is now ready for the insertion of another cartridge.

To use the Magazine: The cut-off lever is pushed backwards and the grip or handle is moved backwards, as in the operation for single-loader, but it now travels 3/8-inch more, and the additional movement brings the breech-block in contact with, and operates, the elevator, and the cartridge resting thereon is raised level with the top of the guide bar and directly in front of the breech-block, by which in its next forward movement the cartridge is pushed into the barrel, and the elevator is pushed down to its former position by the back block, where it is ready to receive another cartridge from the magazine. These movements being repeated, cartridge after cartridge is received from the magazine, raised in front of the breech-block, pushed into the barrel, fired and extracted until the magazine is empty.

Weight of rifle (empty magazine) .... 9 lb. 15.25 oz.

The **Improved Lee Magazine Rifle** comprises the following parts:–

Body & ejector 2
Bolt, pin fixing extractor spring, guide pin trigger bent 3
Extractor, fixing spring, lever, axis pin, cover, screws (2) 7
Striker, trigger bent, fixing pin, stop tube, main-spring 5
Trigger, axis pin, lever, axis pin, spring, holder, fixing pin 7
Trigger guard, screws (2) 3
Magazine, elevating spring, locking bolt, key, screw, table, rivet 7
Cut-off for magazine, levers (2), axis pins (3), bolt, spring, rivet 8

– Furniture & Parts common to other rifles –

Barrel 1
Sight bed, leaf, slide, cap, spring, screws (3), joint pin 9
Stock, nosecap, screw, rod, rod nut, butt swivel, butt plate, screws (2) 8
Band upper, swivel, pin, screws, nut 5
Hand-guard fittings & fixing pin 16
Rod 1

Total 82

Like the Owen Jones, many of the "common parts" refer to those of the .402 Enfield-Martini trials rifle. The report on the Improved Lee rifle continues:

This rifle is fitted with a simple and strong bolt action and differs from most bolt guns in having a powerful extraction, the original mechanism having been improved at the Royal Small Arms Factory. The components of the breech-action are few in number, are easily manufactured, and are not liable to become unserviceable from use.

The magazine which is made to contain five cartridges, is carried filled on the person of the soldier and is only attached to the rifle when required. It is secured by a spring bolt in front of the trigger-guard underneath the shoe [action body]. The cartridges contained in the magazine are fed up into the action by means of a spring at the bottom of the magazine. The magazine must be removed from the rifle for the purpose of re-charging it, but several charged magazines are supplied with each rifle.

Weight of rifle (magazine empty) .... 9 lb. 6 oz.

The **Improved Lee Magazine Rifle, fitted with Bethel-Burton Magazine** comprises the following parts:–

Body, ejector, screw, guide & screw 5
Bolt, pin fixing extractor spring, guide pin for trigger bent 3
Extractor, fixing spring, lever, axis pin, cover, screws (2) 7
Striker, trigger bent, fixing stop, stop tube, main-spring 2
Trigger-guard, screws (3) 4
Trigger, axis pin, lever, axis pin, spring, screw 6
Elevator, axis pin, spring & screw, stop screw 5
Magazine, cover, axis pin, spring & pin, stop, rivets (2), screw 9

– Furniture & Parts common to other rifles –

Barrel 1
Sight bed, leaf, slide, cap, spring, screws (3) joint pin 9
Stock fore-end, nose cap, screw, barrel stop, rod holder, screws (2) 7
Stock butt, & swivel 2
Band upper, swivel, pin, screw, nut 5
Hand-guard fittings, fixing pin 16
Butt plate & screws (2) 3
Rod 1

Total 88

As for the previous two examples, the "common parts" refer to the Enfield-Martini components.

The Small Arms Committee Minutes further describe the Lee-Burton (Improved Lee with Bethel-Burton magazine):

The bolt mechanism of this arm is the same as that of the Improved Lee Magazine Rifle.

The magazine is a hopper placed on the right side of the action and made to contain five cartridges. When not required for use, this hopper lies along the side of the action, the top projecting above the shoe 1 inch. In this position the magazine cannot feed the cartridges it contains into the action.

When required for use, the magazine is pushed upwards until the top projects 2 inches above the shoe, and when so placed it feeds the cartridges which it contains onto an elevator, which raises them in turn in front of the bolt, ready for loading.

On the bolt being drawn back, the empty case is extracted and the elevator raises another cartridge in front of the bolt ready for loading. The magazine cannot be re-charged in either position.

Weight of rifle (magazine empty) .... 9lb. 14 oz.

The terminology "shoe" is used in the prior descriptions of the Lee action; this means the action body or receiver, a carry-over from the Snider era when it was used to define the action body. It originated from Snider's descriptions in the Patent Specifications although the use of this term did not continue for very much longer.

Quite obviously, many of these rifles were contemporaries of the trials .402 Enfield-Martini rifles due to their form of

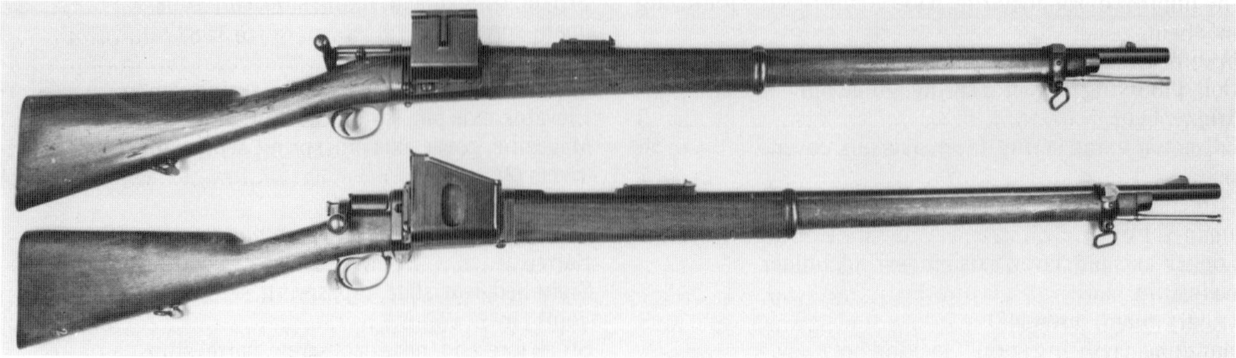

Above & below: Early Bethel Burton rifle (upper) and eventual Improved Lee with Bethel Burton magazine.

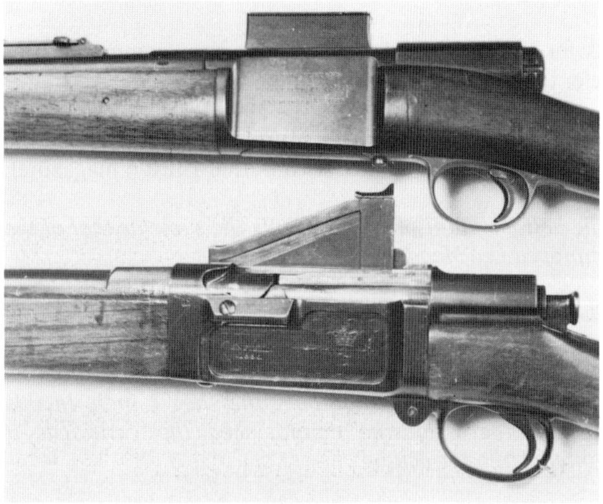

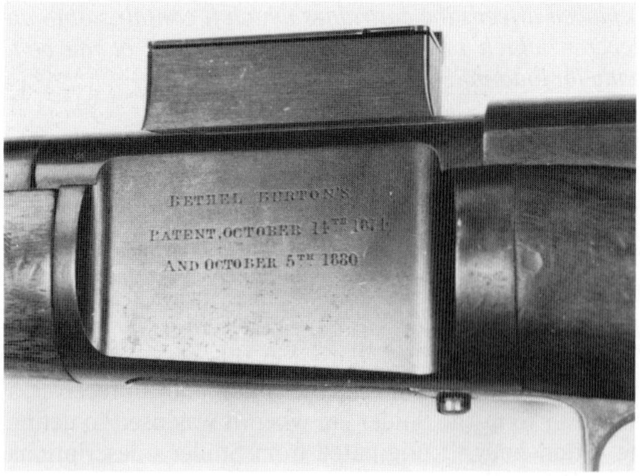

Receiver engraved- BETHEL BURTON'S PATENT, OCTOBER 14th 1874 AND OCTOBER 5TH 1880'.

MOD Pattern Room

stocking-up. The 1st Model Enfield-Martini is readily distinguished by its smaller backsight mounted near the receiver and the provision of a top handguard. It was largely the development of Col. Arbuthnot, Supt. R.S.A.F. A batch of 1,000 of these trials rifles with their characteristic clearance between the barrel and fore-end forward of the lower barrel band was made in early 1886.

The style of stocking-up of trials and experimental rifles can often serve as a good indicator of the origin and production date of that particular arm, especially as many such rifles were not specifically marked. Furniture and fittings like the buttplate, bands and especially the sword bar or mounting stud can also provide clues in this regard. These particular trials rifles have an overall length of about 53-in., approximately 3-in. longer than the following models of .402 trials magazine rifles, which also have a more conventional style of stocking-up and fittings.

Other notable features of the trials Improved Lee are the one-piece stock designs, compared with the Lee-Burton and Owen Jones which utilise the two-piece stock. A magazine cut-off was fitted to the Improved Lee but the cut-off slide and thumb projection is on the left side of the receiver instead of the right as for subsequent models. The term "Improved Lee," from this point on, was used right up until the rifle's eventual adoption, so it may cause some confusion as this nomenclature was applied to all the subsequent U.S. patterns, each incorporating additional refinements.

Comparative firing rapidity trials between Improved Lee, Lee-Burton, Owen Jones and Martini-Henry are of interest (the following figures were for 10 rounds, as single loaders):

Martini-Henry .... 19.3 secs
Owen Jones .... 27.5 secs
Improved Lee with Bethel-Burton magazine ..... 25.3 secs
Improved Lee .... 22.3 secs

The time required to fire 6 rounds; having one cartridge in the barrel and five in the magazine:

Martini-Henry (as a single loader) .... 12.5 secs
Owen Jones .... 4.8 secs
Improved Lee with Bethel-Burton magazine ...... 6.0 secs
Improved Lee .... 5.6 secs

The number of rounds fired in one minute, with magazine only being used:

Martini-Henry (as a single loader) .... 27.0 rounds
Owen Jones .... 29.3 rounds
Improved Lee with Bethel-Burton
magazine .... 24.0 rounds
Improved Lee .... not carried out

It is quite apparent that when magazine fire was used, the rate of the Martini was considerably slower until the other magazines had to be recharged, after which the Martini was at no real disadvantage. The three trials rifles had 5-round magazines, those on the Owen Jones and Lee-Burton being hopper-feed types with loading through the top.

The version of Owen Jones rifle used in these particular trials had been improved late in 1885, giving it some advantage over the Lee rifles, in the Committee's opinion. The mechanism was simplified and magazine remodelled so that by moving it down, it acted as a cut-off for single fire; thus, the elevator (21 parts) was dispensed with.

Therefore, the Owen Jones magazine rifle had evolved through four distinct stages since 1883 and these can be classified as:

1st Type- Sheet metal box magazine in the butt; the breech block action operated by means of a knob in a slot on the right side of the butt and receiver; thumb operated trigger.

2nd Type- Sheet metal box magazine in the butt; the block action operated by a slide underneath the butt, shaped something like the hand grip of the Winchester and Marlin lever action designs but with the bottom flat portion removed; thumb operated trigger at top, rear of action.

3rd Type- Hopper magazine on right side of butt; block action operated by the same type of slide underneath the butt; conventional style trigger underneath the rear of the receiver.

4th Type- Simplified hopper magazine on right side of receiver with the elevator and cut-off switch eliminated; the same action as previous types operated by a slide underneath the butt; conventional style trigger.

From top: 1st, 2nd, 3rd & 4th Types of Owen Jones experimental and trials rifles. *MOD Pattern Room*

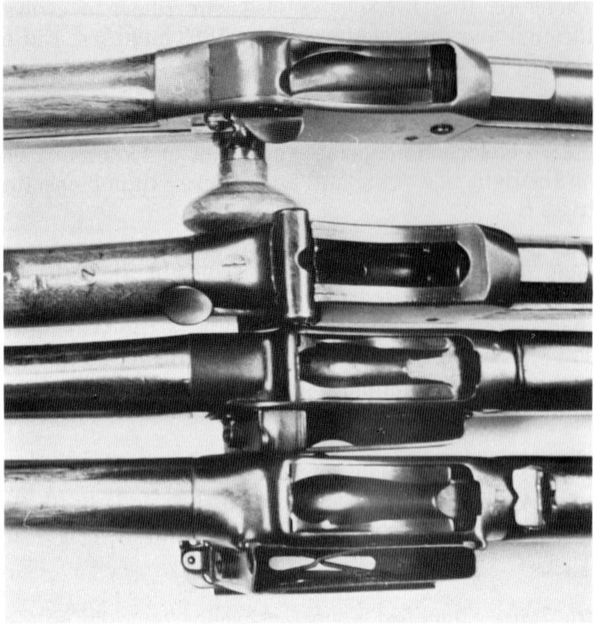

From top: Uppermost action views of 1st, 2nd, 3rd & 4th types of Owen Jones rifles. *MOD Pattern Room*

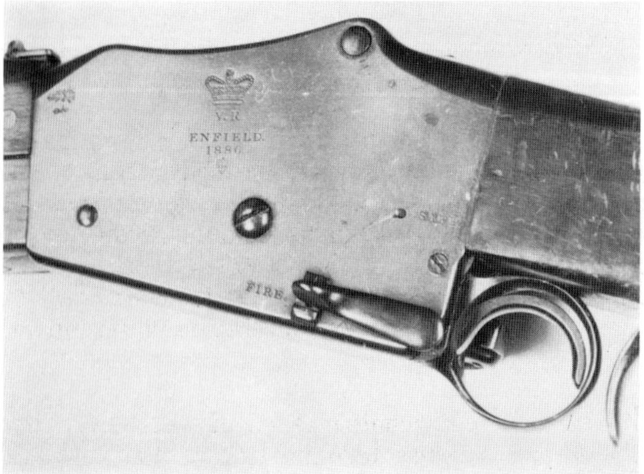

Left view, trials Owen Jones action. *MOD Pattern Room*

The "Improved Owen Jones," classed above as the 4th type, was recommended for troop trials in service, with the required quantity of rifles being set at one hundred. Its main competitor at this time, the Lee-Burton, was also ordered for comparative trials.

Upon receipt of this demand for trials rifles, the R.S.A.F. Superintendent indicated that 100 Lee rifles could be purchased and altered to take the Burton magazine, but the Committee considered that the Owen Jones fulfilled the requirements and so money was appropriated by the Navy and a quantity of 5,000 rifles placed on tentative order. Soon afterwards, this figure was reduced to 2,000 rifles and the Superintendent indicated that before 31st March 1886, no finished arms could possibly be delivered.

The Committee favoured the Owen Jones, notwithstanding the recommendations of the R.S.A.F. Superintendent who considered the Lee to be a better choice and capable of further development; the Owen Jones was much more complicated and expensive to produce. The Navy was quite impatient to have a magazine arm and urgent correspondence ensued between the Director of Naval Ordnance, Director of Artillery, Superintendent R.S.A.F. and the Surveyor General, as to whether the Lee would be tried along with the Owen Jones and about the interpretation of the Committee's recommendations. The Superintendent made it quite clear that if the Owen Jones did not answer well, he would consider himself absolved from any blame in that matter and could not be held responsible in any way.

In order to clarify the issue, the President of the Small Arms Committee was requested to inquire of the Committee if it recommended that both rifles be made for further trials and on 11th November 1886 they replied that after carefully considering the correspondence and the recent success of trials in which the Owen Jones had fired 2,000 rounds in one day without any difficulties, they were in favour of recommending its adoption. However, considering the opinion of the Superintendent, R.S.A.F., they recommended more extended trials between the Owen Jones and the Lee with Bethel-Burton magazine.

His Royal Highness the Field Marshal, Commander-in-Chief, considered that 150 of each type of rifle should be produced for exhaustive trials by troops and he dissolved the Small Arms Committee as their task was considered to have been accomplished. The Navy's order for 2,000 Owen Jones rifles still stood and so that 150 of each type would also be available, the order was placed with R.S.A.F. for 2,150 Owen Jones and 150 Lee-Burton rifles, at a costing of not more than £8 each. A sample Owen Jones fitted with a bayonet was forwarded to the Navy on 1st October 1886.

With the concurrence of H.R.H., Commander-in-Chief, the Secretary of State for War had decided to reconstitute the Special Committee on Small Arms, for the purpose of considering the reports which were made on the 1,000 Enfield-Martini rifles issued for troop trials. The reconstituted Committee comprised of:

President: Maj.-Gen. P. Smith, C.B
Members: Col. H.T. Arbuthnot, R.A. Supt. R.S.A.F. (until 23rd October, 1886)
Col. J.M.G. Tongue, Cmdt. School of Musketry
Col. Sir H. St. J. Halford, 1st Vol. Bn., Leicester Regt.
Col. C.G. Slade, 2nd Bn. Rifle Brigade
Hon. Guy Dawnay
Commander J.E. Meryon, R.N.
Secretary: Lt.-Col. H.H. Oldham

This particular Committee advised the Director of Naval Ordnance that they considered the Owen Jones to be a less-serviceable weapon than the Lee and furthermore inquired as to whether a smaller number of Owen Jones rifles might be ordered and a quantity of Lee-Burton rifles be taken for comparative tests by the Navy. It was pointed out that the Lee-Burton could be made more rapidly than the Owen Jones, and a Lee-Burton was sent to Portsmouth along with the Owen Jones for comparative tests by the Navy; these tests took place in mid-November 1886, under typical conditions as required by the Navy.

From H.M.S. "Excellent" (Naval shore base at Portsmouth), the Captain informed his superiors at the Admiralty that after a series of trials between the two rifles, the Lee was preferred on six counts:

1. The Lee was of much greater simplicity in construction and less likely to get out of order.
2. Several jams occurred with the Owen Jones while none of importance were encountered with the Lee.
3. Spare parts were much more easily replaced on the Lee while the Owen Jones was extremely complicated.
4. The Lee-Burton magazine was much more simple and very easily detached.
5. The Lee extractor was much more powerful.
6. The bolt action was preferred over the block due to its shorter movement; the falling block had the disadvantage of carrying dust or wet into the rifle when the block was lowered.

The new Committee had examined the rifles before it by this time, as well as some more recent inventions including details of a Schulhoff and the Rubini. It was considered that the Rubini had merits which might even make it superior to the Lee and so a specimen was made up in December at Enfield for specific trials.

Though pressed to make a decision so that production of the new magazine rifle could commence, the Committee preferred to delay their final decision until mid-January 1887, but made these observations:

1. The Owen Jones had been eliminated as a result of its trial against the Lee-Burton at H.M.S. "Excellent."
2. New inventions were being considered, namely-
   (a) the Schulhoff system;
   (b) the compound gun, which had a separate magazine that could be attached or removed by an armourer but not taken off by the soldier;
   (c) the Improved Lee, with a quick detachable magazine.
3. The form of the magazine was a separate question as either the Rubini or Lee bolts could be used with any of the magazines.
4. That they be advised as to whether the magazine should be part and parcel of the arm itself, such as the Schulhoff, or a separate part so that one or two magazines would be carried on the person and attached at will.

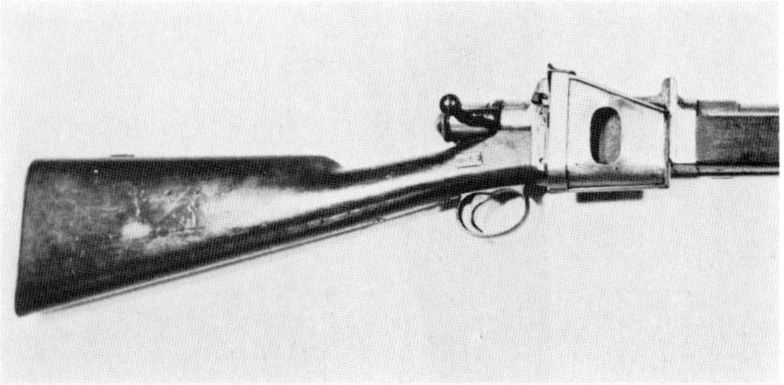

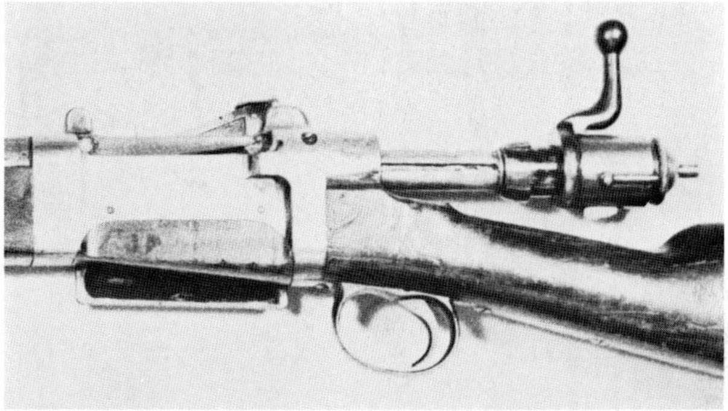

Above & below: Rubini rifle manufactured at Enfield and fitted with hopper magazine. Referred to as a "Compound" gun by Col. C.G. Slade. *MOD Pattern Room*

Consideration at the highest level was required on the last of these points as it involved military tactics and strategies, so a special Committee was organised, the President of which was Major-General Sir Evelyn Wood V.C., along with seven senior officers. Although they stated that they preferred, with one dissension, the permanently attached magazine, they made no binding policy at that stage.

The Small Arms Committee examined a .298-in. Rubini rifle on 2nd December 1886 and this eventually resulted in one of the most important decisions made by this Committee, to adopt a smaller calibre than .402-in. Subsequent trials of the Rubini (Rubin being its Anglicised name) cartridge proved most satisfactory in relation to accuracy, range, penetration, and effect on flesh and bone. Major Rubini was director of the Swiss Laboratory at Thun and he had developed the small calibre metal jacketed bullet in conjunction with a Professor Hebler.

In 1886, smokeless powder was introduced for use in rifles, invented by Monsieur Vielle and adopted by the French for use in the Lebel in 1887. Smokeless powder had a number of advantages; there was less fouling, no smoke to obscure the target or betray the position of the shooter, and higher bullet velocities with generally lower pressures.

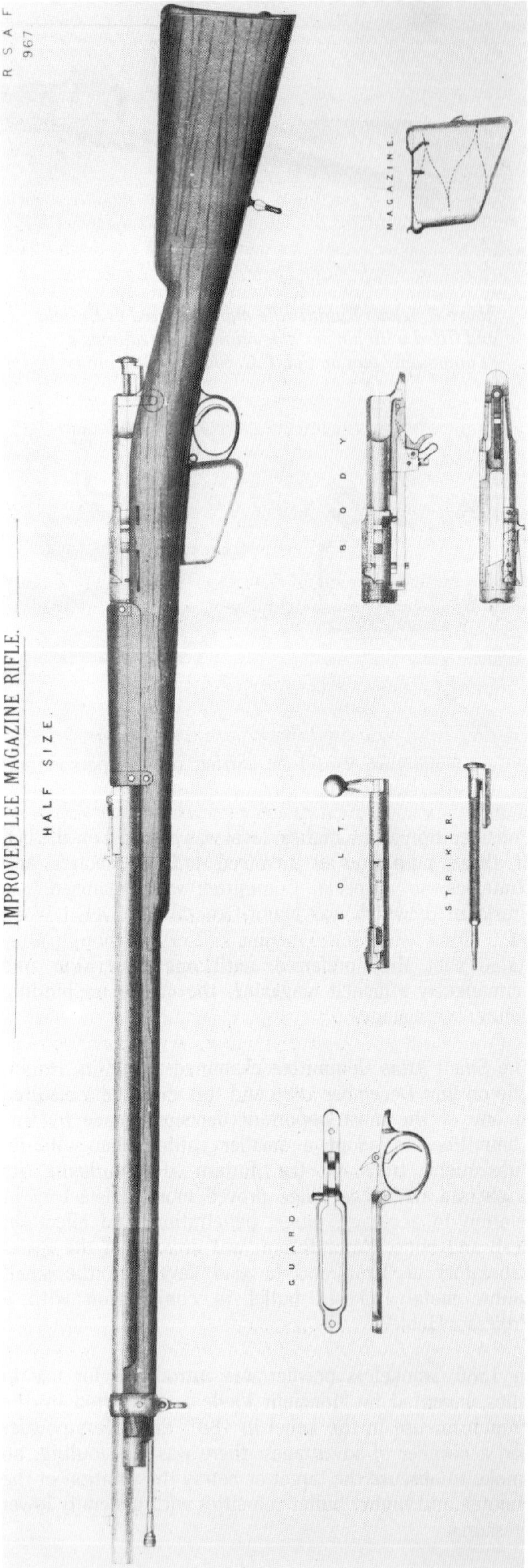
R S A F
967
IMPROVED LEE MAGAZINE RIFLE.
HALF SIZE.
MAGAZINE.
BODY
BOLT.
STRIKER.
GUARD.

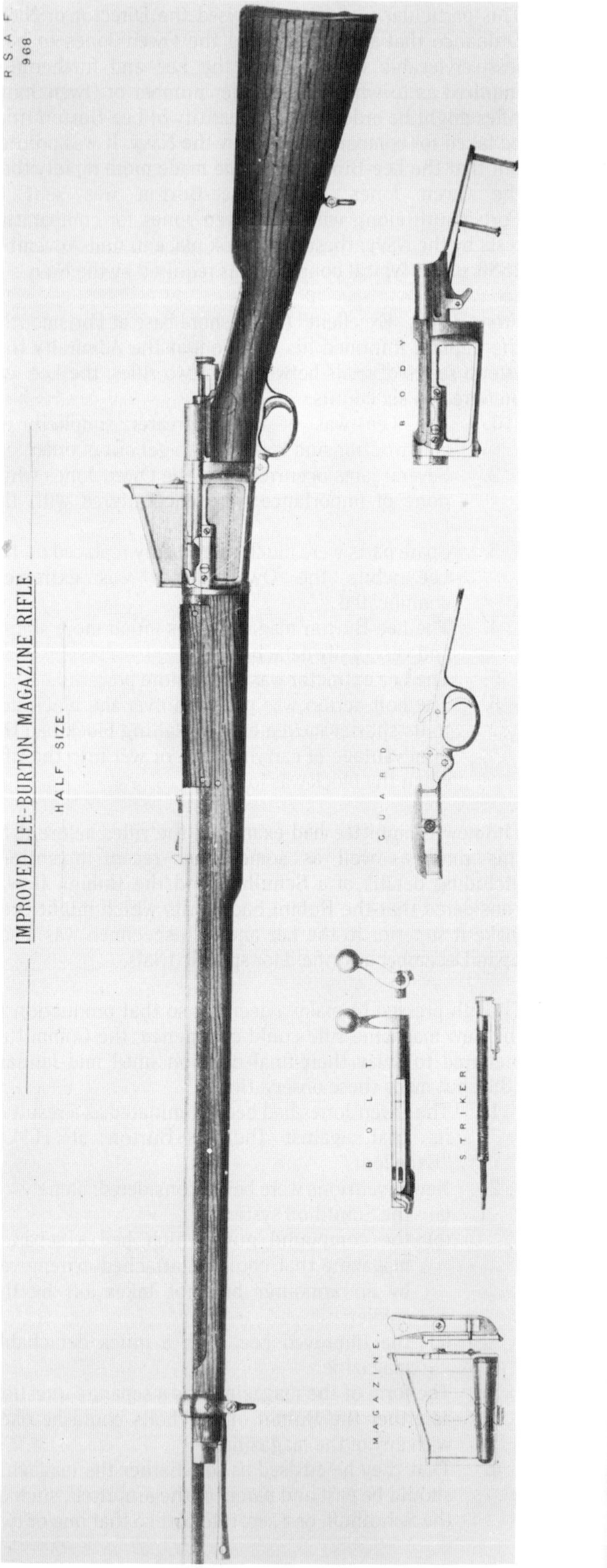
R S A F
968
IMPROVED LEE-BURTON MAGAZINE RIFLE.
HALF SIZE.
BODY
GUARD
BOLT
STRIKER
MAGAZINE.

Experiments with the Rubini bolt at Enfield did not prove to be entirely satisfactory so the Committee recommended that a trial be made between the Lee-Burton and Improved Lee, as the Owen Jones had been dropped from contention. The exclusion of the Rubin bolt from consideration meant that the Lee action was decided upon, but the feed system and other peripherals were yet to be finalised.

Politics and professional jealousies also influenced the decisions made upon a new arm and there was rivalry between various factions which sometimes involved members of the Committee. Some of the Committee reports have appendixes from individual Committee members which were appended because there were differences of opinion on certain points or additional criteria. The following appendix by Col. C.G. Slade is included here because it provides an excellent precis of the position at the time and the standing of other European powers. Slade wrote:

1. It having been definitely decided that the small-arm of the future shall be provided with a magazine, it only now remains to consider the type or system best adapted to the naval and land forces. It is possible that the form of magazine arm best suited to the Army might not find favour with the Royal Navy, and vice versa. If, however, the same ammunition be used, no inconvenience would be likely to arise from having two systems.

2. In a "Progress Report" dated 1st January 1887, the Special Small Arms Committee invited an expression of opinion from the authorities as to which type or system of magazine arms, looking at the question purely from the tactical point of view, would be most suitable for the Service, and added "that such an expression of opinion would materially assist the Committee in coming to a speedy decision."

3. As the tactical question involved in the introduction of a magazine arm is only one of the considerations, a general summary embracing all the information possessed by the Committee, together with the advantages and disadvantages connected with the different types of magazine arms, may be found useful to those to whom the question has been referred to others.

4. It may be accepted that every European Power has either adopted a magazine rifle or is about to do so. From the latest reports received from our Military Attaches at Foreign Courts, the state of the question on the Continent is as follows:-

5. In Germany, the converted Mauser, carrying 10 rounds in a tube under the barrel, has been decided upon. This arm has been before the Small Arms Committee and was rejected; it is heavy and clumsy and the method of carrying the cartridges in the magazine was not approved of. It is probable that two motives influenced the decision of the German military authorities- viz., economy and a desire to be the first in the field. It is believed that 12 Army Corps are already in possession of this arm.

6. The French have carried out exhaustive trials with the "Normale" rifle and 10,000 have been ordered. It is not yet absolutely certain that this arm will be definitely adopted, but it is more than probable.

The calibre of this rifle is .315-in. and the bullet is a compound one, being faced with steel. Some trials have lately been made with bullets of hardened lead. The magazine is in the arm itself, under the barrel. The Committee have not had an opportunity of seeing this gun. Number of cartridges, 10 to 15.

7. The Austrians are still wavering between the Schuloff and the Mannlicher. The Small Arms Committee have seen and rejected the latter; the former is still before them for final trial.

8. The Russians have not yet decided upon any particular system, but retain the Berdan rifle.

9. The Italians have lately discarded the Vetterli in favour of the Vitali, with an overhanging magazine holding 4 or 5 cartridges.

10. The Turks have not yet adopted any system.

11. The Swiss have used the Vetterli for 12 years, but are now trying the Rubini.

12. After protracted inquiry and trials, the field of selection has been narrowed down to three types or systems of magazine arms as the most suitable for final trial.

13. Magazine arms may be classed under three heads:-

(a) Those having the magazine as part and parcel of the arm itself, such as the German Mauser, the Austrian Schuloff, the Vetterli, the new French rifles, and some few others.

(b) Those that have the magazine screwed on or affixed by an armourer as an extra part, and which remains as a fixture when once attached. To this class belong the Bethel Burton, the Mannlicher, and others.

(c) Those that have the magazine or magazines carried on the person of the soldier, and attached at will, as the Improved Lee.

14. The three arms selected by the Small Arms Committee for final trial each represent one of the three types or systems above mentioned, and are:-

(a) The new Schuloff (latest pattern).

(b) A rifle made up at Enfield, which, for want of a better name, may be designated here as the "Compound " gun; it has the Rubini or Lee bolt, whichever may turn out eventually to be the simplest and most serviceable, and is fitted with the Burton-Enfield magazine.

(c) The Improved Lee.

BREECH-LOADING AND MAGAZINE RIFLES SUBMITTED TO THE COMMITTEE

	Action Type	*Magazine Type*	*Result of Trial*	*Remarks*
Bethel Burton magazine rifle	Bolt	Side Hopper	Extraction and ejection unsatisfactory; elevating spoon broke	Improved action proposed, but not yet submitted. Rejected.
Blakeney magazine rifle	No record	No record	No record	None available.
Bullard carbine	Bolt	In tube under barrel	Not tried	Magazine type objected to. Rejected.
Chaffee-Recce	Bolt	Tube in butt	Not satisfactory	Rejected.
Deely-Edge magazine rifle	Lever Action, falling block	Attachable magazine	Not satisfactory	No further trial recommended. Rejected.
Evans magazine rifle	No record	No record	Only drawings submitted	Designed for 2-in. long cartridges, not for service 3 1/4-in. length.
Fosbery magazine rifle	Hinged block	Hopper magazine on both sides of action	Several misfires	Action complicated, rifle heavy. rejected.
Garbe B.L. rifle	Lever at side	None	Not tried, report only submitted	No rifle submitted for trial.
Gardner magazine rifle	Moving barrel	In tube under barrel	Not tried with Gatling cartridge	Not made for Gatling cartridge.
Godsal B.L. rifle	Sliding block	None	Not tried	Not submitted for trial.
Godwin magazine rifle	No record	No record	Only drawings submitted	Not submitted for trial, not suitable for Service.
Grant repeating rifle	No record	No record	Only drawings submitted	Not suitable for service.
Hotchkiss	Bolt	In tube in butt	Not tried	Not suitable for service.
Jarman magazine rifle;				
No. 1	Bolt	In tube under barrel	Not tried	Magazine not approved of. Rejected.
No. 2	Bolt	Detachable, on right side of action	Stood endurance test	Magazine too complicated. Rejected.
No. 3	Bolt	Nordenfelt hopper	Bolt did not act freely	Hoppers too large. Rejected.
Lee magazine rifle	Bolt	Detachable, under action	Stood all tests	Still under trial.
Lorentzen magazine	Bolt	Magazine in the butt	Not tried, not being made for Service cartridge	Clumsy and unfit for the Service. Rejected.
Mannlicher magazine rifle;				
No. 1	Bolt	Treble involving tube in butt	Gave way under trial	Rejected.
No. 2	Bolt	Tube in fore-end	Gave way under trial	Rejected.
No. 3	Bolt	Attachable magazine above action	Gave way under trial	Rejected.
No. 4	Bolt	Attachable magazine under action	Gave way under trial	Rejected.
No. 5	Drawings only	No record	No record	Not presented for trial.
Mauser magazine rifle	Bolt	Tube under fore-end	Not tried	Rejected.
Needham magazine rifle	No record	No record	Only rough drawings submitted	Considered unsuitable for Service.
Nemetz magazine rifle	Bolt	Horizontal, magazine detachable	Not tried	Considered unsuitable for Service.
Owen Jones magazine rifle	Hinged block	Hopper magazine	Stood all tests	Still under trial.
Paulson magazine rifle	No record	No record	Drawings only submitted	Not tested.
Pieri B.L. rifle	Block	None	Not tried	Not suited for Service.
Remington-Lee magazine rifle	Bolt	Detachable magazine under action	Not tried	Not made to take Gatling cartridges.
Schulhoff magazine rifle;				
No. 1	Block & lever	Tube in butt	Failed defective cartridge test	Rejected.
No. 2	Bolt	Tube in fore-end	Failed defective cartridge test	Rejected.
No. 3	Bolt	Tube in butt	Failed defective cartridge test	Rejected.
No. 4	Bolt	Tube in butt	Failed defective cartridge test	Rejected.
No. 5	No record	No record	Cartridges jammed in endurance test	Rejected.
Skerray magazine rifle	No record	No record	Drawings only submitted	Considered unsuitable for Service.
Spencer-Lee magazine rifle	Hinged block	Detachable under action	Failed under endurance test	Rejected.
Spitalki magazine rifle	Bolt	Revolving magazine	No trial	Rejected, too clumsy.
Studer B.L. rifle	Block	None	Failed on being tested	Rejected.
Whitney Arms Company magazine rifle	Bolt	Tube under barrel	No trial	Rejected; not suited for Service.
Winchester magazine rifle	Bolt	Tube under barrel	No trial	Rejected; not suited for Service.

15. ADVANTAGES OF TYPE (a)- The new Schuloff.

(1) The magazine is always at hand, being part of the arm itself.

(2) The soldier can use his rifle either as a single loader, or load through and use the magazine at will.

(3) The magazine can be kept full or partly full.

(4) From one round up to the maximum number contained in the magazine can be fired rapidly or slowly.

(5) The magazine can be loaded easily and quickly.

(6) The number of cartridges contained in the new Schuloff is 10. This arm has not yet gone through the various tests with our ammunition.

DISADVANTAGES OF TYPE (a).

(1) The permanent addition of a magazine adds to the weight of the arm and makes it less handy on all occasions.

(2) It would be difficult to exercise much control over the soldier, as regards using his magazine, when not actually required. It would be a temptation for him to do so.

(3) Once committed to the Schuloff, there would be no changing afterwards.

(4) A breakdown of any sort in the magazine would probably disable the gun.

(5) Cost of manufacture, and difficulty and cost of repair by armourer.

(6) If this arm was adopted as the Service arm and machinery put up for it, it would probably be a long and costly process; it would also eventually necessitate its issue to Auxiliary Forces.

16. ADVANTAGES OF TYPE (b)- The Compound Gun.

(1) The magazine holding 5 cartridges is simple and not likely to get out of order, if it did, it would not affect the arm as a single loader.

(2) The magazine made of steel is not part of the gun, it is not what is called a detachable magazine, i.e. one that can be taken off and put on quickly by the soldier, but it is screwed on at the side of the breech by an armourer, the gun can be used perfectly well without it.

(3) The magazine when not in use rides snugly enough, nearly level with the line of the barrel, and is not inconvenient as regards handling of the arm. When required to be brought into play it is pushed up and secured in that position, it then overhangs the barrel.

(4) The magazine can be kept full or partly full, and the arm can be used when the magazine is affixed, either as a magazine arm, or as a single loader, at will. The Lee bolt or the Rubini bolt, the two best which the Committee have seen up to date, could be used with this arm. The Rubini bolt has not yet stood the various tests with our ammunition; the Lee bolt has stood all the tests.

DISADVANTAGES OF TYPE (b).

(1) As the magazine when once put on by an armourer remains a permanent fixture, it is open to the objections urged against the Schuloff.

(2) It does not seem quite as sound to have a small-arm, the magazine of which is not a part of the weapon, nor yet attachable by the soldier at will.

(3) Those who advocate it say "it would be kept in store and only put on when a regiment took the field." This is not practical. A battalion might receive sudden orders, or have companies on detachment. The question of drill comes in also. Whatever system of magazine arm be adopted, soldiers must be drilled to its use, and made to fire a certain number of rounds annually through it.

(4) The magazine is by no means easy to load quickly at present, as the cartridges are inclined to topple over point downwards.

(5) The magazine holds 5 cartridges only.

17. ADVANTAGES OF TYPE (c)- Improved Lee.

(1) The Lee bolt action is strong and simple, and the rifle with the detachable magazine has stood all the firing tests.

(2) The magazine, holding 5 cartridges, is carried on the soldier; he would, probably, carry two ready loaded.

(3) The magazine is attached as quickly and as easily as loading one cartridge into the chamber.

(4) The rifle can be used as a single loader when the magazine is on.

(5) Except when absolutely required for use, the magazine would not be attached to the rifle, and it would be used as an ordinary single loader.

(6) Almost perfect control would be exercised over the soldier as regards when to use the magazine.

(7) This arm would be easy to manufacture and easy to repair and is a thoroughly sound, serviceable soldier's arm.

DISADVANTAGES OF TYPE (c).

(1) The magazine cannot be loaded when on the rifle.

(2) It would be an extra article for the soldier to carry, he might possibly lose it.

(3) After rapid firing, as a single loader with magazine attached, the latter works loose and is liable to drop out. This latter defect is a mechanical detail and can be overcome.

18. In deciding upon a new arm for the Naval and Land Forces, the magazine part of it is but one factor in the question, and that a minor one. In the first place, the arm must be as serviceable and handy in every respect as the present Service weapon and capable of being used as a single loader or magazine arm at will. Strength to stand the rough usage of service, non-liability to get out of repair, simplicity, convenient weight, lowness of trajectory, accuracy, ease of manufacture and repair, are the most important points. The barrel, rifling, calibre, ammunition and sighting have all been settled.

COMPARATIVE TABLE OF EUROPEAN SMALL ARMS ... 1879

Provided to the Small Arms Committee.

Country & Type of Arm	*Calibre*	*Rifling Grooves, Depth*	*Mechanism, Closing*	*Mechanism, Striking*	*Length of arm*	*Weight of arm*	*Muzzle Velocity*	*Powder Charge*	*Sighted up to*
Austria:-									
Werndl, Patt. 1867-73.	.433-in.	6; .0079-in.	Plug, parallel axis to barrel.	Flat spring.	50.388 in.	9.9 lb.	1430 fps.	61.729 grs.	1400 paces.
Werndl carbine, 1867 & 1873; rifle 1873 cond.	.433-in.	6; .0079-in.	Plug, parallel axis to barrel.	Flat spring.	39.529 in.	9.2 lb.	978 fps.	32.408 grs.	600 paces.
Wanzl (converted).	.5474-in.	4; .0079-in.	Tabatiere opening upwards.	Lock.	52.752 in.	9.5 lb.	1240 fps.	67.594 grs.	900 paces.
Germany:-									
Mauser, 1871.	.433-in.	4; .0118-in.	Bolt.	Spiral spring.	52.356 in.	9.9 lb.	1411 fps.	77.162 grs.	1750 yds.
France:-									
Gras, 1874.	.433-in.	4; .0118-in.	Bolt.	Spiral spring.	51.377 in.	9.3 lb.	1430 fps.	81.021 grs.	2000 yds.
England:-									
Martini-Henry.	.45-in.	7 heptagonal; .0073-in.	Block.	Spiral spring.	46.452 in.	8.8 lb.	1315 fps.	84.878 grs.	1400 yds.
Russia:-									
Berdan II.	.42-in.	6; .0106-in.	Bolt.	Spiral spring.	53.544 in.	9.7 lb.	1430 fps.	77.162 grs.	1500 paces.
Krnka (converted).	.6005-in.	4; .015-in.	Tabatiere opening upwards.	Lock.	53.502 in.	9.9 lb.	1073 fps.	78.242 grs.	800 paces.
Italy:-									
Vetterli.	.4098-in.	4; .0098-in.	Bolt.	Spiral spring.	52.949 in.	9.7 lb.	1427 fps.	61.729 grs.	1110 yds.
Spain:-									
Remington.	.433-in.	6; .0079-in.	Block.	Direct.	51.180 in.	9.0 lb.	1411 fps.	77.162 grs.	1110 yds.
Switzerland:-									
Vetterli repeater.	.4098-in.	4; .0098-in.	Bolt.	Spiral spring.	51.180 in.	10.4 lb.	1417 fps.	57.872 grs.	1110 yds.

Troop Trials model Remington Lee (.43 Spanish) with bolt open. These rifles were fitted with Speed's magazine cut-off at R.S.A.F. Enfield
MOD Pattern Room

19. The Small Arms Committee have only, therefore, at the present moment, two questions before them connected with the new arm- viz., the breech action, and the form or type of magazine.

20. The falling block of the Martini breech-action has up to the present moment, baffled every inventor who has attempted to adapt a magazine attachment to it, and it may almost be accepted that it is a mechanical impossibility to do so. The Martini breech-action is consequently doomed.

Every Continental Power has adopted a bolt action of some sort or other and as this form of action is strong, simple and easy to manufacture, it is not necessary at present to look beyond the Improved Lee or Rubini bolts. The former has stood every test and trial, a thoroughly reliable action.

The latter appears to be even superior to the Lee, but it has not yet stood the severe trials to which the Lee was subjected and to adopt it hastily would be rash.

21. In considering the question of the magazine itself, the following are the main points to be considered:

(a) The tactical aspect. This embraces:

(a1) The employment and use of the magazine on service.

(a2) The number of cartridges that should be carried in the magazine.

(a3) Whether the magazine should be a fixture, a semi-fixture or detachable at will by the firer.

(a4) Rapidity of fire over a single loader.

(a5) Fire tactics and fire discipline.

22. (a1) Rapid firing through the magazine will probably only be resorted to in the final stage of delivering or sustaining an attack. The magazine would prove of advantage to small bodies of cavalry holding a defile or bridge, to the escort and detachment of a machine gun, and to sentries; in the Royal Navy, in boats, in the "tops" and against torpedo boats.

(a2) The number of cartridges that should be carried is a moot point. Prima facie, the greater the number, the more formidable the arm, but the minimum should not be less than five. A large number of cartridges of the Enfield-Martini calibre, and present pattern, would necessitate a clumsy magazine, which would be an awkward addition to the rifle.

23. (a3) The advantages and disadvantages of each of these three systems have been enumerated before and a decision on this point must be arrived at by the value attached to the various arguments for and against.

24. (a4) It has been proved by experiments, that when firing for so short a period even as half a minute, the number of rounds that can be fired from almost any magazine arm, is not greater than can be fired in the same time from any good military single loader.

It follows therefore, that as regards rapidity of fire, a magazine arm is only superior to a single loader to the extent of the number of cartridges that can be fired continuously through the magazine without reloading; and this will depend on the contents of the magazine. Say, from 5 to 10 rounds.

25. (a5) The almost universal conversion of single loaders into magazine arms, or arms capable of being used at will as repeaters, will introduce a new element into fire tactics, but it may be doubted whether any change will consequently take place in these tactics, or if any such change would not be a step in the wrong direction.

Fire discipline, the necessity of controlling and steadying men firing, and thus preventing a wasteful expenditure of ammunition will be of just as much importance as ever.

26. In a circular memorandum dated 2nd November 1886, His Royal Highness the Field-Marshal Commanding in Chief directed that "Independent firing in the attack formation should be discontinued altogether and that in future volley-firing should be employed during all stages of the attack."

It is said that with the introduction of magazine arms a new system of fire discipline will come into play and that the soldier will be taught to load through and fire from his magazine on all occasions. The command "fire one, two or more shots" will be substituted for the caution "fire one, two or more volleys." This would be simply re-introducing "Independent firing" with all its attendant evils.

27. The magazine question is more or less a political one and the general conversion of single loaders into repeaters which is taking place all over Europe is due, not so much to the superiority of the magazine arm as to the feeling that it is necessary to keep abreast of one's neighbours.

28. An examination of the state of affairs both at home and abroad shows clearly that the weapon of the future has yet to be found. The Germans have only adopted the Mauser in order to get a magazine arm as quickly as possible and for economical reasons. The Austrians are taking up the Mannlicher in a half-hearted way and at the last moment may change it.

The French are still carrying on exhaustive trials with the "Normale" which may not turn out satisfactory after all.

The Italians are giving up the Vetterli and substituting the Vitali. The Russians and the Turks are perhaps prudently holding their hands. In England, the delay that has taken place has probably saved the country from prematurely adopting an imperfect weapon.

29. On 1st of the present month the Small Arms Committee reported that "three arms only, each representing different systems, remained for the final trial;" but even since that report was tendered, several new systems have been brought to their notice, the most promising being one submitted by Mr. Morris, which seems simple and has a semi-positive feed. So far as can be judged from a model, this invention is worthy of examination.

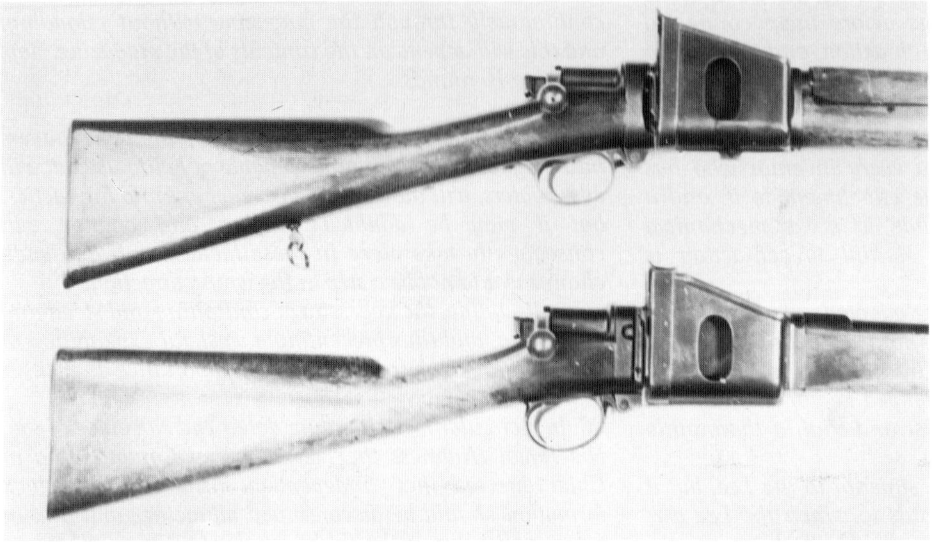

The .402-in. LEE-BURTON. Top specimen is earlier model with handguard and clearanced fore-end. The lower model also has magazine fixture improvements and no lower sling swivel. These also feature Martini style two-piece furniture. MOD Pattern Room

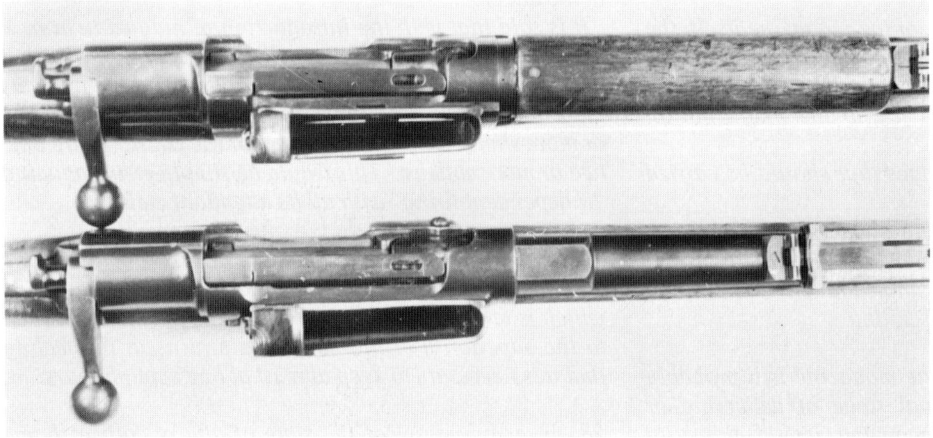

Top view of action. The same Lee bolt system was retained through the Lee-Burton series. MOD Pattern Room

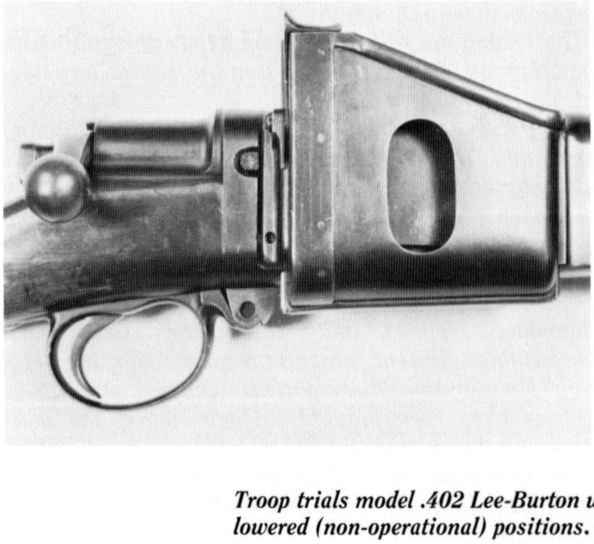

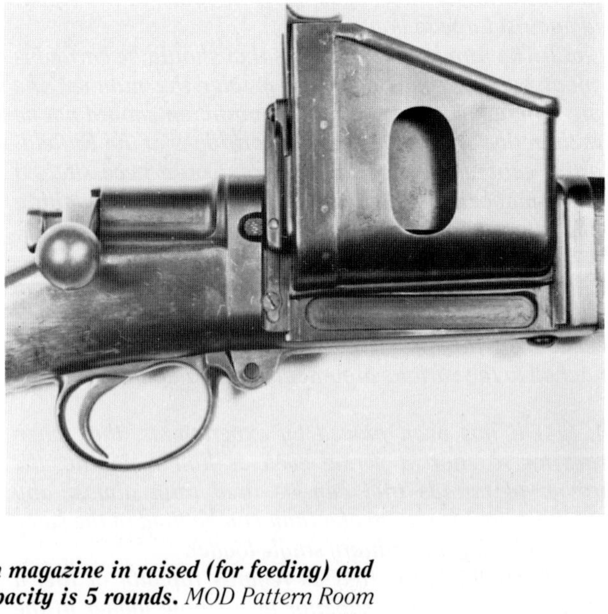

Troop trials model .402 Lee-Burton with magazine in raised (for feeding) and lowered (non-operational) positions. Capacity is 5 rounds. MOD Pattern Room

30. For manufacturing and other reasons, it is no doubt very desirable that a speedy decision should be come to, and the Small Arms Committee have been pressed to render their final report without further delay- say, by the end of the present month.

31. The whole question should be most carefully considered in all aspects, and the peculiar conditions of service of our Army, extending over every quarter of the globe, should be borne in mind.

32. It would be short-sighted to lay stress on the magazine qualities of an arm only, to the exclusion of the other points enumerated in paragraph 19 of this Memorandum, which are far more important; more especially when it is probable that the magazine would seldom or ever be used, whereas the rifle should be at all times a simple, handy and efficient single loader not likely to get out of order.

33. Of the three magazine arms still before the Small Arms Committee for final trials, only one, the Improved Lee, has stood all the tests and could at this moment be safely recommended as a thoroughly serviceable soldier's weapon for present manufacture. Should time admit of it, other systems, including the Morris, would be worthy of trial.

34. The small-arm of the future will be a small bore, probably .300, the bullet steel-cased, and the charge compressed powder or a smokeless chemical compound. The soldier will carry on his person from 120 to 150 rounds, the magazine will be detachable and the soldier will carry two or three, ready loaded, each holding 10 to 15 cartridges. A proportion of the ammunition in the ammunition carts will be ready packed in these little magazines. If a man lose one of his magazines it would be easily replaced. The country that first adopts an arm of this nature will be in advance of other powers.

C. G. Slade, Colonel
Commanding 2nd Battalion Rifle Brigade
Woolwich, 13th January 1887"

This appreciation of the situation by Colonel Slade was certainly far-sighted and while his predictions did not hold true to their full extent upon the Lee-Metford introduction, it certainly can be applied to the contemporary situation.

The time being taken by the Committee to finalise the new magazine rifle was causing concern in some quarters as instructions were issued on 18th February 1887 that the said body report as speedily as possible on the comparative merits of the Lee-Burton and Improved Lee and that their inquiries were not to enter upon points already decided upon, namely the rifling (Metford), the breech bolt (Lee), cartridge and calibre (.402-in.). The last point is surprising in view of the contemporary advances in smaller calibres in Europe, but typical of certain War Office attitudes at the time.

Regarding other designs still under consideration, the Schuloff magazine was rejected on account of its complicated mechanism, weight, expense and the absence of any marked superiority over the Improved Lee. Another late consideration was the Morris magazine but although fifty of the new 1888 Trials Magazine Rifles were assembled with the Morris system, it too was eventually rejected.

A magazine conveyor or carrier was patented by Mr. Morris in 1886 and another invention of his, Patent No. 5786 of April 20th 1887, was the type fitted to the 50 Trials rifles. The box magazine was positioned underneath the receiver and it is basically similar to that on the Lee, except that the magazine platform or elevator was depressed by means of an external projection or knob on the magazine. This enabled the platform to be locked in the depressed position so that rapid loading of the cartridges might be effected.

FIG. 1.

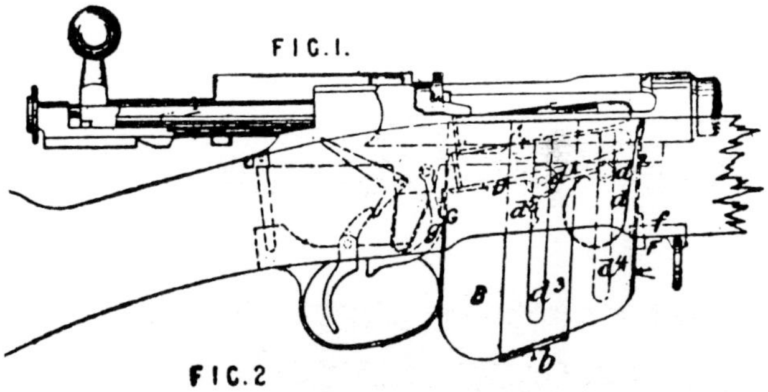

FIG. 2

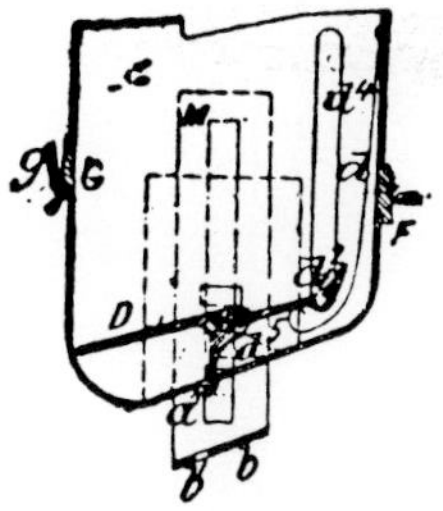

Morris Patent, No. 5786 of 20.4.1887.

The new British agent for the Lee rifle, Capt. E. Palliser, offered to submit the latest pattern Lee rifle with most recent improvements in November 1886 and this model of Lee rifle (still referred to as "Improved Lee") arrived from the United States by mid-January 1887. Sufficient improvements had been effected to warrant the suspension of manufacture of the Lee-Burton pending results of tests with the "new" Improved Lee.

This latest model Lee permitted loading of the magazine on the rifle whereas prior models with detachable magazines required that the magazine be removed to be filled. As this new model was chambered for the Spanish Remington cartridge, it was advised that about three weeks were required to alter it for the British .402 cartridge. By this time, a special magazine cut-off had been developed at Enfield which permitted single loading with the filled magazine being held in reserve.

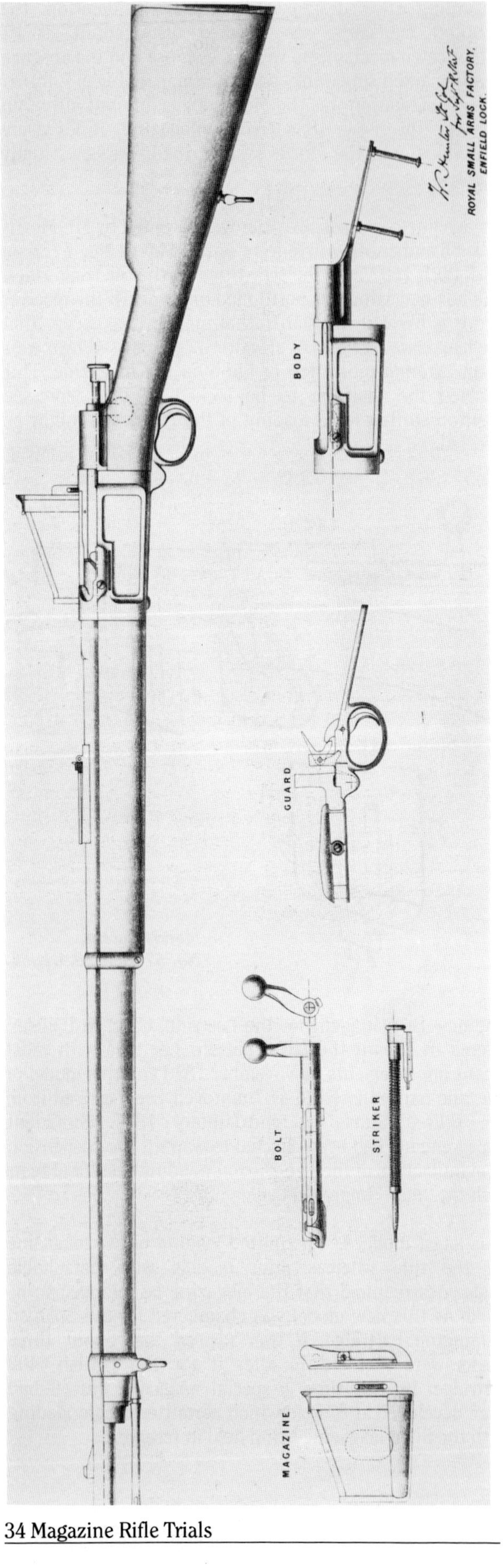

LEE-BURTON MAGAZINE REPEATING RIFLE

WEIGHT, DIMENSIONS, &c.

Length of rifle	*4 ft. 2 ins.*
Average weight of rifle and magazine, empty	*10¼ lbs.*
Length of barrel	*30 inches*
Number of grooves	*7*
Twist of rifling	*1 turn in 15 ins.*
Calibre	*.402 inch*
Ammunition charge	*85 grains*
Bullet weight	*384 grains*

This rifle has a magazine capable of holding five or any less number of cartridges. The magazine is held on the side of the body in such a manner that it can be raised or lowered by a momentary action of the operator; when raised it is in such a position to feed the cartridges into the breech; when lowered it is in such a position that the cartridges may be carried in perfect safety in any position, without the possibility of a cartridge entering the breech. The arm can then be used as a simple breech-loader. The magazine is retained in the raised and lowered positions by a spring acting upon a projection on the magazine; when in the raised position, a point on the spring engages in a recess in the projection, which effectually secures the magazine in position, as it cannot be lowered except by compressing the spring to release it from the recess; this is done by a thumb-piece on the spring.

The cartridges are retained in the magazine when in the firing position, or when carrying the rifle, by means of a spring cap hinged on the rear end of the magazine.

A hole is cut on the outside of the magazine for the purpose of lightening it and to admit of the number of cartridges in the magazine being observed. The cartridges are fed into the breech of the rifle by gravity.

INSTRUCTIONS FOR STRIPPING

1. *Remove magazine.*
2. *Remove magazine fixing spring.*
3. *Remove extractor spring and extractor.*
4. *Remove tang screws.*
5. *Remove guard, elevator and parts attached.*
6. *Remove stock butt.*
7. *Remove bolt.*
8. *Strip guard, removing axis pin, taking out the trigger and trigger lever.*
9. *Remove screw and trigger spring.*
10. *Strip bolt, by pressing lug of stop, brush to right and draw out firing pin, main spring, stop brush, and sear, remove firing pin, unscrew sear, remove it, also stop brush and spring.*

In assembling, the above order is reversed.

At this time there were 300 Lee-Burton rifles in the course of construction and it was inquired if 300 Improved Lee rifles might be speedily made up for comparative trials, but to set up and manufacture such a quantity in such a short space of time was not possible. So, 300 were ordered from the Remington factory in the United States, in 11mm (.433-in.) Spanish calibre, on the undertaking that the rifles would be delivered within three weeks of the order and that 120 rounds of ammunition would be supplied with each rifle. They arrived by April and were then stamped with the War Department mark on the butt; examples are illustrated here and in the first section of Chapter 12.

Enfield production records indicate that a total of 327 Lee-Burton rifles were made at Enfield in the financial year, 1st April 1887 to 31st March 1888. As there were two distinct models, it seems that about two dozen of the initial order of 150 were supplied and the remaining 300-odd were the subsequent model used in the trials against the Improved Lee.

An interesting reference in these production figures is to "25 arms, Enfield, repeating, converted to Lee pattern"; the particular form and calibre of these converted rifles is not listed although they were probably converted from Lee-Burton models and the calibre was likely to have been .402 with ratchet rifling. A box magazine of Lee or Speed design would have been applied. An example of what is considered to be one of these conversion is illustrated here.

Above & below: Lee-Burton conversion to fixed magazine underneath with hinged plate at bottom. *Tower of London*

In April 1887, descriptions of the Lee-Burton and Improved Lee rifles were issued along with details of the forthcoming trials; each rifle was to be issued with 10 rounds of dummy and 100 rounds of ball ammunition. The original drawings, specifications and instructions have been reproduced on the adjacent pages.

These Remington-made Lee rifles feature one-piece stocks and what is considered to be the American style furniture and general appearance. While some Improved Lee rifles were made at Enfield during this period, they may be regarded more as experimental models rather than troop trials types. These particular Enfield-made Improved Lee rifles are also distinctive in that they were stocked up in a similar fashion to the .402 Enfield-Martini series; in a number of respects, this development was parallel and comparable.

The M.O.D. Pattern Room, now housed at Nottingham, has four successive variations of the early trials magazine Lee rifles in their collection; these have been described in "British Rifles – A Catalogue of the Enfield Pattern Room" (published in 1981 by the H.M.S.O.). The custodian, Herbert Woodend, has catalogued and described these particular items, although these four variants were not illustrated in that particular handbook due to space restriction.

The progressive improvement in these models provides a good indication of the U.S. and, to a much lesser degree, the R.S.A.F. development, over the period from 1883 until 1886. In 1886, the three hundred Improved Lee rifles in .43-in. (11mm) Spanish calibre were brought in from the Remington factory for troop trials against the Lee-Burton. Woodend's catalogue titles for each of the following four types have been applied to the following description of refinements made to the original 1879 Model (as submitted for the 1880 trials).

Original U.S. production Remington Lee, the Mod. 1879. Note the position of the bolt handle. *R. Faris*

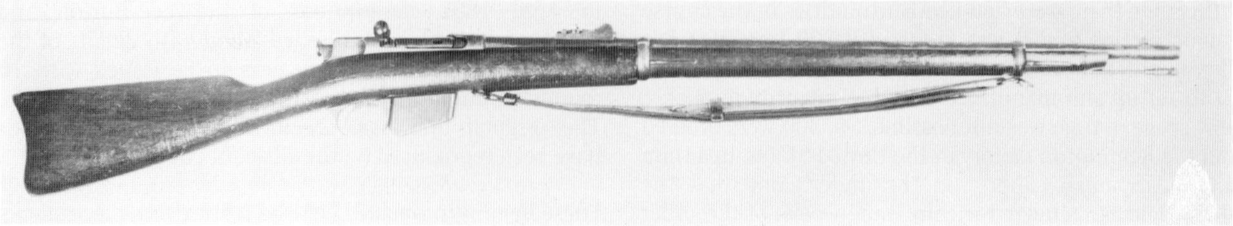

U.S. Model 1879 Lee rifle, manufactured by Remington. *R. Faris*

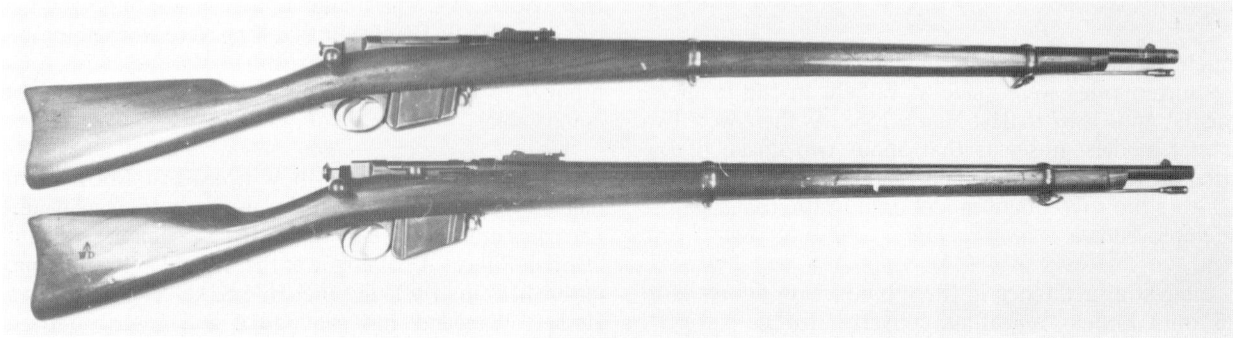

Subsequent Remington Lee rifles imported for evaluation and trials. Improved Mod. 1879 in .45/70 calibre (top) and the eventual troop trials model in .43 Spanish. Note War Department mark on the butt. *MOD Pattern Room*

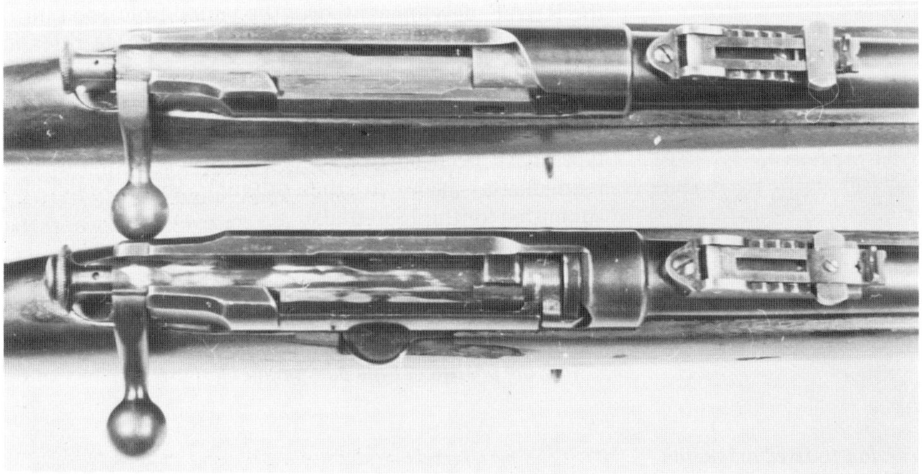

Top: Improved Model 1879 (1882/1885). Bottom: Troop trials (1886) model with Speed cut-off on right side. *MOD Pattern Room*

Troop trials model (.43 Spanish) with bolt open. These were fitted with Speed's magazine cut-off at RSAF Enfield. *MOD Pattern Room*

Rifle, Lee, .45," Model 1879.
Overall Length:- 53.0 in. Weight:- 9 lb. 5 oz.
Barrel Length:- 33.2 in. Year:- 1880.
Martini-Henry type barrel with .450" Gatling chamber.

Bolt action, with rear locking action. The bolt handle stem is positioned closer to the middle of the bolt length. One-piece bolt body with extractor claw formed on rotating collar and retained by a flat spring; the extractor fixes to the locking rib of the bolt body by a hook arrangement. To withdraw the bolt, the extractor spring has to be eased forward until it disengages, when the extractor assembly can be removed and the bolt withdrawn from the action. Some problems were experienced with the extractor design and primary extraction. Five-round detachable magazine, no vertical rib indentations for strengthening.

Rifle, Lee, .45," Model 1879 (Improved).
Overall Length:- 51.7 in. Weight:- 9 lb.
Barrel Length:- 33.1 in. Year:- 1882/1885.
Remington manufacture barrel chambered for .45/70."

Received from America in late 1882. Bolt action, with bolt handle stem at the rear of the bolt body. Improved extractor system, as suggested by Supt. R.S.A.F. Enfield; angle altered in the action so that in the initial turn of the bolt to open the breech, the camming action makes for stronger primary extraction. Magazine can only be charged when removed from the rifle; a hinged plate swings out from the left of the action body when the magazine is removed, to serve as a guide for single loading. Slight modification to the magazine, with two retaining notches on the back strap to enable single loading with the full magazine in reserve when positioned on the second notch; when snapped home to the upper notch, magazine loading is utilised.

Rifle, Improved Lee, .45."
Overall Length:- 52.8 in. Weight:- 8 lb. 14 oz.
Barrel Length:- 33.3 in. Year:- Early 1886.
Enfield manufacture barrel chambered for .45" Gatling.

This particular specimen was produced at R.S.A.F. Enfield and is generally similar to the previous-mentioned example made by Remington. Salient differences are that the box magazine is non-detachable and requires loading through the top of the receiver with the bolt open, there is a magazine cut-off fitted onto the left side of the action body and a rocking arm at the front, left side of the body wall, assists the cartridge extractor on the bolt. The last two improvements were Enfield developments. The stock is a one-piece type, similar to the Remington-produced models and the lines of the Enfield-Martini (relating to the nosecap and clearanced barrel/fore-end arrangement) are quite obvious.

Rifle, Enfield Lee, .4."
Overall Length:- 50.2 in. Weight:- 10 lb. 2 oz.
Barrel Length:- 31.2 in. Year:- 1886.
Enfield manufacture barrel with 7-groove Ratchet rifling, chambered for .402" Enfield-Martini.

The most obvious difference in this prototype model is the two-piece furniture similar to the Martini style, with separate butt and fore-end. It was also produced at the R.S.A.F. and is marked "ENFIELD 1886" on the action body. Whilst the action and bolt is very similar to the previous-listed prototype, the magazine cut-off has been changed to the right side of the action. The internal spring lip which holds the cartridges in the magazine, is positioned in the left wall of the action body.

Woodend, in the Enfield Pattern Room catalogue, indicates that this model was an experiment in economy, constructed primarily to investigate the possibility of using existing stocks of Martini-Henry and Enfield-Martini furniture blanks, should the Lee rifle be adopted. The barrel, fore-end, sights, barrel band and clearing rod are similar to those of the 2nd Pattern Enfield-Martini rifle, although the adapted E.-M. barrel and fore-end required some shortening because of the greater length of the Lee action, compared with that of the Martini. Requirements of a similar overall length governed these relative measurements.

Rifle, Remington Lee, 11mm (.43" Spanish).
Overall Length:- 51.7 in. Weight:- 10 lb. 2 oz.
Barrel Length:- 32.3 in. Year:- 1886.
Remington manufacture, chambered for .43" Spanish.

This was the latest Remington production model in late 1886, submitted to the Small Arms Committee in January of 1887. The fixed magazine permits loading through the top of the open action and this model also incorporates some Remington factory improvements such as the trigger arrangement and the bolt, hence the nomenclature "Remington Lee." The magazine cut-off on the right side of the action, designed by Speed, was installed at R.S.A.F. Enfield.

This troop trials model features certain Remington-designed bolt improvements, namely a detachable bolt head formed with an upright gas check shield, fixed to the bolt body by a simple bayonet lock. In order to adjust the firing pin protrusion and to facilitate the bolt disassembly, a removable collar was threaded to the rear of the firing pin.

Certain modifications were applied to the Remington-produced rifles; some were rebarrelled for the British service or other experimental cartridges and most were fitted with a magazine cut-off which permitted single loading with the full magazine held in reserve. The first with Speed's cut-off had this device fitted into the left side of the receiver (as indicated on the early 1886 model

Enfield-made Lee prototype rifles. Enfield Improved Lee in .45-in. Gatling calibre (top) and the Enfield Lee .402-in. rifle (bottom). MOD Pattern Room

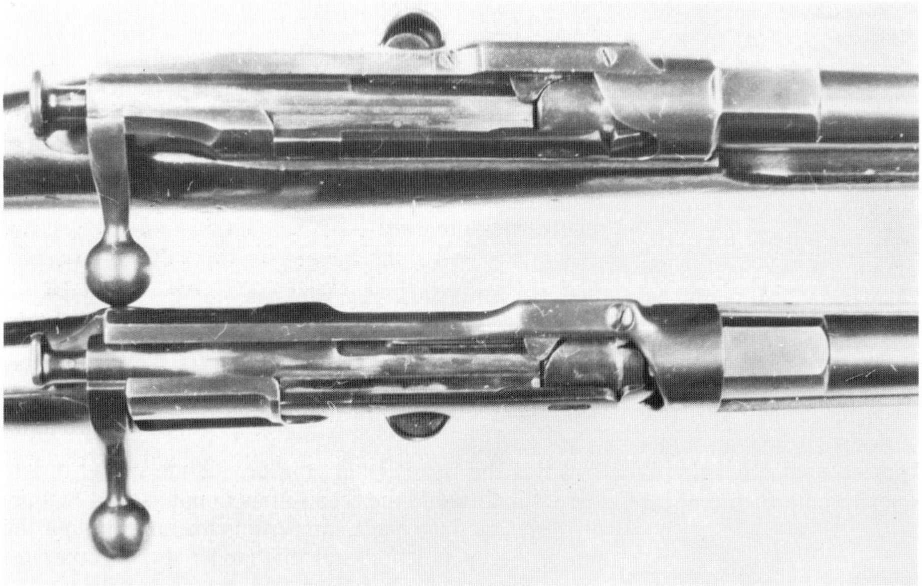

Top action views of the two rifles above. Note the Martini barrel knox form. Speed's magazine cut-off was changed from the left side (on the first model) to the right side of the following and eventual service models.
MOD Pattern Room

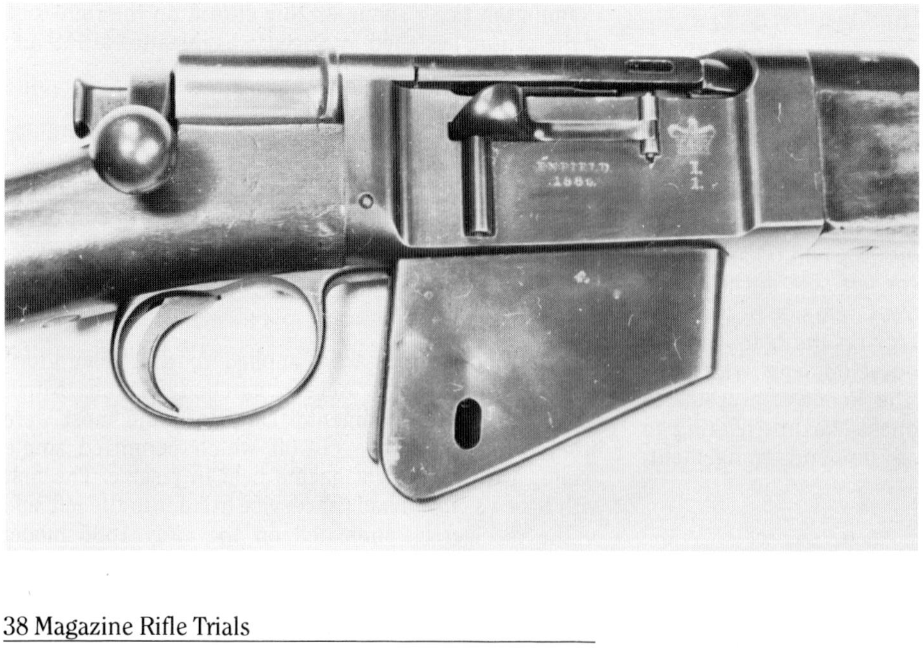

Action markings, .402-in. Enfield Lee rifle, dated 1886. RSAF influences are the square-shaped action and two-piece stock and fore-end configuration, from the Martini era.
MOD Pattern Room

Improved Lee described previously) while the 300 trials rifles had the cut-off mounted onto the right side. These particular modifications were carried out at the Royal Small Arms Factory at Enfield.

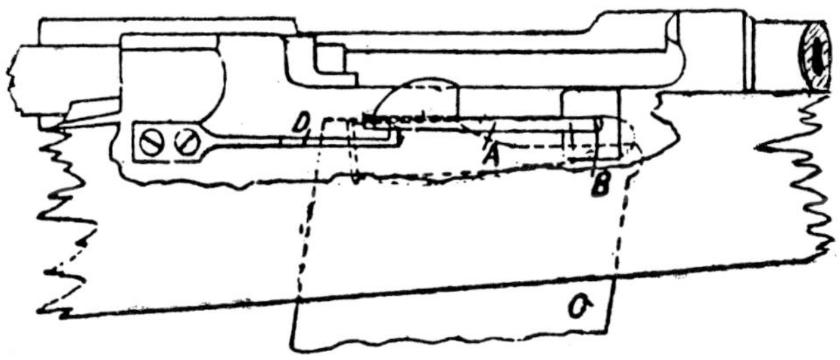

Speed Patent cut-off drawing, No. 6335 of 30.4.1887.

J.J. Speed's patent for the magazine cut-off was No. 6335 of April 30th 1887, although it underwent further modification before its eventual form on the Magazine Lee-Metford service rifle. Another improvement registered with the British Patent Office was No. 11319 of August 18th 1887, by James P. Lee and this relates to the bolt with a moveable head with new form of extractor, gas vents in the receiver and an improved magazine.

Other notable differences include a smaller and neater cocking piece and trigger guard and, compared with some earlier models, a new backsight with longer bed and folding leaf. The backsight slide incorporates a wide, shallow "V" notch and the foresight has a ramp which does not permit the fixing of the usual socket bayonet.

Those Commanding Officers of units undertaking the trials were instructed to make written reports of:

- The handiness of the rifle
- Which magazines were preferred
- The facility as both single and magazine loaders
- Use as both single and magazine loaders
- Cartridge feed in rapid magazine fire
- Which arm performed better as a single loader, magazine rifle, and in rapid magazine fire
- Was any difficulty experienced with extraction in the Lee-Burton
- Was the magazine awkward in drill with arms, or in carrying
- Did the magazine injure clothing or accoutrements.

The results of these user trials were markedly in favour of the Improved Lee and led to the rejection of the Lee-Burton on a number of counts; recent improvements to the U.S. Lee had also rectified a number of the small problems previously encountered. Now that the action and form of the magazine had been decided upon, attention was turned towards the calibre, even though instructions had been issued that this be pursued no further.

Various small bores were under test at this time by some European countries, with considerable successes, and some of these reports were tabled before the Committee. In connection with the calibre comparison, a trial was conducted between the Enfield-Martini and Rubini rifles and results presented on 8th June 1887.

	Enfield-Martini	*Rubini*
Calibre ...	*.405 in.*	*.298 in.*
Rifling grooves;		
Number... ...	*7*	*3*
Form	*Circular*	*Elliptical*
Depth	*.004-in.*	*.005-in.*
Length one turn ...	*16 in.*	*10-1/2in.*
Form of chamber...	*Bottle shaped*	*Straight taper*
Length of barrel ...	*2 ft. 9-1/4in.*	*2 ft. 8 in.*
Bullet (cylindro conoidal)		
Length	*1.274-in.*	*1.187-in.*
Diameter ...	*.402 in.*	*.313 in.*
Weight	*384 grs.*	*216 grs.*
d2/W	*2.945*	*2.9*
Proportion of weight, powder charge to bullet	*1 to 4.5*	*1 to 3*
Powder charge ...	*85 grs.*	*70 grs.*
Weight of 100 cartridges	*10 lb. 10 oz.*	*6 lb. 9 oz.*
Length of cartridge	*3.307 in.*	*3.062 in.*
Muzzle velocity ...	*1,570 fps.*	*1,850 fps.*

Some tests were made on penetration through a variety of materials and animal carcasses so as to ascertain tissue and bone damage; the Rubin cartridge's flatter trajectory, better accuracy and higher velocity confirmed its superiority over the .402 round and the Committee had no choice but to recommend the adoption of the smaller calibre. This decision also sounded the death knell for the Enfield-Martini because the .402 cartridge was not to be adopted, it would not even be suitable as a reserve grade arm. The Enfield-Martini was then converted to the .450 Martini-Henry Mk IV pattern so that the considerable quantity of .402 trials Martini rifles on hand might be employed by Reserve and Colonial forces.

Quite a few difficulties were being encountered by the Royal Laboratory and Royal Gunpowder factory in duplicating the powder used in the Rubin cartridge; some nitro compounds were being tested as well as the black powder. The Duttenhofer compound, which was a nitro-based mixture, was also under consideration.

The Committee had decided upon a barrel by October 1887, in .303-in. calibre with Metford's bore grooving and entry; the Lee bolt action had been decided upon for some time. Joseph Speed, then the Assistant Manager at R.S.A.F. Enfield, also made improvements to the rifle design and he was to play an important role in the early evolution of the service rifle. Speed took out patents on various improvements to the Lee system over the following decade and acknowledgment of his patents may found marked on the commercial or "volunteer" Lee-Metford and Lee-Enfield rifles as these are stamped "LEE-SPEED PATENTS" on the action body and certain other components.

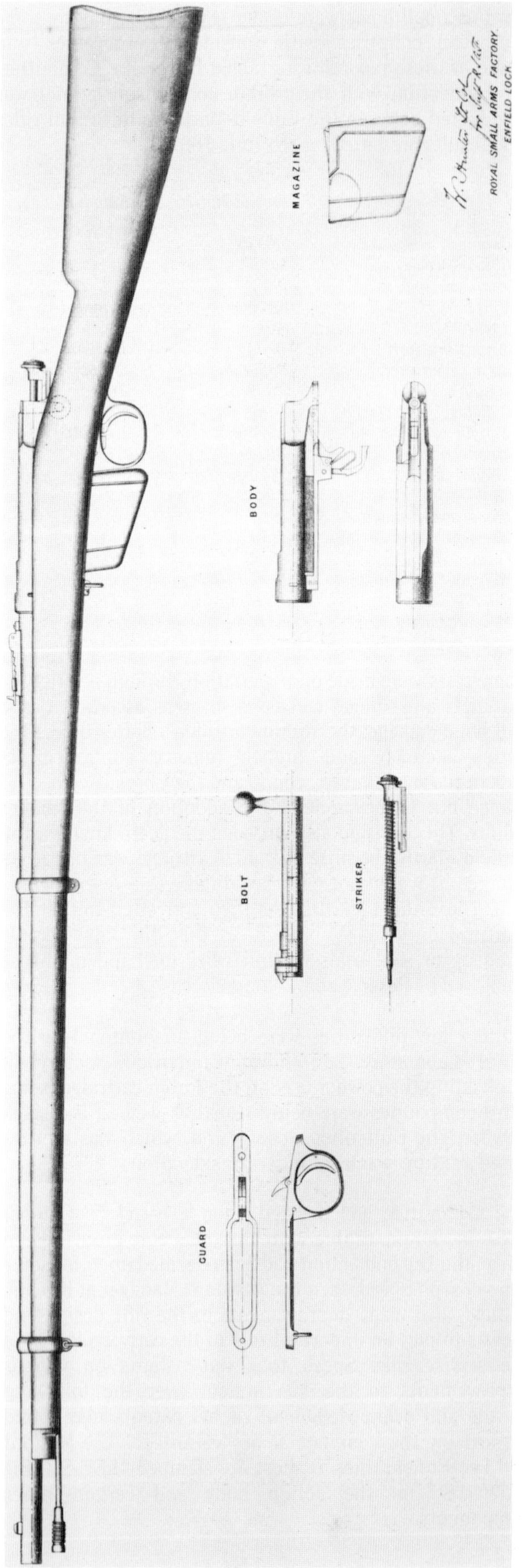

IMPROVED LEE MAGAZINE REPEATING RIFLE

WEIGHT, DIMENSIONS, &c.

Length of rifle	*4 ft. 3½ ins.*
Average weight of rifle and magazine, empty	*10 lbs.*
Length of barrel	*32 inches*
Number of grooves	*5*
Twist of rifling	*1 turn in 20 ins.*
Calibre	*.433 inch*

Ammunition:- Spanish Remington

Ammunition charge	*87 grains*
Bullet weight	*376 grains*

This rifle has a fixed magazine in front of the guard, capable of containing five cartridges. The magazine is loaded from the top, by pressing one cartridge in at a time; these are lifted upwards by the action of a "C" spring, which brings them into position to be pushed home by the forward movement of the bolt; a maintaining spring is fitted on the side of the body for the purpose of keeping the cartridges in the magazine, and also for allowing one cartridge at a time to rise to a convenient position, to be pushed home by the forward movement of the bolt. The right hand side of the body is fitted with a "cut-off" to enable the cartridges in the magazine to be disconnected from the action of the bolt at will, and the rifle used as a single loader.

The bolt is the latest "Improved Lee" with the cam removed from the front to the back of the body, this gives a clear space for loading the magazine from the top, the extractor is brought lower down on the side of the body, forming a guide and steadying the bolt, and allowing it, the extractor, to get in front of the cartridge when feeding from the magazine.

INSTRUCTIONS FOR STRIPPING

1. *Remove keeper screw, depress spring in guard, and take out magazine.*
2. *Remove extractor spring, by forcing it forward, take out extractor and front of bolt.*
3. *Remove bolt.*
4. *Remove tang and guard screws.*
5. *Remove upper and lower bands with cleaning rod.*
6. *Remove stock.*
7. *To take out main spring and striker, compress spring, and unscrew keeper nut, remove cocking piece, and take out striker and main spring.*
8. *To remove "cut-off," drive out axis pin, unscrew spring screws, and remove spring.*

In assembling, the above order is reversed.

The Speed patents incorporated in the Lee-Metford rifle were:

Magazine & cut off .... 6335 of 30.4.1887
Magazine construction, long range
sight & bolt dustcover .... 13335 of 1.10.1887
Magazine spring .... 17944 of 30.12.1887
Magazine construction, handguard
& safety catch .... 15786 of 1.11.1888

Other patents were later taken out on improvements which were utilised in following patterns of the service Lee rifle.

17,944. Speed, J. J. Dec. 30.

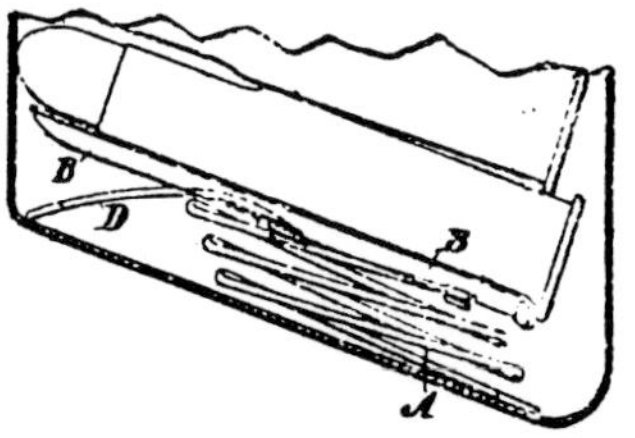

Magazines. — Consists of a cartridge - elevating spring A, the convolutions of which are oblong, and which is so formed that, when compressed (as shown in the Figure), the ends of each of its convolution will be in front of or behind the ends of the convolution above it. A supplementary spring D is formed with or separate from the mainspring to assist in raising the follower B during the discharge of the first two or three cartridges.

13,335. Speed, J. J. Oct. 1.

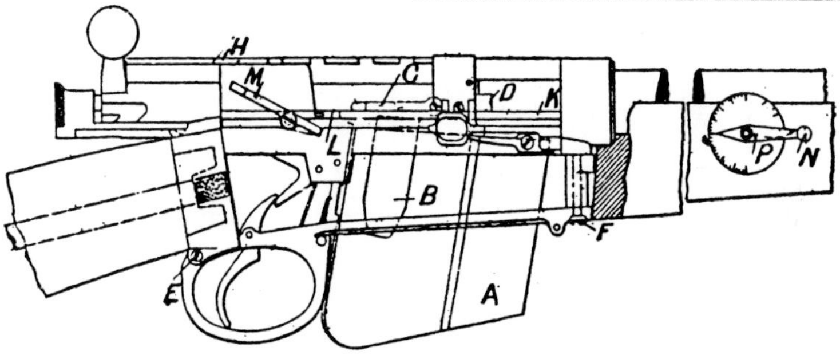

Breech actions, sliding breech - block ; magazines. — The improvements have special reference to the "Lee" magazine rifle. The detachable magazine A is provided with a spring hook B, on one or both sides, the hook projecting over the mouth of the magazine, the object of this hook being to enable the magazine to be loaded in position or when detached. The spring ejector C strikes the cartridge as soon as the former is clear of the breech-bolt. The extractor D is pivoted to the head of the breech-bolt, and is pressed by a spring which fits into a recess in the bolt head. The trigger guard is attached to the shoe by the screws E, F. The head of the bolt has a groove fitting the rib K, which rib is cut away at L to permit the removal of the bolt. M is a piece which fills up the cut-away part of the rib when the gun is in use.

Stocks ; protecting from dust and rain.—The fore-stock is made and attached separate from the butt-stock. H is a curved shield which is attached to the breech-bolt and moves with it.

Sights.—The fore sight N is made adjustable by being carried on the pointer P, which is pivoted to a graduated disc.

15,786. Speed, J. J. Nov. 1.

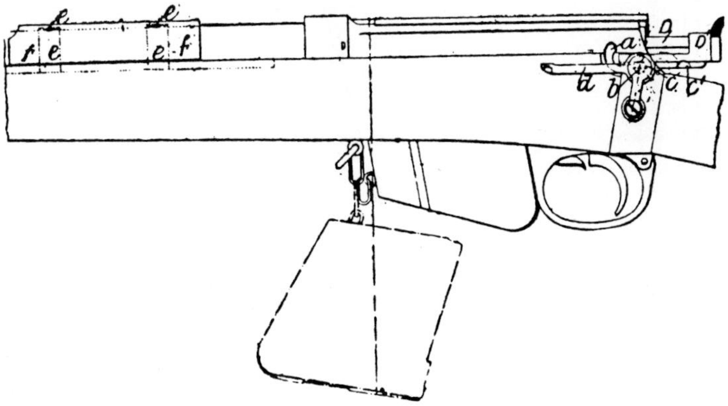

Protecting from heat of barrel. — A hand-guard for the barrel is formed of a wooden shield *f* attached by rivets to the spring clips *e*, which grip the barrel.

Breech actions, sliding breech-block ; magazines.—The gun has a bolt action, the firing-pin D having a cocking-piece D^1 provided with semi-cylindrical slots *c*, c^1. A transverse bolt *b*, operated by the lever *a*, is cut away and, when turned into the position shown in dotted lines, it engages with one of the slots and locks the firing-bolt in its forward or rearward position. The magazine is provided with a link hinge for suspending it from the receiver, and is made to contain a zig-zag column of cartridges. The cartridge carrier has side projections at its forward end engaging in vertical guides in the sides of the magazine and has an upwardly-extending flange at its rear end.

Sights.—The transverse bolt *b* carries a pivoted arm *d*, which, when raised, forms a peep-sight for use with the fore-sight described in Specification No. 13,335, A.D. 1887.]

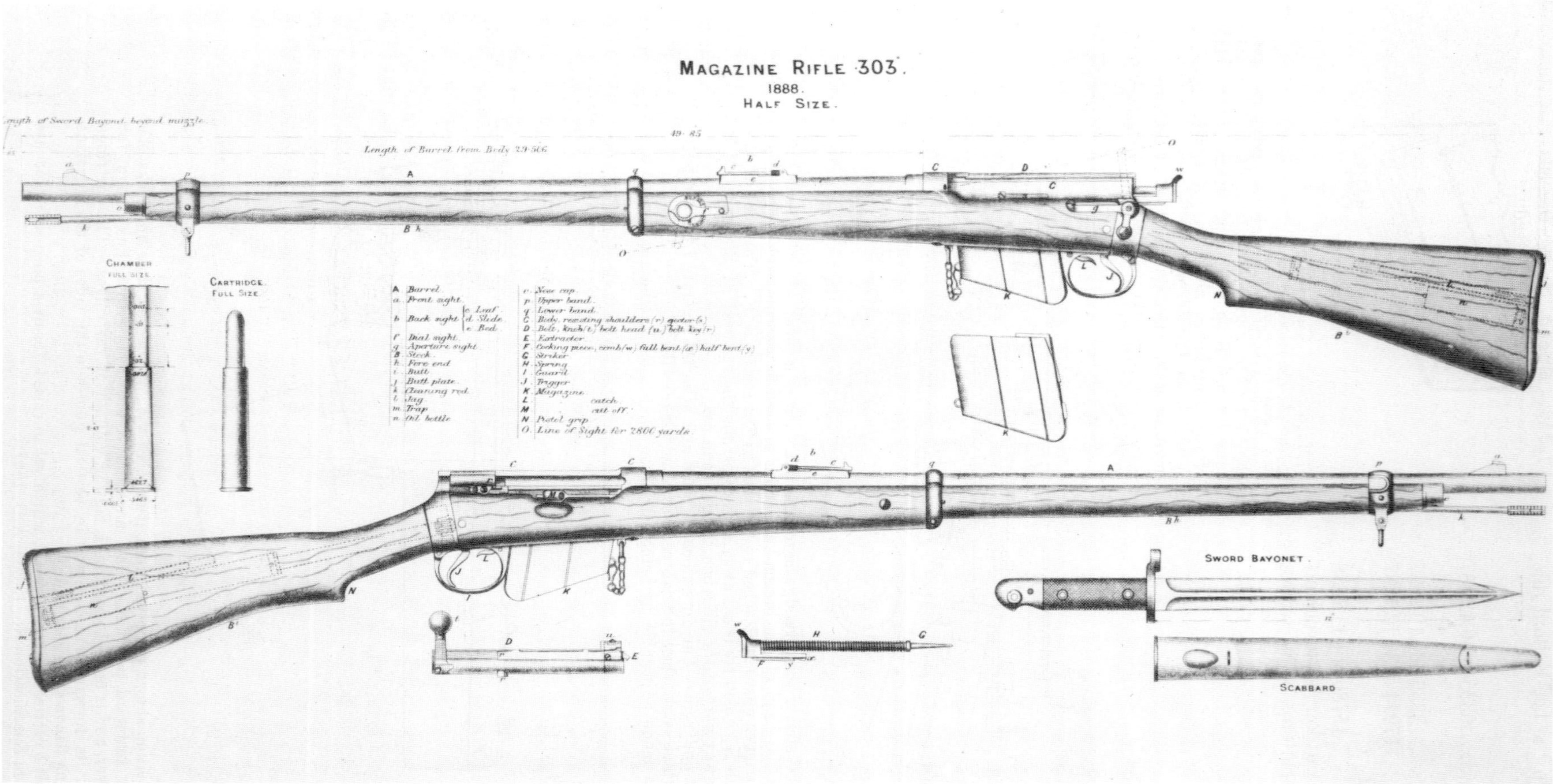

Reduced copy of the .303 Magazine Trials Rifle print as included in the Small Arms Committee reports. This also shows the .303 straight-cased Rubin cartridge and trials bayonet, from which the Patt. 1888 sword bayonet blade form was adapted. There was no safety catch and the magazine was retained by a link chain. The fore-end nosecap and band are Martini style. *Lord Cottesloe*

The 1888 Lee Trials Rifles made at Enfield were provisionally known as the "303-inch Pattern 1888" and incorporated improvements such as the gas escapes and bolt modifications of Lee, as well as certain Speed 1887 patents relating to the magazine, bolt dustcover, dial sight, trigger guard attachment and form of the magazine spring.

Some other improvements made at Enfield include the undercut half-bent, strengthening of the right side of the body, alteration to the comb of the striker, reduction in magazine size and its attachment by a small chain. This was to be the pattern for the 350 troop trials rifles; 50 carbines based on this model were also ordered for trial by cavalry. As the cartridge was still under consideration regarding the shape of the case, the bullet and the propellant, these trials rifles used the rimmed Rubin cartridge.

Action close-up, .303 Trials Rifle. *MOD Pattern Room*

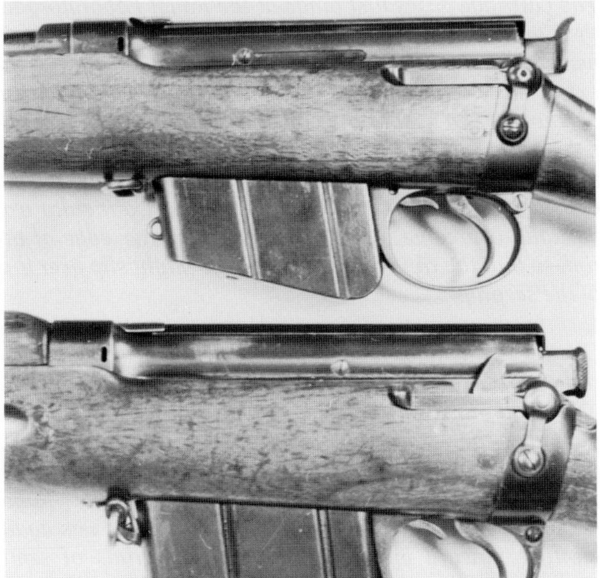

Comparative left view of Patt. 1888 Trials Rifle and ensuing service Mk I Magazine Rifle. *MOD Pattern Room*

Drawings and illustrations of the 1888 trials rifle show the unmistakable lines of the succeeding Lee-Metford and Lee-Enfield rifles, with considerable changes having been made since the .402 and .43 Spanish, Improved Lee models. The Martini style upper band, fore-end cap and bayonet fitting of the previous trials rifles were retained but a new bayonet was introduced for these trials which primarily resulted in the Pattern 1888 sword bayonet. This trials bayonet has a 12-in. blade, double-edged for its full length, and it weighs 1 lb; fixed on the rifle, the combined length was just over 5 ft.

Some other features special to this trials rifle action are the long ejector on the left side of the receiver, and the pivoting bolthead release catch arrangement on the right side of the receiver body.

The butt trap has provision for an oil bottle and jag, and the folding leaf sights are graduated from 200 to 1,600 yds with the volley sights ranged from 1,400 to 2,800 yds. The detachable magazine was linked to the rifle by a short chain, with a 7-round capacity. Each rifle was issued with a spare magazine in a black canvas pouch and for the trials and each rifle was provided with 350 rounds of ball and 17 rounds of dummy ammunition for these trials.

The specifications were given as being:

Length of rifle	49.85 ins.
Length of barrel, from body	29.506 ins.
Calibre	.303 inch
No. of grooves	7
Rifling	Metford
Twist of rifling	1 turn in 10
Average weight	9 lb. 2 oz.

Ball Ammunition:

Charge, Pellet of compressed powder	70 grs.
Bullet, Cased in copper	217 grs.
Weight of cartridge, complete	1 oz.
Weight of 100 rounds	6 lb. 4 oz.

(This black powder ammunition was especially provided for the trials as the pattern for adoption was not yet confirmed.)

A broad cross-section of troops in Britain as well as in Egypt, Canada, Ireland, Bengal, Bombay, Madras and various naval establishments at home and in Malta were issued with these trials rifles. Fifty magazine carbines were also issued for trial at Aldershot, Brighton, Ireland and Hythe, all within the British Isles.

Records show that 387 of these 1888 Magazine Trials Rifles were constructed at Enfield and an additional 50 fitted with the Morris magazine. A total of 51 magazine carbines was also assembled, one retained as a pattern and the remainder issued for trials; but no examples have been examined or reported in recent years to permit examination or provide an illustration.

L-R:
.450 Martini-Henry
.450 Gardner Gatling
.43 Spanish
.402 Early Enfield-Martini
.402 Enfield-Martini
.303 Rubin
.303 Rubin Rimless (from trials Rubini rifle)
.303 Mk I Cordite
M. Golland & H. Woodend

Some unusual conversions were produced, from the Martini-Henry to Magazine Lee; Enfield factory Ordnance Returns describe "2 Rifles, magazine, Patt. 1888, with Martini-Henry stock and action, straight tapered chamber" and "1 Rifle, magazine, Patt. 1888, with Martini-Henry stock and action, chambered for Rubin ringed cartridge." To effect such a conversion would have not have been easy, one would assume certain similarities to the Lee-Burton in the basic action style and other features of the 1888 trials rifle in the stocking-up and furniture, in .303 calibre.

Summaries of early reports issued by the War Office on 21st December 1888 indicate that the new rifle was considered to be handy, the magazine and action were very satisfactory and it was easily cleaned, although some considered the sword bayonet too short. The carbines showed up similarly well and were considered to be an improvement on the Martini-Henry in many ways.

Development of the cartridge was still proceeding and the progress made in this field is best described in this report from the Committee:

The Cartridge Case.

The original Rubin case was a straight taper (sometimes also described as "square neck") with a brass ring at the end for fixing the bullet. This design was retained by the Committee in ordering cartridges from the Royal Laboratory in the first instance. In experiments at Woolwich there were several instances of the ring blowing forward and lodging in the barrel. No such accident had come to the knowledge of the Committee out of many thousands of rounds fired. Nevertheless experiments were made to ascertain whether any danger existed and it was proved that a ring left in the barrel was liable to bulge it.

On 3rd July 1888, a barrel of one of the experimental rifles for trial by troops, burst at the muzzle whilst being tried at Enfield and a case without a ring was found near the firer. On examination, another barrel was found to be bulged. In presence of this defect the Committee endeavoured to obtain a necked case in order to obviate the necessity for a ring. The difficulty lay in getting the powder pellet into the necked case and the Superintendent, Royal Laboratory, brought the Morse cartridge before the Committee.

In this case the base was removable and the pellet could be inserted from the base. There was a difficulty, however, in making the case watertight and the Morse case was abandoned shortly afterwards upon the discovery of a method of necking the case after the insertion of the powder pellet. These cartridges were successfully tried on 15th February 1888 but as the bullet was not held in the case sufficiently tight, the neck was lengthened .1-in on the recommendation of the Committee on 19th July 1888.

The cartridge cases first made at the Royal Laboratory for the Committee were made with rims. The presence of a rim was objected to on two grounds: firstly, on account of the extra size required for the magazine; secondly, on account of the feed of the cartridges out of the magazine, it being found that a jam was sometimes caused by the rim of the top cartridge catching behind that of the second cartridge and carrying it forward. The last difficulty was partially overcome by bevelling off the outer edge of the rim, so that the rim of the top cartridge might slip over that of the second.

On 1st September 1887, the Committee advocated a rimless cartridge, similar to the original Rubin. The objections to it were: firstly, the danger of misfires; secondly, the feed of machine guns. A semi-rimless was therefore recommended, but this was objected to, 25th January, on manufacturing grounds, by the Royal Laboratory. On 26th January 1888, the matter was again brought forward by the Committee and trials of semi-rimless cartridges made at Woolwich were begun on 11th April 1888. The trials of this form of case were perfectly satisfactory in the view of the Committee, but on 1st November 1888, the Superintendent, Royal Small-Arms Factory, as a rifle manufacturer, expressed to the Secretary of State such a strong opinion against the case that it was finally abandoned from that date.

The Primer

The primer first suggested by the Committee, 15th October 1887, was that known as the French cap, which had given great satisfaction to them in the Rubin cartridge. On 20th October 1887, however, Colonel Barlow expressed a wish to make the caps in one piece instead of in three, as in the French cap. This cap was tried by the Committee and it was not strong enough to withstand the back pressure of the powder, the resulting blow-back disabling a large number of mainsprings and jamming the strikers. On 19th May 1888, the Royal Laboratory reported this difficulty overcome.

On 24th May 1888, Rubin ammunition to be used by the troops in trying the experimental small-bore rifle was tested at Enfield and a large proportion of blow-backs resulted. It was found that the Rubin Company had substituted a single for a triple cap. The Royal Laboratory succeeded in recapping the faulty ammunition with Winchester caps.

While the various trials were still under way, further improvements were being made with the rifles, which also incorporated:

- Combination fore-end cap and sword bar and new form of bayonet pommel (resulting in the Patt. 1888) designed by John Rigby;
- Lewes' frontsight and corresponding backsight;
- Simplified cartridge ejector;
- Removal of projecting thumb piece on the cocking piece;
- Slight changes to the magazine which now held 8 rounds;
- Safety catch fitted to left side of receiver, adjacent to the long range backsight;
- Substitution of a link loop for the magazine link chain;
- Fitting a wooden handguard behind the backsight;
- Substitution of the .303-in. cartridge for the Rubin round.

16,321. Rigby, J. Nov. 10.

Bayonets, attaching to gun barrels.—The nose cap D, on the end of the stock, is provided with a piece E, through which the head G' of the cleaning-rod G projects. The sword bayonet C is attached beneath the barrel, as shown, the cross-piece J fitting over the end of the muzzle, while the hilt fits over the projecting piece E, and is secured by a cross-piece on the spring bolt H catching into the recess I.

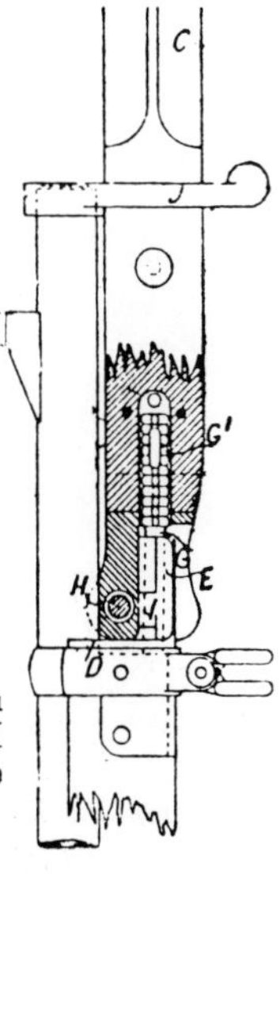

J.J. Speed's designs again figured in a number of these improvements, some of which were registered under Patent No. 15,786 of 1st November 1888. John Rigby's patent of the bayonet attachment was No. 16,321 of 10th November 1888.

A Mauser type action with Pattern 1888 Trials Rifle stock and magazine is in the Pattern Room collection and although no specific documentation is at hand, it appears to have been set up at Enfield at this time. It has certain similarities to the U.S. Model 1903 Springfield even though it predates the Springfield by about 15 years.

This experimental hybrid is in .303-in. calibre but chambered for the Rubin cartridge, sighted for black powder to 1,600 yds with the numbers on the right side of the leaf (like the 1888 Trials Rifles). The example

Above & below: Experimental "Mauser-Metford" rifles. Note the Martini and Metford profile knox forms of the barrels. *MOD Pattern Room*

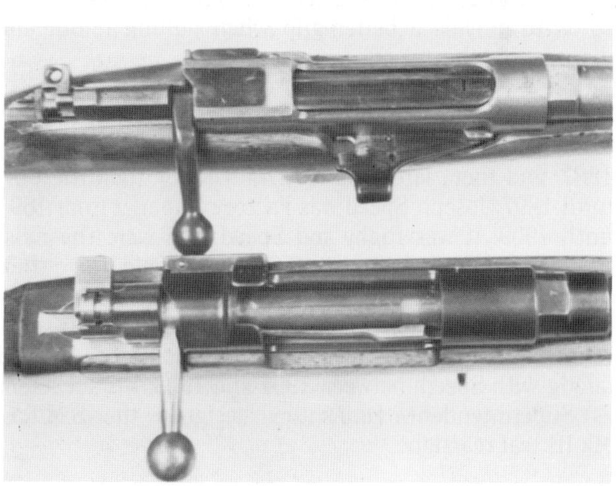

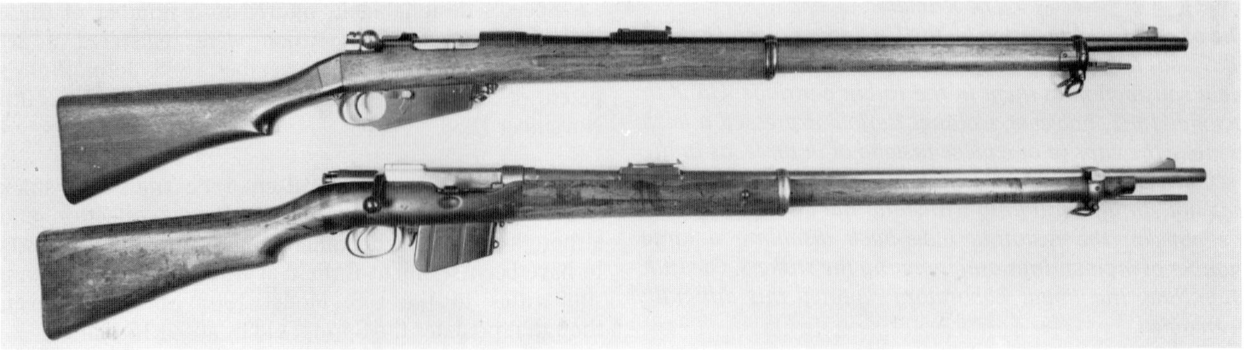

"Mauser-Metford" rifles (in reverse order to the illustrations on the previous page). Note the fore-end and front band assembly types. *MOD Pattern Room*

illustrated incorporates Speed's magazine improvements and cut-off and the long range sights. It has 8-groove Metford rifling (the 1888 Trials rifle has 7 groove) and the knox form is marked "M.M.2" which may be some clue as to its designation (maybe even Mauser-Metford?).

The other "Mauser-Metford" rifle shown has a Mod. 1896 type bolt, two-piece stock and clip-loaded continental style magazine with indicator for the number of rounds. This post-dates the other variant by a few years; it has Lewes sights and other Mk 1 Lee-Metford features.

Prior to the introduction of the Magazine .303 rifle, one of the last experimental models was a modified 1888 Trials rifle which incorporates some of the improvements listed in the previous paragraphs. This particular rifle has an experimental safety catch mounted on both sides of the action and from the shape and mode of operation of the left hand side catch, it is obviously a predecessor of the safety catch on the Mk 1 Magazine Rifle. The safety mechanism on the right side is separate from that on the left and it is automatically disengaged when the bolt is drawn to the rear, working on a double-hinged principle.

One of the 1888 Trials rifles had the magazine cut-off knob modified so as to form a hinged ring which was positioned in a small recess in the fore-stock in its folded position, when not in use. This ring has been chequered so as to provide a better grip when pulling it out and disengaging the cut-off.

Col. Arbuthnot, who had been Superintendent at R.S.A.F. Enfield since 1880, was replaced by Col. King-Harmon in 1887, and then, later in 1887, John Rigby took the post until 1897. Joseph Speed was Factory Manager from 1891 until 1909; it was Rigby and Speed who were the most prominent innovators at R.S.A.F. Enfield for their improvements to the British Service Lee. A later Superintendent, Col. H. S. Watkin, contributed some refinements to the following Short Magazine Lee-Enfield (along with Speed) between 1901 and 1903; his successor as Superintendent, Dr. Ashton, designed the S.M.L.E. Mk III leaf rearsight.

Above & below: Early Magazine Lee-Metford rifle with experimental safety catches. *Tower of London*

Experimental folding magazine cut-off knob on modified 1888 Trials rifle. *Tower of London*

THE MAGAZINE LEE-METFORD

CHAPTER 3

On Sunday, 1st December 1889, the Magazine Rifle Mark I was finally announced in the List of Changes, with an approval date of 22nd December 1888; some of the appurtenances were approved later, on 12th August 1889. It is worthy of note that at first, the official name for the new rifle was not "Lee-Metford," but "Magazine Rifle Mark I"; the authoritative text from the List of Changes reads:

Para 5877:-
Rifle, Magazine (Mark I).
Bottle, Oil (Mark I).
Jag (Mark I).
Scabbard, Sword-Bayonet, Pattern 1888 (Mark I).
Sword-Bayonet, Pattern 1888 (Mark I).
Protector, Front Sight, Rifle, Magazine (Mark I) and Muzzle, Naval.

The above-mentioned articles have been sealed to govern manufacture.

The breech action of the rifle is on the "bolt" system. The cocking piece is so arranged that the rifle can be carried at half cock. Covers are fitted to the bolt and bolt head to protect the action from sand and dirt.

A "safety catch" is fitted on the left side of the body, the pulling back of which, when the rifle is at full cock, prevents any effect being caused by pulling the trigger, and, when springs are "eased," locks the action and prevents the bolt from becoming accidently opened.

Barrel & Rifling:

Length of barrel	*30.2 in.*
Calibre	*.303 in.*
Rifling	*Metford system*
Grooves, Number	*7*
Grooves, Depth	*.004 inch*
Width of Lands	*.023 inch*
Spiral, left hand	*1 turn in 10 ins. or 33 calibres*

Magazine: The magazine consists of a sheet-steel box, inserted from under the body, in front of the trigger guard, through an opening in the body. It is held in position by a spring in the body engaging in a notch in the magazine. It will contain eight cartridges and may be filled when in position in the rifle, or when detached, by inserting the cartridges one by one. A spring at the bottom of the magazine presses upwards a moveable platform, forcing the column of cartridges also upwards.

A "cut-off" is fitted to the right side of the body, which, when pressed inwards, stops the supply of cartridges from the magazine, so that the arm may be used as a single loader. When the "cut-off" is pulled out, the lower edge of the bolt, on its being driven forward, engages the top edge of the uppermost cartridge in the magazine and forces it into the chamber, and so on, until the magazine is emptied.

The magazine can be removed by pressing a small lever inside the trigger guard. One magazine is attached to each rifle, being secured from loss by a chain link; a spare magazine is also issued with each arm.

Stock: The stock, like that of the Martini-Henry, is in two parts, the fore-end and the butt. Under the hinder part of the small of the butt, is a projection forming a so-called "pistol grip." The butt is secured to the body by a "stock bolt." The butt plate is fitted with a trap, giving access to the unoccupied portion of the stock bolt recess, which is arranged to contain an oil bottle and jag.

All rifles of this pattern will have the same length of butt, instead of butts being "long" and "short" as heretofore. The butt is 3/8 inch shorter than that of the "short butt" Martini-Henry rifle.

Nose-cap: This is fitted with a bar on top for attachment of the sword-bayonet underneath the barrel.

Cleaning rod: The cleaning rod is screwed at one end with a left hand thread to fit the jag; when the rod is carried in the rifle this thread is screwed into a hole tapped in the body to receive it, the other end of the rod is recessed and slotted for the convenience of screwing it into the jag or body.

Piling swivel: As the cleaning rod when in its place, does not project sufficiently for use in piling arms, this operation is effected by means of a divided swivel, which is attached to the upper band in addition to the ordinary upper swivel.

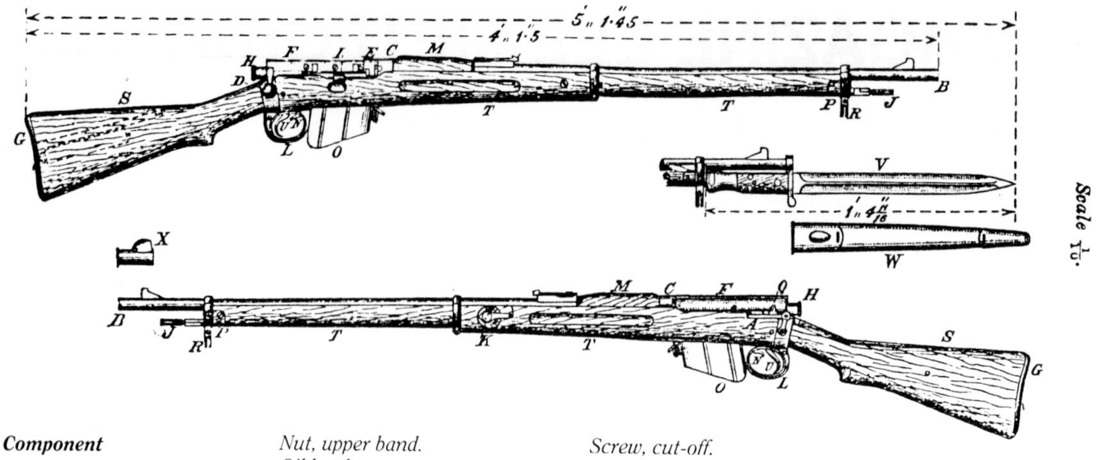

Ref. ***Component***

A. Aperture sight.
B. Barrel.
C. Body.
D. Bolt.
Bolt head.
E. Bolt head cover.
F. Bolt cover.
G. Butt plate.
Butt plate trap.
Band, upper.
Band, lower.
H. Cocking piece.
I. Cut-off.
Collar, guard.
J. Cleaning rod.
K. Dial sight plate.
Dial sight pointer.
Extractor.
L. Guard.
M. Hand-guard.
Jag.
N. Magazine catch.
O. Magazine case.
Magazine platform.
Magazine upper link.
Magazine spring.
P. Nose cap.
Nut, upper band.
Oil bottle.
Pin, bolt head cover.
Pin, sight, axis.
Pin, stop, upper band, and oil bottle.
Pin, lower band.
Pin, trap, axis.
Pin, magazine catch.
Pin, trigger.
Rivets, spring, hand-guard. (4)
Rivets, fore-end.
Spring, extractor.
Spring, cut-off.
Spring, retaining.
Spring, main.
Spring, sear.
Spring, magazine.
Spring, butt-plate.
Spring, dial sight.
Spring, aperture sight.
Spring, sight.
Screw, lower band.
Screw, guard, front.
Screw, guard, back.
Screw, guard, swivel.
Screw, bolt cover. (2)
Screw, extractor.
Screw, cut-off.
Screw, cut-off spring.
Screw, keeper, striker.
Screw, sear.
Screw, sight cap.
Screw, sight bed.
Screw, dial sight, spring.
Screw, dial sight, pivot.
Screw, spring, aperture sight.
Screw, spring, butt.
Screw, spring, sight.
Screw, swivel.
Screw, nose cap.
Screw, ejector.
Screw, bolt head.
Screw, butt plate. (2)
Striker.
Stock bolt.
Q. Safety-bolt.
Sight slide.
Sight leaf.
Sight cap.
Sight, front.
Sight bed.
Sear.
Swivel, guard.
Swivel, band.
R. Swivel, piling.
S. Stock-butt.
T. Stock, fore-end.
U. Trigger.
Washer, dial sight.
Washer, hand guard. (4)
Washer, rivet, fore-end. (4)
V. ***Sword-bayonet, Pattern 1888.***
Blade.
Bolt.
Bolt spring.
Bolt nut.
Crosspiece.
Grips, wood. (2)
Pommel.
Rivets, brass. (2)
Washers, brass, small. (4)
Washers, brass, large. (4)
W. ***Scabbard, sword-bayonet, Pattern 1888.***
Leather, blocked and blackened.
Chape, steel.
Locket, steel.
Locket, springs. (2)
Locket, stud.
Tips, chape.
Mouthpiece.
Laces, iron wire.
X. Protector, front sight.

Hand-guard: A wooden hand-guard is fixed over the breech end of the barrel to protect the hand when the barrel becomes hot. It is held in its place by two steel springs, which clip round the barrel.

Sighting: The rifle is provided with two sets of sights. The foresight and the backsight are fitted in the usual position on the barrel. The foresight is a square block with a vertical cut through it, showing a fine line of sight. Aim is taken by fitting this square in a corresponding square notch in the backsight, so that lines of equal width may be seen on each side of it, and aligning the central line of light on the point to be hit. The lowest or "fixed sight" is that for 300 yards. Using this sight, a head and shoulders can be hit at any distance between 375 and 150 yards, whilst a 6-ft figure can be hit up to about 500 yards, without in either case aiming off the figures. The highest graduation is for 1,900 yards. The rifle is also fitted with extreme range sights. The front sight which is also called the dial sight, is graduated from 1,800 yards up to 3,500 yards. It consists of a bead fixed to a revolving index hand. The index is set to the required distance, which is marked on the edge of the dial plate, and aim is taken by aligning the bead on the object through a circular hole in the "aperture sight" which is raised into a vertical position when in use. Both these sights are on the left side of the rifle.

Jag: The jag is of steel, 7.25 inches long; it is tapped at one end to receive a cleaning rod, and slotted at the other to receive the material used for cleaning.

Oil bottle: The oil bottle is of brass, fitted with a screwed stopper. A pin is attached to the stopper for applying oil to the parts of the action of the rifle.

Sight protector: This is issued for naval service only. It is of brass and forms a shield for the foresight and muzzle.

Sword-bayonet: This is of the form shown in the drawing. It is two-edged and is fitted with wood "grips." It is fixed under the barrel, the bar and the head of the cleaning rod fitting into a recess in the hilt. It is secured by a stud and spring in the usual way.

Scabbard: The scabbard is of black leather, steel mounted, the locket being fitted with springs.

The sword-bayonets and magazines are interchangeable but owing to the necessity for allowing a manufacturing tolerance, some will fit the rifle closer and better than others. For this reason armourers should select those that are most suitable before numbering them.

There are 88 different parts in the Mk I Magazine Rifle for which 950 machines were employed in production; this includes the screws and pins, but excludes the various rivets and appurtenances. First proof of the barrel was made before rifling and the breech was closed by a screwed plug with an axial touch hole and loaded with 110 grains RFG black powder and 350 grain lead slug. The second proof was applied after the rifling was cut and the body and bolt assembled onto it; effectively a proof of the action. A cartridge was used for the second proof, with a pellet and bullet calculated to exert a pressure of 24 tons per sq. inch. A third proof consisted of firing 5 service cartridges from the assembled rifle into a sand butt to ascertain that the mechanism functioned properly. After this, a percentage of the finished rifles were fired at 500 yards in accuracy tests, fitted into a machine rest.

Comparison; 1888 Trials Rifle (top) and early Mk I Magazine Rifle (bottom), left view of action. *MOD Pattern Room*

There are few of the original Mk I Magazine Lee-Metford rifles remaining today in their original form as most were updated to the Mk I* pattern or later converted to various training rifles. Parts special to the Mk I rifle which differ from those on the Mk I* are: body, front and rear sights, safety catch, handguard, butt, mainspring, magazine spring, dial sights and piling swivel. More related details will be found in the specifications of the Mk I* Magazine Lee-Metford rifle, a little further on in this chapter.

Differences may be noticed in the early production Mk I Magazine Rifle handguard; the first type has squared ends at the front of the handguard (also evident in official drawings of the early Magazine Rifle, such as the illustration in the List of Changes), then it was found that the handguard was prone to splitting so the wood thickness was increased. However, the extra height of the handguard due to this increase obscured the sights at minimum ranges, so a sighting groove was cut along the top of the handguard, in line with the sights.

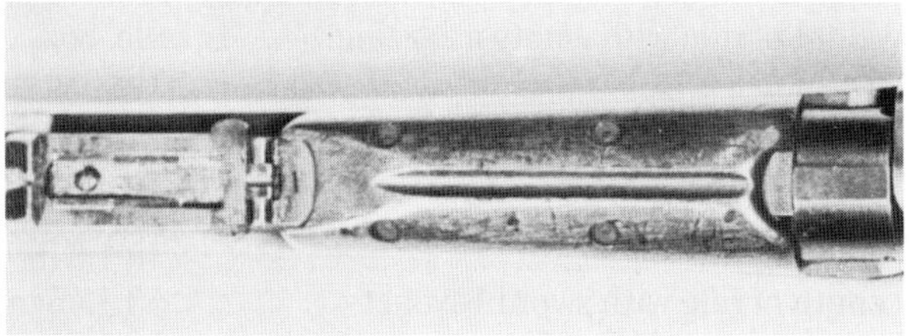

Grooved handguard, early Mk I Rifle. *MOD Pattern Room*

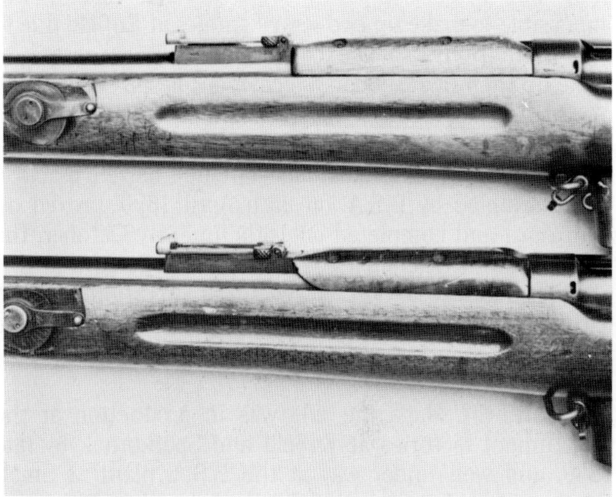

Early Mk I Rifle with square ends at front of handguard, changed to permit removal with fingers. *MOD Pattern Room*

Then it was discovered that the main reason for the handguards cracking was due to the troops removing the guard with the bayonet or metal implement, in order to clean underneath. So, the front edges of the handguards were rounded off to enable its removal by hand and the high curved top of the handguard was smoothed off to the level of the new sighting groove. The front flat spring was also shortened soon afterwards, recorded in the List of Changes § 6760, approved 19th December 1890, with the introduction of the Mk I* model.

Accuracy problems were caused with the original sight graduations and ranges by certain theoretical miscalculations of range scales; originally 500 to 1,900 yards on the backsight and 1,800 to 3,500 yards on the long range dial sight, in anticipation of the smokeless powder development. When the smokeless, cordite loads were introduced in 1891 the true performance was established and these scales altered accordingly. Blackpowder loads were introduced with the new .303 round as an interim measure and the true backsight graduations for black powder were up to 1,600 yards compared with 1,800 yards for cordite loads. Because of the intended use of smokeless propellant cartridges in the Lee-Metford rifles, they were fitted with the 1,800 yard rearsight.

Weights and dimensions of the new rifle were recorded as:

Weight of rifle with empty magazine	9lb. 8oz.
Weight of sword-bayonet	15.5 oz.
Weight of sword-bayonet scabbard	4.75 oz.
Weight of empty magazine	4.75 oz.
Weight of filled magazine	13 oz.
Length of rifle	4ft 1.5 in.
Length of rifle with sword-bayonet	5ft 1.45 in.
Length of sword bayonet	1ft 4.7 in.

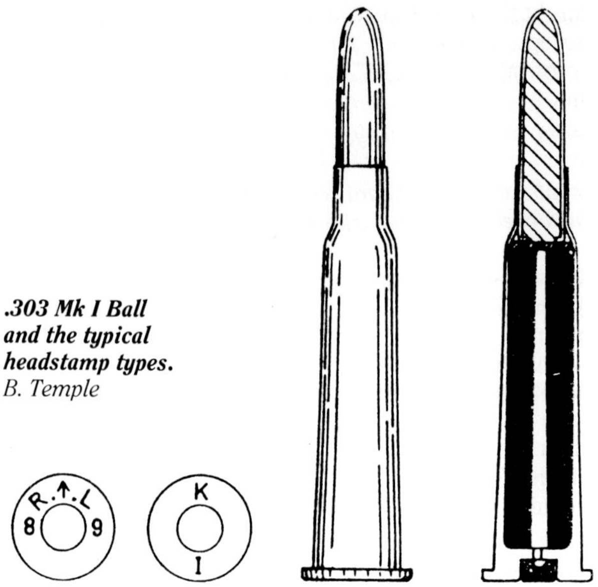

.303 Mk I Ball and the typical headstamp types.
B. Temple

Sample or pattern rifles were forwarded to the Birmingham Small Arms and London Small Arms factories in January 1889 but these companies were advised that it was impossible to make up two sets of gauges at Enfield due to time limitations, so the companies were requested to make up their own sets on a repayment basis. Orders were placed for 200,000 new magazine rifles with B.S.A. and L.S.A. at a price of £5/16/- each for the first 50,000 and a reduced price for the remainder. The first barrelled action was assembled by B.S.A. on 25th April 1890, proved on 30th April and completed on 16th June. By October, the weekly output reached 1,000 and it was ordered that this figure be reduced to 800. The eventual alteration to Mk I* pattern was advised in May 1891, although the definite production changeover date was not so precise.

Therefore, the Magazine rifle was in production at the Government factories at Enfield and Sparkbrook by late 1889, and well under way at the B.S.A. plant at Small Heath and by L.S.A. at the Old Ford, London. The main reason why so few of the new rifles were delivered by the end of 1890 was because minor improvements were still being applied, mostly related to the handguard, and these changes also involved amendments to the patterns.

The new .303 cartridge, Mark I, was announced in the same monthly issue of the List of Changes as the Mk I Magazine Rifle, and approved on 2nd February 1889.

Para 5883:–

Cartridge, S.A. Ball, Magazine Rifle (Mark I)
Solid case, 303-inch.

A pattern of the above-mentioned cartridge for the magazine rifle, L.o.C. 5877, has been sealed to govern supplies.

The case is made of solid-drawn brass. A cap chamber is bored in the base, to receive the cap and anvil, which are also made of brass. A central fire-hole communicates from the cap chamber to the interior of the case.

The charge weighs 71.5 (+ or – 2.5) grains of the ordinary description of gunpowder, and is pressed into the form of a pellet with both ends slightly rounded.

The cartridges are packed, bullet and base alternately in opposite directions, in bundles of 10, in dried brown paper wrappers, tied with string, with a slip of thin paper passing round each cartridge.

Weights &c.

Weight of bullet	*215 grs.*
Weight of case	*164 grs.*
Weight of powder charge	*75.5 grs.*
Weight of bundle of 10	*about 10.5 ozs.*

A small-arm ammunition box will contain 1,100 cartridges, and when so filled its weight is about 85 lb. 8 oz.

A smokeless propellant for the .303 cartridge was not approved until 3rd November 1891 and introduced in the List of Changes, § 6735. A new explosive called "Cordite" was used, developed by Abel, Dewar and Dellner; it consisted of nitro-glycerine, guncotton and mineral jelly. Characteristic of other smokeless powders, it gave less fouling, higher velocities and lower pressures, but it had one major drawback, greater heat which resulted in a reduced barrel life. These ramifications are enlarged upon in the subsequent chapter on the Lee-Enfield.

The Special Committee on Small Arms met on 17th February 1890 for the last time. Col. H.T. Arbuthnot had been replaced as Superintendent R.S.A.F. Enfield by John Rigby. Joseph Speed, responsible for many of the Lee rifle improvements, was appointed Factory Manager soon afterwards. Arbuthnot had disagreed many times with Sir Henry Halford, a prominent Committee member and close friend of William Metford, but Arbuthnot's career came to an hasty conclusion after he was reputedly observed accepting money at the Nordenfelt offices. Nordenfelt was a private contractor engaged in the manufacture of machine guns and other ordnance, often in competition with the Government factory; considering the powerful enemies that Arbuthnot had made, such an embarrassing predicament concluded Arbuthnot's Small Arms Factory career.

Arbuthnot's contribution to the Magazine Rifle is generally all too conveniently ignored; he stood fast in his preference for the Lee when the remainder of the Committee was in favour of adopting the Owen Jones, and he was responsible for the recall of the 1883 Committee to clarify their recommendation. His practical, mechanical ability and foresight at this time was most fortunate indeed although his successor, John Rigby, played a more prominent role.

A safety trigger proposed by Rigby was considered to be worthy of trial, but later rejected. Rigby also devised some special cartridges for lessening the recoil of the service rifle, although in tests it was reported that there was a loss in velocity but no apparent difference in recoil. The Rigby patent acknowledgment is often seen on commercially produced nosecaps for the Lee-Metford and Lee-Enfield rifles.

The Committee, in their final report in February 1890, indicated that two specimen magazine carbines had been submitted, one each for cavalry and artillery, but that no experiments had been conducted to ascertain their suitability. They also stated that the question of propellant and bullet for the .303 cartridge presented grave difficulties and that the erosion and metallic fouling had not been overcome. With ongoing European development in this regard, it was intended to introduce a smokeless propellant along with the new Magazine Rifle, but due to a variety of problems, the black powder load was adopted until such time as a satisfactory smokeless powder could be developed and confirmed as suitable.

In mid-1890, a new "Committee on the Magazine Rifle" was formed and its first meeting held at the War Office on 5th August 1890, with many new members. In fact, only one member of the previous committee, Col. Slade, now Commandant of the School of Musketry, sat on the board. Reports from the new Committee indicate much less disagreement amongst the members, although in fairness to the previous Committees, the new body was not responsible for such wide-ranging decisions because the new rifle and ammunition had already been defined.

Some minor changes to the Magazine Rifle were announced in the List of Changes during this period. It was found that the sharp edges of the cleaning rod could easily score the bore and so it was ordered that the screwed end be rounded off with a file; § 6076 with an approval date of 18th June 1890. On 29th October, the jags were ordered returned to store and pull-throughs issued in their stead.

The spare magazines were also to be returned to store with the jags and the introduction of the pull-through resulted in another small change to the rifle; plugs were ordered to fill the pin hole through the butt, one on each side, after the removal of the oil bottle stop pin. It was later announced that there was no need to shorten the jag hole in the butt by plugging it; this hole was then used to accommodate the weight of the pull-through and was drilled shorter in future rifle manufacture. List of Changes references for these items are §§ 6234 and 6235, approved on 29th October and 3rd October 1890, respectively.

About 13,000 of the new rifles had been produced by November 1889 and the Martini-Henry rifles of the Aldershot division (nine battalions) were exchanged for Lee-Metfords in December 1889, followed by the Guards in February of 1890. The only further progress regarding issue of the new Magazine Rifles was that by November 1890, four rifles were issued per company to the remainder of the Infantry for training purposes and general familiarisation.

Therefore, the new rifle had been on issue to some units for about six months by August 1890 when reports were called in from various Officers Commanding. Only at Hythe, Aldershot and in the Brigade of Guards, was the rifle on general issue, supplied with ball cartridges; those rifles supplied for instructional purposes were not issued with ammunition. Problems were being encountered with some .303 ammunition, that marked "E.B." and made in January 1890 was found to shed its nickel jacket in the barrel and the ammunition on the whole was indifferent, although the Government-made batches were better.

The most objectionable point in the accounts and summaries of the new rifles related to the Lewes sights. Lt.-Col. Lewes of the Northamptonshire Regiment devised these sights which were introduced upon the recommendation of the Committee but the Naval authorities did not approve of them. They only withdrew their objection on the undertaking that the subject would be re-opened when the following Mark of the magazine rifle should be contemplated by the Committee or the War Office.

These particular sights are quite unusual and consist of a block foresight with a slit instead of a blade and this was aligned with a squared "U" backsight. It was complained

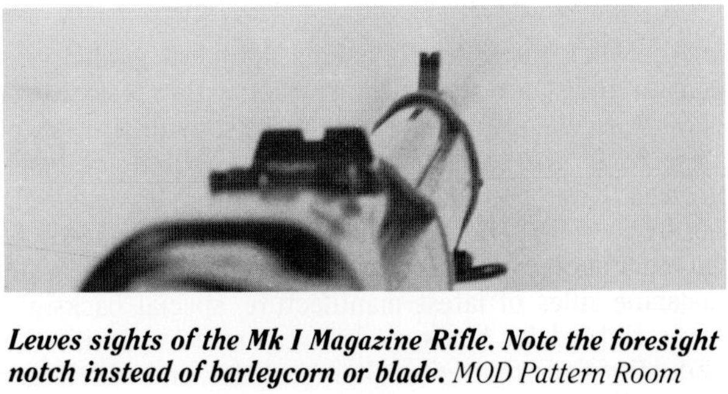

Lewes sights of the Mk I Magazine Rifle. Note the foresight notch instead of barleycorn or blade. *MOD Pattern Room*

that they were very liable to damage and inferior to the system on the Martini-Henry. As a result, and keeping in mind the objection of the Naval authorities, the Committee recommended that the Martini-Henry system be reverted to, with the probability of some slight changes to the form of the backsight notch as well as the barleycorn foresight.

Lewes made very vocal objection to this condemnation and attended a Committee meeting on 11th December 1890 at which he claimed that these sights had not been properly made to his specifications (even though a seal had been attached to a sample rifle with his signature, to indicate that he judged them to be correct) and that he had been misrepresented. However, the Lewes sights were abandoned and he was awarded a royalty of £500 for the use of his patent, No. 14,093 of 1st October 1888.

In view of comparative trials conducted with other more conventional barleycorn type sights, the decision to discard Lewes' sights was fully justified. These tests were conducted at Hythe over three days and reported on 8th October 1890. For the trial, fifty Mk I Magazine Rifles were set up with the Martini-Henry type sights and tested against a similar number of service (Lewes sighted) Magazine Rifles. The trial indicated that the barleycorn system 'was superior and considering the questions of expense, difficulty of manufacture, liability to injury and difficulty of repair, the Lewes sight appeared even less satisfactory.

Other service complaints were made relating to the sharp edges of the backsight cap and also about the graduations, so it was decided to round off the offending edges on the cap and mark the even numbers along one edge of the leaf, with the odd numbers stamped on the other edge.

It was recommended by the Commandant of the School of Musketry that the backsight bed graduations be altered in order that the fixed sight should be at 500 yds; the Committee concurred and decided to incorporate this in the eventual Mk II rifle. Approved on 25th August 1890, although it was pointed out that on the proposed Mk II rifle, the bed had a gently sloping ramp so as to enable ranges from 200 to 500 to be sighted with the leaf in the "down" position, by moving the backsight slide. It was also requested that the backsight notch be more open and shallow than that on the Martini-Henry rifle.

On the Mk I Magazine Rifle, backsight leaf graduations were intended for smokeless powder with a velocity of 2,250 ft./sec. and requests were made for substitution of these backsight leaves with patterns sighted for black powder. It was pointed out also that these leaves would require new caps, but as it had been decided to re-equip the regiments of the Mediterranean garrisons with magazine rifles of latest manufacture, special backsight leaves sighted for blackpowder were made up, with the graduations and figures on alternate sides, and with larger figures. The height of the leaf cap had also been reduced to make sighting correct for the 300 and 400 yard graduations.

The Lee Arms Syndicate in the United States had developed an improved magazine by late 1889, this held ten rounds in a staggered pattern and resulted in a wider but shorter magazine case. This also facilitated easier loading of the magazine, especially when fitted on the rifle and this was the basis for the proposed Mk II model. Lee was awarded a British patent on his new magazine on 15th May, 1889, Pat. No. 8117.

Directives were sent on 31st January 1890, for three rifles to be prepared and fitted with magazines using this new system, for trials. Tests had already been carried out by the Inspection Department, Woolwich and one rifle fitted with such a magazine was being used in proof work. A new magazine spring was also tested, being a flat type in the shape of a "C," the style eventually adopted and carried on through the Magazine Lee-Enfield series. This was another of Lee's improvements, although it was originally submitted to the Committee for the 8-round Mk I magazine and had then to be adapted for the 10-round model.

There had been some trouble with breakages of the service zig-zag coil magazine springs. So a stronger spring with an additional coil was proposed for the 8-round magazine and recommended for immediate adoption for Mk I rifles, approved on 22nd August 1890. From this date, new springs were plunged into oil after tempering, so as to produce a protective coating.

A high breakage rate occurred with the piling swivels and so a strengthened pattern was approved on 28th August 1890. These were still unsatisfactory, because the narrow opening made the unpiling of arms difficult; after 24 piling swivels with wider openings were tested at Hythe, a new pattern was approved on 19th December 1890 and the previous pattern was gradually recalled.

The bolthead was another feature of the new rifle which came in for criticism, in particular those of early manufacture where it appeared that they had not been properly tempered or had been violently abused. The most common fault was breakage of the bolthead shroud or cover. Several modes of bolthead improvements were proposed:

1. By the Director General of Ordnance Factories; strengthened at the dovetail, with the bolthead cover omitted.
2. Sparkbrook pattern, submitted by D.G.O.F. as an alternative to No. 1.
3. By Mr. Tookey, Small Arms Inspection Dept.
4. By Mr. Deeley of Westley Richards & Co.
5. Also by Mr. Deeley of Westley Richards & Co.
6. Todd's pattern submitted by War Office, 29th December 1890.
7. By R.S.A.F. Enfield.

Pattern 1 was that incorporated in the Mk II rifle as first submitted, but the type eventually chosen for the test models was pattern No. 7 which was considered to be the best by far; 100 experimental Mk II rifles currently under manufacture were ordered fitted with these bolts. A transverse pin was still used to secure the bolthead into the bolt body.

It was found that bolthead screws were breaking or loosening in many cases and to remedy this, a screw of mild cast steel, tempered and hardened at the point, was introduced in May 1890; this screw was dispensed with on the Mk II model in production. Bolt cover screws were similarly reported as breaking or working loose but this was rectified by the new type of dustcover proposed for the forthcoming Mk II rifle.

The bolt mainsprings, now better known as firing pin springs or striker springs, were very prone to breakage and many of these were later found to be rusty. It was ordered that they be oiled when issued, as well as on each occasion of their examination, the approval date for this being 18th August 1890. Improved methods of manufacture were later implemented and the mainspring problem was thus generally solved.

Another potential defect had also been found, if the cocking piece were accidentally turned while the bolt was to the rear and an attempt made to close the bolt, a jam would occur. In this instance, the Director-General of Ordnance Factories proposed to cut away a small amount of metal from the body and this was also applied to the proposed Mk II model.

Three bolt variations designated Bolts A, B and C were submitted for testing on 25th and 29th July, indicating that the question of the boltheads must not have been resolved. These three different bolts featured the new boltheads, although pins were still used to secure the bolthead to the bolt body.

Some of the other considered deficiencies of the Mk I rifle related to the fore-end, safety catch and provision for regimental markings. It was felt that the fore-end was not sufficiently full when grasped by the hand during firing, so it was decided to discontinue the wide, horizontal finger groove along each side.

While the jag had been decided against in favour of a pull-through, a new rod was designed so that two might be screwed together for the purpose of driving out cartridge cases in instances when the extractor failed. Thus, its technical name was altered from "cleaning rod" to "clearing rod"; the new rod was applied to the following Mk II and subsequent models. Although the pullthrough had been introduced in 1890, the nomenclature was not changed until early 1894.

Some concerns were also expressed about the safety catch on the Mk I rifle; it was not regarded as a necessity and it was also found that the slots cut in the cocking piece for the catch weakened it, so it was abolished on 31st December 1890. Mark I rifles had the safety catch converted to a plug by removal of the stop pin, lever and half-moon cam; new production Mk I* rifles have no provision for safety catch. The Mk I rifles also have a slight recess in the left side of the butt stock at the socket, to allow for the safety catch thumbpiece when in the engaged position. This also was omitted on the Mk I* version and the proposed Mk II was to have no safety catch as a direct result of the determination.

Previously, unit or regimental marks were stamped or engraved on the steel buttplate of the Mk I rifle but it was found that these markings became obliterated largely as a result of drill work in which the flat of the buttplate constantly came into contact with the ground. It was recommended that a brass disc be inserted in the right side of the butt, upon which these letters and figures might be marked. In principle, this was approved for Mk I arms on 10th August 1890 and a pattern of disc was approved on 30th September 1890. To allow for this, the Mk II rifle was to have a lengthened heel on the top of the buttplate.

Because of the prevalent rusting of barrels at the muzzle, in consequence of soldiers putting in pieces of rag to exclude dust and foreign matter, it was decided to issue sight and muzzle protectors. Approval was given on 3rd January 1891 for a sight cover which fitted over the muzzle and prevented extraneous materials from entering the rifle bore from the muzzle end.

The nomenclature of the Magazine Rifle was changed to "Magazine Rifle, Lee-Metford" on 8th April 1891, in acknowledgment of the contributions of James Lee and William Metford. So it was from this point in time that the rifle was known as the Lee-Metford, in keeping with earlier nomenclature such as Snider-Enfield, and Martini-Henry. This was recorded in § 6476 of the List of Changes also.

The wooden handguards were still being tested as a result of the proposed improvements, and the Director General of Ordnance submitted a modified pattern in which the guard was already in two pieces, divided down the centre and held together by the springs. However, as mentioned previously, this problem was largely overcome by the rounding off of the leading edge at each side to permit removal of the handguard with the fingers. With the approval and introduction of this improved pattern, the modification was applied to all of the Mk I rifles in store, being effected at Weedon. It was also applied to new manufacture and adopted for the following patterns of Magazine Lee-Metford and Lee-Enfield rifles; thus the early style of handguard with squared front ends is quite rare.

Confirmation of the considerable number of alterations made to most of the Mk I rifles may be found on the left

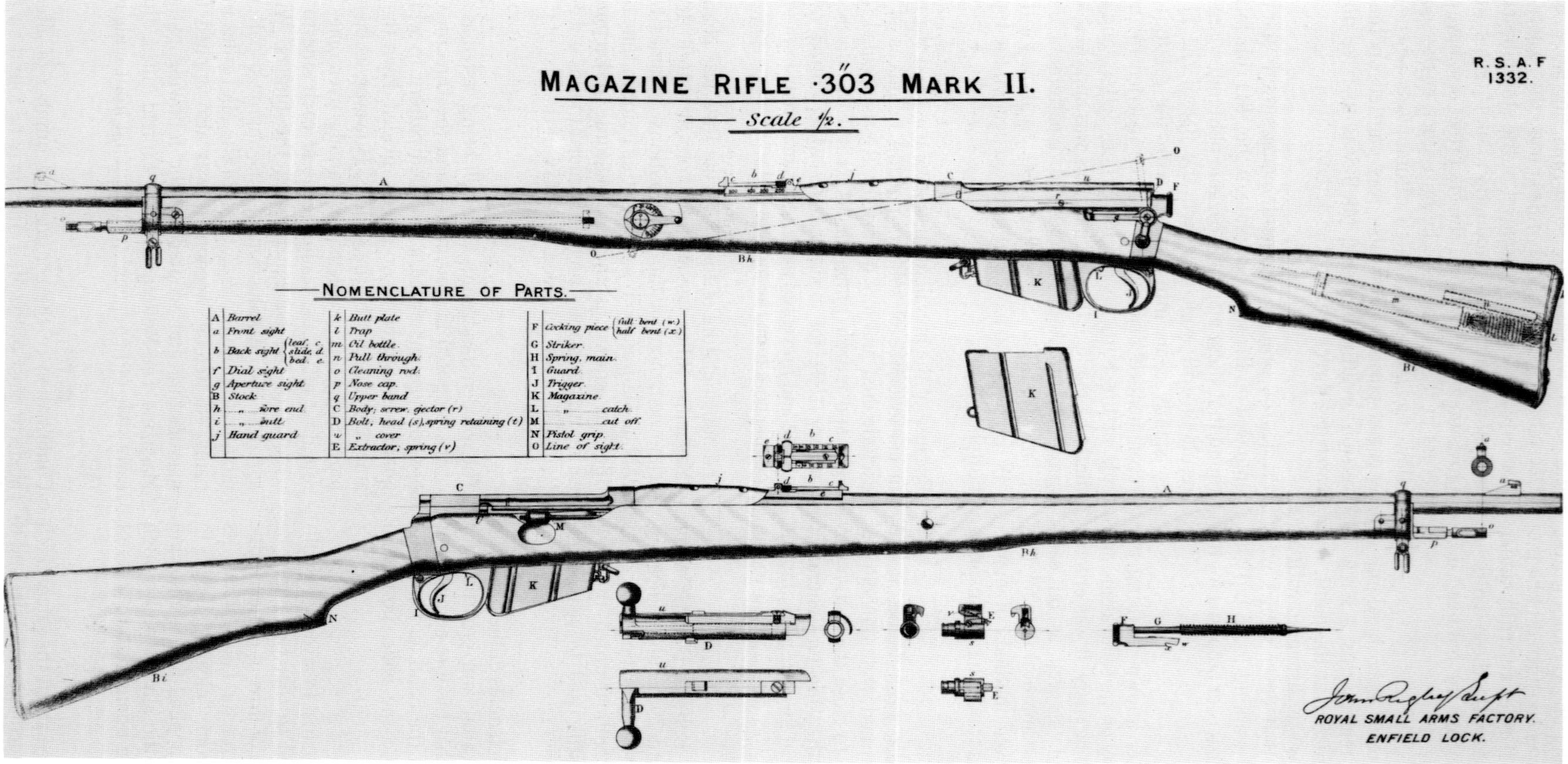

Experimental or Prototype Mark II Magazine Rifle - 1890.
Note some of the Mk I features which were yet to be dispensed with before the eventual introduction of the Mk II model in 1894 (although it was approved earlier in January 1892 and April 1893); the nosecap assembly, recessed stock below the magazine cut-off knob, bolt-head and dustcover. The name "Magazine Lee-Metford" was approved in August 1891. *Lord Cottesloe*

side of the butt socket strap where some inspection stamps and dates of modification are marked. These usually take the form of the last two numbers of the year, e.g. "'91" would indicate 1891, along with the factory inspection stamp. Many of these rifles were later converted to training rifles and thus may be found to have quite a few of these conversion and alteration markings on the left side.

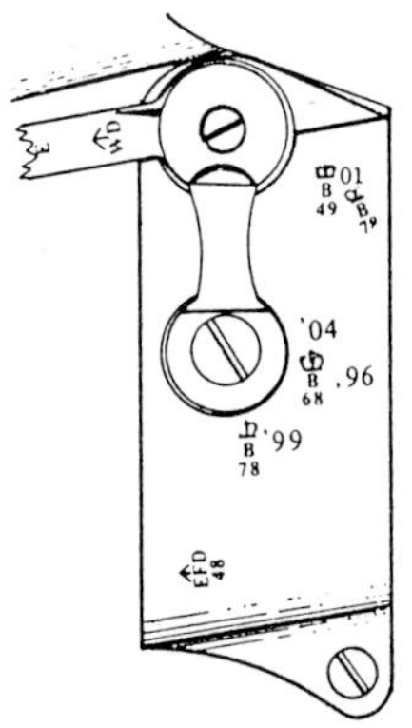

The introduction of a new rifle system resulted in criticism from various quarters; in mid-November 1888, the "Times" made charges against the choice of the new rifle and these were echoed and embellished at later dates. It was claimed that the cost was excessive at £5/10/- each and that the rifle was so condemned that it was to be withdrawn in favour of a new model, the Mk II. Another charge from the "Times" demonstrates the temper of their accusations; that "with such a magazine bolt system, the soldier would not be able to load if firing down a hill."

Many of the Mk I rifles were converted for drill purpose, bayonet practice and small bore training; details of these conversions will be found in the appropriate chapters of this study. On 24th November 1890, the first of the skeleton actions was introduced as an armourers tool. Announced in the List of Changes § 6379 as "Action, Skeleton Mk I" and "Action, Skeleton, Bolt, Breech," they were included with about 30 other items as "Tools, Armourers, Magazine Rifle." Such skeleton actions were used for checking suspect components, as the working and action of such a part put into the action could usually be better observed.

To enable the cut-off spring and its securing screw to be discarded, a modified rifle cut-off was proposed. The new cut-off was slightly thinner and had a small projection to keep it in place whilst in the closed or engaged positions. A slight downward push of the cut-off released it, enabling the cut-off to be swung out and the magazine brought into use. This was not utilised on the Mark I series rifles, but modified slightly and introduced with the Mark II pattern. Similar cut-off plates were used on the following Lee-Enfield and many of the S.M.L.E. rifles.

Numerous small alterations and improvements approved from time to time complicated manufacture and it was not considered desirable to introduce the refinements piecemeal. It was recommended that the Mark II rifle, incorporating these changes, should be considered early in Autumn 1890. Such a rifle was submitted on 17th September 1890 with a reduced diameter barrel and no lower band; it was consequently lighter in weight. Average weight of these particular trials rifles was about 9lb. 1oz., compared with 9lb. 8oz. for the Mark I and 9lb. 4 oz. for the eventual Mark II.

The adjacent drawing of the prototype Mk II rifle (RSAF 1332) shows that the proposed model had a number of innovations although not all were carried through to the second model prototype of which 100 samples were made for trials. Component details and differences with the 2nd model (compared with the Mk I service rifle) were submitted on 17th September and are recorded as being:

Band rear	Omitted.
Barrel	Smaller external diameter to reduce weight.
Body	Modified for 10-rd magazine and omission of safety catch. Cleared in rear to prevent the cocking piece jamming when closing the bolt with the cocking piece in the wrong position.
Bolt	Modified to take new form of bolt-head screw and bolt cover.
Bolt cover	Lengthened in front to form bolthead cover. Holes for bolt cover screws in different positions. Rear turn-over shortened.
Bolt-head cover ...	Omitted.
Bolt-head screw ...	Pin with flat bearing substituted, secured by a keeper screw.
Butt plate	Two alternative buttplates of Delta metal, one of present form, one with heel enlarged for regiment and number.
Cleaning rod	Shortened, head plain, screwed end with Martini-Henry screw. Two when screwed together will answer for driving out jammed cases.
Cocking piece ...	Strengthened by omission of recesses for safety bolt.
Cut-off	Enlarged to suit 10-round magazine.
Handguard	Strengthened owing to thinner barrel.
Magazine	For 10 cartridges, wider and not so deep.
Magazine catch ...	Shortened and more protected.
Magazine link ...	Omitted.
Magazine platform	New pattern for 10-rd magazine, stronger.
Magazine spring ...	Omitted.

Magazine swivel ...	Omitted.
Safety catch	Omitted.
Sharp edges	Removed where possible.
Sear	Freed to prevent jamming.
Sight, fore	Martini-Henry type, barleycorn adjustable and hardened.
Sight, back –	
bed	Altered to read 300 yds with slide at bottom, 400 yds with slide halfway up, and 500 yds. with slide in top position.
cap	Martini-Henry pattern.
leaf	Graduated on alternate sides, figures larger.
slide	Martini-Henry pattern.
Stock, butt	Cleared for alteration to rear of body. Recess for safety catch omitted. Brass disc for regiment and number fitted on right side of toe.
Stock, fore-end ...	Made fuller where gripped by left hand. Grooves omitted. Rod groove omitted.
Swivel, piling	Strengthened pattern recently approved for Mark I rifle.
Weight	With oil bottle and pull through, about 9 lb. 5 oz.

The proposed Mark II rifle is illustrated here and it differs in only a few minor details to the previous listing, the obvious changes are the omission of the lower band, finger grooves and groove underneath the fore-end for the clearing rod. The clearing rod hole has only one opening, in the sword bar projection; the Mk I Lee-Metford front band/nosecap assembly was retained.

The new 10-round magazine also involved widening of the body and the trigger guard as it has a staggered double column of cartridges instead of a single row. New sights, handguard, buttplate, bolt, bolt-head and dustcover were fitted also. One feature which remained on this model but omitted from the eventual Mk II pattern is the scallop out of the right side of the fore-end to enable a better grip on the magazine cut-off knob. A definite step can also be noticed in the backsight ramp at 200 yds. The rearsight leaf was graduated to 1,600 yds and the dial sight plate to 2,800 yds.

For short user trials by the Army and Navy, production of 100 of the proposed new rifles was approved on 9th October 1890. A slightly improved model was submitted for these trials on 24th November 1890 in which the liability to jam had been remedied and the barrel profile considered satisfactory. Although the sights were calibrated for the black powder ammunition (1,600 yds on the leaf and 2,800 yds on the dial plate), 10 of the 100 rifles were sighted with the leaf graduated to 1,800 yds and the dial plate to 2,900 yds, for cordite ammunition.

Allocation of these 100 trials rifles was:–

Aldershot:	52 for B.P. ammunition
	8 for cordite ammunition
School of Musketry, Hythe:	18 for B.P. ammunition
	2 for cordite ammunition
H.M.S. "Excellent":	10 for B.P. ammunition
H.M.S. "Cambridge":	10 for B.P. ammunition

During these tests, some bolts of the trials rifles had the bolt covers work loose; so a screw was used to prevent this occurring and also to stop the bolt-head from falling out when not in the rifle. This was approved on 6th March 1891. The "C" spring for the magazine platform was approved on 14th February 1891 and tests were being made with various modifications to the sights to see if the notch and barleycorn could be fitted to the Mk I rifles already in service or store, as well as those still under manufacture or on order.

Modification of the Lewes sights of the Mk I rifle to barleycorn was shown in RSAF drawing 1830, although the eventual method may well differ in some small respects; the new production sight was approved on 6th July 1891 and the conversion cost estimated at 3/6d for each rifle.

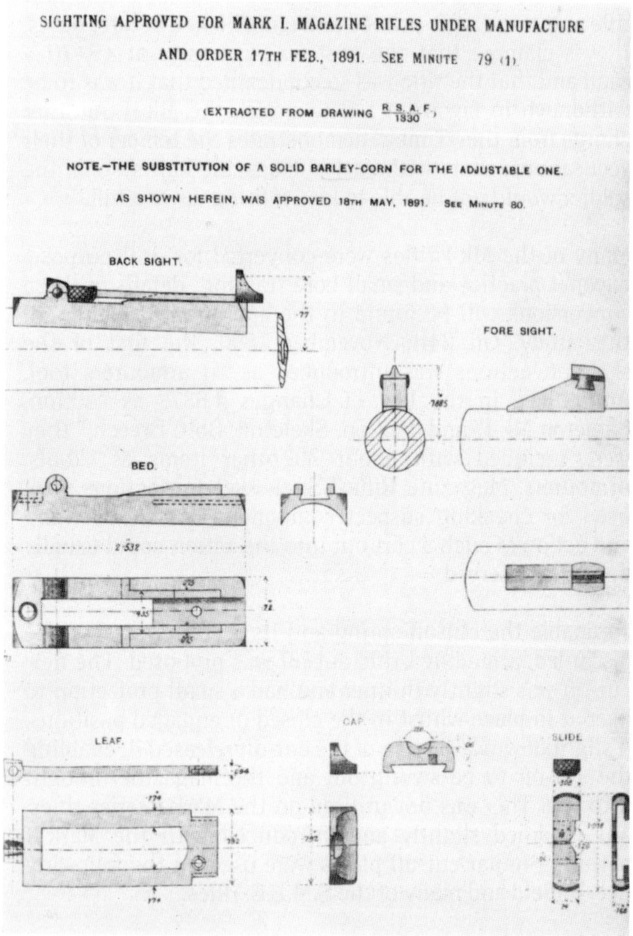

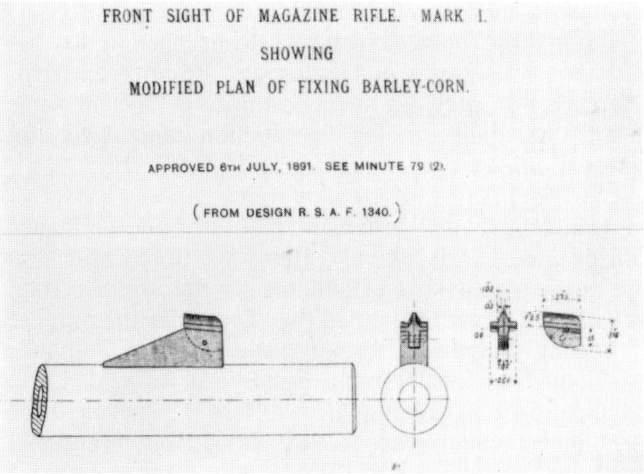

In some of the original RSAF and War Office documentation, the Mk I Improved rifle was referred to as the pattern with barleycorn sights; eventually designated the Mk I* and officially introduced in the List of Changes on 1st July 1892. The Lee-Metford rifle Mk I* and the Mk I* handguard were approved on 19th January 1892 with the following announcement in the List of Changes monthly circulars:–

Para 6760:–

Rifle, Magazine, Lee-Metford (Mark I).*
Rifle, Magazine, Lee-Metford:
Hand-Guard (Mark I).*
Swivels, Piling (Marks II & III).
Disc, Marking, Steel,
for drill arms only.

A pattern of the above-mentioned rifle has been sealed to govern future manufacture of rifles under order and the alteration of Mark I rifles § 5877.

The principal differences between this rifle and Mark I are as follows:

Safety Bolt (or Catch): This is omitted and the cocking piece is without recesses for it. This involves slight alterations in fittings connected with the aperture sight.

Body: Cleared to the rear to prevent jamming of cocking piece when slewed to the right.

Hand-Guard: Of modified pattern. It has the front corners rounded to facilitate removal by hand, and the front spring shortened to grip the barrel less tightly.

Stock, Butt: Alterations to suit the pull-through (See § 6234) are carried out. A leather disc or wad is driven to the bottom of the recess for the oil bottle, to keep the latter from contact with the head of the stock bolt. The oil bottle being shortened, there is now room in the recess for the oil bottle, a piece of flannelette, and the cord of the pull-through.

Disc to Receive Regimental Marks and Numbers: A brass disc is secured by one screw in a recess on the right side flat of the butt to receive regimental marks and numbers. A steel disc (blued) will be employed, instead of the brass disc, in arms issued for drill purposes only (D.P. arms). A pattern of this disc has been separately sealed.

Piling Swivel: Of stronger wire than that previously employed. It is of the same width as the sling swivel, and the opening between the ends is wider than before.

A pattern of this swivel (Mark III) has been separately sealed. Some issues of swivels of a pattern (Mark II) intermediate between Marks I and III have been made; they differ from Mark I only in being of thicker wire.

Mainspring: This consists of 32 coils of .049-in. wire, instead of 39 coils of .040-in. wire, and is 3.25 instead of 5 inches in length.

Magazine Spring: Has four coils instead of three.

Sighting: The front sight is of the "barleycorn" pattern, the barleycorn being set .023 inch to the left of the axis of the barrel to compensate for the tendency of the arm to shoot to the left. In altered rifles the barleycorn is pinned to the original front sight block; in new rifles it is solid. The notches on the back sight are cut at an angle of 90 degrees. The slide is reversible, having one edge straight. It has one vertical white line; no wind-gauge lines.

The graduations are for ammunition giving a muzzle velocity of 2,000 ft/secs instead of 2,200 ft/secs. The leaf is graduated in hundreds of yards up to 1,800 yds, successive hundreds being marked on alternate sides of the slide. Short lines indicating 50 yds are marked on the outer edges.

The Dial Sight: Is graduated from 1,600 yds to 2,900 yds.

The Back Sight Bed: Is arranged for the following sighting;Slide at bottom of leaf, the latter down, 200 yds (lowest sight).
Slide at top of leaf, the latter down, 500 yds (fixed sight). The ramp is made with a gentle slope, so that the slide may be moved to any point between the 200 and 500 yards, and it is graduated in hundreds of yards from 200 to 500.

Sight Protectors: Will be issued with rifles of the pattern § 6256- (Extension of § 5877 to Land Service use).

Dimensions &c.

Length of rifle	*4 ft. 1.5 in.*
Average weight, without sword-bayonet, and the magazine empty	*9 lb. 8.8 oz.*

In the same issue of the List of Changes, a match shooting slide was announced with the official approval date of 13th May 1892. The official introduction and illustration best portray this particular part:–

Para 6762:-

Rifle, Magazine, Lee-Metford: Slide, Sight, Match Shooting.
Mark I and Mark II rifles.*

A pattern of the above-mentioned article has been sealed to govern supplies on repayment for the use, when desired, of officers, non-commissioned officers and men of regiments or corps armed with the Lee-Metford rifle, in match shooting only.

Full size.

It differs from the Service rifle sight slide in the following respects:

It is provided with a step (A) or reading part, which is the same height over the aiming part as the reading part in the present slide is above the bottom of the notch, so that with a full sight the elevations marked on the leaf are available with this slide. It is also marked with vertical lines for wind gauges, which are absent in the slides of Mark I rifles.*

Records indicate that more than 6,000 of these match-shooting slides were made at the Royal Small Arms Factory, Enfield, between 1894 and 1896.

Those rifles of Mark I pattern were virtually all upgraded to Mark I* so that all the Lee-Metford rifles in service would be of the same configuration. This work was undertaken at the various Ordnance factories and the "*" designation stamped on the butt socket after the numeral "I." To ascertain whether a Mk I* rifle was converted, one should inspect the manufacture year; Mk I rifles were pre-1892, while those made from 1892 onwards were Mk I*.

The safety catch operating arm and cam were removed in the Mk I – I* conversion, leaving the safety catch as a plug to fill the hole; on the Mk I* this portion was machined solid with no hole. The cocking piece on the Mk I* has no notches for the safety catch engagement. It may be noticed that the Mk I – I* conversion often has the "*" slightly out of line, although this is not always a sure way to ascertain whether the Mk I* has been a conversion.

After some time in service, it had been found that screwing home the clearing rod of the Mk I* Lee-Metford was liable to damage the front guard screw and so it was instructed that 1.5 turns or 3/32-in. be taken off the threaded end of the rod. This relatively minor modification was officially approved on 26th September 1893.

Particulars of the government factories at Enfield and at Sparkbrook, as well as the contemporary commercial rifle contractors, B.S.A. and L.S.A., are of particular interest at this point in time; the turn of the century probably represents the most active and diverse period for the various rifle factories in London and Birmingham. This reference also provides an opportunity to present some details of their origins, production and size, for comparison.

In 1653, the Long Parliament requested the Ordnance Officer to deal with John and Henry Worth of Enfield, on the northern outskirts of London, for the use of certain mills on the River Lea, called the "Lock," for the purpose of making gunpowder. In 1804, the Board of Ordnance established a Small Arms Factory at Enfield Lock, principally for assembling "Brown Bess" muskets and the ground on which the factory stood was eventually purchased by the Board in 1812.

In 1816 the barrel manufacture was transferred from the Lewisham Royal Manufactory to Enfield and by 1819 there were three recognised Arms Departments; The Tower, Lewisham and Enfield. By 1823, the "Royal Armoury Mills" employed about 42 people and by 1860 there were adjacent cottages for 2 foremen and 60 factory employees; the factory area was then given as 27 acres. A new barrel rolling mill was installed in 1852 and two 46 hp water-wheels had been built; production was rated at 7,000 small arms and 1,500 swords (annually?).

Strikes and high prices demanded by contractors who supplied the bulk of military rifles at this time resulted in small arms manufacture being taken into Government hands under the auspices of the Board of Ordnance. The latest manufacturing techniques in America were investigated and machinery purchased for a new production system; Mr. H. Burton, Master Armourer at Harpers Ferry, U.S.A. was employed as the first factory manager.

Production of service rifles was recorded as being more than 90,000 by the end of the 1860 financial year and for the following year it is reported that 1,700 men were employed at the factory. A separate Inspection Department was formed in 1888 and by 1890 the average annual output was about 60,000 rifles, 5,000 pistols and 8,000 swords or bayonets. A 48-hour working week was commenced in 1894.

By the 1890's, the buildings were single storey with the exception of the stores and some of the older shops, with an available floor space of over 30,000 square feet. Machinery was capable of turning out 2,000 magazine rifles a week, with their bayonets, scabbards and accessories as well as 1,000 spare barrels, 1,000 spare bayonets, 200 cavalry swords or naval cutlasses, 50 lances or boarding pikes and 4 or more complete machine guns. R.S.A.F. Enfield was also involved in some repair work.

The factory lighting was partly gas, partly electric, supplemented by 1,400 Edison Swan lamps. The machinery and dynamos were driven by 1,050 hp steam engines and the gas works mainly provided the lights, though it also supplied the blow pipes for hardening and tempering steel components. Extensive stores of rough gun stocks were kept as the wood was required to be air seasoned for two or three years before use. In the early 1890's, over 3,000 workers were employed at the factory. Finished arms were generally forwarded to Weedon.

The Government purchased the National Arms & Ammunition Co. factory in Montgomery St., Sparkbrook, Birmingham in 1886, although it was not used for the production of complete rifles until 1889. N.A.& A.Co. had been formed in 1872 and applied for a share of the Ordnance contracts, producing some Snider carbines and Martini rifles on government contract and filling Colonial and European orders, however the company ran into severe financial difficulties and was liquidated in 1883. This then became R.S.A.F. Sparkbrook which commenced production with the Lee-Metford rifle and later manufactured the Lee-Enfield rifle and the Mark I S.M.L.E. until it closed early in 1906.

Some Government repair work was effected in Birmingham at the R.S.A.R.F. (Royal Small Arms Repair Factory) Bagot St., Birmingham, but when this was closed in 1894 after more than a century, Government repair work was transferred to Sparkbrook. The Bagot St. establishment was often referred to by the locals and the trade as "the Tower" and in 1914 the premises was taken over by Greener.

The Birmingham Repair mark consists of a Crown over "BR" and is found on many Lee-Metford and Lee-Enfield rifles and carbines, especially on the left side of the receiver butt socket, as well as some bayonets and swords of the period.

BR
40

The Sparkbrook factory, about three-quarters of a mile from B.S.A. Small Heath, was put into general engineering work rather than arms production after its purchase by B.S.A. in 1906 although a few Mark I* S.M.L.E. rifles were produced there in 1906, marked "B.S.A. SPARKBROOK." At the outbreak of the Great War it was producing small engineers tools, motor car bodies and parts for the B.S.A. 13.9 hp motor cars. Car body production was halted and rifle conversion undertaken, most probably the C.L.L.E. Mark I* and S.M.L.E. conversions for .303 Mark VII ammunition. Some parts for the Lewis machine gun were made there also.

B.S.A. factory, circa 1866. Additional factories were later added to the wings. Note the green fields in the foreground.

Another facade of the B.S.A. factory, circa 1911.

The Birmingham Small Arms Co. was formed in 1861 and the erection of works at Small Heath commenced that same year. Production commenced with the Pattern 1853 Enfield rifled musket and the .577 Snider breechloader; B.S.A. became the largest commercial manufacturer in Britain, through two world wars and diversification to produce machine guns and anti-tank rifles, cycles and motor vehicles, tools and machinery. After purchase of the Adderley Park Mills in 1873, it became the Birmingham Small Arms & Metals Co. Ltd. which accounts for the change in manufacture marking to "B.S.A. & M.Co." This was changed back to B.S.A. Co. Ltd. in 1897 and then the firm name became B.S.A. Guns before World War 2.

By the time of the Great War, B.S.A. had four factory sites, at Small Heath, Sparkbrook, Redditch and Coventry Rd., Birmingham. During the 2nd World War the B.S.A. operations were much expanded and another rifle factory built at Shirley, an outer Birmingham suburb, although its main production comprised the No. 4 and No. 5 rifles; the No. 1 rifle was not manufactured there.

London Small Arms Co. had their factory at the Victoria Mills, Old Ford Rd. in London from 1866 until 1921 when they moved to the Albion Works, Ossary Rd. This plant was closed down in 1925 and B.S.A. was left as the sole commercial contractor producing finished S.M.L.E. rifles.

Upon the introduction of the Magazine Lee-Metford rifle, an agreement was reached between B.S.A. and L.S.A. for proportionate shares of the War Office orders for military rifles; such a cartel had been in operation during the Snider and Martini eras. Patterns of the new Lee-Metford rifles were forwarded to these two companies but in January 1889 they were advised that Enfield found it impossible to make up two sets of gauges required for each factory and they were requested to produce the gauges themselves.

These companies were awarded regular contracts with the War Office and also produced the service pattern rifles for Colonial governments and others for the commercial trade. A change in British government policy in 1912 resulted in these companies chiefly supplying India and the Colonies during 1913 and 1914 but during the Great War, production was pushed to the limit on government orders.

Production figures for the Marks I and I* Lee-Metford rifles are not distinguished in the contract records; of the total 360,000 or so Mark I variation Lee-Metford rifles, 188,930 were produced at Enfield, 63,072 at Sparkbrook, 76,500 at B.S.A. and about 40,000 at L.S.A.

Trials of the proposed new Mk II Lee-Metford rifle occurred during March and April 1891 and it was generally well reported upon and widely recognised as a considerable improvement over the Mark I. In the cordite vs black powder comparison, cordite loads were preferred as there was less fouling and they were more pleasant to shoot. Resulting modifications to 50 of these 100 rifles, for further trials were:

- Bolt resisting shoulders modified to reduce friction;
- Altered method of securing the bolt cover, with two systems tried and six rifles altered on each plan;
- Bolt cover made to fit up close to the breech;
- Abolition of the fore-end "scoop" under the cut-off;
- The rear end of the cut-off not left finished as sharp;
- Magazine spring strengthened, friction minimised by reducing surface contact between platform and magazine case;
- Body of the rifle altered to permit cartridges to pass more freely into the chamber;
- Magazine catch modified to allow the magazine to be more easily detached;
- Trigger sear modified to avoid sticking when clogged with dirt;
- The clearing rod strengthened;
- A means of securing the barrel to the fore-end at the point where the lower band was positioned on the Mk I rifle.

Of those rifles tested with the different bolt covers, one pattern had a different retaining spring while the other cover was made to act as the spring itself. The latter type was more successful and it was adopted and used on the ensuing Lee-Metford and Lee-Enfield rifles and carbines. Some rifles with the new pattern bolt cover and also employing a new mode of bolt-head attachment were tested at Hythe, however the method of attachment used on these particular rifles is recorded as being not entirely suitable.

A system for attaching the bolt-head was recommended by Mr. J. Deeley, whereby it was screwed into the bolt; some of Deeley's pattern bolts were already in use for proving ammunition, where they had answered well. Fifty rifles were fitted with the Deeley bolt-head and issued for trial, each with 1,000 rounds of cordite ammunition and distributed to Hythe, Aldershot, Ireland and the Naval establishments of HMS "Excellent" and "Cambridge." Of these rifles, two at Aldershot and four in Ireland were fitted with lower bands.

Penn and Deeley's patent bolt-head proved to be quite satisfactory and it was incorporated in the proposed Mk II rifle. In fact, this method of bolt-head attachment was successful enough to warrant its use on the following family of Lee rifles and carbines until its conclusion with the No. 4 and 7.62mm variants. The application date for this bolt-head patent by Penn and Deeley was 25th November 1890, numbered 19145; it was also associated with an exclusive allocation to the gunmakers Westley Richards.

Following the recommendation of the Committee that the barrel be secured to the fore-end at the point where the lower band was positioned on the Mk I rifle, it had been decided to fit a thin ring around the barrel and secure it to the dial sight screw. This inner band and its function would have been somewhat similar to that of the Mk I and Mk III SMLE rifles, but the idea was discarded for the proposed Mk II Lee-Metford because the lower sling swivel was required for attaching the carrying sling.

The idea of this inner band was to secure the barrel to the fore-end at this point, as the trials Mk II rifle had no lower band which served this purpose on the earlier rifles. The decision to retain the lower band was made on 28th October 1891, so that the sling attachment would be available for arms issued to rifle battalions. Six of the Mk II rifles with the lower band and a butt swivel were tried by some rifle regiments; after their approval it was recommended that this system be universally adapted.

Hitherto, the rifle sling swivels were mounted onto the top band adjacent to the piling swivel, and at the front of the trigger guard next to the magazine link. Even the preceding Martini-Henry, Snider and Enfield long rifles had the sling swivels mounted to the top band and the trigger guard. Rifle Regiments, however, required that the swivels be mounted differently due to the special way in which they slung their arms; with the Enfield and Snider, they were issued with short rifles which had the lower swivel on the butt. When the Martini-Henry rifle was introduced, all the corps armed with rifles were issued with a standard length of arm and so Rifle Regiments had the sling swivels mounted onto the butt and lower band instead of to the trigger guard and the upper band.

The pattern of sling swivels used by the Rifle Regiments simplified the arrangement of piling and sling swivels on the upper band as only the piling swivel required to be attached to the upper band. Because of the convenience of this arrangement, it was introduced for all branches of the Service issued with the rifle and so another aspect was standardised. From this time on, butt swivels were fitted to all new rifles.

An illustration of the pattern rifle which resulted from the trials and subsequent modifications is provided (RSAF 1375) and this model was submitted for approval and adoption as the Mark II Lee-Metford rifle; it will be noticed that the Mk I Lee-Metford nosecap and upper band were still applied along with the earlier sling swivel arrangement, as this was yet to be changed. Requests were made for wind gauge lines on the backsight slide similar to the Martini-Henry and the Committee indicated that they had no objections to having these incorporated.

A small radius was cut at the corner marked "A" after September 1891.

It was reported that a fracture had occurred at the rear of the cocking piece on a proof rifle, at the angle "A" marked in the sketch. This resulted in the approval of a small radius instead of the square cut corner, on 3rd September 1891. Another minor change, approved on 19th January 1892, was the regraduation of the dial plate and backsight due to the flatter trajectory of the bullet with the new cordite load. Graduations were to 1,800 yds on the backsight and 2,900 yds on the dial plate for cordite ammunition.

A report was issued by the Director-General of Ordnance Factories on 2nd February 1891 regarding suggestions by Lieut.-Col. Lewes as to modifications for the Lee-Metford action, which improved the bolt-head and its means of attachment, and had a modified ejector. These were reported as being "ingenious and answering the purpose proposed, but slightly weakened the body and would increase costs." Perhaps some tact and diplomacy was being employed as Lewes was no doubt put out over the rejection of his sights.

But Lewes submitted his rifle again, modified so as to overcome some of these criticisms; the sear and magazine catch, sear spring and retaining spring were attached to the trigger guard instead of the body, so that the parts might be removed for cleaning without removing the fore-end. These points were considered for the 2nd Model prototype of the Mk II rifle which was not yet ready; Lewes was advised on 3rd September 1891 that no further steps would be taken on the matter.

The Commandant of the School of Musketry, on 26th August 1891, suggested that the backsight leaf ladder be nicked at the intermediate 50 yds between every 100 yds graduation; on either the outside or inside of the leaf, which ever was found to be most convenient, similar to that done on the leaf of the latest pattern German service rifle. This proposal was incorporated in the pattern of Mark II rifle submitted by the Adjutant-General on 19th September 1891 and approved by the Secretary of State on 24th September 1891.

A lighter pattern of handguard than those submitted on the trials Mk II rifles was proposed by the Director-General of Ordnance Factories, with more rounded sides so as to reduce the bulk to be grasped when carried at the trail. This was incorporated in the two rifles sent to B.S.A. and L.S.A., as examples of the probable Mark II.

The decision taken to dispense with the sling swivel on the upper band and the introduction of the butt swivel, permitted a stronger piling swivel and a modified upper band. The fore-end cap, sword bar and upper band are all incorporated into the one fitting, a design by Rigby, upon which he took out a patent. This was another innovation of the eventual Mk II rifle which had not been included on either of the two earlier models of Mk II prototypes.

Differences in components between the new magazine rifles, Mark I, Mark I Improved (Mk I*) and Mark II Experimental were listed and described on 6th January 1892. Even at this stage, it is pointed out that the "Experimental Mk II" was still not quite up to the development of the eventual adopted Mark II, the final approval date given for some features being 12th April 1893. The following tabulation not only compares the Experimental Mark II with its immediate predecessor, the Mark I*, but also records dates of approval for the various improvements which consequently caused its upgrading from the original Mark I model.

COMPONENT DIFFERENCES between MAGAZINE LEE-METFORD RIFLES, as at 6th January 1892.

M.L.M. Mk I	*M.L.M. Mk I Improved (Mk I*)*	*M.L.M. Mk II Experimental (at 6/1/1892)*
Barrel:	Sharp angle at chamber end bevelled, 30/11/89. This has been done to all rifles.	Slightly reduced external diameter after first six calibres.
Body:	Cleared in rear to prevent jamming of cocking piece when slewed to the right, 31/12/90. Rifles made subsequent to the approval have profile of body slightly modified. Rifles in store at Weedon have had the bodies cleared there. Rifles in the hands of troops issued prior to the alteration have not yet been altered. Alterations at Weedon were effected between February and July, 1891. Modified for omission of safety bolt, and addition of aperture sight screw. Rifles sent into store at Weedon prior to the approval have had the tooth and arm of the safety bolt removed there; the bolt then acts only as a washer and pin for aperture sight. Rifles in the hands of troops issued prior to alteration have safety bolts.	Modified for 10-rd. magazine. Grooves cut in both sides, to improve cartridge feed from magazine. Resisting shoulders cut with screw pitch, to facilitate opening of bolt after firing. Cam groove enlarged for larger resisting stud of bolt. Socket, cut-off way, and barrel end simplified to cheapen manufacture. Screw thread for rod end not cut.
Bolt:	Rounded at end, to suit rounded angle of cocking piece, 3/9/91. In rifles made subsequent to the approval only.	Front end of cylinder screwed internally with right hand thread for bolt head. Studs in rib modified, and undercut from rear, and a circumferential groove cut on end of bolt for securing the cover. Camming modified, and cam-way cut deeper. Resisting lug made .1-in. longer and .02-in. wider. Resisting shoulders of lug and rib cut with .1-in. screw pitch. Bolt head and bolt cover screw holes omitted.
Bolt head:		Made from solid piece of steel, without dovetail. Tenon screwed externally with right hand thread for screwing into the bolt. Slot in tenon for bolt-head screw omitted.
Bolt head cover:		Omitted.
Bolt cover:		Made thinner and of spring steel, set to spring into its place on the bolt, and modified for new method of attachment. Lengthened in front to form bolt-head cover. Brazed joints in rear turnover done away with. Bolt-head and bolt-cover screws omitted.
Butt plate:		Yellow metal, with heel enlarged for regiment and number.
Putt plate trap:		Made from yellow metal.
Cocking piece:	Recesses for safety bolt omitted, 31/12/90, in rifles made subsequent to approval only. Angle rounded, 3/9/91, in rifles made subsequent to approval only.	Cocking stud enlarged and strengthened to suit modified modified cam in bolt.
Cut-off:		Altered to suit 10-rd. magazine. Made slightly thinner with a projection on it and spring to retain its place when closed. Holes for cut-off spring omitted.
Cleaning rod:	End of screw thread rounded off, 18/6/90, in rods made subsequent to approval. Those in hands of troops have altered by armourer serjeants, LoC 6076.	Shortened. Slot in head omitted. Head fluted longitudinally and tapped to permit two rods to be screwed together. End screwed with right hand screw.
Dial-sight plate:	Graduated for Cordite, 2,000 ft/sec. MV, instead of Nobel, 2,200 ft/sec. The dials of some rifles in store are graduated for Nobel, and some are ungraduated.	
Dial sight pointer:		Bead reduced in size.
Extractor:		Head central with respect to arm, bearing surface of claw increased.
Guard:	Screwed hole for magazine catch securing screw, omitted, 25/3/91, in rifles made subsequent to approval only.	Modified to suit 10-rd magazine.
Hand-guard:	Central depression omitted, height decreased, 26/11/89, but none on issue. Subsequently, front corners rounded to facilitate removal, 13/2/91. Sealed as Mk I*. All in store at Weedon are Mk I*; those rifles in the hands of troops prior to the alteration are the original Mk I.	Thicker, to suit thinner barrel, and exterior rounded.
Jag:	Omitted, 19/3/90, LoC 6234.	
Magazine catch:		Finger grip modified, and chequering omitted.
Magazine case:		Wider and shallower, to hold 10 cartridges in two columns instead of 8 in one.
Magazine platform:		Modified for two columns of cartridges, and "C" spring.
Oil bottle:	Shortened 1-in., 21/7/91, and sealed as Mk II, 20/8/91. Issued with all new rifles, for Mk I pattern in the hands of troops, issued to replace damaged or broken ones.	
Pin, bolt-head cover:		Omitted.
Pin, stop, oil-bottle:	Omitted, 2/5/90. Withdrawn, see LoC 6234.	

M.L.M. Mk I	*M.L.M. Mk I Improved (Mk I*)*	*M.L.M. Mk II Experimental (at 6/1/1892)*
Spring, main:	32 coils, 3.25-in. long, .049-in. wire instead of 39 coils, 5-in. long, .040-in. wire, 23/7/91. In all rifles subsequent to approval. Rifles in store prior to approval and in Mediterranean garrisons, have mainsprings of better wire and fire-blackened. Rifles in the hands of troops at home have the original spring, except where replaced by the better wire or new pattern spring on account of breakages.	
Spring, magazine:	A fourth coil added, 22/8/90; in all rifles except those in the hands of troops at home, for which, however, they are issued to replace damaged or defective ones.	C pattern of ribbon steel, front end bent over to engage front of magazine, rear end turned around and soldered to a pin for attaching to platform.
Spring, butt plate:	Increased in width at the point, 20/2/90, in all rifles made subsequent to approval only.	
Spring, hand-guard (2):	Front spring shortened to grip barrel less tightly, 13/2/91, in rifles made subsequent to the approval only.	Omitted.
Spring, bolt-cover (2):	Made a tight fit, 24/5/90. In all rifles except those in hands of troops at home, for which, however, they are issued to replace damaged or defective ones.	
Screw, cut-off:		Omitted.
Screw, ejector:	Mild-cast steel, hardened at point and tempered, instead of mild steel, case hardened; 24/5/90. In all rifles made subsequent to the approval only.	
Screw, bolt-head:	Mild cast-steel, hardened at point and tempered, instead mild steel, case hardened; 24/5/90. In all rifles except those in the hands of troops at home, for which, however, they are issued to replace damaged and defective ones.	Omitted.
Safety bolt:	Omitted, 31/12/90. Still exists in rifles in the hands of troops at home.	
Sight slide:	V-notch, 90 degrees, reversible, with straight edge; ends lengthened to bear on ramp.	
Sight leaf:	Graduated 2,000 ft/sec. MV, lines and figures on alternate sides, figures larger, short lines denoting intermediate 50 yards, marked on outside edges of sides, sharp edges rounded.	
Sight cap:	V-notch, 90 degrees; sharp edges rounded.	
Sight, front:	Barleycorn. Set .023-in. to the left to counteract a tendency to shoot left.	
Sight bed:	Arranged as follows:- Slide at bottom of leaf, latter down 200 yds, "lowest sight". Slide at top of leaf, latter down 500 yds, "fixed sight". Ramp gently sloping, allowing the slide to move from 200 to 500 yds, and graduated at 200, 300, 400 and 500 yards. Approval dates for the sight changes were:- 17/02/91, 18/5/91, 7/8/91 and 16/11/91. All rifles in store, and in the hands of troops are being altered from the Lewes pattern. Rifles in Mediterranean garrisons have Lewes sighting and are graduated for black powder, with 300 yds "lowest" and 400 yds "fixed" sights. Altered rifles will have the barleycorn let in.	
Sear:		Freed at sides, with knuckle joint at the trigger end.
Swivel, piling:	Strengthened; 28/8/90, sealed as Mk II. 19/12/90, sealed as Mk III.	
Stock, butt:	Stop pin hole and recess for safety bolt omitted. Jag hole reduced in depth, 2/5/90, in rifles made subsequent to approval only. Hole cut for brass marking disc; in all rifles except those in the hands of troops at home.	Smaller in grip, and end modified to fit altered socket. Recess cut for enlarged heel of butt plate. Hole for marking disc omitted.
Stock, fore-end:	Recess for safety bolt omitted, 31/12/90, in all rifles made subsequent to approval.	Altered for 10-rd. magazine. Hand grooves omitted and made fuller where gripped by the left hand. Cut-off recess omitted. Rod groove shortened and prepared for rod stop. Groove cut for elevating aperture sight.
Trigger:		End altered to engage in knuckle joint of sear, and so enable the latter to be worked up if it jams.
	New Components:- Disc, marking, 30/8/91, in all rifles except those in the hands of troops at home. Also screw for. Pull through; Leather, 26/4/90; Whipcord, Mk II, 2/10/90 and 29/10/90; new issue for all rifles. Screw, sight, aperture, 4/3/91; in all new rifles. Sight protector, 3/1/91; new issue for all rifles. Washer, leather, head of stockbolt, 9/9/91.	No marking disc. New Component:- Rod stop.

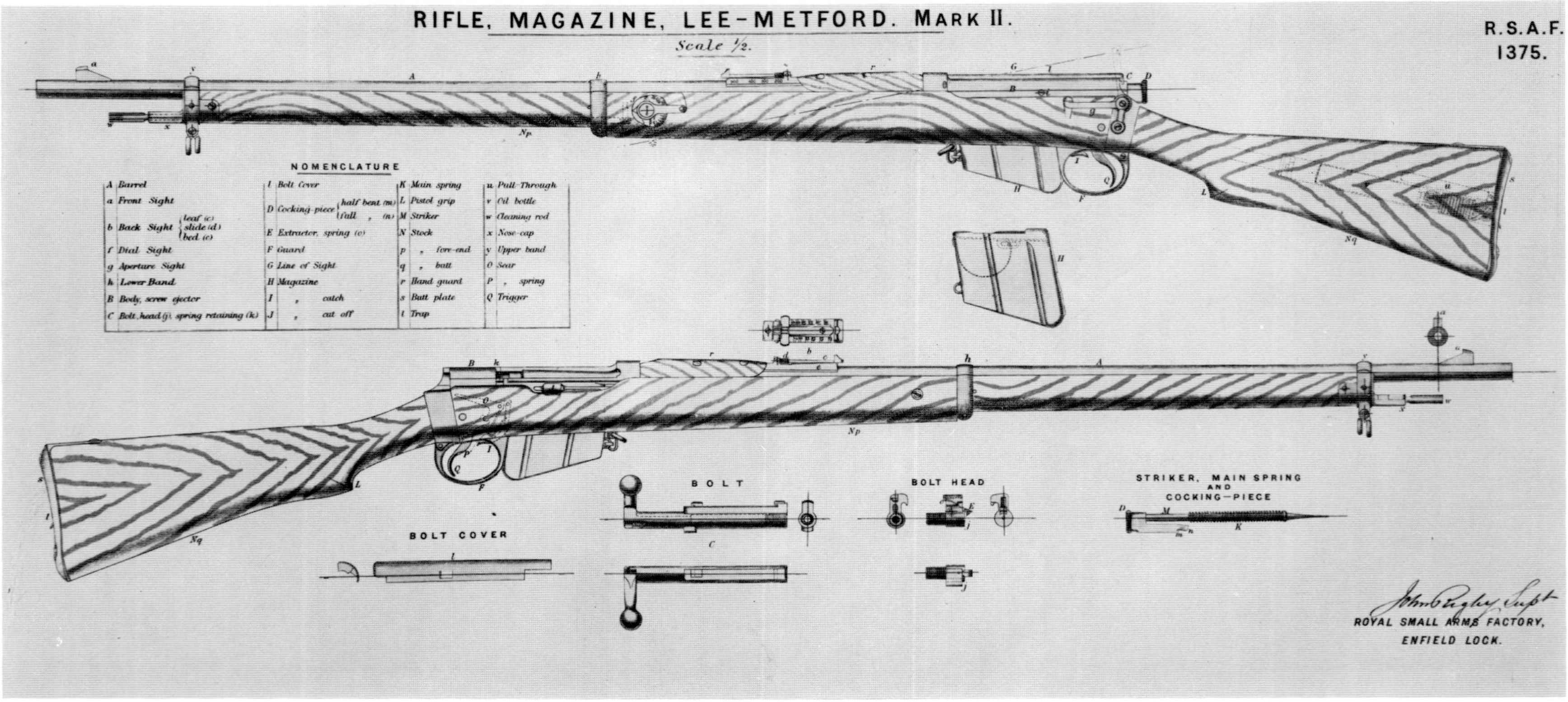

A further drawing of the Mark II model, now referred to as "Lee-Metford". Compared with the drawing a few pages earlier, this model incorporates more of the improvements eventually introduced with the Mark II. Although this particular drawing (RSAF 1375) is dated 1891 and the introduction of the Mark II M.L.M. was some years away, new features such as the Deeley bolthead, spring-steel dustcover and double-row 10-round magazine are evident. Note the earlier nosecap/front band assembly. *Lord Cottesloe*

The Mark II Magazine Lee-Metford Rifle was announced in the List of Changes on 1st May 1894, with the approval dates of 30th January 1892 and 12th April 1893. Some Enfield 1892 dated Mk II M.L.M. rifles have been noted, without serial numbers. Some of these particular examples were later utilised for other trials and fittings, such as sights, rifling, safety-catches. So they can be regarded as pre-production models.

Reproduced hereunder is the particular LoC announcement, the Mk II model was made at Enfield and Sparkbrook, with additional government contracts awarded to the B.S.A. and L.S.A. factories. Production totalled about 252,000 rifles.

Para 7404:-
Rifle, Magazine, Lee-Metford (Mark II).

A pattern of this rifle has been sealed to govern future manufacture. It differs from the Mark I (§ 6760), principally in the following particulars:-*

The barrel is lightened by being reduced in external diameter, except near the breech end.

The body is modified so as to suit the new magazine; grooves are cut in both sides to improve the feed of cartridges from the magazine. The resisting shoulders are cut with a screw pitch to facilitate the opening of the rifle after firing.

The bolt has the front end of the cylinder screwed internally with a right hand thread to receive the stem of the bolt head. The studs on the rib are modified and undercut from the rear and a circumferential groove is cut on the end of the bolt for securing the cover. The resisting lug is made slightly larger and the resisting shoulders of lug and rib are cut with a slight screw pitch to suit those of the body.

The bolt head is made from a solid piece of steel, without dovetail; the tenon or stem is screwed externally with a right hand thread so that it screws into the bolt, and no bolt head screw is required.

The bolt cover is made of spring steel, set to spring into its place on the bolt, so that screws to secure it are not required; it is lengthened in front to form a bolt head cover.

The butt plate and trap are made of yellow metal. The heel of the butt plate is enlarged to admit of regimental marks being engraved thereon.

The cut-off is altered to suit the new magazine. It is made slightly thinner and is furnished with a projection to retain it in its place when closed, the cut-off itself acting as a spring.

The clearing rod is shortened, the slot in the head is omitted and the head is fluted longitudinally. It is tapped at the head and smaller end so as to admit of two rods being screwed together if required to drive out a fired case from the chamber. This is the only purpose for which the rod is to be employed.

The magazine is made wider and shallower. It will hold 10 cartridges, arranged in two columns, instead of 8 in one column.

The magazine spring is of "C" shape, made of ribbon steel, the front end bent over to engage the front of the magazine, the rear turned round and a pin driven through it for attaching it to the platform.

The sight slide is wider and is marked with wind-gauge lines. The sight leaf is graduated from 600 to 1,800 yards by figures on alternate sides of the leaf, with short lines denoting intermediate 50 yards marked on the edges. The sight bed is graduated from 200 to 500 yards and the dial sight from 1,600 to 2,800 yards.

The fore-end of the stock is made fuller at the part gripped by the left hand when in the firing position; the hand grooves are omitted.

The nose-cap is altered to form both nose-cap and upper band. The swivel on the guard is discontinued and the arm is fitted with a butt swivel.

Other less important modifications in various parts have been introduced.

Length of barrel	*30.2 ins.*
Calibre	*.303-in.*
Rifling	*Metford system*
Grooves	*7*
Groove depth	*.004-in.*
Width of lands	*.023-in.*
Spiral, left handed	*1 turn in 10 in. or 33 calibres*
Length of rifle	*4 ft. 1.5 ins.*
Length of rifle with sword bayonet ...	*5 ft. 1.45 ins.*
Length of sword bayonet	*1 ft. 4.7 ins.*
Weight of rifle with magazine (empty) ...	*9 lb. 4 oz.*
Weight of sword bayonet	*15.5 oz.*
Weight of sword bayonet scabbard ...	*4.75 oz.*
Weight of magazine (empty)	*4.75 oz.*
Weight of magazine (filled)	*14.25 oz.*

The following parts are special to Mark II rifles:-

Band, lower.	*Cut-off.*
Barrel.	*Extractor.*
Bead, pointer, dial sight.	*Guard, hand.*
Body.	*Guard, trigger.*
Bed, sight, back.	*Head, bolt, breech.*
Bolt, breech.	*Leaf, backsight.*
Cap, nose (combined with upper band).	*Link, magazine.*
	Magazine.
Cap, sight, back.	*Nut, rod.*
Case, magazine.	*Plate, butt.*
Catch, magazine.	*Plate, sight, dial.*
Cocking piece.	*Platform, magazine.*
Cover, bolt, breech.	*Pointer, sight, dial.*

Rod, cleaning.
Screw, band, lower.
Screw, piling swivel.
Screw, strap, butt-plate.
Sear.
Sight, back.
Sight, dial.
Slide, sight, back.
Spring, guard, hand, lower.
Spring, guard, hand, upper.
Spring, platform, magazine.
Stock, butt.
Stock, fore-end.
Swivel, band, lower.
Swivel, piling.
Trap, butt-plate.
Trigger.

The following parts which are in Mark I rifle are omitted in Mark II:-*

Bolt-head cover.
Magazine, upper link.
Pin, bolt-head cover.
Spring, cut-off.
Screw, bolt cover.
Screw, cut-off spring.
Screw, bolt-head.
Screw, guard, swivel.
Disc, marking.
Screw, disc, marking.
Band, upper, with pin, screw and nut.
Swivel guard.

The remaining parts are interchangeable between Marks I and II rifles.

Major parts listing for Mark II M.L.M. illustrated below:-

A.—Barrel.
B.—Body.
C.—Bolt.
D.—Bolt head.
E.—Hand guard.
F.—Bolt cover.
G.—Butt plate.
H.—Cocking piece.
I.—Cut-off.
J.—Clearing rod.
K.—Dial sight pointer.
L.—Guard.
M.—Magazine catch.
N.—Magazine case.
O.—Magazine platform.
P.—Magazine spring.
Q.—Sight slide.
R.—Stock butt.
S.—Stock fore end.
T.—Trigger.
V.—Rod nut in fore end.

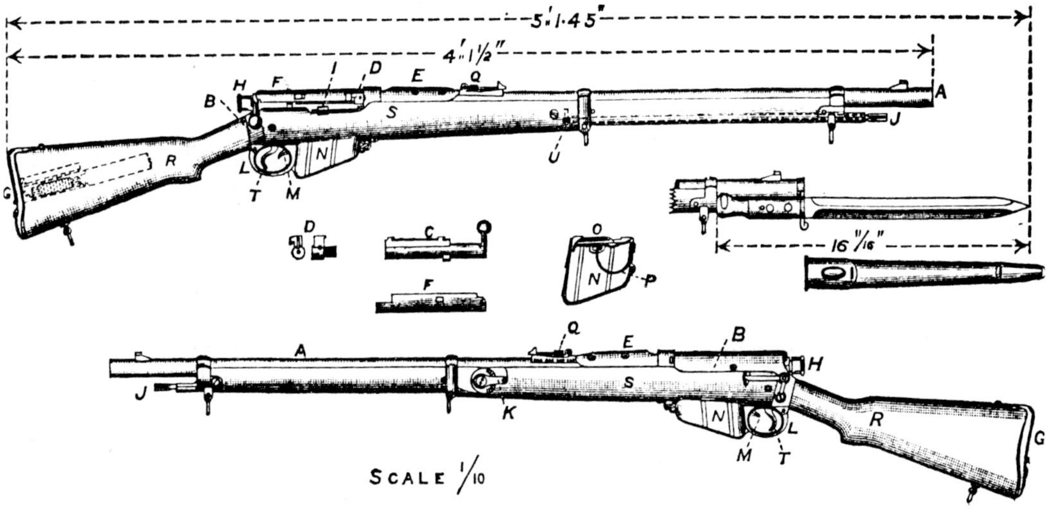

A skeleton action of the Mark II Magazine Lee-Metford was approved on 1st November 1893 and a pattern sealed so as to govern future manufacture. Previously, these short actions had been used by armourers for testing suspect action components and such usage continued with ensuing patterns of the Lee-Enfield although similar variations became increasingly used for instructional purposes as well. The "Action, Gauge" and "Action, Implement" were also announced but these were mainly intended for measuring or working on the rifles.

B.S.A. and L.S.A., the contractors, were advised of the impending change of pattern in 1890 and with the original contract for 200,000 Mark I (and I*) Lee-Metford rifles well under way it was decided that about half should be the new Mark II pattern. Subsequent delays in the Committee's decision on the new model resulted in additional Mk I* rifles being produced and the first Mk II rifles were assembled in October 1892. By March 1894, the weekly rate of production was 600 at B.S.A. and 400 at L.S.A. At this time R.S.A.F. Enfield capacity was reported to be about 2,000 per week although this was not likely to have been the actual output. Another government factory was also in operation at Sparkbrook, in Birmingham.

A Lee-Metford sporting rifle was introduced by B.S.A. in February 1892 which sold for £6/10/-. Later on, a cavalry carbine based on the service pattern but known as the B.S.A. Trade Pattern was also made. The Lee-Metford service rifles made by the contractors were available to the retail trade, listed in 1894 as £4 for the Mk I* model and £5 for the Mk II. While these particular rifles are generally identical to their service contract counterparts, they are marked "LEE-SPEED PATENTS" in place of the usual Government or Ordnance contract markings on the right side of the receiver butt socket.

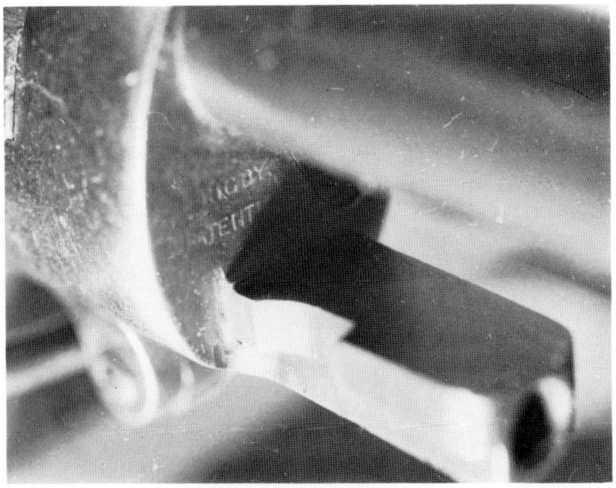

Patent acknowledgement marks found on the commercial or "Volunteer" model Lee-Enfield and Lee-Metford rifles and carbines and sporting models.
Top: The Lee and Speed patent mark on the receiver body.
Top right: Rigby patent mark on the nosecap and sword bayonet bar assembly, marked on the front face.
Right: Deeley's bolthead patent mark, No. 19,145 of 1890.

The British South Africa Co. purchased a few thousand Mk I* and Mk II Lee-Metford rifles between December 1894 and September 1895, along with 380 Lee-Metford sporting models with a flat rib along the top of the barrel. These too were "Volunteer patterns" or commercial production, without the Government markings (Royal cypher and year of manufacture along with the military Mark designation) on the butt socket; only the patent acknowledgment and manufacturer were marked at this position. Many of these rifles were purchased from B.S.A. in haste for rapid consignment to South Africa and on one occasion the B.S.A. factory worked all night to complete such rifles for shipment the next morning.

Many of these rifles were used in the Jameson raid over the New Year period, 1896, as hundreds of rifles were seized and later observed in Pretoria to have the B.S.A. & M. Co. markings, obviously having originated from the British South Africa Co. The B.S.A. and L.S.A. factories also supplied 3,400 rifles to Afghanistan, all of which were the Mk II Lee-Metford pattern. Many others were sold through the trade for use in competitive club shooting.

Whilst the new Mark II rifle rectified most of the considered failings of the Marks I and I*, complaints were still being made about the bolt-head and so on 11th July 1894, case-hardened iron was approved instead of steel (as the metal from which the bolt-heads were to be made). It would appear that the original material was considered to be too hard or brittle.

The safety catch question was raised again and in the preparation of a carbine for mounted troops, the cavalry indicated that such a catch was preferred. Ever since the introduction of the Martini-Henry, there were requests for a half-cock or safety catch; in fact, trials patterns of the Martini-Henry rifle were so fitted. By 1873 it had been discontinued on the Martini-Henry but requests were made for a half-cock on the Martini-Henry cavalry carbine which was introduced in 1877, although no such fitting was provided on the adopted pattern.

The first Lee-Metford had been fitted with a safety locking bolt upon its introduction in 1889, largely as a result of this pressure, only to be discontinued on the Marks I* and II Lee-Metford rifles from 1892. However, when a safety catch was re-introduced in 1894, it remained throughout the series of Lee rifles and is still a feature of service rifles today, although the requirements of modern warfare have dictated a little differently.

A new cocking piece was eventually submitted with the Lee-Metford carbine, with a safety catch attached to the right hand side. The cocking piece was a knurled rim around the back edge, so that it might be easily pulled back from the fired or half cock position with the fingers. This new cocking piece and safety catch necessitated the extension of the bolt and firing pin at the rear by about one inch.

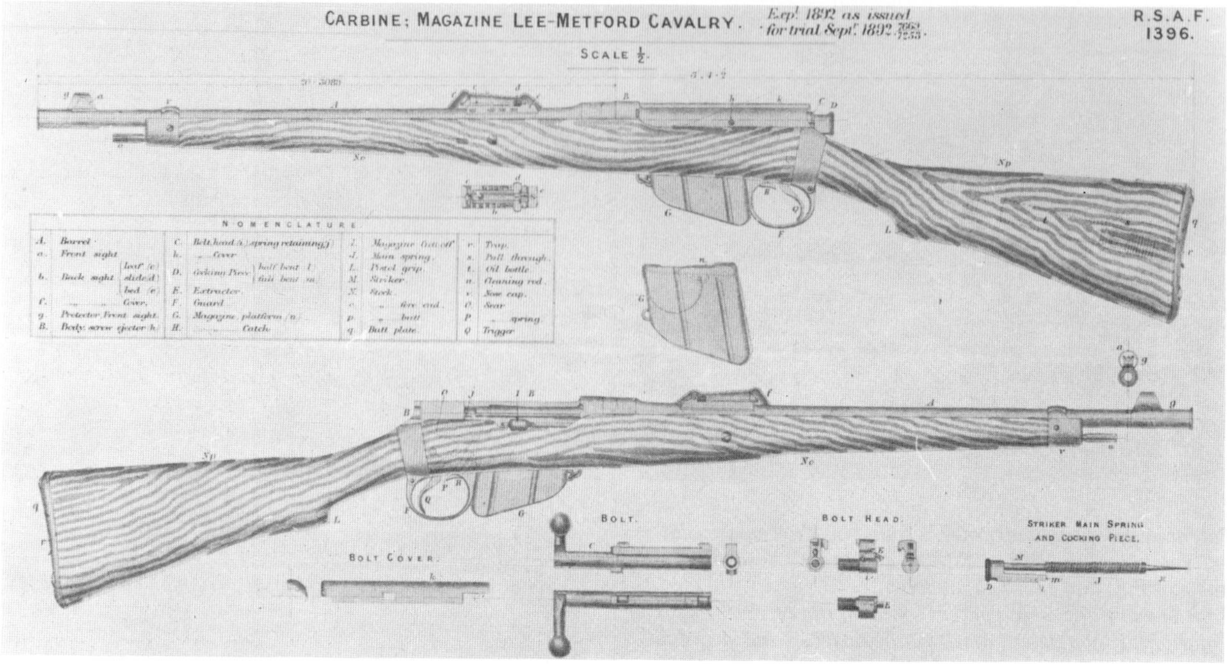

1892 Trials Carbine. Nosecap is .303 Martini Cavalry style. *Lord Cottesloe*

Whilst this carbine was the first to be introduced in a small series of production Magazine Lee carbines, some had been already assembled for trials and consideration; 50 carbines were manufactured and issued to the cavalry in the 1888 trials and in 1889 both an artillery and cavalry model were prepared for submission to the Committee. The basic difference between artillery and cavalry models is that artillery models have a sword bar for attachment of the bayonet while the cavalry models have no such provision. The cavalry carbines have a smoother appearance as they were usually carried in a saddle bucket.

It was complained of the trials Magazine Lee-Metford carbine in March 1889 that the dial sight made it clumsy in the stock and heavy, so it was agreed to use a 2,000 yd. backsight and dispense with the long-range dial sights. Two months later, an improved model without the dial sight and reduced furniture was submitted. It was also decided that there would be no magazine carbine for the Royal Artillery and Royal Engineers as these corps were to have Martini carbines converted to .303 calibre and fitted for the Pattern 1888 sword-bayonet.

The question of the magazine cavalry carbine was not proceeded with because ammunition problems were not yet resolved and with the impending introduction of a new Mark II Lee-Metford rifle, the improvements could also be incorporated in a new carbine. The sample carbine examined in May 1889 was 7 lb. 6.5 oz. (1.5 oz. lighter than the regulation Martini-Henry carbine) and a later model with 3-in. longer fore-end and 2.6-in. longer barrel was reported as being 3.75 lb. heavier than its Martini-Henry counterpart.

In January 1891 a new pattern of magazine carbine was proposed which embodied modifications recommended for the Mark II rifle and on 2nd March 1891, the Director General of Ordnance Factories was requested by the Director of Artillery to prepare such a pattern. The was further delayed, however, because of further improvements being made on the new rifle. A pattern was eventually submitted in February 1892, sighted for cordite and with no cleaning rod.

It was suggested that a rod was required to enable the removal of a possible stuck cartridge case and so a fresh pattern with such a rod was submitted to the Director General in May 1892. Recommendations were made for the production of a trials batch of 100 rifles, but fitted with foresight protector wings. These were forwarded in September, 90 to Aldershot and the remainder to Hythe, along with special saddle buckets. The specifications of these trials carbines are given as:

Overall length	3 ft. 4 in. (40 in.)
Barrel length	1 ft. 9 in. (21 in.)
Rearsight graduations	to 2,000 yds.
Magazine capacity	10 rounds.
Rifling, &c.	7 groove, Metford.

This particular trials model has the Mark II rifle type bolt and dustcover, 10-round staggered row magazine, leather sight protector and rifle style buttplate with butt trap. The nosecap is similar to that on the following .303 Martini-Metford Cavalry Carbine Mk III and Martini-Enfield Cavalry models.

But there were some criticisms and so a new modified pattern was prepared and some of the previous models upgraded. Trials of these 50 improved carbines were conducted in January 1894, with the improvements recorded as being:

- Addition of a wooden handguard.
- Better balanced arm by lightening sections slightly.
- Magazine capacity reduced from 10 to 6 rounds.
- Bolt knob lever bent down because it injured the rider's knee when the carbine was in the bucket.
- Backsight altered as it was previously undersighted.

The reports in March 1894 resulted in further modifications and improvements:

- Fore-end extended to within 1-in. of the muzzle.
- Nosecap altered to incorporate protector wings like those of the German carbine.
- Solid band fitted to fore-end for better support.
- Bolt lever projection reduced (flattened handle knob).
- Fore-end and handguard reduced in size near muzzle.

It was also recommended that the safety catch as approved for the rifle be fitted to the bolt and that the barrel be reduced in length a little.

A pattern was prepared accordingly and submitted on 5th June 1894 and the pattern sealed on 15th June 1894. The Magazine Lee-Metford Carbine Mk I was approved on 29th September 1894 and announced in the List of Changes on 1st April 1895. They were only manufactured at Enfield and the total production was recorded as being 18,686 carbines.

Para 7751:-
Carbine, Magazine, Lee-Metford (Mark I)
With cleaning rod, cavalry only.

A pattern of this carbine has been sealed to govern future manufacture. It differs from the Lee-Metford magazine rifles Mark II (§ 7404) principally in the following particulars:

The bolt has an extension at the rear end, about one inch long, having two grooves for the safety catch. The lever is bent, so that it lies close to the body, and the knob of the lever [bolt handle] is flattened at the top.

The cocking piece is lengthened and fitted with a safety catch, with pin and spring, by means of which the whole action can be locked either at full cock or when the mainspring is eased.

The nose-cap (combined with the upper band) has two wings, to protect the foresight, and is recessed at the back to receive and secure the handguard at that end.

The band, which is oval in form inside, has no shoulders, and passes over both the handguard and fore-end of the stock, and is held in position by means of a band retaining spring, in the underside of the fore-end.

The handguard, which is of wood, extends from the nose-cap to the backsight, and is secured at the backsight and by the band; when it is required to remove it, the band screw is unscrewed, the band forced up to the nosecap, and the leaf of the backsight is raised. The guard can then be taken off.

The butt-plate is of the same size and shape as that of the Lee-Metford magazine rifle Mark I, but is made of delta metal instead of iron.*

The stock butt is smaller all over, except at the socket and butt-plate ends.

The trigger guard has a loop for magazine link across the front instead of at the side; it is also recessed at the front end so as to clear the link and link loop on the magazine when it is placed in the carbine.

The link loop, which is brazed on across the case, is smaller, as it has to pass through the guard.

The magazine is shallower and is made to hold six cartridges. The magazine spring is shorter, in consequence of the magazine case being shallower. It is of similar shape and is fitted to the case and platform in a similar manner to that of the Lee-Metford magazine rifle Mark II.

The Carbine is not provided with the dial or aperture sights.

The backsight is graduated for cordite. The sight leaf is marked from 600 to 2,000 yards and also on alternate sides of the leaf with short lines, denoting intermediate 50 yds. The sight bed is graduated from 200 to 500 yards.

The slide is reversible, the depth of the bar being the same as that of the V; it is marked with a centre line and two wind gauge lines. [The carbine slide is .025-in. narrower than the Lee-Metford rifle slides.]

The carbine has a sling bar let into the right side of the butt, fixed by two screws inserted from the left; a sling loop is swivelled in a lump on the left side of the body and is held in position by a pin driven into a hole in the body.

A recess is provided in the butt, as in the rifle, for the carriage of an oil bottle and a pullthrough.

Length of barrel	*20.75 ins.*
Calibre	*.303-in.*
Rifling	*Metford system*
Grooves	*7*
Groove depth	*.004-in.*
Width of lands	*.023-in.*
Spiral (left handed) ...	*1 turn in 10 ins. or 33 calibres*
Length of carbine	*3 ft 3.9 ins.*
Weight of carbine	*7 lb. 7 oz.*
Weight of magazine, empty	*3.5 oz.*
Weight of magazine, filled	*9.25 oz.*

The following components are special to this carbine:-

Band.
Barrel.
Bar, sling.
Body.
Bolt, breech.
Cap, nose.
Case, magazine.
Catch, safety.
Cocking piece.
Guard, hand.
Guard, trigger.
Link, magazine.
Loop, sling.
Pins, loop, sling.
Pins, catch, safety.
Plate, butt (except pin, spring & trap).
Screws, bar, sling, long.
Screws, bar, sling, short.
Screw, cap, nose.
Screw, keeper, striker.
Spring, band, retaining.
Spring, magazine, platform.
Spring, pin, safety catch.
Stock, butt.
Stock, fore-end.
Striker.

The following components are common to this carbine and the Lee-Metford magazine rifle, Mark II:–

Bolt, stock.
Catch, magazine.
Collar, screw, guard, front.
Cover, bolt, breech.
Extractor.
Cut-off.
Head, bolt, breech.
Pin, axis, sight, back.
Pin, axis, trap, butt plate.
Pin, catch, magazine.
Pin, spring, magazine platform.
Pin, trigger.
Platform, magazine.
Rivets, fore-end.
Screws, band.
Screw, cut-off.

"D" loop sling swivel, originally on all the Magazine Lee-Metford Cavalry Carbines. This was ordered to be removed in March of 1896; few carbines are encountered with this swivel today. *MOD Pattern Room*

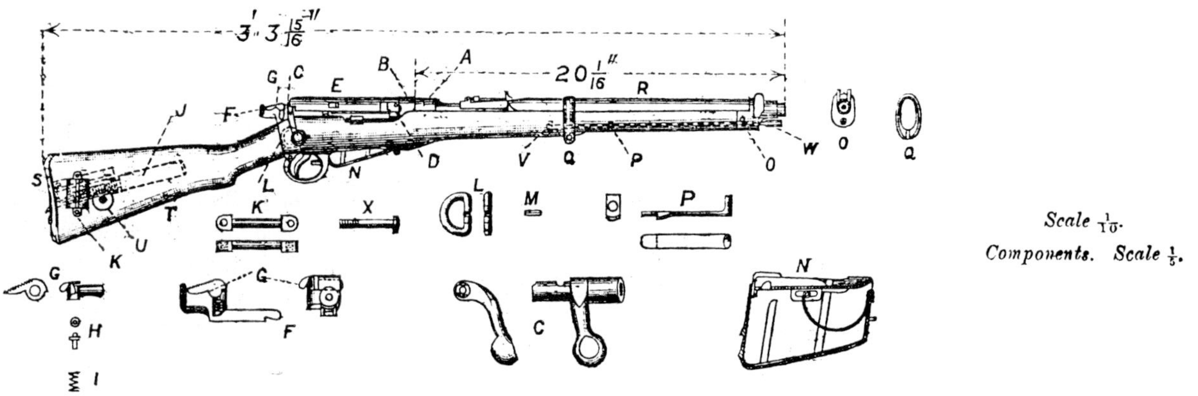

A.—Barrel.
B.—Body.
C.—Bolt.
D.—Bolt head.
E.—Bolt cover.
F.—Cocking piece.
G.—Safety catch.
H.—Safety catch pin.
I.—Safety catch pin spring.
J.—Oil bottle.
K.—Sling bar.
L.—Sling loop.
M.—Sling loop pin.
N.—Magazine.
O.—Nosecap.
P.—Spring, band, retaining.
Q.—Band.
R.—Hand guard.
S.—Butt plate.
T.—Butt stock.
U.—Disc, marking.
V.—Rod nut.
W.—Clearing rod.
X.—Sling bar screw { one short. one long.

A very handy arm, this carbine is ideal for cavalry use and was favoured by mounted troops. It also introduced the bolt-mounted safety catch which then resulted in the advance to Mark II* Lee-Metford rifle, and like the Mk I* Lee-Metford rifle, the carbine utilised a butt marking disc instead of the long buttplate as on the contemporary rifles. Those special fittings for mounted use include the neat nosecap with integral foresight protectors, a recessed sling bar in the butt and a sling loop on the left side of the butt socket. The short 6-round magazine was also an unusual feature.

It was decided to abolish the sling fitting on the left side of the receiver on 6th March 1896. This modification was effected locally, in accordance with these instructions:

The "D" loop and projection are to be sawn off with a frame saw, and the side of the body smoothed with a file and blacked with "pale golden lacquer" and lamp black, mixed so as to form a paste. The body should be slightly warmed to ensure the blacking setting well. A quarter of a pint of golden lacquer and a quarter-pound of lamp black will suffice for 100 carbines.

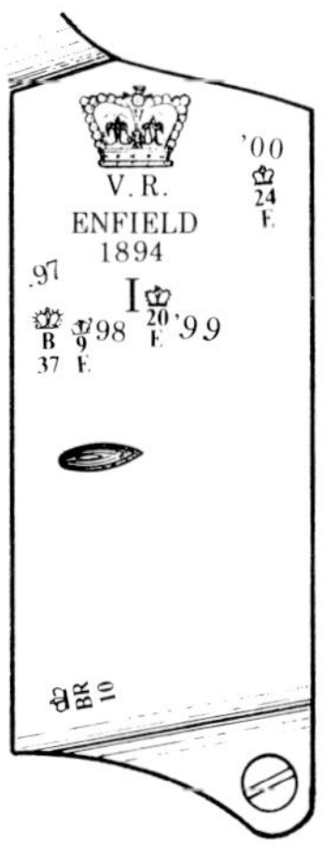

Typical Inspection marks on rifles and carbines. Stamped as subsequent modifications or repairs were effected and usually include the year, eg. '97, '99 or '00 (illustrated).

Another alteration to the carbine approved after the introduction of the Lee-Enfield carbine, was the fitting of a leather backsight cover, secured by a wood screw on each side of the fore-end; a longer one on the right side and a shorter one on the left. This cover was affixed to the left screw and secured over the right screw head when the sight was not in use, this alteration was effected locally by unit armourers.

Although the cavalry carbine was approved in September 1894, the first hundred were manufactured in the fiscal year 1st April 1892 – 31st March 1893, with a further fifty during the following year. These were most likely trials models and probably differ from the subsequent pattern in minor details such as the safety catch or stocking-up. Mass production commenced during the 1894-95 financial year and the service carbines were only produced at Enfield R.S.A.F.

Enfield production registers indicate that during 1895-96, 150 Lee-Metford carbines were supplied new with Morris tubes, there were 5 fitted with special slings, 735 were assembled without sling loops and then 30 without sling loops or sling bars in the butt. The last-mentioned Metford carbines would therefore resemble the Lee-Enfield carbine as introduced during the following year, except for the form of rifling and small differences such as the sights and sight protecting wings. The latter are only likely to be distinguished only after a close examination.

Introduction of the safety catch with the cavalry carbine was then implemented on the magazine rifle and resulted in the advancement of the rifle Mark from II to II*. This was approved on 22nd April 1895 and described in the monthly List of Changes issued on 1st August 1895 as below.

Para 7879:–

Rifle, Magazine, Lee-Metford (Mark II)*
With clearing rod.

A pattern of the above-mentioned rifle has been sealed to govern future manufacture. It differs from the Magazine Lee-Metford rifle, Mark II (§§ 7404 & 7573) in the following particulars:

The bolt has an extension at the rear end, of about 1 inch, in which are two grooves for the safety catch.

The cocking piece is lengthened and is fitted with a safety catch with pin and spring.

The safety catch consists of a transverse bar, capable of being turned by a thumbpiece attached to it. The bar is cylindrical, but has a portion cut away, so that when it is in the position shown in Fig. 1, the firing bolt is free to pass it, but when in the position shown in Fig. 2, the cylinder engages in one or other of the two grooves cut in the firing bolt (according to whether the arm is cocked or otherwise) and locks the action.

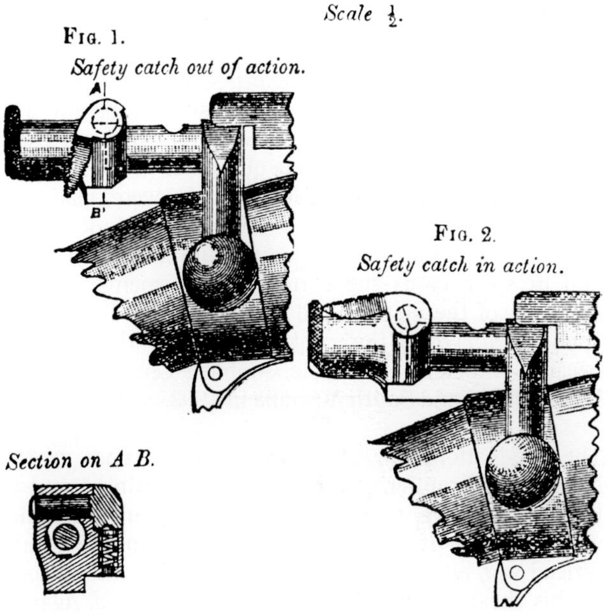

The striker is lengthened to suit the altered cocking piece. The undermentioned parts are interchangeable with those of Magazine Lee-Metford Cavalry Carbine, Mark I (§ 7751), viz.:–

Catch, safety.
Cocking piece.
Pin, catch, safety.
Spring, pin, catch, safety.
Screw, keeper, striker.
Striker.

This model, especially in its original form with Metford barrel, is quite scarce today. With total production recorded as being only 13,479 rifles at the two government factories of Enfield and Sparkbrook, it represents only one in every forty-five of the Marks I, I* and II Lee-Metford rifles. None of the service Mark II* rifles were produced by B.S.A. or L.S.A.

Stronger wire was used for the piling swivels of Marks II and II* Lee-Metford rifles, but whilst they were less likely to bend, they were more prone to fracture at the joint where attached to the screw. So, on 1st July 1895, approval was given for a small radius to be made at the point "A," instead of sharp angles. This pattern swivel was carried over to the Magazine Lee-Enfield rifle series.

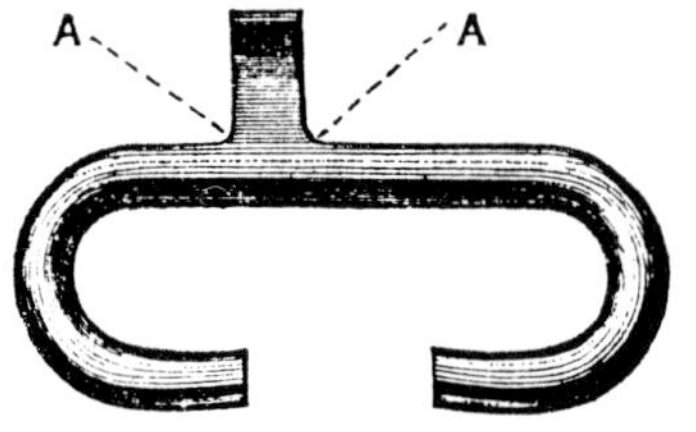

The Magazine Lee-Enfield rifle was introduced a few months afterwards as a result of the poor showing of Metford's rifling when used with the smokeless cordite cartridges. But even after the introduction of the Lee-Enfield in 1895, the Lee-Metford was sill manufactured in various configurations and many were modified and given a new lease of life. Considerable quantities of the standard service configuration Lee-Metford rifle were made on Colonial orders for some years afterwards. Mark I* Lee-Metford rifles of post-1900 manufacture have been noted and in some cases the current improvements such as omission of the clearing rod and groove have also been applied. Not only the contractors filled these orders because Enfield, for example, supplied Lee-Metfords for India in 1901 and South Australia in 1903.

The Lee-Metford saw active service many times, in the North-West Frontier of the Indian sub-continent and in the Sudan, where its record of performance and reliability was excellent. It was used quite widely in the South African war, although there were more Lee-Enfields used in this theatre, especially as the war progressed. Against massed charges of poorly armed tribesmen it performed admirably but against the more experienced and prudent Boers it fared less effectively.

A number of Lee-Metfords were fitted with Enfield-rifled barrels when replacement became necessary; it was noted that Enfield-rifled barrels had a different M.P.I., so on 25th February 1901, orders were issued that their Metford sights be altered to compensate for this. Foresights had been offset to the left to compensate for bullet drift and these foresights were marked with a "*." Such barrels were fitted with central "V" backsight slides and another slide was produced upon which the "V" notch was .003-in. to the left of centre to compensate for drift. As the foresight was not adjustable, approval was given for use of the slide with the offset "V" notch to be replaced with the centrally located pattern in instances where the rifle shot to the left by 5 inches or more. Thus, some of the interchangeability problems between Enfield and Metford rifled barrels were rectified.

Due to the detrimental effect of the cordite propellant on Metford rifled barrels, it was inevitable that replacement barrel supplies would run short, so new barrels with Enfield rifling, but with the Lee-Metford Mark I or I* external profiles, were made up for fitting to the Mark I* Lee-Metford rifles. These were marked "E I*" on the knox form. Replacement of worn Mark II and II* barrels was simple because their barrel profiles are identical to the Lee-Enfield Mk I and I* and they could be readily substituted. Lee-Enfield barrels are marked with an "E" on the knox form, whereas the Metford barrels have no such indication as to the rifling form marked on their barrels.

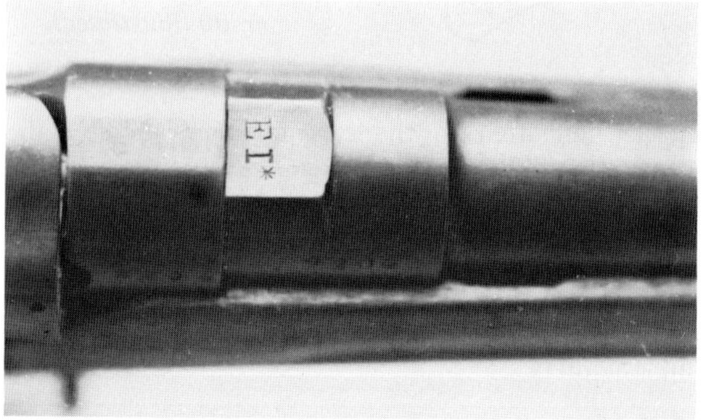

This indicates Enfield rifling in a Metford profile barrel.

Approval was given on 13th January 1902, for Lee-Metford carbines which were fitted with Enfield barrels and with the foresight wings drawn out (to the same height as those on the Lee-Enfield carbines) to be listed as Lee-Enfield Mk I carbines. These Lee-Enfield barrels, too, were marked with an "E" on the knox form to denote the rifling.

A similar change was introduced in nomenclature, where Mark II* Lee-Metford rifles were fitted with Enfield rifled barrels. Where the Mk II* Lee-Metford rifle fore-end was retained, they were to be known as Lee-Enfield Mark I rifles; upon a Lee-Enfield fore-end without provision for the clearing rod being fitted, they became known as the Mark I* Lee-Enfield. Mark designations on the receiver were reputedly altered to conform with the new nomenclature, although no such examples have been noted to date.

Some special models of the Lee-Metford rifle were made up for India in 1904, consisting of otherwise standard Mk I* and Mk II rifles fitted with the S.M.L.E. Mk I pattern of charger-loading guide. To do this, the dustcover and its securing ribs were removed from the bolt and the Mk I S.M.L.E. bolt-head fitted. A corresponding charger tab was fixed onto the left wall of the receiver. In some cases a new S.M.L.E. bolt was required and other small modifications had to be made. It appears that these rifles were converted in Birmingham in 1904, at the Sparkbrook factory. This pattern of Charger-Loader was the basis for the first rifle produced in India at Ishapore, but only samples were made in England. Further details are provided in Chapter 10.

Handguards on the Lee-Enfield rifle were found to be prone to splitting along the wood grain and so a local repair was authorised which has been recorded in the List of Changes § 10763, issued 1st December 1901 and 1st March 1902. The repair consisted of inserting two steel "rivets" of No. 16 WG wire, across the width of the handguard after the split was glued and the guard placed on a mandrel. A somewhat similar method was later used in Australian-made S.M.L.E. handguards, except that threaded brass rod was used instead of the steel wire.

Scale ½.

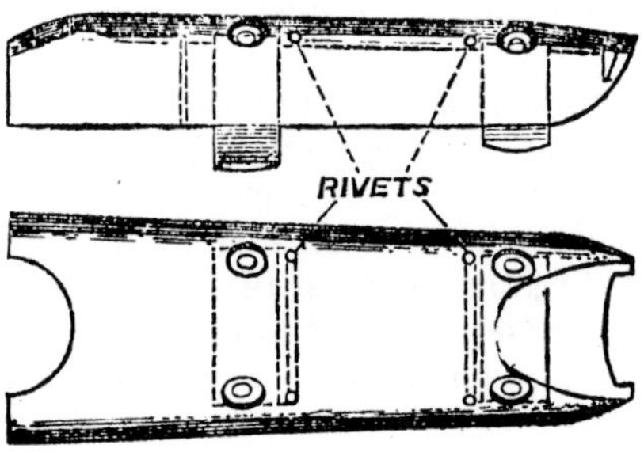

The rivet positioning for repair of Mk I* Lee-Metford handguards was as illustrated; the position as applied to the Marks II and II* rifles is also shown.

Scale ½.
M.L.M. (Marks II., II.*) and M.L.E. rifles.

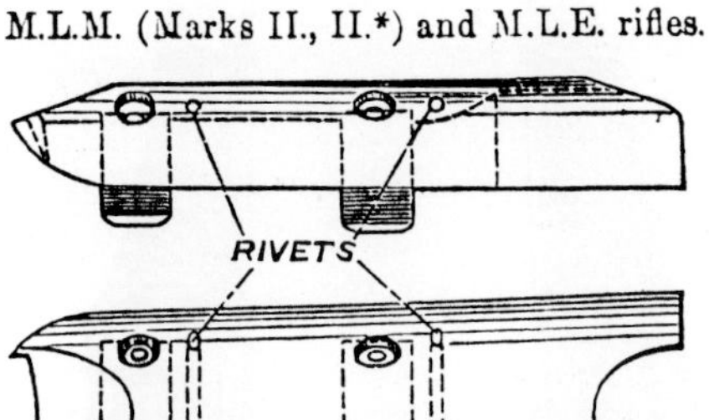

Although the Lee-Metford became a 2nd line issue after the introduction of the Short Lee-Enfield, modifications were still approved which are of interest in this study. On 16th February 1906, approval was given for reduction in weight of the single stage trigger pull-off, from a 6 lb. minimum and 8 lb. maximum, to a 5 lb. minimum and 7 lb. maximum, thus lightening the trigger pull. This was effected by increasing the angle of the cocking piece bent. A description and illustration is to be found in the List of Changes § 13136.

The new Short Magazine Lee-Enfield Mark III rifle introduced a different charger guide in 1907; this formed a bridge across the top of the receiver, and some Mk II Lee-Metford rifles were fitted with this charger bridge, along with a much larger number of Lee-Enfield rifles, to become known as "Charger Loaders." The official designation of the Lee-Metford version was "Rifle, Charger-Loading, Magazine, Lee-Metford Mark II" and it was approved along with its Lee-Enfield counterpart on 1st July 1907. The full text of this announcement is included towards the end of the following chapter.

In addition to the charger-loading bridge, new sights were fitted to the Charger Loading Lee-Metford and Lee-Enfield rifles, to conform more with those on the Mk III S.M.L.E. A blade foresight was utilised, differing from that of the Charger-Loading Lee-Enfield in that it is centrally mounted and marked with the letter "M"; the foresight protector, too, is stamped "M." Some Mk II* Lee-Metford rifles, previously upgraded to Mk I or Mk I* Lee-Enfield rifles, were also converted to Charger Loaders, but where Enfield barrels were fitted, they were made to conform with the C.L.L.E. pattern and described as such. A close examination of the components list in the List of Changes entry will reveal minor differences between the M.L.M. and M.L.E. Charger Loaders; this has also been reproduced in Chapter 4.

The Charger-Loading Lee-Metford conversion was ordered to be discontinued in 1909 and the rifles were to be fitted with the Lee-Enfield bolt with safety catch, to become known as Charger Loading Lee-Enfield Mk I* rifles. Many were refurbished with Enfield barrels. Subsequent upon there being no Metford barrels fitted any longer, in 1910 it was approved that the distinguishing letter "E" be omitted from the knox form of replacement barrels for Lee-Metford and Lee-Enfield rifles and carbines. Enfield-rifled barrels for the Mark I* Lee-Metford rifle (marked "E I*") also had this marking omitted. It was instructed that, in future, any Lee-Metford rifles fitted with Enfield barrels would be referred to as Lee-Enfield rifles, this being marked on the receiver. Similarly, Lee-Metford carbines became known as Lee-Enfield carbines in instances where the Enfield-rifled barrel was fitted.

A special front handguard was approved in 1907 for the Lee-Metford rifle, made of asbestos and supplied through the trade because it was not of government manufacture. This handguard was also approved for use on the Lee-Enfield and Charger Loading rifles. The announcement in the List of Changes reads:

Para 14069:-
Guard, Hand, Front, M.L.M. Rifle
Asbestos; Pat. No. 13754 of 1899 & 6934 of 1907;
also for M.L.E., M.L.M.C.L. & M.L.E.C.L. rifles.

A pattern of the above-mentioned handguard has been sealed to govern supplies for the use of officers, non-commissioned officers and men in voluntary practices.

The handguard is made of unglazed black cotton twill, with a wire frame sewn in, and lined with asbestos; five bands of black webbing are sewn on to fix the handguard to the rifle, the bands being held by spring buttons.

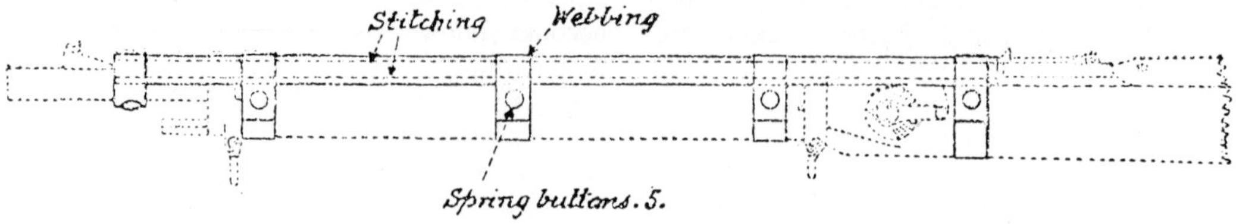

The handguard is fitted on the rifle between the front and back sights, and completely covers the exposed portion of the barrel.

The handguard will not be supplied from store and when required is to be obtained direct from the trade by the purchaser.

Some of the nomenclature changes for components of the Charger Loading Lee-Metford rifle were recorded in the List of Changes, § 14758 in 1909, although by this time the service life of the Lee-Metford was virtually at an end, except for the occasional rifle or Charger-Loading conversion used by Naval forces or reservists.

Another List of Changes entry, § 15304, refers to Naval service of the M.L.M. and M.L.E. rifles in relation to the fitting of spare boltheads. To apply case-hardening it was instructed that a piece of wet rag be wrapped four or five times around the screwed stem of the bolthead, gripped in tongs. Then the face of the bolt-head was heated to a blood red colour in the forge and the part to be hardened was covered with scintilla powder, replaced in the fire so as to not remove the powder and again made red-hot for about one and a half minutes. The red-hot bolt-head was then dipped in cold water, cleaned and adjusted.

It was not until 1926 that the Lee-Metford rifle was declared obsolete, although some C.L.L.M. rifles were again used during World War 2 and components such as bolts and boltheads manufactured by Parker Hale in Birmingham. Even in 1939 and 1940, some M.L.M. Mk II cocking pieces were converted to the flat-sided S.M.L.E. type by grinding and machining, for fitting onto the Navy issue .22RF Short Rifle Mk II.

Above & below: Test sample Service Competition backsight and slide fitted to Mark II M.L.M. rifle. *Ian Skennerton*

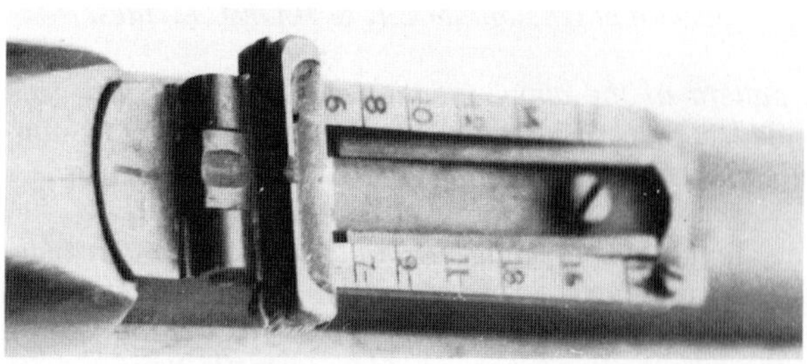

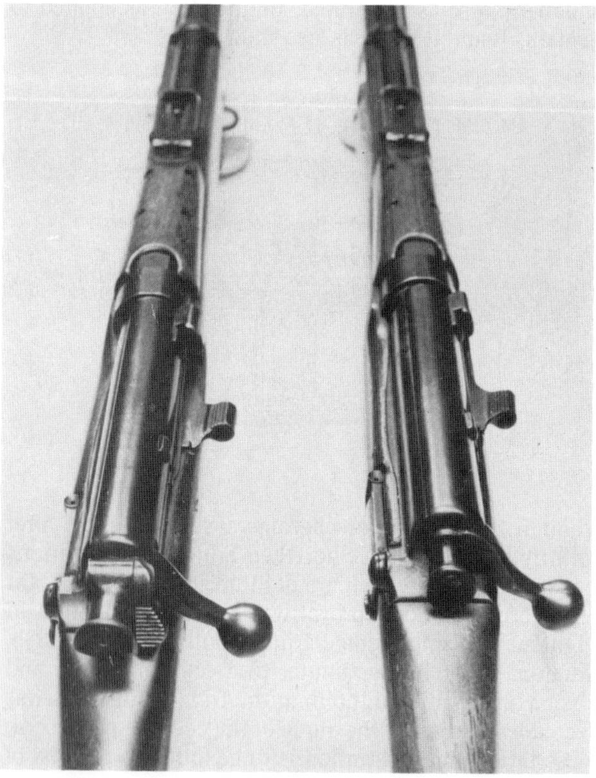

Bolt comparison. Safety catch on Mk II* M.L.M. rifle (left) compared with Mk II M.L.M. model (right).
MOD Pattern Room

THE MAGAZINE LEE-ENFIELD

CHAPTER 4

The War Office, the delegated Small Arms Committee and the associated service manufactories all originally expected that a smokeless propellant cartridge powder would be introduced in conjunction with the Magazine Lee-Metford rifle. But a number of difficulties were being experienced with such propellants so it was decided to use black powder in the interim. Chemists at the War Department and the Royal Laboratory Woolwich were hard pressed to resolve the problems and encountered difficulties in producing a smokeless powder in the short period of time designated.

To summarise the course of events, the early service ball cartridge progression is again outlined. On 20th February 1889, the Mark I .303 ball cartridge was approved; the previous 1888 Trials rifles had been chambered for the Rubin cartridge. There was 71.5 grains of pelletised black powder in the Mark I cartridge to propel a 215 grain bullet, the primer was a Boxer design with an integrated anvil. A Mark II version was approved on 19th July 1890, the difference mainly consisting of a change in the bullet.

A smokeless cartridge, the Cordite Mark I was initially approved on 3rd November 1891 and it underwent some minor changes within a short space of time until the Mark II Cordite round was approved on 17th July 1893. The Cordite Mk I case was Boxer-primed whereas the Cordite Mk II used the larger Berdan primer. Cordite rounds can usually be identified by the letter "C" along with the Mark number on the case headstamp, whereas the black powder loads had no such distinguishing letter along with the Mark number. The most conspicuous exception to this occurred with the first supply of cordite cartridges which had varnished cases; the headstamp used the letter "C" without any distinguishing numeral or Mark number.

These early .303-in. cartridge details have been provided as a background to the introduction of Enfield rifling. For comprehensive, illustrated descriptions of the considerable variety of British .303 ball, blank, dummy, incendiary, tracer and armour-piercing cartridges, Barry Temple's "Identification Manual on the .303 British Service Cartridge" (Vols. 1 – 5) is highly recommended.

"Cordite" was so named because it resembles strands or cords; composed of 58% nitro-glycerine, 37% gun cotton and 5% mineral jelly. It was much more stable than black powder and produced little or no fouling. Its main disadvantage was the greater heat generated, which, in

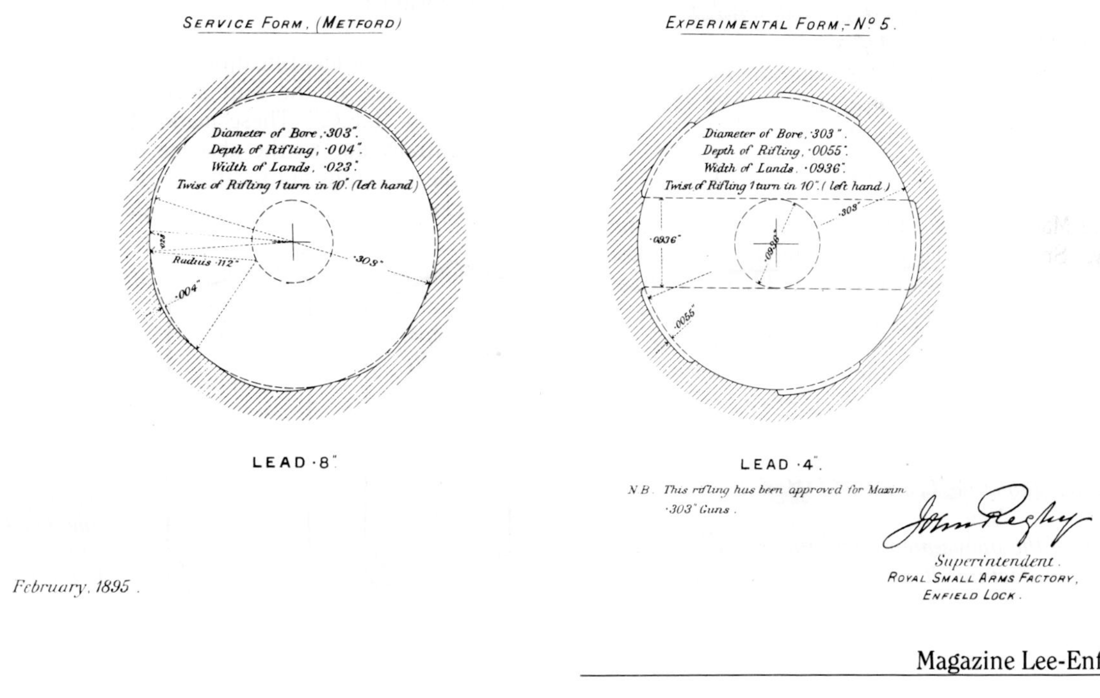

The clearing rods were officially abolished for all .303-in. rifles and carbines on 11th and 19th May 1899 and although this action may seem to be of scant significance, it resulted in an advancement of nomenclature of the Lee-Enfield rifle and carbine from Mark I to Mark I*. So, the new rifle became the Magazine Lee-Enfield Rifle Mark I* and the new carbine, the Magazine Lee-Enfield Carbine Mark I*. The only differences are associated with the absence of clearing rod and its provision in the fore-end.

The .303-inch Martini-Enfield carbines are also referred to in the official announcement in the List of Changes and this notice reads:

Para 9700:-
Carbines, Artillery, Martini-Enfield, Marks I and II**
Carbines, Cavalry-
*Magazine Lee-Enfield, Mark I**
*Martini-Enfield, Mark I**
*Rifle, Magazine, Lee-Enfield, Mark I**

Abolition of clearing rod from all .303-in. rifles and carbines.

It has been decided that the use of clearing rods with .303-inch rifles and carbines shall be discontinued. The rods will become obsolete and will be dealt with accordingly.

Patterns of the above-mentioned rifles and carbines have been approved to govern future manufacture. They differ from the previous pattern only in the omission of the clearing rod and the hole, groove and rod nut in fore-end, and hole in nose cap. In the case of the Magazine Cavalry Carbine, the hole for clearing rod in the band, retaining spring, is also omitted.

This particular designation advance was marked on the side of the butt socket, stamped after the Mark numeral "I" which then becomes "I*." As the rod hole in the carbine is not obvious except for the hole in the nosecap, it is difficult to distinguish a Mk I from Mk I* Lee-Enfield carbine without inspecting either the markings or the nosecap. Some Mark I carbines were upgraded by fitting the Mk I* nosecap. On the rifle, the long, open groove underneath the forward portion of the fore-end was deleted on the Mk I* and there is no corresponding hole in the nosecap/sword bar. Rods were also withdrawn from earlier model rifles and carbines in service.

Regarding production of the new, upgraded models without clearing rod provision, some 590,000 Mk I* rifles were made by the two government factories and the two contractors while 26,647 Mk I* cavalry carbines were manufactured at Enfield. Production of the Mk I* variant was nearly double that of its Mk I predecessor.

Some interesting anomalies may be encountered regarding this change, usually in contract rifles or carbines made for Indian, colonial or volunteer orders. With the abolition of clearing rods, some Lee-Metford rifles of various Mark configurations (i.e. I*, II & II*) were made without provision for the clearing rods. This can be readily certified by checking the year of manufacture (post 1899) marked on the receiver body.

With some of the commercial or contract production, there are instances of official modifications and upgrades being left for some time before being implemented; this seems to have occurred more often with L.S.A. rather than B.S.A. It was obviously not as practical or economical for these non-government factories to alter machinery with as much rapidity as the Enfield or Sparkbrook factories. Similar anomalies may be encountered in some Mark I and I* S.M.L.E. rifles by L.S.A.

In January 1900, a new Small Arms Committee was formed, consisting of the Superintendents and Chief Inspectors at Enfield and Woolwich, the Commandant at Hythe, District Inspector of Musketry from Aldershot, a representative from the Navy, and later in the year, one from the National Rifle Association. The war in South Africa was still in progress and service equipment was under review in light of battle experience as complaints were forthcoming from the field. Some fresh complaints were levelled at the Lee-Metford and Lee-Enfield rifles and carbines, mostly through combat experience and comparison with the 7mm Mauser Model 1896 in the proficient hands of the burghers.

It has been mentioned in the second chapter (on the trials) that there were claims of bias on the earlier Committees, so care had been taken in the structure and membership of the new Committee in that departments of both manufacture and inspection of arms and ammunition were represented, in addition to the Army, Navy, and soon afterwards, the National Rifle Association. The President was considered to be impartial as he had no commitments in the fields of manufacture, design or inspection. The constitution of this Committee was approved by the Secretary of State on 18th January 1900 and the care in selection of members was largely due to the significance of the their assignment.

There had been a series of complaints from South Africa about the rifle sighting, duly substantiated by tests with new production rifles after their delivery to store in England; orders were given that sighting of rifles under manufacture be suspended, as of 22nd January 1900. An emergency alteration was put into operation, consisting of inserting backsight leaves with the notch .03-in. to the left of the centre line, so as to maintain a rapid supply to the troops in South Africa.

An altered foresight was recommended for rifles under manufacture, similar to that of the Mark I* Lee-Metford rifle converted from the Lewes system. Another recommendation was that the amount of deflection be altered to .02-in. to the left of the axis of the bore, instead of .05-in. as was the current practice. However, both these recommendations were held over, pending further trials results.

LEE-ENFIELD CARBINES

Left side view of Lee-Enfield Cavalry Carbine, marked with the Royal Cypher over "ENFIELD", year of manufacture and "L.E.C." over "I", the Mark number. This indicates a Mark I Lee-Enfield Carbine. Note the short, 6-round magazine. *MOD Pattern Room*

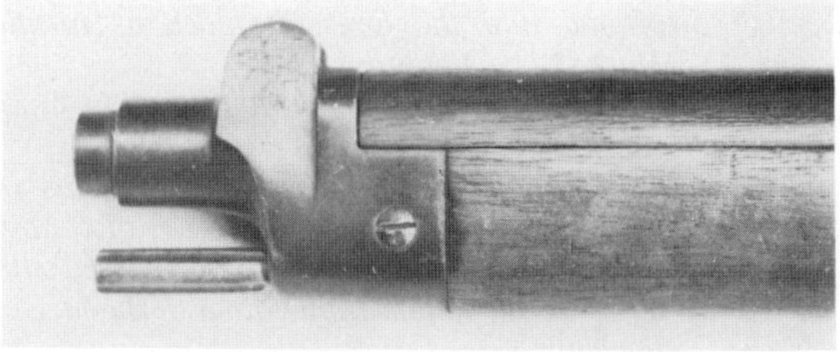

The Lee-Enfield Carbine has slightly higher foresight wings than its Lee-Metford forebear. *MOD Pattern Room*

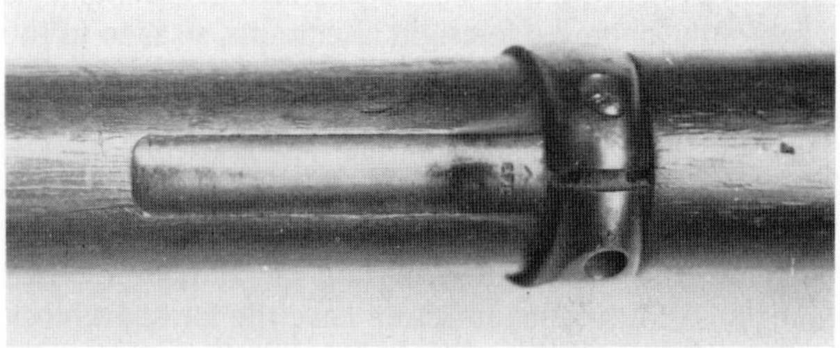

Spring band retainer on the underside of Lee-Metford and Lee-Enfield Cavalry Carbine fore-ends. *MOD Pattern Room*

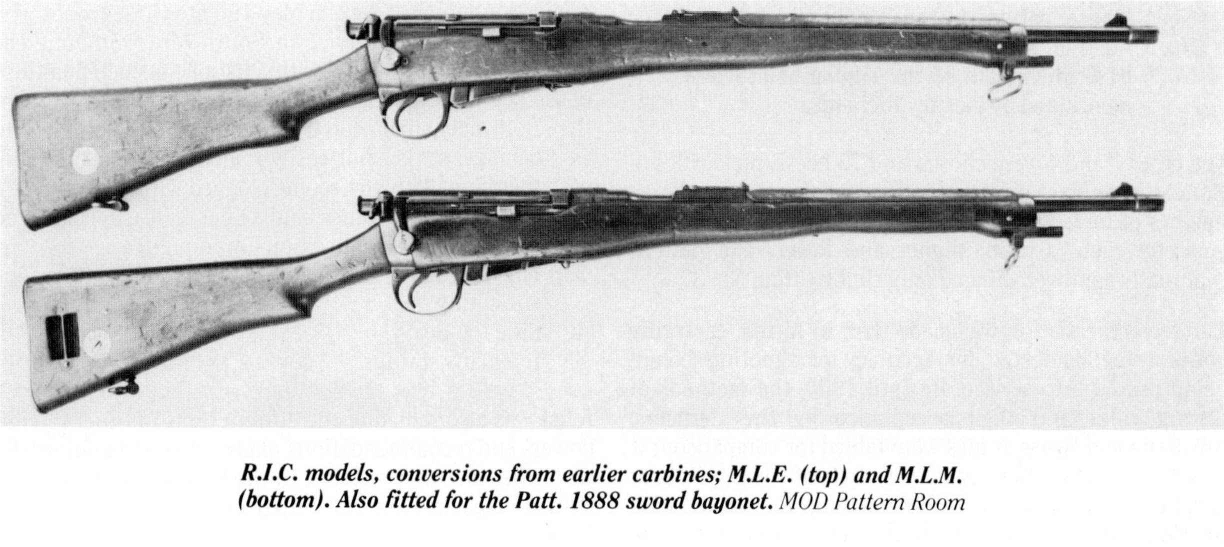

R.I.C. models, conversions from earlier carbines; M.L.E. (top) and M.L.M. (bottom). Also fitted for the Patt. 1888 sword bayonet. *MOD Pattern Room*

Ten rifles each, of Sparkbrook, B.S.A. and Enfield make, and of as early manufacture as was available, were tested against ten each of current production from Sparkbrook and Enfield. The new Sparkbrook rifles were found to have a deflection of 20 inches to the right at 200 yards while the others had a mean error of 6 to 12 inches at the same distance. All rifles were found to be undersighted; the Enfield and Sparkbrook rifles were 70 yards short at 600 yards and the B.S.A. rifles were 50 yards short at that distance. The ammunition used in these tests was Mark II, of Royal Laboratory manufacture.

The sighting and accuracy difference was highlighted between the groups of rifles from the different factories and it was realised that a fixed foresight would be impractical because of this variance. So, a moveable foresight was recommended for adoption, to permit correction and calibration.

There was no means of lateral adjustment on the Lee-Metford or Lee-Enfield rearsight, so a moveable foresight was the only alternative and such a foresight would require some form of sight protector. A fixed foresight protector was proposed in October 1898 and tests by both the Army and Navy proved satisfactory, resulting in its recommendation for adoption. This pattern appears to have been a hood type secured by a split pin, while a similar model secured by a screw was being tested at Aldershot by the Navy. It was suggested that advantage be taken of the approved foresight alterations, so as to fit the fixed foresight protectors at the same time.

The Enfield Royal Small Arms Factory submitted a drawing and details for a moveable barleycorn sight, fitted into a dovetail on the sight block, made of steel and hardened and tempered by blazing off. It was proposed that rifles of future manufacture have the foresight block soldered on centrally, and the adjustable foresight moved to the left. For the actual form of the foresight, six patterns were submitted:

Nos. 1 – 3, by the Superintendent, R.S.A.F.
No. 4, Mauser pattern, by the Commandant, Hythe.
No. 5, Austrian Mannlicher, similar to No. 4.
No. 6, by Commandant, Hythe, similar to a pattern mentioned by Westley Richards.

The types 3 and 4 were chosen and 75 Lee-Enfield rifles of Enfield manufacture were allocated for trials. Three heights of foresights were proposed; the standard height and one each, .015-in. higher and lower. The pattern eventually approved differed only slightly from No. 3.

The recommendation was made that, in future, each rifle and carbine be tested for accuracy of shooting before being passed into service. In April 1900, the methods of testing rifles and their acceptance by the Germans, Austrians and Swiss armies were tabled for comparison; it was noted that in these countries, each rifle was tested for accuracy and not just a small percentage as was the case in England. On these grounds, and considering the sighting defects which had occurred largely due to the lack of individual testing, the Committee decided that after manufacture, the shooting and accuracy tests would be conducted with all rifles and carbines.

The uneven bearing of the fore-end and also on the shoulder parts of the breech bolt was another cause of inaccuracy. Tests were conducted but it was discovered that generally the effects were not sufficient to warrant any further experiments, although a little more clearance was allowed between the fore-end and the barrel so that slight movement was permitted. A clearance of .01-inch was approved and this necessitated enlargement of the barrel hole in the nosecap and at the backsight in carbines. It was also decided to pay more attention to the bearing of resistance shoulders on the bolts and bodies during manufacture of Lee-Enfield rifles and carbines.

The sighting problems were considered resolved upon the implementation of the following instructions, issued in the List of Changes § 10393, and approved on 22nd October 1900:

Method 1. *For future manufacture:–*
Barleycorn foresight to be solid and fixed .02-in. to the left of the axis of the bore.

Method 2. *For rifles sent for repair to Birmingham:–*
Removable barleycorn foresight to be fitted .02-in. to the left of the axis of the bore and pinned into the foresight block as in the Mark I Lee-Metford.*

Method 3. *For rifles in the hands of troops and in Army store:–*
The V notches in the slides and caps of backsights to be .03-in. to the left of centre; rifles altered by this method to be marked with a "" at the bottom right corner of the face of the leaf.*

Rifles altered by either Method 1 or 2 were marked with a "" on the right side of the foresight block.*

A questionnaire was sent in May 1900 to those troops who had been on active service in South Africa; in June the Committee was presented with a list of considered failings of the Lee-Enfield. These were:

1. Bolt mechanism too complicated, bolthead likely to damage, striker not easily removed for cleaning and therefore likely to become clogged with oil.
2. Magazine too flimsy and exposed, easily damaged and could not be filled quickly enough.
3. The magazine catch and trigger were activated by one spring.
4. Rifle too heavy.
5. Too many springs.

A list was also tabled of the current arms of other foreign powers and recommendations made for a comparative trial with rifles on other systems. In the South African War, the British magazine rifle was pitted against the Mauser 1896 Model (essentially the Mod. 1893, although Boer Mausers

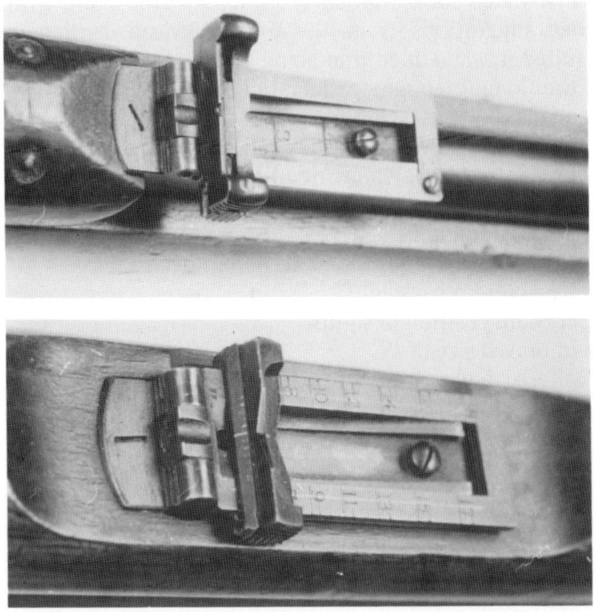

Above & below: 1900 and 1901 trials adjustable rifle backsights and foresights. *MOD Pattern Room*

are marked as Model 1896 and Model 1895) in most capable hands and few systems in use by any contemporary power would not have been found wanting in some respect.

The German Mauser rifle, the principle arm of the Boers during the South African conflict, could be clip loaded, had an integral magazine and was capable of excellent accuracy, complemented of course by the skill, marksmanship and fieldcraft of the burgher. Evidently, the considered failings Nos. 1 & 2 (previously listed) were influenced by a comparison with the Boer Mauser. The lighter double pull-off of the Mauser trigger was also preferred to the heavy, single stage Lee-Enfield trigger.

Looking at the overall view, the Committee considered that the following points should be decided upon:

1. Was .303-in. the ideal calibre? (The Boer M'96 Mauser was 7mm).
2. Should the magazine rifle be used as a single loader or solely as a magazine loader?
3. Was an auto-loading rifle desirable?
4. Was a rifle with a shorter barrel and longer bayonet desirable, with a view to one arm being used by all the services?

A new pattern backsight was recommended in January 1901 for trial, this was a much finer type and capable of more minute adjustment. Troops at Hythe, Salisbury and H.M.S. "Excellent" tested these rifles along with the hooded adjustable foresights. Modification of sights to this pattern required a new ramp, dovetailed and screwed to the original bed, and the fitting of a new leaf.

The Committee instructed that a rifle and carbine be prepared: a Lee-Enfield rifle with hood, new pattern removable sight, new pattern backsight and slide, and a longer handguard; and a Lee-Enfield carbine with a hood, bead foresight and new pattern backsight.

NOSE-CAP (WITH SIGHT PROTECTOR)

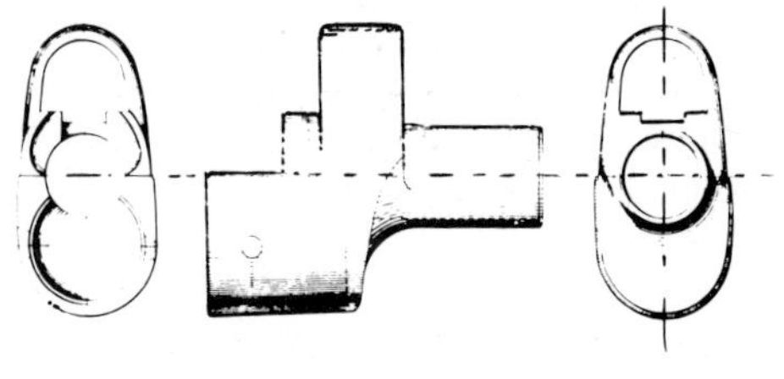

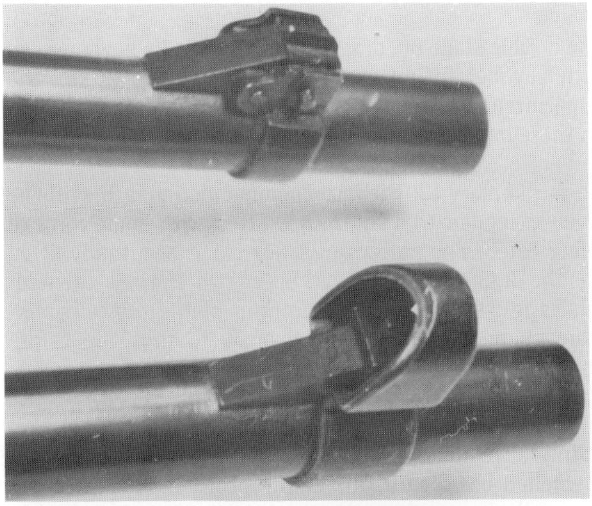

Trials carbine foresight protector, incorporated in the nosecap assembly. Note special foresight. *MOD Pattern Room*

On 2nd October 1900, these patterns were submitted before the Committee.

A few examples of the rifles and carbines set up at Enfield in an attempt to remedy these sight problems are presently in the Pattern Room collection, listed in their Catalogue, numbers RB214, RB217 and RB218. Some were specially marked, like the 1900 Cavalry Carbine "A Mk I*" which represents the final development of experimental sights and work with Westley Richards. With a new sight incorporating a windgauge, the backsight is adjustable with knurled knobs on each side, somewhat similar to that on the Charger-Loading Lee-Enfield rifle; the bead foresight is also adjustable in a dovetail on the sight block. This particular model was provisionally approved in 1901 and production specifications finalised in 1902, but not proceeded with due to the introduction of the Short Lee-Enfield rifle.

Adjustable rearsight on trials M.L.E. Mk I* Cavalry Carbine. Windage adjustment is on the slide. *MOD Pattern Room*

Other examples of the sighting trials rifles, such as RB217 and RB218 at the Pattern Room feature moveable foresights in a dovetail slot, windage adjustment on the backsight, improved backsight notches and frontsight profiles. Some also have different handguards and foresight protectors.

During October, further experiments were carried out with sights by the 3rd Dragoon Guards at Curragh, comprising of Lee-Enfield carbines fitted with four types of adjustable foresights:

Service barleycorn: Result; satisfactory.
U.S. Springfield: Result; clumsy and unsatisfactory.
Bead: Result; satisfactory.
Bead foresight with nosecap and hood combined, and broadened "V" on the cap and backsight:
Result; good, recommended for adoption.

The only real complaint was that the foresight hoods were found to scrape on the leather saddle buckets, so a smaller foresight hood was tried. This was found to be firm in the Mk III carbine bucket but would not fit the Mk I carbine bucket, M.H. Converted. The new parts proposed for the carbine were; nosecap with hood, leaf sight with wind slide, and moveable dovetail bead slide, at a total cost estimated at about 4/10d.

Also related to proposed improvements during this period after the turn of the century, a few different types of experimental rifles were fabricated with and without the long handguard (another Mauser influence) and with minor variations in foresights and nosecaps, but with the same type of backsights. All were fitted with adjustable foresights and this necessitated an enlargement of the foresight slot in the nosecap, which provides a feature for ready identification of these particular test rifles.

The requests for a lighter, more handy arm along with other suggestions from the South African battlefield resulted in a modified rifle being prepared and submitted to the Committee for consideration. Some of the features were:

Light barrel, by turning down the diameter.
Freedom of barrel (more free-floating).
Lightening of butt.
Long handguard.
Nosecap with improved mode of fixing bayonet.
New pattern backsight.
Forward position of sights.
Improved size of "V."
Altered drag pull-off.
Bead foresight.
Height of backsight and foresight above barrel axis.
Loading by chargers.
Polished wood furniture.

Designated the "No. 1 Modified Rifle," it weighed 8 lb. 6 oz. compared with 9 lb. 4 oz. for the service rifle. The sights incorporated the previous-listed recommendations while the new design of nosecap and bayonet mount removed the strain on the barrel when the bayonet was fixed, but entailed a new pattern of handle for the bayonet. No safety catch was fitted although the Committee considered one to be desirable.

The double stage or "drag pull" trigger and a new nosecap with bayonet fitting were designed by Col. Watkin, the Superintendent at R.S.A.F. Enfield; he took out patents on these improvements, No. 14,163, dated July 11th 1901.

On 7th January 1901 it was recommended that this rifle be given extended trial by troops, fitted with some form of safety catch. The pattern of safety catch eventually fitted was the design of Watkin and Speed, which they patented; this is illustrated at the beginning of the next chapter. More changes were yet to be made to this rifle before undergoing troop trials.

Further alterations were also recommended, consisting mainly of relatively minor sight modifications; fitting an enamelled bead sight, the addition of two new parts for the pull-off, moving the backsight assembly further forward to the lower band, and the incorporation of Mr. Punchard's invention for rapid filling of the magazine. A rifle with these modifications was then exhibited to the Committee.

There had been many requests for different lengths of butt for the service rifle since the introduction of the Lee-Metford, this was considered by some officers to be most important for the soldier's shooting. In September 1901, the Committee finally approved the introduction of short and long butts for the carbine and rifle and ordered that the distinguishing letters "L" or "S" be stamped on the appropriate butts, ½-in. in front of the tang of the buttplate. Short butts are ½-in. shorter than the normal model and the long butts are ½-in. longer.

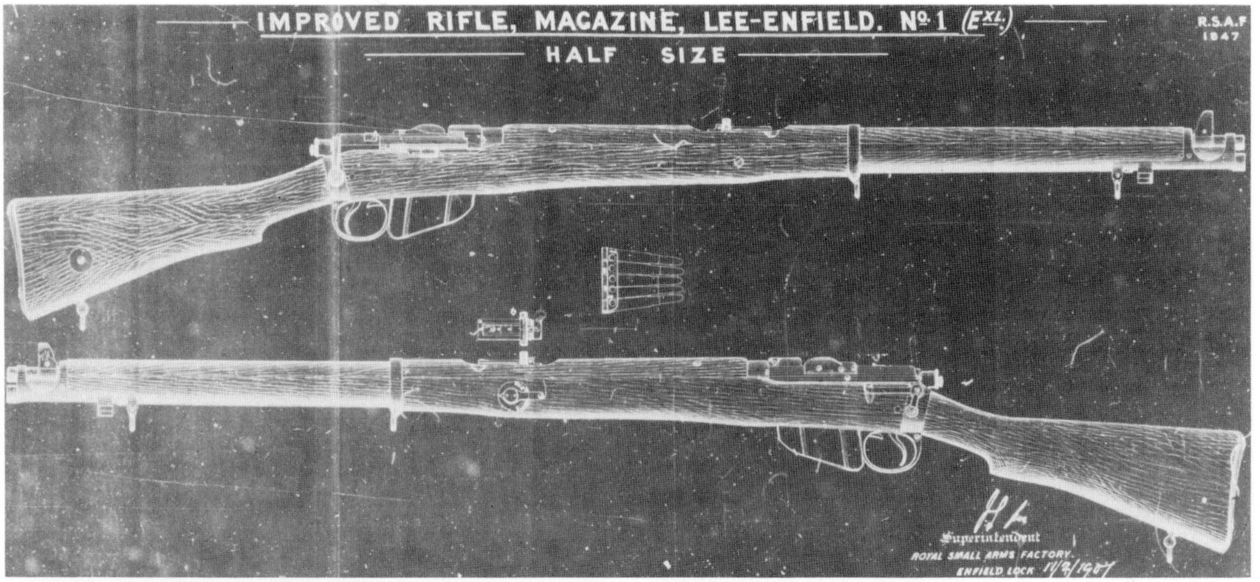

Whilst this 1901 experimental has charger-loading, sight, nosecap and bayonet stud features of the eventual Mk I SMLE, it still has the 30.2-in. long barrel. *MOD Pattern Room*

14,163. Watkin, H. S. S. July 11. *1901*

Firing - appliances generally applicable.—The trigger *a*, Fig. 1, is provided with two shoulders *c*, *e* which act successively on the sear arm *d* so as to give a "drag pull."

Sword bayonets, attaching to gun barrels.—A metal sleeve engages over the barrel and fore-end, and is connected thereto by screws *m*, *n*, Fig. 2. It is provided with a stud *g* to receive the hole in the cross *h* of the sword bayonet, with the usual sword bar *k*, and with a hole *p* for the barrel.

Sights.—The bayonet carrier is also formed with a hood *q* for protecting the fore-sight.

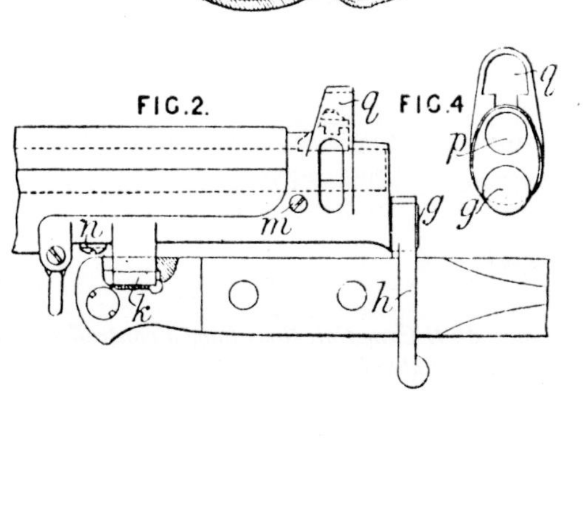

14,162. Watkin, H. S. S., and **Speed, J. J.** July 11. *1901*

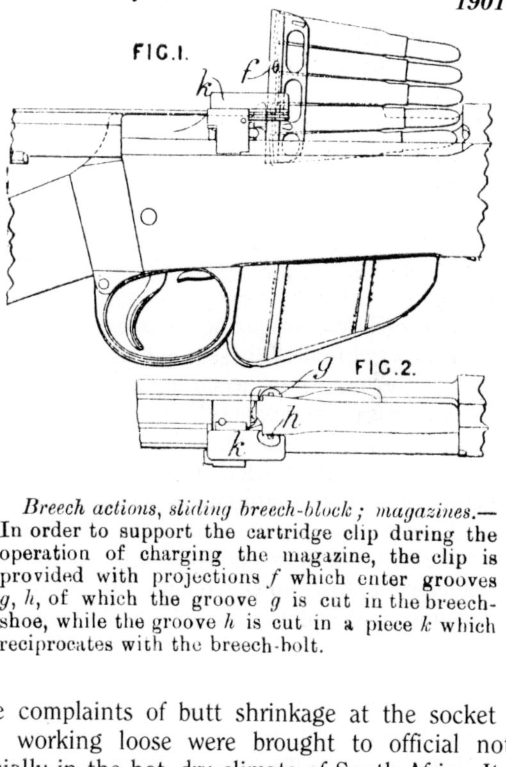

Breech actions, sliding breech-block; magazines.—In order to support the cartridge clip during the operation of charging the magazine, the clip is provided with projections *f* which enter grooves *g*, *h*, of which the groove *g* is cut in the breech-shoe, while the groove *h* is cut in a piece *k* which reciprocates with the breech-bolt.

Some complaints of butt shrinkage at the socket and butts working loose were brought to official notice, especially in the hot, dry climate of South Africa. It was noted that once the butts had worked loose, the stockbolt could not be readily re-tightened because a special long screwdriver was required.

There were two ways in which these problem could be readily corrected and so the following modifications were introduced. Firstly, the butt sockets were soaked in benzole and paraffin wax then compressed to size; these butts were marked with the letter "P" on the right side near the socket. Secondly, the threaded portion of the stockbolt which protruded through the back of the receiver was squared off after being lengthened slightly and a plate was inletted into the back of the fore-wood into which the tip of the stockbolt was fitted; stocks and stockbolts thus modified were designated "No. 2." The small plate let into the back of the fore-end was called a "keeper plate." Where this was fitted, the butt was marked "2"; in cases where both modifications were employed the butt was marked "P 2," on the small of the butt.

The stockbolt hole was deepened ¼-in. or if the stock was loose in the butt socket, the stock was shortened by a similar amount and new shoulders were formed to bear on the butt socket. These modifications permitted the squared end of the stockbolt to extend beyond the forward face of the butt socket.

Such butts with compressed sockets were produced for the Lee-Enfield, especially those rifles destined for Colonial service in warmer climates. Fewer of the squared stockbolts and keeper-fitted fore-ends were made for the Lee-Enfield although this mode of keeping the butt firm and securing the stockbolt was later applied to the S.M.L.E.

There were two patterns of the new modified rifle tried at the School of Musketry, Hythe, with quite favourable reports. The first type, designated the "No. 1 Modified Rifle with modifications as suggested by the Committee," had no butt trap and was provided with a charger-loading system designed by Speed and Watkin (and eventually adopted on the Mark I Short Magazine Lee-Enfield); it was well accepted.

A Punchard (Harris) magazine charger operated by a thumbpiece was fitted to the second type, in which the cartridges were "poured in" when the lever was depressed. No butt trap was fitted to this rifle either. The new pattern rearsight was found to be more vulnerable due to its increased size over the service model and there were also complaints of heat haze, although this was later explained as being vapour from wax underneath the handguard. Both these patterns were still fitted with the full-length, 30.2-in. barrel and to further reduce the weight it was recommended that longitudinal holes be drilled inside the butt.

The Harris patents were Nos. 10,239 and 16,284 of 1900 and Nos. 14,640 and 17,870 of 1901. These feature external control of the magazine platform, as the platform and spring are attached to the rifle body instead of the magazine case. The controlling lever is usually mounted on the left side of the fore-end. The Ross (Pat. No. 20,644 of 15th October 1901) and the Edwards (Pat. Nos. 4415, 6062, 9014 & 13,124 of 1901) systems were also tested. These were generally more simple than the Harris and the illustrated rifle shows a Ross system with controlling levers on both sides. These systems required a smaller capacity magazine, some of which were fixed, and others such as that illustrated, have a detachable case.

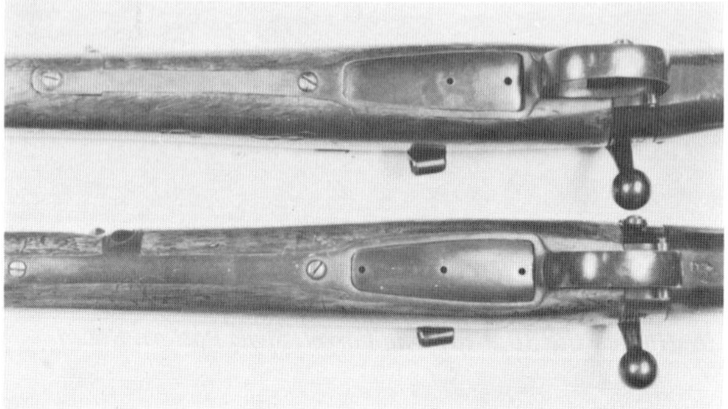

These trials rifles feature "controllable platform" loading systems, the magazine lifter being depressed by an external lever on the left side of the receiver. *MOD Pattern Room*

Trials were undertaken with these patent magazine and loading systems; the Harris type in December 1901 and again in 1902, and the Edwards patent in April 1902, but apart from certain mechanical faults, the loose cartridges were generally found to spill easily on the ground whilst being loaded. The success of 5-round charger clips precluded this and when it was decided in May 1902 that charger loading would be adopted with the new service short rifle, further consideration of these patent magazine systems was no longer necessary.

Such a controllable platform was used on the Mk I and Mk II series of Canadian Ross rifles and another feature of these experimental rifles, similar to the Ross, is the foresight hood. Sir Charles Ross brought his rifle over to England to have the War Office recommend it for trials which he hoped would result in its adoption for British service, but this was rejected.

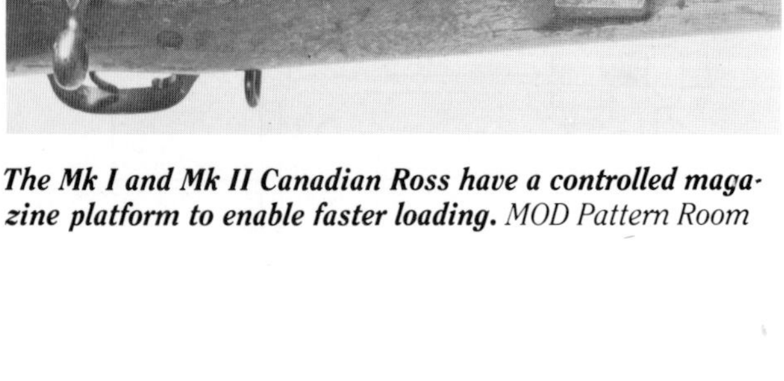

The Mk I and Mk II Canadian Ross have a controlled magazine platform to enable faster loading. *MOD Pattern Room*

In Canada, successful trials with the Ross had been conducted in Quebec in August 1901; then in April of 1902, the Canadian Department of Militia and Defence received the sealed pattern of the Mark I Ross to govern its manufacture there. Although the Marks I and II Ross rifles used the externally controlled magazine platform, on the Mark III rifle this feature was omitted. The controllable platform and foresight cover indicate some interchange of ideas between the Ross rifle and the Lee-Enfield experimental models.

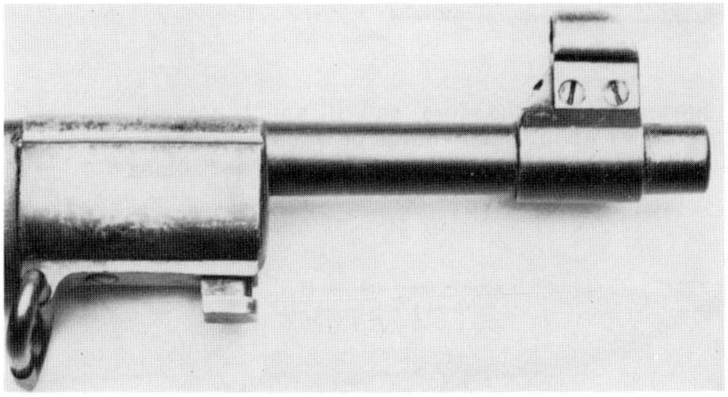

Ross rifle foresight hood protector. *MOD Pattern Room*

Another experimental Lee-Enfield rifle with further modifications was tested on 18th, 19th and 21st December 1901, with these additional alterations:

The barrel freed in the nosecap.
Altered pull-off.
Backsight moved forward and modified.
Foresight and backsight raised by .1-inch.
Holes drilled in the butt for weight reduction.
Bead sight altered.
Chargers improved to prevent lateral play of cartridges.

These modifications were reported upon after further testing, the only general complaints now being that the backsight centre was too small, the backsight was unsatisfactory, and of the modified rifles, those with the drilled (lightened) butts were found to be heavier! It should be realised that the tests and trials of various batches of rifles often occurred simultaneously, so that complaints were often rectified and the resulting improved rifles tested again whilst the original trials were still in progress.

There was still some development being carried on with the carbines and it was decided to incorporate the alterations which had thus far been reasonably well received, in a short rifle, so as to enable one length of rifle to be used for the various services. Requests had been made from South Africa for a longer-barrelled arm for the cavalry and Field Marshal Lord Roberts pressed for the adoption of a shortened rifle for both the infantry and cavalry.

The general complaint about the carbines related to their inferior accuracy, to which its shorter sighting radius contributed significantly. The war in South Africa demanded quick and accurate shooting at ranges which differed from those in previous wars; the Boers had no desire for close-quarter or hand to hand combat. In this respect, the weight and unwieldiness of the long rifle was also a disadvantage in quick movement from cover to cover. The extra reach with the sword bayonet was infrequently employed.

In due course, an experimental rifle was shortened by five inches to see if handling and balance was suitable and if such a length would be acceptable to both infantry and cavalry units. The illustrated rifle was one of the previous experimental models of 1901 with an adjustable foresight and hood, an experimental backsight and a controllable platform magazine system with two controlling levers. The rearsight was moved forward to the lower band as ordered in mid-December, hence the gap between the handguard and backsight. After this, the barrel and fore-end were shortened by 5 inches and the nosecap and foresight refitted. This particular test model can therefore be classified as the first prototype Short, Magazine Lee-Enfield rifle.

The following, obvious step was to shorten the Enfield Modified Rifle by some five inches, which is essentially the Short Lee-Enfield trials rifle; descriptions of this trials rifle and its subsequent development will be found at the beginning of the next chapter.

While the Short Magazine Lee-Enfield rifle was adopted in 1902, some special Magazine Lee-Enfield rifles and carbines were yet to be introduced. A more novel method of shortening the Lee-Enfield to render it more portable, especially for mounted infantry, was the application of a

Prototype Shortened Lee-Enfield; with controllable platform loading, adjustable backsight and foresight hood, and 25.2-in barrel. *MOD Pattern Room*

25,483. Rimington-Wilson, R. H. R. Dec. 23 1899.

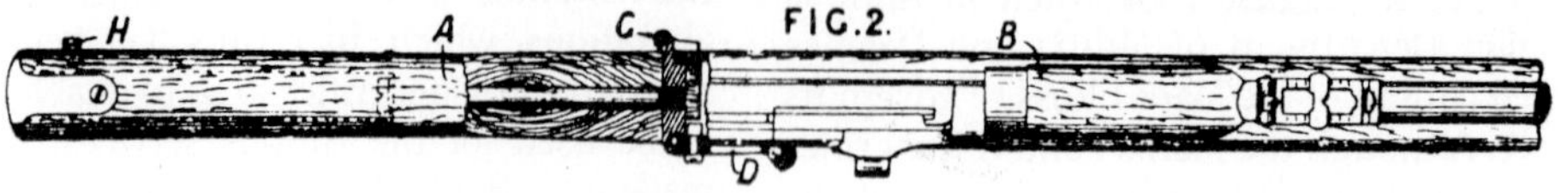

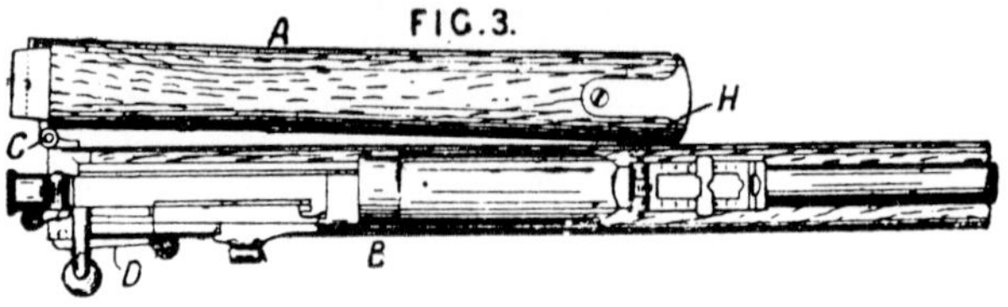

Stocks.—To facilitate transport, the stock A is hinged at C to the fore end B, and is secured, when in position for use, by the catch D which either turns upon a horizontal pivot, as shown, or upon a vertical one. A buffer button H prevents jar to the gun when the stock is folded smartly back.

Rimington-Wilson modified Magazine Lee-Enfield with butt extended and in the firing position. *MOD Pattern Room*

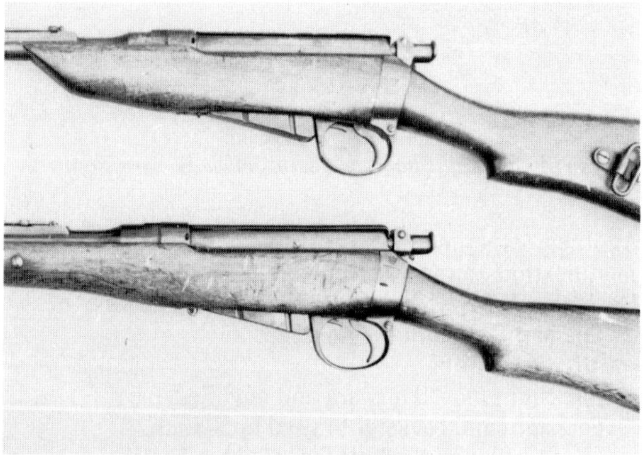

Boer War souvenir: captured Lee-Enfield carbine (top) compared with service issue model (bottom). *J. Tenniswood*

folding butt. Although this was more of a commercial application, it was put up as a solution to one of the main problems experienced by mounted infantry. The illustrated specimen was the subject of a patent taken out by R. Rimington-Wilson on 23rd December 1899.

During the South African war, a handy method often applied for lightening rifles and carbines was simply to "sporterise" them by removing the front portion of the fore-end, forward of the lower band. Quite a few of these unofficially modified Boer War rifles and carbines have been noted and it was not uncommon for Boers to similarly alter captured Lee-Metfords and Lee-Enfields when using them during the later phases of the war.

It is quite obvious that such practices were unlikely to have occurred with the traditional British regiments but it seems to have been carried out in certain Colonial mounted units, particularly among the Australian and South African volunteers. Of course the facility to fix a bayonet was removed when the nosecap and fore-end were dispensed with but the modus operandi of these forces called for little use of the bayonet, except perhaps as a convenient utensil and tool.

Such a cavalry carbine is illustrated here, in which the fore-end has been considerably shortened to reduce weight. A Mauser lower sling swivel has been screwed onto the left side of the butt, probably for attachment to a thong so as to prevent its loss when being used on horseback. This is likely to have been a captured British carbine, modified and used by Boer forces.

Col. Lockyer, Chief Inspector of Small Arms, also put forward a shortened rifle design of which there were two basic types:

"A" Pattern with a heavy barrel and
"B" Pattern with a light barrel.

Essentially a shortened Lee-Enfield rifle, the "A" pattern merely had a Lee-Enfield barrel shortened by 5 inches while on the "B" pattern it was turned down to a thinner external diameter. The Lockyer experimental rifles are illustrated and were contemporary with the 1901 trials

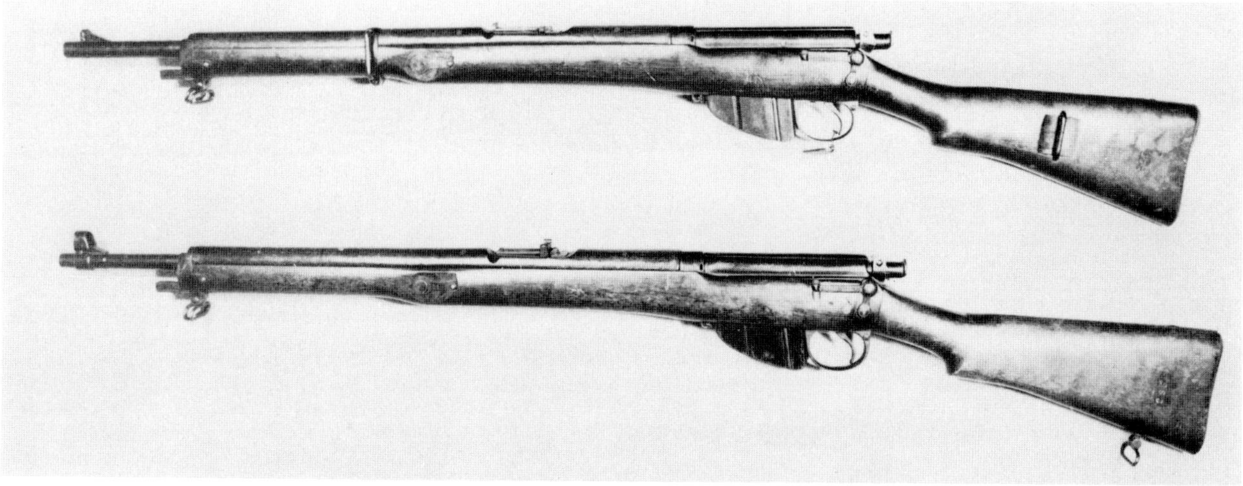

Above & below: Lockyer 1901 Carbines. "A" (upper) and "B" (lower) types differ mainly in the sights and sling fittings, although the "B" pattern has a lightweight, turned-down barrel. *Imperial War Museum*

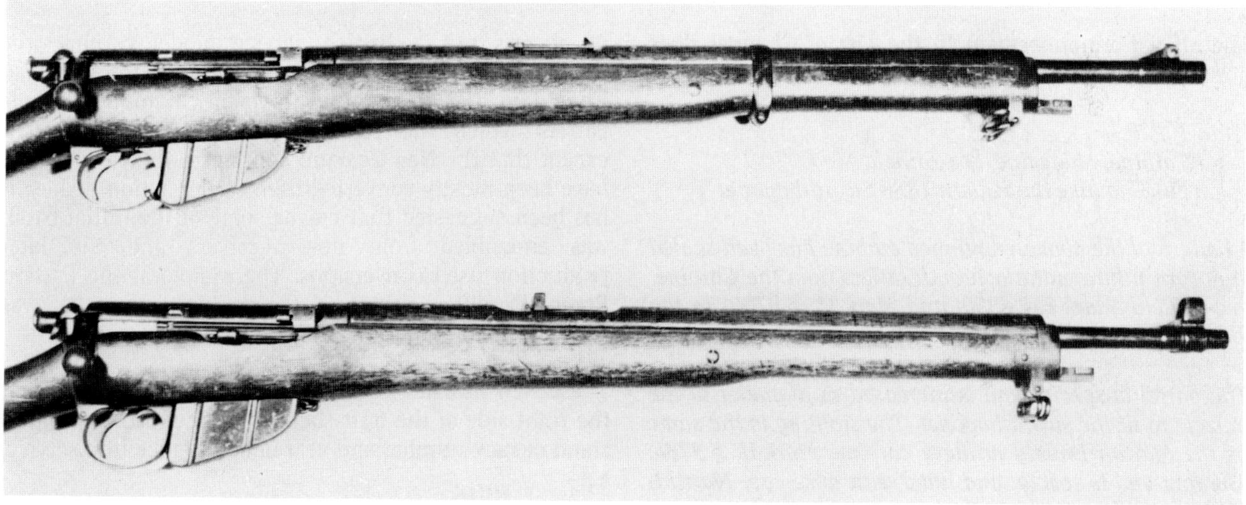

rifles. One feature similar to the trials rifles is the adjustable foresight and provision for a protecting hood.

These Lockyer barrels are 24-ins. long, making the overall lengths a little over 37-ins., the full-length handguards are fitted. It will be noticed that one of the "A" type carbines is fitted with a sling bar on the left side, similar to the Mod. 1896 Mauser carbine, while the bottom one, a "B" pattern, has experimental sights identical to those on the 1901 trials rifles previously described.

The sights on these Lockyer carbines are graduated to 500 yds on the bed and 2,000 yds on the leaf ladder. As can be seen from the illustrations, some models were fitted with a band while others were not. There was no provision for the fitting of a clearing rod as the requirement had been cancelled a few years previous. Only a few of these carbines were assembled for restricted trials, most probably at the Enfield factory.

A new carbine was approved in 1900 which bears some resemblance to the Lockyer patterns and is a little unusual in that it was not adopted for general British service yet promulgated in the List of Changes; the only batches assembled were for supply to a Colonial government.

This particular carbine is fitted for the Pattern 1888 sword bayonet, it has a heavier barrel with similar muzzle diameter to the Lee-Enfield rifle, a special fore-end fitted with no lower band, and a barrel and fore-end shape quite similar to those of the Martini-Enfield Artillery Carbine Mk III (which was approved on 12th July 1899). A pattern arm for this Lee-Enfield carbine was made at Enfield to guide any future manufacture, dated 1900 and serial numbered "1." Like the previous service carbines, the manufacture and identification markings were stamped on the left side of the receiver butt socket.

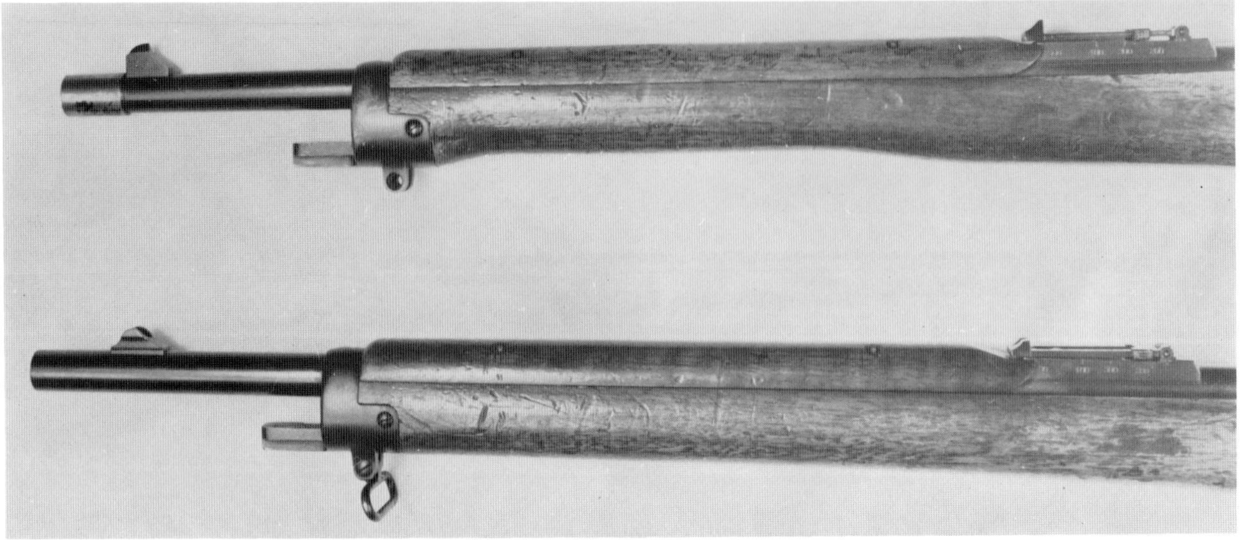

Fore-end profile comparison of the R.I.C. (top) and N.Z. Pattern (bottom) Lee-Enfield Carbines. Both fix the Patt. '88 sword bayonet. *MOD Pattern Room*

The official announcement in the List of Changes does not have a date of approval, but it reads:

Para 10220:–
Carbine, Magazine, Lee-Enfield
Fitted to take the Pattern 1888 Sword-bayonet

A pattern of the above-mentioned carbine has been sealed to govern future manufacture. It differs from the Carbine, Lee-Enfield, Mark I (§ 8390) and Mark I (§ 9700) in the following manner:*

The barrel is special and is increased in diameter at the muzzle to fit the sword-bayonet. The sighting is the same as the Martini-Enfield artillery carbine, Mark II, § 9786. The fore-end is special and fitted with nose-cap, Martini-Enfield artillery carbine. The handguard is also special and fitted with two springs.

Length of barrel	*21 ins.*
Calibre	*.303 inch.*
Rifling	*Enfield system.*
Grooves, number	*5.*
Grooves, depth	*.005 inch.*
Width of lands	*.0936 inch.*
Spiral, left handed ...	*...1 turn in 10 ins. or 33 calibres.*
Sighting system ...	*V notch and barleycorn foresight.*

The foresight is fitted with wings for its protection.

Length of carbine, without sword bayonet	*3 ft. 4.25 ins.*
Length of carbine with sword bayonet	*4 ft. 4.25 ins.*
Weight of carbine	*7 lb. 8 oz.*
Weight of carbine with sword bayonet	*8 lb. 7.5 oz.*

As it was not a British service model carbine, the production records indicate that only 1,500 were made at Enfield in 1901-03. These are identical to the sample pattern which was sealed, and mentioned previously, except that the New Zealand contract carbines appear to have been mostly converted from earlier action bodies; it has been suggested that maybe some of the initial batch was assembled from new receivers and the later production used conversions. The sample at the Pattern Room also differs slightly in the markings, as an "A" was added to the designation "L.E.C.."

These New Zealand issue carbines were marked "N↑Z" on the right side of the butt socket along with the carbine's stand or rack number and year of acceptance into service, e.g.-

$$\frac{1008}{1903}$$

Maj. E.G.B. Reynolds, in his book "The Lee-Enfield Rifle," mentions that this particular carbine order was for New South Wales but this is incorrect as the entire batch of 1,500 carbines was for New Zealand service. The first thousand was produced in 1901-02 with the remainder made in 1903.

In October, 1902, an altered backsight was approved for service carbines fitted for the Pattern 1888 sword bayonet; it has a narrower slide and lower sight cap. Details are given in the List of Changes reference § 11807 and the lower right corner of the leaf was marked "E.C./ 88" so as to distinguish it from the earlier carbine backsights.

A common repair problem with the Lee-Metford and Lee-Enfield rifles was damaged handguards; they were liable to split longitudinally and so a repair was authorised which involved gluing the split and placing the handguard on a mandrel and drilling two holes. These holes were drilled across the width of the guard, as per the illustration.

Scale ½.

M.L.M. (Marks II., II.*) and M.L.E. rifles.

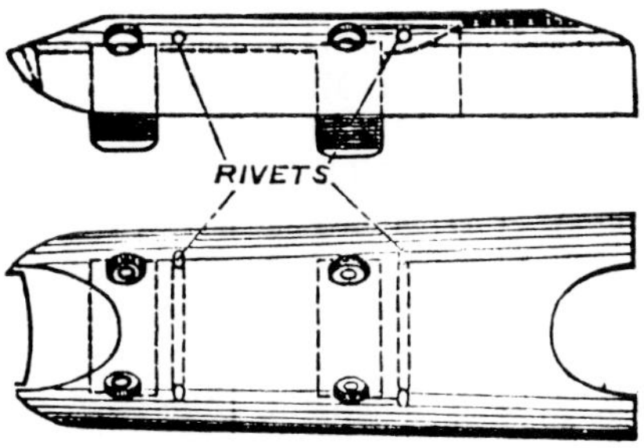

Described in the List of Changes § 11034, a length of No. 16 WG wire was then inserted into the hole so as to reinforce the guard. This repair was also carried out on the Lee-Metford and Martini-Enfield rifles although in each instance the wire rivets were positioned differently. A somewhat similar method for repair or reinforcement of the S.M.L.E. rifle was later employed in Australia, except that threaded brass rod was employed rather than wire.

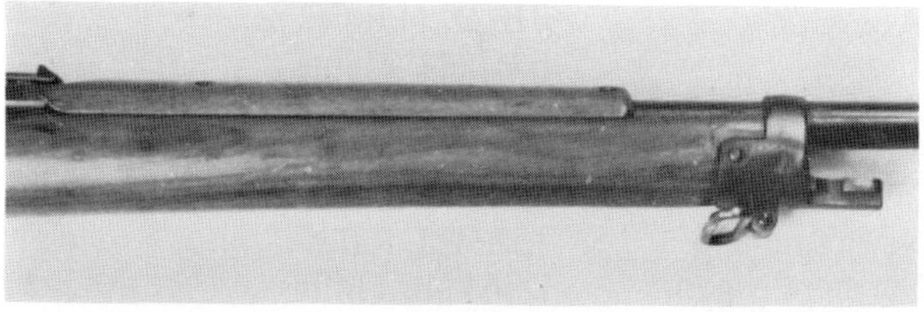

Shortened M.L.E. Carbine handguard; a common Colonial repair when the front end splintered. *I. Skennerton*

Enfield production records show that in the 1902-03 fiscal year, 432 Enfield-made Mark I* Lee-Enfield rifles were stocked up with American Whitewood fore-ends and handguards and a further 6 with special butts and buttplates. The following year, 9,919 Mark I* Lee-Enfield "modified" rifles were fitted with the S.M.L.E. pattern butt. Another 31,000 Mark I* Magazine Lee-Enfield rifles were altered to the "modified pattern" at 3/9d each. How these rifles were "modified" is uncertain but they were standard length Lee-Enfield rifles with the service pattern sights. In 1904-05, a further 80 Mark I* Lee-Enfield rifles were listed as being fitted with the S.M.L.E. butt. This butt was the Mk I S.M.L.E. pattern which has a sheet steel buttplate with no trap.

Some special carbines were undergoing evaluation at Hythe during this period and although they were not Lee-Enfield models, mention is warranted here as they may well have influenced some thinking on the short rifle and also incorporate some advanced design concepts. These were the "Thorneycroft" carbines which have in more recent times become known as "bullpup" types, where the trigger assembly and right hand grip are forward of the receiver group, with the magazine positioned in or under the butt.

This design effectively shortens the overall length by about 7 inches and reduces the weight by up to 10%. Some of the later British experimentals were of the "bullpup" design, namely the SREM of 1944, the EM2 of 1948 and the recent 4.85mm and 5.56mm IW and SA 80; the last model has now been adopted for British service as the L85A1. Thorneycroft took out a patent on his ideas, No. 14,622 of 18th July 1901.

The design concept is quite obvious from the illustration; the trigger assembly is forward of the breech instead of behind or beneath it. This results in savings in weight and overall length. The first model of the Thorneycroft weighed 7 lb. 8 oz., even though it was fitted with a 30.2 in. Lee-Enfield barrel, it was 7.5 in. shorter in overall length. In the 1902 evaluation trials at Hythe, the main complaints were of excessive recoil, awkward handling and of fired cases striking the fingers during ejection. However, the essential design was considered to have merit so further improved models were submitted. After more testing at Hythe in 1903, it was decided to suspend further tests as the basic faults of the first model rifle still remained.

Although the Short Magazine Lee-Enfield rifle was introduced in 1902-03, it did not see the demise of the Magazine Lee-Enfield development because another carbine was yet to be introduced and special conversion programs carried out with the Lee-Enfield rifles so as to upgrade them for issue to reservists and Naval forces.

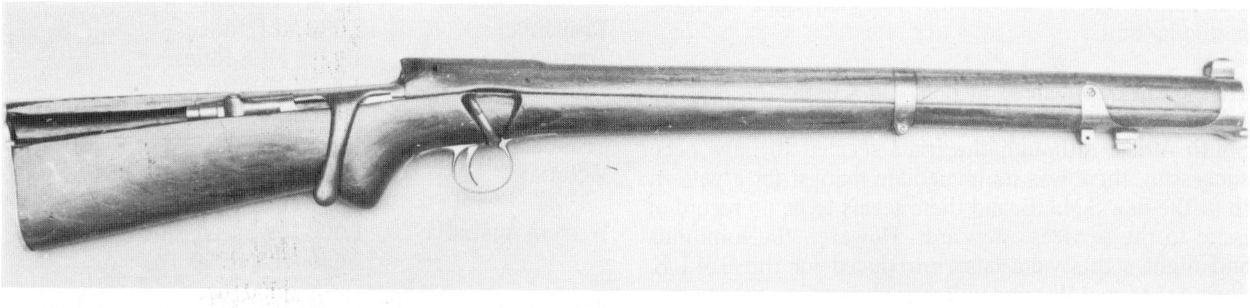

Thorneycroft carbine; forerunner of the "Bullpup" designs. *MOD Pattern Room*

This next carbine was made especially for the Royal Irish Constabulary, a conversion from the Lee-Enfield cavalry model, although a few used the Lee-Metford carbine as the basis for conversion. An order was initially placed for 10,000 of these carbines in 1905 and typical of previous R.I.C. issue carbines such as the Snider and Martini, they were 2nd line arms converted from earlier, obsolete models. No announcement of the R.I.C. carbine was made in the List of Changes and this carbine has often been confused with the Lee-Enfield carbine made for New Zealand as both were made to fix the Pattern 1888 bayonet. See page 79 for illustration of the R.I.C. models.

This model is quite distinctive due to the bayonet fitting and the shape of the fore-wood which was modified from the cavalry model; on close examination the fore-end will be noticed to have sections of wood inletted and glued into place. Butt discs are generally stamped "R.I.C." along with the issue or rack number.

It had been officially declared that Lee-Metford rifles fitted with Enfield barrels would become known as Lee-Enfield rifles, in the instance of the Mk II Lee-Metford, the bodies are recorded as having been re-stamped "L.E. I" for the new designation. On 16th March 1904, approval was given for these particular rifles to be redesignated "Mark II Lee-Metford" again and the "L.E. I" marking cancelled. This was due to the different pattern of bolt, without the safety catch of the Mark II* Lee-Metford and Marks I and I* Lee-Enfield. Similarly, with the Lee-Metford carbine, where the Enfield barrel was fitted, it became known as a Lee-Enfield carbine.

So as to maintain a trigger pressure similar to that of the new Short Lee-Enfield, a modification was approved on 16th February 1906 and announced in the List of Changes § 13136, lightening the pull by 1 lb. to a new low of 5 lb. and high of 7 lb. The Lee-Enfield has a single stage trigger while the Short rifle has a double stage pull. This lightened trigger pull was also applied to the Lee-Metford.

A very minor modification approved at the same time is of interest as it details an unusual way of slinging the rifle. It appears that Mounted Infantry and Imperial Yeomanry slung their rifles from the nosecap swivel, of which the securing screw was found to work loose. To prevent this, the threaded end of the screw was riveted over slightly. This alteration was applied to Mark II and II* Lee-Metford and Mark I and I* Lee-Enfield rifles on issue to these particular units.

Trials were commenced in April 1903 upon the issue of 1,000 sets of night sights, 950 of which were sent to South Africa. Although the trials seemed to have been successful, there was no immediate request for a pattern to fit the new S.M.L.E. and there seems to be no record of issue to the services afterwards. However, the luminous and night sights were later introduced for the S.M.L.E. during the Great War and quite widely used.

The Lee-Metford and Lee-Enfield rifles were issued on a wide scale throughout the British Colonies and Dependencies by this time. A return of the "Strength of Land Forces for the year ending 31st December 1904" for all the Colonies (excepting India) indicates the extent of these issues:–

Barbados:	100 MLM & MLE Rifles
	342 MLE & ME Carbines, for Police
Basutoland:	103 MLE Rifles
Bechuanaland:	196 MLE Rifles
British Central Africa:	6 MLE & MLM Rifles
British Guiana:	600 MLE Rifles
British Somaliland:	1,881 Rifles (includes Martini?)
	329 Carbines (includes Martini?)
Canada:	164 MLM Rifles
	53 MLM Carbines
	30,276 MLE Rifles
	4,276 MLE Carbines
Cape Colony:	23,179 MLM & MLE Rifles
	214 MLM & MLE Carbines
Ceylon:	130 MLE Carbines
	2,400 MLE & ME Rifles
Falkland Is.:	100 MLE Rifles
Fiji:	150 MLE Rifles
Hong Kong Volunteers:	82 MLE Rifles, with 316 MLE Rifles on order
Natal & Zululand:	5,035 MLE Rifles
	316 MLM Rifles
	6 MLM Carbines
New South Wales:	15,040 MLE & ME Rifles
New Zealand:	19,985 MLE Rifles
	1,471 MLE Carbines
Newfoundland:	132 MLE Rifles
Northern Nigeria:	2,363 MLE Rifles
	9 MLM Rifles
	1,000 MLE Carbines
Queensland:	3,774 MLE Rifles
	62 MLM Rifles
Rhodesia, N.W.:	17 MLM Rifles
Rhodesia, Southern:	2,380 Rifles (includes Martini?)
	214 Carbines (includes Martini?)
St. Helena:	150 Rifles Mk I (MLE or MLM?)
Straits Settlements-	
Malay States Guides:	800 MLE Rifles
	20 MLE Carbines
Singapore Vol. Corps:	522 MLE Rifles
	127 MLE Carbines
Penang Vol. Corps:	200 MLE & ME Rifles
South Australia:	2,819 MLE Rifles
Tasmania:	2,296 MLE Rifles
Transvaal:	13,229 MLE Rifles
	100 MLM Rifles
	823 Carbines (includes Martini?)
Uganda:	100 MLE Rifles
Victoria:	2 MLM Rifles
	7,547 MLE Rifles
Western Australia:	1,000 Lee Speed Rifles (trade)
	2,650 MLE Rifles

[Abbreviations- MLM = Magazine Lee-Metford ME = Martini-Enfield MLE = Magazine Lee-Enfield]

The Martini-Enfield, Martini-Metford and Martini-Henry rifles and carbines represent the bulk of the arms issued throughout the Colonies around the turn of the century and there were still many Sniders and non-standard arms such as Winchesters and Swinburn Martinis in service. At this time the new Short Magazine Lee-Enfield had just been introduced and so more of the superseded Lee-Enfield and Lee-Metford arms were sold, loaned or gifted to the various Colonies. This does not infer that the Colonies were usually equipped with 2nd line or reserve grade arms; the more affluent and defence-conscious governments such as those in Canada and Australasia ordered the latest patterns.

Introduction of charger-loading with the Mark I Short Magazine Lee-Enfield rifle saw 9 long rifles converted to this system at Sparkbrook in 1904-05, comprising 3 each of Mark I* Lee-Metford, Mark II Lee-Metford and Lee-Enfield rifles. Probably a feasibility study, the Mark I S.M.L.E. charger loading system consists of a charger guide on the left side of the receiver and a sliding guide on the right side at the top of the bolthead. A stop nib was affixed to the right side of the receiver. India adopted this charger-loading conversion of the Lee-Enfield rifle and the nine converted rifles may well have been modified as a result of a request from India. The India Pattern charger-loading conversions were carried out at Ishapore and details are provided in Chapter 10.

A special backsight slide with windgauge was approved for Martini-Enfield, Lee-Metford and Lee-Enfield rifles on 16th May 1906, although it was principally used on the Lee-Enfield rifles. It was not, strictly speaking, a service issue as it was supplied on a repayment basis, for use in voluntary practice. Promulgated in the List of Changes § 13314, the official description reads:

Demands for such numbers of these slides as may be required should be put forward in the usual way by officers commanding.

The slide consists of an elevation slide, fitted with a gib spring at the side to tension the slide on the leaf; and a wind gauge which is dovetailed in the elevation slide, and is capable of lateral adjustment for ¼-in. to the right or left. A transverse spring is fitted within the dovetail; this spring limits the movement and regulates the tension of the wind-gauge slide.

Full size.

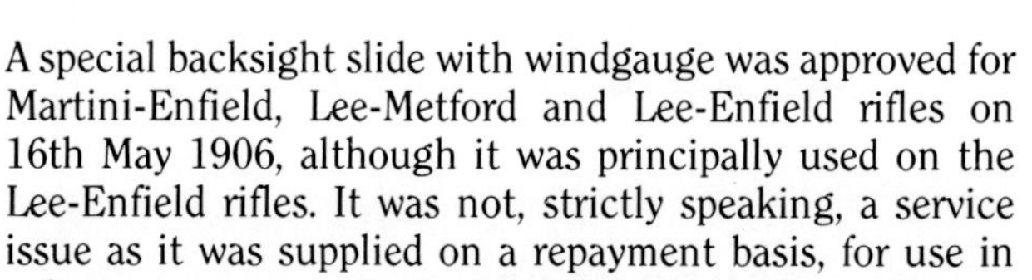

The wind gauge is provided with two V notches, one for use at ranges from 200 to 400 yards with the leaf down, and the other for use at ranges from 500 to 1,800 yards with the leaf raised. It is also marked with centre lines under each V-notch, which act as pointers to the wind gauge scale on the back edge of the elevation slide. This scale has divisions of .0388-inch on the back edge, marked by alternate long and short lines, each division representing a wind allowance of 6 inches for each 100 yards the firer is distant from the target.

To fit this slide, the ordinary cap screw, cap and slide must be removed from the leaf and the wind-gauge slide put on in its place.

To remove the wind gauge from the elevation slide, move the wind gauge to the extreme right, then with a piece of wire of a suitable size inserted in the hole in the top of the wind gauge, depress the wind-gauge spring, at the same time pressing the wind gauge to the right.

The pattern introduced by § 6762 is hereby declared obsolete.

This slide is also quite similar to the experimental pattern used in the 1901 trials.

The Mark III Short Magazine Lee-Enfield rifle introduced a new charger guide in 1907, this formed a bridge across the top of the receiver and on 8th March, a pattern of M.L.E. rifle modified to this form of charger loading was decided upon. This particular conversion also incorporated improved sights and was approved on 1st July 1907, being announced in the monthly List of Changes.

Para 13992:-

Rifle, Charger Loading, Magazine, Lee-Metford (Mk II)
Converted to charger loading from Rifle, Magazine, Lee-Metford, Mk II.

Rifle, Charger Loading, Magazine, Lee-Enfield Mk I)*
Converted to charger loading from Rifles, Magazine, Lee-Enfield Marks I and I, and Lee-Metford, Mk II*.*

Patterns of the above-mentioned rifles have been approved to govern conversion.

The conversion consists generally in fitting the rifle with a bridge charger guide and a new magazine for charger loading, an adjustable blade foresight with fixed protector, and a new backsight leaf with a slide having a windgauge, with U-notches, adjustable laterally by a screw having its head on the left, and a clamping nut on the right for fixing the slide on the leaf to prevent movement of the slide when firing. The bridge charger guide and the magazine are similar to those of the Rifle, Short, M.L.E., Mark III, § 13853.

Blade, foresight: Foresights with a straight edged blade, in five heights, will be used, according to requirements, to enable variations in shooting to be corrected. They will be

CHARGER LOADING LEE-ENFIELD.

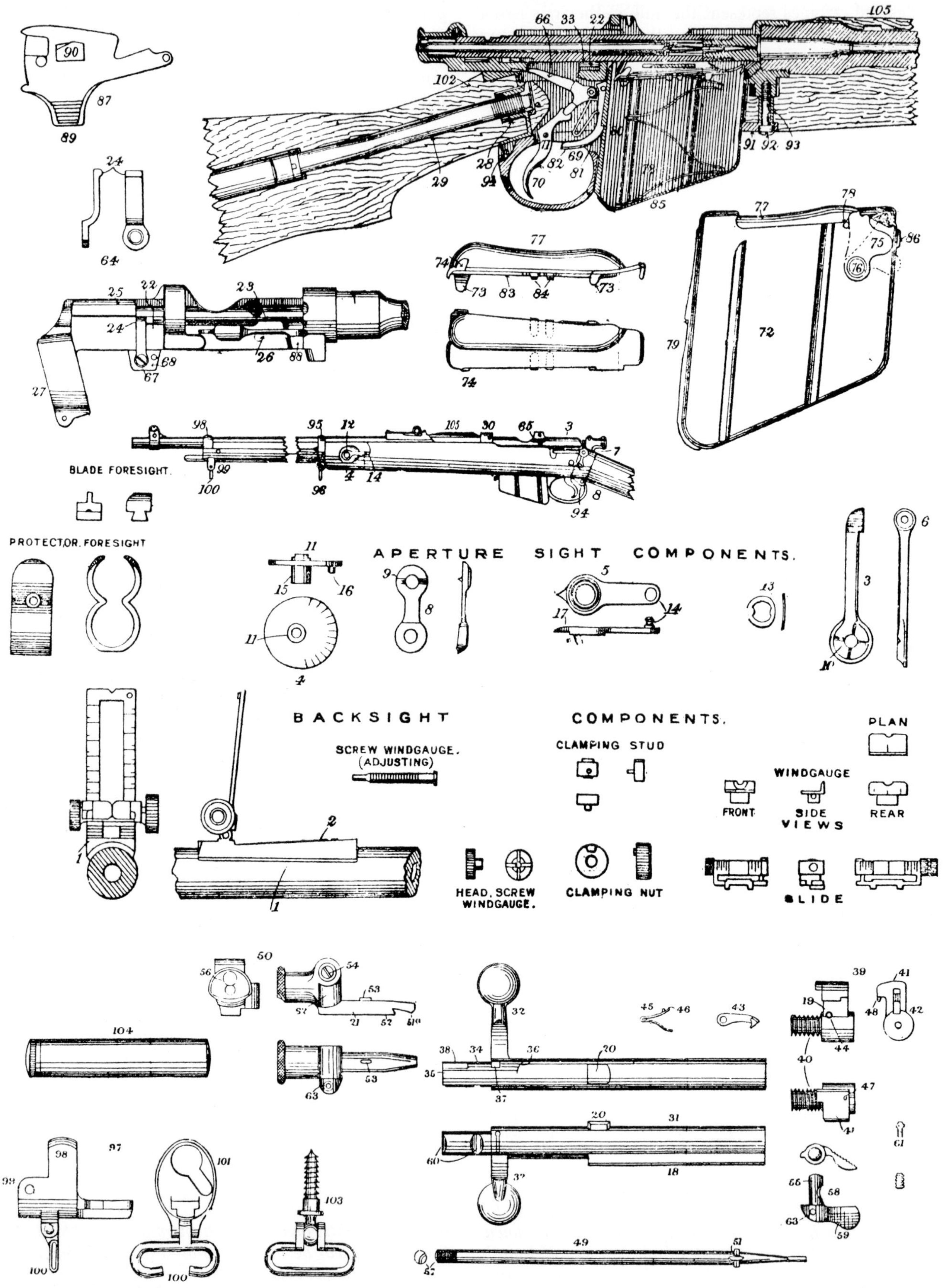

PLATE VI.—NAMES OF THE PARTS OF RIFLES, CHARGER-LOADING M.L.E., REFERRED TO IN PLATES VII TO X.

1. Backsight bed.
2. ,, ,, ramp.
3. Aperture sight.
4. Dial sight plate.
5. ,, ,, pointer.
6. Aperture sight peep hole.
7. ,, ,, pivot screw.
8. ,, ,, spring.
9. ,, ,, spring ribs.
10. ,, sight cross-cut notches.
11. Pivot aperture sight pointer.
12. ,, screw.
13. Spring disc.
14. Bead pointer.
15. Screwed boss, fixing dial sight.
16. Centring pin dial sight.
17. Index point of pointer.
18. Bolt rib.
19. Extractor seating.
20. Bolt lug.
21. Tongue of cocking-piece.
22. Recess for lug of bolt.
23. Body bead.
24. Retaining catch.
25. Resistance shoulder.
26. Cut-off slot.
27. Body socket.
28. ,, ,, boss.
29. Stock bolt.
30. Gas escape hole.
31. Bolt breech.
32. Lever bolt breech.
33. Lug seating.
34. Long cam groove.
35. Short ,, ,,
36. Cam-shaped face.
37., 38. Cam grooves separating studs.
39. Bolt head.
40. ,, ,, tenon.
41. ,, ,, projection.
42. Extractor slot.
43. Extractor.
44. ,, axis screw.
45. ,, spring.
46. ,, ,, pin.
47. ,, ,, ,, hole.
48. Bolt head hook.
49. Striker.
50. Cocking piece.
51. Striker collar.
51A. Cocking-piece full bent.
52. ,, ,, half ,,
54. ,, ,, seating for safety catch.
55. Safety catch.
56. Striker keeper-screw.
57. Recess in striker for the keeper-screw.
58. Safety catch stem.
59. Safety catch finger piece.
60. ,, ,, locking grooves.
61. ,, ,, plunger.
62 Seating for plunger and spiral spring of safety catch.
63. Nipple holes for plunger of safety catch.
64. Retaining catch spring.
65. Ejector.
66. Sear.
67. ,, screw.
68. ,, seating in body.
69. ,, spring.
70. Trigger.
71. ,, axis pin.
72. Magazine box or case.
73., 74. ,, platform guides.
75. ,, stop clip.
76. ,, ,, ,, stud.
77. ,, platform.
78. ,, clip stop.
79. ,, tooth.
80. ,, catch tooth.
81. ,, catch.
82. ,, ,, finger piece.
83. Magazine platform plate.
84. ,, ,, ,, tongues.
85. ,, ,, spring.
86. ,, ,, auxiliary spring.
87. Cut-off.
88. ,, screw.
89. ,, thumbpiece.
90. ,, catch.
91. Guard trigger.
92 Screw guard front.
93. ,, ,, collar.
94. ,, ,, ,, rear.
95. Lower band.
96. Sling swivel.
97. Nose-cap.
98. ,, band.
99. ,, fixing screw.
100. Piling swivel.
101. Diagonal slot in nose cap to clear the fore-sight block when assembling, &c.
102. Stock bolt, plate keeper.
103. Stock butt swivel.
104. Oil bottle, Mark III (latest Mark IV).

marked on the left top side with the figures 933, 948, 963, 978, and 993 respectively, representing their heights in decimal parts of an inch, from the axis of the barrel. The blades will be of two patterns:– those for the L.M. Mark II (which have the foresight block central) will be marked with the letter "M," and those for the L.E., Mark 1 (which have the foresight block .027-inch to the left) with the letter "E" after the figure.*

Rifles, C.L., M.L.E., Mark I, converted from Rifles, M.L.M., Mark II*, will be fitted with blades, foresight, and protectors, foresight, of the pattern fitted to Rifles, C.L., M.L.M., Mark II, marked with the letter "M," except where fitted with barrels having Enfield rifling, which are marked with the letter "E" on the knox form. Rifles C.L., M.L.M. Mark II, fitted with barrels having Enfield rifling, which are marked with the letter "E" on the knox form, will be fitted with blades, foresight, and protectors, foresight, of the pattern fitted to Rifles C.L., M.L.E., Mark I*, marked with the letter "E."*

Leaf, backsight: The leaf, backsight is stronger and the elevation lines are further from the axis than in the case of the old leaf removed on conversion, and is without a cap and screw. A stop screw for the slide is fitted at the top corner on the right hand side.

Slide, backsight: The slide, backsight is fitted with a wind gauge, adjustable laterally by a screw having a head fixed by a pin on the left hand side. A clamping nut is fitted to the slide on right side, which, when screwed up, presses a clamping stud on to the edge of the leaf for the purpose of fixing the slide in any position on the leaf; a friction spring is fitted in the leaf slot on the left side, and presses on the side of the leaf; an extension of this spring acts in the notches cut at right angles on the inside face of the head of the windgauge screw.

The slide is marked with the scale of eight divisions on the front and rear faces, four on the right and four on the left, for the adjustment of the windgauge. Each division represents 6 inches on the target per 100 yards. The slide also has a platinum centre line on the rear face.

The windgauge is fitted into the slide and held and positioned by the windgauge screw, the necessary friction being given by a flat bowed spring fitted in the slide underneath the windgauge. The windgauge has two sighting bars with U notches and platinum centre lines, one for use with leaf down, and one for use when leaf is raised. Lines are marked on the front and rear faces to correspond with the inner lines on the slide for centreing the windgauge.

The windgauge screw head is notched at quarter turns to receive the extension to the slide spring. Each quarter turn represents 1 inch on the target per 100 yards, and six quarter turns one division of the scale on the slide.

Protectors, foresight: The sight protector is fitted on the barrel and foresight block, and is fixed by a screw. The protectors are of two patterns; one for the Rifle, C.L., M.L.M., marked with the letter "M," and one for Rifle, C.L., M.L.E., Mark I, marked with the letter "E."*

(Note: See blade, foresight, for remarks as to difference of sight protectors and blades when Rifles, C.L. M.L.E. Mark I are fitted with barrels having Metford rifling, and Rifles, C.L., M.L.M., Mark II, are fitted with barrels having Enfield rifling.)*

Average weight of rifles . . . . . . 9 lb. 5 oz.

The following is a list of components, showing which are special to these rifles and which are common to other arms. Those which are common to other arms are denoted by the word "Common" and their interchangeability is as stated in the "Priced Vocabulary of Stores," &c.

COMPONENT	*SPECIAL or COMMON*
Bands, lower, M.L.M.R., II.	*Common.*
Barrels, M.L.M.C.L.R., II.	*Special to M.L.M.C.L.R., II, but will be found in M.L.E. C.L.R., I*, converted from M.L.M.R., II*.*
Barrels, M.L.E.C.L.R., I. #*	*Special to M.L.E.C.L.R., I*, but will be found in M.L.M. C.L.R., II that are re-barrelled with barrels having Enfield rifling.*

COMPONENT	SPECIAL or COMMON
Beads, pointer, dial sight, M.L.M.R., II.	Common.
Beds, backsight, M.L.M. C.L.R., II.	Special to M.L.M.C.L.R., II, and M.L.E.C.L.R., I*.
Blades, foresight, M.L.M. C.L.R., II.	Special to M.L.M.C.L.R., II, but will be found in M.L.E. C.L.R., I*, fitted with barrels having Metford rifling.
Blades, foresight, M.L.E. C.L.R., I*.	Special to M.L.E.C.L.R., I*, but will be found in M.L.M. C.L.R., II, fitted with barrels having Enfield rifling.
Bodies, M.L.M.C.L.R., II.	Special to M.L.M.C.L.R., II. and M.L.E.C.L.R., I*.
Bolts, breech, M.L.M.C.L.R., II.	Special to M.L.M.C.L.R., II.
Bolts, breech, M.L.E.C.L.R., I*.	Special to M.L.E.C.L.R., I*.
Bolts, stock, M.L.E., No. 2.	Common.
Caps, nose, M.L.E.R.	Common.
Cases, magazine. M.L.E.R.S., I, No. 2.	Common.
Catches, magazine, M.L.E.R., II.	Common.
Catches, safety, M.L.M.C.	Common. (for M.L.E.C.L.R., I* only).
Clips, stop, magazine, M.L.E.R.S., I.	Common.
Cocking pieces, M.L.M.R., II.	Common. (for M.L.M.C.L.R., II only).
Cocking pieces, M.L.M.C.	Common. (for M.L.E.C.L.R., I* only).
Collars, 'screw, guard, trigger, front, M.L.M.	Common.
Cut-offs, M.L.E.R.S., I.	Common. (new manufacture cut-offs, i.e. not converted, are M.L.E.R.S., I* pattern).
Extractors, M.L.E.R.S., I.	Common.
Guards, hand, M.L.E.R.S., I.	Common.
Guards, trigger, M.L.M. C.L.R., II.	Special to M.L.M.C.L.R., II, and M.L.E.C.L.R., I*.
Heads, breech bolt, M.L.M. C.L.R., II.	Special to M.L.M.C.L.R., II, and M.L.E.C.L.R., I*.
Heads, screw, windgauge, backsight, M.L.M.C.L.R., II.	Special to M.L.M.C.L.R., II, and M.L.E.C.L.R., I*.
Leaves, backsight, M.L.M.C.L.R., II.	Special to M.L.M.C.L.R., II, and M.L.E.C.L.R., I*.
Nut, clamping, slide, backsight, M.L.M.C.L.R., II.	Special to M.L.M.C.L.R., II, and M.L.E.C.L.R., I*.
Pins, axis, backsight, M.L.M.R.	Common.
Pins, axis, trap, buttplate, M.L.M.	Common.
Pins, catch, magazine, M.L.M.	Common.
Pins, catch, safety, M.L.M.C.	Common. (for M.L.E.C.L.R., I* only).
Pins, fixing, head, screw, windgauge, M.L.M.C.L.R., II.	Special to M.L.M.C.L.R., II, and M.L.E.C.L.R., I*.
Pins, stop, band, lower, M.L.M.R.	Common.
Pins, trigger, M.L.M.	Common.
Plates, butt, M.L.M.R., II.	Common.
Plates, keeper, stockbolt, M.L.E.R.S., I.	Common.
Plates, sight, dial, M.L.M.R., II.	Common.
Platforms, magazine, M.L.E.R.S., I.	Common.
Pointers, sight, dial, M.L.M.R., II.	Common.
Protectors, foresight, M.L.M.C.L.R., II.	Special to M.L.M.C.L.R. II, will be found on M.L.E. C.L.R., I*, fitted with barrels having Metford rifling.

COMPONENT	SPECIAL or COMMON
Protectors, foresight, M.L.E.C.L.R., I*.	Special to M.L.E.C.L.R. I*, will be found on M.L.M. C.L.R., II, fitted with barrels having Enfield rifling.
Rivets, fore-end, M.L.M.	Common.
Rivets, spring, handguard, M.L.M.	Common.
Screws, band, lower, M.L.M.R., II.	Common.
Screws, bed, backsight, M.L.M.R.	Common.
Screws, cap, nose, M.L.M.R.	Common.
Screws, cut-off, M.L.M.	Common.
Screws, ejector, M.L.M.	Common.
Screws, extractor, M.L.M.	Common.
Screws, guard, trigger, M.L.M., back.	Common.
Screws, guard, trigger, M.L.M., front.	Common.
Screws, keeper, striker, M.L.M.R.	Common. (for M.L.M.C.L.R., II only).
Screws, keeper, striker, M.L.M.C.	Common. (for M.L.E.C.L.R., I* only).
Screws, plate, butt, iron.	Common.
Screws, protectors, foresight, M.L.M.C.L.R., II.	Special to M.L.M.C.L.R., II, and M.L.E.C.L.R., I*.
Screws, sear, M.L.M.	Common.
Screws, sight, aperture, M.L.M.R.	Common.
Screws, sight, dial, M.L.M., fixing.	Common.
Screws, sight, dial, M.L.M., pivot.	Common.
Screws, spring, sight, aperture, M.L.M.	Common.
Screws, spring, sight, back,	Common.
Screws, spring, trap, butt plate, M.L.M.	Common.
Screws, stop, slide, backsight, M.L.M.C.L.R., II.	Special to M.L.M.C.L.R. II, and M.L.E.C.L.R., I*.
Screws, strap, butt plate, iron.	Common.
Screws, swivel, piling, M.L.M.R., II.	Common.
Screws, windgauge, M.L.M. C.L.R., II.	Special to M.L.M.C.L.R., II, and M.L.E.C.L.R., I*.
Sears, M.L.M.R., II.	Common.
Sights, aperture, M.L.M.R.	Common.
Slides, backsight, M.L.M. C.L.R., II.	Special to M.L.M.C.L.R., II, and M.L.E.C.L.R., I*.
Springs, auxiliary, platform, magazine, M.L.E.R.S., I, No. 2.	Common.
Springs, extractor, M.L.M.	Common.
Springs, guard, hand, M.L.M.R., II, lower.	Common.
Springs, guard, hand, M.L.M.R., II, upper.	Common.
Springs, main, M.L.M.	Common.
Springs, pin, catch, safety, M.L.M.C.	Common. (for M.L.E.C.L.R., I*).
Springs, platform, magazine, M.L.E.R.S., I.	Common.
Springs, retaining, breech bolt, M.L.M.	Common.
Springs, sear, M.L.M.	Common.
Springs, sight, aperture, M.L.M.	Common.
Springs, sight, back, M.L.M.R.	Common.
Springs, sight, dial, M.L.M.	Common.
Springs, slide, backsight, M.L.M. C.L.R., II.	Special to M.L.M.C.L.R., II, and M.L.E.C.L.R., I*.
Springs, trap, butt plate, M.L.M.	Common.
Springs, windgauge, backsight, M.L.M.C.L.R., II.	Special to M.L.M.C.L.R., II, and M.L.E.C.L.R., I*.
Stocks, butt, M.L.M.R., II, No. 2 (long, normal or short).	Common.

COMPONENT	*SPECIAL or COMMON*
Stocks, fore-end, M.L.M. C.L.R., II.	*Special to M.L.M.C.L.R., II, and M.L.E.C.L.R. I*.*
Strikers, M.L.M.R.	*Common. (for M.L.M.C.L.R., II, only).*
Strikers, M.L.M.C.	*Common. (for M.L.E.C.L.R., I*, only).*
Studs, clamping, slide, back-sight, M.L.M.C.L.R., II.	*Special to M.L.M.C.L.R., II, and M.L.E.C.L.R., I*.*
Studs, clip, stop, magazine, M.L.E.R.S., I.	*Common.*
Swivels, band, M.L.M.R., II.	*Common.*
Swivels, butt, M.L.M.R.	*Common.*
Swivels, piling, M.L.M.R. II.	*Common.*
Traps, butt plate, M.L.M.R., II.	*Common.*
Triggers, M.L.M.R., II.	*Common.*
Wads, bolt, stock, M.L.M.	*Common.*
Washers, bolt, stock.	*Common.*
Washers, guard, hand, or rivet, fore-end, M.L.M.	*Common.*
Washers, screw, fixing, dial sight, M.L.M.R.	*Common.*
Windgauges, backsight, M.L.M. C.L.R., II.	*Special to M.L.M.C.L.R., II, and M.L.E.C.L.R., I*.*

\# *When issued with bodies in exchange for M.L.M.C.L.R., II, the marking must be changed on the lines laid down in LoC § 12184.*

The specifications were officially given as being:

Weight, with oil bottle & pullthrough	*9 lb. 5 oz.*
Weight of sword bayonet, without scabbard	*15.5 oz.*
Length of rifle without bayonet	*4 ft. 1.5 in.*
Length of rifle with bayonet	*5 ft. 1.5 in.*
Length of barrel	*2 ft. 6.09 in.*
Calibre	*.303 in.*
Rifling system	*Enfield.*
Twist of rifling	*Uniform, left handed, 1 turn in 10 in. or 33 calibres.*
Muzzle velocity (about) ...	*Mk VI ammunition - 2,060 fps. Mk VII ammunition - 2,440 fps.*
Number of grooves	*5.*
Depth of grooves	*.0065 in.*
Width of lands	*.0936 in.*

The conversions to the new charger-loading system were effected at R.S.A.F. Enfield, as well as by the trade at B.S.A. Co., L.S.A. Co. and Vickers. Enfield conversions were effected in 1908, 1909, 1910 and 1913, with the major share being converted in 1909 and 1910. More than 100,000 were done at Enfield and double this quantity converted by B.S.A., L.S.A. and Vickers and completed by June 1911. The Vickers conversions are marked "V.S.M." for the factory name, which indicates Vickers Sons & Maxim.

Enfield was the sole government factory engaged in the manufacture of small arms by this time because R.S.A.F. Sparkbrook, previously employed in the manufacture of the Lee-Metford, Lee-Enfield and Short Lee-Enfield rifles, had been sold to B.S.A. Co. in 1906. The Enfield factory, about 10 miles north of central London, was the principal Government manufactory from 1860 until October 1988 when the Conservative government sold off Royal Ordnance Factories to British Aerospace. The land at Enfield was then re-allocated for housing and residential estates and the newly-privatised manufacturing concern merged with the R.O.F. establishment at Nottingham.

The primary commercial or non-government contractors were the Birmingham Small Arms Co. at Small Heath, Birmingham and the London Small Arms Co. at Victoria Mills, Old Ford Rd., in the London inner city suburb of Tower Hamlets, E2 or E3.

The Birmingham Small Arms factory commenced producing service rifles in 1861 with the Enfield rifled-musket at their Small Heath plant; B.S.A. became the largest commercial manufacturer of small arms in Britain. Production continued through two World Wars and B.S.A. set up a number of subsidiary factories, especially during the 1939-45 war, one of the largest being at Shirley on the outskirts of Birmingham. Shirley commenced production in 1941 with the No. 4 rifle. It has only been in the last decade that the B.S.A. factory has been forced to close due to the restricted commercial markets and government contracts.

London Small Arms Co. commenced rifle production at the Victoria Mill, Old Ford in East London in 1866 and continued in this location until 1920. Like its Birmingham competitor, L.S.A. manufactured the percussion Enfield, Snider, Martini, Lee-Metford, Lee-Enfield and Short Lee-Enfield rifles but it never matched the capacity or output of B.S.A. In 1921 the address was changed to Albion Works, Ossary Rd., and production ceased in 1925.

The National Arms & Ammunition Co., set up in Birmingham in 1872, was forced to close due to financial difficulties in 1883. The Government bought the factory for £55,000 in 1886 and spent a further £30,000 on plant; this became the Royal Small Arms Factory, Sparkbrook. During the early 1890's, Sparkbrook was turning out rifles at between 2/- and 2/6d each cheaper than Enfield. In 1906 it was closed down and bought by B.S.A. but only a few rifles were turned out by "B.S.A. – Sparkbrook" as the factory was mainly used for engine and motor vehicle production.

Vickers was another traditional British armaments manufacturer although the company commenced operation in the latter part of the 19th century and was more involved in the manufacture of machine guns and the larger ordnance. Their production was also more diversified and later extended to naval and air force equipment and armaments. The company underwent a number of changes during its service contract life, so the appropriate manufacturer could be Vickers Sons & Maxim (V.S.M.), Vickers Armstrong at Bath, or Vickers Ltd. at Crayford, Kent, depending upon the period of time, the conversion or manufacture, or the arm or component.

A special asbestos front handguard was approved for use with the Lee-Enfield and Charger-Loading Lee-Enfield rifles on 21st October 1907 and published in the List of Changes § 14069. Details and an illustration of this item will be found towards the end of the preceding chapter as it was initially approved for fitting to the Lee-Metford rifle.

Approval was given for the conversion of Lee-Enfield and Lee-Metford rifles to the small-bore aiming-tube system in 1907. The .22 rimfire cartridge was used and the rifles were marked "A.T." on the action body, bolt, butt and fore-end; these trainers are more fully described in Chapter 7. Mention of this conversion has been made here because components are sometimes encountered marked "A.T."; complete A.T. rifles with the tubes removed or the bolthead changed back to the .303 pattern have also been noted.

On 5th March, 1909, an altered magazine cut-off was approved for the Mark III Short Lee-Enfield and its use was extended to the Lee-Enfield and Charger-Loading rifles. The particular introduction was announced in the List of Changes § 14639 and a further description is provided in the ensuing chapter. Another small alteration applied to Lee-Metford and Lee-Enfield rifles in service during this time was the rounding off of the edge on the bolt face. Approved on 4th May 1909, it was officially announced in § 14710 of the List of Changes.

A somewhat unusual Lee-Enfield was approved for India on 1st December 1909, and it differs from the usual service model in that it was a single-loader. Therefore the magazine and cut-off were not fitted and special filling pieces were used to block up the magazine way. A sheet metal loading platform was provided for positioning the cartridge prior to its being fed into the chamber.

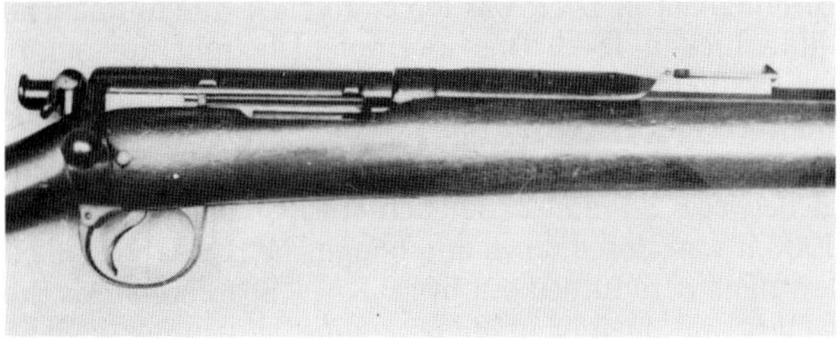

The Single Loader M.L.E. as approved for India. Note the modified magazine floorplate. *MOD Pattern Room*

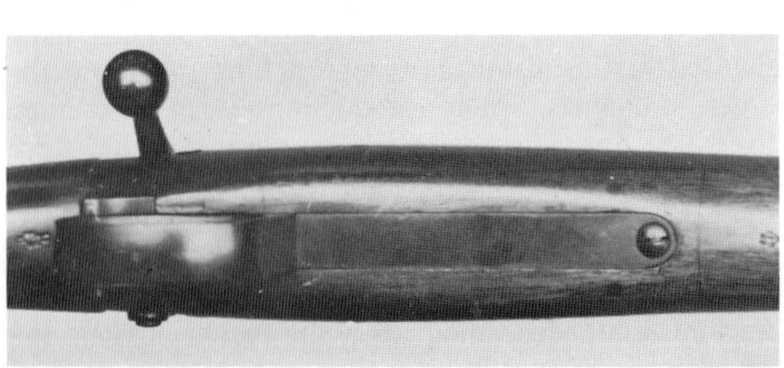

But for the missing magazine and cut-off, this conversion is only obvious upon examination of the trigger guard or floorplate, or when the bolt is drawn back, as this rifle is identical to the service model in all other respects. It was approved on 31st August for issue to Frontier Levies in India and 2,000 were converted at Enfield during 1911 and 1912.

India also introduced further conversions of the Lee-Enfield carbine; the first was a model which fixed the Pattern 1888 sword-bayonet and was approved in India in 1904, the second consisted of upgrading its sighting in 1924 in order that Mark VII ammunition could be used. Indian production at Ishapore is described in Chapter 10.

The Birmingham Small Arms Co. Ltd. 1909 Catalogue illustrates and describes the military, target match and sporting rifles as well as military and sporting carbines, sights and accessories. Giving B.S.A. addresses of "Small Arms Factory, Small Heath, Birmingham" and "6 Great Winchester Street, London E.C.," the charger-loading conversion of Lee-Enfield rifle is described along with commercial variants of the M.L.E. with half-stock fore-ends, smaller capacity magazines, express rearsights with folding leaves and adjustable foresights.

The half-stocked "trade pattern" sporters have 24.5-in. barrels and circular sling loops, some have the military style safety catches as on the M.L.E. and M.L.M. Mk II*, while others have no safety catches at all, as on the M.L.M. Mk II. Another variation sometimes encountered on these commercial patterns is the B.S.A. Patent Locking Bolt, which is a sliding safety catch at the top of the wrist and operated by the thumb; it is marked "B.S.A. Pat.."

Wind-gauge backsight slides for the Lee-Metford and Lee-Enfield ladder sights were offered in five different patterns, along with B.S.A. asbestos packed bands and nosecaps to better bed and accommodate the heat expansion of the barrel.

Other variations were illustrated and described, "For Officer's Use" and "Trade Pattern" with and without provision for bayonet, chequering of the fore-end and wrist, the B.S.A. Patent safety catch, and carbine or rifle style backsight. The Officer pattern has a 19-in. barrel whereas the Trade pattern has a 21-in. barrel. Overall lengths are listed in catalogues at 38 in. and 40 in. respectively.

On 11th October 1910, the case hardening of spare bolt-heads was approved for the M.L.M. and M.L.E. rifles in Naval service, published in the List of Changes, § 15304. A piece of wet rag was wrapped four or five times around the screwed stem of the bolt-head, gripped in tongs, and heated in the forge until blood red. The portion to be hardened was covered with scintilla powder and replaced in the fire so as not to remove the powder, heated to red-hot again for about one and a half minutes and then dipped in cold water, cleaned and adjusted.

A sporting model with chequering and grip safety. *MOD Pattern Room*

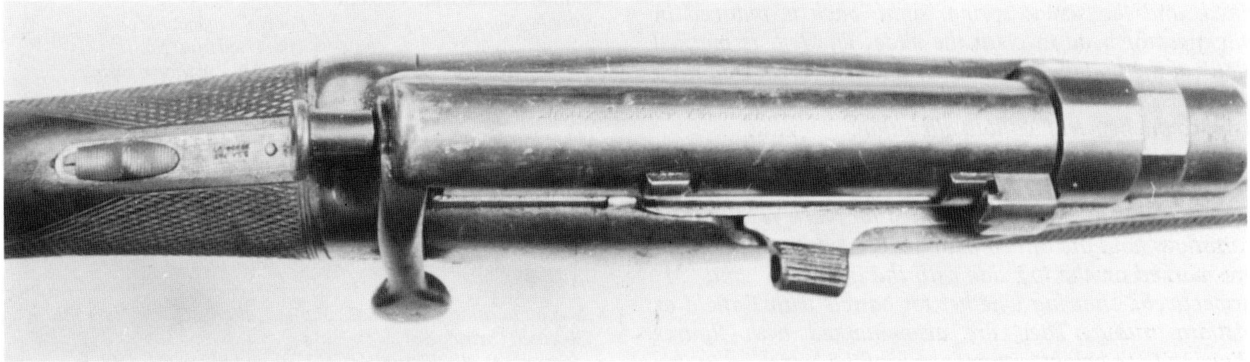

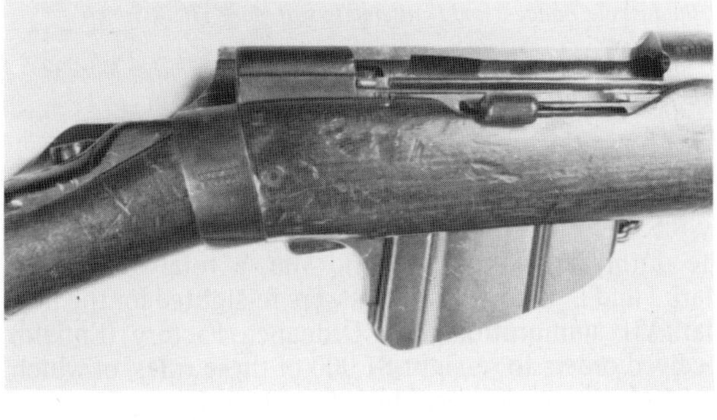

Thumb trigger model (bolt removed). Some were considered for sporting and Rifle Club competition. *MOD Pattern Room*

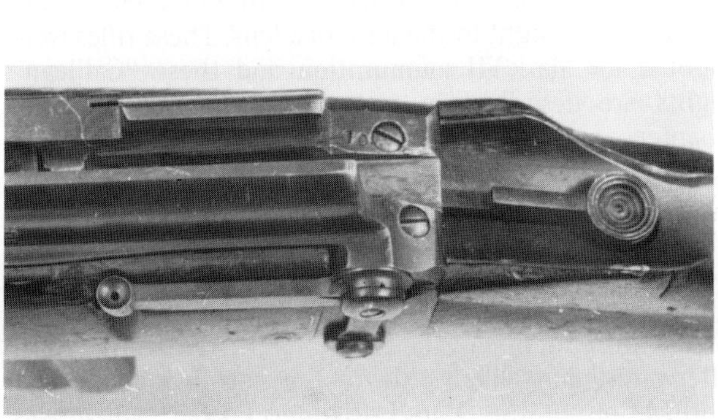

Regarding the introduction of the Charger Loading Lee-Enfield Mk I* in 1907, an unusual situation in the lineage of the British rifle occurred in that the Mk I Charger Loading Lee-Enfield is not mentioned officially until 1914. There was no prior official approval or introduction recorded, possibly because early conversions were not for general issue but intended for certain Naval units.

The charger bridge on the Mk I C.L.L.E. is fitted like that of the Mk I* model, although the sighting is not as elaborate and it conforms more with the standard Lee-Enfield type, without windage adjustment or locking wheels on the slide. Most Mk I charger-loaders were converted for use by the Navy and the first official reference in 1914 describes sighting and magazine alterations to accommodate the Mark VII ball ammunition.

The particular modifications for Mk VII ammunition consisted of:

1. Reduction of height in the backsight bed ramp.
2. Fitting a new backsight leaf graduated to 1,900 yds., marked "C.L." at the bottom right hand corner.
3. V-type notch on the backsight, cut centrally.
4. Rifles with .05-in. offset barleycorns were fitted with adjustable barleycorns set .02-in. left of the bore axis.
5. The dial sight plate had the graduations milled off and was recalibrated from 1,700 to 2,700 yds; these plates were stamped "C.L."

The introduction of the Mk VII cartridge also resulted in similar modification to the Mark I* Charger-Loader and consisted of fitting regraduated backsights and dial plates as well as new foresights. This alteration was officially promulgated in § 17011 of the List of Changes:–

Para 17011:–

*Rifle, Charger-Loading, M.L.E., Mark I**
Alteration to suit Mark VII .303-inch cartridges.

Consequent on the decision to alter rifles of the abovementioned pattern to suit the Cartridges, S.A., ball, .303-inch, Mark VII (§ 15629), the following alterations to components of such rifles in the hands of troops and in store will be made:–

*Backsight: Ramps of bed reduced to give lower elevations, and new leaf graduated for elevations up to 1,900 yards fitted, and the screw, spring, sight, back is reduced in thickness of head to clear the slide. The leaf is marked "I*2" at the bottom right corner. The slide recovered from the leaf graduated for Mark VI .303-inch cartridges is fitted to the new leaf without alteration.*

Blades, foresight: The blades, foresight, are of six lower heights than those for rifles sighted for Mark VI .303-inch cartridges, and are of two patterns, to denote which, they are marked on the top side with the letters "E" and "M" respectively, showing whether for barrels with Enfield or Metford rifling. They are also marked with figures showing their heights from the axis of the barrel.

Dial sight: The graduations for Mark VI .303-inch cartridges are milled off and the plate re-graduated for elevations from 1,700 to 2,700 yards. The plate is marked "C.L." on the face.

Case, magazine; Platform, magazine; Spring, auxiliary, platform, magazine; and Body: These components are altered as shown in § 15638 for Rifles, Short, M.L.E., altered to suit Mark VII .303-inch cartridges.

List of components: The following is a list of altered and new components. Components of new patterns to replace altered components, as announced in § 15638, have been included:

COMPONENT	*DETAIL*
#Barrels, with body, M.L.E. C.L.R., Mark I, No. 2.*	*With sights assembled with .849-in. blade, foresight, but without protector, foresight and screw. Marked H.V. on barrel in front of sight bed.*
#Barrels, without body, M.L.E. C.L.R., Mark I, No. 2.*	*As above.*
#Blades, foresight, M.L.E.C.L.R., Mark I - .819-inch .834-inch .849-inch .864-inch .879-inch*	*For rifles with barrels having Enfield rifling.*
.894-inch...	*For rifles sighted for Mark VII .303-inch cartridges.*
#Blades, foresight, M.L.M.C.L.R., Mark II - .819-inch .834-inch .849-inch .864-inch .879-inch	*For rifles with barrels having Metford rifling.*
.894-inch...	*For rifles sighted for Mark VII .303-inch cartridges.*
#Bodies, M.L.E.C.L.R., Mark I, No. 2.*	*Recognisable from No. I by polished bullet lead.*
>Cases, magazine, M.L.E.R.S., Mark I– No. 3. ...	*With stop slip and stud, lip, and rib, spring. Converted from Nos. 1 & 2.*
No. 4. ...	*When stock of No. 3 is used up.*
#Leaves, backsight, M.L.E.C.L.R., Mark I, No. 2.*	*Unassembled. Marked "I*2" at right bottom corner.*
+Plates, sight, dial, M.L.E. C.L.R., Mark I.	*Marked "C.L." on face. Not issued separately.*
>Platforms, magazine, M.L.E.R.S., Mark I - No. 2. ...	*For No. 3 magazine case; converted from No. 1.*
No. 3. ...	*With No. 2 platform spring; for No. 4 magazine case.*
†Screw, spring, sight, back, M.L.E.C.L.R., Mark I.*	*[no details]*
>Springs, auxiliary, platform, magazine, M.L.E.R.S., Mark I - No. 3. ...	*For No. 3 magazine case; converted from Nos. 1 & 2.*
No. 4. ...	*For No. 4 magazine case.*

 # Special to rifles, C.L., M.L.E., Mark I, altered to suit Mark VII .303-inch cartridges.*

> Common to all Rifles, Short, M.L.E., and Rifles, C.L., M.L.E., altered to suit Mark VII .303 cartridges.

+ Common to all Rifles, C.L., M.L.E., Marks I (not yet announced in the List of Changes) and I altered to suit Mark VII .303-inch cartridges.*

† Suitable for M.L.E.C.L. rifles, Mark I, sighted for either Mark VI or VII .303-inch cartridges.*

This re-sighting of Charger-Loading rifles was carried out by R.S.A.F. Enfield, B.S.A. Co. and Westley Richards. After the outbreak of the 1st World War, a total of 251,985 Mark I and I* Charger-Loaders were re-sighted for the new Mark VII ammunition. The Ordnance Factory (Enfield) received orders to re-sight 81,000 of these rifles, of which 80,559 had been delivered by 1st April 1917.

During 1915, 225 Charger-Loading Lee-Enfield Mark I* rifles were fitted with special optical sights by B.S.A., comprising an aperture backsight with lens eyepiece and telescopic foresight in the form of a lens. These rifles were sighted for Mark VII ammunition and these "Galilean" sights are described in more detail in the following chapter.

Even though approval was given for the alteration of the Mk I C.L.L.E. to suit Mk VII ammunition on 1st September, it was one month before similar approval was given for the Mk I* C.L.L.E. Also, the announcement and official introduction of the Mk I C.L.L.E. was delayed until afterwards, most likely due to the non-introduction or notification of the Mark I Charger Loader which was

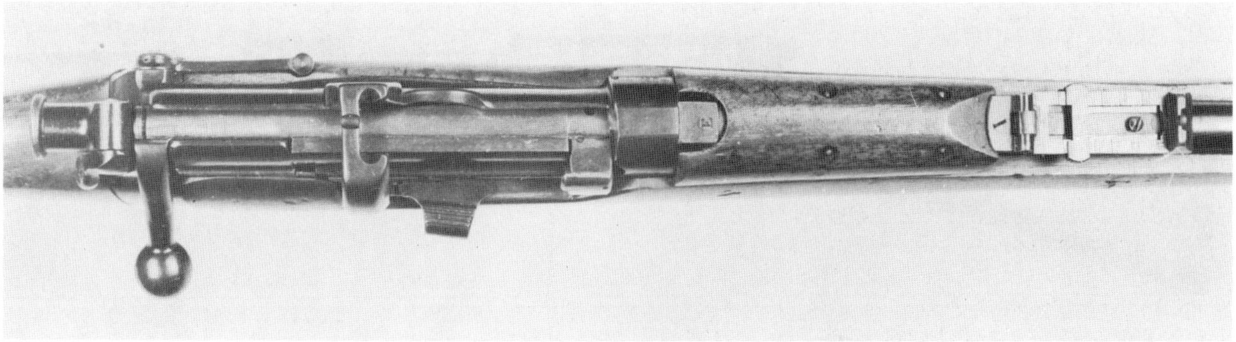

Sighting modifications to the C.L.L.E. Mk I for Mk VII ammunition. Note sighting groove in charger bridge. *MOD Pattern Room*

finally officially approved on 17th December 1914. This late List of Changes announcement reads:

Para 17041:–
Rifle, Charger-Loading, M.L.E., Mark I

1. Alteration to suit Mark VII .303-inch Cartridges
2. This pattern not previously announced in List of Changes

1. Consequent on the decision to alter rifles of the above-mentioned pattern to suit the Cartridge, S.A. ball, .303-inch, Mark VII (§ 15629), the following alterations to components of such rifles in store will be made.

Backsight: Ramps of bed reduced to give lower elevations, and new leaf, with central V notches in cap and slide, graduated for elevations up to 1,900 yards, fitted. The leaf is marked "C.L." at the right bottom corner. The slide recovered from the leaf graduated for Mark VI .303-inch cartridges is fitted to the new leaf, provided it has a central V notch.

Foresight: Such rifles as have the foresight barleycorn placed .05-inch left of the axis of the bore are fitted, while undergoing alteration, with a removable barleycorn, .02-inch left of the axis of the bore (i.e. Barleycorn, foresight, M.L.E. rifle) and a fixing pin (i.e. Pin, barleycorn, foresight, M.L.M. rifle) of the same pattern as those used in method 2 of § 10393, so that only leaves with central V notches in cap and slide, as before described, are required.

Dial sight: The dial sight is altered and marked in the same way as that of the Mark I rifle, described in § 17011.*

Case, magazine; Platform, magazine; Spring, auxiliary, platform, magazine; and Body: These components are altered as shown in § 15638 for Rifles, Short, M.L.E., altered to suit Mark VII .303-inch cartridges.

List of components: The following is a list of altered and new components. Components of new patterns to replace altered components, as announced § 15638, have been included:

COMPONENT	*DETAIL*
#Barrels, with body, M.L.E.C.L. rifle, Mark I - No. 2.	*With sights assembled. Marked H.V. on barrel in front of sight bed.*
#Barrels, without body, M.L.E.C.L. rifle, Mark I - No. 2.	*As above.*
#Bodies, M.L.E.C.L. rifle, Mark I - No. 2.	*Recognisable from No. 1 by the polished bullet lead.*
>Cases, magazine, M.L.E. rifle, short - No. 3.	*With stop clip and stud, lip and rib spring. Converted from Nos. 1 & 2.*
No. 4.	*When stock of No. 3 is used up.*
#Leaves, backsight, M.L.E.C.L. rifle, Mark I.	*With cap, screw and slide. Marked "C.L." at right bottom corner.*
+Plates, sight, dial, M.L.E.C.L. rifle, Mark I.	*Marked "C.L." on face. Not issued separately.*
>Platforms, magazine, M.L.E. rifle, short - No. 2.	*For No. 3 case. Converted from Nos. 1 & 2.*
No. 3.	*With No. 2 platform spring; for No. 4 magazine case.*
>Springs, auxiliary, platform, magazine, M.L.E. rifle, short - No. 3.	*For No. 3 magazine case. Converted from Nos. 1 & 2.*
No. 4.	*For No. 4 magazine case.*

 # Special to rifles, C.L., M.L.E., Mark I, altered to suit Mark VII .303-inch cartridges.
> Common to all Rifles, Short, M.L.E., and Rifles, C.L., M.L.E., altered to suit Mark VII .303 cartridges.
+ Common to all Rifles, C.L., M.L.E., Marks I and I, altered to suit Mark VII .303-inch cartridges.*

2. This pattern rifle has not previously been announced in List of Changes. It was converted on the same lines and from the same patterns of rifles as the Mark I rifle (§ 13992) except that the sights already on the rifle before conversion (i.e. a solid unadjustable foresight and a backsight without a windgauge as in M.L.M. and M.L.E. rifles) were left untouched by the conversion, and no foresight protector and screw were fitted. The body, also, differs from the Mark I* body in having a groove cut through the bridge charger guide to clear the line of sight at the lower elevations.*

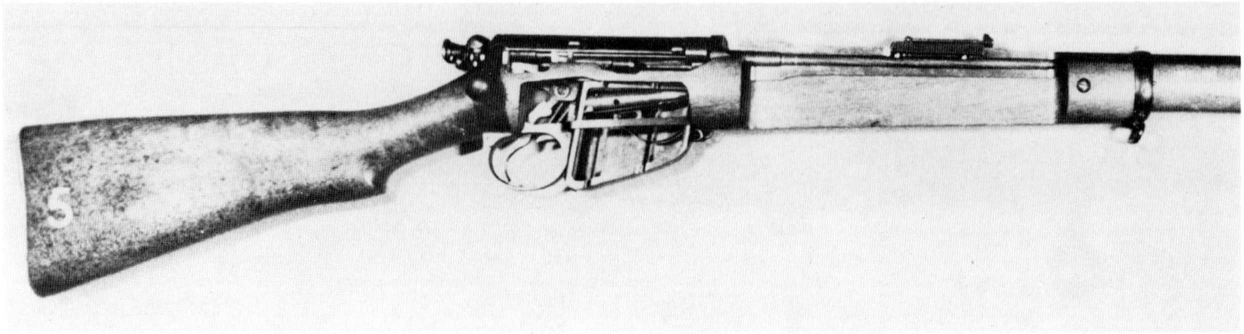

Full-length sectionised M.L.E. rifle. *Imperial War Museum*

The Mark I and I* charger-loading models saw the eclipse of the Magazine Lee-Enfield's service career although some were continued with as reserve grade and training rifles until the 1939-45 war when additional barrels, bolts and boltheads had to be produced by Parker Hale and B.S.A. due to the scarcity of some spare parts. The C.L.L.E. was often referred to as the "Territorial Pattern" because of its extensive reserve role.

The U.S. Remington factory's commercial models of Lee sporting rifles also provide an interesting comparison with the British service development and some parallels can be drawn. Their 1904-1905 Catalogue describes the mechanism as being similar to the Remington-Lee U.S. Navy Model Rifle and lists the available calibres as 6mm US Navy, 30-30, 30-40 US Gov't., 7mm, 7.65mm, .32 and .35 Special, 32-40, .405W, 38-55 HP smokeless, 38-72 and .303 British. An Express rifle version was also available in .44-77 bottle-necked, .43 Spanish, .45-70 or .45-90.

The sales brochures list the following features of the military bolt system magazine rifle in its sporting rifle configuration:

(a) By half cocking the piece, the mechanism becomes locked, and in this condition its safety from premature discharge and the impossibility of accidentally opening the gun makes the Remington-Lee particularly desirable for carrying on horseback or on the trail.

(b) Each rifle furnished with four magazines holding five cartridges each, which will enable the shooter to discharge twenty shots in an incredibly short space of time and without stopping to replenish the magazines.

(c) By pressing the magazine catch, in front of the trigger, the magazine and contents can be instantly removed without the possibility of a premature explosion, which frequently occurs where the cartridges have to be worked out by the manipulation of the action.

Four magazines were provided with each rifle, which had a round barrel, open sporting sights and provision for the mounting of Lyman sights. The list price was $25.00¢ and a special model with 26-in. half-octagon barrel, full pistol grip with rubber cap, selected English walnut stock, fine checkering, adjusted trigger pull, fine finish metal work and fittings and special Lyman sights was listed at $60.00¢.

Specifications are given as:–

Weight:	From 7 to 7.25 pounds.
Magazine:	5 rounds (4 rounds for .32-40 & .38-55)
Barrel Length:	24-in., 26-in. (standard) or 28-in.

The larger Express model in .43, .44 and .45 calibres was listed at $20.00¢ and weighed a little more at 8.5 pounds; it was also fitted with the optional buck-horn sights.

These rifles can also be considered as contemporaries of the B.S.A. trade pattern models from the B.S.A. 1909 Catalogue, described and illustrated a few pages earlier. It makes an interesting comparison of the marketplaces both sides of the Atlantic; many of the B.S.A. rifles were sent to South Africa.

Even the British B.S.A. 1926 catalogue illustrates and lists the Magazine Lee-Enfield half-stocked sporting rifles; Metford rifling was also still available. At this comparatively late date, these rifles were in M.L.E. configuration, with or without the safety catch on the cocking piece, and without the charger loading facility. They were available in .303, 7mm and 8mm calibres, with 5-round or 10-round magazines and with express-type sights and deluxe chequered fore-end and buttstock if so desired.

The prices range between £5/12/6d for the "Trade Pattern" with military sights to £9/-/- for the "High Velocity Sporting Rifle" with chequered stock, horn fore-end tip, pistol grip and express sights. By comparison, a selected quality, government pattern Mark III S.M.L.E. by B.S.A. was quoted in the same catalogue at £8/5/-. The Pattern 1907 sword-bayonet with hooked quillon was £1/2/6d.

Sporter model Lee-Metford and Lee-Enfield rifles of this era were often taken out on safari and game expeditions to the colonies, especially in Africa. The .303 sporter was frequently used as a general purpose camp gun, supplementary to the big game rifle.

A PROGRESSION OF EARLY BRITISH SERVICE LEE'S.

Top to bottom:

Magazine Lee-Metford Rifle, Mark I
Magazine Lee-Metford Cavalry Carbine, Mark I
Magazine Lee-Enfield Rifle, Mark I
Short, Magazine Lee-Enfield Rifle, Mark I

MOD Pattern Room

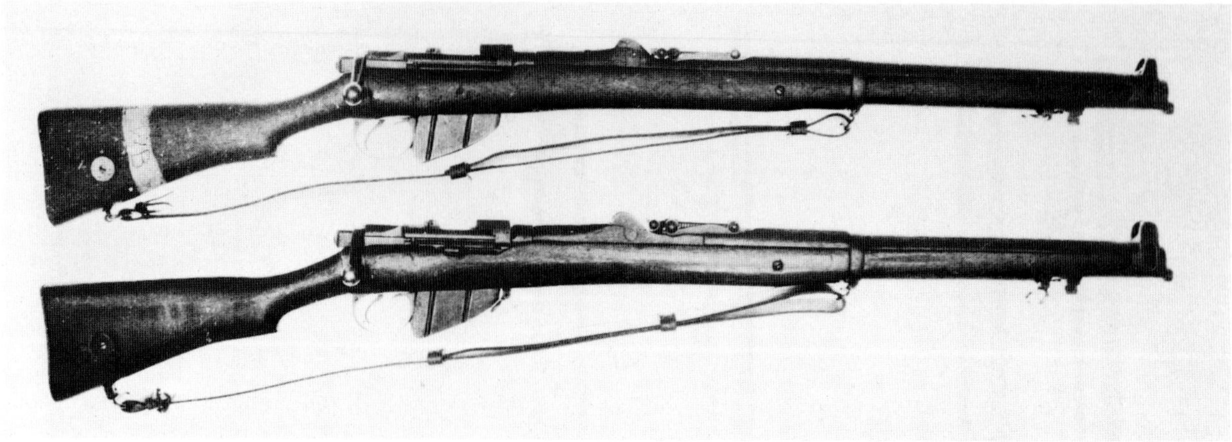

The 1902 troop trials Short Rifle "A" Pattern (top) compared with the eventual service Mark I Short, Magazine Lee-Enfield. R. Faris

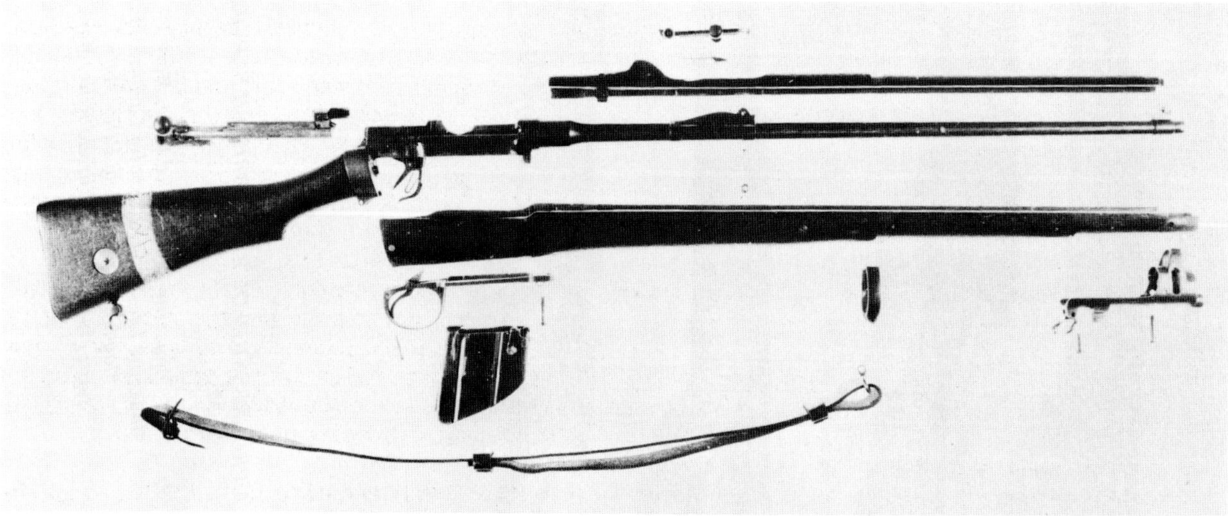

Above & below: The "A" Pattern trials Short Rifle stripped. Note the one-piece handguard, Watkin and Speed backsight, sliding charger guide on the bolthead and the enclosed hood over the foresight, incorporated in the nosecap. R. Faris

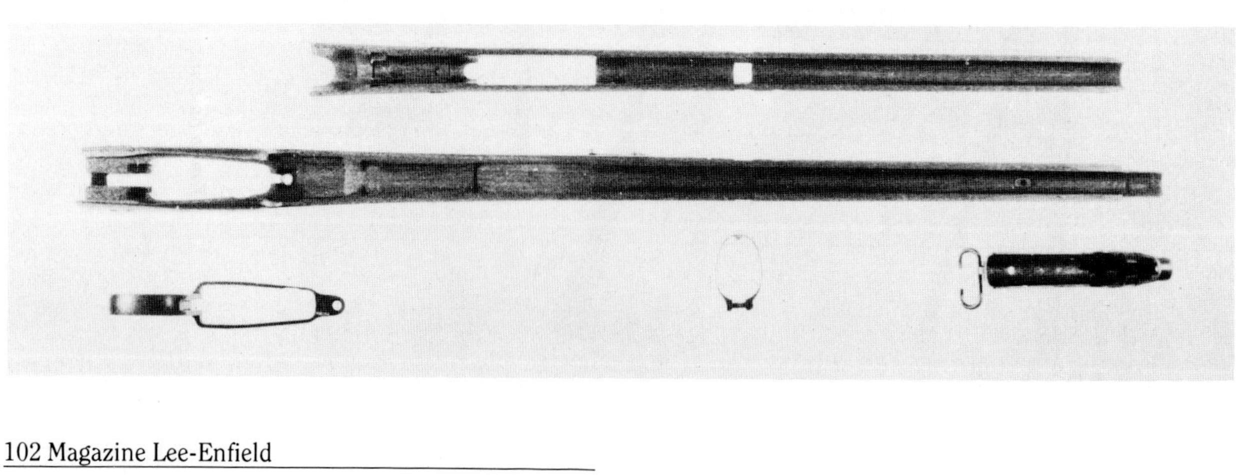

A SHORT RIFLE, THE S.M.L.E.

CHAPTER 5

In the later months of 1901, some 1,055 "Shortened Enfield Modified Rifles" were fabricated at the Royal Small Arms Factory, Enfield; these were based upon the No. 1 Modified Rifle shortened and also incorporated further, more recent improvements. Trials of the new short rifle were carried out by the Army, Navy and Royal Marines early in 1902, which generally resulted in recommendations for its universal adoption, that is to say, a uniform length rifle for all services and forces.

The new trials rifle was about 4.5 inches longer than the cavalry carbine and it weighed only 8 oz. more. The cavalry was quite pleased with the handling of the new rifle and they reported more accurate shooting than with their cavalry carbines. For the other branches of the Services, including the Navy, it weighed 1 lb. 8 oz. less than the long rifle, and was 5 inches shorter and considerably more handy. A nominal one thousand of these rifles were issued for extensive troop trials.

Some minor changes were implemented as the troop trials progressed, as was often the case when such trials took a considerable length of time, as reports from the various units came to hand. The trials rifles were fitted with a bead foresight but the form of barleycorn already in service was preferred; the foresight hood was not generally liked. The cavalry was reported as not having favoured the Lee-Metford and Lee-Enfield cavalry carbine sight protectors as previously on issue.

This Shortened Rifle was provided with a long type of handguard which was well received; Speed and Watkin's new safety catch was also welcomed. Aluminium and brass buttplates were tried but the Committee recommended a thin sheet-steel pattern. Previously, steel had not been favoured due to its tendency to rust and there was no trap fitted for access to the stockbolt hole in the butt. There was no magazine cut-off, although it had been recommended that provision be made for one in the event of its later being found to be necessary.

So as to provide a better evaluation and suitable comparison with the service rifles and carbine, this thousand Shortened Enfield Modified rifles underwent trials alongside their current service counterparts. Those units which carried out the trials were the Royal Naval School of Gunnery at Sheerness, H.M.S. "Excellent," 3rd Provisional Regiment of Hussars, 4th Provisional Regiment of Dragoons, H.Q. Cavalry Brigade, 3rd Bn. Manchester Regiment, 2nd Bn. Highland Light Infantry, 2nd Infantry Brigade, 4th Bn. Middlesex Regiment, 1st Infantry Brigade, 1st Army Corps, 21st Lancers, 2nd Bn. Liverpool Regiment, Board of Officers at the Curragh, 4th Bn. Manchester Regiment, 1st Bn. Yorkshire Light Infantry, 4th Bn. Lancashire Fusileers, H.Q. South-Eastern District and the Provisional Bn. (10th) at Shorncliffe.

A further quick-filling system for the magazine was tested against the charger-loading system, this controllable platform being of Edward's design (see the previous chapter for descriptions of the Punchard or Harris, Ross and Edwards quick-filling systems); the charger-loading system was preferred. An improved, tightened charger guide was used in October 1902 although the first model was found to develop excessive play.

The hooded sight protectors were considered unsatisfactory and so it was proposed that the tops be cut off the hoods. So fifty rifles with hood tops removed were issued to various cavalry regiments for comparison against a similar number of earlier trials models with the hoods, but fitted with barleycorn front sights.

Regarding the backsight, two patterns had been issued for the 1,000 trials rifles, differing again from the experimental pattern for the 1901 long rifles (illustrated in the previous chapter). The 1901 type was found to be too vulnerable to damage and so the two new models, designated "A" and "B" were tried, with approximately 500 rifles fitted with each type of sight.

Type "A" was not unlike the eventual Mark I S.M.L.E. pattern, with ivory knobs on the slide and very fine adjustment for range and windage. The hinge pin was to the front and so the sight hinged forward as elevation was increased. Markings were in increments of 100 yards and numbered from 200 to 2,000 yds. The sight was another Speed and Watkin patent.

The "B" pattern was modified from the early 1901 experimental model using a more conventional type of leaf with a clamping bar and a release catch on the left hand side. Like the early pattern sight, the leaf had two

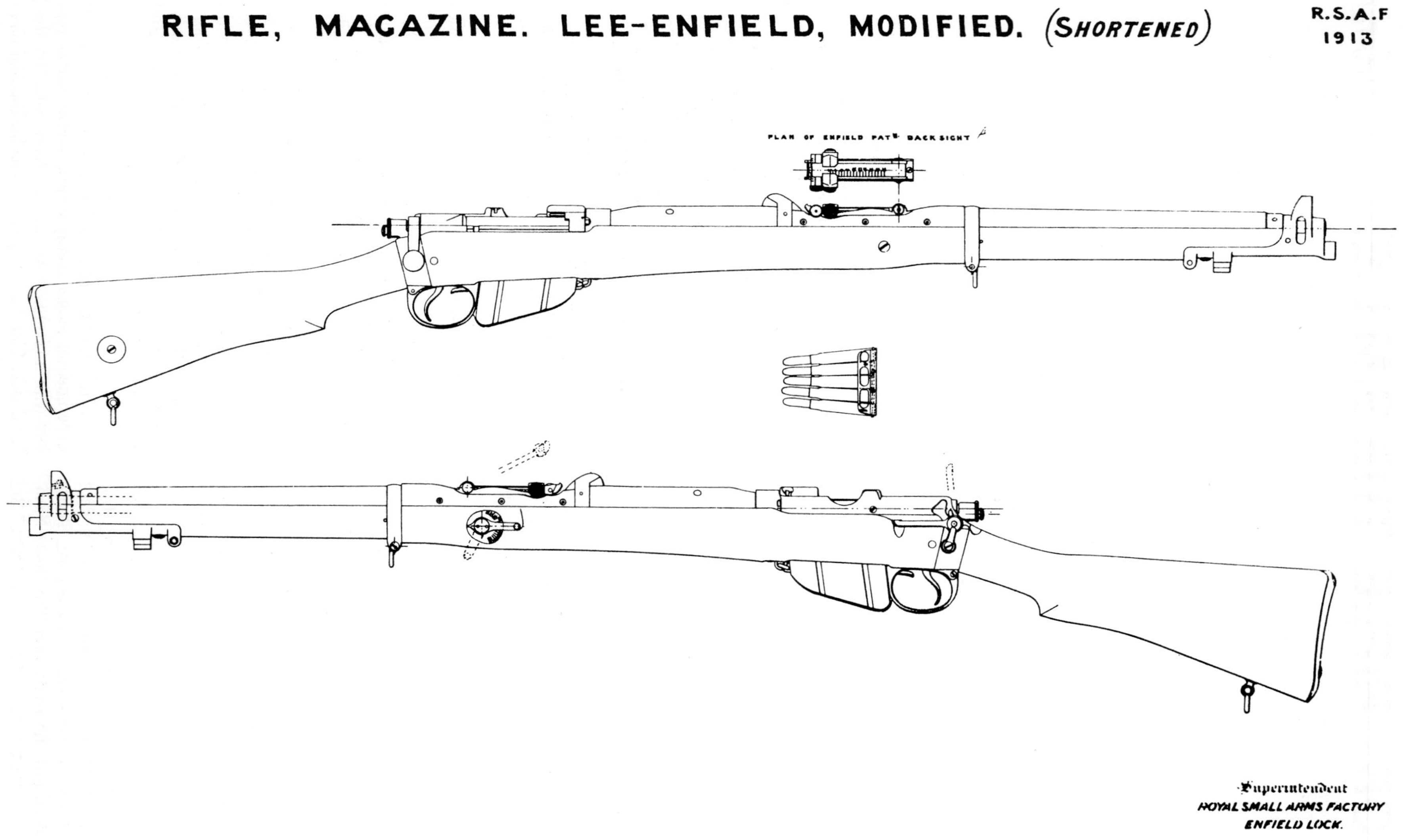

1902 Trials Rifle Type "A". This has the Watkin and Speed backsight. The main differences to the adopted Mark I Short Rifle are found in the nosecap with fully enclosed hood, different form of backsight protector mounting, and the differently constructed handguard. MOD Pattern Room positions, folded down for ranges to 500 yards and raised vertically for ranges from 500 to 2,000 yards. The lateral adjustment or windage was made by a wheel at the rear of the bed. This pattern was found to become very hot to touch after rapid fire and the method of clamping the bar at set ranges was unsatisfactory as it often moved when the rifle was fired.

6743. Watkin, H. S. S., and **Speed, J. J.** March 23. ***1903***

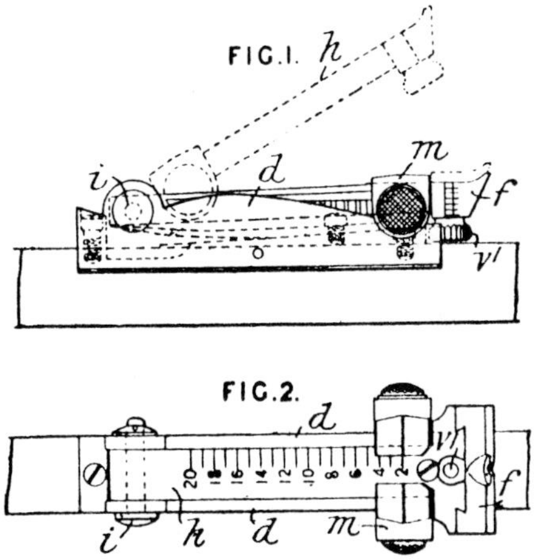

Sights.—Relates to back sights, in which the leaf is gradually elevated for range by a slide which travels down the leaf and along the top of a curved bed. Fig. 1 shows a side elevation, and Fig. 2 a plan. The leaf *h* is hinged at *i*, and consists of a graduated bar having teeth along its edges to engage with spring catches on the slide *m*, which rests on the top surfaces of the curved bars *d*. The leaf bar *h* is graduated for hundreds of yards range, the necessary adjustments for intermediate ranges being effected by the screw v^1 which moves the notched bar *f* on the bar *h*.

6744. Watkin, H. S. S., and **Speed, J. J.** March 23. ***1903***

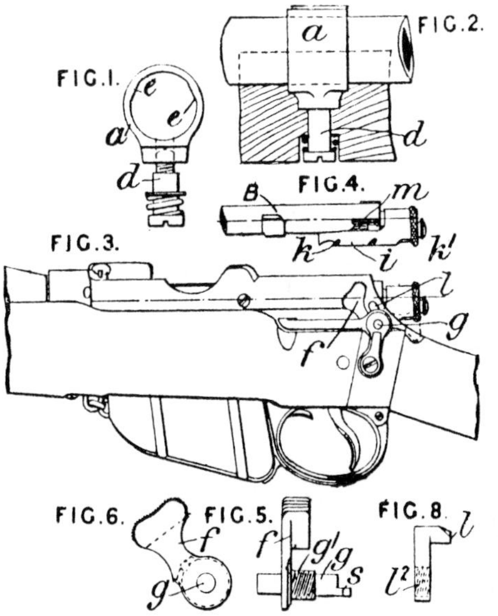

Barrels ; stocks.—Relates to means for attaching the barrel to the stock, and consists in the employment of rings *a*, Figs. 1, 2, passing over the barrel and having flats *e* to grip the barrel as soon as the rings are drawn towards the stock by the spring-supported screws *d*.

Breech actions, sliding breech-block.—Relates to safety devices for bolt guns. Fig. 3 shows the gun in elevation, Fig. 4 the breech-bolt and hammer detached, Figs. 5, 6, the hammer safety, and Fig. 8 the bolt safety. The pin *g*, provided with the thumb lever *f*, is cut away at its inner end *s*, so that when in the position shown in Fig. 3 it clears the hammer *i*, but when the lever *f* is pulled rearwards the part *s* engages in one of the slots *k*, k^1 of the hammer, and locks it against movement. The rocking of the lever *f* also acts by a quick thread g^1 on a nut l^2 so as to carry the pin *l* into the recess *m* of the breech-bolt B to lock it against rotation.

Further improvements incorporated in the new rifle were designs of Speed and Watkin and they took out patents on the safety catch and inner barrel band. This particular patent is No. 6744 of 23rd March 1903.

A drop in muzzle velocity was observed with the shorter barrels so certain modifications were made to increase velocity to over 2,000 ft/sec. again. This was achieved by using progressive depth rifling, becoming deeper towards the muzzle; this form of rifling was eventually adopted with the Mark I S.M.L.E. and the depth of the rifling was .0065-in. at the muzzle and .005-in. at the breech end.

Three different lengths of butt were issued, short, normal and long; the short type marked with an "S" and the long with an "L." There was no bolt dustcover and the bolt had a turned-down knob. Two lengths of bayonet were tried, a 12-in. blade model which became the Pattern 1903 and a 17-in. blade type that was double-edged like the Pattern 1888. The shorter pattern was preferred. The cavalry was also issued with experimental rifle buckets for trial with the new rifle.

Due to the duration and diversity of the trials, other units also participated; the Plymouth Division of the Royal Marine Artillery, and the Royal Marine Light Infantry at Chatham, Deal and Portsmouth. The trials were considered to be quite successful and for the first time in Britain's military history, the cavalry, artillery, infantry, rifle regiments, engineers, navy and marines were to have a standard rifle. At various times each of these corps had their own specific arm and the simplification in manufacture, supply and spare parts was a tremendous gain, especially when compared with the situation some forty years previous.

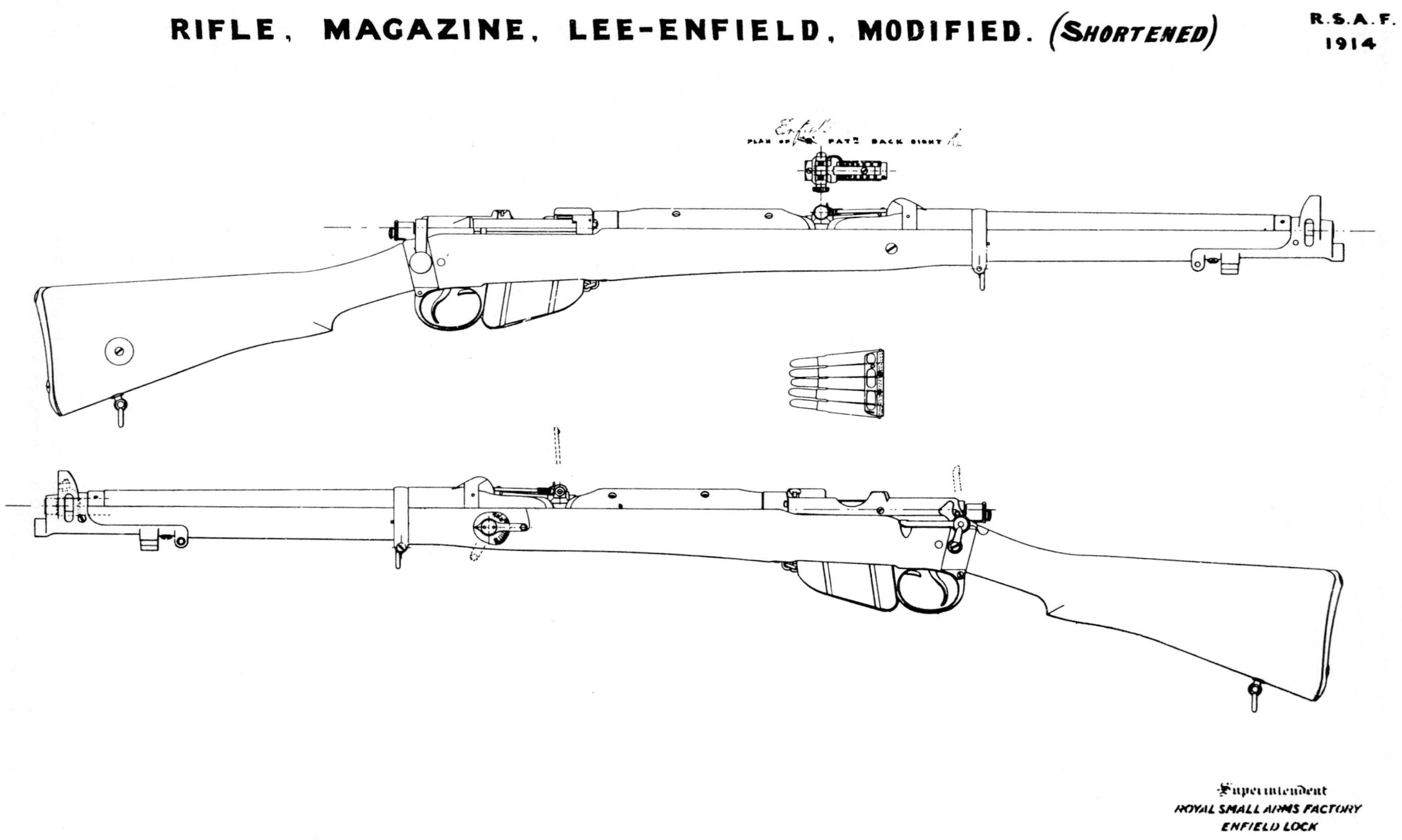

1902 Trials Rifle Type "B". Similar to the previously-illustrated "A" Pattern but for the backsight and handguards. On this drawing, the backsight description has been changed a little so as to conform more correctly with that of the "A" Pattern drawing. MOD Pattern Room

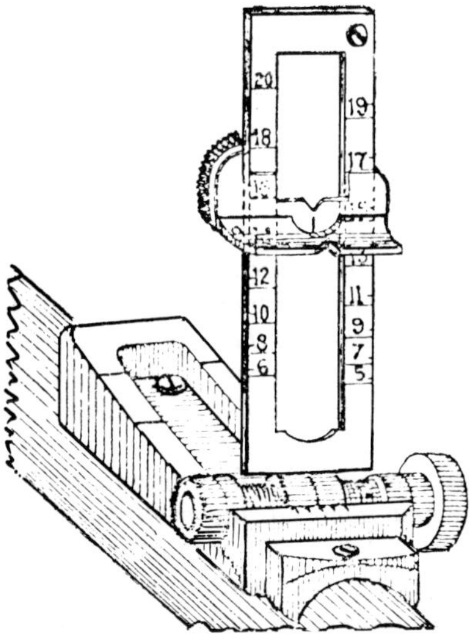

"B" Pattern backsight in the raised position.

There was much adverse criticism of the proposed short rifle, not, as in the past, from a small vocal element of disappointed inventors or journalists looking for headlines, but from the strongly-entrenched Rifle Association fraternity and the gun trade. Their main concern was that the shorter and lighter barrel would not be as accurate, the sighting radius too short, and the recoil considerably increased, however, the Small Arms Committee would not compromise its decision. Results from the trials indicated that the accuracy was as good as that of the long rifle and that the proposed rifle had better handling qualities.

A modified trials rifle was submitted to the Committee on 15th December 1902 incorporating certain improvements, namely:

An adjustable barleycorn foresight,
A deeper magazine,
Tighter charger guides,
Provision for a magazine cut-off,
A steel sheet-metal buttplate,
An adjusting screw fitted to the elevation slide,
The windgauge omitted,
The top cut out of the foresight protector hood,
Modified handguards and backsight protector.

This was the rifle which was to be adopted as the S.M.L.E.

Very few of these trials rifles have survived as most of them were converted to S.M.L.E. .22RF Aiming Tube Mark I rifles in late 1906 and early 1907.

The Short, Magazine Lee-Enfield Rifle Mark I is 5 ins. shorter and 1.25 lb. lighter than the Magazine Lee-Enfield and it was first approved on 23rd December 1902 and promulgated in the List of Changes, § 11715.

Para 11715:-

Rifle, Short, Magazine, Lee-Enfield (Mark I).

A pattern of the above-mentioned rifle has been sealed to govern future manufacture. The magazine will hold 10 rounds, and is filled by cartridges carried in chargers (§ 11753); guides to hold the latter are provided on the bolt head and body, while the five cartridges held in the charger are swept out of it by the thumb into the magazine. The bolt and cocking piece may be locked in the "full cock" and "fired" positions by a safety catch, and locking bolt, situated on the left side of the body, between it and the aperture sight; they are held in position by the aperture sight spring.

Barrel: The barrel is smaller in diameter externally, and 5 inches shorter than that of the L.E. rifle; it is fitted with a band which carries the block foresight, this is keyed and pinned to the barrel, and is dovetailed to carry an adjustable barleycorn foresight (three heights of which will be provided, known as high, normal and low, and marked "H," "N" and "L" respectively), to enable variations in shooting to be corrected before the rifle is issued to the Service. The heights of the high and low barleycorns differ from the normal by .015-inch.

Backsight: The backsight is fitted with a leaf pivoted to the bed at the front end; at the rear end is a cap in which the V is cut. Elevation is given by moving the slide, which rests on curved ramps on each side of the bed. The leaf is graduated by lines for every 100 yards from 200 to 2,000, the even numbers being marked by figures. The slide can be set at any 100 yards graduation or intermediate 50 yards elevation, and is held in position by means of catches engaging in notches on each side of the leaf. To set the slide these catches are disengaged from the notches by pressing the bone studs at each side of the slide. The cap is joined to the leaf by a vertical dovetail, and it can be given a fine adjustment for intermediate ranges between the 50 yards elevations afforded by the slide by means of a vertical screw underneath the cap of the backsight; a smaller vernier scale divided to give a vertical movement of .0106 inch being provided on the left edge of the cap and leaf. Each division on the vernier represents 2 inches elevation per 100 yards. By raising the slide to its highest limit, 2,050 yards elevation can be obtained. The dial sight for long ranges is graduated from 1,600 to 2,800 yards, marked "L.E.S.."

The following particulars show the principal differences in detail between the short rifle and the M.L.M. and M.L.E. rifles:-

Body: The body is made with charger guide on the left to receive the charger by which the magazine is loaded, and a stop on the right which forces the charger guide on the bolt head forward when the bolt is drawn fully back; it is also arranged to receive the locking bolt and safety catch to lock the bolt and cocking piece; the cut-off slot is left in the body for the insertion of cut-off if required. The left side of the body is cut away to afford clearance for the thumb of the right hand when pressing the cartridges from the charger into the magazine. A cut-off will be supplied for naval service only.

Rifle, short, magazine, Lee-Enfield. **(Mark I.)**

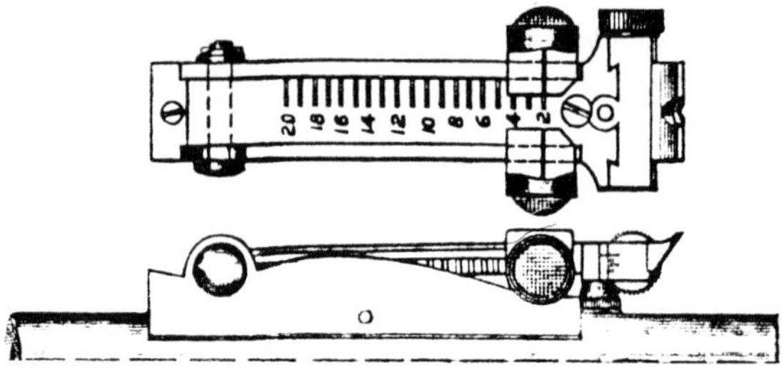

Views of backsight.
Scale ½.

Pin, fixing, bed, sight, back.
Pin, fixing, block, band, foresight.
Pin, fixing, stud, head, catch, slide, sight, back. (2)
Pin, fixing, washer, pin, axis, sight, back.
Pin, joint, band, outer.
Plate, dial, sight.
Plate, keeper, screw, wind-gauge.
Platform, magazine.
Screw, dial sight, fixing.
Screw, fine adjustment, sight leaf.
Screw, guard, back.
Screw, nut, keeper, striker.
Screw, protector, sight, side, handguard, rear.
Screw, spring, sight, aperture.
Screw, spring, sight, back.
Screw, stop, charger guide.
Screw, wind-gauge.
Screw, band, inner.
Screws, band, outer, and swivels, butt & piling. (3)
Screws, bed, back; back & front.
Screws, keeper, fine adjustment, sight, back, and handguard, cap and sight protector, top. (4)
Screws, nosecap; Back & front.
Sear.
Slide, sight, back.
Slide, sight, back, fine adjustment.
Slide, sight, back, wind-gauge.
Spring, auxiliary, platform, magazine.
Spring, nut, keeper, striker.
Spring, platform, magazine.
Spring, screw, band, inner.
Spring, screw, wind-gauge.
Spring, sight, aperture.
Spring, sight, back.
Springs, catch, slide, sight leaf. (2)
Stem, swivel, butt.
Stocks, butt; Long, Normal & Short.
Stock, fore-end.
Striker.
Stud, clip, stop, magazine.
Trigger.
Washer, pin, axis, sight, back.
Washer, rivet, fore-end; and spring, handguard. (4)
Washer, spring, band, inner.

** In rifles for Naval Service only.*

Only two of these rifles were manufactured without windgauge and they would have been pattern arms. The same financial year, 1st April 1903 to 31st March 1904, only 961 S.M.L.E. rifles were made at Enfield, these being fitted with the windgauge. Production began in earnest the following year, when over 67,000 were made at Enfield and nearly 10,000 at Sparkbrook. This same year (1904-05), nine long rifles were converted to charger-loading with the addition of a guide on the left side of the receiver and another on the bolthead, in a manner similar to that of the Mk I S.M.L.E.; one of these patterns was later adopted for India.

Some difficulties had been experienced with consistent accuracy of the new production rifles and the advent of the new Short rifle did not rectify the problem as reports of the "wild shooting" persisted. Variable factors were closely investigated and many experiments effected but no real success was achieved in the short term. It was found that a short lead in the rifling resulted in better accuracy and a new type of rifling was proposed by Capt. Wallace, described as "Metford-Enfield" rifling. Reports of trials with this rifling were presented on 29th August 1904 because 12 barrels had been rifled on this system. Of these, 7 were used by troops for comparative trials regarding cleaning and 3 at Hythe to test accuracy as well as facility of cleaning.

These barrels were found to foul less and were also more easily cleaned while those with the short lead were noticeably more accurate. Further trials were ordered using 20 service barrels with the short lead and 20 with Wallace's rifling and the short lead, but the Wallace rifling was found to have no marked advantage over the service type, so no more tests were conducted. The shorter lead, though, was adopted later and even in 1908 it was applied to new Martini-Metford, Martini-Enfield and Magazine Lee-Metford and Lee-Enfield rifle and carbine barrels; noted in the List of Changes, § 14234.

Some 60 rifles were fitted with a V-notch on the backsight similar to the U.S. Krag rifle and this shape was later modified so as to have parallel sides with a shallow V at the bottom. The U-notch sight was eventually introduced with the Mark III S.M.L.E. rifle in 1907.

Although the shorter lead improved accuracy somewhat, there were still problems with reliable accuracy and it was discovered that relieving the last 14 inches of the bore often caused this upset because the bullet was not always evenly supported by the rifling just prior to its leaving the muzzle. So the Committee approved parallel rifling for the full length of the bore which meant that the lapping of the last 14 inches was discontinued. Progressive depth of rifling, however, continued in production until 1917.

After the 1,000 trials short rifles were returned to Enfield, a report was prepared on their condition. It was found that the safety catches suffered nigh on a 10% casualty rate and this was considered to be the weakest part of the new rifle. A comparatively small number of striker (firing pin) retaining screws had become broken or worked loose and some of the magazine auxiliary springs were defective or had become loose. Some fore-ends were split and quite a few had lost rivets and washers but as these rifles had in many cases been deliberately worked hard, the mortality rate of components was not considered to be excessive, except perhaps for the safety catch. Most of the barrels were noticeably worn although this was expected as most other wear was not regarded as being abnormal.

The first 300 of the new service rifles at Hythe were reported upon and the main complaints were:

The handguard was too easily displaced,
The striker keeper nut broke or was damaged on 27 rifles,
Shooting was not satisfactory, largely due to the poor sighting in,
Some screws, mainly the outer band screw and backsight fine adjustment screw, worked loose.

These faults were remedied by official modifications applied to subsequent models; this was largely the reason for the "double introduction" of the Mark I S.M.L.E. in the List of Changes.

Proposals were made regarding the conversion of the outdated Lee-Metford Mark I*, the Lee-Metford Mark II and II* and eventually the Lee-Enfield rifles to the Short Lee-Enfield pattern so as to utilise those outmoded rifles still in service or returned to store. The true chronological progression of these conversions is not readily ascertained by their sequence in List of Changes paragraph entries, it appears that these were later arranged so as to keep the announcements of new and converted rifles in order; the various approval dates also indicate this.

The first official introduction of such a conversion appears in the List of Changes § 11948, with an approval date of 2nd November 1903.

Para 11948:-
Rifle, Short, Magazine, Lee-Enfield, Converted (Mark I)
Converted from Rifle, Magazine, Lee-Metford, Mark I.*

A pattern of the above-mentioned rifle has been sealed to govern conversion as may be ordered. The conversion is made from rifles, M.L.M., Mark I, and consists generally in fitting the rifles with new sights, shorter and lighter barrels, and adapting them for the magazine to be filled from a charger, § 11753.*

The arm is fitted to carry the sword-bayonet, pattern 1903, §§ 11716 and 11717, on the nosecap. The barrel and sights are identical with those described in § 11947.

Original pattern Short Rifle Mk I Cond. MOD Pattern Room

Body: The body of the rifle is altered to receive the magazine, rifle, short, M.L.E., Mark I, the charger and the new locking and safety bolts.

Bolt: The bolt has the cover stops and a portion of the top of the rib removed, the cover screw holes filled in, and hole bored for screw, keeper, screw head, bolt breech.

Bolt head: A slot is cut in the tang of the bolt head, which acts as a key when stripping and assembling the striker and cocking piece; the screw, keeper, screw bolt head and screw, bolt head, first being removed.

Dial sight plate: The dial sight plate has the original lining removed, and is relined and marked.

Dial sight pointer: The dial sight pointer is fitted with a new bead.

Trigger guard: The guard is expanded to receive the magazine, Rifle, short, M.L.E., Mark I. A piece is brazed in the magazine slot for magazine seating, the swivel boss is removed, and a new link loop is fitted to receive link, Rifle M.L.M. Mark II; it is also lightened.

Stock, butt: The stock butt is reduced externally, and bored longitudinally with three holes for lightness; it is arranged to take the butt plate and stock bolt, Rifle, Short, M.L.E. Mark I. When the stock butt is small at the socket end, the shoulders and end are cut back 3/16 inch, this shortens the converted rifle, as compared with the new, by about 1/4 inch. In packing converted rifles, an additional packing piece, 1/4 inch thick, will be required, inserted at the end of the chest behind the bridge fittings.

Stock, fore-end: The stock, fore-end, has a liner fitted and glued in; it is opened to receive the altered body; the clearing rod slot and the band pin hole are filled up. It is arranged to take the stock bolt keeper plate.

All dimensions, weights, &c., are the same as for the Rifle, Short, M.L.E., Mark I, § 11947 and Converted Mark II, § 11949. The following components are special to this rifle:

Body +	*Bolt head*	*Stock, fore-end +*
Bolt +	*Stock, butt +*	*Screw, bolt head*

(+ Components marked thus are converted from Rifle, M.L.M., Mark I.)*

All other components are common to Rifle, Short, M.L.E., Mark I, § 11947 and Rifle, Short, M.L.E. Converted, Mark II, § 11949.

This conversion involved a number of changes as the Mark I series Lee-Metford has a single-row magazine which results in a narrower opening and recess in the lower part of the receiver body. The bolt and its dustcover are also different, attached by screws; these also resulted in a more complex conversion. This particular conversion never went into full production as only nine examples are recorded as having been converted at Enfield; it was declared obsolete in 1906.

This conversion is particularly distinctive because the original fore-end was retained, with the finger grooves and oval-shaped recess for the magazine cut-off thumbgrip; a close examination of the receiver body and bolt reveals the conversion due to the filled and brazed screw holes and fabricated charger guide. The specimen examined has the original manufacture markings on the right side of the receiver butt socket while the left side is marked "ENFIELD 1904" and "Sht. L.E." and "Cond. I." No cut-off is fitted, in keeping with the original Mark I S.M.L.E. specifications. A sheet-steel buttplate was also provided.

The similar upgrading of the Marks II and II* Lee-Metford and Marks I and I* Lee-Enfield rifles to Mark I S.M.L.E. pattern was effected by fitting the new short barrels, adapting the charger system and modifying the wood furniture. This particular adaption was designated the Converted Mark II and it was first approved on 16th January 1903 with later alterations approved on 14th October and 6th November 1903. This is of particular interest because its approval was prior to the re-sealing of the Mark I S.M.L.E. pattern, which occurred on 1st January 1904.

The announcement in the List of Changes reads:

Para 11949:–

Rifle, Short, Magazine Lee-Enfield, Converted (Mark II)
Converted from Rifles, Magazine L.E., Marks I and I, and Magazine L.M., Marks II and II*.*

A pattern of the above-mentioned rifle has been sealed to govern conversion as may be ordered. The rifle is very similar to the Rifle, Short, Magazine, Lee-Enfield, § 11947.

The conversion is made from Rifles, M.L.E., Marks I and I, and M.L.M. Marks II and II*, and consists generally in fitting the rifles with new sights, shorter and lighter barrels, and adapting them for the magazine to be filled from a charger, § 11753. The arm is fitted to carry the sword-bayonet, pattern 1903, §§ 11716 and 11717, on the nosecap.*

The barrel, sights, handguards and nosecap are identical with those described in § 11947.

Body: The body of the rifle is altered to receive the charger and the new locking and safety bolts.

Bolt: The bolt has the cover lugs and the extension on the rear end for safety catch arrangement removed, and striker hole bushed. In the M.L.M. Mark II rifles, the cover lugs only have been removed (as there is no safety catch provision).

Bolt head: The bolt head is altered to receive the charger guide, and has a slot cut in the screwed end of the bolt head which acts as a key when stripping and assembling the striker and cocking piece.

Guard: The trigger guard has a slot cut in the back to lighten it, and is freed at the front to clear the stop clip of the magazine.

Stock fore-end: The stock fore-end has a liner fitted and glued in; the swivel slot and lower band pin hole are filled up, and the fore-end reduced externally; and an extra stop pin for the outer band is fitted.

Stock butt: The stock butt is reduced externally, bored longitudinally with four holes for lightness, the stock bolt hole deepened, and recess at socket end cut to clear locking bolt, and recess for butt plate strap filled up, it is also fitted with a marking disc. The butt plate is of sheet steel without a trap for the oil bottle, which is not carried in the butt.

When the stock butt is small at the socket end, the shoulder and end are cut back 3/16 inch. This shortens the converted rifle as compared with the new by about 1/4 inch. In packing converted rifles an additional packing piece 1/4 inch thick will be required, inserted at the end of the chest behind the bridge fittings.

All dimensions, weights, &c., are the same as for the Rifle, Short, M.L.E., Mark I, § 11947.

All components are interchangeable with the new short rifle, M.L.E., with the following exceptions:

Screw, guard, back,	*Spring, sight, aperture, and*
Washers, rivet, fore-end (2),	*Screw, spring, sight, aperture,*

which are the same as for the Rifle, M.L.E., Mark I.*

Provision was made for a magazine cut-off in the new short rifle but none was fitted. Some of the Converted Mark II rifles were assembled for the Navy who requested that the cut-off be fitted. The cut-off was approved for Naval service on 12th August 1903 but this particular cut-off differs from that of the Magazine Lee-Enfield in that a recess is cut out of the left side near the front, to allow for the stop clip on the magazine and the left edge of the cut-off is turned up. This particular cut-off was also fitted to Naval issue Mark I S.M.L.E. rifles.

A slight alteration to the form of this cut-off was approved on 18th February 1904 and announced in the List of Changes § 12129 as "Cut-off, rifle, short, M.L.E. Mark I and Converted Mark II." The text continues:

A pattern of the above-mentioned component has been sealed in substitution of that described in § 11850. It differs in that the clearance of the magazine stop lip at front has been increased, and the lead for cartridge head increased at rear.

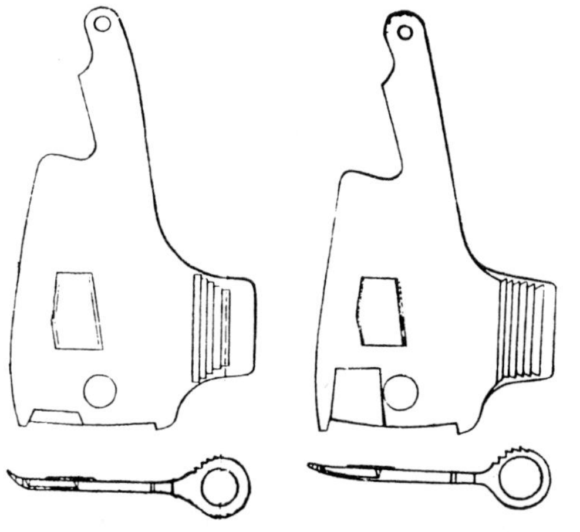

The left cut-off was that introduced § 11850, that on the right is per § 12129, in 1903 and 1904 respectively, for Naval service only. Fitting of the magazine cut-off was later extended to the Land Service as well.

This was only approved for Naval service and the use of the cut-off was not extended to Land forces until 1906. From 1906 the cut-off remained on the rifle until the middle of the 1st World War when it was omitted as an economy measure. After the Great War it was again incorporated in the new service rifles; the Mark V in 1922 and Mark VI in 1929 although it was omitted on the .276 (Pattern 1913) trials rifle in 1913.

Except for the more apparent difference in the cut-off for the Marks I and I* M.L.M. rifle, at a glance the following cut-offs for M.L.M. and M.L.E. rifles as well as the various Marks of S.M.L.E. rifle up to No. 1 Mk VI and trials No. 4, appear to be similar. However, a close examination will reveal minor differences between the various models; such differences may be noted in details such as the stop nibs, chamfer along the back edge and shape of the cut-out for the forward right hand magazine lip.

By 1904, virtually all world powers had adopted clip or charger-loaded small-bore magazine rifles which could be categorised into the Mauser or Mauser-type, Mannlicher or "Other." While this last class includes the British Lee-Enfield, French Lebel, Russian Nagant and U.S. or Norwegian Krag, the Mauser was the most popular, indicated in the following table by countries of usage.

MAUSER RIFLES

	Pattern	Calibre (mm.)	Calibre (in.)
Argentine	1891	7.65 mm.	.301 inch.
Belgium	1889	7.65 mm.	.301 inch.
Bolivia	1891	7.65 mm.	.301 inch.
Brazil	1904	7 mm.	.276 inch.
Chile	1904	7 mm.	.276 inch.
China	1893	7 mm.	.276 inch.
Columbia	1891	7 mm.	.276 inch.
Ecuador	1891	7.65 mm.	.301 inch.
Germany	1898	7.92 mm.	.311 inch.
Luxembourg	1896	6.5 mm.	.256 inch.
Mexico	1902	7 mm.	.276 inch.
Peru	1891	7.65 mm.	.301 inch.
Portugal	1904	6.5 mm.	.256 inch.
Persia	Various old patterns.		
Spain	1896	7 mm.	.276 inch.
Sweden	1906	6.5 mm.	.256 inch.
Turkey	1905	7.65 mm.	.301 inch.
	And other patterns.		
Uruguay	1895	7 mm	.276 inch.
YugoSlavia	1899	7 mm	.276 inch.

MAUSER-TYPE RIFLES

Italy (Carcano)	1891	6.5 mm	.256 inch
Japan (Arisaka)	T'38	6.5 mm	.256 inch
U.S.A. (Springfield)	1903	7.62 mm	.300 inch

Even after the initial issues of the Mark I S.M.L.E. at home and abroad, few of the units would have had sufficient time to become familiar with the new rifles before the following questionnaire was prepared and forwarded by the War Office.

1. Do the foresight wings shut out too much light from the foresight?
2. Is the barleycorn satisfactory as to shape and adjustability?
3. Is the "V" backsight notch the best shape and size to suit the barleycorn?
4. Is the windgauge advantageous for field service purposes?
5. Is a fine adjustment on the backsight necessary?
6. Is the long handguard necessary?
7. Is the safety catch necessary?
8. Is the double pull-off preferred to the dead pull?
9. Does the charger loading meet requirements?
10. Does the slight increase in recoil affect shooting?
11. Is the rifle considered superior to the long rifle in handiness, accuracy and general utility?
12. Is the rifle preferred for snap shooting?
13. Is the decrease in weight considered important as regards marches, &c., and lessening fatigue to the firer in continuous fire?
14. Does the rifle fairly sustain accuracy after considerable use?

The reports were then summarised and Lord Kitchener and his staff prepared the following answers:

1. The foresight wings are not a disadvantage and thoroughly protect the foresight.
2. A sharper sight, such as on the Mauser, is preferred.
3. The "V" of the backsight is too narrow at the top.
4. The wind-gauge is not necessary for field service, but also not disadvantageous, and could be useful in the hands of trained marksmen.
5. The same remarks apply for the fine adjustment.
6. The long handguard is necessary; preventing burnt hands on the barrel after rapid firing and lessening mirage.
7. The safety on the long rifle is preferable to that on the short rifle.
8. The double pull is superior when troops become used to it.
9. One charger full (5 rounds) is all that can be got into the magazine comfortably, there does not seem to be any necessity for the magazine to hold more than five rounds.
   If a shorter magazine were implemented, the rifle could be carried better at the slope and the trail.
10. Men will get used to it, after firing a few rounds the increase in recoil compared with the long rifle is hardly noticed.
11. For Cavalry and for short men like the Gurkhas, it is excellent compared with the long rifle. It is considered that it will be generally preferred for handiness and there have been less complaints with the short rifle than any new rifle previously introduced into the Service.
12. The small "V" in the backsight is a disadvantage in snap shooting and though the short rifle is much handier, it seems a little heavy at the muzzle.
13. No opportunities have arisen to answer these questions.
14. Same as 13.

In addition, it was felt that a cut-off would be necessary, the oil bottle and pull-through should be carried in a butt trap, a bolt cover was necessary, some rifles should be made with a left hand action to suit men who shoot from the left shoulder, and some complaints were made from India as to breakages of a number of components due to poor quality of the metal.

As a result of this report magazine cut-offs were re-issued, ordered on 25th October 1906, along with the recommendation for re-introduction of a butt trap in the next pattern of rifle, the Mark I* S.M.L.E. The requests for a left-handed action and bolt dustcover were not pursued.

To ascertain the advantages, if any, of the short barrel over the long one, 48 rifles were prepared for trial at Hythe, consisting of:

12 latest pattern short rifles,
12 Magazine Lee-Enfield Mark I* rifles,
12 Magazine Lee-Enfield rifles with short rifle sights, and adapted for charger loading,
12 Magazine Lee-Enfield rifles with short barrels, and adapted for charger loading.

These comparative trials were conducted between March and May 1905 and the purpose was to compare the velocity, accuracy, systems of sighting and speed of loading of each pattern of rifle, as well as relative fatigue of the soldiers after continuous fire and long marches.

It was found that the long (M.L.E.) rifles fitted with short barrels were the most accurate and the short rifles were the worst, although the difference was only marginal. The new sighting system proved to be superior to the old pattern but the foresight protector and coarse foresight of the short rifle gave poorer results at the falling targets and in poor light. Some barrels were tested with the shorter lead to the rifling, and deeper grooves, and these proved to be more accurate.

New patterns of the S.M.L.E. Mark I and Mark II Converted were sealed on 26th September 1906, and again announced and updated in the List of Changes, § 13509, due to the many small changes that had been applied since the initial introduction of these rifles. The following improvements were incorporated over the original patterns sealed in 1903:

1. Five different heights of barleycorn foresight were available, marked on the left side, .93, .945, .96, .975 and .99; these being the relative heights above the axis of the bore.
2. The rearsight, with new fine adjustment screw with larger milled head and a washer and two spiral springs to tension the fine adjustment. The top edge of the wind gauge was hardened to help prevent damage.
3. A double spring was fitted to the rear handguard, secured by two rivets. On the earlier pattern, these rivets were positioned at each side of the bore line; on the later pattern with double spring, the rivets are in line with the bore.
4. A new trigger guard with sling swivel forward of the magazine was fitted, the magazine link loop being omitted from the trigger guard and magazine. This was to be the No. 1 magazine.
5. A stronger firing pin keeper screw and nut were fitted, with a rope thread.
6. A stud and spring were fitted to the fore-end which act on the barrel from underneath the nosecap. These stocks were marked "S" below the nosecap. The rivet and two washers in the fore-end were replaced by a screwed pin.
7. 1/8-inch shorter sling swivels were fitted.
8. A piling swivel was fitted to the nosecap, except for those rifles issued to the cavalry.

The approval dates for these improvements and List of Changes paragraph reference numbers (§) were:–

Hardened top edge of wind gauge; 24.02.1905, § 13509
New fine adjustment springs in sight; 02.03.1905, § 13509
Vertical cuts at end of rear handguard; 14.02.1905, § 13509
Double spring for rear handguard; 14.02.1905, § 13509

Sling swivel lug for trigger guard; 30.09.1905, § 13509
Omission of magazine link loop; 30.09.1905, § 13509
Stronger keeper nut for striker; 03.11.1905, § 13509
Modified magazine platform spring; 25.01.1905, § 13509
Fore-end recess for barrel spring; 10.04.1905, § 13509
Screwed pin replacing fore-end rivets; 24.10.1905, § 13509
Piling swivel for rifles (not Cavalry); 30.09.1905, § 13509
Shorter sling swivels, Land service; 26.09.1905, § 12992
Sharp edges on body removed; 09.05.1906, § 13313

In essence, the new sealed patterns served to "tidy up paperwork" as most of these changes had been incorporated into new production or effected by armourers. These were embodied in the new S.M.L.E. Mark I* which was approved on 27th March 1906. The announcement in the List of Changes was:

Para 13577:–

Rifle, Short, Magazine, Lee-Enfield. (Mark I).*

A pattern of the above-mentioned rifle has been approved to govern present manufacture and alteration of Mark I rifles as may be ordered.

It differs from the Rifle, Short, M.L.E., Mark I, § 13509, in the following particulars:

The buttplate is made of gun metal and is fitted with a trap for insertion of the oil bottle and pullthrough in the butt.

The stock, butt, is recessed to receive a small bracket fixed by two screws, to carry the sling swivel, and also for the gunmetal butt plate. The division between the two upper lightening holes is cut away to allow of the insertion of the pullthrough weight.

A leather wad is inserted on the top of the stock bolt to act as a buffer for the oil bottle.

A magazine case (No. 2) which is deeper at front to facilitate loading, is fitted, and an auxiliary spring (No. 2) having a straight instead of curved end, is fitted to the magazine case.

The magazine case and auxiliary spring will be marked with the figure "2" for purposes of identification.

To prevent the swivel screws working loose by the movement of the swivels, the swivel screws are bored longitudinally at the screwed ends to enable the end of the screw, when screwed home, to be expanded by the armourer, with the centre punch issued as part of the foresight adjusting tools.

A keeper screw for striker, which can be turned by a coin, is fitted to the cocking piece. This screw replaces the nut, screw and spring, keeper striker.

The sharp corners of various components have been rounded to prevent injury to equipment and clothing.

The following is a list of new and additional components:

COMPONENT	*REMARKS*
Bracket, swivel, butt. ...	*Common to Rifles, short, M.L.E., Marks I* and Converted II*.*
Screws, bracket, swivel, butt (2 of).	*Same as Screw, strap, butt plate, iron.*
Butt plate.	*Special.*
Butt plate trap.	*Special.*
Spring, trap, butt plate. ...	*Special.*
Screw, spring, trap, butt plate.	*Special.*
Pin, axis, trap, butt plate.	*Special.*
# Magazine case, No. 2. ...	*Interchangeable, all short rifles.*
# Spring, auxiliary, platform, magazine, No. 2.	*For magazine case, No. 2.*
Wad, stock bolt.	*Same as Wad, stock bolt, M.L.M.*
Stock, butt.	*Special.*
# Screw, swivel	*Interchangeable, all short rifles.*
# Screw, keeper, striker. ...	*Common to all Rifles, short, M.L.E. Marks I* and Converted II*. Will be fitted to all short rifles locally as stocks of nuts, screws and springs are used up.*

(# Rifles, short, M.L.E., Marks I and Converted II* of early manufacture were not fitted with these components.)*

The principal advances therefore were the gunmetal or brass style buttplate with trap, lower sling swivel fitted into a plate inletted and held into the butt by two screws (not just secured by the swivel/screw combination as previously), the deeper No. 2 magazine, and the earlier method of locking the striker unit by a spring plunger was replaced by a simple keeper screw at the rear of the cocking piece. Some of these new features remained characteristics of the S.M.L.E. rifle throughout its production life.

Some notations on these changes introduced in LoC § 13577 are of interest in respect of the nomenclature of component parts; the gun metal buttplate was to become referred to as the No. 1 brass buttplate (see § 13742) and the deeper magazine case, the No. 2 case. The butt trap of this buttplate is removable in the same manner as the later No. 4 rifles. Other notes on these changes are that the Mark I* S.M.L.E. rifles made at B.S.A.-Sparkbrook had No. 1 magazine cases without the loop and early Mk I* rifles made at the same B.S.A.-owned plant (ex-Government) still had no keeper nut on the cocking piece for the striker or firing pin.

According to the Enfield Pattern Book, the Converted Mark II* S.M.L.E. was approved on 15th March, nearly a fortnight before the Mark I* S.M.L.E. and consisted of upgrading Lee-Metford Marks II and II* and Lee-Enfield Marks I and I* to the S.M.L.E. Mark I* pattern. Its introduction in the List of Changes puts the approval date of the Converted Mark II* at 9th July. It should be pointed out that the margin of the original List of Changes announcement also records the individual approval dates for various items incorporated in the introduction, so there may well have been some confusion here. Other references may also be found in the original LoC margins, such as the numbers of relevant War Office papers.

Para 13578:–
Rifle, Short, Magazine Lee-Enfield, Converted (Mark II)*
Converted from Rifles, Magazine Lee-Enfield Marks I and I, and Magazine Lee-Metford Marks II and II*.*

A pattern of the above-mentioned rifle has been approved to govern future manufacture. It differs from the Rifle, Short, M.L.E., Converted, Mark II (§ 13509) in the following particulars:

The stock, butt is recessed to receive a small bracket fixed by two screws, to carry the sling swivel. The stock bolt hole is of a depth suitable for Stock bolt No. 2, M.L.E. rifle, which will be used. The division between the two upper lightening holes is cut away to allow of the insertion of the pullthrough weight.

The Butt plate, M.L.M. rifle Mark II having a strap at heel for regimental marks, and a trap for the insertion of the oil bottle and pullthrough in the butt, is retained. The Wad, stock bolt, M.L.M. is retained.

The Screw, strap, butt plate, iron is retained. The Marking disc and screw are omitted.

A magazine case, No. 2, which is deeper at front to facilitate loading, is fitted and an auxiliary spring (No. 2) having a straight instead of a curved end, is fitted to the magazine case. The magazine case and auxiliary spring will be marked with the figure "2" for purposes of identification.

To prevent the swivel screws working loose by the movement of the swivels, the swivel screws are bored longitudinally at the screwed end to enable the end of the screw, when screwed home, to be expanded by the armourer with the centre punch issued as part of the foresight adjusting tools.

A keeper screw for striker, which can be turned by a coin, is fitted to the cocking piece. This screw replaces the nut, screw and spring, keeper striker.

The sharp corners of various components have been rounded to prevent injury to equipment and clothing.

See footnote (#) § 13577 for Rifle, Short, M.L.E., Mark I, as regards certain components not being fitted to Converted Mark II* rifles.*

Some observations on this particular List of Changes entry are that the initial paragraph "has been approved to govern future manufacture" is not consistent with general announcements for converted rifles which usually read "has been approved to govern future conversion," and the classification of Rifles, M.L.E. Marks I and I* and M.L.M. Marks II and II* for conversion should be amended to include M.L.E. carbines also, as some carbine bodies have been noted utilised for this particular conversion.

Skeleton actions of the Mark I S.M.L.E. and Converted Marks I and II were approved on 27th February 1904 and described in the List of Changes:

Para 12220:–
Action, Skeleton, Rifle, Short, M.L.E. (Mark I)
Also Rifles, Short, M.L.E., Converted Marks I and II.

A pattern of the above-mentioned article has been sealed to govern future manufacture.

The action of the Rifle, Short, M.L.E. Converted, Mark I differs slightly in detail and the components are not interchangeable with Rifles, Short, M.L.E., Mark I and Converted Mark II, but the principle of construction being the same, this pattern of skeleton action adequately illustrates the mechanism of any pattern of short rifle.

Skeleton actions for Naval service differ only from those for Land Service in being fitted with a cut-off.

Some skeleton Lee-Enfield actions were also updated by conversion to the S.M.L.E. pattern, with approval dates given as 18th March and 11th May 1904.

Para 12221:–
Action, Skeleton, Rifle, Short, M.L.E. Converted (Mark I)
Converted from Action, Skeleton, Rifle, M.L.E.

A pattern of the above-mentioned conversion has been sealed to govern conversion as they may be required. The conversion consists in altering the body, bolt, guard and stock-butt, and fitting the action with short rifle components.

The body differs slightly from the Action, Skeleton, Rifle, Short, M.L.E. (Mark I) described in § 12220 in the following manner: the body is converted and the clearance on left side of body left for the thumb, when loading with the charger, is shallower.

In all other particulars the converted skeleton action is similar to the new. Skeleton actions for Naval Service differ only from those for Land Service in being fitted with a cut-off.

The coin slotted keeper screw for the firing pin or striker as introduced on the Mark I* and Converted Mark II* short rifles was approved for fitting to all models of the Short Magazine Lee-Enfield on 17th August 1906; this was effected by removing the nut, screw and spring from the cocking piece and tapping the existing keeper hole for the new screw.

Another alteration also approved included fitting of the smaller swivels to butt and lower band. This swivel is

1/8-inch shorter than the M.L.M. Mark II pattern previously used, approved for Land Service on 26th September 1905; it was mainly intended to make the rifle fit more easily into the saddle bucket. On 9th May 1906, approval was given for the removal of sharp edges on the charger guide on the left side of the body, as well as the charger guide stop on the right side; these had been found to injure the bandolier and clothing when the rifle was carried at the slope.

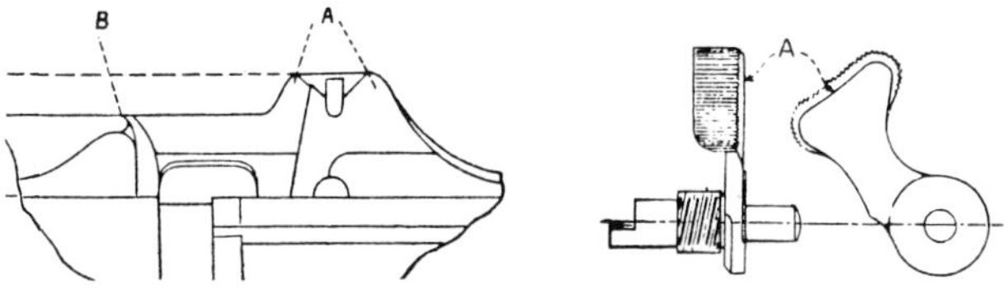

Removal of sharp edges on the Mark I S.M.L.E. as per LoC §§ 13313, 13549 at the points marked "A" and "B".

Early in 1906, the government factory at Sparkbrook was closed. Production records indicate that the Mark I* S.M.L.E. rifle was not manufactured by R.S.A.F. Sparkbrook, and that the Mark I version was the last type made there. The main government factory was at Enfield Lock, and B.S.A. and L.S.A. produced arms on contract for the government. The Sparkbrook factory was sold to B.S.A. and although it is recorded that no rifles were made there, B.S.A. must have carried out production for a short time at least, because some Mark I* S.M.L.E. rifles have been noted with the manufacturer's marking "B.S.A.-SPARKBROOK." However, from 1907, the Sparkbrook factory was not used for arms production until the 1st World War and even then it was mainly for machine gun parts.

BSA-Sparkbrook receiver designation marking on some Mark I* S.M.L.E. rifles, 1906 - 1907.

In January 1907, some comparative tests were undertaken with the Mark I* S.M.L.E., the Canadian Ross and the Model 1903 Springfield. These rifles underwent accuracy tests at a variety of ranges, rapid firing, snap shooting and handling, and sand tests; it was found that the S.M.L.E. and Ross had comparable accuracy and grouping capabilities, better than the Springfield, and the S.M.L.E. was the only rifle to function properly while sand and foreign matter was poured over the action. Minor damage was sustained to the Ross and Springfield actions during these tests, from which the S.M.L.E. emerged unscathed, apparently the most practical service rifle.

By this time, the Mark I S.M.L.E. had been in the hands of some troops for a number of years, at home and overseas, and various recommendations had been made regarding possible improvements. Paramount criticism related to the charger-loading fittings; it was found that the charger guide on the bolthead soon developed excessive play and even though a tightened variation had already been introduced, it was still not considered entirely satisfactory.

In February 1906, two patterns of charger loading attachments were put up for consideration; one by B.S.A. and the other designed by the Chief Inspector of Small Arms. Experiments had been under way at Enfield with a solid, one-piece type which formed a bridge across the receiver; this was the C.I.S.A. pattern. Some rifles were fitted with this charger bridge to test strength and ease of loading and it was proven satisfactory. As a result, trials were recommended at Hythe in April 1906, after which the C.I.S.A. pattern of charger loading bridge was chosen for the new rifle. From the accompanying illustration, it will be noticed that the charger bridge has a slightly more squared appearance than that on the eventual Mark III S.M.L.E.

Prototype Mark I S.M.L.E. rifle with experimental charger bridge fitting. This is more squared in its shape and the bolt has the later style of bolthead. *Tower of London*

As a result of complaints made about the sighting, the Committee tested and approved a blade foresight to replace the barleycorn, with a U-notch in the rearsight instead of a V. The foresight protectors were also widened to admit more light onto the foresight. These changes were the result of the questionnaire and report sent back by Lord Kitchener and his staff after the first issues of the Mark I S.M.L.E.

In 1905, Col. Watkin was succeeded by Dr. Ashton as Superintendent at R.S.A.F. Enfield, although Joseph Speed was still the factory manager. The new backsight for the proposed rifle, which eventually became known as the Mark III S.M.L.E., was designed by Dr. Ashton and Joseph Speed; a patent was taken out on this backsight, No. 4776 of 27th February 1906.

4776. Ashton, H. T., and **Speed, J. J.** Feb. 27.

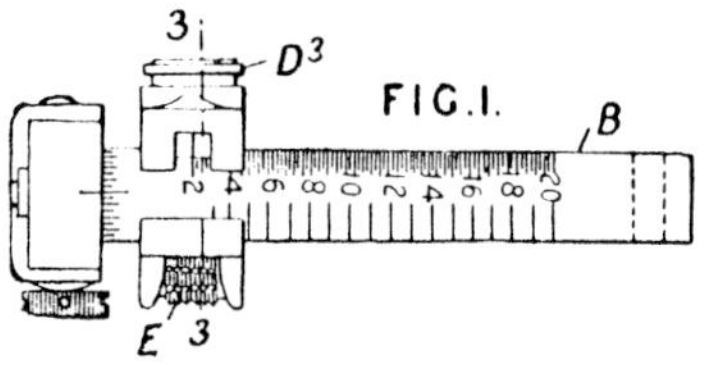

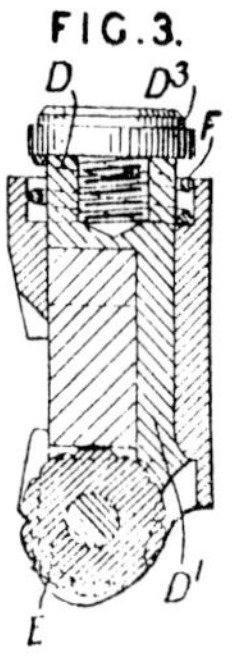

Sights.— Relates to improvements in backsights of the kind described in Specification No. 6743, A.D. 1903. Fig. 1 is a plan and Fig. 3 a section on the line 3–3, Fig. 1, The slide on the leaf B carries a worm E engaging with teeth cut on the side of the leaf. Fine adjustments are made by turning the worm E, which is provided with longitudinal cuts to serve as graduation marks and to afford a hold for the fingers. The worm is disengaged, to allow coarse adjustment by sliding to be made, by pushing in against the action of the spring F the head D^3 of the sliding carrier D, D^1 in which the worm E is mounted.

Some of the improvements to the Lee-Enfield were still relative to the original complaints made from the war in South Africa, regarding the sighting and charger-loading capacity. The Short Rifle in 1902 was of course a result of these investigations but the sighting and charger-loading facilities of the Mark I S.M.L.E. were not entirely successful, resulting in the Mark III S.M.L.E. Rifle in early 1907.

In the South African campaigns, it has been quite widely accepted that the Lee-Enfield was distinctly inferior to the Boer Mauser, but this was not necessarily so. Its coarse sights and lack of charger loading, along with some improperly sighted rifles due to lack of proper testing and calibration, were the main shortcomings; the Lee-Enfield was still quite highly regarded, even by the Boers who used it quite extensively in the latter stages of the war. The marksmanship and fieldcraft of the average British soldier was markedly inferior to that of the burgher and this important difference was too often the main reason for the improperly attributed superiority of Mauser over Lee-Enfield; that many of the Boers preferred the Lee-Enfield is an indication of its true measure.

Success of the new fixed bridge charger guide and improved sights resulted in the decision to incorporate these features on a small number of rifles for trial and six of these were completed by 19th September 1906 and sent to Aldershot. Favourable reports resulted in the Committee recommending its adoption as a new pattern of rifle. Because the Short Lee-Enfield had already proceeded through two marks, the proposed rifle was to be allocated the ensuing numeral, "III."

The new, Short, Magazine, Lee-Enfield Mark III rifle was approved on 26th January 1907.

Para 13853:-

Rifles, Short, Magazine, Lee-Enfield (Mark III)

A pattern of the above-mentioned rifle has been approved to govern future manufacture. It differs from the Rifles, Short, M.L.E. Marks I and I, described in §§ 11947, 13509 and 13577, in the following particulars:*

Blade foresight: Foresights with a straight edged blade in five heights instead of a barleycorn will be used, according to requirements, to enable variations in shooting to be corrected. They will be marked on the top left side with the figures 0, 015, 03, 045 and 06 respectively, representing 1.0 inch, 1.015 inches, 1.03 inches, 1.045 inches and 1.06 inches, their heights from the axis of the barrel.

Backsight: The bed is made wider to strengthen the axis joint, and the front part is made tubular so as to encircle the barrel, and is fixed to the barrel by a crosspin and the point of the spring screw to prevent its working loose.

The leaf, which is made to turn over on to the handguard and rebound, is graduated on the top left side by lines representing every 25 yards between 200 and 2,000 yards in addition to the lines on the right for hundreds of yards, as in the Mark I rifle.

The slide may be set at any elevation, above or below the elevation at which the slide was last set, in multiples of 50 yards, by pressing the catch on the left side and releasing a fine-adjustment worm wheel engaged in thread notches on the right side of the leaf, thus enabling the slide to be moved quickly along the leaf. The worm wheel is pivoted in the right side of the catch, and may be rotated in either direction at right angles to the leaf, this movement providing fine-adjustment for the slide. The periphery of the worm wheel is divided by 10 thumbnail notches, the distance between each notch representing 5 yards in range, i.e., five notches 25 yards, or one division on the left side of the leaf, and one complete revolution 50 yards.

The wind gauge is fitted directly on the rear end of the leaf, and is held in position by the wind-gauge screw. A

U-notch instead of a V-notch, is cut in the top edge, and the face is roughened to prevent the light being reflected from it. Bright centre lines are marked on the face of the wind gauge for aiming, and on the top as a centre line for use in conjunction with the wind-gauge scale on the leaf.

The wind-gauge scale is marked with divisions representing the same amount of deviation on the target (6 inches per 100 yards) as those on the Mark I rifle. Each quarter turn of the wind-gauge screw represents 1 inch deviation per 100 yards, and at each quarter turn a friction spring engages in a nick inside the head of the screw, thus checking its rotation.

Body: The body is fitted with a charger bridge guide; the slots for the charger stops being sloped in front so that if a charger is left in the charger guide after pressing the cartridges into the magazine, the act of closing the bolt ejects the empty charger.

Bolt-head: The charger guide and the slide for the charger guide are omitted.

Butt-plate: The trap of the butt-plate is fitted with an axis pin in a similar manner to that of the Rifle, M.L.M., Mark II, and the trap spring is tapered in thickness. This butt-plate is similar to the No. 2 butt-plate for Mark I rifles announced in § 13742.*

Band, inner: The inner band is 1 inch nearer the breech end of the barrel, and is bored larger to suit the larger diameter of the barrel at that position.

Cut-off: The cut-off is strengthened at the joint end and is similar to that for Mark I rifles announced in § 13742.*

Locking bolt: The locking bolt is slightly modified to suit the stop pins which have been coned to ensure smoother working of the locking bolt.

Nosecap: The nosecap is lightened and the shape of the foresight protecting wings altered to better enable the object to be seen in moving practice, and to admit more light to the foresight.

Handguard, rear: The rear handguard is modified by the removal of the sight protector.

Handguard, front: The front handguard is increased in thickness in front of the outer band to strengthen it.

Stock, fore-end: The position of the inner band seating is moved 1 inch nearer to the breech to strengthen the fore-end; and a new form of backsight protector is fitted to the fore-end and fixed by a screw and nut.

Spring, sight, aperture: The aperture sight spring is recessed for the head of the screw, which is rounded to prevent injury to clothing and accoutrements.

Weight of rifle .... 8 lb. 10.5 oz.

The following is a list of components, showing which are special to this rifle and which are common to other "marks" of short rifles:

COMPONENT	*SPECIAL or COMMON*
Bands, inner	*Common to Marks III and converted IV.*
Bands, outer	*Common to all Marks.*
Blades, foresight	*Common to Marks III and converted IV.*
Barrel	*Common to Marks III and converted IV.*
Bead, pointer, dial sight ...	*Common to all Marks.*
Bed, backsight	*Common to Marks III and converted IV.*
Block, band, foresight ...	*Common to all Marks.*
Body	*Special to Mark III.*
Bolts, breech	*Common to all Marks.*
Bolts, locking	*Common to Marks III and converted IV.*
Bolts, stock	*Common to Marks I, I*, converted II, III.*
Bracket, swivel, butt ...	*Common to Marks I*, converted II*, III and converted IV.*
Caps, handguard, front ...	*Common to all marks.*
Caps, nose	*Common to Marks III and converted IV, but may be used to replace nosecaps fitted to other Marks.*
Case, magazine	*Common to all Marks.*
Catches, magazine ...	*Common to all Marks.*
Catches, safety, bolt, locking	*Common to all Marks.*
Catches, slide, backsight...	*Common to Marks III and converted IV.*
Clip, stop, magazine ...	*Common to all Marks.*
Cocking piece	*Common to all Marks.*
Collar, screw, front, trigger guard	*Common to all Marks.*
Cut-off	*Common to all Marks.*
Disc, marking, butt ...	*Common to Marks I, I*, converted II and III.*
Extractor	*Common to all Marks.*
Guard, hand, front ...	*Common to Marks III and converted IV.*
Guard, hand, rear ...	*Common to Marks III and converted IV.*
Guard, trigger	*Common to all Marks.*
Head, breech bolt ...	*Common to Marks III and converted IV.*
Head, screw, wind gauge ...	*Common to Marks III and converted IV.*
Key, block, band, foresight.	*Common to all Marks.*
Leaf, backsight	*Common to Marks III and converted IV.*
Nut, screw, protector, backsight	*Common to Marks III and converted IV.*
Nut, screw, back, nosecap...	*Common to all Marks.*
Pin, axis, backsight ...	*Common to Marks III and converted IV.*
Pin, axis, trap, butt plate	*Common to Marks I, I* and converted II fitted with No. 2 butt plate, and III.*
Pin, axis, worm, fine adjustment	*Common to Marks I, I* and converted II fitted with No. 2 butt plate, and III.*
Pin, axis, catch, slide, backsight	*Common to Marks III and converted IV.*
Pin, catch, magazine ...	*Common to all Marks.*
Pin, fixing, bed, backsight	*Common to Marks III and converted IV.*
Pin, fixing, block, band, foresight	*Common to all Marks.*
Pin, fixing, head, screw, wind gauge	*Common to Marks III and converted IV.*
Pin, fixing, washer, pin, axis, backsight	*Common to all Marks.*
Pin, joint, band, outer ...	*Common to all Marks.*
Pin, screwed, fore-end ...	*Common to Marks I, I* and III.*
Pin, stop, band, outer ...	*Common to all Marks. Same as "Pin, stop, band, lower, M.L.M. rifle".*
Pin, stop, bolt, locking ...	*Common to Marks III and converted IV.*
Pin, trigger	*Common to all Marks.*
Plate, butt	*Marks I* and III, and also I and converted II when fitted with No. 2 brass buttplates.*
Plate, keeper. stockbolt ...	*Common to all Marks.*
Plate, sight, dial ...	*Common to all Marks.*
Platform, magazine ...	*Common to all Marks.*
Pointer, sight, dial ...	*Common to all Marks.*
Protector, sight, back ...	*Common to Marks III and converted IV.*
Rivets, spring, handguard...	*Common to all Marks.*

COMPONENT	SPECIAL or COMMON
Screw, band, inner ...	Common to all Marks.
Screw, cap, nose, back ...	Common to all Marks.
Screw, cap, nose, front ...	Common to all Marks.
Screw, catch, slide, backsight	Common to Marks III and converted IV.
Screw, cut-off	Common to all Marks.
Screw, disc, marking, butt..	Common to Marks I, I*, converted II and III.
Screw, ejector	Common to all Marks.
Screw, extractor ...	Common to all Marks.
Screw, guard, trigger, back	Common to Marks I, I* and III.
Screw, guard, trigger, front	Common to all Marks.
Screw, handguard cap ...	Common to all Marks. Same as "Screw, keeper, fine-adjustment, and handguard cap, &c."
Screw, keeper, striker ...	Common to all Marks.
Screw, plate, butt ...	Common to all Marks.
Screw, protector, backsight.	Common to Marks III and converted IV.
Screw, sear	Common to all Marks.
Screw, sight, dial, fixing ...	Common to all Marks.
Screw, sight, dial, pivot ...	Common to all Marks.
Screw, spring, sight, aperture	Common to Marks I, I* and III.
Screw, spring, trap, butt-plate	Common to Marks I* and III, and I and converted II when fitted with brass butt-plates.
Screw, spring, sight, back..	Common to Marks III and converted IV.
Screw, swivel, bracket, butt	Common to Marks I* and converted II*, III and converted IV. Same as "Screw, strap, buttplate, iron."
Screw, swivel	Common to all Marks.
Screw, wind gauge, backsight	Common to Marks III and converted IV.
Sear	Common to all Marks.
Sight, aperture	Common to all Marks.
Slide, backsight ...	Common to Marks III and converted IV.
Spring, auxiliary, platform magazine	Common to all Marks fitted with No. 2 magazine.
Spring, catch, slide, sight, back	Common to Marks III and converted IV.
Spring, extractor ...	Common to all Marks.
Spring, guard, hand, rear, double	Common to all Marks.
Spring, main	Common to all Marks.
Spring, platform, magazine..	Common to all Marks.
Spring, retaining breech bolt	Common to all Marks.
Spring, screw, band, inner..	Common to all Marks.
Spring, screw, wind gauge, back sight	Common to Marks III and converted IV.
Spring, sear	Common to all Marks.
Spring, sight, aperture ...	Special to Mark III, but may be used to replace similar spring in Mark I and I*.
Spring, sight, back ...	Common to Mark III and converted IV, but may be used to replace similar spring in all other Marks.
Spring, sight, dial ...	Common to all Marks.
Spring, trap, butt-plate ...	Common to Marks I* and III, and I and converted II when fitted with brass buttplates.
Spring, stud, fore-end ...	Common to all Marks.
Spring, wind gauge, backsight	Common to Marks III and converted IV.
Stock, butt (long, normal or short)	Common to Marks III and converted IV.
Stock, fore-end	Special to Mark III, may be used for converted Mark IV if required.
Striker	Common to all Marks.
Stud, clip, stop, magazine..	Common to all Marks.
Stud, fore-end	Common to all Marks.
Swivel, band, &c... ...	Common to all Marks.
Swivel, piling	Common to all Marks.
Trap, butt-plate	Common to Marks I, I* and converted II fitted with No. 2 butt-plate, and III.
Trigger	Common to all Marks.
Wad, bolt, stock ...	Common to all Marks.
Washer, bolt, stock ...	Common to all Marks.
Washer, guard, hand ...	Common to all Marks.
Washer, pin, axis, backsight	Common to all Marks.
Washer, screw, fixing, dial sight (and nut, screw, protector, backsight)	Common to all Marks, but in Marks III and converted IV, one additional for "Washer, nut, screw, protector, backsight".
Washer, spring, band, inner.	Common to all Marks.
Wind gauge, backsight ...	Common to Marks III and converted IV.
Worm, fine-adjustment, backsight	Common to Marks III and converted IV.

It will be noted that in this List of Changes announcement, the Locking Bolt is referred to rather than the Safety Catch; in general usage, the locking bolt is technically that portion of the safety operated by the thumb, acting upon the smaller catch which locks the cocking piece and bolt. However there is some confusion in later List of Changes components tabulation between the safety catch and the locking bolt. Some other minor inconsistencies and possible oversight becomes obvious from this LoC entry, in the parts and classifications; such as the Swivels, band being "Common to all Marks," Trigger guard back screws being "Common to Marks I, I* and III," and the purported difference between "Pins, axis, worm, fine-adjustment" and "Pins, axis, catch, slide, backsight."

As occurred with the Marks I and I* S.M.L.E. rifles, conversions were also sealed to govern the upgrading of earlier Lee-Metford and Lee-Enfield rifles. Following the numerical sequence, the new conversion then became the Mark IV and was approved on 17th June 1907 as the Short, Magazine Lee-Enfield Converted Mark IV; it was announced in the List of Changes, § 13854 as hereunder:

Para 13854:–
Rifle, Short, Magazine, Lee-Enfield, Converted (Mark IV)
Converted from Rifle, M.L.E. Marks I and I; and M.L.M. Marks II and II*.*

A pattern of the above-mentioned rifle has been approved to govern future conversion. This conversion differs from the converted Mark II described in § 13578 in that it embodies the special features of the Mark III described in § 13853.*

Its weight is 8 lb. 14.5 oz. and it is the same as the Mark III arm in all details, the following components excepted:

COMPONENT	REMARKS
Body	Special to Converted Mark IV.
Bolt, stock	Same as M.L.E. No. 2.
Cut-off	Converted as described in § 12129; and common to all short rifles, vide § 13612.
Disc, marking, butt ...	Omitted from Converted Mark IV as the butt plate has a strap on which the marking is placed.
Pin, axis, trap, butt plate.	Same as M.L.M.
Plate, butt	Same as M.L.M.R. Mark II.
Rivet, fore-end ...	Same as M.L.M. Not removed from fore-end on conversion.
Screw, disc, marking, butt..	Omitted from Converted Mark IV.
Screw, guard, trigger, back.	Same as M.L.M.

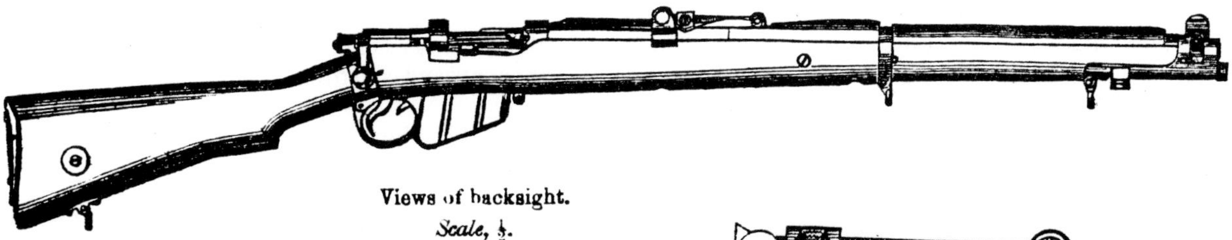

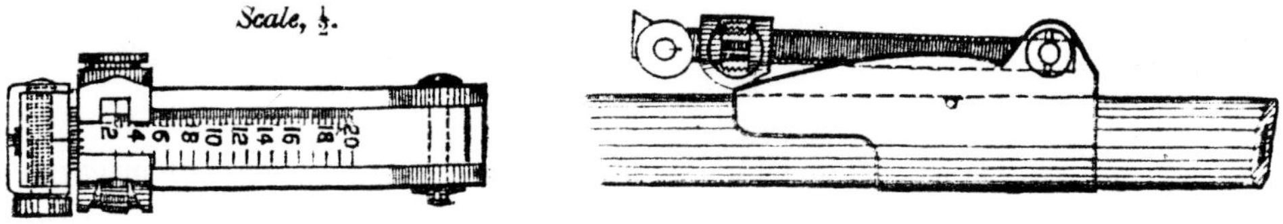

COMPONENT	*REMARKS*
Screw, spring, trap, butt plate	*Same as M.L.M.*
Screw, spring, sight, aperture	*Same as M.L.M.*
Screw, strap, butt plate, iron	*This screw is also used for Bracket, swivel, butt, in all short rifles fitted with that component.*
Spring, trap, butt plate ...	*Same as M.L.M.*
Spring, sight, aperture ...	*Same as M.L.M.*
Stock, butt ...	*Same as M.L.E.R.S., Converted Mark II*.*
Stock, fore-end ...	*Special to Converted Mark IV, in that it retains the rivet and two washers on conversion, instead of being fitted with a screwed pin as in the case of the Mk III; but interchangeability of fore-ends is not affected.*
Trap, butt-plate	*Same as M.L.M.R., Mark II.*
Washer, rivet, fore-end	*Same as M.L.M. Not removed from fore-end on conversion.*

In 1907, when the Mark III S.M.L.E. was introduced, a modified safety catch was tested incorporating a press stud on the right side of the receiver, which prevented the catch from being accidentally moved forward and therefore disengaged. The catch on the left side, although different in shape to the type on the Mark III, is identical in operation to the conventional service pattern except that the button on the right side must be depressed before the safety catch could be moved forwards. A small groove and stop are evident in the left side of the receiver, to act as a guide for a ball bearing behind the safety catch. The illustrated rifle has no unusual receiver markings that would indicate special application or assembly; it is otherwise identical to the Mark III rifle.

This safety catch was a design of Captain Wallace, put before the Small Arms Committee in May 1907; trials were conducted by the end of the year. It was Wallace who proposed the Metford form of rifling for the short rifle in 1904 which performed creditably in trials. About a dozen of the new Mark III rifles were modified with this safety catch, of which five were forwarded to H.M.S. "Excellent," five to the 16th Lancers at Aldershot, and one to the C.I.S.A.

These Wallace safety catch trials rifles are dated 1907 on the receiver and the catch was found to be quite satisfactory in operation but difficult to clean when sand or foreign matter lodged behind it. Another disadvantage was that it weakened the receiver slightly although the main consideration was that such a change in the safety mechanism would delay the introduction of the new mark of rifle. As a result, Wallace's design was decided against and the question of a safety catch improvement not pursued any further.

Above & below: Wallace design safety catch applied to some test rifles in late 1907. *Imperial War Museum*

Further tests were made with Wallace's modified Metford rifling, featuring shallow grooves; although it was found to be quite satisfactory there were insufficient advantages to warrant changing from the Enfield type. Nine different rates of twist were tested in 1906 and 1907, varying from 1 turn in 15 inches to 1 in 6 inches, although the standard rate of twist, 1 in 10 inches, still proved to be the most suitable and so no changes to the current rifling were recommended.

Interchangeability of components was a feature of the British service rifle from the mid-19th century. This means that component parts were made to readily interchange between rifles of the same type, without hand finishing or special fitting. However, it was pointed out that bedding, and therefore the shooting, differed from rifle to rifle, so to avoid accidental interchange, the fore-end and nosecap were numbered to match the serial number of each particular rifle.

Due to head spacing and sighting requirements, the bolts, barrels and rearsights were already serial numbered to the receiver, and had been since the introduction of the Lee-Metford. This numbering of the receiver, barrel, bolt, rear sight, nosecap and fore-end was continued throughout the Mark III and Mark III* series in British, Australian and Indian production, although in Indian service this numbering was extended back to those Magazine Lee-Metford and Lee-Enfield rifles and carbines still on issue. Numbering of the fore-end and nosecap in British service was approved on 21st October and 22nd November, 1907.

Skeleton actions of the Mark III and Converted Mark IV Short Lee-Enfields were approved on 21st May 1908. When the skeleton actions were first introduced in the Martini rifle and carbine era, they were made up from selected parts because they were designed as armourer's tools but by this time many of the components were reject components and the skeleton action were used in a slightly different role. Viewing slots were cut into certain parts, like the receiver and bolt, and a shortened barrel was fitted, without a fore-end; this meant that operating parts in the action could be observed. To test suspect components, it merely required substitution of the respective part with the suspect one and the movement could be observed through the viewing slot.

From about 1900 their role changed, so that by the time of the Great War many reject parts were used and the action was mostly intended as a demonstration or educational aid. During both World Wars, the conversion to skeleton actions was largely effected by the trade and so minor variations may be encountered in the form and position of the viewing slots and cuts. Skeleton Mark III and III* actions were also assembled in India and Australia, and again, variations may be noticed in the form and mode of sectionizing. More details on the skeleton actions will be found towards the end of Chapter 12.

As the Mark III (and subsequent Mark III*) S.M.L.E. represents the largest production of the Lee-Enfield rifle and was also the prime model manufactured in India and Australia, a closer examination is made of the specifications, components and construction of this model.

The Short Magazine Lee-Enfield Rifle Marks III and IV.

[Information regarding points in which Marks I, I*, II and II* differ from Marks III and IV, is given in squared brackets.]

Weight of rifle with magazine empty,	
Mark III, about	8 lbs. 10.5 oz.
Mark IV, about	8 lbs. 14.5 oz.
[Mark I & II series, about	8 lbs. 2.5 oz.]
Weight of sword-bayonet, no scabbard, about	1 lb. 0.5 oz.
Length of butt, long, about	1 ft. 1.8 ins.
Length of butt, normal, about	1 ft. 1.3 ins.
Length of butt, short, about	1 ft. 0.8 ins.
Length of rifle with normal length butt	3 ft. 8.5 ins.
Length of rifle with normal length butt, with sword-bayonet, about	5 ft. 2 ins.
Length of barrel, about	1 ft. 1.2 ins.
Calibre	.303 inch.
Rifling system	Enfield.
Twist of rifling:	Left hand, 1 turn in 10 ins. or 33 calibres.
Muzzle velocity, Mk VI ball, about ...	2,060 ft/sec.
Mk VII ball, about ...	2,440 ft/sec.
Number of grooves	5.
Depth of grooves at muzzle	.0065 in.
Depth of grooves at breech, to within 14 in. of muzzle	.00575 in.
Width of lands	.0936 in.
Sighting system	Adjustable blade foresight, radial backsight with vertical adjustment and windgauge.
[Sighting system	Adjustable barleycorn fore-sight, radial backsight with vertical adjustment and windgauge.]
Method of loading	Charger holding 5 rounds.

The Short Magazine Lee-Enfield is constructed on the bolt system, the breech being closed by a bolt worked by a lever on the right side. The bolt contains the mainspring and striker.

The striker is inserted from the front of the bolt; its rear end passes through the mainspring and screws into the cocking-piece, thus connecting the cocking piece and bolt. The front of the mainspring bears against the collar on the striker and the rear and against the back of the bolt chamber.

The bolt-head is a separate component. It carries the extractor, extractor screw and a V-shaped spring and is connected with the bolt by means of a screwed tenon, in which is cut a slot which acts as a key for removing the striker from the cocking piece. A hook on the right of the bolt-head, which engages with a rib on the body, secures the bolt-head when the bolt is rotated.

[The head has a slide for the bolt-head charger guide.]

On the left of the body, about a cartridge length from the mouth of the chamber, the end of the ejector screw projects into the bolt way. In front of the ejector screw, the left side of the body is cut away to afford clearance for the thumb, when loading with the charger. Immediately in rear of the thumb clearance is the bridge charger guide.

[There is also a raised portion in rear of the thumb clearance, in which a slot is cut. The guides into which the charger is inserted when loading are formed by the slot cut in the raised portion of the body behind the thumb clearance and the slot in the bolt-head charger guide.]

A cut-off which consists of a flat piece of steel hinged to the right side of the body is provided and is moved by means of a thumb piece.

On the left side of the body a socket is arranged to receive the safety catch, which consists of a small transverse locking bolt and pin. The former is threaded at the part where the pin works. On the thumb piece being forced back, it engages in one of the recesses cut in the cocking piece, which is then locked. The pin is brought into action by means of the screw on the locking-bolt; its front end engages in the short cam groove at the rear end of the bolt, and thus prevents the rotation of the bolt.

There are two holes or gas escapes in the body, one on either side of the breech, to facilitate a lateral escape of gas in case of a burst cartridge; there is also a third hole or gas escape in the bolt-head, in case of a blow-back of gas from a defective cap.

The magazine is made from sheet steel and holds 10 cartridges. The cartridges rest on a platform which is actuated by a spring made from ribbon steel. To afford a smooth bearing for the platform an auxiliary spring is fixed by a lip and stud to the front of the magazine. There are four patterns of magazines; Nos. 1 and 2 for use with Mk VI ammunition and Nos. 3 and 4 for use with Mk VI or VII ammunition. Nos. 1 and 2 magazines are fitted on the right side with a stop clip and on the left side with a fixed lip to retain the platform in its place and keep the top cartridge in position. For these purposes, No. 3 magazine has a stop clip on the right side and a spring lip on the left side, and No. 4 magazine is provided with fixed lips on either side. The magazine can be released by pressing the catch which protrudes through the trigger guard in front of the trigger.

The upper band and nose-cap are formed in one piece, with an extension in front, on which the ring of the cross-piece of the bayonet fits, and a bar underneath the rear end to hold the pommel of the bayonet. Lugs are provided to carry a swivel if required. Wings are attached to protect the foresight. The outer band carries the band swivel. The inner band, by which also the barrel is attached to the fore-end, is one inch in rear of the outer band. It is fixed by a screw passing through the fore-end, round which is a spring to maintain the necessary pressure against the barrel.

[The inner band is immediately under the outer band.]

Sling swivels are mounted onto a bracket at the toe of the butt and onto the outer band; referred to as lower and upper swivels respectively. These swivels have the securing lug in an off-centre position and these are mounted on the rifle such that the longer portion of the swivel is towards the right hand side (the side on which the bolt handle is positioned).

The barrel is supported by the fore-end, except from about half-inch in rear of the inner band recess to the knox-form.

[The fore-end is freed from the barrel throughout, except for about half-inch in front and half-inch to the rear of the inner band, and under the knox-form at the breech-end.]

The butt and fore-end are separate. The butt has a butt trap cover with spring to enable an oil bottle and pull-through to be carried in the stock bolt hole. In the butt, four holes are bored longitudinally for lightness, it has a pistol grip and a brass butt plate fixed by two screws; on the right side is a brass disc (Mark II* and IV, strap on butt plate), on which the regiment, number and date of issue are marked. In front is the butt swivel stem to which the swivel is attached. Butts are issued in three lengths, one half-inch longer, and one half-inch shorter than the normal, marked respectively on the heel L and S; the normal bears no mark. The rear end of the body is formed with a socket into which the butt fits, and is held in position by the stock bolt, which is squared at the front end to fit the keeper plate in the fore-end, thus preventing the stock bolt from turning, and the butt from becoming loose. When stripping the rifle, the fore-end must first be removed before turning the stock bolt.

The barrel is screwed into the body and secured to the fore-end by the inner band, nose-cap, outer band and trigger guard.

A wooden handguard completely covers the barrel. It is in two pieces, being divided at the centre of the backsight bed; the front portion is held by the outer band and by a metal cap which fits into a recess between the nosecap and the barrel; the rear portion is held by springs gripping the barrel. Both portions rest on the fore-end and are quite free of the barrel.

[The backsight protector is attached to the rear handguard.]

The foresight is of blade pattern. For Mk VI ammunition it

is made in five heights (marked –015, 0, 015, 03 and 045 respectively) and for Mk VII ammunition in seven heights (marked –06, –045, –03, –015, 0, 015 and 03 respectively) to allow any deviation from the normal in the shooting of individual rifles to be corrected. It is adjustable laterally. It is dovetailed into the foresight block band, which is secured to the barrel by means of a key and pin.

[The foresight is of barleycorn pattern and for Mk VI ammunition is made in five heights (marked .93, .945, .96, .975 and .99 respectively). For Mk VII ammunition it is made in six heights (marked .9, .915, .93, .945, .96 and .975 respectively).]

The backsight is fitted with a leaf and has a curved ramp on each side of the bed on which the slide rests and by means of which elevation is obtained. The backsight protector is level with the rear end of the leaf and is fixed to the fore-end with a screw and nut.

The leaf is pivoted to the front end of the bed and is graduated on the right by lines for every hundred yards from 200 to 2,000 yards, the even numbers only being marked by figures. The slide can be adjusted for every 50 yards and is retained in position by a worm wheel, which engages in a rack on the right side of the leaf. The wheel is disengaged from the rack by pressing the stud on the left inwards. The lines on the left of the leaf give graduations for 25 yards.

[The slide can be adjusted for every 50 yards and is retained in position by means of one of two catches engaging in a notch on the right side of the leaf. These catches are disengaged from the notches by pressing the bone studs on each side of the slide. A vertical fine adjustment and a windgauge are fitted to the leaf.]

Small increases and decreases of elevation can be obtained by rotating the worm wheel, the periphery of which is divided by thumb-nail notches into 10 parts, each part giving a difference of 5 yards range. One complete revolution of the wheel thus corresponds to a difference of 50 yards.

The windgauge is adjusted by means of a screw with milled head on the right. It is provided with a scale, each division on which represents 6 inches per 100 yards. Between the screw head and the windgauge is a small friction spring which prevents accidental movement of the windgauge. This spring gives a click at each quarter revolution of the screw, and 6 clicks, or one and a half revolutions of the milled head, are equivalent to one division on the scale.

[The fine adjustment is joined to the leaf by a dovetail and can be used for obtaining any additional elevation required between the 50 yards elevation given on the leaf. A scale is provided on the left edge of the fine adjustment and leaf. The windgauge is fitted in the vertical fine adjustment and is moved by a screw and provided with a scale. A thumb-nail slot is cut in the top front edge of the fine adjustment as a guide for centring the windgauge. Each division on the two scales represents 6 inches per 100 yards.]

On the left side of the rifle, long range sights are provided for ranges from 1,600 to 2,800 yards. These sights consist of an aperture sight pivoted to the rear of the body and a dial sight, with pointer, secured by means of a screw to the fore-end in rear of the outer band. The ranges are marked on the dial plate.

Action of the mechanism, &c.

On raising the bolt lever the bolt is rotated to the left, thereby forcing the stud on the cocking piece to move backward from the long to the short groove in the rear of the bolt; this action withdraws the striker about 1/8-inch. At the same time, a steel lug on the underside of the bolt works down an inclined slot on the left side of the body, withdrawing the bolt about 1/8th of an inch and effecting primary extraction.

[On drawing back the bolt, the charger guide on the bolt head, coming in contact with the stop on the top right side of the body, is held stationary in position to receive the charger.]

The charger containing 5 cartridges is placed between the guides, and the cartridges are forced into the magazine by the thumb.

On pushing the bolt forward, the charger is thrown out and the full bent of the cocking piece is brought against the nose of the sear. The cocking-piece and striker are thus held stationary whilst the bolt travels forward, the mainspring being compressed between the collar of the striker and the rear end of the mainspring chamber in the bolt.

During the forward movement, the lower part of the bolt head engages behind the upper part of the base of the top cartridge in the magazine and pushes the cartridge into the chamber.

On turning the bolt to the right, the breech is finally closed by the rib on the bolt working over the resisting shoulder on the right side of the body; at the same time the lug on the bolt works into the recess cut on the left side of the body.

On pressing the finger, when the rifle is as full cock, the two ribs on the trigger bear in succession on the lower arm of the sear and produce a double-pull off, the first bringing the nose of the sear to the bottom of the bent of the cocking-piece; the mainspring then carries the striker forward, exploding the charge.

The shock of the discharge is taken equally on either side of the body; on the right by the bolt rib bearing against the resisting shoulder, and on the left, by the bolt lug bearing against the rear wall of the recess in the body.

If the bolt has not been properly turned over when the trigger is pressed, the stud on the cocking piece either causes the breech to close automatically by striking against the rounded corner of the division between the two grooves of the bolt, causing the bolt to turn down and the breech to close, or else the stud on the cocking-piece strikes fully against division between the two grooves, and prevents the striker flying forward. If then the bolt is closed by hand, the whole action becomes locked as the sear nose is engaged by the half-bent, which is undercut, whilst the cocking-piece stud travels halfway down the longer groove; consequently the trigger cannot be pressed, nor can the bolt be rotated until the action is placed at full cock by drawing back the cocking piece.

Upon opening and drawing back the bolt, the cartridge case is drawn out of the chamber by the extractor, and it is ejected when its base comes in contact with the ejector screw.

The safety-catch may be used when the cocking-piece is either at full-cock or in the fired position. In neither case can the cocking-piece be moved backward or forward, nor the bolt rotated. Care should be taken to see that the bolt lever is in its lowest position before moving the safety catch, as if it is slightly raised when the safety catch is moved there is a possibility of the aperture sight spring being strained and weakened.

To strip the bolt: Remove the extractor spring, extractor screw and extractor. Remove the striker keeper screw and see that the stud on cocking-piece is in the long cam.

Unscrew the bolt-head and remove the striker, mainspring and cocking piece. If the striker is so tight in the cocking piece that the bolt-head cannot be moved by hand, place the head in the bolt hole of the body and turn the bolt about a quarter of a turn over to the left. If it is found that the striker is not released by unscrewing the bolt-head to the full extent, draw back the cocking-piece until the stud is clear of the end of the bolt and turn it to the right, so that the stud rests on the rear end of the bolt; screw home the bolt-head, then replace the cocking-piece, with the stud in the long cam, and unscrew the bolt-head.

[Remove the charger guide stop screw and charger guide. Remove the extractor spring, extractor screw and extractor. Raise the striker keeper nut, if fitted, out of the locking position, turn it slightly, or remove the striker keeper screw, and see that the stud on the cocking piece is in the long cam. Remove the striker as in Mark III. Replace the striker keeper nut in locking position and unscrew keeper nut screw, and remove screw, spring and nut.]

To assemble the bolt: Place the mainspring, striker and bolt-head in the bolt and screw home the bolt-head for about six turns; then place the cocking-piece on the bolt, so that the stud is in the long cam, and screw home the bolt-head; draw back the cocking-piece until the stud is clear of the end of the bolt and turn it until the stud rests on the rear end of the bolt. Unscrew the bolt-head about six turns, turn the cocking piece so that the stud moves into the long cam. Screw home the striker until the end is flush with the rear end of the cocking piece, and the keeper-screw recess in the striker is in the correct position; replace the keeper screw, move the cocking-piece to the short cam and screw home the bolt-head, being careful to see that it is screwed fully home before replacing the bolt in the rifle. If it is found impossible to screw home the striker in the cocking piece by hand, extra leverage may be obtained by placing the bolt-head in the bolt hole of the body. Replace the extractor, extractor screw and extractor spring.

[Replace in the cocking piece, the keeper-nut, spring and screw (if fitted) and lift the nut out of the locking position. Replace the mainspring, striker and bolt-head as for Mark III and screw the striker home till the keeper-nut (or screw) recess is in the correct position. Place the keeper-nut in the locking position (or replace the keeper-screw). Screw home the bolt head, as for Mark III. Replace the extractor, extractor screw, extractor spring, charger guide and stop screw.]

To remove the magazine platform: Nos. 1, 2 and 3 magazines, push the stop clip forward to the disengaged position shown in the dotted lines, slightly depress the rear end of the platform at A and lift the front end at B with the fingers, at the same time pressing the platform to the right side and upwards; then pull the platform forward with a slight wriggling lateral movement (taking care to keep the rear side-ears C above the ribs of the case D) and lift the platform with its spring out of the magazine. Under no circumstances must leverage be applied to the lip of the platform at the front end F to force it out. If force is found necessary, a knife blade may be placed on top of the auxiliary spring at the front of the magazine at G, with the point resting under the front of the magazine at the left side of the lip H and a slight upward pressure will release the platform.

※ No. 3 MAGAZINE.

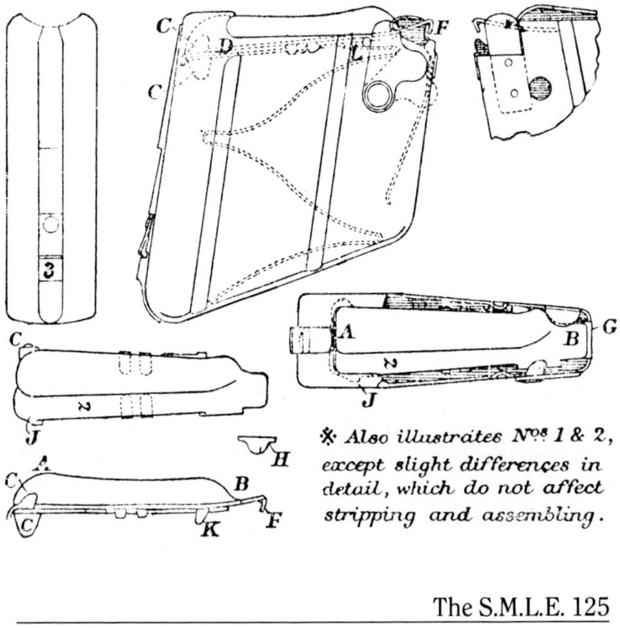

※ *Also illustrates Nos 1 & 2, except slight differences in detail, which do not affect stripping and assembling.*

No. 4 magazine, depress the rear end of the platform at N as far as possible, at the same time holding up the front end; then pull the front end at O towards the rear end of the case – until it passes under the front lips QQ and between the inner forward ribs RR in the case, through which it is sprung. The front end of the platform should then rise up out of the case. Then tilt the rear end of the platform N sideways, the left side S uppermost, and draw forward out of the case.

To assemble the magazine platform: Nos. 1, 2 and 3 magazines, place the spring in the case and, tilting the right side of the platform down slightly at J, press the rear end of the platform into the case just above the ribs D, then see that the front ears K are just in front of the ribs of the case at L, slightly depress the rear end of the platform at A, and press the front end down with a sharp pressure at M, then push back the stop clip against the stop.

Nº 4 MAGAZINE.

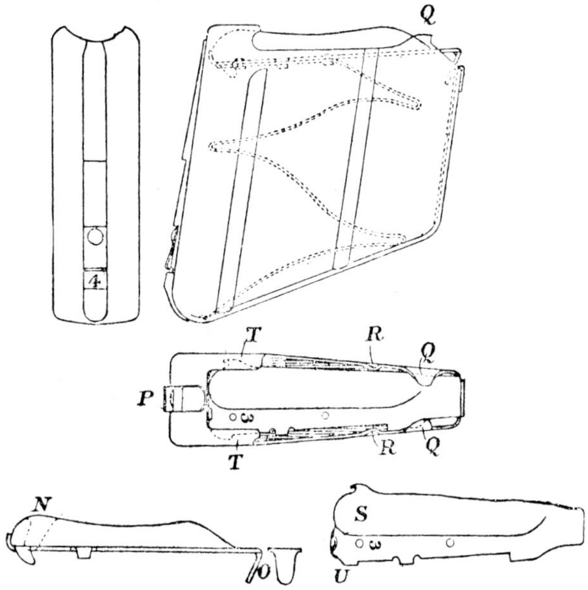

No. 4 magazine, insert the rear end N of the platform in front of the rear lips TT on the case, tilting it sideways so that the right side U enters first, then depress the rear end until the front end is below the level of the front lips QQ on the case, then press forward, forcing the front end through the internal ribs RR in the case.

Names of the Parts of Rifle, Short M.L.E., Mark III.

1.	*Blade foresight.*
2.	*Foresight block.*
3.	*Band foresight block.*
4.	*Key foresight block.*
5.	*Crosspin foresight block.*
5A.	*Backsight bed.*
6.	*Backsight bed crosspin.*
6A.	*Backsight bed sight spring screw.*
7.	*Backsight leaf.*
8.	*Backsight slide.*
9.	*Backsight slide catch.*
10.	*Backsight fine adjustment worm wheel.*
10A.	*Windgauge.*
10B.	*Windgauge screw.*
11.	*Backsight ramps.*
12.	*Seating for safety catch.*
13.	*Safety catch.*
14.	*Locking bolt stem.*
15.	*Bolt.*
16.	*Bolt head.*
17.	*Striker.*
18.	*Cocking piece.*
19.	*Striker collar with stud.*
20.	*Bolt head tenon.*
21.	*Cocking-piece locking recess.*
22.	*Locking bolt.*
23.	*Locking bolt flat.*
24.	*Locking bolt thumbpiece.*
25.	*Locking bolt aperture sight stem.*
26.	*Locking bolt stop pin recess.*
27.	*Locking bolt safety catch stem.*
28.	*Locking bolt safety catch arm.*
29.	*Locking bolt screw threads.*
30.	*Locking bolt seating.*
31.	*Bolt cam grooves.*
32.	*Sear.*
33.	*Sear seating.*
34.	*Sear spring.*
35.	*Magazine catch.*
36.	*Full bent of cocking piece.*
37.	*Short arm of sear.*
38.	*Trigger rib.*
39.	*Trigger rib.*
40.	*Trigger.*
41.	*Trigger axis pin.*
41A.	*Magazine case.*
41B.	*Magazine platform spring.*
41C.	*Magazine auxiliary spring.*
42.	*Guard trigger.*
43.	*Stock fore-end.*
44.	*Spring and stud fore-end.*
45.	*Protector backsight.*
46.	*Handguard front and rear.*
47.	*Spring handguard rear.*
48.	*Lower band groove.*
49.	*Lower band.*
50.	*Nosecap.*
51.	*Foresight protector.*
52.	*Sword bar.*
53.	*Boss for ring of sword bayonet crosspiece.*
54.	*Swivel seating.*
55.	*Swivel piling.*
56.	*Nosecap barrel opening.*
57.	*Inner band.*
58.	*Inner band screw.*
59.	*Inner band screw spring.*
60.	*Butt sling swivel.*
61.	*Sword bayonet, pattern '07.*
62.	*Bridge charger guide.*
63.	*Cut-off.*

An "Outline of Manufacture and Inspection of the Service Rifle" from the 1909 Text Book of Small Arms provides an enlightening view of the Mark III rifle production:–

Specification and Drawing.

The design of the rifle having been decided upon, a specification and drawings are prepared to govern manufacture and inspection.

Whilst the object of the manufacturer is to make a serviceable weapon, well up to the specification, he has always to do so at as low a cost as possible. For this reason an independent inspection is given on the rifle and the points that are consistently before the department charged with this duty are:

1. *Accuracy in shooting.*
2. *Safety and wearing qualities.*
3. *Interchangeability of components.*

Material.

The specification gives particulars as to the material to be employed for each separate component. These materials are steel, iron, wood and brass.

Steel.

Steel, the most important of the materials with which we are concerned, is a variety of iron containing carbon to the extent of from 0.05 to 1.5 per cent. Other substances are also present in varying proportions, but the carbon is the most important ingredient, and it is to the presence of the carbon that steel owes its most valuable property, viz., that by certain heat treatments and methods of cooling it can be brought to any desired state of hardness or

SHORT RIFLE, MAGAZINE LEE-ENFIELD (MARK III).

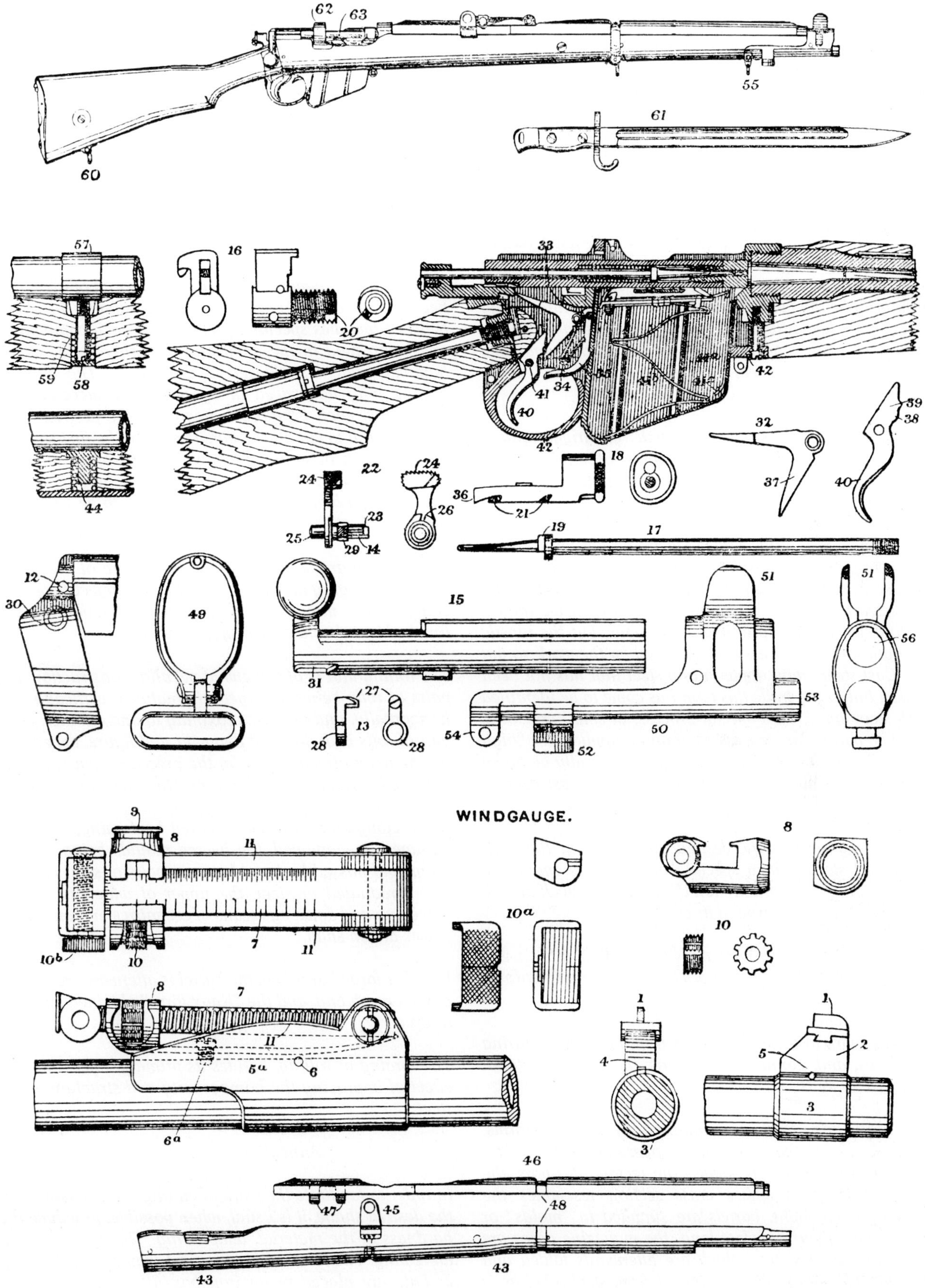

softness. Steel may be classified as "mild" or "high carbon," according as the amount of contained carbon is below or above about 0.3 per cent. As the property of hardening increases, up to a certain point, with the percentage of contained carbon, a "high carbon" steel is employed for such important components as the bolt, cocking piece and sear, which have to resist considerable pressures, or have wearing surfaces. When the component is required for work of a springy nature, such as the cut-off, retaining spring for bolt-head, or sear spring, a special quality "high carbon" steel is employed, as this quality of spring cannot be imparted to a mild steel. For the barrel either "high carbon" or "mild" steel may be used, but, in any case, the steel is of a high quality and greater strength is required of it in its mechanical tests than would ordinarily be the case with mild steel. The manufacturing processes, the difficulties of boring and drilling an accurate "tube," have to be kept in view at the same time as the qualities that will give safety to the barrel under firing pressures.

For the minor components, such as trigger-guard, nosecap, screws, &c., a lower grade of mild steel is employed, which, though of ample strength, is cheaper and easier to work.

The terms "cast" and "forged steel" are apt to be rather confusing. All steel is ordinarily cast into an ingot and by "forged steel" is meant that it has subsequently been subjected to forging. A forging is more expensive than a casting, involving more time and labour, but the tensile strength and soundness are increased by the work so put in. "Cast steel" may apply to any steel that has not been forged, but in the trade this term has come to be identified with steel made, by a special process, in small crucibles. As this particular process is mainly applied to "high carbon" steels of good quality, it may generally be taken that this is the class of steel referred to when "cast steel" is mentioned.

Where toughness is the desired quality, as in the bolt-head and stock-bolt, a special "gun-iron" is employed. This contains less carbon than the mildest of steel, in fact, approaches wrought iron in its composition.

For any particular component the following points have to be borne in mind in deciding which is the most suitable class of steel to use:
(a) The function of the component in the finished rifle.
(b) The work to be performed on it in the manufacturing processes, and, conjointly with this, the cost of material and wear of tools.

The iron and steel are obtained in various forms, depending on the subsequent operations. Thus, for example, the bodies are made from rectangular bars, the trigger from bars whose cross section resembles the finished component, barrels are supplied in "molds" or round bars about one foot long, the magazine cases are stamped out of sheets which are afterwards folded and brazed, the smaller pins are made from steel wire, as is also the main spring, whilst the springs from the rear handguard and magazine platform are made from "ribbon" spring steel.

An Inspector may take a sample of material at any time and subject it to chemical analysis and mechanical tests.

Manufacture.

The principal manufacturing processes are forging and machining, the former being employed as a preliminary to the latter in those cases where the form of the raw material necessitates it.

Drop forging is generally employed. One end of the bar is heated to a cherry red, and brought roughly to shape between two dies specially prepared for each component, one being held in the anvil and the other in the steam hammer, which is allowed to drop on it. The finished forging may be arrived at by one or a succession of operations, each bringing it nearer to the desired form. Whilst hot, it is cut off the bar. These are the first steps in the formation of such components as the body and bolt, and their object is to save the immense amount of cutting which would be required were the article to be cut out of solid.

The preliminary stages in the manufacture of the barrel are either forging or rolling, or both. The object here is to elongate the mould to the approximate length of the finished barrel, and also to shape the exterior.

The rolling mill is of somewhat peculiar construction, ten pairs of rolls being arranged alternatively vertically and horizontally, and in close proximity to one another; each pair of rolls is of different diameter and figure, depending on the work already done by the preceding pairs, and so arranged that the elongation is practically continuous.

If the elongation is to be obtained by forging, a "Ryder hammer" is employed, the heated moulds being held in tongs and passed by hand between pairs of semi-circular dies (graduated in size), the upper of which is rapidly moved up and down by means of eccentrics or cranks on a power-driven shaft.

Whether forged or rolled, the barrel is afterwards re-heated at the breech end, and the "Knox form" obtained by drop-forging.

The forging or rolling, besides its primary object, has the effect of improving the homogeneity and structure of the steel.

Annealing and Pickling.

In order to facilitate the work of cutting the component to the desired shape, it is usual, when possible, to reduce the hardness of the material, temporarily. This is effected by annealing. The finished forgings, for example for the body or bolt, are placed in an iron box filled with vegetable charcoal, and heated in a muffle for several hours, after which the furnace is allowed to completely cool down. During this process the interior is softened, but a hard scale forms on the surface and this is removed by "pickling," i.e., the forgings are immersed for a short time in a solution of sulphuric acid and water. This dissolves the hard scale and the component is then ready for machining.

Machining.

This is a general term and comprises all the various operations which are wholly or partially mechanically performed, and have for their object the reduction of the components to their correct shape and size. In practically every case this is effected by cutting, but there are several methods of applying the cutting tool and the differences between them are indicated in the names of the different operations. The more important of these are turning, drilling, boring, slotting, milling and shaping.

Drilling consists in cutting a hole by means of a drill, the end of which is shaped to form the cutting edge or edges.

Boring consists in enlarging a circular hole by means of a boring bar, the sides of which form cutting edges, or from the sides of which separate cutters project.

In milling, a hardened steel wheel with a series of cutting edges on its rim, or on one or both of its sides, is revolved, and the work is automatically fed forward or rotated as the cutting progresses.

In shaping, the work is held stationary whilst the cut is being effected by the tool. This is given a reciprocating motion, but cuts only during the forward stroke.

Slotting is somewhat similar to shaping, but the tool is given a vertical motion, and the process is used for enlarging a circular hole (already drilled) in an required direction by traversing the work.

It is obvious that there must be relative motion between the tool and the work and, further, that one or both of them must be moved to give the requisite feed to bring or keep the tool in contact with a fresh surface or portion of the work. Much ingenuity has been expended on the design of machines to perform these motions automatically or semi-automatically, enabling large numbers of components to be made with a minimum of labour. It should be understood that the "cut" or amount removed is necessarily small and that, therefore, a large number of varied operations have to be performed on each component before it is brought to the correct size and dimensions. It is one of the chief objects of the factory staff to so arrange the sequence and number of these distinct operations so as to keep down expenses.

It is not supposed that a machine once adjusted to perform a certain operation is capable of repeating the same indefinitely, and reproducing in large numbers components in any particular stage, all absolutely identical. The wear of the cutters and fixings proceeds slowly but surely and it becomes necessary to fix manufacturing limits within which the serviceability of the article is not appreciably affected. These limits which are generally within a few thousandths of an inch, vary with the importance of the component and the stage at which it has arrived, and it will be understood that in some instances no toleration (within the powers of measuring) is permissible.

Gauging.

These considerations bring us to the necessity of gauging. This commences in the machine room, where the machine hands are provided with gauges by which they can check the work being produced, and bring to notice the necessity for re-adjustments of the machine, or for tool-sharpening. Such work is performed by qualified fitters and the gauges by which they are guided are of a higher order of perfection than those used by the machine hands. The machined components are viewed and gauged again by an independent staff.

In designing a gauge for a component it is most important to consider with what other component it interworks or comes in contact, and what parts of each must have a definite size to ensure ease of assembly and interchange-ability. Following from this it will frequently be found that a gauge, or a portion of it, resembles one or more of the components adjacent to that, shape and dimensions of which it is designed to check. The making of gauges calls for exceptional skill and their high price is commensurate with the degree of accuracy required. A special set of gauges, known as standards or reference gauges, are kept by the Inspection Department and used with the greatest reservation for the sole purpose of checking the work gauges and preventing the latter being used when wear or other causes have rendered them unreliable. Thus it may be seen that the whole question of gauging forms an intimate link between the factory and inspection staffs.

Barrel.

As the barrel is the most important component in the rifle and as some of the operations in the making of it are peculiar to it, they are worthy of more detailed investigation. The barrel is subjected to inspection at various stages and the specification requirements at each stage serve as convenient points from which to review the methods employed to arrive thereat. Starting then with the forged or rolled barrel, it is first inspected in the "fine bored stage." It is viewed, and must take a plug .3025-inch in diameter; the exterior must be concentric with the bore and must be generally of correct dimensions. The bore must be straight. Briefly, the processes the barrel has undergone are rough turning of the exterior, drilling, finished turning, polishing and boring. As has always been indicated, it is the ever-present concern of the manufacturer to arrive at the desired end with the minimum of expense, and hence it by no means follows

On display at R.S.A.F. Enfield: Some of the gauges applied for the manufacture and inspection of the Mark III S.M.L.E. rifle. *MOD Pattern Room*

that the procedure in different factories as regards the sequence and number of operations are identical. Thus, for instance, the drilling may be started from both ends simultaneously, and be carried out in several distinct operations, or may proceed from one end to the other without any break, and be completed in one operation. The drilling bits and boring bars slide horizontally along the bed of the machine whilst the barrel revolves. They are bored with a fine hole longitudinally, through which a cooling lubricant is pumped, and this washes out the "swarf" or cuttings from the bore. The important considerations are that the bore must be straight and concentric with the exterior; it is the business of the factory, knowing how far they can trust their own methods and machines, to take steps, such as intermediate viewing and setting, as shall ensure this.

The test for straightness depends upon the light and shade effects which may be seen upon the bright surface of the smooth bore. A line of shadow is thrown down from the muzzle to the breech, and this line must remain straight while the barrel is slowly revolved. Any bend or curvature in the line of shadow indicates a corresponding want of straightness at that point in the bore. A more modern test depends on the multiple reflections of the muzzle, which appear in a smooth straight bore as concentric rings. By suitable regulation of the amount of light which is permitted to strike the muzzle, this can be made an exceedingly delicate test, and the slightest want of concentricity is an indication that some correction is needed. This is obtained by setting.

Barrel-setting was formerly looked upon as the work of a highly-skilled specialist, but, owing to improved mechanical methods, the degree of skill required is now considerably diminished. Having located the error, the older method (which is still largely in use) was to strike the barrel a transverse blow with a copper or brass hammer in the required direction, estimating the required strength of the blow. But, in conjunction with the system of testing by concentricity of rings, means are provided whereby the barrel can be mechanically bent to the necessary extent to overcome the former want of straightness. To test the concentricity of the bore and the exterior, the barrel is spun on plugs entering the bore.

The barrel is next inspected in the "rifled stage." It is viewed, and must be free from greys and scratches; the depth and pitch of the rifling are tested. The important feature of the rifling machine is the method of regulating the cut; this may be taken approximately as one-thousandth of an inch, thus it is clear that the rifling of each groove required, say, five to seven cuts to be made upon it. The cutter box is drawn through the barrel by means of a rod, A, the tooth, B, of the cutter taking out a shaving in its passage. The cut is regulated by the rod, C, for when this is screwed outwards, the spring, D, in the cutter box bears on the cutter, and the slope, E, on the cutter rides up the slope in the cutter box, and consequently the cutting tooth, B, protrudes to a greater extent. The rod, A, is revolved mechanically to give the necessary pitch to the rifling.

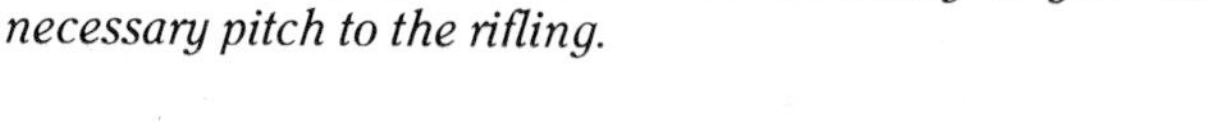

Rifling cutter box.

In the "first chambered" stage the cartridge chamber has been bored to within .002-inch of the finished size, the extractor way has been cut, and minor machining operations, preliminary to the fitting of the sights, have been performed. When next submitted for inspection the assembled sight bed and block band foresight have been fitted, and the dimensions, fitting and positioning of these are tested.

Proof.

At this stage the particular barrel, body and bolt, which will afterwards form part of the same rifle, are first brought together, and, being numbered, the action is completed with "slave components" which are kept for this purpose, the backsight is stripped and the barrel proved. This is carried out with a proof cartridge, giving pressure considerably above that to which the rifle is subjected, and it is fired under conditions ensuring safety to the firer in the event of any portion failing to stand the test. The high pressure has the effect of setting up the chamber and the recoil shoulders of the bolt and body. It is to allow for the consequent enlargement that the first chambering is left two-thousandths of an inch small.

Incidentally, it may be noted that at this firing proof an intimate relationship is established between the recoil shoulders of the bolt and body, and the fact that this affects the shooting of the weapon is the reason for taking pains to ensure that rifles always have their own bolt.

After proof, the barrel, body, bolt and bolt-head are examined to see that they are sound and they are then marked with the proof mark.

The barrel next comes forward in the "browned and finished chambered" stage. Browning will be considered together with the other finishing operations. Since the proof the chamber has been brought to finished dimensions, and it is now gauged. In addition the bore is spun, for straightness, on a rod of which only two small portions bear, namely, at the breech end and the centre of the bore; any want of straightness is thus made apparent by a lateral movement of the muzzle end. In addition, the bore must be parallel from end to end and take a .303 or .304 plug and reject a .305 plug. The barrel and body (which has been browned) are then assembled and submitted in the "breeched up" stage; the relative positions of the flat on the barrel and the body are tested, and the barrel and body, supported on a plug in the muzzle and another in the boltway of the body, are spun to ensure their axes being in alignment. Finally, the sights are completely assembled and the barrel is placed in a sight-testing machine which enables the height of the leaf above the axis of the barrel to be checked. This is done for 200, 500, 1,000 and 1,500 yards elevation and intermediate heights are checked by a profile gauge, which is applied to the ramps and checks their figure. The general working and fit of the slide and wind-gauge are also tested.

There is one other operation which is peculiar to barrel work, namely, "lapping." This is effected by placing a stopping of tow and a rod with a jagged end a few inches inside the muzzle, and pouring in molten lead. This sets in the grooves of the rifling and adheres to the rod, by means of which it can be worked up and down the bore, the rod being given a twist so as to follow the rifling. By the addition of a paste and oil and emery powder, which is carried by the lead plug, any small imperfections which may be present can be removed. This process must be used with discrimination as it is clear that unlimited lapping will enlarge the bore considerably.

This completes the manufacture and inspection of the barrel as a separate component.

Finishing Operations.

This term includes operations such as polishing and bobbing, which serve to give a finish to machining instructions; the treatment of steel to give it certain qualities, and also to the preparation of the surface to avoid rusting and to give it a dull appearance.

Polishing is principally used for the barrel and is effected in this case by rapidly moving the barrel between spring clams faced with emery and oil. Bobbing consists of fine-grinding and polishing with rapidly-revolving emery wheels; the work, if cylindrical, is revolved also and is traversed to and fro in front of the bob.

The treatment of the steel has in general two objects, namely, to harden it or give it elasticity. Hardening may be used:

(1) To restore the tensile strength to a cast steel that has been softened for machining purposes;

(2) To give a hard surface to resist wear in cases where moving parts come in contact;

(3) As a preliminary operation to tempering in spring-making.

Hardening is invariably effected by heating the component (generally to about 740 degrees C, a red heat) and then cooling it more or less suddenly, or, as it is termed, "quenching" it, in water or oil. This, however, leaves the steel so brittle that small components can be snapped between the fingers and, therefore, for whichever the above reasons hardening has been used, it is generally modified by subsequent tempering. This consists in again raising the temperature of the steel so to partially soften it, and then allowing it to cool, or quenching it. The amount of temper to be given is governed by the purpose the component is to serve and is regulated by the temperature to which it is raised, and the rate of cooling. In tempering springs, the temperature employed is about 290 degrees C. A bath of nitre or lead is sometimes employed as a tempering medium.

Case hardening is used to give a hard surface, where required, to those components made of iron or mild steel, which do not contain sufficient carbon to enable them to be hardened as has already been described. In these cases the parts are packed up in an iron box with bone-dust or shreds of leather (which forms the source of supply of the carbon necessary to effect the process), the lid of the box is sealed with fireclay and the whole subjected to lengthy heating at considerable temperature. They are then quenched in oil or water.

The process used for protecting the surface or for giving it a dull appearance are oil blacking, browning and blueing.

Oil blacking is effected by heating the component and then dipping it in oil and allowing the oil to drain off; this leaves a black carbonaceous deposit on the surface. A similar process, blazing off, which derives its name from the fact that the oil is ignited by the heat of the component. These methods are almost a part of hardening and tempering, to which they frequently form the sequel, the heating required for the former processes serving also for the latter, whilst the dipping in oil in the latter serves the double purpose of quenching the part and blacking it. Oil blacking and brazing off can, however, be carried out with mild steel and these processes are used for nearly all the smaller components.

Browning may similarly follow the hardening and tempering processes, or it may be applied direct. It is a chemical process by which the surface of the steel is given a dull, dark appearance by oxidation produced by acids used in the mixture.

Blueing is limited to a few components, chiefly to be found on the dial and back sights. It is effected by oxidation of the surface by heat. The part to be treated is heated in charcoal or in a bath of nitre and manganese dioxide.

Stocks.

Walnut is the wood most commonly used for stocks; the wood is received from contractors roughly cut to form, and, after examination, it is carefully stored to season for about three years.

Amongst the machining operations, the turning of the butt and fore-end on Blanchard lathes is particularly interesting; in this machine the wood and an iron dummy shaped to represent the butt or fore-end, as the case may be, revolve side by side at the same rate. A wheel is kept in contact with the dummy, and by the protrusions or recesses thereon is moved away from or allowed to approach the axis of rotation. The cutters are carried on the rim of a similar wheel, which is rapidly revolved, and both this and the guiding wheel being supported in the same frame (which travels laterally) the cutters are constrained to follow the movements of the guiding wheel, and thus give to the wood a shape similar to that of the dummy.

A similar arrangement is used for glass-papering the parts after the finished turning, but in this case the cutters are replaced by layers of felt and glass-paper glued on to the rim of the wheel. The finished butt and fore-end are both soaked in linseed oil for half-an-hour. The inspection of these components is left to the finished stage, when, in addition to the usual points, the quantity and condition of the wood are looked to.

Finished View.

The assembly of the components to form the finished rifle calls for no special remarks.

The points of importance in the final inspection of the rifle are:
(1) To see that all components are complete, and are marked with a view mark and number and proof mark where applicable.
(2) To see that the various parts fit and bear correctly, and in this connection the fit of the barrel, fore-end and nose-cap, or "stocking up," is of special moment; the object is to control the vibrations of the barrel on firing, and to secure a common basis from which the sighting of the rifle has been carried out. Any difference in this particular must affect the accuracy of the weapon and is therefore carefully looked to.
(3) The working of the action is examined and the magazine, charger guide and extraction tested with the aid of dummy cartridges and fired cases.
(4) The weight of pull-off and certain springs are tested.
(5) The general finish is examined, the aperture and dial sights are tested and the working of the slide and wind-gauge of the backsight looked to.
If satisfactory, the rifle receives the finished view mark and is then sent to the range for testing.

Testing.

All rifles are fired at 100 feet and 10 per cent are also fired at 600 yards, any which give doubtful results at the shorter range being included among those tested at the longer. In both cases the rifle is held in a mechanical rest, designed to approximate the conditions under which it is ordinarily fired as regards points of support, recoil, &c. Means are provided for laying the rifle by hand-wheels, and with the aid of a special telescope laid on the blade of the foresight and bearing on the upper surface of the wind-gauge, a true, full, regulation sight is obtained. After laying and before firing, the telescope is carefully removed.

The following is the procedure in the firing test at 100 feet range: trial shots are test fired, and, if necessary, the fore-sight blade is adjusted laterally (within limits), or it may be replaced by a lower or higher one as required. Five rounds are then fired and four shots out of the five must come within a rectangle 1 inch broad and 1.5 inches high. Failing this, the rifle is not accepted. At 600 yards, nine out of a group of ten shots fired must lie within a two-foot circle.

After passing a satisfactory test the blade of the fore-sight is fixed by the top edge of the front of the stool of the block band fore-sight being punched into a recess cut in the fore-sight. The rifles are once more given a general inspection and are then packed and passed into store.

With the closure of the government Sparkbrook factory in 1906, the Ordnance work was wholly taken over by Enfield, by now the only Government Small Arms Factory. Despite the closure of Sparkbrook and removal of repair facilities, there was only a slight increase in staff at Enfield. A table of employees at Enfield in 1906-07 not only provides an indication of the size of the establishment, but also a proportion between the various machine shops, tradesmen and associated facilities.

	1st Apr. 1906	31st Mar. 1907
Tool Room	159	95
Machine Gun (.303 Maxim conversion)	61	46
Examination	91	114
Bayonet	42	38
Barrel	137	207
Repair (transferred from Birmingham)	–	53
Foundry	26	24
Millwrights, Labour, &c.	108	109
Gas Works	7	5
Stocking	136	114
Carpenters	20	15
Smithy	70	39
Large Room	289	356
Sighting	235	242
Magazine	61	72
Browning	39	43
Hardening	48	48
Assembly	114	102
Screw	54	50
Polishing	35	45
Grinding	7	2
Writers	24	26
Miscellaneous	32	34
Stores	46	45
Accounts	39	41
TOTALS	...1,880 ...	... 2,085

Approval was given for trials to be conducted to test endurance and wear and tear of barrels in 1908, incorporating alloys of Chrome Vanadium, Manganese, Nickel and Titanium. These tests were quite extensive and not completed until 1911. It was found that machining the barrels made of harder alloys was much more complicated and there was no appreciable improvement in accuracy. Barrel life was not prolonged as much as was first considered likely and upon considering the additional cost and machining problems, it was decided to continue with the original formulations.

Special markings were used on components which, although rejected for general service, were still suitable for roles such as skeleton actions, aiming tube rifles, repairs and conversions. These marks were:–

A. T.	Aiming Tube.
[ ]	Repairs.
[C]	Conversions.
S.K. A.	Skeleton Actions.

A heavier extractor spring was approved for rifles Mark III and Converted Mark IV on 9th April 1908; the tension was increased from between 4.5 and 5.5 lb. to between 7 and 9 lb. The heavier tension spring was marked with the number "3" on the inside of the longer end in order that it might be readily identified.

Sound suppressors or "silencers" were brought to the notice of the Committee; some other nations were conducting small scale tests to ascertain their feasibility at about this time. One of the best was Sir Hiram Maxim's design which he patented in 1908, No. 6680, and the Small Arms Committee considered that it had no military application given the current circumstances and requirements. Silencing devices of this period were generally the "add on" types which fixed onto the muzzle.

Another device brought to the notice of the Small Arms Committee was Major Muller's Fire Control which permitted a rifle to be fired upon reaching a predetermined angle of elevation; a sample of this interesting device is held in the Imperial War Museum collection and is illustrated here along with copies of Muller's patents dated 1907, 1908 and 1910. The last one most resembles the mechanism in the butt of the rifle illustrated.

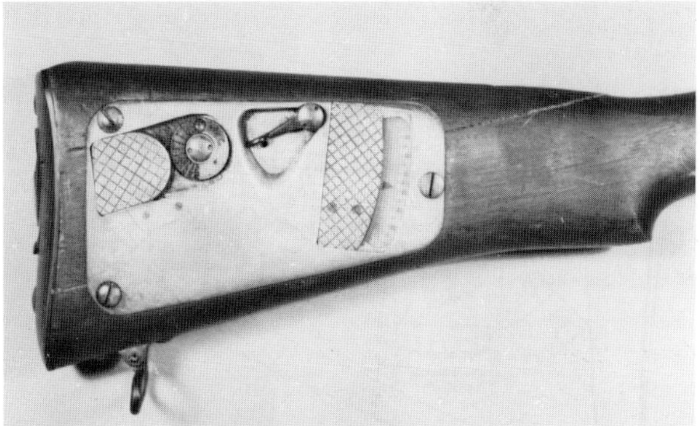

Muller Fire Control which fired the rifle upon reaching a predetermined degree of elevation. The patent drawings are shown on the next page. *Imperial War Museum*

Similar devices have been noted on U.S. and German service rifles held in official collections, indicating the likelihood of Muller's fire control being put up for consideration by the various Ordnance boards and military authorities in Europe and the United States.

6358. Müller, A. March 23, 1907.

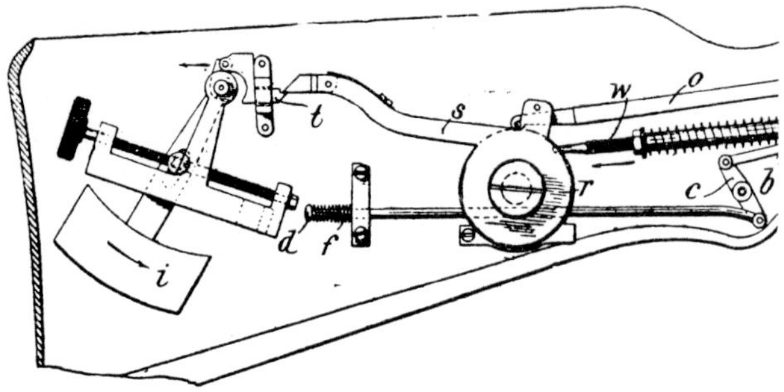

9252. Müller, A. April 25, 1908.

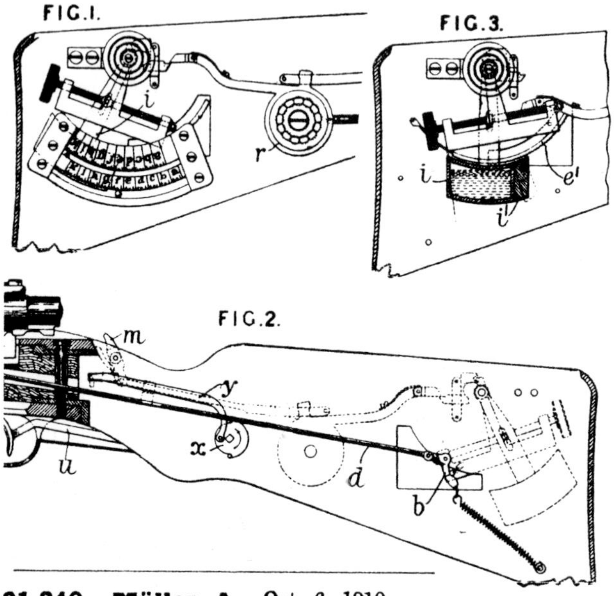

21,340. Müller, A. Oct. 6, 1910.

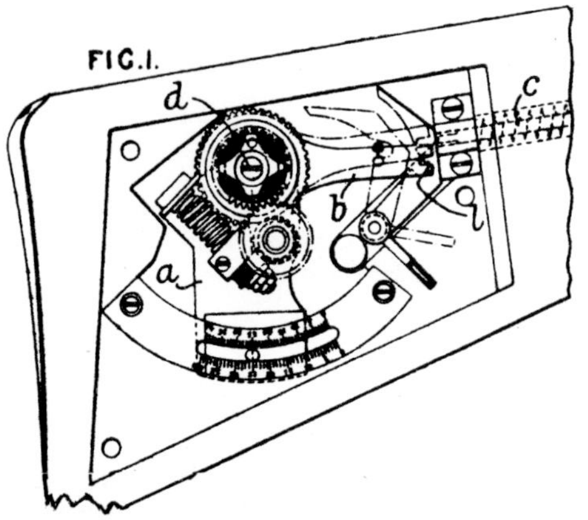

Some more converted S.M.L.E. rifles for the Navy were introduced in the Converted series; these were the Marks I**, Converted II** and Converted II***. The official approval dates are recorded as 4th January, 25th June and 5th July respectively, and the alterations were effected at the Royal Naval Ordnance Depots (R.N.O.D.) at Chatham, Portsmouth and Plymouth.

These particular conversions are of special interest because they were not intended to duplicate other recently introduced patterns; in some minor respects they do not resemble any other models. The earlier Converted Mk II conforms with the Mk I S.M.L.E. and the Converted Mk II* is like the Mk I* S.M.L.E.; the Converted Marks I**, II** and II*** have distinctive sight protector ears on the nosecap to accommodate the Mk III S.M.L.E. foresight, but with the earlier Mk I S.M.L.E. type of charger guide mounted onto the bolthead.

The official announcement of these Naval conversions was made in the List of Changes:

Para 14936:–

*Rifle, Short, Magazine Lee-Enfield, Mk I***
Converted from Rifle, Short, M.L.E., Mk I.

*Rifle, Short, Magazine Lee-Enfield, Cond. Mk II***
Converted from Rifle, Short, M.L.E., Cond. Mk II.

*Rifle, Short, Magazine Lee-Enfield, Cond. Mk II****
Converted from Rifle, Short, M.L.E., Cond. Mk II.*

Patterns of the above-mentioned rifles have been sealed to record the above-mentioned conversion. The conversion (which is carried out in Naval Ordnance Depots, Chatham, Portsmouth and Plymouth, in accordance with Admiralty letters of 4th January 1908 and of 1st July 1908) consists in fitting a Blade, foresight, M.L.E. rifle, short, Mark III (§ 13853) and a higher windgauge with top edge hardened, and with a U-notch to the backsight to correspond. The windgauge is special to the above-mentioned rifles.

In addition to the above-mentioned conversion, the following additions and alterations have been made:–

§ 13313. Body: Sharp edges on charger guides on left side and on charger guide stop on right side have been removed to prevent damage to clothing and equipment.

§ 13509. Backsight: The rear end of the sight leaf is bored to receive two spiral springs to tension the fine adjustment, and a new screw, fine-adjustment, having a larger head and finer milling, is fitted, having a tension spring under the head to prevent the screw jarring loose on firing the rifle.

Guard, hand, rear: The rear handguard is fitted with a double spring, fixed by two longer rivets placed in a central position in the length of the spring.

Guard, trigger: The trigger guard is provided with a lug at the front to carry a sling swivel and screw when required in that position, and the magazine link loop is removed.

Link, magazine: The magazine link has been removed.

Magazine: The magazine link loop has been removed.

Spring, platform, magazine: A retaining nib has been riveted to the spring to prevent any movement of the platform on the spring.

Stock, fore-end: The stock, fore-end has been recessed to receive a stud and spring to centre the barrel in the nosecap. It is also suitably recessed for the double spring of the handguard.

Plate, butt: The Plate, butt, M.L.E. rifle, short, Mark I, No. 2 (§ 13742) has been fitted and a wad, stockbolt, is inserted in the stock, butt, except in the case of the converted Mark II***, which is already fitted with the M.L.M. rifle, Mark II, buttplate and a wad inserted before conversion from Mark II*.*

*Stock, butt: The stock, butt, has been suitably recessed to receive the butt plate, except in the case of the converted Mark II***, which was already fitted as described above.*

*§ 13511. Screw, keeper, striker: A keeper screw (which can be turned by a coin) for retaining the striker in the cocking piece, has been fitted in place of the nut, screw, and spring, keeper, striker, except in the case of the converted Mark II***, which was already fitted before conversion from Mark II*.*

*§ 13549. Bolt, locking: The sharp edges of the locking bolt have been removed to prevent damage to clothing and equipment, except in the case of the converted Mark II***, which was already so altered before conversion from Mark II*.*

*§ 13577. Screws, swivel: To prevent the swivel screws working loose, they are bored longitudinally at the screwed end to enable the end of the screw, when screwed home, to be expanded by the armourer, with the centre punch issued as part of the foresight adjusting tools, except in the case of the converted Mark II***, which was already so altered before conversion from Mark II*.*

§ 13724. Guide, charger, No. 2, and screw: The No. 2 charger guide and screw have been fitted. [This No. 2 pattern differs from the No. 1 in that it has a larger and stronger screw and is only interchangeable with the No. 1 pattern when fitted with the No. 2 screw.]

§ 14070. Nosecap and stock, fore-end: The nosecap and stock, fore-end have been numbered to correspond with the barrel number.

Head, breech, bolt: The edge of the front face of the bolt-head has been rounded in accordance with the instructions given in § 14710.

All spare parts in store at Naval Ordnance depots are being altered as affected by the above-mentioned modifications. When the conversion described above has been carried out, the marking on the body and butt of the rifles will be amended accordingly.

Regarding the markings which indicate that the conversion was effected, it only consisted of the additional stars or asterisks being stamped on the action body; in the case of the Mark I** they are found on the right side of the body while those on the Converted Marks II** and II*** are located on the left side because the original Mk II series conversion markings were stamped on the left side of the butt socket. Naval issue rifles were usually marked with an "N" on the left side of the butt socket and as most of these rifles were intended for Naval service, the "N" marking is usually encountered on these particular conversions.

In 1912, most of these rifles were fitted with charger-loading bridges and had certain modifications effected at R.S.A.F. Enfield so as to bring them more in line with the S.M.L.E. Mark III, along with modified, drawn-out nosecap sight protectors which make them quite distinctive. Details of this modification will be found in the following pages, near particulars of the changes to sights for the .303 Mk VII High Velocity ammunition.

In 1908 or 1909, the shape of the charger bridges on new production Mark III and Converted Mark IV rifles was altered slightly; the front face on the right-hand side was more deeply bevelled. This was not specifically mentioned in the List of Changes or Army Circulars as it was an internal manufacturing amendment.

Other minor changes applied to the Mark III and Converted Mark IV rifles were the introduction of the magazine cut-off Mark III and the lightened spring for the fore-end stud. The new cut-off was approved on 5th March 1909 and it was also applied to other short rifles and the charger-loading long rifles. Difficulties had been experienced with excessive wear and improper function of the stop nibs, so the rear stop nibs were strengthened and the inner and outer stop nibs had their angles altered. The weaker fore-end stud spring was required as a vertical "stringing out" of shots was found to occur due to an overstrong spring.

LoC § 14639. Increased clearance in magazine case at A to allow for the new magazine cut-off.

Yet another change implemented in manufacture during this period was the omission of the Pin, stop, outer band, which was approved on 10th November 1909. This thin, transverse pin had been inserted just forward of the lower band, since the time of the Martini, so as to prevent the band moving forward under firing recoil.

Some small changes were also effected to the rearsight of the Mark III and Converted Mark IV rifles in late 1909. The first comprises merely of the fitting of washers to take up the backlash of the windgauge screw, announced in the List of Changes, § 14937. The next alteration was a change in the manufacture of the fine adjustment screw and windgauge head; the List of Changes announcement reads:

Para 15077:–

Rifles, Short, M.L.E. Mark III, Converted Mark IV, and Short, .22-inch, R.F. Mark I.

1. Alteration of Worm, fine adjustment, backsight.
2. Fitting of larger Head, screw, windgauge, backsight.

1. In future manufacture the thumb-nail notches in the worm, fine-adjustment, backsight, of the above-mentioned rifles will be "V"-shaped instead of rectangular, and will extend to three-fourths of the depth of the thread of the worm instead of to the bottom of the thread.

2. In future manufacture the head, screw, windgauge, backsight, will be .095 inch larger in diameter and the fixing pin will be correspondingly longer. The larger head and longer pin will be described as "No. 2."

When fitting the No. 2 head and fixing pin to backsights with slides of early manufacture the following instructions will be observed by armourers:

A radial groove is to be filed in the right rear side of the slide, at "A," to give clearance to the No. 2 head. The depth to which the groove is to be filed will be determined by trying the slide on the sight leaf (assembled with wind-gauge and screw with the No. 2 head) until it can be set just below the 200 yards position.

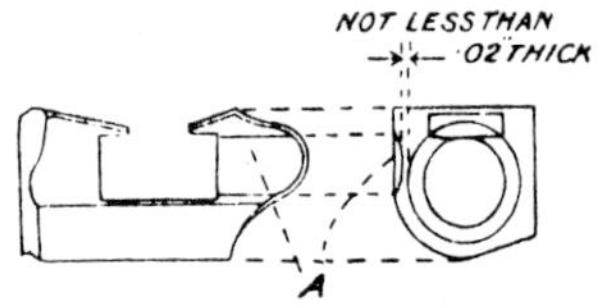

After adjusting the slide to suit the No. 2 head, it will be blacked in accordance with the instructions given on page 52, "Instructions for Armourers, 1904," for the temporary blacking of sights.

This change in the fine adjustment screw and wind-gauge screw head was similarly applied to the .22 Short Rifle Mark I as indicated in the LoC announcement. Although this .22 rimfire trainer bears little resemblance to the Mark III rifle, the sighting is intended to be a duplication and so the Mark III rifle backsight is fitted, necessitating these modifications.

The next series of modifications occurred as a result of the introduction of an improved service cartridge, the Mark VII ball on 3rd November 1910. The main differences between the Mk VI and VII are:

	Mark VI	Mark VII
Bullet:	215 gn round nose	174 gn pointed
Charge:	30 – 32.5 gn Cordite	37 gn M.D.T. (Modified Cordite)
Muzzle Velocity:	2,060 fps.	2,440 fps.

The higher velocity and lighter projectile of the Mk VII round produced a flatter trajectory and so all service rifles required sight recalibration. To hold and feed the differently shaped projectile also required minor changes to the magazine case and lips.

The first detailed official announcement of the changes required due to the new cartridge were made in the List of Changes issued on 1st October 1911, having been approved on 6th July 1911. This involved changes in new production rifles as well as the conversion of arms in store and in the hands of troops. The announcement reads:

Para 15638:–

Rifles, Short, M.L.E.-
Marks I, I and III.*
Converted Marks II, II and IV.*

1. Rifles in hands of troops and in store will be altered to suit Mark VII .303-inch cartridges.

Rifles, Short, M.L.E.-
Marks III and Converted Mark IV.

2. Future manufacture and conversion to be made to suit Mark VII .303-inch cartridges.

Consequent on the adoption of the "Cartridge. S.A., Ball, .303-inch, Mark VII" (§ 15629), the following alterations to components of rifles of the above-mentioned patterns in the hands of troop and in store, and the following change in pattern of components in rifles of future manufacture and conversion have been made.

1. Rifles in the hands of troops and in store.

Bed and spring, backsight: Ramps reduced to give lower elevations, and spring shortened and bevelled at rear end to clear slide in lowest position. Existing springs held as spare, in charge of units or in store will be similarly shortened and bevelled when being fitted to rifles sighted for Mark VII ammunition.

Screw, spring, sight, back: In Marks I and I and converted Marks II and II* rifles a new screw (No. 2) for the backsight spring with shorter head to clear the groove in the slide, is fitted.*

Slide, backsight; Marks I and I and converted Marks II and II* rifles: The slide is reversed on the sight leaf. It is altered to suit the reduced ramps of the backsight bed by bevelling the rear of the underside to form a new ramp seating; recesses are cut right and left of the rear of the slide to form clearances for the knuckles of the sight bed joint when the slide is set at extreme elevations, and a groove is cut through the underside to clear the reduced head of the spring. It is marked "2" on the original ramp seating.*

Mark III and Converted Mark IV rifles: Ramp seating is bevelled at front to clear ramps of sight bed and groove cut on underside to clear head of spring screw. It is marked "2" on the underside.

Barleycorns and blades, foresight: Two barleycorns, foresight, .015-inch and .03-inch lower than the existing lowest height, and three blades, foresight, .015-inch, .03-inch and .045-inch lower than the existing lowest height, have been introduced for use as required.

The barleycorns, foresight, are marked on the top left side with the figures "9" and "915" while the blades, foresight are marked "-06," "-045" and "-03," representing respectively .9-inch and .915-inch, .94-inch, .955-inch and .97-inch in height from the axis of the barrel.

The .99-inch barleycorn, foresight, and the 1.045-inch blade, foresight will not be used in rifles sighted for Mark VII ammunition.

Dial sight: The existing graduations are milled off, and the plate re-graduated, except where the dial plate has already been rendered too thin by previous conversion, when a new plate is fitted. The dial plate is marked "L.E.S. 2" on the face.

Cases, magazine: The No. 1 and 2 cases have been altered by removing a portion of, and riveting a lip to, the front left side, and by fitting a spring on the back rib to hold the front end of the magazine up to position. Cases of either pattern so altered are marked with the figure "3" on the back rib.

Platform, magazine: The platform is reduced on the front left side to allow it to clear the lip when stripping and assembling. It is marked with the figure "2" on the top.

Springs, auxiliary, platform magazine: The set of the No. 1 and 2 springs is altered to improve the feeding-up of the cartridges from the magazine to the chamber. Springs of either pattern so altered are marked with the figure "3."

Body: Clearance is cut in the magazine way for the magazine lip, and the cartridge lead is polished and prominent corners rounded.

2. Rifles of future manufacture and conversion.

Sights: The sights are as in 1.

Case, magazine: A case designed to suit the Mark VII cartridge is fitted. It differs from the altered case described in 1, in having a fixed lip, instead of a stop clip on the right front side, and a fixed lip on the left front side, both lips being formed integral with the case. It is marked with the figure "4" on the back rib.

Platform magazine: The platform is without a bottom plate, and is narrower at the front end than previous patterns to facilitate stripping and assembling. It is marked with the figure "3" on top.

Spring, platform, magazine: The platform spring is attached to the platform by two rivets. It is marked with the figure "2" on the top bend.

Spring, auxiliary, platform, magazine: The auxiliary spring of new pattern is marked with the figure "4."

Body: The body is as described at 1, but with the sharp edges of the cartridge leads removed.

Marking of rifles: 1 and 2. Rifles fitted suitable for Mark VII cartridges are marked "H.V." on the barrel, immediately behind the backsight.

List of components: The following is a list of the altered and new components, showing their interchangeability:

COMPONENT	DETAILS
Barleycorns, foresight, M.L.E. rifle, short, Mark I:	And I*, and converted Marks II and II*.
.9-inch & .915-inch.	For rifles sighted for Mark VII ammunition.
#Barrels with body, M.L.E. rifle, short:	Sighted for Mark VII ammunition.
Mark I and I*.	With sights assembled (with .945-inch barleycorn foresight), inner band, and pins, stop, bolt, locking.
Converted, Mark II.	As above.
Mark III.	With sights assembled (with 1.0-inch blade, foresight), inner band, and pins, stop, bolt, locking.
Blades, foresight, M.L.E. rifle, short, Mark III and IV: .94-inch, .955-inch and .97-inch.	For rifles sighted for Mark VII ammunition.
Cases, magazine, M.L.E. rifle, short, Mark I: No. 3.	With stop clip and stud, lip, and rib spring; converted from Nos. 1 and 2. For all rifles, S.M.L.E., sighted for Mark VII ammunition.
No. 4.	For all rifles, S.M.L.E., sighted for Mark VII ammunition.
Lips, magazine, M.L.E. rifle, short, Mark I.	For No. 3 magazine case. Not issued separately.
Plates, sight, dial, M.L.E. rifle, short, Mark III, No. 2.	All rifles, short, M.L.E. sighted for Mark VII ammunition.
Platforms, magazine, M.L.E. rifle, short, Mark I: No. 2.	For No. 3 magazine case; converted from No. 1.
No. 3.	With No. 2 platform spring; for No. 4 magazine case.
Rivets, lip, magazine, M.L.E. rifle, short, Mark I.	And for rivets, spring, platform, magazine; for No. 3 magazine case and No. 2 platform spring. Not issued separately.
Rivets, spring, rib, magazine, M.L.E. rifle, short, Mark I.	For Nos. 3 and 4 magazine cases. Not issued separately.
Screws, spring, sight, back, M.L.E. rifle, short, Mark I, No. 2.	And I* and converted Marks II and II*. For rifles sighted for Mark VII ammunition. Suitable to replace the No. 1 screw in other rifles.
Slides, backsight, M.L.E. rifle short: +Mark I, No. 2.	And I* and converted Marks II and II*. For rifles sighted for either Mark VI or Mark VII ammunition.
Mark III, No. 2.	And IV. For rifles sighted for Mark VII ammunition.
Springs, auxiliary, platform, magazine, M.L.E. rifle, short, Mark I: No. 3.	For No. 3 magazine case; converted from Nos. 1 & 2.
No. 4.	For No. 4 magazine case.

COMPONENT	*DETAILS*
Springs, platform, magazine, M.L.E. rifle, short, Mark I, No. 2.	*For No. 3 magazine platform. Not issued separately.*
Springs, rib, magazine, M.L.E. rifle, short, Mark I.	*For Nos. 3 and 4 magazine cases. Not issued separately.*
Springs, sight, back, M.L.E. rifle, short:	*Existing patterns altered to suit rifles sighted for Mark VII ammunition; also suitable when altered for rifles not so sighted.*
Mark I.	*And I*, and converted Marks II and II*.*
Mark III.	*And IV; also suitable for all other marks.*

 # The Marks I and III barrels with body are suitable for converted rifles (the Mark I for converted Marks II and II, and the Mark III for converted Mark IV) if the Springs and Screws, sight aperture, of the converted rifles are exchanged for those of short rifle pattern.*

+ When used for rifles sighted for Mark VI ammunition, this slide will not be placed in the reversed position.

Rifles in Naval Service will not be altered until the Mark VI ammunition is available.

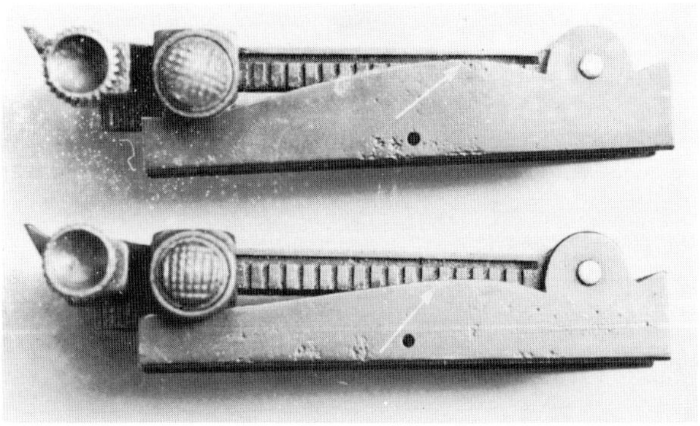

Modified sight bed (lower) for the Mk VII ammunition. Note the difference in ramp profiles. *John Beards*

Sighting changes made to the various rifles for higher velocity Mk VII ammunition involved the reduction in height of the backsight bed ramp as the flatter trajectory did not require such a high curve. On close examination the difference in height of the sight bed ramps can readily determine whether it was intended for Mk VI or VII ammunition. The marking of the dial sight plate, "L.E.S. 2" instead of "L.E.S." and the "H.V." mark on the barrel makes this conversion relatively easy to confirm. Even towards the end of the Great War in 1917, records indicate that sizeable quantities of the S.M.L.E. and C.L.L.E. rifles were being re-sighted for Mk VII ammunition at the Ordnance Factory.

Obviously it was some time before the new ammunition became available in sufficient quantity to meet all service requirements and so it was the Navy that had to wait until full and adequate supplies became obtainable. The minor changes made to the receiver body were because of the new pointed projectile which sometimes caused the second last round in the magazine to feed high and the sharp bullet point often became caught on the lip and did not feed into the chamber.

The approval for alteration of the Mark I Charger Loading Lee-Enfield for Mk VII ammunition was not given until 1st September 1914, and for the Mk I* Charger Loading Lee-Enfield it was 2nd October 1914; most likely due to the rifle shortages at the beginning of the Great War. Those modifications made to rifle bodies for the new ammunition were carried out by R.S.A.F. Enfield, B.S.A. Co. and Westley Richards. Details of the conversion applied to the Charger-Loading rifles will be found near the end of Chapter 4.

The next rifles in line for upgrading were the naval S.M.L.E. Mark I**, Converted Mark II** and Converted Mark II***. On 29th May and 12th August 1912, approval was given for the pattern at the Naval Ordnance Depot, Portsmouth, to be resealed in view of the following alterations which were to be carried out at Naval Ordnance Depots or the Small Arms Factory at Enfield.

These alterations comprised of:

- Removing the charger guide on the bolthead and left side of the body;
- Fitting a charger bridge guide onto the body;
- Straightening the incurving wings of the nosecap;
- Cutting a clearance in the fore-end for charger bridge;
- Recessing the fore-end for the fore-end stud and spring, located underneath the nosecap.

Because all of the Mark I**, Converted Mark II** and Converted Mark II*** rifles were to be upgraded, no change in their nomenclature was to be made.

Not very long after the Mark VII ammunition was issued, it was complained that the sharper recoil caused certain screws on the rifle to loosen; namely the nosecap screw, sear screw and front trigger guard screw. On the Marks I and I* and Converted II and II* rifles, the backsight bed screws were also found to be liable to loosen and on the Mark III rifle, the sight spring screw was similarly affected. As a result, with an approval date of 27th March 1912, the screws were ordered to be tightened up and the metal around the screw heads centre-punched so as to prevent further loosening up. This was to be carried out on rifles in future production as well as those arms in store and in the hands of troops.

Centre-punching was effected as per the accompanying drawings:

GUARD TRIGGER AND SCREW GUARD FRONT

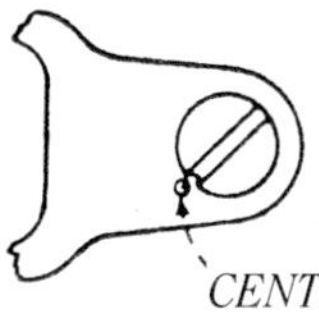

CENTRE PUNCHED

SPRING RETAINING AND SCREW SEAR

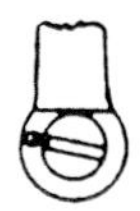

SIGHT BED AND SCREW

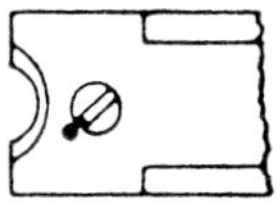

A relatively minor alteration was approved soon afterwards which none the less altered the external appearance of rifles in service with certain military units; the removal of the piling swivel on the nosecap of rifles carried in saddle buckets or in covers on limbers. The securing screw was put back into the nosecap.

The ensuing S.M.L.E. conversion introduced was the Mark I*** for Land Service and it consists of a S.M.L.E. Mark I* modified for the Mk VII ammunition. Approved on 22nd April 1914, a windgauge with "U" sighting notch was fitted, as well as a blade foresight in place of the barleycorn type. The windgauge is the same as that fitted to rifles in Naval Service and the foresight blade is the pattern for the Mark III S.M.L.E. for Mark VII ammunition. Two additional asterisks or stars were stamped after the original "I*" designation on the right side of the receiver at the butt socket. As these were done by hand with individual stamps, they usually don't line up neatly.

The Mark I*** Short Magazine Lee-Enfield was the last of the factory conversions approved for the S.M.L.E. Although some Mk III rifles were later upgraded to Mk III* pattern, it was not a specific conversion and not always effected in the same way; many such upgraded rifles were not marked with the additional "*" after the original designation as it was often performed by armourers.

The preceding sequence and form of the Converted rifles may seem complex and confusing at first but it follows quite an ordered progression. A table is provided in order that these might be more readily understood and recognised.

Rifle, SMLE	*Converted from:*	*Upgraded to:*
Mk I**.	SMLE Mk I.	Some features of SMLE Mk III.
Mk I***.	SMLE Mk I* & Mk I**.	Mk VII ammunition sighting.
Cond Mk I. ...	MLM Mk I*.	SMLE Mk I.
Cond Mk II. ...	MLM Mks II & II* & MLE Mks I & I*.	SMLE Mk I.
Cond Mk II*. ...	MLM Mks II & II* & MLE Mks I & I*.	SMLE Mk I*.
Cond Mk II**. ...	SMLE Cond Mk II	Some features of SMLE Mk III.
Cond Mk II***. ...	SMLE Cond Mk II*	Some features of SMLE Mk III.
Cond Mk IV. ...	MLM Mks II & II*, & MLE Mks I & I*.	SMLE Mk III.

The conversion markings were made on the left side of the receiver butt socket (except for Marks I** & I***) and these rifles will be noted to have the original rifle designation, factory and production year on the right side of the butt socket, with the appropriate conversion designation, factory and conversion year on the left side. The exceptions, Marks I** and I*** have no specific conversion markings except for the additional stars or asterisks stamped after the original mark designation on the right side of the butt socket.

Some S.M.L.E. Mark I**, Converted Mark II** and Converted Mark II*** rifles were transferred from Naval to Land Service and other alterations approved on 18th August 1915. These were announced in the List of Changes:

Para 17464:-
Rifles, Short, Magazine, Lee-Enfield-
*Mark I***
*Converted Mark II***
*Converted Mark II***.*

1. *Change of service letter on transfer of a number to Land Service.*
2. *Alteration to suit Mark VII .303-inch cartridges.*
3. *Some Mark II** and II*** not fitted with the charger bridge guide.*

1. A number of the above-mentioned rifles (§§ 14936, 16091) having been transferred to Land Service, the service letter will be as now shown. [This does not apply to any markings on the rifle.]

2. These rifles are being altered to suit Mark VII .303-inch cartridges in accordance with § 15638, as for Marks I and I, and Converted Marks II and II*, the special windgauge (now common to Mark I***, § 16910) and blade, foresight, being retained.*

*3. A few Marks II** and II*** are not fitted with the bridge charger guide, and differ from the remainder in being fitted with:*
Bodies, M.L.E.R.S., Converted, Mark II.
Gauges, charger, head, breech bolt, M.L.E.R.S., Mark I, No. 2.
Heads, breech bolt, M.L.E.R.S., Mark I.
Screws, stop, charger guide, M.L.E.R.S., Mark I, No. 2.

*(Note- Mark I** rifles, without charger bridge guides, on being altered to suit Mark VII .303-inch cartridges, become Mark I***, § 16910.)*

This publication is official acknowledgment of probably the only significant muddle in a series of otherwise orderly conversions. All of the Converted Mark II** and Mark II*** rifles were supposed to have been modified with the charger bridge but it is obvious that some were not so modified and so these rifles could be encountered with the form of charger-loading provision of either the

S.M.L.E. Mark I or Mark III series. However, the nomenclature remained the same, whether fitted with the bolthead charger guide or bridge charger guide. This complication occurred because of the Naval conversions being effected outside of the usual R.S.A.F. Ordnance parameters.

Another similar anomaly which may be encountered is the Mark I*** S.M.L.E. fitted with charger bridge guide; according to the official documentation, the Mark I*** retains the earlier type of charger-loading facility, yet specimens of the Mark I*** rifle have been noted with the charger bridge. Whether this was done as a wartime contingency or whether some Mark I*** rifles were included with batches of the Converted Marks II** and II*** rifles when they were so fitted, is not certain.

Although it is often accepted that the Short Magazine Lee-Enfield is inferior to the Mauser system in some respects, it is still one of the most "soldier-proof" bolt action rifles ever designed. It showed up better in reliability and adverse condition trials and is renowned for the operating speed of the bolt, especially in well-trained hands. In 1912, some trials were conducted at Hythe against the German service rifle and it was found that about 14 or 15 rounds per minute could be fired from the German rifle, compared with 28 for the S.M.L.E.

When the Imperial German forces came up against the cream of Britain's infantry on the Western Front in 1914, in some cases they attributed the large volume of fire directed against them to a high issue scale of light machine guns amongst their British counterparts; in reality it was only the proficient manipulation of the S.M.L.E. rifle bolt.

The question of a new rifle had been raised in 1910 and the Small Arms Committee was directed to investigate the matter, deciding later that same year on certain features for the proposed new rifle. This eventually became the .276 Pattern 1913 trials rifle and was eventually brought into service as the .303 Pattern 1914. Had it not been for the outbreak of World War 1, the history and lineage of the Lee-Enfield rifle would have been quite different.

Requirements for such a new rifle included:
- Front locking bolt action,
- A rimless cartridge,
- Aperture sights,
- One-piece stock,
- and heavier barrel.

Whilst these were departures from the S.M.L.E. configuration, the Committee still wished to retain the following S.M.L.E. features:
- A similar weight, length and recoil,
- Full length handguard,
- Charger loading,
- and the 10-round magazine.

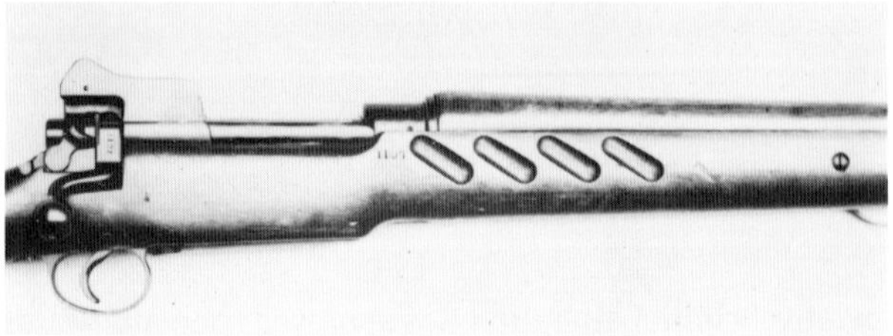

Pattern 1913 Trial Rifle. *MOD Pattern Room*

Development of the Pattern 1913, Pattern 1914 and U.S. Model 1917 rifles is described in Skennerton's "The U.S. Enfield." However, a summary is provided here due to its relevance to this part of the Lee-Enfield story.

In excess of a thousand .276 trials rifles were produced at Enfield during late 1912 and early 1913 for issue at home and abroad; these are known as the Pattern 1913. Problems were encountered with extraction, excessive muzzle blast, heat and higher pressures, mostly due to the high-velocity cartridge. Other considered failings were the smaller capacity magazine, weak front handguard and loose safety catch lever.

A few Improved Models were built up as a result of the various criticisms but the outbreak of the Great War resulted in the suspension of further development. Production of this rifle was eventually commenced in 1916 in the United States, in .303-inch calibre, as the Pattern 1914. In the United States, the P'14 and M'17 are often referred to as the "Enfield model."

The Pattern 1914 rifle was manufactured under contract by Winchester and the two Remington factories at Ilion and Eddystone, so three different manufacture markings may be encountered. Upon completion of the British contracts, the design was modified a little and produced in .30-06 calibre as the U.S. Model of 1917.

The beginning of the Great War changed the course of the Pattern 1913 rifle development; so the future of the Lee-Enfield rifle was also altered because it is probable that the Mark III S.M.L.E. would have been replaced in service by the new .276-inch rifle. Instead, the service life of the S.M.L.E. was about to begin in earnest, to prove its reliability in many theatres of war.

Dummy S.M.L.E. rifles were assembled for training purposes, because of the wartime shortages. A number of various patterns were utilised for this purpose, many made locally, so some variation may be encountered depending on when and where they were produced. Similar style dummy rifles were made during the 2nd World War, and also in Canada, Australia, India and other countries of the Empire and the following Commonwealth. Some are merely wooden silhouettes of the original rifle, other variations may have additional original style fittings.

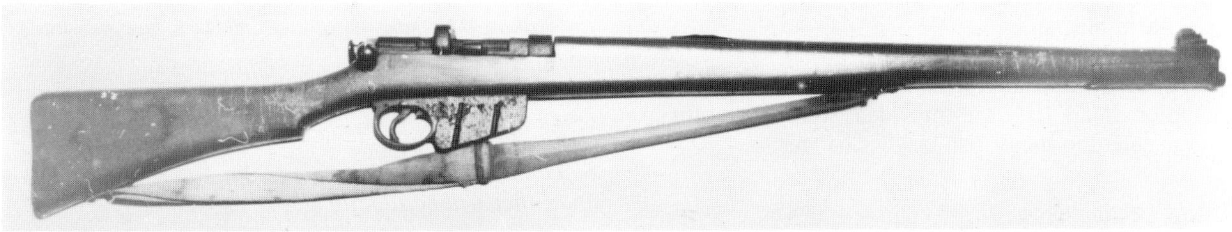

Dummy SMLE Trainer, with one-piece cast metal action, trigger and magazine. *School of Infantry Museum*

The example illustrated here has a London maker's stamp on the butt, "H.H. SUMMERS, 23 CHEAPSIDE, E.C.." It has some metal components and sling fittings, permitting more practice other than just parade-ground drill. Metal casting has produced a real-size bolt handle and the magazine cut-off is also worthy of note; other external details of the action are quite realistic. It is recorded as being used for the preliminary training of the huge new armies hastily raised in 1914-15.

Many conversions to "D.P." or "Drill Purpose" were effected during the war, generally using downgraded or reject rifles and components. Actions of these rifles could still be manipulated and operated but were not intended for the firing of ball ammunition. Many such conversions were also applied to earlier patterns and obsolete models; by mid-1917, some 173,975 M.L.E. rifles had been done at the Ordnance Factories and Depots. The D.P. conversion usually only involved the stamping of these letters on the action and various other parts after that particular rifle had been downgraded.

The use of a particular model of rifle in any major conflict extending over a few years usually results in a number of improvements and the introduction of various fittings, accessories and specialist devices; the S.M.L.E. had more than a fair share. These include grenade dischargers, clinometers, wirebreakers and cutters, optical and telescopic sights, large capacity magazines, night sights and remote firing devices and other such wartime combat requirements.

Sundry economies were introduced into the manufacturing procedure and the following deletions were recommended to be applied to the sealed pattern of the Mark III S.M.L.E.:

Lightening hole in the nosecap boss,
Ejector screw and its hole in the action body, moving the ejector recess shoulder forward slightly,
Lightening recess in the backsight protector,
Lightening groove in the triggerguard, and
Lightening holes in the butt.

The above alterations were calculated to increase the weight by nearly 2 ozs. but with a saving of 9 men and 10 machines per rifle in manufacture. It was also pointed out that the omission of the long range volley sights and the windgauge would save 1.5 ozs., which would make the nett all-up weight about the same. Such a rifle with these "economies" was forwarded to the School of Musketry for a report and the resulting recommendation was that the modifications be incorporated into production. It should be pointed out that there was consideration of omission of the magazine cut-off at this stage also.

The omission of the magazine cut-off in the Mark III rifle was first instituted by B.S.A. in September of 1915, when certain sections at their factory were being reorganised. Although permission was received from the War Office for this alteration to the specifications, the Inspectorate and some other departments were not advised and the B.S.A. management was strongly censured for not proceeding through the proper channels. In fact, they were first ordered to revert to the original pattern or else the rifles would be failed on inspection, but the omission of the magazine cut-off was approved soon afterwards, although it was only as a temporary proposal at first.

The magazine cut-off was still considered a requirement for mounted forces and certain other units and it was re-introduced to a limited extent after the war was over. Post-war development of the S.M.L.E., the Marks V and VI, incorporate the cut-off, and even the trials No. 4 Mark I in 1933 was so furnished.

The Mark III* Short Magazine Lee-Enfield was officially approved on 2nd January 1916, but while the Mark III* usually has no provision for the magazine cut-off, the omission of the cut-off was not necessarily intended to be a feature of the Mark III*. The Mk III* model was produced in Australia and India during the 2nd World War until as late as the 1950s'; in Britain the Mk III* represents the major production for the latter part of the

Great War as well as a sizeable percentage of the World War 2 rifle numbers.

Features of these changes to the Mark III and the new production Mark III* are provided from the official announcement in the British List of Changes:-

Para 17622:-

Rifle, Short, M.L.E., Mark III.

1. Modifications.

Rifle, Short, M.L.E., Mark III, without cut-off.*

2. Introduction.

1. In future manufacture, rifles of the above-mentioned pattern may embody any of the following modifications:

(a) The omission of long range sights (dial and aperture) from stock, fore-end and body. In this case the stock, fore-end, being without recesses for the dial sight and screw and the aperture sight and a washer being employed to replace the base of the aperture sight as a seating for the active end of the aperture sight spring.

(b) The wind-gauge on the back sight leaf being replaced by a fixed cap held to the leaf by a screw.

(c) The lug on the striker collar being omitted, and grooves being cut through the collar to enable the Tool, removing striker, pattern B, to be used for the removal of the striker from, and its replacement in, the bolt.

(d) The swivel lugs being omitted from the underside of the front end of the guard, and a wire loop being fitted on the left side of that position for the attachment of the breech cover.

2. A pattern of this rifle has been approved to govern manufacture of the rifles required without cut-offs.

It differs from the Mark III rifle in that the body is not slotted and drilled to receive the cut-off and screw, but, like the Mark III, may embody the modifications mentioned at 1 above.

The following is a list of the new components necessitated by the changes described above:

COMPONENT	*DETAIL*
Barrels, with body, M.L.E. rifle, short, Mark III.*	*With sights assembled (with 1.0-in. blade, foresight), inner band and pins, stop, bolt, locking; sighted for Mark VII ammunition.*
Bodies, M.L.E., rifle, short, Mark III.*	*With pins, stop, bolt, locking.*
Caps, backsight, M.L.E. rifle, short, Mark III.	*And III*; for No. 2 leaf.*
Leaves, backsight, M.L.E. rifle, short, Mark III, No. 2.	*And III*; assembled with cap and screw.*
Screws, cap, backsight, M.L.E. rifle, short, Mark III.	*And III*; for No. 2 leaf.*
Stocks, fore-end, M.L.E. rifle, short, Mark III, No. 2.	*And III*; not recessed for dial sight.*
Strikers, M.L.E. rifle, short, Mark I, No. 2.	*All M.L.E. short rifles.*
Washer, spring, sight, aperture, M.L.E. rifle, short, Mark III.	*For III and III* rifles without aperture sight.*

When fitted with stock, fore-end, No. 2 to rifles with cut-off fitted, they will be slightly altered as follows:

The upward extension, on the right side of the stock, which covers the position of the cut-off slot in the body, will be cut away, and an angular recess will be cut to half the depth of the cut-off log recess to clear the cut-off joint when the cut-off is opened.

When fitting these stocks to rifles fitted with aperture sights, the aperture sight will be replaced by the Washer, spring, sight, aperture, but should this component not be available, the pillar of the aperture sight will be filed off at the base, and the base disc used as a washer under the spring.

Considerable numbers of the Mark III rifles were later upgraded to Mark III* pattern by alteration of some or all of these features. Some were factory conversions in which a new fore-end was fitted and the windage adjustment and magazine cut-off removed, in the course of which the body was usually stamped with a "*" after the Mark III designation on the receiver markings; other conversions merely consisted of the particular items being removed when the rifle was sent to armourers for repair. In the case of unserviceable parts on the long range sight, the pointer was removed and the stud on the plate ground off so that the bare dial plate remained in the fore-end to plug the hole.

The cut-off still seems to have been a requirement for some forces, such as the Navy, because some 1917 Enfield Mk III* rifles have been noted with cut-off slots and the "*" suffix cancelled, with the cut-off plates added again.

As a result of the abolition of the long range sights, a similar alteration was approved for earlier model S.M.L.E. rifles and Charger Loading Lee-Enfields, on 2nd May 1916, in that it was instructed that particular unserviceable components not be replaced. The following was ordered:

1. If either the bead, the pointer, the spring, or the pivot screw of the dial sight becomes unserviceable, the whole of these components will be removed and returned to store, the dial plate and fixing screw being left on the rifle for the purpose of covering the hole in the fore-end. The aperture sight will be left on the rifle.

2. If either the dial plate or the fixing screws becomes unserviceable, the whole of the dial sight components will be removed and returned to store; the hole in the fore-end being plugged.

3. If the aperture sight becomes unserviceable, the pillar to be filed off at the base, and the base disc used as a washer

British Service Type 30 (top) and Type 38 Arisaka Rifles. *I. Skennerton*

under the spring. In the case of deficiency the Washer, spring, sight, aperture, M.L.E.R.S., Mark III (§ 17622) may be employed to replace the aperture sight.

The desperate shortage of Britain's service rifles resulted in the acquisition and subsequent introduction into service of some rifles of her allies. List of Changes § 17213, approved on 24th February 1915, announced the "Rifle, magazine, .256-inch, Patterns 1900 and 1907" and the "Carbine, magazine, .256-inch, Pattern 1907" along with the bayonets and appurtenances. This was the Japanese Arisaka rifle and carbine, of which 150,000 were brought in during 1914-15 and initially introduced for Land Service. On 15th June 1915, their use was extended to the Navy. The Pattern 1900 and 1907 are better known as the Arisaka Types 30 and 38 and those used in British service were stamped with typical British issue markings, usually on the butt or upper tang of the action. These rifles generally have a deep blued finish on the metal.

The British service designations are a curious anomaly; the Japanese designations are Meiji 30 and Meiji 38 for the rifles and Meiji 38 for the carbine; the conversions of these years to the western calendar works out at 1897 and 1905 respectively as the Meiji calendar commenced with the year one in 1868.

The rifle of another ally, Canada, was approved for British service on 21st October 1915. This was the Mark III Ross, with which many of the Canadians had been armed when they went into the trenches in Europe, to find that the complicated mechanism and fine machining was extremely liable to stoppages, largely due to the mud and some unsuitable (non-Canadian) ammunition. Under such circumstances, of course, the S.M.L.E. had few peers.

Two models of the Ross were used officially in the British Service, the Mark IIIA and IIIB. Rifles were ordered direct from the Ross Rifle factory in Quebec and the contracts were not cancelled by the British government until March 1917 by which time the Pattern 1914 rifle was in plentiful supply from the United States. The figure of Ross rifles supplied by Canada is a little confusing; a total order of 118,000 has been mentioned although contracts appear to have been for the supply of 100,000 and only 66,590 are recorded as being delivered by mid-1916.

The Ross and Arisaka rifles were not front line issues, but relegated to training, guard duty, home defence and some Naval service. Except for its sniping function, the Pattern 1914, too, was relegated to a 2nd line role. Many other arms were used, virtually whatever was available in quantity. The Navy used some .44/40 Winchester carbines and the introduction of .44/40 ball and dummy ammunition into service was recorded in the List of Changes, § 19506. This all served to release larger quantities of S.M.L.E. rifles for active service.

A more unusual rifle for British service was the .303 Bannerman, a U.S. hybrid. Francis Bannerman & Co. of 499 – 501 Broadway, New York City, had purchased reject "brittle" Springfield actions from the U.S. government and then modified, rebuilt and assembled rifles for sale on the surplus market. A contribution towards Britain's war effort, 1,000 rifles with bayonets, cartridge belts and accessories were provided; the patriotic Bannerman family originated from Scotland.

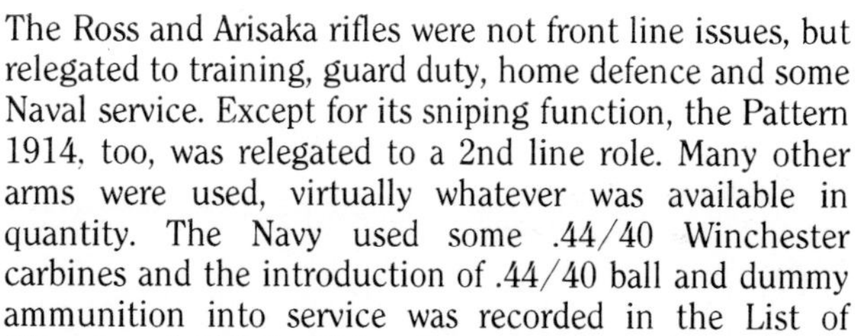

U.S. Bannerman action; note the name, logo and .303 designation on the receiver ring. *MOD Pattern Room*

.303 Bannerman Rifle, a conversion of the U.S. Springfield. *MOD Pattern Room*

The early Springfield actions had been re-heat treated by R. F. Sedgley & Co., Philadelphia, which rectified most of the action faults. The stocks are Springfield, the barrels, sights and trigger guards are from the Krag, handguards and followers from the Mauser, and the rifle finished off with a dummy magazine floorplate. They are marked on the receiver with the Bannerman crest and "BANNERMAN" in an arc. However, it was found that these rifles would not feed the rimmed .303 cartridge properly so they were reputedly all stamped "DP" and not even used for firing practice.

The new trench warfare in Europe also brought about some unusual applications of the S.M.L.E.; periscope rifles, rifle batteries, night sights and sniping rifles, to mention a few examples. The periscope rifle comprises a jig with special levers and mirrors which enable the rifle to be worked remotely so that the operator was not exposed to enemy fire whilst aiming and manipulating the rifle. Sighting was effected through the mirrors of an attached periscope.

In numerous cases the trenches of opposing forces were quite close and exposure above the parapet to fire at the enemy often proved fatal; it was quite common for periscopes to have the top lenses shot out. There were several unofficial variations of locally made periscope rifles and rifle batteries and many such ingenious ideas would have been the product of idle hours in an all-too familiar dug-out.

Rifle batteries were rifles set up in fixed positions so as to bring fire upon a certain point of the enemy defences, say, a damaged parapet, exposed communication trench or latrine. Thus, at any time, day or night, even if the target were obscured by fog, smoke or darkness, rifles could be fired merely by pulling a lever or line from safe cover.

Records indicate that some patterns were tested at Hythe in October 1915 but were not entirely satisfactory. A sample pattern by Platt Bros., Oldham, was successfully tested at Hythe in November 1915 and an initial order for 250 units placed with Platts. Another 250 were ordered from S.H. Heyward & Co. Ltd. in January 1916. The official pattern was approved on 27th April 1916, consisting of jigs fitted inside a box to house three S.M.L.E. rifles, onto which hardened steel plate could be installed for protection against enemy fire. Such a battery is described in the List of Changes, § 17750.

Other interesting Great War adaptions of the S.M.L.E. rifle include mounting on aircraft, although such applications were usually done locally and not necessarily according to any official or set-down procedure. A 1915 photograph of a 5 Sqn. Bristol Scout B shows a Mk III S.M.L.E. rifle stripped of its butt and secured to the plane's fuselage at such an angle that its firing forward would just miss the propellor blades in front. From the cockpit, the pilot could manipulate the bolt and trigger and roughly sight along the line of fire. No doubt there were other such applications by naval, air, tank and ground forces during this period, warranted by the new mode of warfare, tactics and armaments.

The night sights were quite widely used in the new trench warfare and one type officially introduced in late 1916 was designated "Sights, luminous, No. 1 Mk I." This comprises a fold-down leaf behind the foresight with a luminous bar to replace the blade, and a cap which screwed over the backsight with a luminous bar on either side. Daylight firing required only to fold down the moveable blade held in position by a spring. More technical details may be found in the List of Changes §§ 17920 and 19952. Similar types of night sights were commonly used on the various machine guns also.

J.E. Martin of Glasgow patented a form of luminous foresight for the Short Magazine Lee-Enfield rifle, pat. 23,433 (1914), accepted 3rd January 1916. This sight, however, is different to the ensuing military pattern as it positioned behind the foresight only and had no complementing luminous backsight.

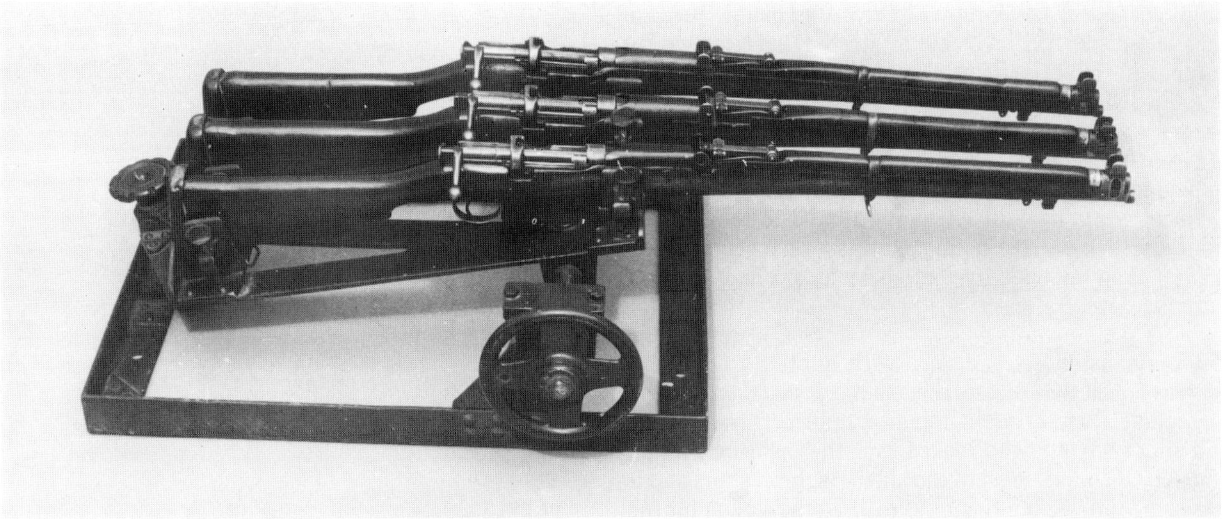

British Rifle Battery with three S.M.L.E. Mk III* rifles clamped in position. Armoured shields were often mounted in front. *Imperial War Museum*

LUMINOUS FORE-SIGHT FOR THE LEE-ENFIELD RIFLE.

23,433 (1914). J. E. Martin, Glasgow. The inventor claims: " A sighting device consisting of a plate provided with means for attachment to a rifle, the plate presenting a luminous screen of substantial U-shape, the base of the U being formed with lateral extensions serving to guide the vision of the marksman with respect to the rear sight and the gap between the limbs of the U serving to guide the vision of the marksman towards the object."

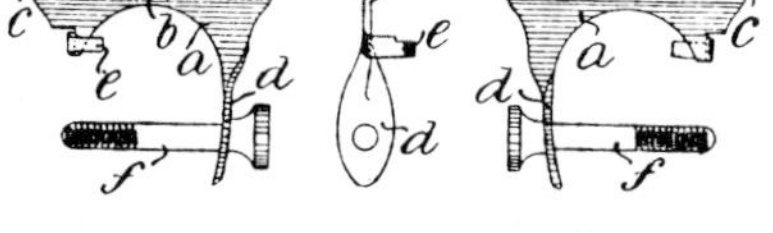

Pat. No. 23,423 of 1914. Accepted 3rd January 1916.

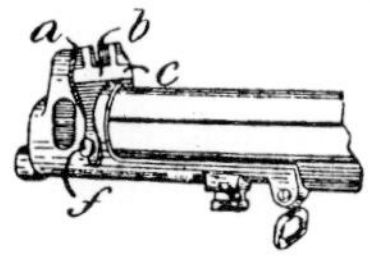

The sight illustrated is designed for the Short L.-E. British Service Pattern Rifle. The plate *a* is fitted preferably rearwardly of the fore-sight and it is, as shown, provided with a luminous screen *b* and the lateral extensions *c*. The gap between the upstanding luminous extensions constitutes the foresight. The plate *a* may be clipped on to the wings of the foresight protector but preferably the downwardly projecting limb *d* is arranged partly to surround the barrel whilst the tongue *e* engages with the foresight protector. The limb *d* receives a screw threaded pin *f* which takes the place of the ordinary nose cap screw. Accepted January 3, 1916.

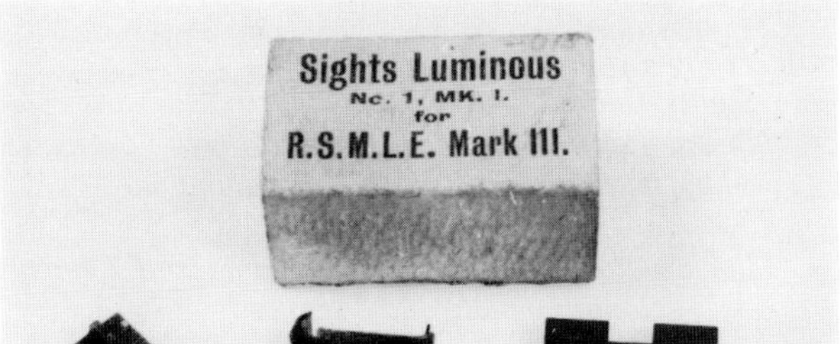

S.M.L.E. Luminous Sights. The spring (middle) fits around the foresight assembly. *MOD Pattern Room*

Until the Great War, mortars and trench howitzers utilised various mechanisms to ignite the bomb propellant charge; in some cases these were merely cut-down rifle actions like the the Mk III S.M.L.E. illustrated here, attached to a 2-in. Trench Howitzer Mk I. A special .303 blank cartridge was fired in the rifle action to initiate the mortar bomb propellant. Britain's introduction of the Stokes mortar markedly increased the rate of fire because the separate initiating cartridge and mechanism was no longer required. The initiating charge was inserted into the base of the mortar bomb and ignited by a pin inside the mortar, after the bomb was dropped down the muzzle.

Trench mortar with S.M.L.E. action initiator. This fires a special blank cartridge to ignite the mortar propellant charge. *School of Infantry Museum*

Sniping came of age during the Great War, mainly because the telescope had become a more reliable piece of equipment by this time. The progress of sniping was regulated by the development of optics and it was the Germans who first appeared to appreciate the possibilities of exerting control over an area by "picking off' individuals and restricting movement by this means. The Germans sent some 20,000 sniping units to the western front in 1914-15, consisting of Zeiss, Goerz and Luxor 2.75 or 3-power telescopes fitted by a variety of mounts onto the Gew 98 Mauser.

The Germans chose marksmen with hunting experience, such as Forest Guards, and the usual practice was for them to seclude themselves in ruins or upturned tree trunks and shoot anyone who looked over the opposing trench parapet. Of course, most of the kills were head shots and it was some time before the Allies realised the reason for the "sudden plague of head wounds." Although this loss may not have been enormous compared with other casualty figures in trench warfare, it severely restricted movement and lowered morale. The Lee-Enfield rifle with its standard sights was no match for these experts.

To answer this threat in the short term, many British units privately acquired sporting rifles, especially Mannlichers and Mausers fitted with telescopic sights as well as some captured German sniper rifles taken in trench raids. At first the role of the British sniper was to locate and neutralise his German counterpart but as the war progressed and better equipment became available, he was used more to minimise movement in certain areas or to locate and silence enemy machine gun positions. There were even instances of individual snipers putting artillery pieces out of action by picking off the crew as they manned their gun and towards the end of the war armour-piercing ammunition was issued to snipers for putting enemy machine guns and gun sights out of commission.

The early sporting style sniper rifles presented problems regarding ammunition supply and replacement of parts; by the summer of 1915, small numbers of S.M.L.E. rifles fitted with telescopic sights began to reach the front. One of the main objects of the telescopic sight was to make the target more distinct because many objects could not be distinguished with the naked eye but required viewing through binoculars or a telescope. If a rifle were brought to bear upon such a target as a loophole or observation post, the object could not be adequately surveyed over the conventional open sights.

Initially, Rifle Club-type optical sights were introduced; these generally comprise a lens to the left or in front of the foresight, lined up with a lens at the rear, thus effecting a simple Galilean telescope with no protective or focusing tube. Four main types were used, the Martin, Lattey, Gibbs and the Neill (which was also known as the Barnett or Ulster sight).

OPTICAL SIGHTS FOR RIFLES.

1,850 (1915). T. Caldwell, Belfast. The backsight clearing lens of this plain Galilean optical combination is fitted in a mounting which is attached to the side of the body of the Lee-Enfield rifle in the place ordinarily occupied by the long range aperture stem. The fore-sight carrying the magnifying lens is mounted on the side of the nose cap on a line laterally with the ordinary fore-sight. The backsight lens carrier may be adjusted so that the sights may be zeroed and a movable aperture is provided in front of the lens so that it may be quickly adjusted for shooting at either 200 or 500 yards.

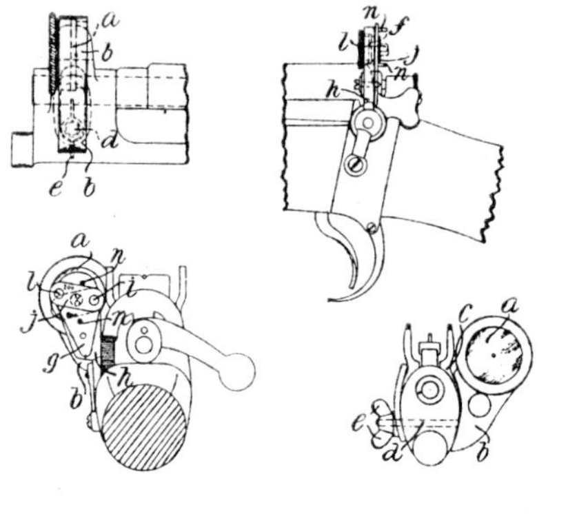

Patent No. 1850 of 1915.

The Martin sight was used with B.S.A. Co.'s aperture backsight which fits in place of the long range aperture sight. The Neill sight, too, consists of a lens mounted to the left side of the nosecap, with backsight that replaced the long range aperture, but it has limited lateral adjustment. Captain Lattey's was the most widely used and it utilises the original rifle foresight and backsight with lenses placed in front of them. The Gibbs sight is very well made and this design fixes the front lens at the rear of the foresight, with a red pointer painted on it. A small number of B.S.A. sights were also used.

Other such sights were purchased privately, including 1,400 of the Latteys. Government contracts were placed for about 7,000 Lattey sights, 4,000 Neill sights, 575 Martin sights and 80 B.S.A. patterns. The prices at the time, which had a considerable bearing on the relative quantities purchased, were:

Martin sight	about £3/-/-
Neill sight	about £1/15/-
Lattey sight	7/6d.
Gibbs sight	between £4 – £5.

Such optical sights were officially approved on 28th September 1915, in two variations; the Lattey and the Neill. The descriptions in the List of Changes are:–

Para 17556:–

Sights, Optical (Lattey), Mark I
Rifles, short, M.L.E., Mark III and Converted Mark IV, fitted with adjustable windgauge.

Back: with lens, lens retaining plate and 2 screws; clamping plate and 2 screws.

Front: with lens, lens retaining plate and 3 screws; clamping lug and screw.

Introduction.

Lattey sights fitted. *L. Walsh*

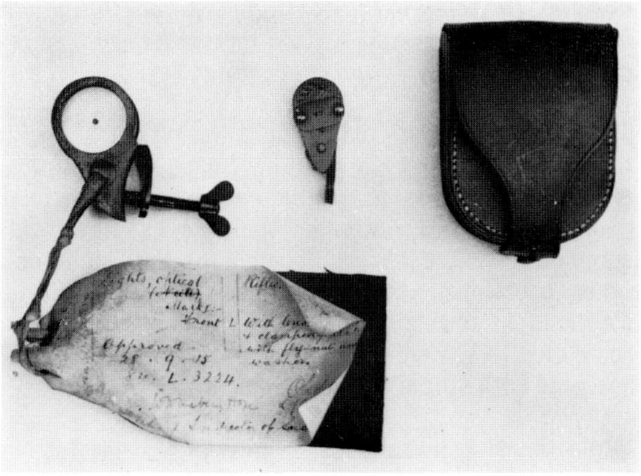

Neill sights, the Sealed Pattern *MOD Pattern Room*

Patterns of the above-mentioned sights have been approved to govern supplies as may be ordered.

Backsight: The backsight consists of a frame with lens fitted in the rear face, retained by a plate fixed by two screws. A hook shoulder is formed on the front face and an adjustable hook clamping plate fitted on the bottom, fixed by two screws.

The backsight is fitted on the rear face of the adjustable wind-gauge of the backsight of Rifles, Short, Mark III, and Converted Mark IV, in the following manner:

Partially unscrew the bottom clamping plate screws and hook the shoulder, on the front face of the frame, over the top edge of the windgauge, and press the hook of the bottom clamping plate up in front of the bottom edge of the windgauge, and tighten up the two clamping plate screws.

Frontsight: The frontsight consists of a frame with a lens fitted in the front face, retained by a plate fixed with three screws. A hook clamping lug with screw is fitted in the left side, and a horse-shoe shaped opening formed in the bottom of the frame.

The frontsight is fitted on the nosecap of the rifle, in front of the sight protector, in the following manner:

With the clamping lug and screw on the left side apply the horse-shoe shaped bottom portion of the sight to the front oval-formed portion of the nosecap; if the sight will not pass over the nosecap adjust by filing the inside sides of the horse-shoe until the sights can be pressed on, then turn the clamping lug into the lightening slots below the sight protector, and tighten up the sight by means of the screw.

The following List of Changes reference was for the Neill sight.

Para 17577:-

Sights, Optical (Neill), Mark I
Rifles, short, M.L.E.

Back:- with lens and bush, and aperture plate with fixing screw, two washers, and elevation plate.

Front:- with lens and bush, and clamping plate with fly-nut and washer.

Introduction.

Patterns of the above-mentioned sights have been approved to govern supplies as may be ordered.

Backsight: The backsight consists of a frame into which a bush, fitted with a lens, is screwed, and an adjustable aperture plate with a pivoted elevation plate fitted on the rear face and fixed by a screw and two washers.

The backsight is fitted on the left side of the rifle under the spring of the long range aperture sight, and replaces the long range aperture sight. To allow the backsight to fold down horizontally when not in use, a suitable clearance will be cut in the stock, fore-end.

A few backsights of early supply were graduated for 200 and 500 yards elevation only, and a single aperture bored in the aperture plate. Later supplies are graduated for 200, 400, 500 and 600 yards elevation, and four apertures are bored in the aperture plate.

Frontsight: The frontsight consists of a frame into which a bush, fitted with a lens, is screwed, and is provided with a screw, fixed in the bottom end, for attaching the sight to the rifle by means of a clamping plate, fly-nut, and washers.

The frontsight is fitted on the left side of the rifle by boring a hole through the wood of the fore-end at the bottom end of the lightening slots below the sight protector of the nosecap. The screw of the frontsight is passed through this hole, and the frontsight is fixed in position by means of the clamping plate, fly-nut, and washer.

The name of the Neill optical sight was officially changed to Barnett in 1916; there must have been some dispute as to whose invention the sight was and Barnett's claim seems to have been stronger than Neill's. However, in common reference it was still called the Neill sight, or on other occasions, the Ulster. This particular sight is identical to one for which a patent was taken out, accredited to T. Caldwell of Belfast, Pat. no. 1850 (1915), accepted 20th January 1916.

While the Martin sight had been approved on 1st May, 1915, prior to the Neill and Lattey, it was not announced in the List of Changes until later. It was often the case that items were officially approved after they had already been introduced into service, especially during extended hostilities.

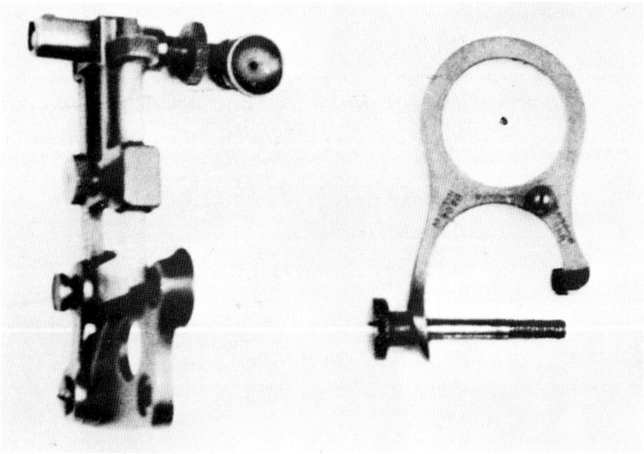

Martin sights. The rear lens is incorporated in the aperture backsight. *MOD Pattern Room*

The official description in the List of Changes reads:–

Para 17757:–

Sights, Optical (Martin), Mark I
Rifles, short, M.L.E.

Back & Front.

Introduction.

Patterns of the above-mentioned sights have been approved to govern supplies as may be ordered.

Back sight: The backsight consists of a frame, in which the aperture sight with lens is slideably fitted and adjusted vertically and laterally. It is fixed to the left rear end of the body, and replaces the aperture sight (or washer) of the rifle, being tensioned by the aperture sight spring.

There are slight variations in certain details of some of the back sights issued.

Front sight: The frontsight consists of a frame, with a lens, fitted in the front face, retained by a plate fixed by four screws. A long arm is formed on the left side, through which a fixing screw passes, and a hook on the right side engages in the lightening recesses on the right side of the nosecap.

In fitting the sight, the front nosecap screw of the rifle is removed, the hook of the front sight placed in the lightening recesses on the right side and the long arm on the left side of the nosecap, the sight bridging the nosecap behind the sight protector. The fixing screw is then inserted and screwed home, and the screw at the rear of the sight adjusted to prevent longitudinal movement.

A general instruction for fixing and adjusting the sights on the rifle is issued in each box with the sights.

These Galilean sights had some distinct disadvantages:

1. The field of view was very small, about one degree or only 5 to 7 feet per 100 yards, with a magnification of 2 to 3 times; so difficulty was experienced in picking up and identifying an object quickly.
2. As the telescope formed by the two lenses was not enclosed, light often interfered with perfect vision and dirt or dust readily accumulated on the lens surfaces.
3. The sights were vulnerable to damage and working loose.
4. Because of the mounting systems, sights were not always securely mounted and therefore less reliable and accurate.

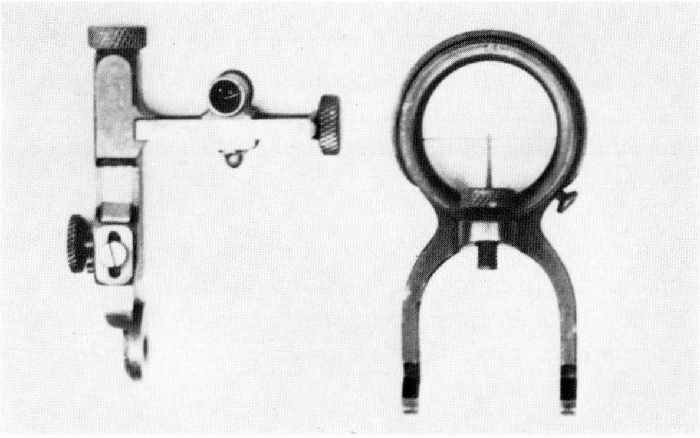

Gibbs sights. Note the post & crosswire reticule. *HBSA*

The first contracts were signed with English gunmakers for the fitting of telescopic sights to S.M.L.E. rifles in 1915 and by early 1916 they were coming into general use at the front. After this time, the use of Galilean sights diminished. Some Mark I* Charger Loading Lee-Enfield rifles were set up by B.S.A. in March 1915 with optical sights, the first batch consisted of 25 rifles and it was followed by 200 more. These rifles were most probably fitted with the B.S.A. pattern optical sight and used by the Territorial reserves.

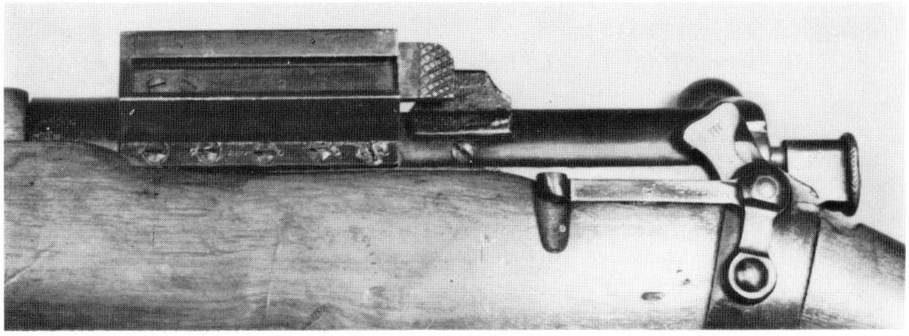

Periscopic Prism Co. dovetail mount patented Feb. 1915 by Rolfe-Martin.
J. Alley

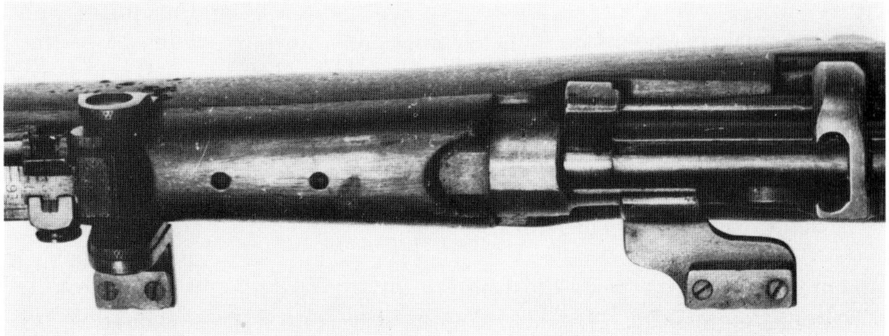

Whitehead mounts for the long Winchester telescopic sight.
R. Etherington

Holland & Holland mounts, used with Aldis telescopic sights.
R. Faris

A few variations of telescopic sights and mounts were used on the S.M.L.E. Mark III, which made possible about one dozen combinations. Most of these had offset-mounted telescopes so as to enable charger loading and use of the open sights, although this did limit the effectiveness of the telescopic sight to a certain degree. After complaints from the battlefront about the unsuitability of the side-mounted scopes, some models were fitted over the bore. The three main manufacturers of telescopic sights were Aldis Bros., Periscopic Prism Co. Ltd. and Winchester, although smaller quantities of Watts & Son, Evans, Standly, Gibbs, Fidgelands, Jeffries and Zeiss telescopes were also used.

There was no lateral (windage) adjustment on some of the earlier telescopes, so this adjustment was effected on the mounts. The focus slide with attached dust cover plate operating in a helical slot at the top of the tube is typical of many early telescopes and although this telescope production extended over a period of only two years, it will be found that some technical particulars such as the lens arrangement, magnification, field of view, and range graduations, were modified and altered.

Although some of the earlier telescopes were calibrated to 800 yards, the usual graduations were in 100 yard increments to 600 yards. Telescopic sights used in the Service were usually engraved with ordnance or acceptance stamps, most commonly the broad arrow. Until 1918, there was no official or specific pattern of sight and by this time the Pattern 1914 rifle was favoured for sniping; there was no official approved pattern of telescopic sight for the S.M.L.E.

The adaption and fitting of these sights to rifles was undertaken by the trade: the Aldis Bros., Periscopic Prism Co., Holland & Holland, James Purdey & Sons, J. Rigby & Co., W. Jeffrey & Co., J. Bartle & Co., D. Frazer & Co., Whitehead Bros. and William Evans. The Periscopic Prism Co. later fitted most of the Pattern 1914 .303 rifles with the Pattern 1918 telescopic sight and it was Holland & Holland who set up nearly all of the No. 4 (T) equipments during the 1939-45 war.

The Great War S.M.L.E. sniping rifles or telescopes were not officially announced in the List of Changes although mention is made of some of the accessories in Stores Vocabularies and the List of Changes; items such as rubber eye-cups "for use on all telescopes except Winchester." The Government contracts for the fitting of telescopes to S.M.L.E. rifles resulted in a total of 9,788 sniper rifles being supplied and passed by the Chief

Inspector of Small Arms at Enfield by the end of the war. This total was composed of:

- 4,830 Periscopic Prism Co. sights fitted by P.P.Co.,
- 3,196 Aldis Bros. sights fitted by various gunmakers,
- 908 Winchester sights fitted by Whitehead Bros.,
- 855 Assorted other sights, fitted by various gunmakers.

Aldis sights were classified into four different patterns while the P.P.Co. sights differed only in minor variations such as the solid or split drum. The Winchester sights were the A5 and B4 models.

Those gunmakers who set up Aldis telescopes on a variety of side and overbore mounts were Purdey, Holland & Holland, Atkin, Bartle and Aldis Bros. The assorted sights set up on various mounts by other gunsmiths include:

Watts sights set up by Rigby, Frazer and Woodward,
Evans and Goerz sights set up by Evans, and
Jeffrey sights set up by Jeffrey.

Comprehensive lists and illustrations of the sight variations, gunmakers, types of mounts and contract quantities will be found in Chapter 3 of Skennerton's "The British Sniper" along with the Specifications for fitting the telescopic sights onto the Mark III Short Lee-Enfield Rifle.

Most of the sights used were the Periscopic Prism Co. pattern; P.P.Co. is more renowned for the Patt. 1918 telescope which was fitted to the Pattern 1914 rifle to become the No. 3 Mk I* (T) sniper rifle; the Patt. 1918 scope was also that applied to the Australian S.M.L.E. sniper rifle towards the end of World War 2. Periscopic Prism Co. Ltd. of 23 Prince of Wales Crescent, Kentish Town, London, and A.B. Rolfe-Martin patented a mount for their telescopes in February 1915; it consists of a mount with female dovetail screwed and soldered to the left side of the rifle receiver body. A long male dovetail is incorporated on the base of the telescope rings in the upper mount, held in position by a long leaf spring and chequered thumb catch on the lower mount.

Periscopic Prism Co. was awarded sizeable contracts for unit sights for aircraft machine guns as well as rifle telescopes but delivery dates of these top priority items fell far behind schedule due to poor organisation and some financial problems, so the Government eventually paid out the shareholders and it was taken over by the Ministry of Munitions. The telescope illustrated is typical of P.P.Co. production for the S.M.L.E., over 12 inches long with a field of view of 9.5 degrees and an eye relief of 3 inches. The drum is ranged from 100 to 600 yards, in 100-yard increments.

Aldis telescopes were made by the Aldis Bros. of Birmingham, and are inscribed as such on the steel tube. Their four patterns used have a magnification rated at about 2.5 power and a 19mm objective lens, although there were small differences between some models. The range drum is usually marked from 1 to 6, indicating 100 to 600 yards. Rifle serial numbers were engraved on the telescope tubes and some of these sights were later fitted onto the No. 3 (Pattern 1914) sniper rifles early in World War 2.

Like the P.P.Co. telescopic sight, the Aldis has a focusing clamp screw and dust cover slide operating in a helical slot forward of the range drum; the range drum has a clamping screw mounted at the front on the Models 1 and 2, and at the rear of the Models 3 and 4. Lateral adjustment is made by altering the prism cell with a special tool or a coin and there are divisions marked on the prism cell adaptor to indicate adjustment after removal of the protecting cap and loosening the three small set screws.

The slender Winchester telescope is better known in its sniping role with the U.S. Model 1903 Springfield. The two models used in service with the S.M.L.E. were the A5 and B4; unlike the British-made rifle telescopes which usually have a picket and crosshair reticule, the Winchester has cross hairs with no picket. The A5 model was rated at 5-power magnification and the B4 rated at 4-power.

The calibration and alignment of the Winchester sights was quite different to their British-made counterparts and there is no range drum on the tube. Due to the higher magnification, long and slender tube and relatively small lenses they have a smaller field of view and poorer light gathering qualities. This sight was primarily a target-type scope pressed into military service whereas most of the British telescopic sights were specially designed for their service role. Winchester scopes were mostly mounted overbore on two piece mounts, by Whitehead Bros.; the Canadians were issued with the Winchester sight as well, mounted on the Ross and some Patt. 1914 rifles.

The Watts and Sons telescope is quite scarce and the only example noted is 10.5 inches long with no provision for lateral adjustment; this particular museum specimen has no graduations on the range drum, which would indicate that it was not brought into service. It is considered that those issued in service were calibrated to 600 yards on the drum scale. Another telescope set up in small quantities was the Evans.

The new Pattern 1914 rifle, produced in the United States, became available from late in 1916. It proved to be more accurate than the S.M.L.E., especially at the shorter ranges up to 300 yards, distances at which much of the sniping was done. To permit more accurate elevation, a fine vertical adjustment backsight was introduced onto some Winchester-made rifles and hen designated the Pattern 1914 Mark I* (F). This was approved on 22nd November 1917 and was successful enough warrant general preference over the scope-sighted S.M.L.E. sniper models.

Soon afterwards, a new telescopic sight was introduced and mated to the Pattern 1914 rifle to produce a reliable

Right (from top): Fidgelands, Watts, Periscopic Prism Co. and Winchester sights.

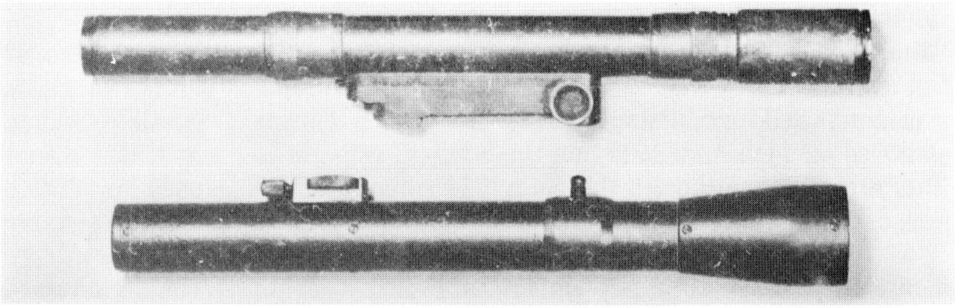

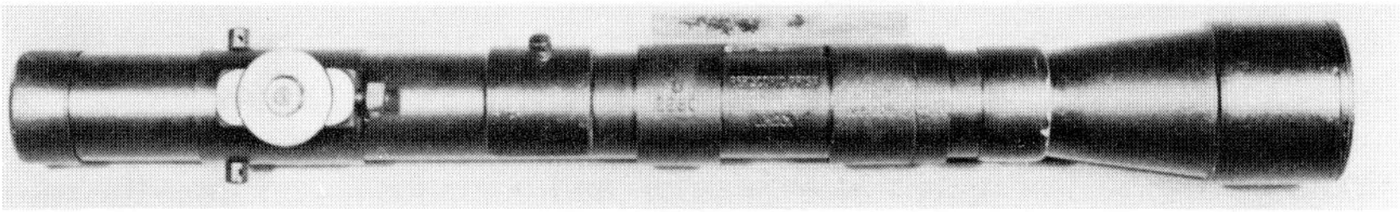

Below (from top):
P.P. Co. scope, P.P. Co. mounts; Aldis scope, Purdey mounts; Winchester scope, Whitehead mounts.
J. Alley, MOD Pattern Room, and R. Etherington

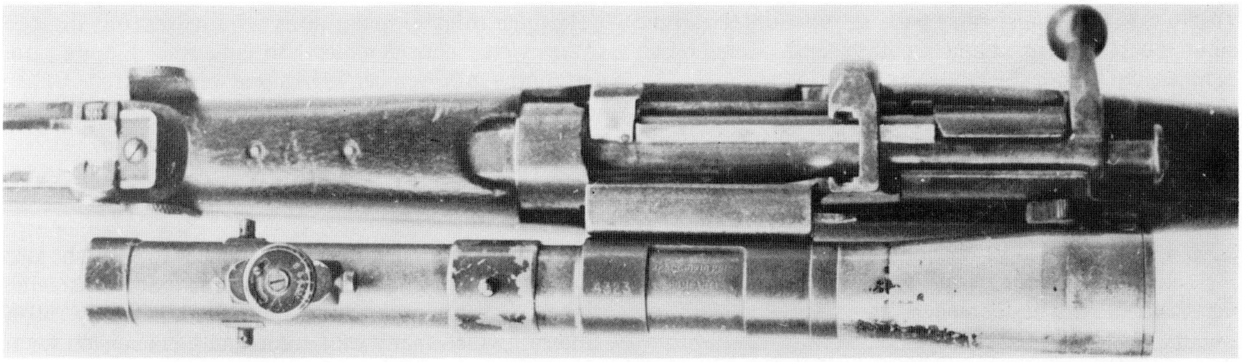

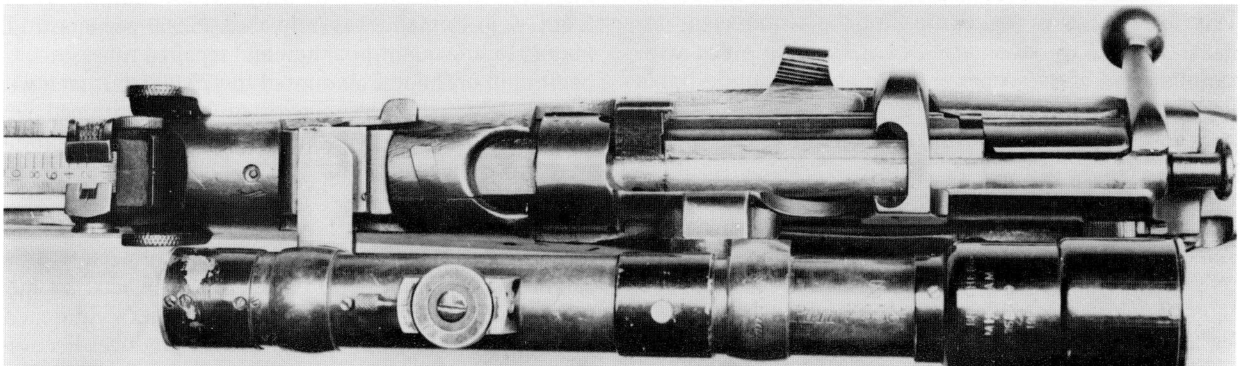

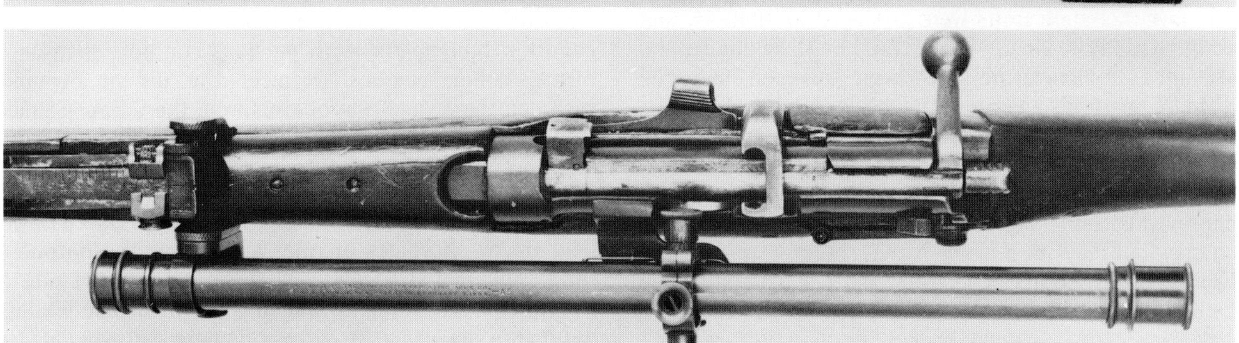

sniping rifle unit. The telescope was designated the Pattern 1918 and made at the Periscopic Prism Co. factory, now run by the Ministry of Munitions. The "Rifle, magazine, .303-inch, Pattern 1914 Mark I* W (T)" was approved on 11th April 1918 and the S.M.L.E. sniper rifle thus superseded, although the Patt. 1914 sniper rifle did not see enough active service to prove itself properly.

It was not until 30th September 1921 that "Sights telescopic for Short Magazine Lee-Enfield; Holsters, leather, for telescopic sights; and Eyepieces, rubber, sights telescopic, S.A., No. 1" were declared obsolete. It was instructed that the S.M.L.E. rifles fitted with telescopic sights be disposed of by return to Weedon and demands issued for the Pattern '14 sniper to replace them. The S.M.L.E. sniper were then stripped of components.

Equipment for snipers had been investigated by the Small Arms Committee early in 1921; this resulted in the return order for the S.M.L.E. snipers and their breaking up. A requirement was decided upon for about 3,000 sniping rifles at an issue of 8 per battalion. During the war, nearly 10,000 S.M.L.E. sniper rifles were set up with telescopic sights but the peacetime requirements were much reduced.

An ordnance stocktake of S.M.L.E. sniper rifles in store at Weedon in July 1921 lists:

745 fitted with P.P.Co. telescope and mounts;
99 with Aldis telescopes and mounts;
41 Aldis telescopes on Holland & Holland mounts;
106 Aldis telescopes on Purdey mounts;
and 13 Winchester models.

Periscopic Prism Co. set up 2,000 new Patt. '14 rifles with Patt. '18 scopes in 1918, short of the requirement for 3,000 units. The issue of S.M.L.E. sniping rifles with various types of telescopes and mounts was not desirable for the supply consideration, so Territorial Forces were issued with the Patt. '14 (F) rifles until telescope fitted models became available. A further 79 scope-mounted Patt. '14 rifles were set up by B.S.A. by 1938 and the serviceable Aldis sights from the broken-up S.M.L.E. snipers kept in store.

Upon the outbreak of World War 2, these Aldis (and a few P.P.Co.) sights were set up on Patt. '14 rifles in an offset position to the left of the bore, by Alexander Martin, Glasgow; 421 such units were produced in 1940-41, and designated "Rifle No. 3 Mk I* (T) A." In 1926 the general ordnance nomenclature had been changed and the S.M.L.E. patterns became the Rifle No. 1 while the Patt. '14 became the Rifle No. 3.

Late in 1944, Australia introduced the No. 1 rifle with an improved Pattern 1918 telescopic sight mounted overbore on crawfoot mounts, a heavy barrel, special furniture and two different heights of mounts. Further details will be found in Chapter 9.

With particular shortages of spare parts for the earlier Marks of the S.M.L.E. rifle during wartime, approval was given for certain Mk III rifle parts to be fitted where possible. Such an instance is the Mark I S.M.L.E. locking bolt, where approval was given for the Mark III bolt to be used on 26th April 1916; it was necessary to fit the Mark III stop pin as well for precise operation.

The S.M.L.E. cocking piece was next in line for modification and a new pattern was approved on 5th August 1916 on which the knurled head was removed, more metal left above the striker hole, and three finger grooves cut on each side so as to afford a grip when cocking the rifle without manipulating the bolt. This new pattern was approved as a spare for all rifles fitted with the earlier cocking piece, although patterns with knurled or enlarged heads were reintroduced with the Mark VI and early No. 4 rifles, only to be again discontinued when a similar manufacturing expediency was reintroduced during World War 2. This change was also implemented in the Indian and Australian factories.

The quantities of small arms required during the Great War were considerable; British and Empire troop strengths in November 1918 alone were listed as:

British	3,563,466
Home Volunteer	248,444
Indian Native	654,021
Colonial	525,760

Over 1.25 million rifles were issued to British forces during 1914-15 and this figure had increased considerably by 1918. At the outbreak of war, Britain had only 750,000 rifles of which many were undergoing conversion for the Mark VII ammunition. In July, 1914, output of the S.M.L.E. in England was only about 8,000 per month. By May 1918, the quantity of new and repaired rifles since the beginning of the war amounted to 475,000, but men were still undergoing training without being equipped with rifles. Wooden blanks fashioned into basic rifle profiles were quite widely used for drill and basic handling during both major conflicts.

Battlefield rifle losses were far in excess of what was first calculated as all casualties had to be replaced by men, fully equipped, with rifles; of all the soldier's equipment, this article took by far the longest to produce. Recovery rate of rifles from the battlefield was much lower than expected due to the different type of warfare and conditions, compared with previous conflicts such as in Africa. Other avenues to meet the urgent demands included the acquisition of arms from the U.S.A., Canada, Australia and Japan and negotiations were commenced to arrange contract manufacture of rifles in the U.S.A.

Training of the new Kitchener armies in Spring 1915 was held up by shortages of rifles and it was the output of rifles from British factories that determined the rate at which divisions could be put into the field. The wastage figures of small arms were calculated at 8% at the beginning of the war, based on experiences from the Boer War. Such figures were grossly under-estimated during the earlier months of the Great War although when the front lines became more static, the wastage figures were reduced. The other major factor which had an adverse effect on these figures was the increased use of artillery compared with prior conflicts.

British rifle production figures between 1912 and 1915 show how dramatically the output was increased:

	New	*Repaired*	*Total*
1912-13	103,000	81,000	184,000
1913-14	108,000	50,000	158,000
1914-15	418,000	491,000	909,000

The wartime production of the British factories, as well as those imports from North America, comprised:

	Enfield	*B.S.A.*	*L.S.A.*	*UK SMLE Totals*	*Canada Ross*	*USA P'14*
Aug-Dec.						
1914	51,576	56,416	12,101	120,093	-	-
1915	271,856	275,927	65,678	613,461	2,650	-
1916	418,283	435,212	99,433	852,928	33,476	282,495
1917	640,113	468,447	97,012	1,205,572	88,360	835,355
1918	626,330	345,732	89,990	1,062,052	-	-
Totals:	2,008,158	1,581,854	364,214	3,854,106	118,486	1,117,850

Note:- The Enfield figures include those rifles from the "Peddled Scheme" (S.S.A. and N.R.F. receivers and other peddled components assembled at Enfield during 1917-1918). The S.S.A. actions numbered only 279 by Dec. 1916 but totalled some 250,000 by June 1918 when it was taken over by the Ministry of Munitions as National Rifle Factory No. 1. The N.R.F. production is unsure, but considerably less than S.S.A.; these receiver groups were included in the Enfield figures of the peddled scheme.

This table is interesting on a number of counts; until 1917, production at B.S.A. was larger than at Enfield and during 1917 and 1918 the "peddled" rifles were included in the R.S.A.F. returns, boosting the figures by about 300,000. Proportionate capacity of B.S.A. and L.S.A. is evident; L.S.A. production was hampered by labour problems and obsolete machinery. Particularly during 1917, it was found hard to retain skilled labour because of enlistment schemes to which many of these workmen responded.

The production capacity of the Remington and Winchester factories in making 1.25 million rifles in well under two years, commencing with a new design and involving the tooling-up and setting up much of the new machinery, gives and indication of the industrial might of the United States. Australia and India were also producing the S.M.L.E. and although these countries had to meet large orders to equip their own forces, India sent 50,000 rifles and Australia, 30,000 to the threatened Motherland.

Alternative rifle factories and methods of production were investigated in Britain in an urgent effort to meet the huge demands. The Vickers factory at Crayford undertook to produce the Pattern 1914 rifle and although a sample was produced, it was not entirely satisfactory. Nevertheless, an order was placed for 100,000 rifles and bayonets with the rifles costed at £5/14/- each with deliveries at a rate of 2,000 per week commencing on 31st July 1915, to increase to 3,000 per week by 27th November 1915. Difficulties in setting up production resulted in the cancellation of the contract and eventually only a small quantity of bayonets was produced in 1917. Vickers produced mostly artillery and machine guns and acquired a name for late deliveries.

Some minor rifle components such as screws and springs were supplied by various specialist firms before the outbreak of the war and this was expanded into a "peddled scheme" whereby most of the rifle components were made by private industry and then assembled in the government workshops or factories. Late in 1914, the Standard Small Arms Co. was established in Birmingham with the assistance of a £60,000 Government grant to manufacture the S.M.L.E. rifle. A contract was signed with the Ministry on 23rd January 1915.

The S.S.A. Co. was set up by two Birmingham gunmakers, Johnstone and Petersen, and a London businessman, Mr. Waring. The address was listed at 8 Lench St. in Birmingham, one of the old Hollis premises. After more than a year, it became evident that the complete, finished S.M.L.E. rifles would never be produced there, even though some workmen had been trained at Enfield. This was mainly due to problems with the management, a scarcity of skilled labour and the shortage of certain machinery. In fact, within two years of being set up, only five or six thousand receivers and some barrels and trigger guards had been turned out.

By early 1917, the production of receivers, bolts, trigger guards and barrels had begun at S.S.A. in earnest. But management was still considered to be the main problem, so the Ministry of Munitions decided to buy the Company and commenced protracted negotiations that eventually resulted in the acquisition of the whole concern. In June 1918, the S.S.A. Co. became National Rifle Factory No. 1. A new Greener factory was taken over to produce stocks and barrels as the National Rifle Factory No. 2.

S.S.A. & N.R.F. marks on Peddled Scheme receivers. Note adjacent Enfield inspection marks. *J. Alley, B. Labudda*

It has been recently suggested that N.R.F. could indicate "National Revolver Factory" but the Ministry of Munitions records describe the National Rifle Factory, takeover of Standard Small Arms, and its operation and management in some detail. It has also been pointed out that the "NRF" stamping found on the 1918 SMLE receivers would be unlikely to indicate production at a revolver factory.

After it became evident in 1915 that Standard Small Arms would not be able to produce completed rifles, it was decided to assimilate S.S.A. Co. into a "peddled scheme" whereby they would produce receivers, barrels, bolts and trigger guards, while other specialist firms manufactured additional components which could then be assembled at the Royal Small Arms Factory. Late in 1915, the bayonet shop at Enfield was relocated to Sheffield so that this rifle assembly and fitting could be better undertaken at Enfield.

The first batch of 279 peddled rifles was delivered in September 1916 and by the end of 1916 the total amounted to 13,590. By 1st April 1917, S.S.A. Co. had produced 117,952 bodies, increasing to 232,416 sets (bodies, trigger guards and possibly bolts) and 55,548 barrels by 1st April 1918. Therefore, total production figures of the S.S.A. Co. actions exceed 250,000 to which must be added about 6 months production as the National Rifle Factory.

Unlike their Enfield, B.S.A. and L.S.A. counterparts, those receivers produced by the Standard Small Arms Co. and National Rifle Factory are not marked as such on the right side of the receiver butt socket along with the Royal cypher; they have no factory or manufacturer's name in the usual location.

The illustrated receiver markings are typical of the peddled rifles, with the unusual cypher, no factory name and 1916, 1917 or 1918 date. These receivers have the manufacturer's mark on the left side of the receiver to the rear of the safety catch; marked "S.S.A." or "N.R.F.."

Typical action markings of Peddled Scheme rifles.

An example has also been noted with "B.S.A." in the same position, on a 1917-dated Peddled Scheme rifle; some unfinished action bodies were also utilised from B.S.A. production and it is possible that some Enfield-made bodies may have been used in the peddled scheme as well. Peddled scheme receivers have R.S.A.F. Enfield inspection markings.

Components for the peddled rifles, as well as spare parts for rifles from other factories, were produced by the various contractors and some parts were also made for earlier patterns of rifles such as the Mark I S.M.L.E. and Lee-Enfield and Lee-Metford models.

These contractors and the component parts they supplied were:–

Abingdon Ecco Ltd., Tyseley, Birmingham	Safety bolts & catches.
T. Adams, 41 Avenue Rd., Blackheath, Birmingham	Striker forgings.
Anthony & Sons, 7 Unett St., Birmingham	Swivels.
Armourduct Manufacturing Co. Ltd., Witton, Birmingham	Backsight beds.
Joseph B. Ashbury, Lench St., Birmingham	Cocking Pieces.
Automatic Standard Screw Co. Ltd., Charles St., Halifax	Screws.
W. Baker, 87 Snow Hill, Birmingham	Extractors.
R.P. Bannerman & Sons, Ringslade Rd., Wood Green, London	Swivels, Mark I S.M.L.E.
Barker Bros., Unity Works, Constitution Hill, Birmingham	Outer Bands.
Baxter's Bolt, Screw & Rivet Works, Parade, Birmingham	Stockbolts, Marks I and III S.M.L.E.
Bellamy & Smith, St. Mary's Row, Birmingham	M.L.M. Breechbolt retaining springs.
Bell Punch & Printing Co., Tabernacle St., Finsbury Sq., London	Nosecap screws, fore-end studs.
Birmingham Small Arms Co. Ltd., Small Heath, Birmingham	Complete Mark III & III* rifles, most components.
C.G. Bonehill, Belmont Row, Birmingham	Barrel repairs, foresight bands, skeleton actions.
Boss & Co., 13 Dover St., Piccadilly, London W1	Cocking pieces.
Thos, Bradley, Hareford St., Birmingham	Charger guide rivets.
Joseph Brazier & Sons, Brick Kiln St., Wolverhampton	Extractor springs.
British Gun Barrel Co. Ltd., Halesowen, Birmingham	Barrels.
British United Shoe Machinery Co. Ltd., Union Works, Leicester	Backsight beds, triggers, keeper plates.
Burney & Blackburne Ltd., Tongham, Surrey	Stockbolts, converted from old pattern.
Burton, Delingpole & Co., Moseley St., Birmingham	Stockbolts.
(taken over from Standard Small Arms Co.)	
Butler & Spragg Ltd., Cambridge St., Birmingham	Screws (most rifle types).
Chubb & Sons Lock & Safe Co. Ltd., 128 Queen Victoria St., London	Extractor springs, piling swivels.
Criterion Stampings Ltd., Criterion Works, Willenhall	Swivels, cocking pieces, locking bolts.
Davis & Timmins Ltd., 34 York Rd., Kings Cross, London	Screws.
Dunford & Elliot (Sheffield) Ltd., Bath St., Birmingham	Barrels, and repair of.
Eagle Engineering Co., Ltd., Warwick	Trigger springs.
Erebus Manufacturing Co. Ltd., Willenhall, Staffordshire	Backsight protectors.
Field Rifle Co., 58 Cambridge St., Birmingham	Outer bands.
Fletcher, Biggin & Co., Murray Works, 28 Sidney St., Sheffield	Breech bolt retaining springs.
Gimson & Co., Vulcan Rd., Leicester	Triggers.
T. Glover & Co. Ltd., Gothic Works, Angel Rd., Edmonton, London	Backsight protectors, swivels, magazines.
Grant Negros & Co. Ltd., Grantype Works, Dog Lane, Willesden	Backsight beds, cocking pieces, magazine catches.

W.W. Greener, St. Mary's Square, Birmingham	Barrels, Sight beds, breech bolts, butt plates, extractor springs, fore-ends, handguards, backsight slides, and rifle reconditioning.
Guest, Keen & Nettlefolds Ltd., Broad St., Birmingham	Butt plate and marking disc screws.
Herberts (Nottingham) Ltd., Chelsea St., New Basford, Nottingham	Sears, extractor springs, sear springs, sling swivels.
L. Herve, 90 Clerkenwell Rd., London EC1	Fine adjustment screws for Mark I S.M.L.E.
Holland & Holland Ltd., 98 New Bond St., London W1	Cocking pieces.
Holloway & Naughton, 10 Vesey St., Birmingham	Extractor springs.
Jones' Sewing Machine Co. Ltd., Guide Bridge, Manchester	Backsight beds.
Job Lee, Premier Engineering & Tool Works, Kettering	Cocking pieces.
Linley & Co., Nechells Park Works, Eliot St., Birmingham	Magazines, buttplates.
Linotype & Machinery Ltd., 188 Fleet St., London EC4	Backsight beds.
London Small Arms Co. Ltd., Victoria Park Mills, Old Ford, London	Complete rifles, most component parts.
Maxfield & Co., 175 Clifton Rd., Aston, Birmingham	Firing pins, triggers.
Metropolitan Galvanising Co. Ltd., 143 Cannon St., London EC4	Steel marking disc screws.
Mitchell Manufacturing Co., Park St., Birmingham	Swivel brackets, outer bands.
M. Myers & Sons Ltd., Charlotte St., Birmingham	Washers.
Nicholls & Lewis Ltd., 16 Princip St., Birmingham	Backsight protectors.
C. Osborne & Co. Ltd., 12-14 Whittall St., Birmingham	Triggers.
A.G. Parker & Co. Ltd., Whittall St., Birmingham	Barrels, .22RF conversions & magazines.
Perkins, Joseph & Son Ltd., Astwood Bank, Redditch	Safety catches, extractors.
Prescot Clock and Mechanism Co., Prescot, Lancashire	Extractors, pins.
Pulvermann, Martin & Co., 31 Minories, London EC1	Swivel brackets, keeper plates, nuts, collars.
E.J. Riley Ltd., Accrington	Stock butts.
F.C. Scott, 65 Princip St., Birmingham	Breech bolt heads.
Singer Manufacturing Co., Gordon St., Glasgow	Inner & outer bands, foresight blades & bands, safety catches, swivel brackets, extractors, trigger guards, swivels, springs.
Singer Sewing Machine Co. Ltd., Clydebank, Scotland	Swivels, band & piling.
Solenoid Engineering Co., 13 City Rd., London EC1	Backsight protectors.
Standard Rotary Machine Co. Ltd., Rushden, Northants	Magazine catches.
Standard Small Arms Co. Ltd., Lench St., Birmingham	Action bodies, barrels, backsight beds, foresight block bands, bolts, boltheads, trigger guards, charger guides.
Standard Woodwork Manufacturing Co. Ltd., Park Royal, Willesden	Handguards, S.M.L.E. Marks I & III, and M.L.M.
Herbert Terry & Sons Ltd., Redditch	Magazine platform springs.
Timson, Bullock, Barber Ltd., Katesby Rd., Kettering	Backsight beds.
Vickers Ltd., Vickers House, Broadway, Westminster, London	Backsight protectors, barrels, butt swivels, brackets, swivels.
T. Watson, 73 Gurney St., Walworth, London SE	Safety catches, rifle gauges.
Westley Richards & Co. Ltd., Bournbrook, Selly Oak, Birmingham	Barrels, foresights, rifle repairs.
Wolseley Sheep Shearing Machine Co. Ltd., Alma St., Birmingham	Springs.
Wright & Co., 29 Shadwell St., Birmingham	(order transferred to W.W. Greener.)

Other related suppliers (mostly from the Birmingham area) and the items they manufactured were:–

Abingdon Ecco Ltd., Tyseley, Birmingham	Patt. '07 bayonet locking bolts & scabbard locket studs.
Addressograph Co. Ltd., Willesden Green, London	S.M.L.E. luminous front sights.
Thomas A. Ashton Ltd., Sidney St., Sheffield	Patt. '07 scabbards & leathers.
Automatic Standard Screw Co. Ltd., Charles St., Halifax	Patt. '07 & '88 bayonet nuts, rivets & washers.
Baker & Finnemore Ltd., 199 Newhall St., Birmingham	Cartridge chargers & clips.
R.P. Bannerman & Sons, Ringslade Rd., Wood Green, London	S.M.L.E. Rifle manufacturing gauges.
G. Beaton & Sons. Ltd., Holland Park, London	Fencing Muskets.
J. & J. Bent, Charlotte Rd., Stirchley, Birmingham	Oil bottle Mk IV.
E. Bibb & Sons, 157 Moseley Rd., Birmingham	Cartridge chargers.
Birkenhead Educational Dept., Town Hall, Birkenhead	Pullthrough assembly.
Birmingham Metals & Munitions Co., Adderley Pk. Mills, Birmingham	Pullthroughs.
F.C. Blake, Station Ave., Kew Gardens, London	S.M.L.E. Rifle Mark III gauges.
C. Brandauer & Co., 401 New John St., West Birmingham	Cartridge chargers & repair of.
W.H. Briscoe & Co. Ltd., 51 Park St., Birmingham	Oil bottle for rifles.
British Machine Co., West Parade Works, Halifax	S.M.L.E. Rifle manufacturing gauges.
British Stamped Metal Ceiling Co., 97 Victoria St., London EC4	Cartridge chargers & repair of.
W.N. Brunton & Sons, Wire Mills, Musselburgh, Scotland	.303 rifle cleaning rods.
H.M. Budgett, Crown Work, Chelmsford	S.M.L.E. manufacturing gauges.
Bulpitt & Son Ltd., 157 Camden St., Birmingham	Cartridge chargers & repair of.
G.G. Bussey & Co., Museum Works, Rye Lane, Peckham, London	Leather for Patt. '03 & '07 scabbards.
J.A. Chapman, Woodside Lane, Sheffield	Patt. '07 bayonets, scabbards, spares.
James Clews & Sons Ltd., 194 Aston Rd., Birmingham	Muzzle protectors for rifles.
Cogswell & Harrison, Gillingham St., London SW1	Fencing musket parts.
Cook & Co., 18 Exchange St., Manchester	Cartridge chargers.
H.S. Cooke & Co., 116 Northwood St., Birmingham	Oil bottle Mk IV, pullthrough weights.
A. Cooper, 91 Ashton St., Birmingham	Patt. '07 scabbards, parts & repairs.
Alfred Darling & Sons, 25 Pitching Rise, Brighton	S.M.L.E. Inspection gauges.
Davis & Mawson, Plume St., Ashton, Birmingham	Oil bottle Mk IV.
Davis & Timmins Ltd.,	Patt. '87 & '88 bayonet screws & nuts, S.M.L.E. luminous sight clamp screws.
E.R.O. Davis, Leicester Municipal Arts School, Leicester	Pull through parts.
Decimals Ltd., Grange Rd., Selly Oak, Birmingham	Wire cutters, S.A. & parts.
Dingley Bros., Viaduct Works, Coventry	S.M.L.E. rifle gauges.
Eagle Engineering Co. Ltd., Warwick	Sniperscope rods, periscope rests.
E.L. Tool & Gauge Co., 240 Aston Rd., Birmingham	S.M.L.E. rifle gauges.
Foster Accessories Ltd., 21 Milk St., Birmingham	Implement action P'14, S.M.L.E. foresight protectors, cleaning rods & tools.

Gabriel & Co., 4 & 5 AB Row, Birmingham	Oil bottle Mk IV.
Francis Gilbert, Canal St., Birmingham	Reflectors, mirror & stripping tools.
J. Glanville, 1 & 2 Coburn Rd., Bow, London	Rifle rests.
Gramophone Co. Ltd., Hayes, Middlesex	Cartridge chargers & repair of.
Great Eastern Railway, C/- Mr. Hill, Stratford	Trigger pull testers.
S. Hall & Sons Ltd., Vanguard Works, Hay Mills, Birmingham	Oil bottle Mk IV.
Halladays Ltd., 12-18 Geach St., Summer Lane, Birmingham	Oil bottle Mk IV.
Hancock & Corfield, Imperial Works, Morden Rd., Mitcham	Cartridge chargers & repair of.
Harcourts Ltd., 222 Moseley St., Birmingham	Oil bottle Mk IV.
S. Heath & Sons Ltd., Leopold St., Birmingham	Oil bottle Mk IV.
Hepburn, Gale & Ross, Grange Rd., Bermondsey, London SE1	Patt. '07 scabbards, Patt. '88 repair.
Herberts (Nottingham) Ltd., Chelsea St., New Basford, Nottingham	Wire breakers, for attachment to bayonets.
Adam Higler Ltd., 75A Camden Rd., London NW1	Gauges for rifle parts.
Hirst Bros. & Co., Roscoe St., Oldham	Luminous backsight attachments.
Holland & Holland, 98 New Bond St., London W1	Gauges for rifle parts.
J.T. Huxtable & Co., 8-9 Gresse Buildings, Stephen St., London W1	Patt. '03 scabbard tags.
Ingram & Kemp Ltd., Newtown Row, Birmingham	Pullthrough Mk IV.
H. Jenkins & Sons Ltd., Unity Works, Vittoria St., Birmingham	Oilbottles & pullthrough weights.
Larkins & Bailey, 26-31 Eyre Street Hill, Hatton Garden, London	Rifle gauges.
J. Legge & Co. Ltd., Colonial Works, Willenhall	Wire breakers, No. 1 Mk II.
Lightwood & Son Ltd., Price St., St. Mary's, Birmingham	Pullthrough Mk IV & cleaning rods, S.M.L.E. & M.L.M. tools and assorted parts.
Messenger & Co. Ltd., Cumberland Rd., Loughborough	Rifle rests Mk II.
Metropolitan Fire Brigade, Southwark Bridge Rd., London SE1	Pullthrough assembly.
Midland Railway Company	S.M.L.E. rifle gauges.
John Mitchell, Moland St., Birmingham	Cartridge chargers & repair of.
Wm. Mitchell (Pens) Ltd., Bearwood Rd., Smethwick	Cartridge chargers.
R. Mole & Sons, Granville St., Birmingham	Patt. '07 bayonets & parts, also wire breaker bolts & nuts.
R.W. Munro Ltd., 103 Cornwall Rd., South Tottenham, London	Gauges for rifles.
M. Myers & Sons Ltd., Charlotte St., Birmingham	Muzzle protectors, cartridge chargers.
Newall Engineering Co., Blackhorse Lane, Walthamstowe	Gauges for rifle parts.
Northampton Polytechnic Institute, Clerkenwell, London EC	S.M.L.E. rifle gauges.
North Eastern Railway Company	S.M.L.E. rifle parts gauges.
Wm. Notting, Bowling Green Lane, Farringdon Rd., London EC1	S.M.L.E. Inspection gauges.
J. Pace & Sons, Union St., Willenhall	Patt. '88 bayonet bolt springs.
Parkinson & W.B. Cowan Ltd., Bell Barn Rd., Birmingham	Cartridge chargers.
Perforated Music Co. Ltd., 197 City Rd., London EC1	S.M.L.E. rifle gauges.
Perry & Co. Ltd. , 36 Lancaster St., Birmingham	Cartridge chargers.
Pitter's Ventilating & Eng. Co., Upper Market St., Woolwich	Gauges for rifle parts.
E.A.N. Pochin, Westbrook, Chelmsford	S.M.L.E. rifle gauges.
E. Pollard & Co. Ltd., 29 Clerkenwell Rd., London EC1	Rifle rests, aiming.
C.H. Pugh Ltd., Whitworth Works, Tilton Rd., Birmingham	Wirecutters & breakers, cartridge chargers.
James Purdey & Sons, 57 South Audley St., London W1	Muzzle protectors.
Reid Bros. Engineers Ltd., 12 Wharf St., City Rd., London N1	Armourers gauges, plugs & tools.
Rudge, Whitworth Ltd., Crow Lane, Coventry	Wirebreakers Nos. 1, 2 & 3 Mark I, cartridge chargers.
Sanderson Bros. & Neubold Ltd., Newhall Rd., Sheffield	Patt. '07 bayonets.
Edwin Showell & Sons, Stirchley Brass Foundry, Birmingham	Oil bottle & pullthrough weights.
Siemens Bros. & Co., Woolwich, London SE	Rifle aiming rests.
Sperryn & Co. Ltd., Moorsom St. Works, Birmingham	Oil bottles.
J. Stone & Co. Ltd., Oceanic House, 1A Cockspur St., London SW1	Striker removal tools.
H.W. Taylor, 30 Graham St., St. Paul's, Birmingham	Cartridge chargers.
Herbert Terry & Sons Ltd., Redditch	Patt. '88 bayonet bolt springs, grip washers.
S. Thomas & Sons Ltd., British Mills, Redditch	Cartridge chargers & repair of.
Tool Makers & Light Machinery Ltd., Lower Oxgate Rd., Cricklewood	Gauges for rifles & parts.
W. Travers, Tempest St., Wolverhampton	S.M.L.E. rifle gauges.
Tunstall Optical Co., Leslie Park Rd., East Croydon	Gauges for rifle parts.
H. & T. Vaughan, Standard Works, Wood St., Willenhall, Staffs.	Cartridge chargers.
Vellacott & Co., 14 Kirby St., Hatton Gardens, London EC1	S.M.L.E. gauges & inner bands.
Vernier Manufacturing Co., 4 Station Bldgs., Haggerston, London	Gauge plates for S.M.L.E. rifle.
Vining Tractor & Motor Manufacturing Co. Ltd., Shepherds Bush	Gauges for rifle parts.
Voucher Co., Auto Tool Works, Bridge St., Walsall	Patt. '07 bayonet mandril & tools.
T. Watson, 73 Gurney St., Walworth, London SE	Luminous sight attachments.
Webley & Scott Ltd., 81-91 Weaman St., Birmingham	S.M.L.E. & M.L.M. rifle & armourers gauges.
W.R. Welton, 40 Gauden Rd., London SW4	S.M.L.E. gauges.
Western Electric Co. Ltd., North Woolwich, London	Oil bottle Mk IV.
Charles A. White , Aston Technical School, Aston, Birmingham	S.M.L.E. gauges.
John White & Sons Ltd., Park Road Tannery, Bingley, York	Patt. '07 scabbards.
Wilcox & Gibbs Ltd., 20 Fore St., London EC2	S.M.L.E. gauges.
Wilkinson Sword Co., 53 Pall Mall, London SW1	Patt. '07 bayonets, Patt. '03 bolts.
Wolseley Motors Ltd., Gatliff Rd., Edbury Bridge Rd., Pimlico	Rifle & parts gauges.

Some businesses were nationalised by the Government in instances where manufacturing or business difficulties were being experienced in fields critical to armaments production. Standard Small Arms Co. of Lench St., Birmingham became the National Rifle Factory No. 1 in June 1918; a new Greener factory in Birmingham was taken over as the National Rifle Factory No. 2; Periscopic Prism Co. of 23 Prince of Wales Cres., Kentish Town, London was appropriated by the Ministry of Munitions by 1918 and Wolseley Motors Ltd. of Gatliff Rd., Elbury Bridge Rd., Pimlico, London SW1 was taken over as the National Gauge Factory from 20th January 1917.

Britain's arms production was hampered considerably by enemy action during the war, R.S.A.F. Enfield and L.S.A. were bombed on a number of occasions and S.S.A. also had its production diminished due to the air raids. Some of the contractors suffered severe damage; the British Gun Barrel Co. was burned out and as they supplied 700 barrels per week to L.S.A., the London Small Arms production was affected for some time. Most of these air attacks occurred during 1917.

Some other furniture and fitting concessions may also have applied during the war as some L.S.A. Mk III rifles have been observed with original, matching number fore-ends which were converted from the S.M.L.E. Mk I* pattern.

Another wartime measure was the introduction of an emergency class of rifle marked "E.Y."; generally worn or downgraded examples beyond local repair. Early in December 1918, approval was given for some of these rifles to be reinforced by wire binding and the insertion of a wood screw through the fore-end, 1.5-in. in front of the trigger guard, so that the cup discharger might be used for firing grenades. There is some question about the E.Y. origin; it seems that the official approval in late 1918 was belated as it was already being applied. Also debatable is that the origin of the term "E.Y." represents the initials of its "inventor," an Edward Yule.

At first the binding consisted of whipcord or wire flex but copper wire was later utilised. This binding was usually done at two positions on the fore-end, around the fore-wood and rear handguard behind the backsight, and about half-way between the nosecap and outer band. This binding served to minimise flying splinters in the event of a "blow up" caused by pressure in the barrel as well as to stiffen the rifle furniture.

Officially at least, these strengthened E.Y. rifles were intended for use at Army Instruction Schools and the standard issue rifle was intended to be used with the discharger cup in the field. However, as the battalion rifle bombers were a specialist group, it is quite likely that they too reinforced their rifles in a similar manner, for use in this capacity.

Some new rifles were strengthened in this way in Indian and Australian service; in addition to the more usual E.Y.'s, the downgraded and obsolete rifles such as the Mark I series and Converted Marks II and IV's. Many of these saw active service in this form during the 2nd World War, the grenade most commonly used with this combination being the No. 36M or Mills bomb. It was also frequent practice for Indian service rifles to have the fore-end reinforced with the transverse wood screw just forward of the front trigger guard screw. The particular rifle grenade attachments and cups are described in Chapter 11.

Late in the war, on 8th June 1918, a "Bantam" butt was approved, a half-inch shorter than the short butt. This was

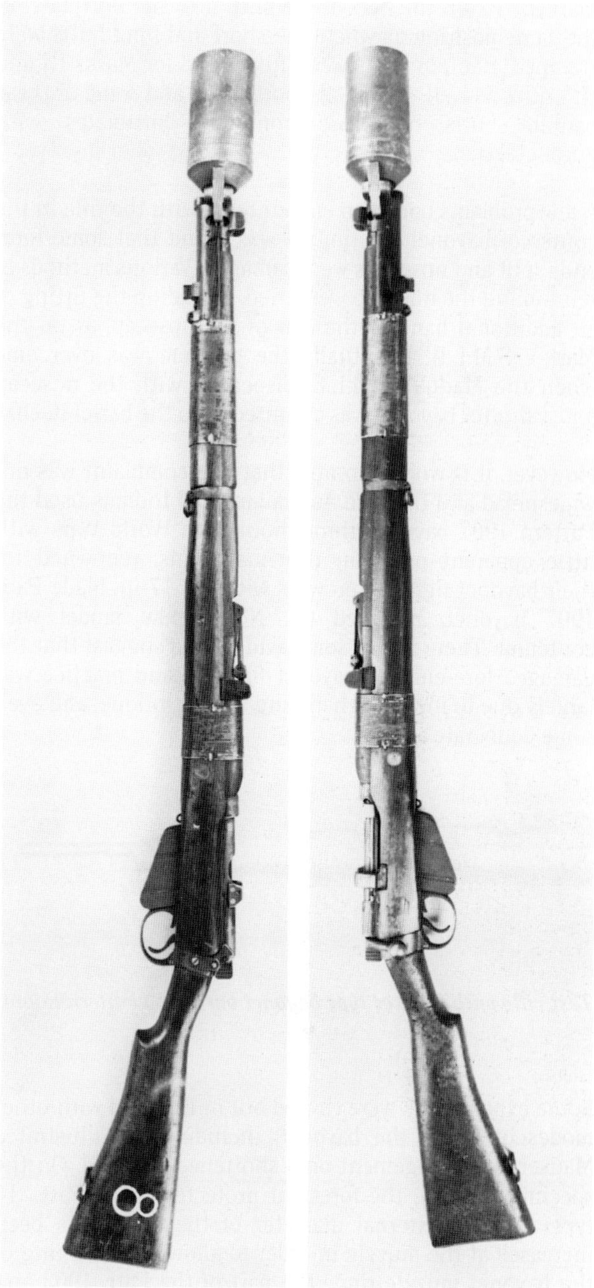

Strengthened (wire bound) Grenade Discharger rifles, often converted from E.Y. models. These often have a reinforcing bolt through the fore-end. *I. Skennerton*

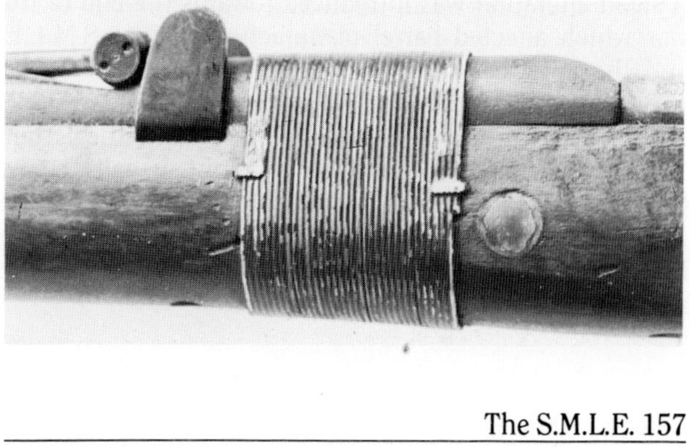

marked "B" on the heel of the butt, near the buttplate, in the same position as where the short and long butts were stamped. Bantam butts were introduced for Marks III and III* rifles as well as the .22 Short Rifles and some of these bantam butts had cast iron/steel buttplates with gunmetal traps.

A few problems had been encountered with the rifle in the course of bayonet fighting; it was found that some fore-ends split and nosecaps were damaged. Various methods of reinforcing the nosecap were tried, including the fitting of an additional band at the rear of the nosecap, as on the Mark V S.M.L.E. Eventually the problem was overcome when the Mark VI S.M.L.E. dispensed with the nosecap and a shorter bayonet was mounted onto the barrel itself.

However, it is worthy to note that this complaint was not widespread and that the Australians and Indians used the Pattern 1907 bayonet throughout both World Wars with little apparent problem; the Australians, renowned for their bayonet fighting prowess with the 17-in. blade Patt. 1907 bayonet, regarded the No. 4 spike model with contempt. There is also some evidence to suggest that the damaged fore-ends in bayonet fighting and practice was largely due to incorrect handling and technique, and even some deliberate abuse.

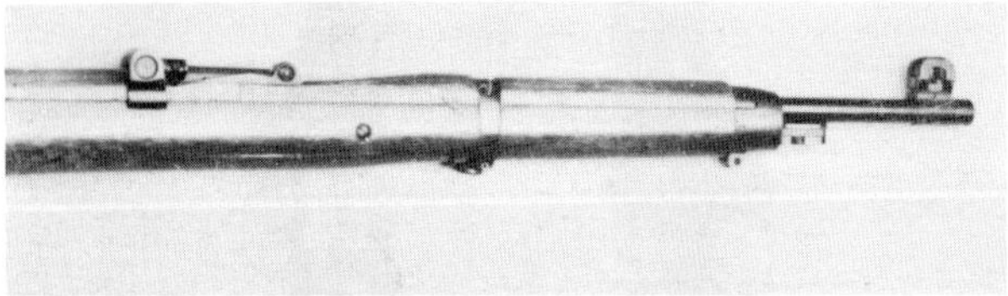

Test rifle with Mauser type bayonet bar. *MOD Pattern Room*

Some experiments were carried out in England with other modes of fixing the bayonet, including the illustrated Mauser bar arrangement on a shortened fore-end. On the specimen shown, the foresight protector is the Patt. '14 type and the external diameter of the barrel has been increased at the muzzle in order to allow for the fixing of the bayonet muzzle ring. The part of the Patt. 1903 and 1907 bayonets commonly referred to as the "muzzle ring" really attaches to the bayonet boss on the nosecap rather than around the muzzle, so this is technically an incorrect term when applied to these bayonets.

A small alteration was introduced towards the end of the war which affected barrel manufacture for the S.M.L.E. and the Patt. 1914 sniper rifle; the modification of the small cone for the chamber, which was increased by .002-inch. In the instance of the S.M.L.E. barrels, the letters "S.C." were marked behind the rearsight, usually below the "H.V." marking which indicates sighting for the Mark VII ammunition. On the Patt. '14 sniper rifle, the barrel was marked "S.C." on the portion exposed in the nosecap opening.

Another economy applied after the introduction of the Mark III* S.M.L.E. was the omission of the butt marking disc and screw. In future, any regimental markings or rack numbers were to be made in the wood of the butt itself. The omission of the butt marking disc is a feature of most Mark III* rifles but it should be realised that many of these changes; the omission of the magazine cut-off, long range sights, backsight windage adjustment and change to the later pattern of cocking piece, were introduced independently. In many rifles the marking disc hole already machined in the butt was filled with a glued wooden disc.

The alterations made from the pre-war S.M.L.E. Mark III to what was the Mark III* by the end of 1918 were really a process of evolution; it is not uncommon to find some features on these rifles which do not normally belong to that particular Mark or star variation. The most often encountered anomaly in this respect relates to the magazine cut-off or provision for same in the receiver and the fore-end; some of the Mark III features were brought back into production after the war was over and similar discrepancies may be encountered in Australian and Indian production, especially during the late 1930's and early 1940's.

A smaller piling swivel was approved on 7th December 1920 and the previous pattern which had endured since the Mark II Lee-Metford rifle was declared obsolete for future manufacture. The smaller pattern was less likely to foul on rifle buckets or in covers on limbers, although the earlier pattern again appears on later Indian and Australian-made rifles. It must also be considered that many of these small components were easily exchanged or altered by unit armourers and such interchange often occurred in normal service.

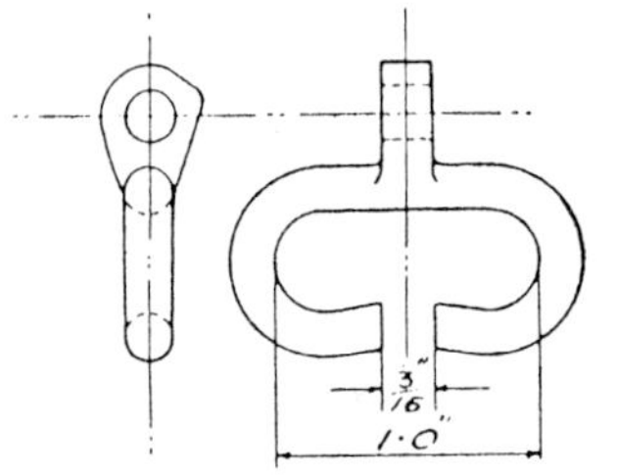

Piling swivel, narrow type, approved on 7/12/1920.

Another small change was the alteration of the transverse nosecap screw, from the ¼-inch diameter head type which is shaped to fit more-or-less flush with the nosecap, to the No. 2 type which has a larger head and is squared off at the back of the screw head so that it stands more proud of the nosecap. This was mainly intended for securing a muzzle protector. Upon the introduction of the new screw, the previous pattern was designated the No. 1 and declared obsolete. The No. 2 screw was also fitted to the Mark V S.M.L.E. This is another instance of a change not being implemented in Australian and Indian production, probably because such a muzzle protector was not on issue there.

Some wartime modifications introduced to increase production were only optional changes and even the introduction of the Mark III* was considered to be a temporary measure. Such changes included the omission of the long range sights, a modified cocking piece, windgauge on the backsight and swivel lugs on the trigger guard. This is why it is often possible to encounter rifles made after 1916 with such Mark III features, even though the Mark III* which incorporated these economies was introduced in early 1916. Such uncharacteristic features are more likely to occur in examples produced by the Enfield or L.S.A. factories.

The following changes were approved to govern future manufacture on 30th June 1920 and 10th January 1921:

1. Omission of long range sights;
2. Use of a T-shaped cap to replace the rearsight windage;
3. Omission of the lug on the front of the striker collar, and grooves cut through the collar of the striker;
4. Omission of the swivel lugs at the front of the trigger guard, and the fitting of a wire loop in its place, for the purpose of securing a fabric breech cover.

Because these changes had already been described and implemented, this is yet another case of "tidying up the paperwork" afterwards.

Complaints were made that the sear screw in the action body was found to work loose and so in 1923, approval was given to cut a V-notch in the end of the screw and centre punch the metal around the screw hole after the screw had been tightened up.

A member of the New Zealand Expeditionary Force, Lance-Corporal Schofield, invented a rotary type 20-round magazine which fitted alongside the standard magazine and operated independently of it. Some models were assembled and tested but while it operated satisfactorily and was considered to have merit, the matter was pursued no further.

A large capacity 20-round box magazine was approved for the S.M.L.E. and a pattern sealed on 25th April 1918 to govern manufacture. This was a more conventional type compared with the Schofield and may have been inspired by the large capacity German box magazine. Because of the slightly conical shape of the .303 cartridge, it is moderately curved and has strengthening ribs on each side which result in an appearance similar to the Bren LMG magazine. Both 15 and 20-round capacities were fabricated for trials but the 20-round pattern was manufactured in quantity. The main reason that larger capacity magazine was requested was because it had been found that mud was transmitted into the magazines via the chargers and it was considered that a large capacity magazine would not have to be filled as often.

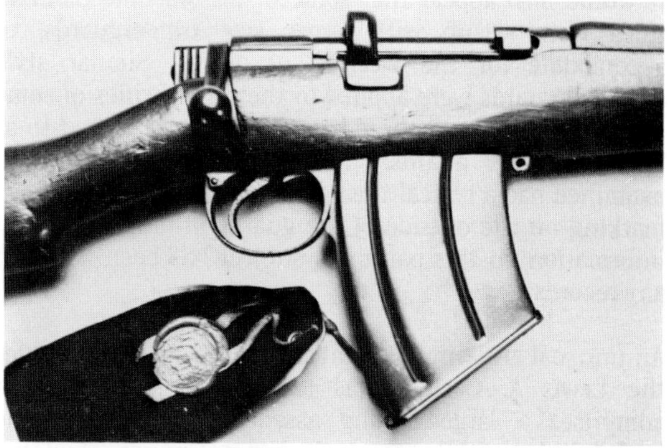

Magazine, Mark I No. 4, a 20-round model.
MOD Pattern Room

This 20-round model was adopted as the "Magazine, Mark I No. 4" and an initial order placed for 100,000. By the end of August 1918, some 120,000 had been issued in France and a further 30,000 sent to store at Weedon. Orders totalled 200,000 and the contracts were completed by early September 1918. The most common contractor's name encountered on this magazine is Linley & Co., although very few seem to have survived. A special leather pouch was also approved for this magazine.

The Mark I No. 4 magazine was intended to be carried in reserve and attached to the rifle after the usual 10-round magazine was exhausted, thus avoiding its reloading by chargers. Should the 20-round magazine be exhausted, it would be reloaded by means of chargers in the conventional manner. However, it was found that the last five rounds inserted by charger were considered to be detrimental to the balance of the rifle, as well as for the grip in bayonet fighting.

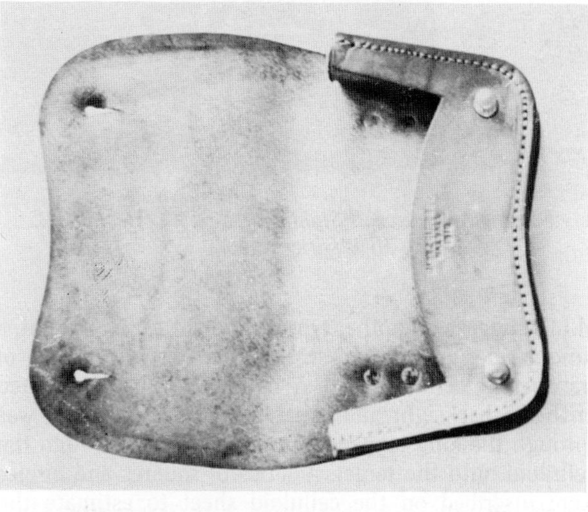

Leather Carrying Case for the 20-round Magazine.
Carl Woods

It would also appear that some of the wartime S.M.L.E. rifles were set-up with larger size triggerguards, to accomodate for the wearing of gloves. Similar style enlarged guards were applied to the service rifles of some other countries, and these have often been referred to as "winter trigger guards." The specimen that has been examined had a typical Great War contract manufacturer's marking on the outside of the guard, although no other information on this particular subject has been noted in any records.

An unusual anti-aircraft sight which somewhat resembles the Lewis A.-A. sight was tested in March 1916; it comprises a large folding assembly attached to the backsight bed, folding forward when not in use. The long, aperture rearsight is mounted in place of the original long range aperture sight. This particular design was forwarded by B.S.A., designed for use at 1,000 yards with an airplane speed of 120 mph. Of course these two variables could not be guaranteed and the sight was liable to damage, so it was considered unsuitable for the Royal Flying Corps or Naval Service for which it had been intended.

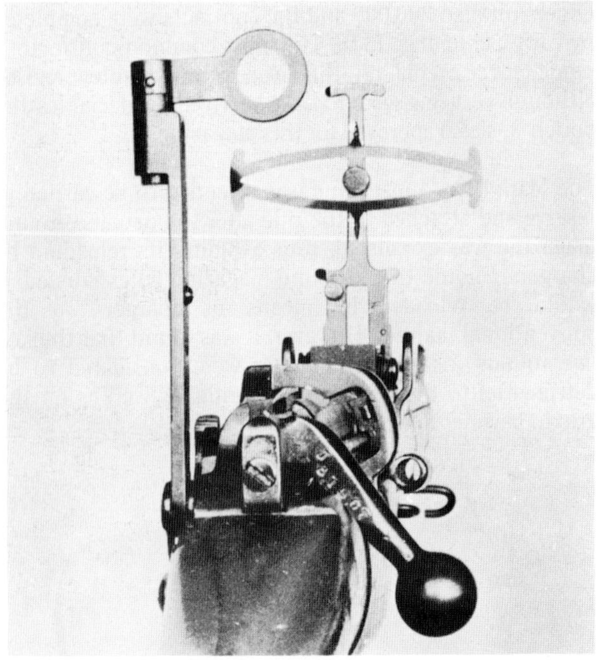

Folding Anti-aircraft Sights fitted to Mk III S.M.L.E.
MOD Pattern Room

M.I.D. suggested another type of rifle anti-aircraft sight, a unique design which featured a square plate of transparent celluloid with a handle attached, mounted onto the backsight and offset to the left. Sighting was through the long range aperture and thence through the celluloid onto the target. A series of squares and circles were inscribed on the celluloid sheet to estimate the required lead and elevation and although this might appear to work in theory, it was considered too fragile and impractical for service use.

The Armistice saw the termination of large-scale S.M.L.E. production in England. B.S.A. Co. produced the rifle from the mid-1930's until the 2nd World War; in Australia and India it was manufactured until after the 2nd World War. It still proved to be an excellent combat rifle in the hands of Australian, New Zealand, Indian and South African forces during 1939-45 and it remained the main issue British rifle until the Allied landings in Europe in 1944 so it can justifiably be claimed as the standard issue rifle for most of the 2nd World War also.

The No. 1 Mark III* is probably the best known S.M.L.E. and as it also represents the main issue pattern during the major portion of the 2nd World War in the European and Far Eastern theatres of war, some official 1942 textbook details of its specifications, mechanism and stripping procedures are reproduced here for general information and for those enthusiasts who wish to maintain or service such rifles.

The No. 1 MARK III* RIFLE

GENERAL DETAILS

Weight (without bayonet) ...	*8 lb. 10.5 oz.*
Type of Ammunition Used ...	*All types of Mk VII.*
Bore of Barrel	*0.303-inch.*
Type of Rifling	*Concentric, left hand.*
Number of Grooves	*Five.*
Twist	*1 turn in 10 inches.*
Muzzle Velocity	*2,440 ft. per second.*
Method of Loading	*Charger with 5 rounds.*
Capacity of Magazine	*10 rounds.*
Weight of Pull-off	*5-6 lbs.*
Type of Sights	*Open "U" and blade foresight.*
Range of Sight	*200-2,000 yards by 5-yard intervals.*
Extreme Range	*3,480 yards at 33 degrees elevation.*

STRIPPING

The service rifle may be stripped by armament personnel only. Before stripping is commenced always see that the rifle is unloaded.

Sequence of Stripping

(a) Remove screws, nose cap front and rear, and nosecap.
(b) Remove outer band screw, swivel lower band and front hand guard.
(c) Remove inner band screw and spring.
(d) Remove magazine.
(e) Remove screws, trigger guard, front and back.
(f) Remove trigger guard, trigger axis pin and trigger.
(g) Raise back sight leaf and remove rear hand guard, lower and back sight.
(h) Remove fore-end.

(i) Remove butt plate screws, butt plate, wad, stock bolt and butt safety catch.
(j) Remove sear spring.
(k) Remove sear axis screw, bolt retaining spring and sear.
(l) Remove magazine catch pin and catch.
(m) Remove sear spring.
(n) To assemble, reverse the above order.

NOTE 1: Check the protrusion of the stock bolt with the gauge before assembling the fore-end.

NOTE 2: See that the sides of the faces of the end of the stock bolt are vertical and will engage correctly with the keeper plate in the fore-end, before assembling the fore-end to the body.

STRIPPING OF COMPONENT PARTS

The Magazine

(a) Depress the rear end of the magazine platform as far as it will go, at the same time holding up the front.
(b) Push the front end towards the rear of the case, until it passes under the front lips. The front end of the platform should then rise out of the case.
(c) Tilt the rear end of the platform sideways, left side uppermost, and draw it forward out of the case.
(d) Remove the platform auxiliary spring.

The Bolt

(a) Remove the extractor spring, extractor screw and extractor.
(b) Remove the striker keeper screw, see that the stud of the cocking piece is in the long cam groove, unscrew the bolt head, unscrew the striker and remove the main spring.

The Back Sight

(a) Remove the pin fixing washer, washer axis pin and leaf.
(b) Press the screw slide catch and remove the spring and catch.
(c) Remove the cap screw and cap.
(d) Remove the screw and spring from the bed.

NOTE: The sight bed is part of the barrel and is not to be removed.

Gauges Used in Assembly

(a) .064 and .074 armourers' dummies to test the minimum and maximum distance between face of bolt head and end of chamber.
(b) Gauge stock bolt. A "go" and "no go" gauge for the protrusion of the stock bolt.
(c) Gauges striker point .040 and .042 to test the protrusion of the striker point.

MECHANISM

The following explanation of the mechanism is based on the assumption that the bolt is fully to the rear and the cartridges have just been pressed into the magazine.

Cocking of the Bolt

On pushing the bolt forward the charger is thrown out and the full bent of the cocking piece is brought against the nose of the sear. The cocking piece and striker are held stationary by this means, whilst the bolt continues forward, the main spring is compressed. During the forward movement of the bolt, the lower part of the bolt engages behind the base of the top round in the magazine and pushes it into the chamber. On turning the bolt to the right, the breech is finally closed by the solid rib on the bolt travelling over the resistance shoulder on the right side of the body, and, at the same time, the solid lug moves into the recess cut in the left side, and the extractor moves out and over the rim of the cartridge ready for extraction.

Release of the Striker

On pressing the trigger, the lower rib first comes in contact with the short arm of the sear and forces the long arm of the sear to come close to the edge of the cocking piece. So far, the pull on the trigger has been light, as the rib is close to the pivot and great leverage is obtained; this is the first half of the pull-off. Continued pressure on the trigger will bring the upper rib against the short arm of the sear. This action requires a greater pressure. The action is more rapid so that the end of the long arm of the sear is drawn smartly off the full bent of the cocking piece. This is the second half of the pull-off. The striker is then free to fly forward under the action of its main-spring and strike the cap of the cartridge.

Force of Recoil

The shock of discharge is taken equally on either side of the body and by the resistance shoulder and solid lug recess.

Unlocking the Bolt

On raising the bolt lever [handle], the bolt is rotated to the left, thus forcing the stud on the cocking piece to move backwards from the long to the short groove in the rear end of the bolt. This action withdraws the striker 1/8-inch. On completion of this movement, the solid lug on the underside of the bolt works down its inclined slot, and thus withdraws the bolt about 1/8 inch and so effects primary extraction.

Extraction and Ejection

Further backward movement of the bolt withdraws the empty case out of the chamber. The case is held by the extractor, which retains it on the face of the bolt until the base of the cartridge strikes the protruding ejector screw on the left side of the bolt-way. The empty case is then ejected out through the right side of the body.

Notes on the Action

If the bolt has not been properly turned over when the trigger is pressed, one of two things will occur:
(a) The stud on the cocking piece causes the breech to close automatically by striking against the division between the two grooves of the bolt, causing the bolt to turn down and the breech to close.

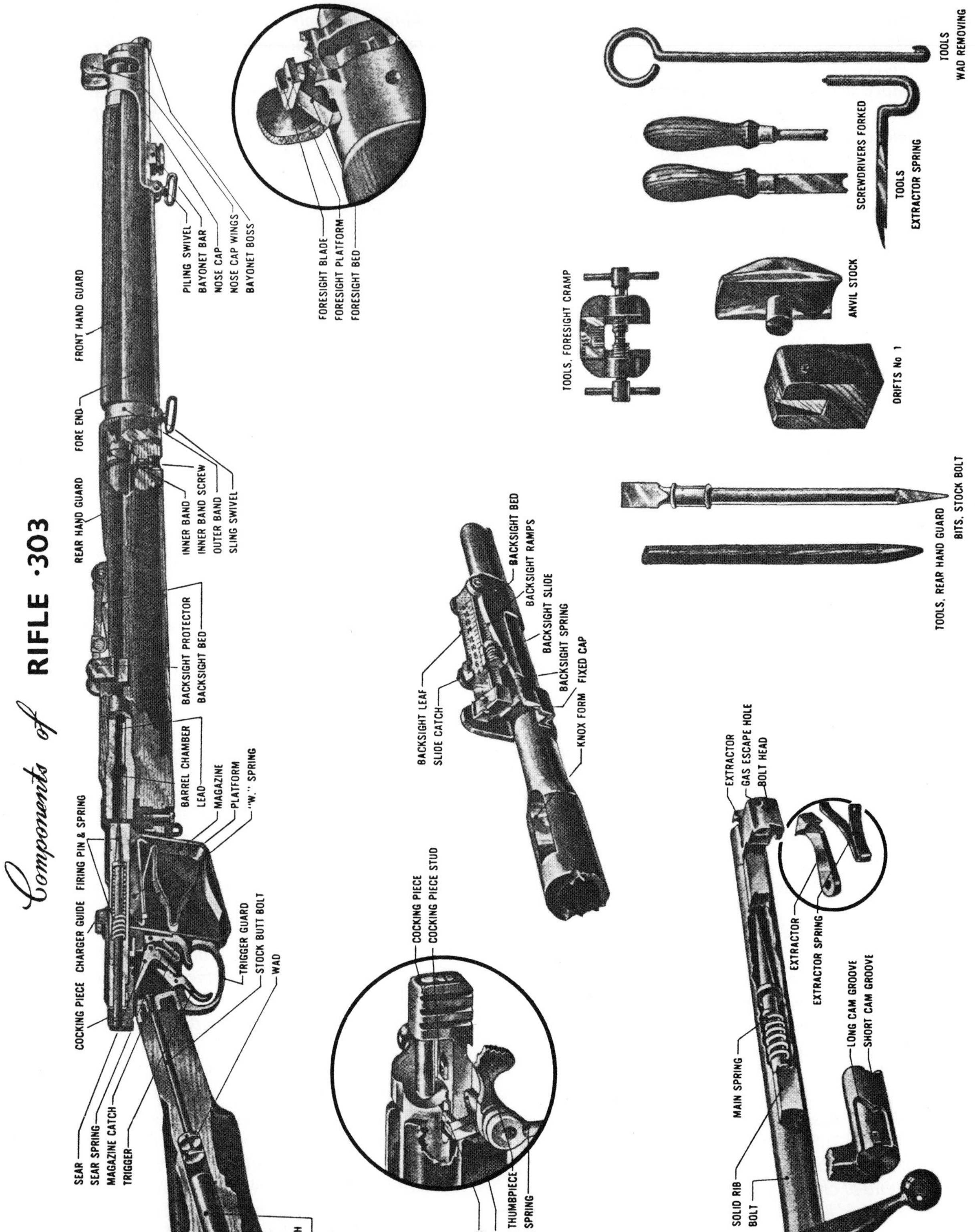
Components of RIFLE ·303
PILING SWIVEL
BAYONET BAR
NOSE CAP
NOSE CAP WINGS
BAYONET BOSS
FRONT HAND GUARD
FORE END
REAR HAND GUARD
INNER BAND
INNER BAND SCREW
OUTER BAND
SLING SWIVEL
BACKSIGHT PROTECTOR
BACKSIGHT BED
BARREL CHAMBER
LEAD
MAGAZINE
PLATFORM
"W." SPRING
COCKING PIECE
CHARGER GUIDE
FIRING PIN & SPRING
TRIGGER GUARD
STOCK BUTT BOLT
WAD
SEAR
SEAR SPRING
MAGAZINE CATCH
TRIGGER
PULL THROUGH
OIL BOTTLE
SLING SWIVEL
BUTT PLATE
BUTT TRAP
BUTT
FORESIGHT BLADE
FORESIGHT PLATFORM
FORESIGHT BED
BACKSIGHT BED
BACKSIGHT RAMPS
BACKSIGHT SLIDE
BACKSIGHT SPRING
BACKSIGHT LEAF
SLIDE CATCH
KNOX FORM
FIXED CAP
COCKING PIECE
COCKING PIECE STUD
SAFETY CATCH
LOCKING BOLT
SAFETY CATCH THUMBPIECE
SAFETY CATCH SPRING
TOOLS WAD REMOVING
SCREWDRIVERS FORKED
TOOLS EXTRACTOR SPRING
TOOLS, FORESIGHT CRAMP
ANVIL STOCK
DRIFTS No 1
TOOLS, REAR HAND GUARD
BITS, STOCK BOLT
EXTRACTOR
GAS ESCAPE HOLE
BOLT HEAD
EXTRACTOR
EXTRACTOR SPRING
LONG CAM GROOVE
SHORT CAM GROOVE
MAIN SPRING
SOLID RIB
BOLT
BOLT LEVER
STRIKER
COCKING PIECE

(b) The stud on the cocking piece strikes fully against the division between the two grooves and prevents the striker flying forward. If the bolt is then closed by hand, the whole action becomes locked, as the sear nose engages in the half bent which is undercut, and the stud of the cocking piece has moved half way down the long groove. The result is that the trigger cannot be pressed until the action is placed at full cock by drawing back the cocking piece by hand.

Action of the Locking Bolt Assembly

The locking bolt and locking bolt safety catch may be applied when the cocking piece is either at full cock or in the fired position. The locking bolt when applied in these positions prevents the cocking piece from moving backwards or forwards, as the safety catch in either position engages in the short groove, thus preventing the bolt from being rotated.

Safety Devices

(a) If the bolt has not been turned sufficiently in the loading action the stud in the cocking piece enters the short groove and so prevents the striker from flying forward.

(b) In the half-cocked position, the pressure on the trigger will not release the sear from the half bent on the cocking piece.

(c) The definite locking of the bolt before the round is fired which is effected by means of the resistance lug and solid rib.

(d) Gas escape holes allow gas to escape when a defective round is fired.

- *(i) Gas escape hole in the bolt head which allows the escape of gas in the event of a pierced cap.*
- *(ii) Gas escape hole in the left side of the breech which allows escape of gas in the event of a split case.*

Loading

(a) Turn the safety catch to the forward position with the thumb of the right hand.

(b) Grasp the knob of the bolt with the forefinger and thumb of the right hand; turn it sharply upward, and draw back the bolt to its full extent.

(c) Take a charger between the thumb and first two fingers of the right hand, and place it vertically in the charger guide.

(d) Place the ball of the thumb immediately in front of the charger on the top round, and, hooking the forefinger under the magazine, force the cartridge down with a firm but continuous pressure until the top cartridge is engaged in the magazine. A further charger may be inserted if necessary.

(e) Force the bolt sharply home, and turn the bolt lever [handle] as far down as it will go.

(f) Turn the safety catch fully to the rear.

Firing

(a) Turn the safety catch forward with the thumb of the right hand.

(b) Take the standard aim.

(c) Press the trigger.

Unloading

(a) Turn the safety catch to the forward position.

(b) Draw the bolt back to its fullest extent.

(c) Work the bolt backwards and forwards, without turning the bolt lever downwards, until the cartridges are removed from the magazine and chamber.

(d) Close the breech and press the trigger.

(e) Apply the safety catch.

STOPPAGES

No Round Enters the Chamber

(a) Draw bolt back and press round down into the magazine and release sharply.

(b) Tap bottom of the magazine.

0Cause: Magazine platform stuck.

Rifle Half-cocked

Pull back cocking piece by hand.

Cause: Faulty loading action.

Empty Case remaining in Bolt-way

Remove by turning rifle over to right.

Cause: Weak extractor spring or faulty ejector screw.

Treatment: Return rifle to armoury for adjustment.

Misfire

Reload; if stoppage is repeated, then:-

Cause: Faulty bolt or insufficient striker protrusion.

Empty Case Remaining in Chamber

Reload and fire.

Cause: Faulty or broken extractor and/or spring (if stoppage repeats).

Treatment: Return rifle to armoury for adjustment.

CARE AND CLEANING

General Notes

(a) Anti-freeze oil only should be used.

(b) Care should be taken to remove dirt, etc., from the gas escape holes and crevices.

(c) Chamber and ammunition should always be free from oil.

(d) The furniture should be well rubbed with oil to prevent the ingress of water.

(e) If possible, a rifle should be allotted to each serviceman under a sergeant and therefore the cleanliness and care of the rifle is the responsibility of the individual soldier who has the rifle on his charge.

INSPECTION OF THE RIFLE

This inspection is carried out half-yearly on all rifles held on charge by the unit, and covers the following:

(a) Examine all parts for security.

(b) Strip, clean and examine the arm.

(c) Weigh all springs.

(d) Gauge the striker protrusion.

(e) Gauge the bolt head with the .064 and .074 gauges.

(f) Check the action for correct functioning.
(g) Check the operation of the bolt and locking catch.

REPAIRS AND ADJUSTMENTS

(a) After any repair or adjustments made on the rifle, tests applicable to the particular part should be carried out before the rifle is regarded as serviceable.
(b) When fitting spare bolts, fore-ends, stocks, nose caps, etc., to rifles, the components must be marked with the rifle number.

To Increase or Decrease the Weight of the Pull-off

(a) To increase weight, reduce the angle which the full bent makes with the bottom of the cocking piece with an oil stone, i.e., metal is removed from top of sear.
(b) To decrease the weight, increase the angle which the full bent makes with the bottom of the cocking piece, i.e., metal is removed from the bottom of the sear.

Fitting a New Bolt Head

These heads are usually supplied longer to allow for fitting.
(a) Assemble new bolt head to bolt, and insert in body.
(b) Insert .064" gauge in chamber and see if bolt goes right home. If it does not, metal is removed from the head until it does. Very fine emery cloth is used in the fitment, ensuring that the face is kept square.
(c) Remove .064" gauge and insert the .074" gauge and see bolt does not close. A new bolt head is fitted when the bolt closes over the .074" gauge.
(d) The face of the bolt head is then rounded off to a radius up to .02-inch.
(e) Test the striker protrusion.

NOTE: It is often possible to exchange one bolt head with another and so obtain two serviceable rifles.

To Fit a New Striker

(a) Strip the bolt.
(b) Assemble new striker and components of bolt. Place stud of cocking piece in long cam groove.
(c) Gauge the radius and protrusions of the striker. Adjust this until correct, by stoning.

Fitting a New Ejector Screw

Remove old screw and insert a new one.

NOTE: The point of the screw in the bolt-way should just touch the bolt in its passage but not retard it.

Fitting a New Fore-end

Generally carried out a Ordnance Depots.
(a) Remove old fore-end and strip from it the keeper plate, nose cap screw nut, collar, backsight protector and nut, and assemble these components to a new fore-end.
(b) Test for fit with the new fore-end, the nose cap and front hand guard. Remove wood as required.
(c) Coat the barrel at the inner band lightly with lamp black and oil, and replace. Assemble the fore-end. Replace the trigger guard and screw. Test for springing and adjust the fore-end as necessary. Test pull-off to see if double pull is present. If incorrect, alter the trigger guard seating as required.
(d) Remove the trigger guard and screws, and the fore-end. Examine the barrel groove for seating of the barrel at the reinforce, and for a distance of ½-inch above and below the inner band.
(e) Lubricate the fore-end before assembling and test the protrusion of the stock bolt.

NOTE: On removal and assembly of fore-end, the use of the No. 1 drift is important to prevent any damage being done.

WEIGHING OF SPRINGS

General Notes

The rifle should be placed in a vice fitted with protective clams for the weighing of all springs so that the rear hand guard may first of all be removed without damage. The rifle must be cleaned, the magazine in position, and the rifle fully cocked. If any spring is overweight, look for burrs, dirt, etc. Then retest. (The tester trigger is to be used for weighing).

Weighing the Pull-off

(a) Place roller of tester trigger over trigger and pull diagonally over small of butt steadily. When trigger begins to move, the weight should read from 3 to 4 lbs.
(b) The pull is then continued until the rifle fires and then the weight should read between 5 to 6 lbs.

Weighing the Main Spring

(a) With the striker in the fired position, place the tag of the tester over the cocking piece thumb grip and pull steadily. The cocking piece should move at 7 to 9 lbs.
(b) Cock the action and apply the tester trigger as in sub-para (a). The cocking piece should move at 14 to 16 lbs.

Weighing the Bolt Head Retaining Spring

(a) Draw back the bolt until the head is over the retaining spring. Place claws of tag over the bolt head and pull vertically upwards.
(b) The bolt head should release at 10 to 16 lbs.
NOTE: If the weight is outside these limits, examine the sear screw to see if it is tight.

Weighing the Extractor Spring

With the bolt head upright in the bolt-way or taken out of the rifle, place the tip of the tag of the tester under the hook of the extractor. When it moves, the weight should read (with a No. 3 spring fitted) from 7 to 9 lbs.

Weighing the Butt Trap Spring

Place the tip of the tag under the rib of the butt trap and pull diagonally up. The trap should move at 2 to 3 lbs.

GENERAL NOTES ON HANDLING THE RIFLE

(a) Never keep rag in the muzzle.
(b) The bolt is only to be removed for cleaning and storage purposes.
(c) Only trained armament personnel are permitted to make examinations or to strip parts from a rifle.

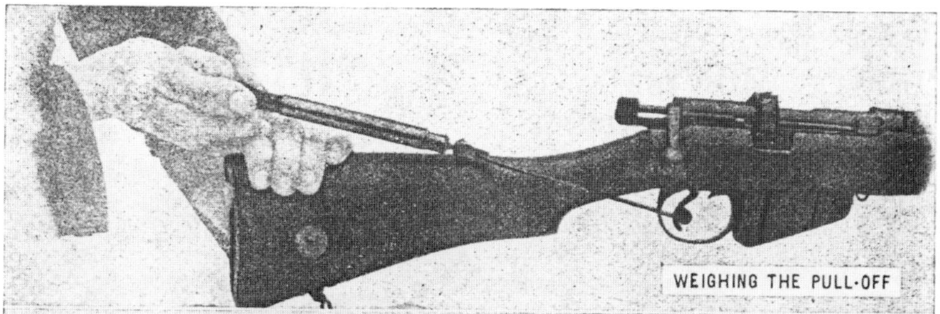

WEIGHING THE PULL-OFF

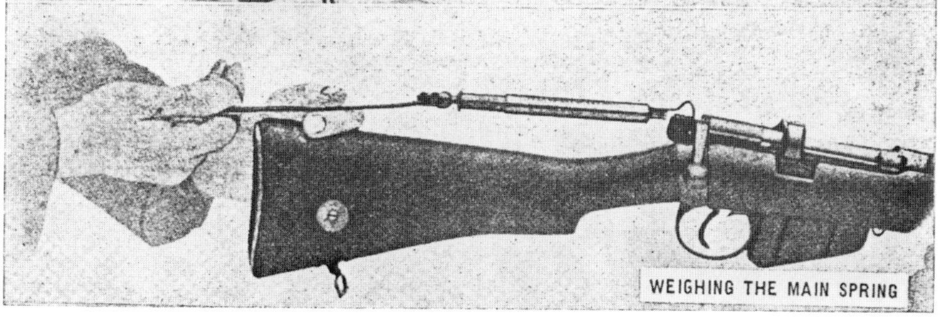

WEIGHING THE MAIN SPRING

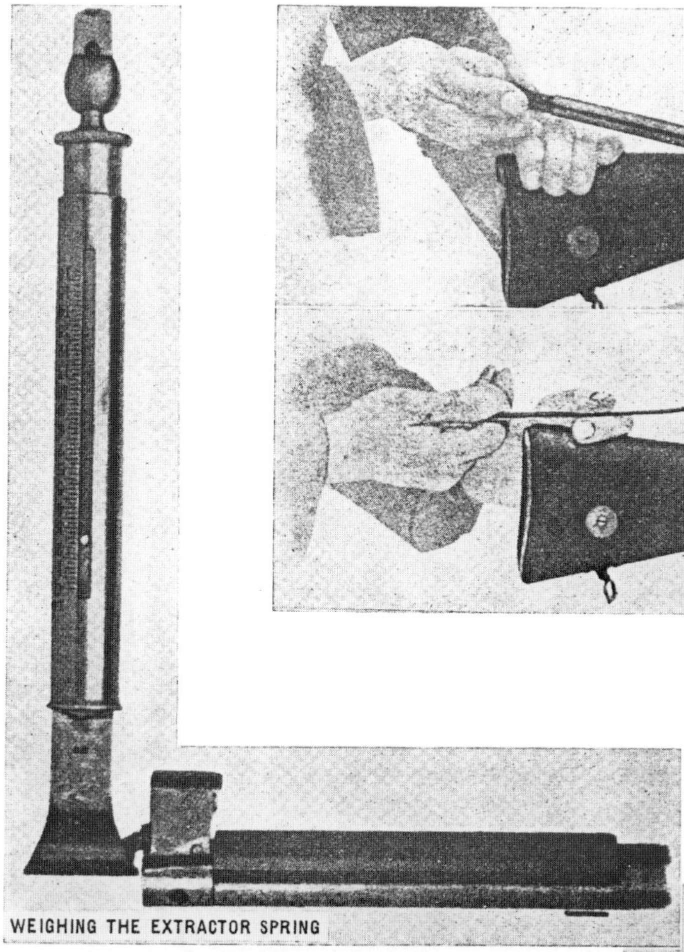

WEIGHING THE EXTRACTOR SPRING

WEIGHING OF SPRINGS

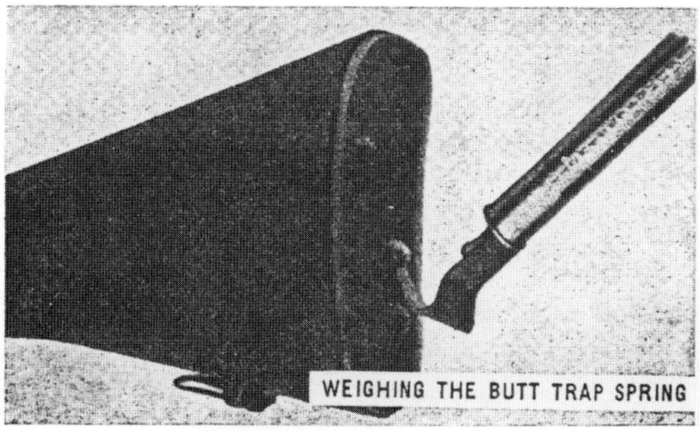

WEIGHING THE BUTT TRAP SPRING

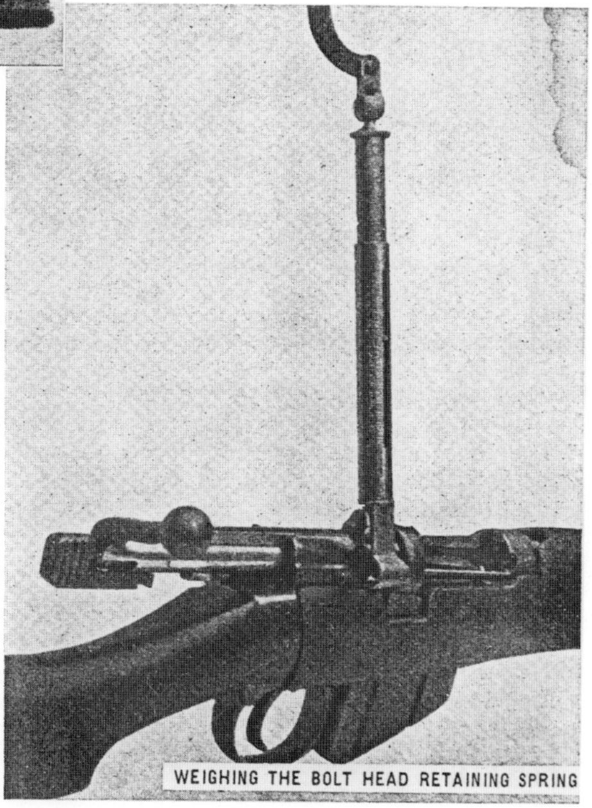

WEIGHING THE BOLT HEAD RETAINING SPRING

Mechanism of
RIFLE ·303

COCKED . . .

LOWER RIB OF TRIGGER IN CONTACT WITH THE SEAR. FULL BENT OF COCKING PIECE HELD ON NOSE OF SEAR. MAINSPRING COMPRESSED.

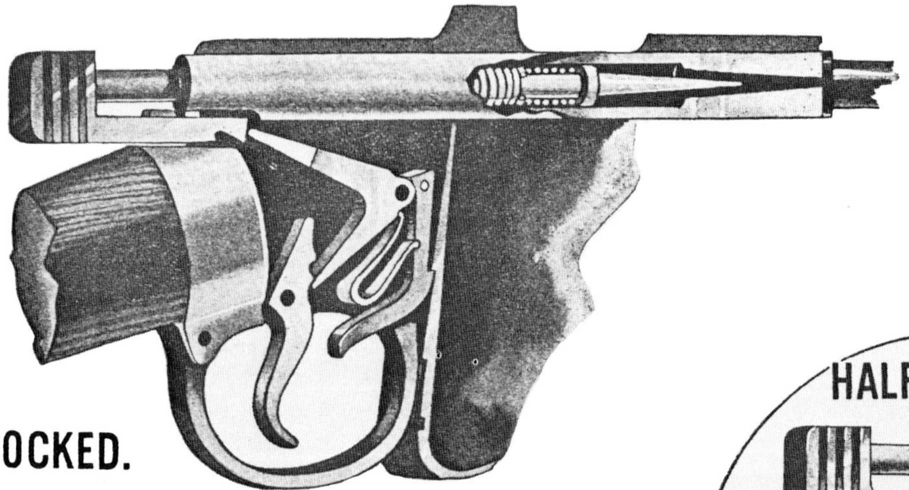

COCKED.

HALF COCKED . . .

NOSE OF THE SEAR ENGAGED IN THE HALF BENT, WHICH IS UNDERCUT, THEREBY LOCKING THE MECHANISM.

FIRED.

FIRED . . .

UPPER RIB OF TRIGGER ENGAGED WITH SEAR. STRIKER CARRIED FORWARD BY MAINSPRING.

Aiming

The point of aim is the lower portion of the target appearing to rest on the blade of the fore sight. The fore sight is to be centrally aligned to the U-notch of the back sight.

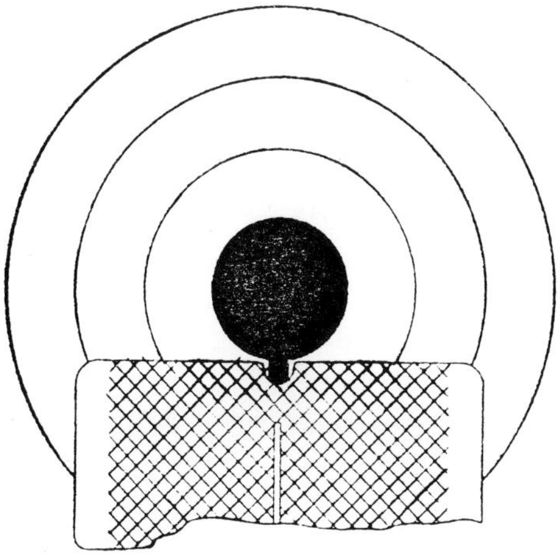

POINT OF AIM

Trigger Pressing

(a) Breathing is to be restricted while aim is being taken.
(b) Two distinct pressures are to be taken. The first when the rifle is brought into the position for aiming, the second when the sights are aligned on the target.
(c) The pull should be taken across the small of the butt. The pressure must be in the nature of a gentle squeeze.

Trials were conducted with aperture sights on the Mark III rifle in 1911, resulting in their recommendation for the next service rifle. During the Great War, 20 S.M.L.E. rifles fitted with aperture sights were tested in France and found to be quite satisfactory, so after the war an order was placed at Enfield for new Mark III rifles fitted with aperture sights, at a rate of 750 per week. In 1921, three such rifles were used for a short period in parade ground drill to ascertain the suitability and sturdiness of these sights and found to be satisfactory, so more were forwarded to cavalry units for similar tests. These trials rifles had minor differences in the form of the backsight but the type on rifle no. 4350 was chosen as the pattern for future manufacture. No upper band was fitted at the base of the nosecap as yet, but these were the basis for the Mark V S.M.L.E. model.

A patent on an aperture sight modification for the S.M.L.E. was taken out by B.S.A. on 30th January 1913, No. 2440. This particular system required that the bolthead be modified so as to clear the bridge piece of the sight base in order that the bolt could be withdrawn and removed from the rifle. This fitting was considered more for a military application than for competitive shooting, for which established types of detachable aperture backsights with fine adjustment were readily available.

2440. Birmingham Small Arms Co., and **Norman, G.** Jan. 30. ***1913.***

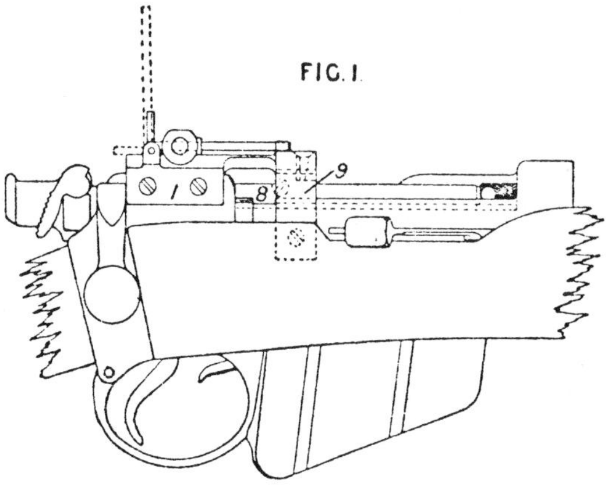

A strengthening outer band was approved on 5th January 1921 for attachment to the nosecap. Six rifles fitted with such a band at the rear of the nosecap were assembled and sent to Hythe for limited trials. Tests were considered quite successful and on 29th June 1921 it was announced that this fitting was approved. These six rifles can be considered prototypes of the Mark V although they were just referred to as Mark III rifles fitted with aperture sights.

By the end of the Great War, a change in concept of the rifle and its role had been brought about by experience in a new type of warfare. Extreme long range and volley firing were no longer requirements and a smaller calibre, high velocity round with corresponding flatter trajectory and increased range were no longer quite as important. New participants had been introduced into battle which also changed the role of the rifle; the tank, hand grenade, aircraft, and new applications of the machine gun. Trench warfare introduced a new perspective to battle, with many terrible consequences.

Reliability under the most adverse conditions was of paramount importance for a military rifle; the more accurate types such as the Pattern 1914 and Ross were excellent at the rifle range and even in troop trials but in the mud of the trenches they were inferior to the S.M.L.E. Something of a scandal had been caused in Canada by reports of despairing Canadian troops in the trenches having to kick open the bolts of their Ross rifles in order to operate them.

Some special cartridges had been developed during the war; armour piercing, tracer and incendiary, and although these were generally used in machine guns, they were also fired from the rifle on certain occasions. For these particular rounds, the .303-inch projectile was considered preferable to the smaller bullets because its larger mass permitted more of the special components to be incorporated in the bullet. Considering these and the previous-mentioned points, moves towards the .256 or .276-in. calibres were, for the moment anyway, abandoned.

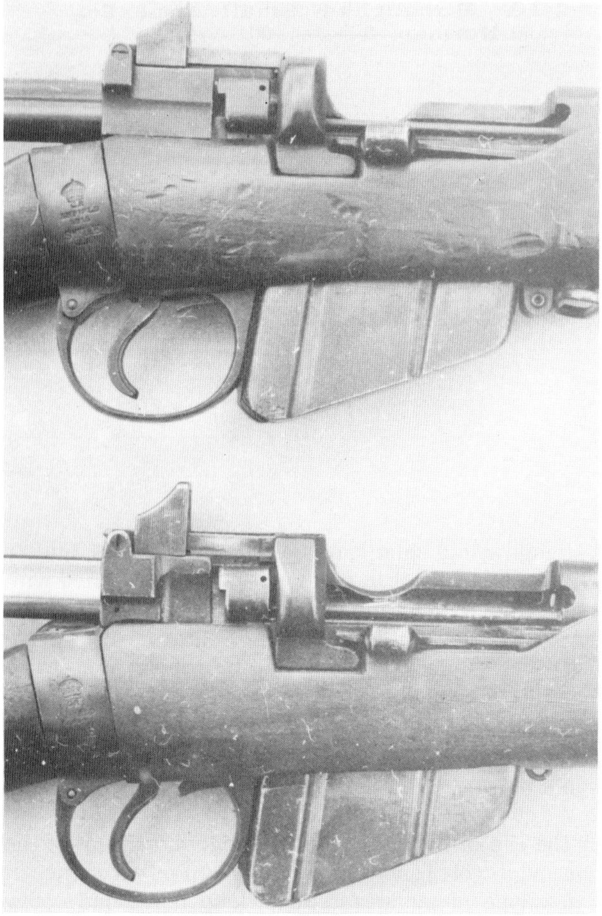

Old Photographs which compare the Trials Mk V (top) and the eventual production Mark V (bottom) rifles.

The new direction was towards firepower and trials with self-loading rifles were conducted by the Small Arms Committee to determine whether any were suitable for extensive trials with a view to adoption. Meanwhile, the Mark III S.M.L.E. fitted with aperture sights had become referred to as the Mark V and the possibility of converting existing stores of Mark III rifles to the Mark V pattern was being investigated.

The new sight had to be brazed on but it was found that the conversion to Mark V was not a very economical proposition. Those components requiring conversion were the action body or receiver, barrel, fore-end, handguards and nosecap and it was costed at £2/16/8d for these modifications to be effected on the Mark III rifle, to bring it up to the new pattern.

Regarding the magazine rifle, the following points were put up for investigation and consideration by the Small Arms Committee:

Provision for rifle grenades;
Possibility of an aperture sight not liable to damage;
Use of rustless metal;
Retention of the magazine cut-off;
An improved bolt action;
A shortened bayonet.

Quite obviously, most of these points were the result of recent experiences in the adverse conditions of the Great War trenches.

Three of these considerations had a bearing on the new Mark V, of which about 20,000 were made in a limited production run during 1922-24. These rifles were manufactured at R.S.A.F. Enfield and although a pattern was sealed to govern production, the Mark V was not officially approved, announced or introduced in the List of Changes. Production was commenced in November 1922 and the receivers are dated between 1922 and 1924.

The Mark V model differs from the Mark III S.M.L.E. in the following respects:

1. A folding aperture backsight is fitted behind the charger bridge, graduated from 200 to 1,450 yards. In the folded down position, a battle aperture sight is provided and the leaf is raised for the longer ranges using an adjustable slide moving in 50 yard increments;
2. The long range dial sight and flip-up aperture have been omitted;
3. An upper band is fitted at the rear of the nosecap, onto which the piling swivel is mounted and both items are secured to the nosecap by the same screw;

Above & below: Mark V S.M.L.E. with backsight raised. Also note the "V"-marked safety catch, band at the rear of the nosecap, and large-headed nosecap screw. *MOD Pattern Room*

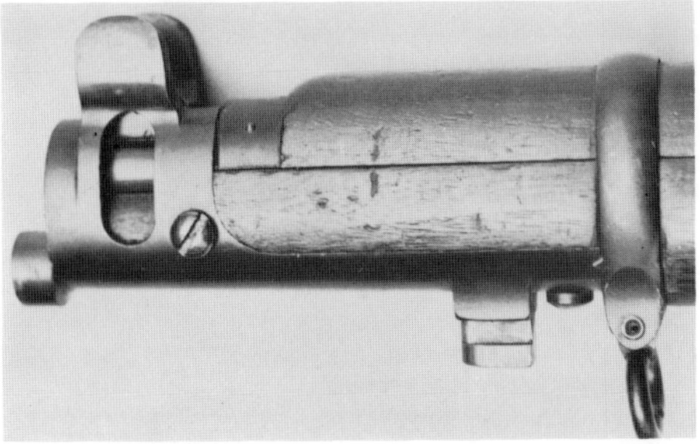

4. A full-length handguard is fitted in two pieces and there is no backsight mounted onto the barrel;
5. The nosecap screw has a large head, slotted for a coin;
6. The magazine cut-off has been retained but there is no spotting or production hole;
7. The swivel mount on the trigger is replaced by a wire loop, as on the Mark III* rifle;
8. The cocking piece is similar to that on the Mark III*, with three vertical grooves milled into each side and no projecting rim or collar at the rear;
9. A differently shaped safety catch is fitted, with angular grooves on the forward edge; it is marked "V" to indicate Mark V.

As well as providing an updated model for manufacture, the new Mark V was intended as a pattern for upgrading Mark III rifles. The two main improvements were the aperture sights which also provided a longer sighting radius, and the reinforced nosecap for bayonet fighting.

Sights on the early production Mark V rifles were found to be unsatisfactory because the backsight slide was discovered to "climb" during firing, due to the recoil. To remedy this, stronger springs were fitted and the nib was re-designed so that the setting of the slide at the various ranges was more positive. Another variation observed in some early backsights is that the graduations are to 1500 yards, made possible by a different indexing position on the backsight slide.

As happened with previous new rifles, the Mark V was sent to Bisley for a showing and tests on the mounds at the National Rifle Association meeting; the consensus of opinion of the service and civilian shooters was that:

The rifle was not properly zeroed;
The trigger pressure was too heavy;
The rifle did not shoot consistently;
Rearsight adjustment in 100 yard steps was not fine enough;
Aperture size of the backsight was not suitable;
The backsight would be prone to damage;
"Aiming off" was more difficult with the aperture sight than with the "U" notch backsight of the Mark III;
The aperture sight was generally approved of;
Disappointment that a heavier barrel was not incorporated.

When the Small Arms Committee was constituted over 20 years earlier, the development of the rifle was their main concern but with the recent changes and developments in strategies and armaments, it became obvious that the rifle often required special and separate consideration and so it was decided to set up a Rifle Sub-Committee.

This new Rifle Sub-Committee recommended a new design on 14th December 1923, which resulted in an experimental model which incorporated:

1. Heavy barrel with the same profile as the M.L.E. rifle, but shortened to 25.2 ins. and fitted with lugs near the muzzle for securing a bayonet or discharger cup.
2. Reduced body weight with modified Mark V type sight bracket and charger guide.
3. Two-piece firing pin with shorter cocking piece to avoid injury to the hand in bayonet fighting.
4. A more compact bolthead with single arm spring, controlled in the body by an internal rib and claw and not protruding over the side of the body.
5. Modified fore-wood and handguard, so that the muzzle protruded a few inches.
6. A folding aperture backsight like the Mark V, only smaller, less liable to damage and with finer vertical adjustments.
7. A new keeper plate with "ears" to strengthen the fore-end.

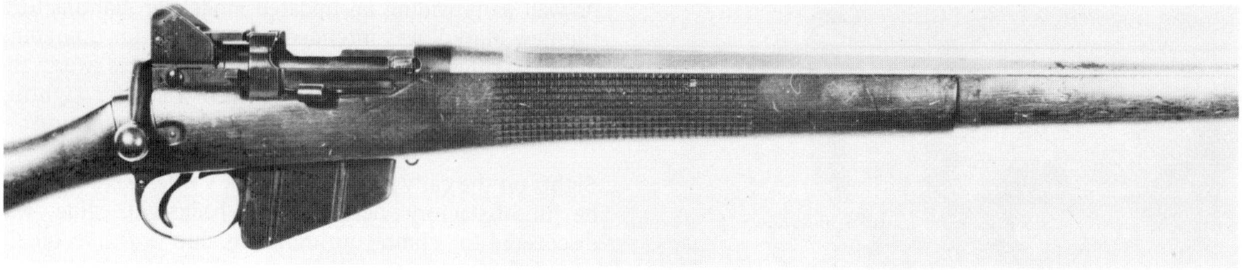

Above & below: 1924 Experimental rifle with handguard removed, showing the special barrel with wide, longitudinal flutes. R. Faris

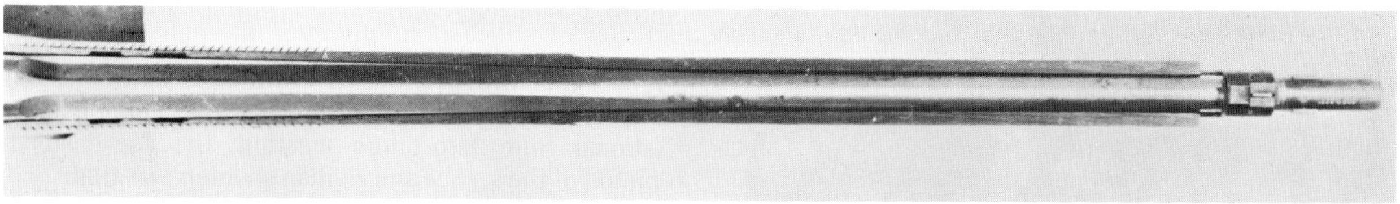

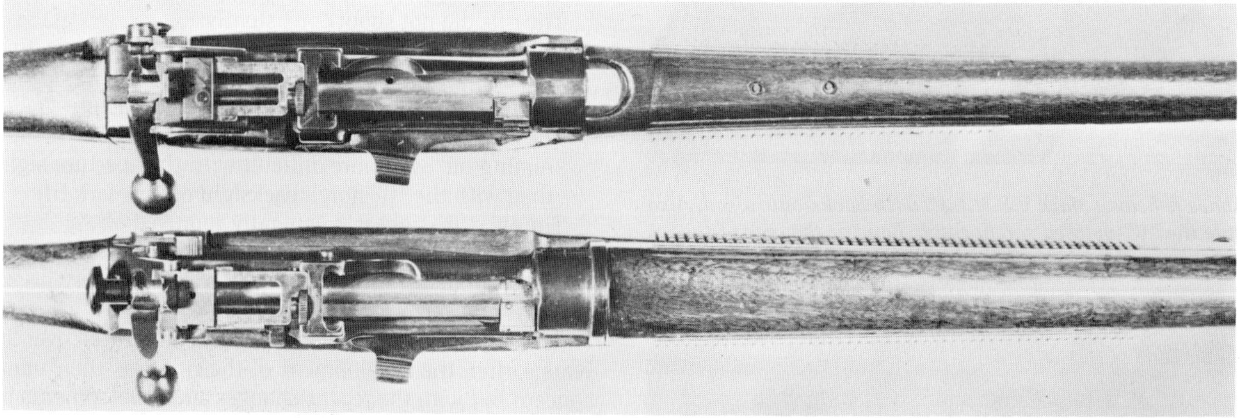

Above & below: Comparison photos of 1924 Experimental rifle with No. 1 Mk VI, with 1924 rifle at top in each case. Note the shortened SMLE-style nosecap, SMLE safety and the different bolt of the 1924 Experimental. R. Faris

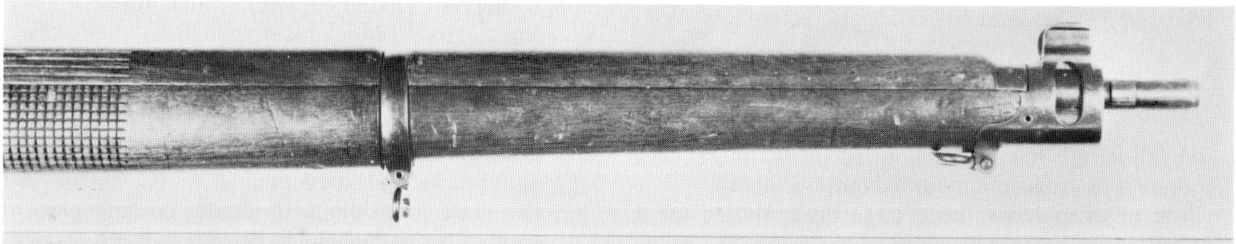

8. An undercut foresight blade for improved definition, with the adjustment effected by a special key.

Two of these rifles were exhibited at the Hythe Small Arms School on 31st July 1924, before students at the Senior Officers School.

This experimental model is quite special, with certain features similar to the Mark V and the following Mark VI S.M.L.E. because it represents a link between the two. The nosecap was like that on the Mark V, but shortened, with the barrel protruding beyond the nosecap; the handguards and style of sights are similar to the Mark V although the backsight and protectors are differently constructed. A battle aperture is provided and the sight is graduated to 1,300 yards, bearing some resemblance to the eventual No. 4 type.

The action body is still like the Marks III and V, with a rounded wall on the left side and circular thumb cut for charger-loading; the charger guide and backsight protecting ears are quite robust. The cocking piece has been specially modified so as to reduce its rearward length and there is a special release catch for the bolthead, secured at the back of the receiver by a special screw.

This prototype is actually the first experimental model of the Mark VI and eventual No. 4 rifles, with its heavy barrel protruding through the nosecap and foresight protector. However, the action body or receiver is much lighter and some problems were encountered with its accuracy which resulted in the recommendation that it be strengthened, as in the following design. The particular trials on these rifles were conducted at Hythe in mid-1924.

The relative success of this new pattern was responsible for the demise of the Mark V and so the plans to convert the previous patterns of S.M.L.E. rifles were also abandoned. Considerable problems were caused in this particular proposed conversion, mainly because the sights were on two separate components; the backsight on the receiver body and the foresight on the barrel, so accurate alignment was almost unobtainable in normal run of work. Where this alignment had to be made in new production rifles it presented less of a problem than with matching older and, all too often, worn parts.

Further modifications were recommended as a result of the Hythe tests, these included:

1. A modified aperture backsight with rapid screw elevation and a "click-stop" device.
2. Chequering of fore-end and buttplate to improve grip.
3. Alteration of the bolthead so that it could be more easily released by the fingers.
4. Provision for attachment of the Pattern 1907 bayonet.

Points numbers 2 and 3 were incorporated in the next prototype but it had already been decided that the long-bladed bayonet was unsatisfactory as a weapon or utility tool and that a much shorter and lighter form of bayonet should be used. The longer, heavier bayonet also required a more substantial mounting which was not in the best interests of weight reduction and the longer blade had poorer penetration and withdrawal qualities.

The proposed lighter bayonet had no "handle" or crossguard and it was intended for use solely on the rifle. A spike, about 7.5 ins. long and cruciform in cross-section, replaced the blade. It was intended that the lugs on the barrel for securing the bayonet be also utilised to fix a grenade launching cup. This bayonet was eventually introduced as the No. 4 Mk I, with relatively minor modifications.

In 1926, six prototype rifles were assembled incorporating the aforementioned suggestions as well as some other features which facilitated more easier production. The illustrated rifle is one of these prototypes and features a strengthened body with integral charger guide, smaller improved bolthead, redesigned and lightened bolt, heavier barrel, rectangular undercut foresight with lateral adjustment, and a chequered fore-end and buttplate. The new, stronger rearsight was less liable to damage, with a quick acting elevation screw which incorporated a click device representing one click for each 25 yards at 500 yards. This was later designated the "A Pattern."

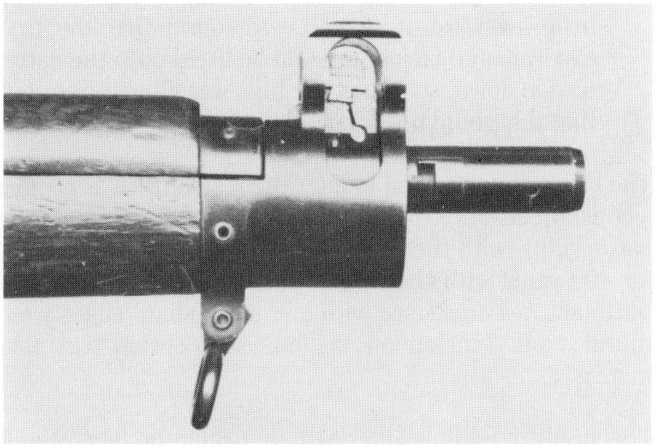

Lightened nosecap of "A" Pattern rifle. *MOD Pattern Room*

During range tests at Enfield, it was found that better accuracy was obtained with the bayonet fixed and after the barrel had been warmed up by a few shots. In July 1926, four of these rifles were sent to the Small Arms School at Hythe and the reports were quite favourable. More trials were then carried out in comparison with the Mark III S.M.L.E. and although the new rifle was generally well reported upon, a number of mostly minor defects showed up, so a few changes were recommended.

Results from these trials as well as opinions from the various establishments where they had been tested, resulted in consideration of the following points by the Committee:–

1. The flared and rounded button at the rear of the cocking piece was approved of as it prevented injury to the hand in bayonet fighting, which was its intention.
2. The bolt and bolthead "worked" more smoothly owing to the new design.
3. The fore-end chequering was approved of.
4. The buttplate chequering was considered unnecessary.
5. The butt trap spring was insufficiently strong and could be opened accidentally.
6. The pistol grip was thought to be positioned too far from the trigger.
7. The rearsight elevation adjustment was thought to be too loose and the click settings were intended for graduation at 500 yards, therefore at other ranges the adjustment was inconsistent. It was considered that each click should alter the elevation by the same amount, at least to 600 yards. Some of the markings on the sight were not definite enough.
8. Requests were made for the foresight bed and sight protectors to be made narrower, in order that the "Degree Method of Indication and Recognition of Targets" could be applied.
9. The back of the cut-off when in the disengaged position, was found to be too sharp.
10. Care had to be taken with the protruding muzzle, that obstructions were not picked up.
11. The bayonet was found to be a little difficult to fix and the balance of the rifle with bayonet fixed was a little awkward at first. It was found that the first experimental bayonets could be fitted onto the barrel upside down, so one of the lugs was altered in order that this could not occur.

The Committee decided to alter the rearsight so as to give a constant advance of 2 inches per 100 yards and have the battle sight (with the leaf in the folded down position) set for 400 yards with the bayonet fixed. Chequering on the buttplate was dispensed with and some sharp edges were rounded off. Tension on the butt trap spring was also increased.

The stores of military rifles held by Britain on 31st March 1924 comprised:

Rifle	*Home*	*Abroad*	*Total*
SMLE Mk III	207,629	45,769	252,998
SMLE Mk III*	1,328,572	5,293	1,333,865
SMLE Mk I & Conversions	90,194	924	91,118
Patt. 1914	764,865	77	764,942
Patt. 1914 (F) & (T)	3,362	159	3,521
SMLE D.P., all Marks	28,510	1,109	29,619
SMLE E.Y., all Marks	68,043	1,935	69,978
Rifle .22 Sht, All Patterns.	10,588	951	11,539

This put the total on hand at more than 2.5 million rifles, and the table also demonstrates the relative proportions of the various rifles at the time.

More experiments were carried out with different rates of twist in rifling in 1926 and the Mark III* was used with special barrels. One of these particular rifles is in the Pattern Room collection and the barrel is marked on the knox form "EXPL SPIRAL ONE TURN IN 8 INCH." Normal twist for the .303 rifle is 1 turn in 10 inches and the more rapid twist is usually required to stabilise heavier bullets. It may well be that experimental ammunition with different propellants and/or different weight projectiles was being tested.

The nomenclature of British service small arms and some other equipment was changed and updated on 31st May 1926 so as to simplify the ever-increasing number or marks and such variation in the war material. The List of Changes announcements, §§ A1635-A1690, run to many pages and an update is also provided in this LoC reference, as to the items then classed as being obsolete, obsolescent, omitted from Vocabulary or with any changes in designation, along with the appropriate cross-reference paragraphs in previous List of Changes entries.

Such changes applicable to this study are:

Old Designation	*New Designation*
Rifle, Short, MLE Mk III	Rifle No. 1 Mk III
Rifle, Short, MLE Mk III*	Rifle No. 1 Mk III*
Rifle, D.P., Short MLE	Rifle No. 1 D.P.
Rifle, E.Y., Short MLE	Rifle No. 1 E.Y.
Rifle, Short, .22 RF Mk IV	Rifle No. 2 Mk IV*
Rifle, Magazine, .303-in Patt '14:	
Mk I* W w/ fine adjustment sight	Rifle No. 3 Mk I* (F)
Mk I* W (T)	Rifle No. 3 Mk I* (T)
Rifle, Short, .22in RF:	
Sight auxiliary, aperture, Mk I	Sight auxiliary, Mk I

The following are listed as being obsolete:

Rifles, Charger-loading, M.L.E.
Rifles, M.L.E.
Rifles, M.L.M.
Rifles, Long, .22-in. RF Mk II

Those rifles omitted from Vocabulary but retained in store for possible future requirements were:

Rifles, Short, M.L.E.:-
Mark I
Mark I*
Mark I***
Converted Mark II
Converted Mark II*
Converted Mark IV

The Rifle, Short, .22-in RF Patt '14 No. 2 was noted as having been omitted from Vocabulary, to be converted to Rifle, Short, .22-in RF Mark IV*.

Prior to this announcement, in 1925 the London Small Arms factory at Albion Works, Ossary Rd. closed down leaving B.S.A. as the only other contractor capable of filling the service orders for rifles. Therefore, by the 1930's, B.S.A. Co. at Small Heath was the only factory in

Britain producing the No. 1 rifle, though these were mostly foreign orders. When the 1939-45 war broke out, B.S.A. was the sole plant that could produce the Short Lee-Enfield.

Large quantities of the No. 1 Marks III and III* rifles were produced by B.S.A. for the trade and foreign governments. A typical, but in some ways slightly unusual example, is the order for Siam on which a tiger head and Siamese script was marked on the right side of the butt socket in place of the usual British factory and designation markings. The accompanying Pattern 1907 bayonets were similarly marked on the blade ricasso.

Siam ordered 10,000 S.M.L.E. rifles from B.S.A. complete with bayonets and slings in January 1920, to a total value of £76,625. The order was completed by December of the same year and the inscription in Siamese (Thai) reads "Mark Rama VI BE 2462." BE 2462 indicates Buddhist era 2462 (AD 1919 on the European calendar). Other markings on these rifles in Siamese script on the receiver, bolt, nosecap boss and bayonet are serial numbers; the transcription in western numerals is:

๑	๒	๓	๔	๕	๖	๗	๘	๙	๐
1	2	3	4	5	6	7	8	9	0

Siamese SMLE & P'07 bayonet markings. The script number on the bayonet ricasso reads "6801" in western numbers.

King Vajiravualh reigned from 1910 to 1925 under the title of Rama VI; he had been educated in England and was a graduate of Sandhurst. Upon completion of a musketry course at Hythe he was attached to the Royal Durham Light Infantry for a short time. The tiger is representative of the Wild Tiger Corps which the king founded in 1911 as a volunteer force largely composed of public servants and also included the royal bodyguard.

Even in the late 1920's, the Lee-Enfield magazine cut-off was undergoing improvement. Enfield Design Department drawings DD(E) 936 show an experimental cut-off made from pressed sheet steel instead of the usual cast type, and the strengthened section at the pivot screw axis hole has the top piece brazed or welded on. The first drawing is dated 2nd May 1928 with a revision on 2nd November 1928. This improved cut-off is very similar to that used on the No. 1 Mk VI rifle soon afterwards.

The inter-war era saw most powers looking towards updating their small arms, but while considerable advances were made in some fields such as machine guns, most countries commenced the 2nd World War with the same rifle systems, and in many cases, the same rifles as used in the 1914-18 war. The notable exception was the United States where the self-loading M1 Garand was adopted although it must be pointed out that the United States was not plunged into the war as early as Europe and had more time for preparation.

The proposed No. 1 Mark VI rifle was almost ready for trial by troops by late 1930 and facilities at Enfield were readied for the commencement of production of a limited number of rifles for these trials. The Superintendent R.S.A.F. advised that the first 25 Mark VI rifles could be sent to Hythe by the end of 1930 and that the first 500 were expected to be completed by 1st May 1931 with a further 500 by mid-June. These troop trials were conducted by the Small Arms School, 3rd Carabineers and 2nd Bn. West Yorkshire Regiment during 1931. Delays had been caused by inconsistent shooting which was traced to certain machining operations on the receiver, this resulted in improper breeching up.

Further trials were commenced at Aldershot, Eastern and Northern Commands and in Egypt in 1932, to be completed by October of 1933. Arrangements were also made for some trials to be carried out in India and the factory at Ishapore made moves towards future production of the new rifle. Although it was not specified, it would appear that the 1932-33 trials used No. 4 rifles.

The production years marked on rifle receivers of this initial production run are 1929 and 1930 although the rifles were not completely assembled until late in 1930 and early in 1931. This was often the case, where the year marked on the receiver or action body referred to when that component or group of components was finished and not necessarily when the rifle was fully assembled. An examination of the year marked on the barrel, receiver and butt of any rifle prior to the No. 4 can prove to be an enlightening exercise in this respect.

The serial numbers of the No. 1 Mk VI rifles commence with the letter "A" and they appear to have run from A0001 to about A1000. The Mk VI has the receiver designation markings on the left side of the receiver at the butt socket, in a similar position to the Lee-Metford and Lee-Enfield carbines or the Converted Marks II and IV series of S.M.L.E. rifles. These No. 1 Mk VI trials rifles were later referred to as the "B Pattern." The A Pattern was the prototype with chequered buttplate, of which only six were produced.

The first 535 Mk VI rifles passed inspection by 31st March 1931 with the remaining 490 being passed by 15th June 1931, a total of 1,025 rifles. It is also recorded that 1,050 spike bayonets were made for these trials rifles; 584 of which passed inspection by 30th April 1931 and 466 by 16th June 1931.

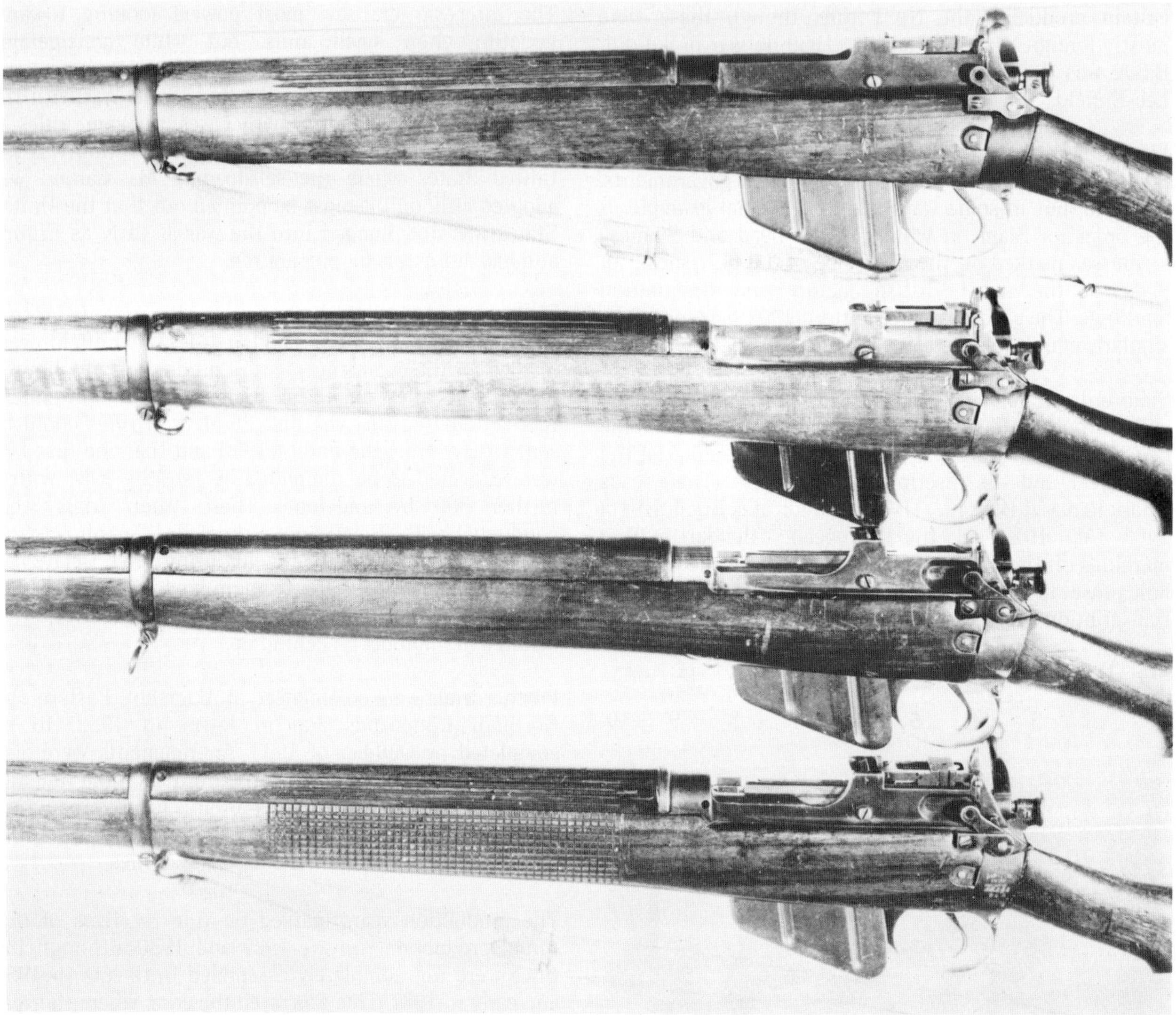

From top: Early No. 4 rifle and intermediate experimental models, to the No. 1 Mk VI (bottom), in reverse chronological sequence. The No. 1 Mk VI type receiver with plain wooden furniture (2nd from bottom) has a missing rearsight. Note the differences in the receiver walls. *R. Faris*

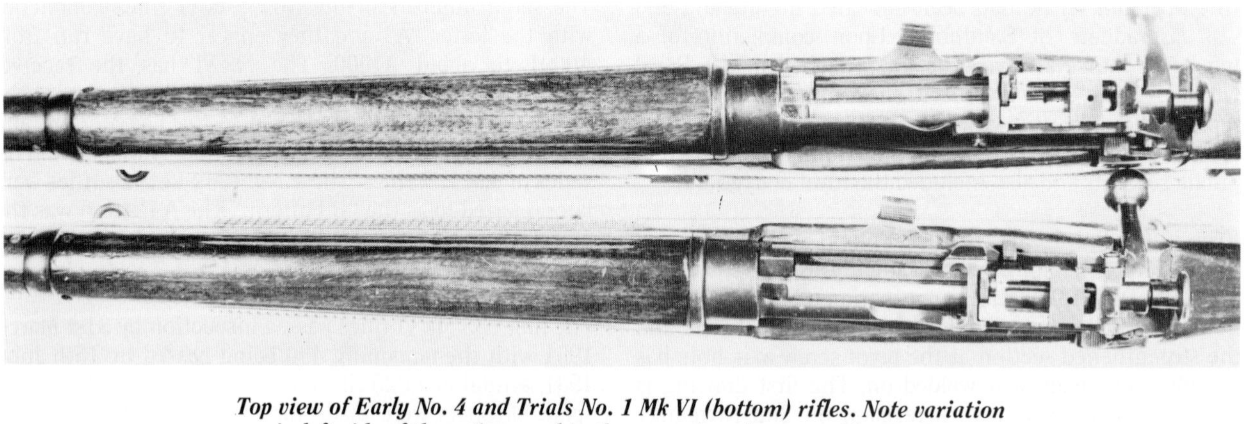

Top view of Early No. 4 and Trials No. 1 Mk VI (bottom) rifles. Note variation in left side of the actions and in the magazine charger guides. *R. Faris*

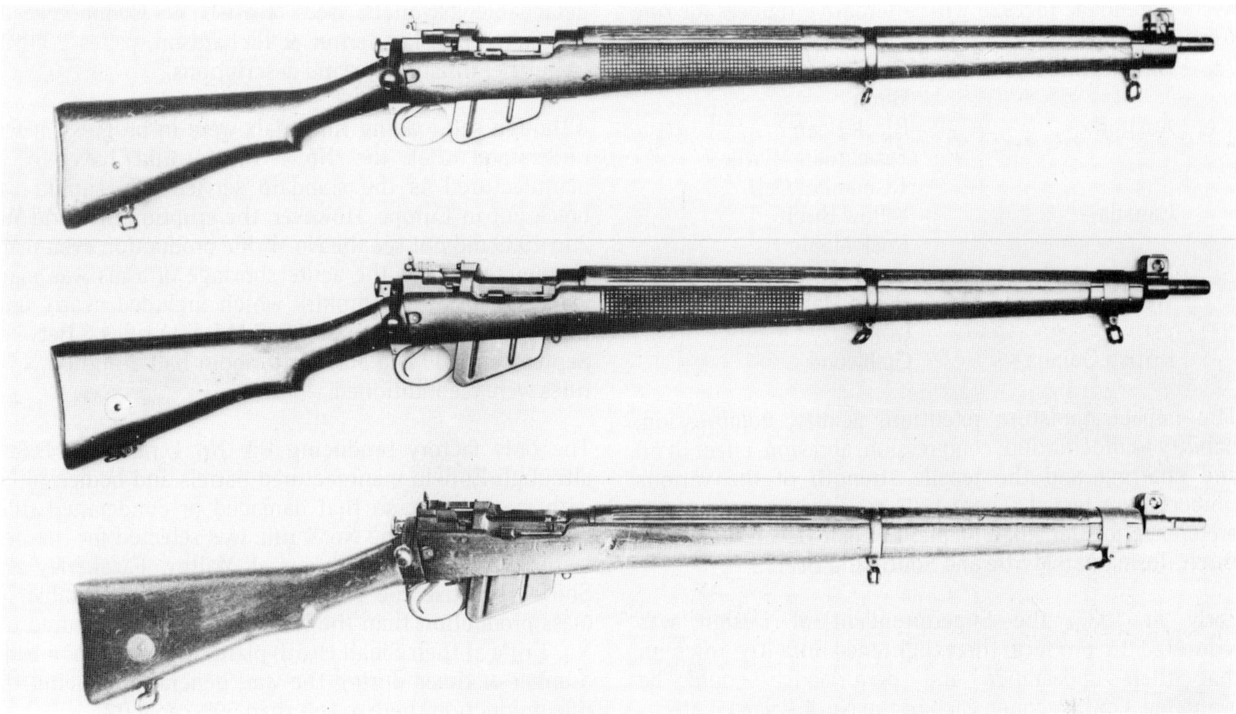

From top: No. 1 Mk VI "A" and "B" Models, and the troop trials No. 4 rifle (also classified as "B" Model due to action similarities). MOD Pattern Room

In mid-1931, the new rifle was redesignated the No. 4 Mark I, although the No. 4 trials rifles are different in that they have a stronger receiver with a higher left wall for the boltway; another obvious feature of the trials No. 4 was the absence of fore-end chequering. About 2,500 of the No. 4 trials rifles were produced by 1933 and these are marked in the same position on the receiver butt socket as the No. 1 Mk VI. Again, the serial numbers commenced at A0001; further details of the No. 4 trials rifles will be found at the beginning of the following chapter.

Two more models of the Mark VI / No. 4 trials rifles were proposed, designated the C and D patterns.

The Model C featured:

- Light buttplate of aluminium alloy
- Strengthened backsight elevator screw
- Strengthened foresight protector
- 12-inch blade bayonet
- Improved backsight leaf
- Altered profile of the fore-end
- Sling swivel bracket mounted on the guard.

The Model D was to incorporate:

- New bolthead retaining catch
- Modified backsight
- New cut-off
- Strengthened right rib on the body
- Increased resistance bearing surfaces.

It was proposed to assemble twelve Model C and 2 Model D rifles but only six of the Model C were built up and the Model D was eventually abandoned because it involved too many major changes. Later on, 50 more of the Model C rifles were fabricated for trial as well. By 22nd May 1935, these had been completed and undergone accuracy tests, ready for trial. Because the self-loading trials were in progress it was generally considered that there was no need to continue with the No. 4 rifle tests, but instructions were given that twelve rifles be sent to Hythe for certain trials on 20th August. History has proven this to be a far-sighted decision.

Therefore, the quantities of these trials rifles are recorded as being:

A Pattern (No. 1 Mk VI)	6
B Pattern (No. 1 Mk VI)	1,025
B Pattern (No. 4 Mk I)	about 2,500
C Pattern (No. 4 Mk I)	56
D Pattern	abandoned

Further development which was carried out on the trials No. 4 rifles is described in the following chapter. In the official documentation, there is little specific reference to No. 1 Mark VI, the Mark VI or No. 4; rather, the Patterns or Models A, B, C and D, although in some instances no type differentiation is made at all. This has resulted in some confusion because the reports from various tests and trials as well as recommendations for improvements and changes become very easily confused.

Tests were made in 1928 with alternative timbers for rifle furniture. These were:

United Kingdom –	Beech
	Birch
Australia –	Black Bean
	Tasmanian Myrtle
	Tasmanian Oak
Canada –	Yellow Birch
	Rock Maple
New Zealand –	Southland Beech
West Africa –	Benin Mahogany
	Iroko
British Guiana –	Crabwood

The impact, moisture retention, acidity, compression, liability to indentation, compression, abrasion, effect of oil and charring and the tensile strength of the various timbers were tested. These tests resulted in an order of preference for Canadian Maple, English Beech, Canadian Birch, Tasmanian Myrtle and Southland Beech.

Early in 1935, the Superintendent of Design was requested to produce three lightened infantry rifles so that their appearance and performance could be evaluated. One specimen each of the No. 4 (trials), Pattern 1914 and No. 1 Mk III rifles was produced for examination by the Committee. While the other two rifles were ordered to be fitted with bayonets, the No. 1 Mark III was not.

Other requirements were:
- A 5-round magazine
- Wooden furniture reduced to "handguard proportions"
- Barrels shortened, possibly to 20 inches
- Weight of about 7.5 lb. or less, if possible
- No provision for the firing of grenades.

The three light infantry rifles were converted from the particular standard patterns and it was inquired if an aperture sight could be fitted, so long as it was practical, although it was the regular open sights which were fitted to the No. 1 version. Light alloy parts were desired for components such as the trigger guard, nosecap, bands, butt plate and oil bottle, to further reduce the weight.

For this particular conversion, a No. 1 Mk III* rifle was used, the barrel shortened to 21.7 inches, the fore-stock "sporterised" by removing the forward portion, the sight leaf lightened and a 5-round magazine fitted; the all-up weight was 7 lb. 10 oz. One each of the shortened No. 1 and No. 4 rifles was produced for examination at the Small Arms Committee meeting on 27th March 1935 but there was no further development in this direction for some time.

These lightened rifle conversions were undertaken by the trade and by the end of March the modified rifles had been submitted. Another conversion of the Pattern 1914 model was requested from Soley Arms Co.; the suppliers of these conversions were B.S.A. and Soley. A few various prototypes were produced, mostly with folding or detachable bayonets; see "British & Commonwealth Bayonets" by Skennerton & Richardson, pages 208-213 for further illustrations and descriptions.

Although self-loading rifle trials were in progress, it was understood that the No. 4 rifle would have to be manufactured as the standard service rifle should war break out in Europe. However, the eruption of World War 2 in 1939 did not see the No. 4 rifle production even ready to commence and the acute shortage of arms was made worse by losses at Dunkirk which included nearly one-third of a million No. 1 (S.M.L.E.) rifles. Between September 1939 and May 1940, about half-a-million No. 1 rifles were reconditioned.

The only factory producing the No. 1 rifle was B.S.A., although Enfield manufactured barrels and bodies, along with many parts, so that damaged or condemned rifles could be rebuilt. The No. 4 rifle was selected for the new factories under construction at Maltby, Fazakerley and Shirley because the No. 4 was much more suitable for mass production than the No. 1. B.S.A. manufactured the No. 1 rifle at their Small Heath plant, which was bombed a number of times during the war, generally reducing the rifle production there to less than 50% capacity.

The first of these Birmingham air raids occurred in August 1940, when the barrel mill and adjacent shops were destroyed, halting production. Severe damage was again inflicted during November, after which some of the vital machinery was dispersed and certain operations carried out elsewhere. Large quantities of components were being produced throughout Britain by private contractors, especially from late 1941.

Many of the B.S.A. made No. 1 rifles made in the late 1930's and early part of the war have "L" prefix serial numbers and Small Heath is recorded as having ceased No. 1 rifle production in October-November 1943. R.S.A.F. continued to produce components, especially furniture, and the lighter-coloured beech timber was used for fore-ends and handguards of many rifles during this period.

No. 1 rifle production, comprising parts assembly and utilisation of many cannibalised parts, was continued at "Dispersal Factories" after the S.M.L.E. plant at Small Heath closed down in late 1943, although this scheme appears to have been in operation well before this. From the examination of some of these rifles, the dispersal factories commenced operation as early as 1940 or 1941 and dates up to 1944 have been noted.

The action markings of these dispersal factory rifles are different in that they have no factory name, the year and rifle designation markings were often made with individual stamps which do not align properly. These rifles were assembled from spare parts wherever possible, many having been stripped from condemned or damaged rifles, although some components like the beech furniture were new.

G. R.
B.
1944
S^{HT} L.E.
III*

Left: Typical Dispersal Factory cypher and factory markings. Below: Similar action marks with later "FTR" marking. *MOD Pattern Room*

Such an example inspected at the M.O.D. Pattern Room and illustrated here, has a 4-groove barrel, B.S.A.produced sight components, Australian-made Queensland Maple butt from a Lithgow Mk III S.M.L.E. and new beech fore-wood and handguards. It is considered that these 1941-44 vintage rifles were assembled in Birmingham and that some of the B.S.A. Small Heath machinery was used. Four groove rifling was approved as a simplification measure in March of 1942 because B.S.A. complained that 2-groove rifling (offered as an alternative in 1941) was not as practicable due to the arrangement of their plant.

Birmingham Small Arms' other rifle factory was at Shirley, although none of the No. 1 components were produced there. Fortunately, the Shirley plant did not suffer the bombing raids endured by Small Heath. B.S.A. employed a total of 42 factories during the war; at Redditch, Mansfield, Leicester, Tipton, Dudley, Bromsgrove, Newcastle, Stoke, Corsham, Ruislip and a number of Birmingham suburbs. These various factories produced the Besa machine gun, Boys anti-tank rifle, Browning machine gun, Hispano Suiza, Oerlikon, Sten machine carbine, Bren machine gun parts, fuzes and high velocity and rocket projectiles. Some rifle components were also manufactured at the other B.S.A. factories.

The figures for No. 1 rifle deliveries into store during 1940-1941 were:-

1940	May	9,618	
	June	7,387	
	July	8,722	
	August	8,919	- 1st air raids, B.S.A. Small Heath.
	September	8,676	
	October	6,197	
	November	5,003	- Production severely hampered by air raids; B.S.A. machinery was dispersed, "peddled scheme" production of the No.1 continued.
	December	4,549	
1941	January	3,311	
	February	1,983	
	March	3,160	
	April	2,387	
	May	5,493	
	June	4,009	- plus Maltby's first No.4 rifles.
	July	2,915	- and Fazakerley's first production.
	August	4,900	- and Shirley's first production.
	September	3,503	- factories producing the No.4 rifle reduced call on No.1 production.
	October	4,169	
	November	2,638	
	December	3,575	

Before mid-1941, none of the No. 4 rifles were available but after this the No. 1 rifle figures were increasingly augmented by the No. 4 rifles from Maltby, Fazakerley and the B.S.A. Shirley factories. R.S.A.F. Enfield and B.S.A. Small Heath produced quite a few separate No. 1 receiver bodies and bodies with barrels attached, these were then assembled using components from down-graded or unserviceable rifles at other ordnance depots and by the gun trade.

Enfield drawings and sketches indicate further concessions in the production of No. 1 parts and accessories:-

07.08.40	Modification to buttstock to accept No. 4 buttplate.
07.08.40	Modification to No. 4 buttplate to suit No. 1 stock.
07.07.41	Experimental sight for use with No. 64 grenade.
11.07.41	Axis pin, backsight, No. 1 rifle.
15.07.41	Experimental sight for use with No. 68 grenade.
20.10.41	Proposed backsight catch.
22.10.41	Proposed amendment to Cocking piece "B."

Repairs to the No. 1 and No. 3 rifles were allocated to many renowned British gunsmiths in addition to Enfield and B.S.A.; many such repairs involved complete rebuilds.

Assembly of the No. 1 rifle from cannibalised parts was carried out by Cogswell & Harrison, Holland & Holland and Westley Richards; general repairs were effected by Boss & Co., Parker-Hale, Skimin & Woods, Purdey, Greener, W. Darlow, Tranter Bros., Wilkes, Midland Gun Co. and other such members of the trade.

These gunmakers were also engaged in the assembly and repair of machine guns and pistols as well as manufacture of certain components. Cogswell & Harrison made 4-inch smoke dischargers as did Westley Richards, A.T. Ralphs and B.S.A. The Lanchester machine carbines were assembled by W.W. Greener and Boss & Co. and most of the gunsmiths made some S.M.L.E. and Sten gun components. Greener made the No. 1 Mk III signal pistol and B.S.A. produced the Boys anti-tank rifle and the Browning and Besa machine guns.

No. 1 rifle actions were also utilised for making up smoke dischargers. These were introduced from 1936, the "Discharger, Smoke Generator, 4-in., No. 1 Mk I" employed the .303-in. Martini-Enfield action, the No. 2 Mk I variant used an S.M.L.E. action, the No. 3 Mk I model was on a Martini action made up as new, and the No. 4 Mk I type applied a P'14 rifle action. Such a discharger is illustrated in the Indian chapter, in 2½-in. configuration.

Sizeable quantities of parts for the No. 1, No. 3 and No. 4 rifles were produced by private industry, particularly the No. 4, and then these parts were forwarded to the Ordnance Factories or stores for assembly, to effect repairs, or to increase reserves. British production of the No. 4 was augmented when supplies from Stevens-Savage and S.A.L. (Long Branch) in North America also became available from 1942.

The 2nd World War British manufacture of the No. 1 rifle components is interesting because many of these parts were marked with the wartime codes. The makers' codes during the 1939-45 war were:–

Manufacturer	*Code*	*Component/s*
Aero Zip Fasteners, Southampton Row, London WC1	S312	Washers, locking bolts.
Albion Pressed Metal Co., Birmingham	M3	Magazines, safety catch springs.
Anstey & Wilson Ltd., Kenyon St., Birmingham	M8	Magazine platforms.
Barlock Typewriter Co., Nottingham	M625	Sears.
B.S.A. Guns Ltd., Small Heath, Birmingham	M47A	Most components, assembly of complete rifles.
T. Blakemore Ltd., Walsall	M32	Extractor springs.
J. Boss & Co., Albemarle St., London W1	S156	Extractors, boltheads, handguards. Repairs.
Butler, Spragg & Co. Ltd., Birmingham	M52	Swivel screws.
Carr & Smiths, Birmingham	M54	Extractor & backsight springs.
Chiswell Wire Co., Sandown Rd., Watford	S169	Springs.
Cogswell & Harrison, Park Rd., East Acton, London W3	S171	Backsight slides, foresight blades.
Cox & Co., Watford-By-Pass, Watford	S177	Safety catch components.
W. Darlow, Bedford	-	General repairs, tubing of No. 1 rifle to .22RF.
Dean Bros. Ltd., Nottingham	M627	Magazine cases, springs.
Economic Stampings Co., Leicester	M605	Extractors.
Electrical Engineering & Plastics Ltd.	-	Extractors.
Elkington & Co. Ltd., Birmingham	M78	Bands, safety catches.
Freeborn Powers Converters Ltd., Rye Rd., Hoddesden	S368	Breech bolts (contract possibly cancelled).
G. E. Fulton & Sons	-	Sight beds, foresight blades.
W. W. Greener, Birmingham	M94	Buttplates, strikers, extractor springs.
Guest, Keen & Nettlefold Ltd., Birmingham	M99	Screws, butt marking discs.
Hanwell Engineering Co. Ltd., Northampton	M62	Boltheads.
John Harper & Co. Ltd., Willenhall	M107	Protectors.
W. A. Hudsons Ltd., Curtain Rd., London EC2	S221	Safety catch components.
Jones Sewing Machine Co. Ltd., Manchester	N36	Boltheads, bolts.
Kings Langley Engineering Co., Kings Langley, Herts.	S60	Safety catch components.
F. W. Kubach, Sylvan Rd., London SE19	S374	Safety catch components.
Midland Gun Co., Birmingham	M170	Miscellaneous repairs & parts.
N. K. Watch Case Manufacturing Co. Ltd.	-	Backsight pins.
R. B. Page & Co., High St., Kingston-on-Thames, Surrey	S91	Foresight blades, strikers, conversion to .22RF.
A. J. Parker, Birmingham	M182	Safety catch washers, springs, magazine platforms.
Parker Hale Ltd., Birmingham	M192	Breech bolts, safety catches, nosecaps, magazines, cocking pieces, extractors, foresight blades, .22RF boltheads.
Premier Screw & Repetition Co., Leicester	-	Screws, pins, rivets.
Premier Spring Co., Redditch	M200	Mainsprings, nosecap screws.
Prince-Smith & Stells Ltd., Keighley	N56	Magazine platforms.
A. T. Ralphs (N.C.H.) Ltd., Leicester	M612	Strikers, pins.
Redfern, Stevens Ltd., Birmingham	M205	Washers, pins.
Restwell Springs Co., Darwen, Lancashire	N165	Mainsprings, fore-end stud springs.
Roneo Ltd., Romford, Essex	S262	Cap screws.
George Salter & Co. Ltd., West Bromwich	M215	Butt trap springs.
Sanbra Ltd., Birmingham	M216	Buttplates.
Sanghams Weston	-	Inner bands.
Tom Senior, Atlas Works, Liversidge, Yorks.	N63	Backsight leaves.
Singer Manufacturing Co. Ltd., Clydebank, Scotland	N67	Markings discs, washers, rivets.
E. Smith & Sons	-	Extractor springs.
Southern Railways, Eastleigh, Hants.	S360	Inner band screws.
Spartan Engineering & Mechanical Co. Ltd., London	S273	Cocking pieces, backsight caps.
Streetly Tools & Stampings, Streetly, Birmingham	M241	Brackets.
Strong & Co.	-	Backsight axis pins, screws.
Tambis Engineering Co., Northolt	-	Backsight components.

Manufacturer	*Code*	*Component/s*
Herbert Terry & Sons Ltd., Redditch	M245	Washers.
Tapley & Co.	-	Magazine stop clips.
Tranter Bros., Birmingham	M248	Miscellaneous repairs & parts.
Trevor Stampings Ltd., Welwyn	S114	Magazine platforms, backsight caps, nosecaps.
W. Trueman & Sons, Birmingham	M249	Stockbolt plates & washers, extractors.
United Spring Co. Ltd., Smethwick	M252	Extractor springs.
Viners Ltd., Sheffield	N79	Sear springs.
Walker & Hall Ltd., Sheffield	N80	Butt trap springs, sling swivels.
Ward & Sons, Birmingham	M262	Washers.
W. J. Wild Ltd., Birmingham	M269	Stock bolt washers.
S. Wright & Sons, Birmingham	M276	Springs.
Zipp Fasteners Co., Edmonton	S300	Washers.

English gun trade members who undertook No. 1 rifle repairs and reconditioning, but did not manufacture components were:-

Holland & Holland, 98 New Bond St., London W1	S51	Also repaired No. 3 & Ross rifles, made No. 4 parts.
J. Purdey & Son, 57-58 South Audley St., London W1	S388	Also repaired No. 3 rifles.
Skimin & Woods, Birmingham	M244	Also repaired No. 3 rifles.
Westley Richards, Bournebrook	M268	Also repaired No. 3 rifles.

Birmingham Small Arms (B.S.A. Guns) had a number of factories engaged in small arms production, as mentioned before, although the No. 1 rifle was produced at the Small Heath plant. Some of the other B.S.A. factories provided parts for the No. 1 also, the various B.S.A. factory codes were:

B.S.A. Small Heath	M47A
B.S.A. Redditch	M47B
B.S.A. Shirley	M47C
B.S.A. Leicester	M615
B.S.A. Mansfield	M616

At this time, different forms of stocking up were tried; whether this was due to shortages of suitable timber or because it reduced the curing and seasoning time required before timber could be cut for furniture is not clear, although it was probably a combination of both. Laminated timbers with resin bonding were tried, giving an appearance like plywood, similar to the wartime German Mauser rifles. Comparative tests were also conducted with the No. 4 rifle in 1943.

Special Bakelite-impregnated furniture was tested in 1934 with 20 rifles and if similar tests carried out in Australia are any indication, it was found to be unsatisfactory due to a tendency to warp, and fracture under pressure; this also affected the rifle accuracy considerably. The reason for a Bakelite finish appears to have been to prevent saturation of the rifle furniture by gas in chemical warfare.

A "Freezing" anti-rust finish was applied to the rifle for tests also; the metal components which were usually finished blued are a grey colour as an experimental zinc-based finish was used. The particular specimen referred to is in the Pattern Room collection and no other information is at hand; some test details and data on this 1918 vintage No. 1 Mk III* rifle is available, although it probably dates from the post-1930's.

An interesting variant of the Lee-Enfield is this particular model used for testing steel helmets. Probably one of a kind, it is a .5-inch calibre smoothbore which fired a short pistol-type cartridge designated the ".500 Shrapnel Ball" and it was loaded with a spherical lead ball for testing helmet quality. This particular rifle was converted from a Magazine Lee-Enfield, fitted with a charger bridge and shortened to an overall length of 35 inches by a 15.5-in. barrel. An aperture sight is mounted on the charger bridge and it is considered that this specimen is of late 2nd World War vintage.

.50 Helmet Testing Rifle *MOD Pattern Room*

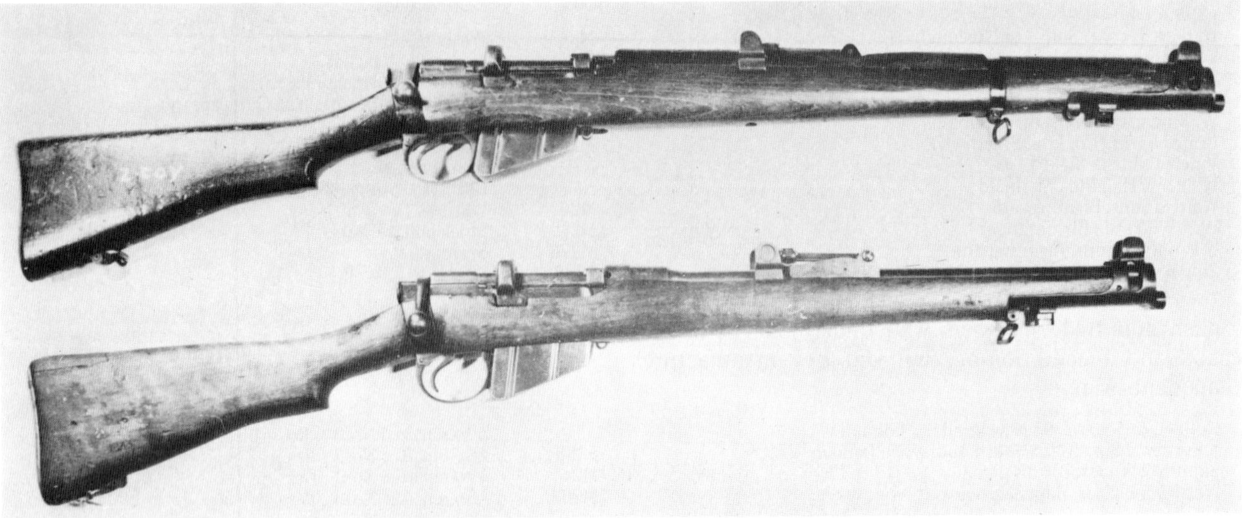

Shortened Mk III* rifles, with 20.2-in. barrel (top) and 18.2-in barrel with no handguards fitted. *J. Alley*

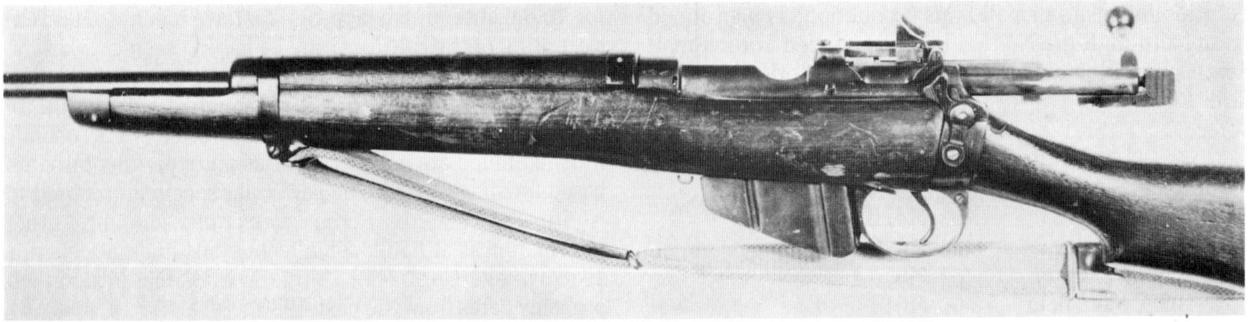

Mark V S.M.L.E. in a carbine configuration. *R. Faris*

But for the "jungle" carbines, there was no further important development of Britain's No. 1 rifle after the introduction of the No. 4. India and Australia continued production of the No. 1 until some time after the 2nd World War and some new models were developed for specific roles, particularly jungle warfare. These versions of the No. 1 and No. 4 were developed in 1943 and the particular acceptance and endurance trials were effected in July 1943.

Some shortened S.M.L.E. rifles were modified in Britain from the standard service pattern during the war by shortening, some have been noted marked "A.I.F." and there has been some conjecture of an Australian link (Australian Imperial Forces) but this is not likely. These rifles are similar in their configuration to the 1944 Australian shortened and lightened rifles, merely the No. 1 rifle shortened, but the Australian models have dual-range aperture sight on the charger bridge. The Australian trials rifles are described in Chapter 9.

Minor variants have been observed amongst these rifles but generally the barrel length is 20.2 inches with a few having an 18.2-in. barrel. Rifles of assorted origin and vintage were utilised and Ishapore, Lithgow and British receivers have been noted. Some have had the handguards removed and the sharp edges of the top of the fore-end rounded off, although all seem to have retained the original type of backsight.

These were most probably modified in 1943 when carbines were being considered for jungle fighting and as India and Australia only produced the No. 1 rifle, Britain's lightened and shortened No. 4 would not be of any benefit, hence the modifications to the No. 1. Eventually, a half-stocked version was put up for adoption.

Because the jungle carbine variant of the No. 4 became the No. 5, so the No. 1 rifle altered to a similar configuration was designated the No. 6, initially for production at the Ishapore factory. It was also adopted in Australia and orders were eventually placed with Lithgow for 40,000 although it never went into production there and only the prototypes and trials No. 6 rifles were made in Australia. Examples of British Mark III and Mark V S.M.L.E. rifles converted to "No. 6 Mark I" are retained in the M.O.D. Pattern Room collection and these are fitted with the No. 5 rifle flash hider assembly, buttplate and flip-up aperture sight.

An example of such a No. 1 Mark III lightened pattern and its component parts has been examined closely and this reveals that a batch of a dozen or more Mark III* rifles, mostly of Great War vintage, were drawn from store, converted and renumbered, probably at R.S.A.F. Enfield. Serial numbers appear to have been commenced with A100 and the conversion was made to specification, drawing D.D.(E) 3710. The rearsight consists of an L-type aperture mounted onto the new-made charger guide. Lightening cuts have been made in the trigger sear, barrel knox form, trigger guard and breech bolt and the original No. 1 butt, fore-end and barrel band have been converted in accordance with the proposal. The buttplate and flash eliminator are early No. 5 prototypes made at Enfield.

Although the Indian and Australian models were fitted with the No. 5 type flash eliminator, their sword bar or bayonet attachment stud has a squared bar for attaching a modified Pattern 1907 bayonet whereas the British No. 5 pattern has a rounded sword bar for the No. 5 knife bayonet. Descriptions of the No. 6 rifle development will be found in Chapters 9 and 10.

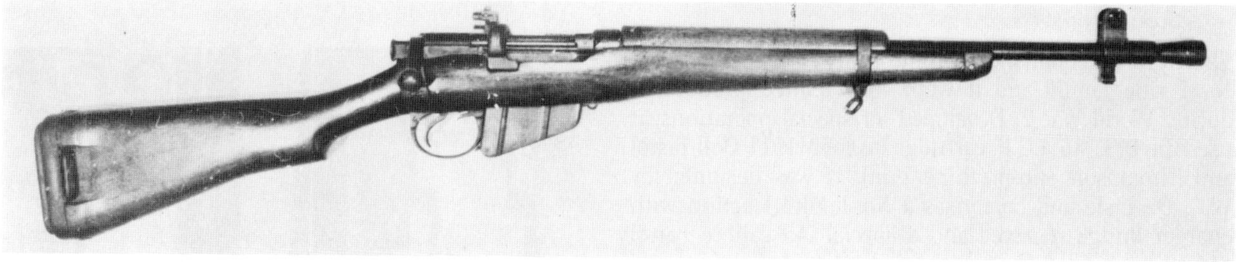

Enfield Prototype No. 1 Mark III Lightened Pattern *R. Etherington*

Above & below: Close-up views of the modified Mk III action and furniture. In the butt, the original sling swivel mount and marking disc holes have been filled and a new section added for the No. 5 type buttplate. *R. Etherington*

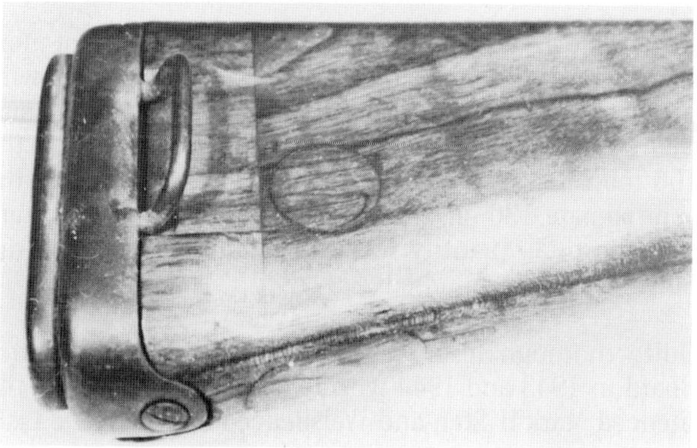

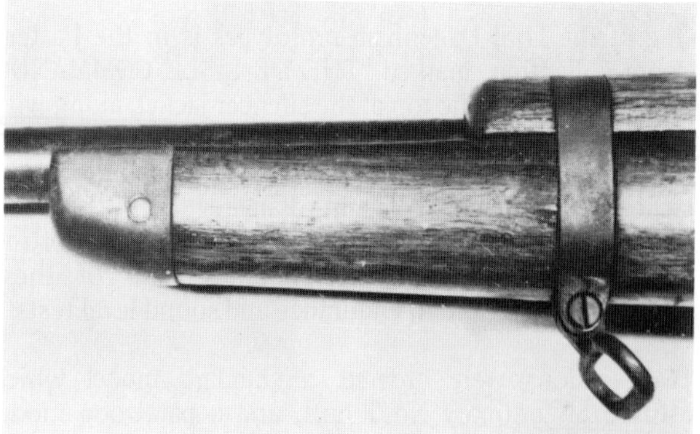

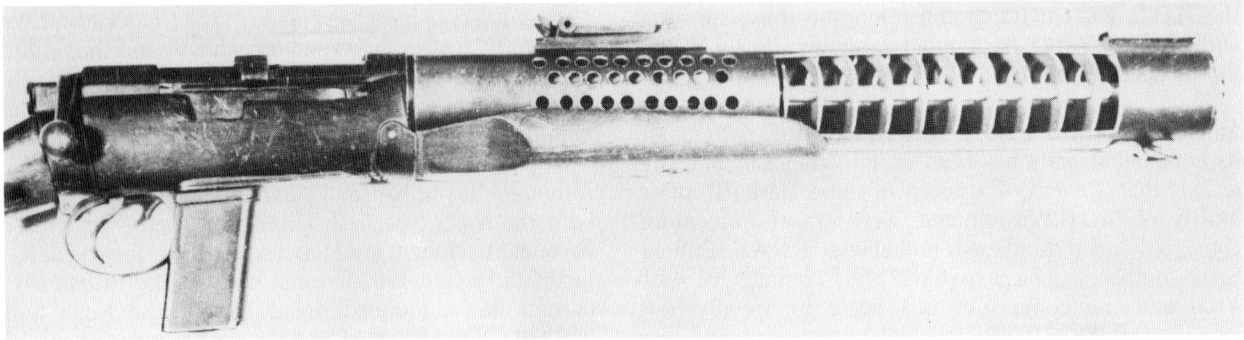

Sectionised De Lisle Commando Carbine showing the silencer section and its baffles. *School of Infantry Museum, Warminster*

The De Lisle commando carbine is also a member of the No. 1 rifle family and it went into limited production during World War 2. Developed for special operations, it uses the U.S. .45 A.C.P. cartridge (as for M1911 Colt pistol and Thompson sub-machine gun). It was designed by W.G. De Lisle and comprises a No. 1 Mk III action with charger bridge removed and a special .45 calibre barrel fitted, extending back into the original boltway and requiring a shortened bolt. A jacket which contains the silencing baffles extends forward of the breech for about 15 inches.

Three silenced rifles were proposed to the Ordnance Board in 1943 by De Lisle, two in .22 RF calibre and one in .45 ACP. The .22 rifles were an auto-loader and a machine carbine which featured a waxed paper belt, but neither of these designs filled any requirement so they were not proceeded with. A small number of .45 prototypes was built up by De Lisle at the Ford Dagenham factory, assisted by Major Sir Donald Campbell who was responsible for organising "practical field tests" in special operations on the occupied French coast.

It is recorded that after successful trials on the Campbell estate, the carbine was offered to the Senior Officer to shoot and he shot at some ducks feeding on the other side of the lake, purportedly some 400 yards distant. One duck was shot dead through the neck and upon recovery of the "target" the officer coolly remarked, "I'm rather out of practice, so perhaps it's hardly fair to blame the gun. Actually I aimed for its head."

One of the Ford Dagenham prototypes is in the Pattern Room collection, marked "THE DE LISLE COMMANDO CARBINE" along the top of the silencer jacket, along with the serial number. A 1910 vintage Mark III S.M.L.E. action was used for this particular conversion and the silencer jacket is slightly smaller in diameter than the later production model and it has an American style sporting rear sight with four range notches. Two of these prototypes were used in endurance and sound level tests.

Two versions were ordered, a standard model which utilised an ordinary No. 1 butt, and a paratroop model

Commando Carbine sights (left) and prototype (right). The 600-yd. sights are from the 9mm Lanchester SMG.

with pistol grip and folding stock. Only 106 carbines are recorded as being produced by Sterling Armament Corporation at Dagenham although initial orders were for 500 carbines, numbers 451-500 were to be the airborne models with the folding buttstock. This was then increased to a total of 600 carbines but Ordnance Board reports indicate that production was terminated at 106 standard models; only prototypes of the folding stock model were assembled.

Other official records put the total production at not more than 130 and some specimens have been noted with Sterling marked serial numbers in excess of this. It may well be that the production run did extend beyond the 106 indicated in the Ordnance Board report, or even the approximate 130 on production records, but it is certain that the total quantity was still relatively small and did not exceed a few hundred.

Quite thorough tests were conducted by the Ordnance Board in 1943 and 1944 with two prototypes, against the silenced Mark II Sten and Welsilenced Sten. The De Lisle carbines fared quite well in both the endurance and sound level tests and later on, two of the Sterling made De Lisle carbines were used for further trials. The order with Sterling was placed for Combined Operations and was cancelled after the mode of warfare changed following the landings in Europe in 1944. The commando carbines were later used quite successfully in the Far East in jungle fighting.

Official specifications of the Commando carbines were given in November 1945 as:–

Weight	8 lb. 4 oz. 7 lb. 2.5 oz. (Airborne model)
Length	2 ft. 11.25 in. 2 ft. 1 in. (Airborne model)
Barrel Length ...	7.25 in.
Sight Radius	12.25 in.
Sights	Blade & V sights, luminous sights for night work.
Silencer	Integral, 13 baffles, made of duralium.
Distance from muzzle to end of silencer ...	10.5 in.
Ammunition	.45-in. Colt auto.
Muzzle Velocity ...	830 ft/sec.
Effective Range ...	275 yards.

Some of the carbines were fitted with luminous night sights by Holland & Holland. A comprehensive description of these carbines along with more illustrations and drawings may be found in "De Lisle's Commando Carbine" by Skennerton.

Another unusual modification of the S.M.L.E. rifle of World War 2 vintage was the Charlton, a gas-operated conversion to light machine gun. The concept of a gas operated Lee-Enfield was not new; Sir A.T. Dawson and Sir G.T. Buckham secured such a patent, No. 128,395 in 1918.

Two other British gas operated conversions of the S.M.L.E. which warrant mention are the Howell and the Howard Francis but these were never proceeded with beyond the prototype stage whereas the New Zealand Charlton was produced in sufficient quantity to be issued to troops, although it is not likely that it ever saw action against hostile forces. Two other similar conversions of the S.M.L.E. were the South African Reider and the Australian Ekins.

Philip Charlton was a New Zealander and his patents for the gas-operated automatic system were taken out in November 1941. At the outbreak of World War 2, New Zealand was desperately short of light machine guns and although production of the Bren began in Australia, Britain and Canada before the war, there were insufficient supplies for New Zealand's forces. Against this background the Charlton was born and it underwent N.Z. Army tests on 7th November 1941.

At first, obsolete Lee-Metford rifles were intended to be the basis for this conversion but then Lee-Enfield rifles were used when no more Lee-Metfords could be obtained. The No. 1 rifles were in too short supply at this time to be considered as a basis for conversion in New Zealand; problems were encountered with the Lee-Metfords and Lee-Enfields due to the age and wear of the actions, some of which were more than 50 years old.

The New Zealand production Charlton outwardly resembles a typical light machine gun with an exposed barrel and cooling fins and a bipod which is attached just forward of the gas take-off, although the exposed mechanism on the right side of the action appears to be somewhat vulnerable and even clumsy. It was intended for use from the hip, shoulder and bipod, primarily as a semi-automatic arm with a full automatic capability in reserve. Pistol grips were fitted for firing from the shoulder and in the hip firing position it was intended that the leg of the bipod be grasped with the left hand.

128,395. Dawson, Sir A. T., and Buckham, Sir G. T. June 19, 1918.

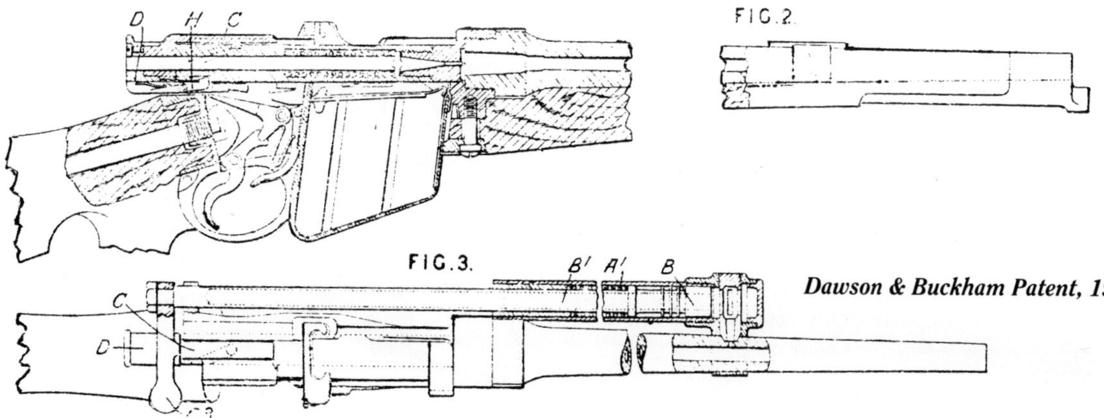

Dawson & Buckham Patent, 1918.

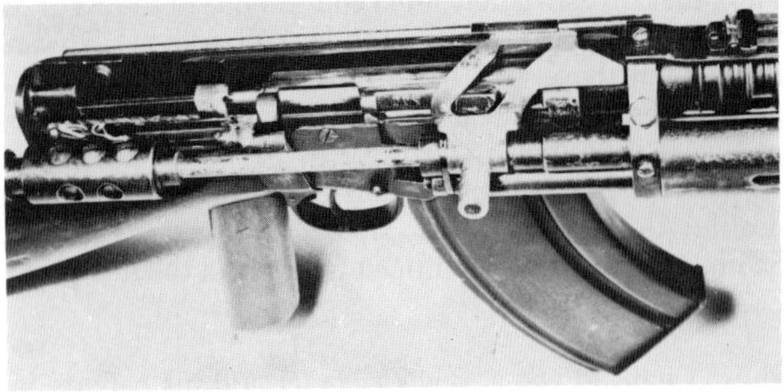

View of the N.Z. Charlton mechanism. *Imperial War Museum*

The first batch of rifles converted on the production line system was composed of 237 Lee-Metfords, most likely Mark I* models. No more Lee-Metford rifles were available so Lee-Enfield rifles were then used which necessitated some changes in the jigs and components. The rate of conversion was intended to be as high as 200 per month but by July 1942, only 431 were done, comprising 237 M.L.M. and 194 M.L.E. rifles. Most of these had previously been downgraded for drill purpose only and many new parts, including barrels, were often required.

A wide variation in hardness of components and receivers was encountered due to their age and so spot softening and grinding was often required. Australia's commitments in the war against Japan meant that certain No. 1 rifle components for use in the Charlton were not available and thus many of the old parts had to be heat treated and rejuvenated, or new ones produced in New Zealand.

Parts made by the N.Z. sub-contractors were:

Cams	Machine Tools Ltd., Wellington
Springs	N. W. Thomas & Co., Wellington
Cooling fins	Precision Engineering Co., Wellington
Trigger sideplates	Precision Engineering Co., Wellington
Sideplates	Motor Mowers Ltd., Hastings
Cocking pieces	Motor Mowers Ltd., Hastings

Most of the Charlton rifles used the Lee-Metford/Lee-Enfield backsight but some of the later models utilised backsights from No. 1 Mark III and III* rifles. The sights were attached to a strap which was secured to the barrel over the sheet metal cooling fins. Either the standard S.M.L.E. rifle magazine or a modified Bren version could be fitted.

Specifications were given as:–

Weight:	Approx. 16 lb. 8 oz.
Rate of Fire:	30 aimed shots per minute, repetition 700 round per minute, automatic
Feed:	10-round magazine, or 30-round modified Bren magazine.

The Charlton conversions were assembled in Charlton's Motor Workshop in Hastings, Hawkes Bay and it is recorded that the contract for 1,500 guns was completed.

Some improvements to the conversion system are recorded in August 1942 but after the war the Charlton automatic rifles were housed in the Palmerston North Showground services stores complex and destroyed in a fire.

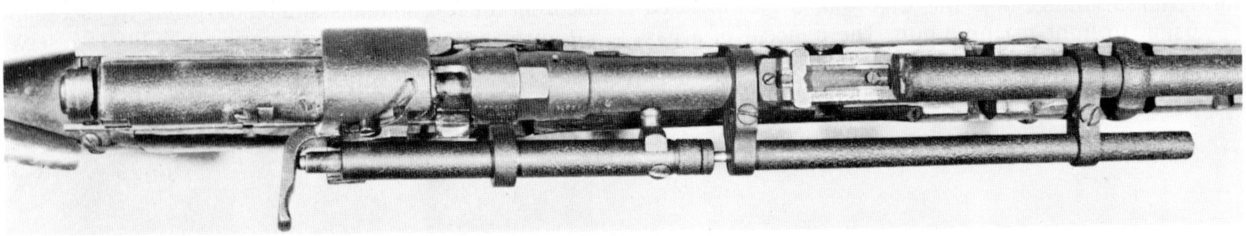

Above & below: Prototype Gas-Operated Lee-Enfield Conversion, probably the Howell pattern. A Mk I* M.L.M. has been used in this case. *Imperial War Museum*

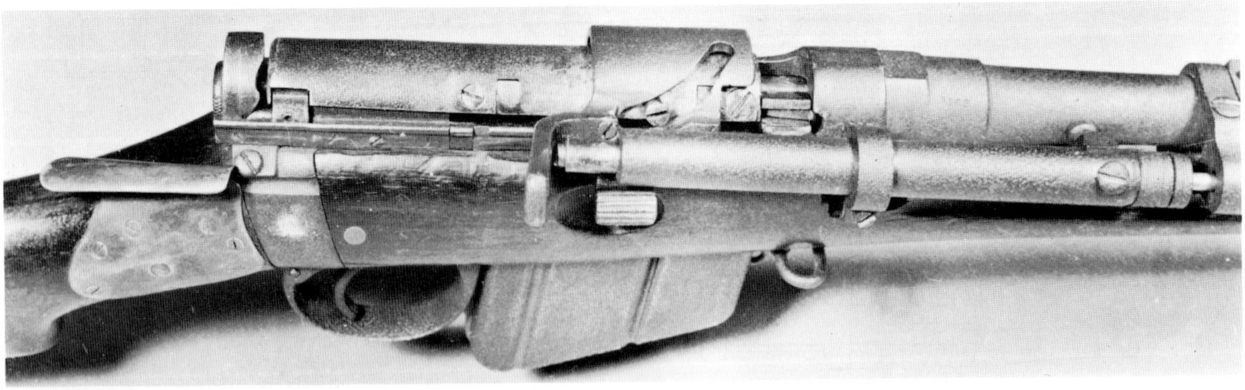

The same building which housed the Charltons also held large quantities of clothing, boots and webbing equipment, along with the stores records. It is reported that there were massive discrepancies in the stores. After the buildings burned down, any checks were impossible and although a Court of Inquiry was convened, nobody was charged due to the lack of evidence.

Another variation of the Charlton design was made in Australia by Electrolux Pty. Ltd., Victoria; instead of using the Lee-Metford or Lee-Enfield as a basis for conversion, the current production No. 1 Mk III* rifle was utilised and so the conversion was considerably less complicated. Original furniture was retained as was the magazine, sights and nosecap but sheet metal guards cover the mechanism on the right side of the rifle so that, from the right side at least, it bears little resemblance to the New Zealand production model.

On the left side of the Electrolux model, a somewhat similar type of selector switch is mounted, the rear pistol grip also is reminiscent of the New Zealand version. The sight arrangement is quite different, consisting of a simple aperture at the rear, precluding any alteration for different ranges.

The 1944 issue of Australian Munitions Digest listed Project No. A93 as "Conversion of .303 Rifles to Charlton M.G." by the Electrolux company at Alexandra Ave., South Yarra, Melbourne, but the project was discontinued by June 1944. It would appear that only one or two prototypes were assembled.

The British Howell predated the Charlton by more than 25 years as it was designed during the Great War; it is probably the first of the seriously considered gas-operated conversions. Although the system of operation is somewhat similar, the Howell is entirely different in layout and external appearance. Despite its ungainliness, it is a relatively simple conversion with some commendable features. The Howell retains the S.M.L.E. full stocked configuration but has a distinctive metal guard on the right side above the tubular steel pistol grip which appears to serve as a guard for the hand, protecting it from the reciprocating parts along the right side of the rifle.

In 1937 and 1940, Mr. N. Howell offered the Ordnance Board a proposal for the "self-loading conversion of hand operated rifles" particularly intended for anti-aircraft use by the Home Guard. No further attention was to paid to this idea and it was the same design as that previously described from the Great War.

Although there were a few other inventors who designed gas-operated adaptions of the Lee-Enfield, some of which were brought to the notice of the Committee, only the Howell and Howard Francis were assembled in England for serious tests and trials. Both were regarded as emergency conversions for home defence rather than systems for possible development of new rifles.

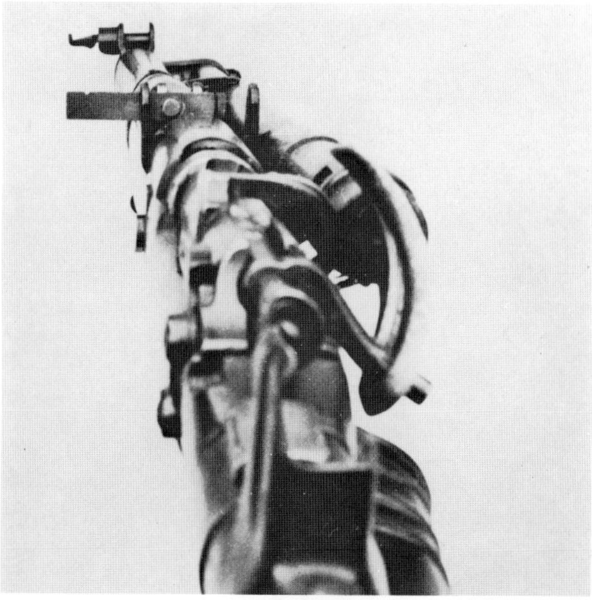

Firer's View of the Howell. *MOD Pattern Room*

While the Howell, Charlton, Reider and Ekins were .303 calibre rifles, the Howard Francis was different in concept as it was a 7.63mm self-loading carbine. Developed in England late in 1943, it was chambered for the 7.63mm Mauser pistol cartridge and was stocked like a "sporter" rifle with a pistol magazine projecting below the normal magazine position. It is a simple blowback design and uses the original .303 barrel with the chamber plugged and reamed out for the new cartridge; the bore size was suitable for the new round, so the barrel requirements remained the same.

The Howard Francis specifications record the overall length as 32 inches with the barrel shortened to 12.75 inches. The magazine has a 12-round capacity and the aperture sights on the receiver give a sighting radius of 13.9 inches; its weight was listed as 8 lb. 1.5 oz. The original trigger mechanism was removed and replaced by a re-designed assembly. Although the accuracy was quite reasonable, trouble was encountered with the feed due to the magazine platform and lips. The project was not proceeded with any further because there was no requirement for such a carbine at that stage of the war.

The South African Reider was brought to the notice of the Ordnance Board when Dr. Van der Bijl, South Africa's Director General of War Supplies advised of a device for adapting the Lee-Enfield rifle to a light machine gun and a sample with particulars of the conversion was conveyed to England in June 1941.

Only 18 S.M.L.E. rifles were converted by Reider, a Cape Town radio and television experimenter and inventor who died in 1954. The work was done in 1940-41 by Reider, the only drawback was recorded as being overheating after firing 100 rounds. There were some differences in the

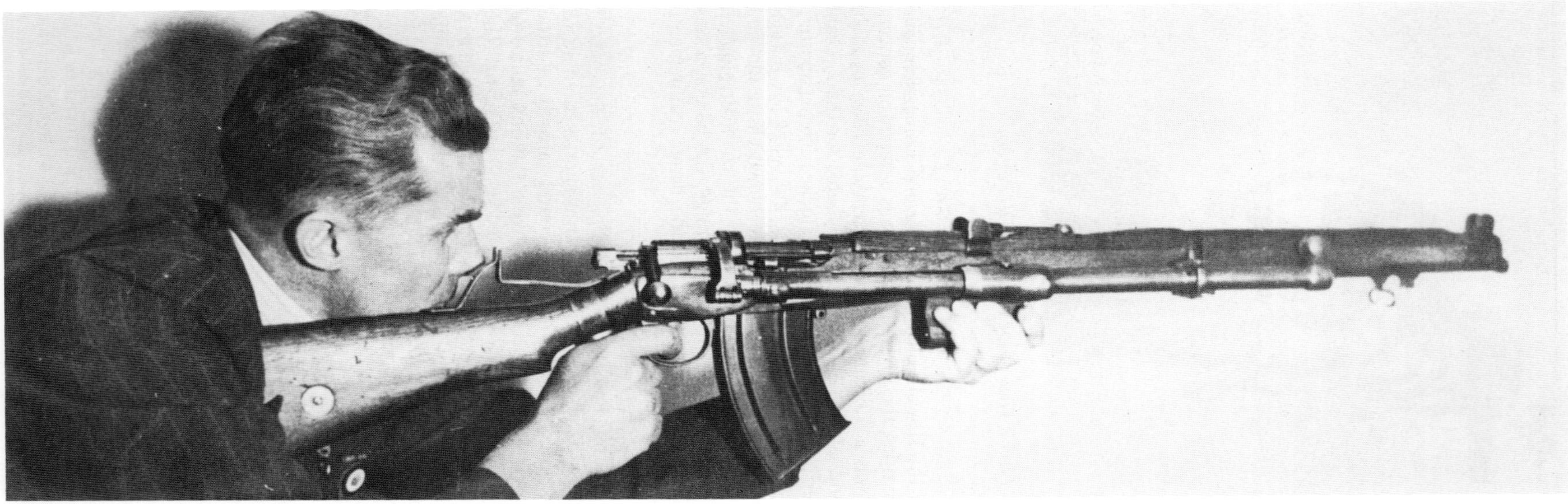

Mr. H.J.R. Reider demonstrating his Automatic Conversion; the modified Bren magazine increased its capacity to 30 rounds compared with the conventional 10-round S.M.L.E. magazine. Historical Firearms Society of South Africa

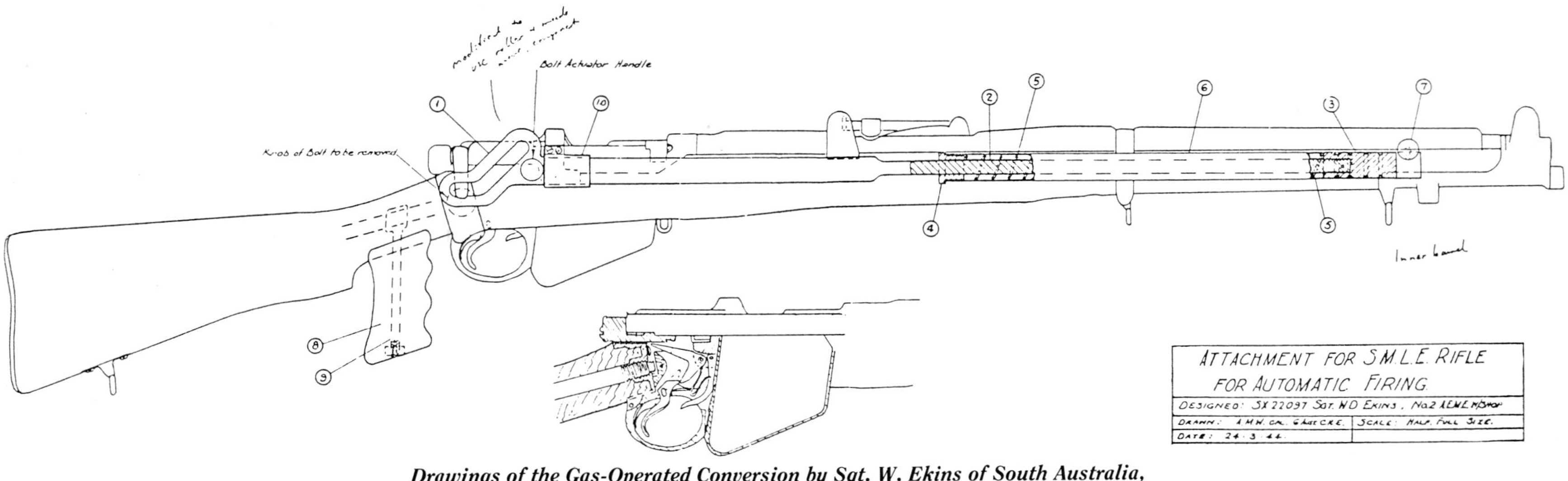

Drawings of the Gas-Operated Conversion by Sgt. W. Ekins of South Australia, dated 24.3.1944. It is similar to the Howell and Reider conversions, with a gas piston acting on a cam to operate the bolt. Australian War Memorial

conversions as his "new plan" was implemented in August 1941. Three rifles were sent overseas for assessment and the Reider was fitted with a 25 or 30-round magazine with a cyclic rate of fire estimated at more than 250 rpm; the cost of each adaptor was estimated at about £2/-/-, with both repetition and fully automatic fire possible.

The South African Reider being tested, mounted on a tripod. *Historical Firearms Society of South Africa*

The original S.M.L.E. rifle on which the design evolved was serial no. M 45374, it was presented to Reider by the Admiralty in 1944 and is now housed in the Military Museum at the Castle, Cape Town. Rifles nos. 8134, B 92462, B 92868, F 59817, G 47054, N 8109, S 48250, Z 36139 and Z 39133 were converted to the original plan while nos. F 26776, F 53065, F 59724, F 45222, F 68561, H 43785, H 86891 and Z 50675 appear to have been converted to the improved plan. Those rifles sent overseas were nos. 8134, F 26776 and F 59724.

Australia also had its share of such inventors because a design by Sgt. W. Ekins is recorded as being considered, somewhat similar to the Howell. Drawings dated March 1944 are titled "Attachment for S.M.L.E. Rifle for Automatic Firing" and refer to the No. 2 A.E.M.E. workshop. A common feature of the Reider, Charlton, Howell and Ekins designs is the outboard gas piston assembly on the right side of the gun, acting directly upon the breech bolt or the bolt handle.

Another conversion of the Short Lee-Enfield rifle is the shortened Mark V, a take-down variant attributed to South Africa and reputedly modified by Greener in England for the South African Police. This reduces the arm to "saddle bag" proportions as its 14-in. barrel permits a taken-down length of only 21 inches with an assembled length of less than 33 inches.

The No. 1 Mark III and III* rifles, as well as the complementing skeleton actions and the No. 3 Mark I* (T) and (T) A were declared obsolete in British service in 1947; for Land Service on 22nd March, the Air Service on 13th February and in Naval Service on 13th March. However, the venerable S.M.L.E. was reinstated as a current store again in 1954 in its No. 1 Mk III* configuration. It was still in use with Australian and Indian forces when the 7.62mm F.N. F.A.L. rifle came onto general issue by the 1960's.

Take-down Conversion of the Mark V S.M.L.E. *R. Faris*

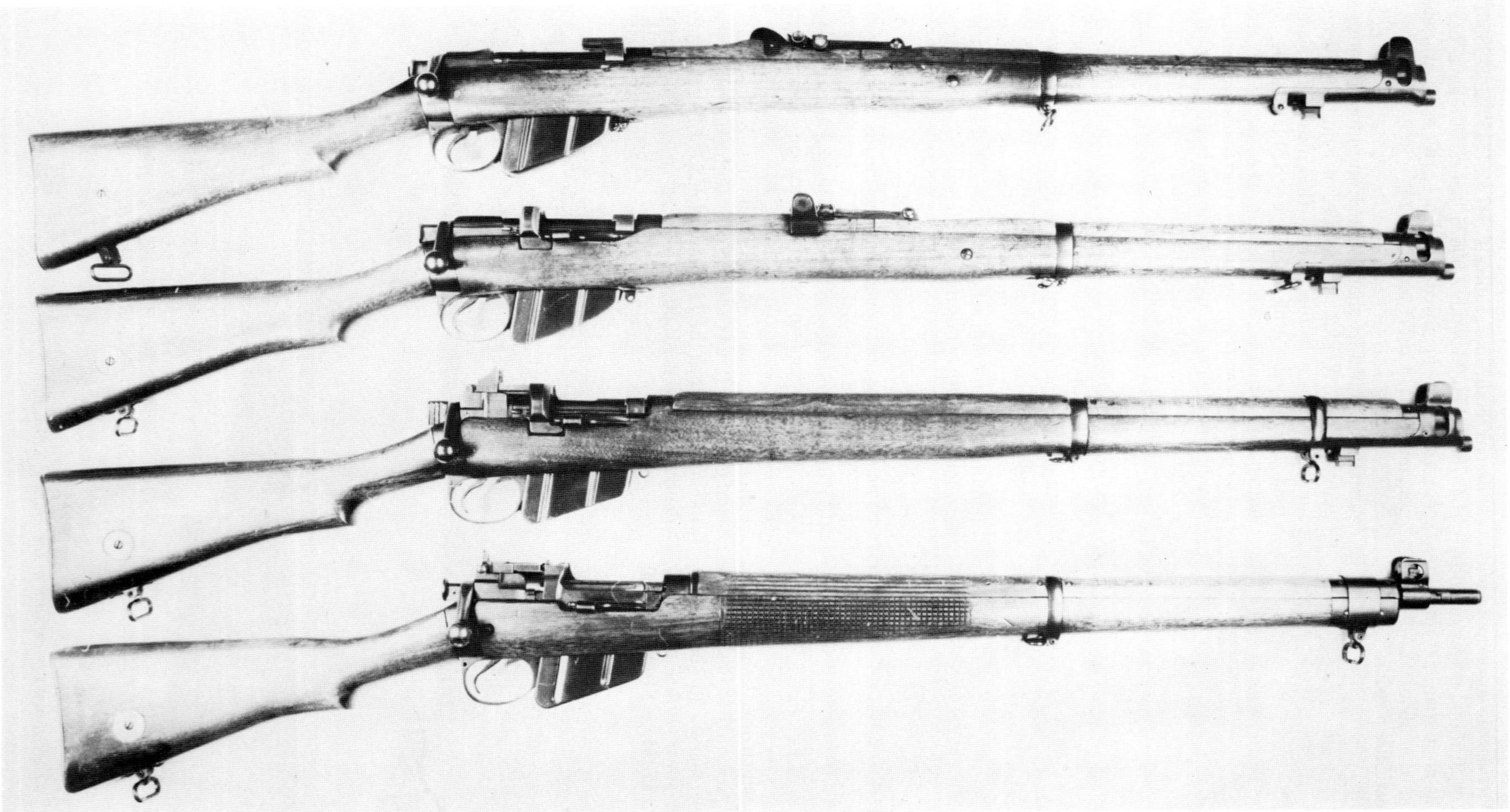

DEVELOPMENT OF THE NO. 1 RIFLE

Top to bottom:

Short, Magazine Lee-Enfield Mark I
Short, Magazine Lee-Enfield Mark III
Short, Magazine Lee-Enfield Mark V
Rifle No. 1 Mark VI

MOD Pattern Room

It may be noticed that some of the parts and fittings of Indian and Australian service No. 1 rifles are different, not just in the shape or markings due to manufacture in these countries, but in subsequent alterations and modifications. For example, many Australian nosecaps later had "windows" cut through the left and right foresight protector ears; such details will be found in Chapters 9 and 10.

Indonesia refurbished some S.M.L.E. rifles and these have special markings; all the original manufacture and inspection marks appear to have been machined or ground off and new markings placed on the right side of the receiver butt socket in place of the original ones. Most of these particular rifles originally appear to have been Australian production.

A new serial number range was used on these Indonesian rifles, e.g. 0-002878, stamped on the top of the receiver ring, barrel knox form, underneath or on the side of the nosecap and on the bolt, all in different positions to those normally found on British, Indian and Australian service rifles. These rifles were brought into Indonesian service in the mid-1950's and some No. 4 rifles were similarly refurbished and issued. The No. 1 and No. 4 rifles were not manufactured in South-East Asia.

Another S.M.L.E. adaptation emanating from this part of the world is the modified nosecap illustrated here with extended sword bar to fix the Arisaka bayonet, probably effected post-war when there was an abundance of the Japanese bayonets. This conversion seems to have been done in Indonesia.

Extended sword-bayonet bar on No. 1 rifle nosecap, to enable fitting of Arisaka bayonet; Indonesia. *J. Alley*

"Winter" trigger guard, for use with gloved hands. *R. Faris*

An enlarged trigger guard has been mentioned during the Great War narrative; the example here is considered to be later due to its fabricated construction. Another "winter trigger" has been noted although this was a conventional machined type similar to the service trigger guard, only larger. The specimen illustrated is probably later and maybe it was not British made, as no such references have been observed in War Office or other such official papers. Some crudely modified S.M.L.E. trigger guards have been noted from the Great War period, where the back part has been hammered forward so that there is a larger space in front, just behind the magazine, presumably for better accommodating a gloved hand. It could well be that some enlarged trigger guards were made extra-service.

Modified handguards; the thin extensions on each side, at the backsight bed base, have been trimmed off. *J. Alley*

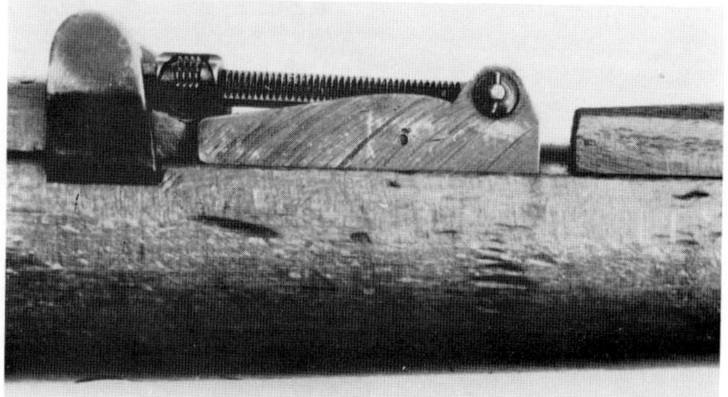

Modified handguards; note the roughly machined backsight bed and WW2 beech furniture. *B. Labudda*

Some ex-service No. 1 rifles have been recently noted with modified handguards at the position of the backsight bed. This appears to have been done after the thin extension portions have become broken or cracked and these have then been removed from the front and rear handguards, exposing the backsight bed. Many Indian service rifle were so modified and it appears that this adjustment has also been applied to No. 1 rifles in other countries.

The bipod illustrated below has not been noted in any of the official documentation although a monopod and bipod were tested with the No. 4 rifle during the war; these are illustrated in the following chapter. Likewise, the carrying handle on the example shown here has not been noted in War Office or Ministry of Defence papers, so both these items may well have emanated from some foreign service.

Canada did not manufacture the S.M.L.E. No. 1 rifle, though it appears that some replacement .303 No. 1 rifle barrels were produced at Long Branch.

In general terms, the S.M.L.E. (No. 1) action was not sufficiently strong for use with the 7.62mm NATO cartridge, although tests were carried out in this form in Britain, India and Australia. Some 7.62mm barrels were considered for both the No. 1 and No. 3 rifles in November 1959 and five for each type were made at Enfield by May of the following year, although no results of any consequence were reported.

In Australia, trials were carried out with reinforced .303 No. 6 rifles chambered for the new 7.62mm round although these were not satisfactory or even successful; details and illustrations will be found in Chapter 9, the Australian section. In India the problem of tensile strength was overcome by making the receiver bodies from higher grade steel, these particular No. 1 rifles, designated the 7.62mm Rifle 2A, are described in Chapter 10. In British service, it was only the No. 4 action which was utilised in the 7.62 x 54mm role.

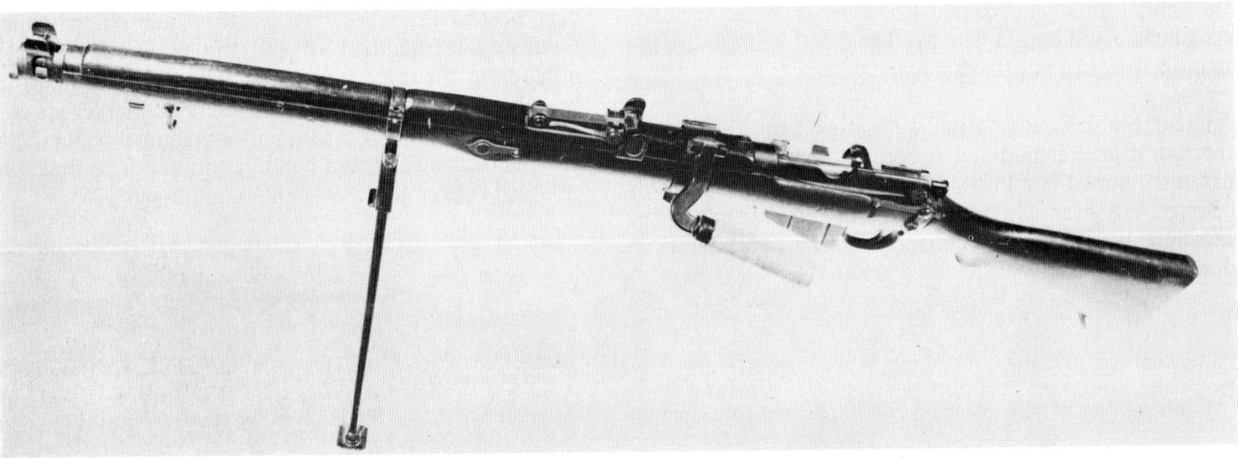

Bipod and carry handle fitted, the right leg partly obscured. *R. Faris*

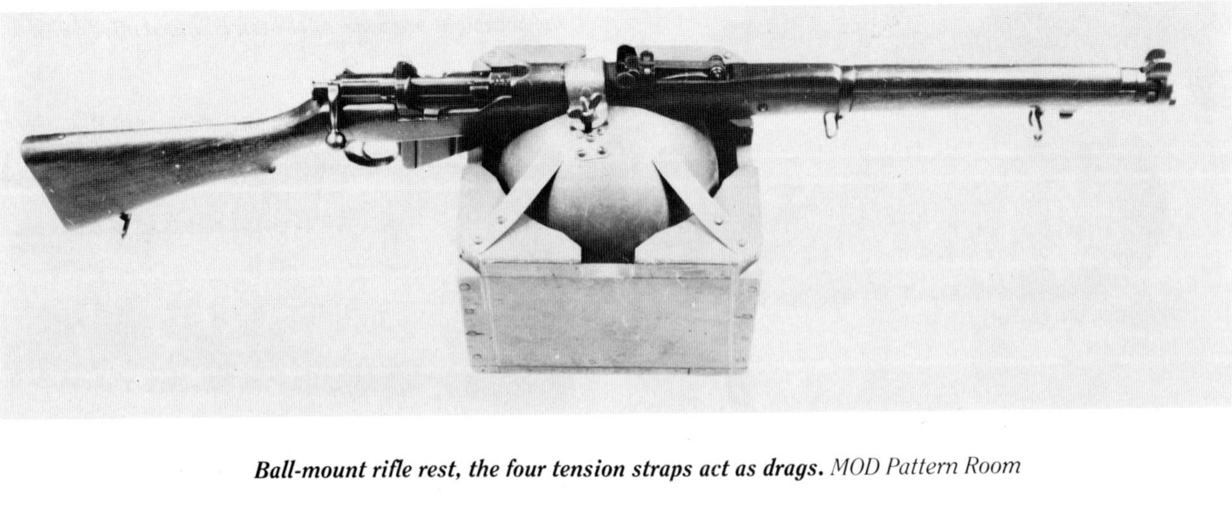

Ball-mount rifle rest, the four tension straps act as drags. *MOD Pattern Room*

RIFLES NO. 4 & NO. 5

The No. 4 series represents one of the most commonly encountered models and largest production runs of the Lee-Enfield rifle although very few were actually manufactured at R.S.A.F. Enfield. Until the 2nd World War, the Royal Small Arms Factory at Enfield was responsible for producing most of the Lee-Metford, Lee-Enfield and Short Lee-Enfield series; other rifles which belong to the No. 4 "family" are the No. 5 (also known as the Jungle Carbine) and the Nos. 7, 8 and 9 which are .22 RF trainers. The No. 4 issues also overlapped the 7.62mm NATO era because the L8 series and heavy barrel, accurized L39 and L42 rifles were developed and introduced by the early 1970's.

It is also worthy of note that the No. 4 was manufactured in larger quantities overseas than at home. Nearly 2.2 million were produced in Canada by Small Arms Limited at Long Branch, and in the United States at the Stevens-Savage factories in Massachusetts. In England, the wartime No. 4 was made in three newly-established factories constructed especially for this rifle, at R.O.F. Fazakerley, R.O.F. Maltby and B.S.A. Shirley. Two were government Royal Ordnance Factories while Shirley was set up by B.S.A. Guns. During the Great War, the "peddled scheme" incorporated the production of millions of component parts by contractors, the 2nd World War No. 4 manufacture employed a similar system but on a broader scale.

Rifle No. 1 Mk VI was re-designated the No. 4 Mk I in mid-1931. No. 4 trials rifles have many features similar to the Mk VI; magazine cut-off, waisted foresight protector, button-headed cocking piece and bronze coloured buttplate. But the distinguishing differences which are classified as No. 4 rather than No. 1 Mk VI are the higher, flat left wall of the receiver, absence of the safety catch recess in the receiver, and omission of fore-end chequering. Trials No. 4 rifles are marked with the cypher, year of manufacture and rifle designation, on the left side of the receiver butt socket, like the No. 1 Mk VI. Typical No. 4 trials rifle markings are:

G R
ENFIELD
1931
No 4 Mk I
A 0485

The nomenclature change from No. 1 Mk VI to No. 4 Mk I was effected during 1931, but some of the earliest of the No. 4 marked trials rifle may be found to be dated 1930 and most components but for the receiver were commenced in manufacture as parts for the Mk VI. An interesting consideration is that despite the significant alterations to the receiver body, the trials No. 1 Mk VI and this early model of the No. 4 Mk I were still officially referred to as the Pattern "B." The "A" pattern was the first of the No. 1 Mk VI rifles with chequered buttplate and other relatively minor differences while the following "C" Pattern incorporated some improvements which were used in the eventual No. 4 rifle. Further details of the "A," "B" and "C" Patterns will be found towards the end of the previous chapter.

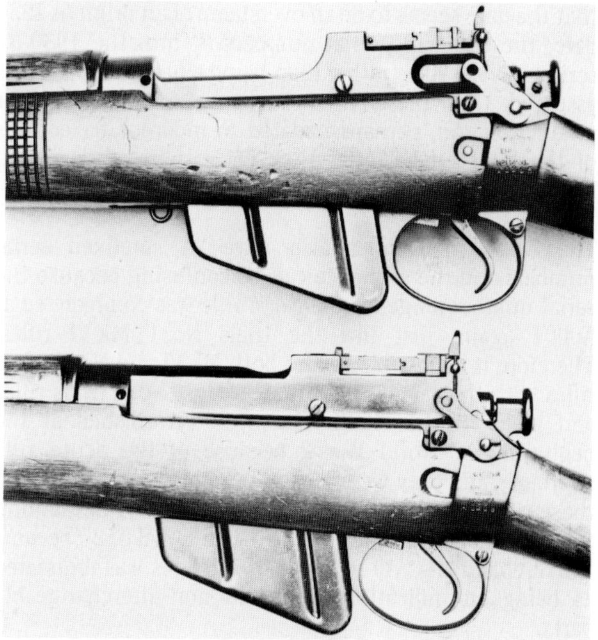

No. 1 Mk VI (top) and trials No. 4 Mk I (bottom) receivers.
MOD Pattern Room

After the batches of No. 1 Mk VI and No. 4 Mk I trials rifles were produced, some more rifles were ordered so as "to keep rifle work at R.S.A.F. alive and to perfect certain manufacturing processes." On these basis, a further 500 rifles were ordered on 4th April 1932 and these incorporated:

Minor improvements to the cocking piece, striker and keeper screw, and buttplate;
Plunger instead of a ball on the rearsight, along with a new type of spring;
New backsight with 6 clicks instead of 4 per segment of adjustment,
Higher quality steel for the breechbolt.

These were modelled on the "C" pattern and very similar to the eventual No. 4 Mark I service rifle.

A quantity of these No. 4 rifles had already been produced and according to a report in 1933, they were in addition to those proposed for the 1933-34 financial year. This makes a total of 2,479 No. 4 trials rifles. It is pointed out that this figure does not include the nominal 1,000 No. 1 Mk VI rifles manufactured in 1929-30 because by 1933, the distinction had been made between the No. 4 and Mk VI patterns of the new rifles.

Perhaps a better perspective of the situation would be to consider the Mk VI as a troop trials rifle whereas the No. 4 better addressed the development of appropriate manufacturing techniques for the new style of rifle; this was one of the major aims of the No. 4, to facilitate the more modern procedures in production which involved less specialist procedures.

On some of the 1933-dated No. 4 rifles it has been noted that the date seems to be an overstamp of an original 1931 date; these specific serial numbers fit into the 1930-31 series of No. 4 rifles rather than being a new group, so it is assumed that these examples were from the earlier production run, perhaps updated to incorporate features of the later model, after the "C" pattern as described previously.

The No. 4 trials rifles also have "A" prefixed serial numbers and the sequence can be confusing because the serial number range for the No. 4 rifle was commenced at A0001 again, just like the trials No. 1 Mk VI rifles. Therefore it is possible to find both Mk VI and No. 4 trials rifles with the same serial number. Most No. 4 trials rifles and many of the Mk VI rifles were refurbished at the beginning of World War 2 because of the acute rifle shortages and they were mostly renumbered. On many of these rifles the renumbering consisted merely of marking the "A" prefix after the number, so rifle "A2312" became "2312A" and the "A" suffix on No. 4 rifles was registered as being an indication of certain non-interchangeable parts.

These ex-trials No. 4 rifles which were taken into service during the wartime years, marked with the "A" suffix to the serial number, were finally ordered to be withdrawn and scrapped in the mid-1950's. A few years earlier, in January 1951, the E.M.E. Regulations announced that "Rifles No. 4 Mk 1 which bear the suffix letter 'A' to the serial number, e.g. 444A, are fitted with components which are not generally interchangeable, and care must be taken when changing such components. When a replacement is required and a spare component will not fit, the rifle will be exchanged."

On most trials rifles the re-numbering is visible because part or all of the original serial number can be identified. Some discrepancies may also be encountered in receiver bodies because of different machining operations; a few unfinished receivers or receivers from other trials batches were built up also and included in this re-fitted series. Due to the severe rifle shortages, especially after the losses at Dunkirk and upon the British evacuation from Europe, all suitable actions were utilised and some of these receivers do not have the original style of trials markings; some with a more rough finish also are not up to the standard of the original trials rifles.

Some of these early No. 4 Mk I rifles were produced as factory exercises in manufacture and not intended for troop trials. A group of rifles considered in this category is the "No. 4 Mk I EXP" series, this designation being marked on the brass butt disc with no designation, factory or year of manufacture being stamped on the receiver. Such a specimen, No. 37 is in the Pattern Room collection and Nos. 42, 45 and 63 have also been noted in other museum and private collections, so this production run was of a significant size. These particular examples are considered to be of post-1933 vintage, probably made during 1934 or 1935, and like the other trials No. 4 rifles, were produced at R.S.A.F. Enfield.

A large portion of the original trials No. 4 rebuilds were later converted to sniper models; factory and contract records indicate that about 2,000 of the No. 1 Mk VI and No. 4 Mk I rifles were rebuilt as No. 4 rifles during the early years of World War 2, along with 1,400 as No. 4 Mk I (T) sniper rifles. These particular sniper rifle conversions utilised only the No. 4 receivers because the No. I Mk VI is not suitable for positioning the scope mount pads due to the curved shape and form of the lower wall on the left side of the receiver. The 2,000 trials rifle rebuilds to a No. 4 configuration included most of the trials No. 1 Mk VI rifles on hand as well as the remainder of the No. 4 trials models.

During the extended troop trials which commenced in 1932, accuracy proved to be inconsistent with some early production rifles but this seems to have been rectified after the 1932 trials. Twelve rifles were sent to Hythe in 1935 for comparison tests with twelve No. 1 Mk III rifles and the No. 4 proved to be more accurate in deliberate and rapid fire, snap shooting and fire with movement, with a margin of improvement ranging between 2% and 20%. The aperture sights were considered excellent, earlier aperture sights such as those on the Mark V proved to be liable to damage. The new sight was strongly constructed and quite sturdy in the both folded down and elevated positions; it is also considerably smaller and therefore less liable to damage.

The new bayonet for the proposed No. 4 rifle was found to effect shooting considerably more than the Pattern 1907 model fitted onto the No. 1 rifle, because it mounted directly onto the barrel and was supported by it entirely. When this new bayonet was proposed, it was thought that the heavier barrel of the No. 4 rifle, coupled with the smaller bayonet, would not affect the shooting very much at all.

Moving the bayonet from the barrel onto a separate fixing point was one of the features of the change from Magazine Lee-Enfield to Short Lee-Enfield in 1902 and while it had the advantage of removing the weight of the bayonet from the barrel, the separate nosecap/bayonet attachment resulted in a significant weight increase, particularly at the muzzle, making the rifle "nose heavy" for quick pointing and snap shooting.

The short, cruciform blade of the new bayonet did not affect the handling as much when fixed. Other complaints levelled against the Pattern 1907 bayonet were that the long, broad blade glinted even in moonlight or when Very lights were fired, and it often bent or broke. It was not ideal for penetration or withdrawal and had to be frequently sharpened.

The No. 4 bayonet did not have a "handle" and could not be used as a knife of short-sword and was really only a poker, although it was pointed out that the Pattern 1907 was not suitable for use as a knife, unless for clearing brush or chopping; it was even ineffective for cutting up duckboards and ammunition boxes. At last, it was generally realised that the length of reach with a bayonet was not as important as the handiness and that a longer bayonet was often more clumsy in bayonet fighting. For this reason, the Pattern 1903 had been widely preferred to the Pattern 1907 for bayonet work.

At this point the practice of marking the receiver with the Royal cypher above the rifle designation, factory and year of manufacture was discontinued. The cypher or Royal sign is composed of the crown and reigning monarch's initial followed by "R"; so Victoria Regina was abbreviated to "V.R." and Queen Victoria's cypher appears on Lee-Metford and Lee-Enfield rifles and carbines until she was succeeded by King Edward VII in 1901.

V. R.

E. R.

G. R.

The initials in the cypher were then changed to "E.R." for Edward Rex. Edward reigned until 1910 when George ascended the throne and the initials then changed to "G.R." indicating George Rex. King George reigned until 1936, by which time the use of the Royal cypher had been dispensed with on small arms, except for some late production B.S.A. No. 1 rifles which continued from the late 1930's until the middle of World War 2. Birmingham dispersal factory No. 1 rifles will also be noted with the cypher. These cypher initials were also incorporated in various proof marks which can assist in the dating of certain components.

In the Royal cypher, the shape of the crowns will be noticed to differ and these have been classified as "King's crown" and "Queen's crown"; the Queen's crown is more rounded at the top right and left sides. Less obvious, but nonetheless clear and consistent variations will be noted between the crowns of Edward and George, as well as the designs used by the various factories. The Sparkbrook and S.S.A. cyphers in particular are quite different.

Appropriate years of reigning British monarchs were:

Victoria	1837 - 1901
Edward VII	1901 - 1910
George V	1910 - 1936
Edward VIII	1936
George VI	1936 - 1952

Indian production rifles used a modified cypher until the declaration as a republic in 1950, when the new national symbol call the "ashoka" was used on No. 1 rifles in place of the Royal cypher.

Australian rifles used no cypher on the receiver butt socket at all, just a shield design on early production Mark III rifles and after that, only the lettering to indicate the factory, year and rifle designation. Further details of the Indian and Australian markings will be found in Chapters 9 and 10. Like the British wartime No. 4 rifle, those manufactured in North America did not use the cypher.

A new lightened rifle was proposed for infantry early in 1935 when it was ordered that user trials be conducted with shortened No. 4 Mk I, No. 3 and No. 1 Mk III models so as to indicate the relative appearance and performance which could be expected. The requirement was for a rifle weighing 7 lb. 8 oz. or less and a standard spike bayonet was to be fitted to the No. 4 rifle, a special folding spike for the No. 3 and no bayonet at all for the No. 1; a description will also be found in the previous chapter.

The ensuing modified No. 4 weighed 7 lb. 14.25 oz. and the barrel was shortened by 3.6 ins. to about 21.5 ins. long. An aluminium buttplate was utilised along with a 5-round magazine to keep weight at a minimum; dural pressings were also intended for some components. In a report forwarded with the rifle, it was noted that "we are inclined to think that if a high standard of accuracy is essential, a nominal one-piece stock and a front locking action are an almost essential basis." An allegory may be made drawn with the wartime Canadian Long Branch lightweight No. 4 model.

Attempts were made to reduce the weight of the receiver with a series of experimental rifles designated the LT1 and LT2. Drawings for these rifle modifications are dated July 1937 and the differences between the models were in the lightening cuts on the receiver. LT1 had "rotary cuts"

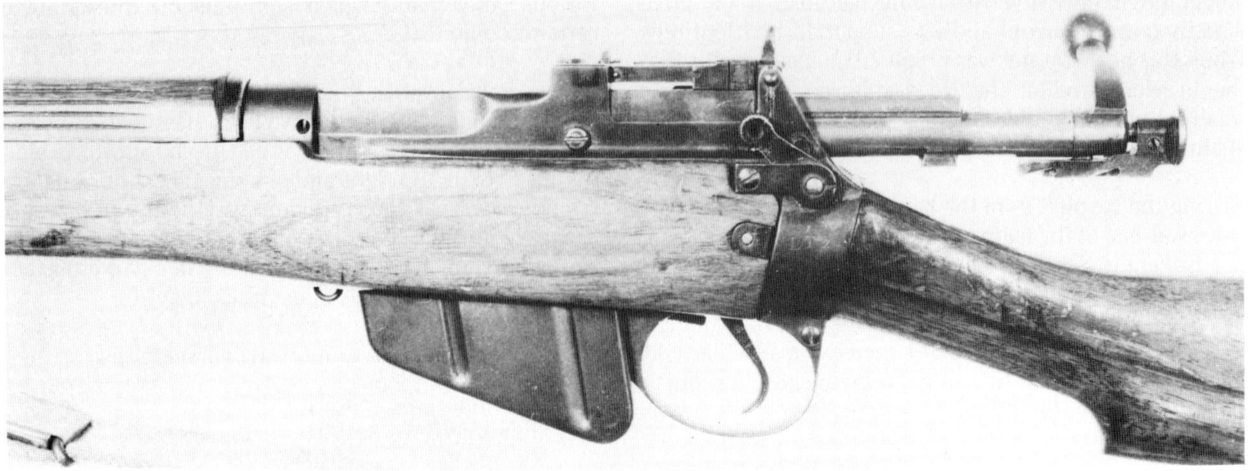

Above & below: 1937 LT2 Experimental rifle (engraved on left side of butt socket). An attempt to reduce receiver weight, using unfinished production trials body. *R. Faris*

whereas LT2 had "reverse cuts and straight cuts." These "cuts" refer to machined-off portions of the receiver body.

Evident in the accompanying illustration, the left wall of the action has been reduced in thickness and an attempt made to equalise the mass of metal on each side, upon which the bolt locking lugs were supported. These attempts were not successful in that consistent shooting could not be obtained from these rifles. The Canadians incorporated a similar style of modifications in their lightened rifle series during World War 2, (mentioned a few paragraphs earlier, in reference to the lightened infantry model in 1935).

Due to the unstable situation in Europe during the late 1930's and the dearth of service rifle production in Britain but for a token amount by B.S.A., efforts were made to get production of the No. 4 under way. One of the intentions when the No. 1 Mk VI was first produced for trials, was to design a rifle better suited to modern production techniques and the No. 4 exploited this as much as possible. At first, it was complained that the old hands at Enfield R.S.A.F. could not get used to the new techniques in manufacture of the Mark VI as it relied much less upon individual skills of the craftsmen than previous rifles. Even in the late 1930's, many of the traditional gunmaking routines were still employed in Government ordnance factories.

In June 1939 it was realised that Europe would inevitably be plunged into war again and a decision had to be made regarding production of a general service arm. The self-loading rifle was still under development and the choice lay between the No. 1 and No. 4 rifles. The former was ill-suited to mass production and so the latter was chosen. Of all the Lee-Enfield designs, the No. 4 was preferred because of its heavy barrel, aperture sighting and general suitability for mass production. Orders were then given for the preparation of new rifle factories to manufacture the No. 4, the decision to produce the No. 4 rifle was made in October 1939 although it was about three years before rifles were being turned out.

But the No. 4 was not to be made at Enfield; two new Royal Ordnance Factories were set up, at Fazakerley near Liverpool in Merseyside, and at Maltby, about 10 miles north-east of Sheffield in South Yorkshire. Birmingham Small Arms, already at Small Heath in Birmingham proper, established another plant to produce the No. 4 rifle at 85 Factory, Marshall Lake Rd., Shirley, an outer Birmingham suburb. These three factories commenced production in 1941.

Fazakerley is on the outskirts of Liverpool, not far from the Aintree Racecourse. In preparation for building the factory, the first sod was turned in May 1940; small arms production commenced in June 1941. Its output represents half the total of British production of small arms during World War 2 and at its peak, employed 12,000 people, 70% of which were women workers. It continued as a major arms producer and repair factory after the end of the war, into the 7.62mm era.

The Maltby factory was in operation in mid-1940, about one year ahead of Fazakerley, although it never had the capacity or manufacturing facilities of Fazakerley. Its main role was co-ordination and assembly of the rifle components produced by the various sub-contractors.

At the beginning of the war, R.S.A.F. Enfield was arranged so as to manufacture machine guns and pistols, the B.S.A. factory at Small Heath concentrated upon machine guns and anti-tank rifles. Some No. 1 Mark III and III* rifles were still being produced at Small Heath and Enfield was still making receivers, barrels and furniture for the No. 1 rifle. Between September 1939 and May 1940, some 500,000 Mark III and III* rifles underwent factory reconditioning and repair.

Upon the realisation in 1939 that the forthcoming war would be a protracted struggle, arrangements were set up to produce the No. 4 rifle in North America; this is described in Chapter 8. In 1927, India had decided to adopt the No. 1 Mk VI and purchased machinery, the specifications and tooling and started the manufacture of some components. In 1930 the project was suspended because the War Office categorically stated that this rifle would not be adopted and so India utilised the machinery in other production and disposed of the tooling and components. So, at the onset of the 2nd World War, India had little choice but to continue with No. 1 rifle manufacture.

The No. 4 Mark I rifle, along with three variations of the No. 4 bayonet, was approved on 15th November 1939 and 13th February 1941. The official announcement in the List of Changes was:-

Para B4737:-

Rifles, No. 4 Mk. I
With long butt
With normal butt
With short butt

Accessories for the Rifle
Bayonets, No. 4, for Rifles, No. 4 Mk I
Mk I
Mk II
*Mk II**
Scabbards, Bayonet, No. 4 Mk I
Tools, Removing, Foresight Screw for Rifles, No. 4 Mk I
Tools, Spring, Extractor, No. 2 for Rifles, No. 4 Mk I
Gauges, Armourers
Striker point, No. 4, for Rifles No. 4 Mk I

1. Introduction.

Tools, Spring, Extractor, No. 1

2. Nomenclature.

Bayonets, No. 4 Mk I

3. Obsolescent.

1. Approval has been given for the manufacture of the above-mentioned rifle and accessories. A brief description is given hereunder:

The Rifle, No. 4 Mk I is similar to the No. 1 rifle, differing in the following respects:

The only radical departure in design is in the method of sighting, an aperture backsight being hinged at the rear of the body similar to that on the No. 3 rifle.

The heavy nosecap is abolished, the barrel is exposed approximately 3-in. at the muzzle end, and all screw threads are to standard specifications. Due to manufacturing differences, the majority of the components are not interchangeable with those of the No. 1 rifle.

Briefly the differences in the principal parts are as follows:-

The barrel is the same length, but the external diameter is larger. Solid lugs are provided at the muzzle end for fixing the bayonet and foresight band block.

The lower band is approximately .5-in. wide and is similar to the existing band on the No. 3 rifle. An alternative design is also approved which differs only in that the lugs, which house the clamping screw, are manufactured separately and welded to the band.

The upper band is approximately 1-in. wide and hinged at the top. It is assembled, at the front end, over the front handguard and stock fore-end. An alternative design is also approved, which differs only in that the lugs, which house the clamping screw, are manufactured separately and welded to the band. This alternative band is not hinged.

The foresight band block is slotted from the forward end of the dovetail to the foresight protector screw hole to act as a clamp for the foresight blade.

The foresight blade screw is formed to take a special tool to prevent unauthorised adjustment.

The bolt ribway, or bolt head guide, is formed with an inner undercut ledge which ensures more effective control of the bolt head. The sides of the charger guide are formed solid with the body and connected by a bridge piece. Side ears are formed at the rear end and are drilled transversely for the backsight axis pin. A "T"- shaped guide groove is cut on the right side to house a new type of bolt head catch.

The breech bolt rib is wider and has a lightening groove cut longitudinally. The bolt head tenon hole has a plain portion for approximately .4-inch.

The breech bolt head wing is shorter and grooved to engage the undercut ledge on the body. Four differing lengths are provided, Nos. 0, 1, 2 and 3, progressing by

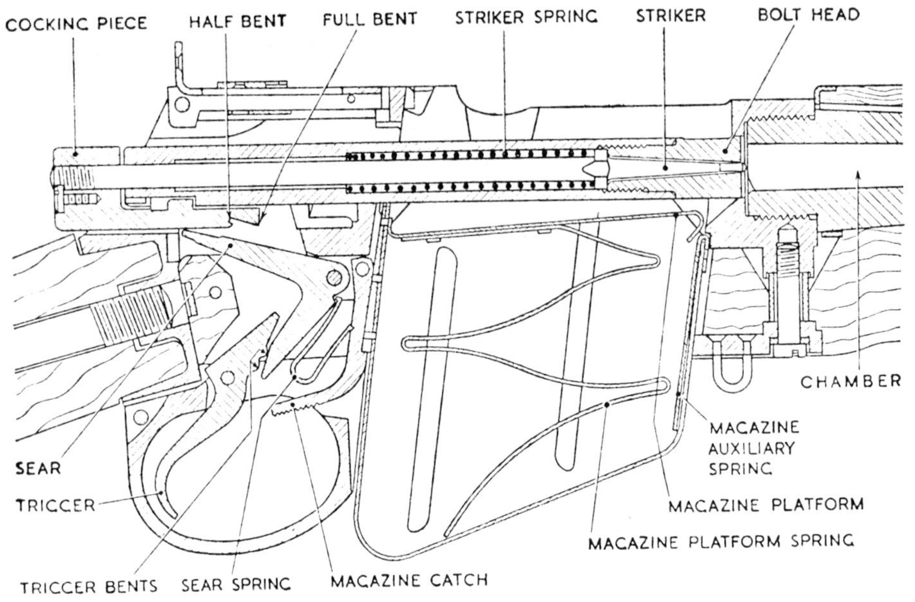

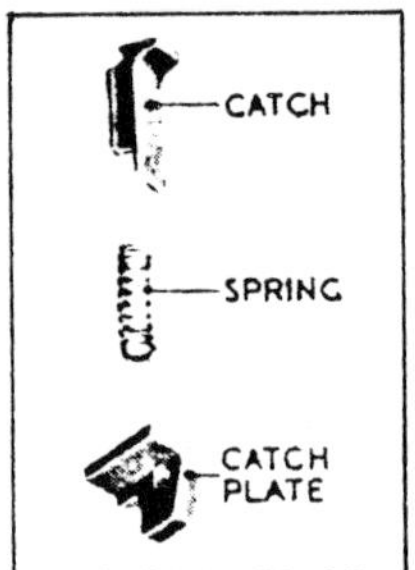

COMPONENTS OF THE BOLT HEAD CATCH

RIFLE No. 4 Mk I

(A) RIFLE FIRED

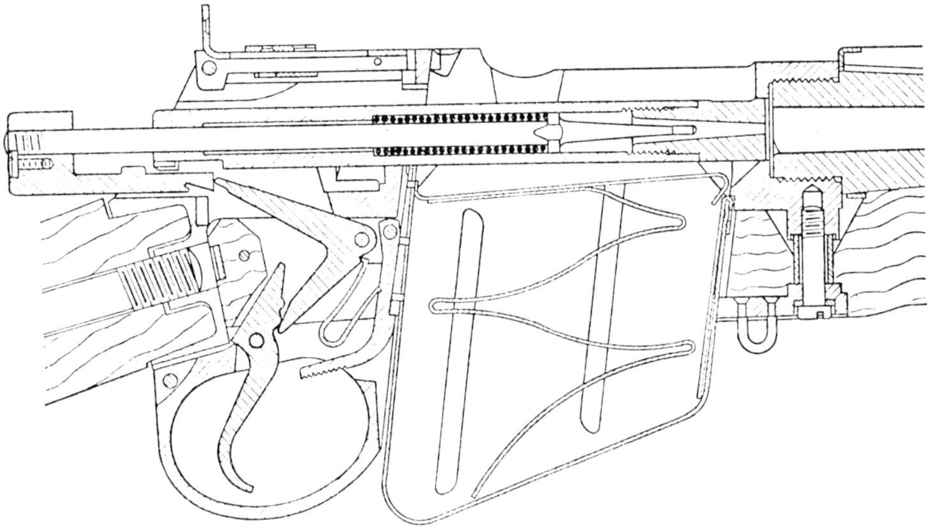

RIFLE No. 4 Mk I

(B) RIFLE COCKED

increments of 0.003-in. respectively, for convenience of head space adjustment.

The breech bolt catch is "T"-shaped and has a thumb-piece at the top formed with a tongue to engage the breech bolt head groove. The rear face has a radial groove to house a coil spring.

The breech bolt head catch plate is "L"-shaped, the long arm having a tapped hole. It is retained to the body by the magazine catch screw.

The backsight consists of a leaf, slide, nut, adjusting screw, retaining pin, plunger and spring.

The leaf is graduated from 200 to 1,300 yards by increments of 50 yards. A battle sight is provided on the rear face for use when the leaf is down. The leaf is retained in the upright position by a plunger and spring located in the body.

The slide, finely chequered on the rear face, is controlled by the adjusting screw which passes through a detachable nut. Rotating the screw clockwise raises the slide.

The adjusting screw is approximately 1.75-in. long and has a fixed knurled head. It is retained in the leaf by a lateral pin.

The plunger and spring housed at the top of the leaf, sets on serrations beneath the head of the adjusting screw to form a clicking device.

The butt plate strap is wider and the rear face is deeply recessed to enable the trap to be opened. The inner face is recessed in lieu of the hole for the trap axis pin.

The butt plate trap differs in that the lip is omitted and it is provided with a transverse pin. The pin rotates in the butt plate recess and the trap is retained by the spring.

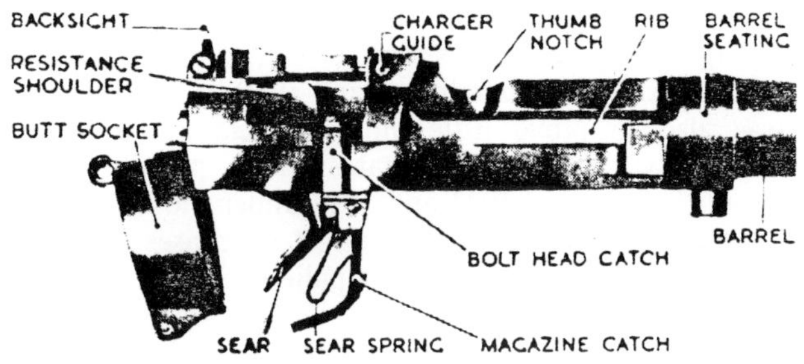

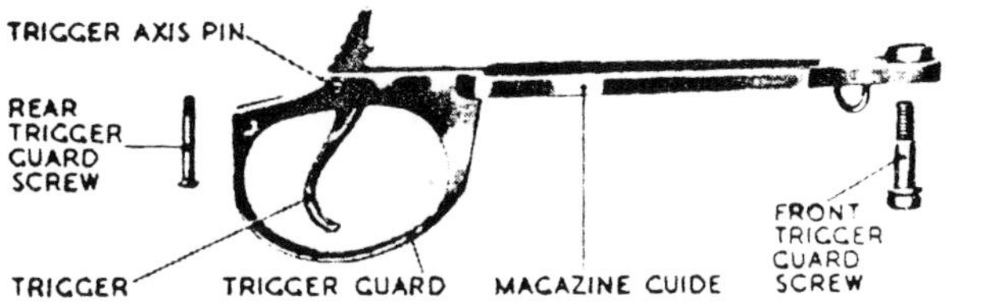

The cocking piece, Mk I, has a dome-shaped head. The Mk II is of a different design to facilitate ease of manufacture. No more Mk I pattern will be manufactured and all replacements will be of Mk II pattern.

The locking bolt thumb-piece is longer and has a radial head.

The locking bolt spring has a square boss with projections for positioning in a groove in the left of the body forward of the butt socket.

The magazine catch tail is altered in set and projects at a different angle into the bow of the trigger guard.

The magazine case is modified to suit its central location in the body. The rib on the rear face does not extend beyond the catch ledge and the rib spring is omitted.

The magazine platform is altered in contour to facilitate manufacture and can be identified by the sharper taper and wider turned-over lip at the forward end.

The magazine auxiliary spring is of approximately the same length, but the turned-over end is not a full radius. It has a round hole, in lieu of a slot, for fitting the rib of the case.

Note: The magazine case, platform and auxiliary spring are not numbered.

The foresight protector is similar to that fitted on the No. 3 rifle. It is assembled around the foresight band block, and partially engages with the lugs of the barrel for the foresight band block, a transverse screw locking it in position. An alternative design is approved, differing mainly in manufacture. A plain hole is bored eliminating the interruptions for the lugs on the barrel. The outside is plain in lieu of the shaped design.

The fore-end stock is provided with a cap, approximately 2.25-in. long, at the forward end and a tie-plate is fitted in lieu of the screwed pin. The barrel groove is recessed for lightening purposes.

The front and rear handguards are similar to those fitted on the No. 3 rifle. The front guard is fitted with a cap, approximately 2.25-in. long, and two liners. The rear hand guard has one liner and is retained by a ring encircling the rear end of the barrel.

The stock bolt spring washer replaces the keeper plate, and is inserted between the head of the stock bolt and steel washer.

General Particulars

Bore, Rifling & Chamber	
Bore	*.303-in. nominal.*
Rifling Form	*Concentric.*
No. of grooves	*5.*
Direction of grooves	*Left hand.*
Pitch	*1 turn in 10-in.*
Width of grooves	*.936-in.*
Depth of grooves	*.005-in.*
Chamber	*for .303-in. Service ammunition.*
Weight of rifle	
Without bayonet	*9 lb. 1 oz.*
With bayonet	*9 lb. 8 oz.*
Length	
Without bayonet	*44.43-in.*
With bayonet	*52.79-in.*
Capacity of magazine	*10 rounds.*
Sight radius at 200 yards	*28.72-in.*

Accessories

A brief description of each is given hereunder.

The Bayonet, No. 4 fits over the muzzle end of the rifle barrel and consists of a blade, catch, plunger and spring.

The blade is approximately 8-in. long. The Mk I is cruciform in section, while the Mk II and II are circular. All three taper to the point and are provided with a barrel socket at the rear end. The barrel socket is offset approximately 1-in. from the blade and is approximately 2-in. long, being bored to assemble over the muzzle. The rear of the socket is recessed internally to engage the two lugs on the rifle barrel, drilled and slotted externally to house the spring and catch respectively.*

The catch is rectangular in shape having a radial face and lip. It is drilled longitudinally to house the tail of the plunger and the rear end is serrated to assist manipulation.

The plunger is cylindrical, approximately 1-in. long. It is reduced in diameter at one end and bored from the opposite end to accommodate the coil spring.

The main difference between the Mk II and II is in manufacture only. The Mk II is made in one piece whereas the Mk II* is manufactured in two parts, known* *as the spike and socket, which are finally brazed together, and are not issued separately.*

The Scabbard, bayonet, No. 4 Mk I is a tapered steel tube, approximately 8-in. long, with a detachable cylindrical steel or zinc base alloy mouth-piece and a tip in the form of a knob. A spring, to hold the bayonet in position, is held between the mouth-piece and tube by the two mouth-piece screws. A leather liner is positioned in the lower extremity of the tube to cushion the bayonet, and a fixed stud is provided for the attachment of the scabbard to the frog.

The Tool, removing, foresight screw, approximately 7.25-in. long is of steel rod with a wood handle. The end is slotted to accommodate the projections on the head of the foresight blade screw.

The Tool, spring, extractor, No. 2 is similar in shape to the No. 1 pattern but differs in that the stripping pin is smaller in diameter and the assembling end has a radius in lieu of a lip.

The Gauge, striker point, No. 3 is of steel for checking, by armourers, the protrusion of the striker when assembled.

2. Consequent upon the introduction of Tools, spring, extractor, No. 2, the nomenclature of the existing Tools, spring, extractor will now read as shown.

3. Consequent upon the introduction of Bayonets, No. 4 Mks II and II, the Mk I pattern is hereby declared obsolescent. Existing stocks will be used up.*

In the original General Particulars reproduced in the previous pages from the List of Changes entries, there was a mistake regarding the rifling (right hand twist was specified); this should have been "left hand twist." The bayonet blade length was wrong, too, being listed at 9-in. instead of 8-in. However, both these particulars have been corrected in the LoC reference reproduced here.

Troops were being issued with the No. 1 rifle during the early part of the war and it was still in production in Birmingham. It was not until June 1941 that the first No. 4 rifles were completed at Maltby, 25 in number, and the following month Fazakerley turned out its first production batch of 300. The next month, August, B.S.A. Shirley produced its first 300.

The No. 4 rifle was first issued to front line troops the following year; they were generally armed with the No. 1 rifle. Home Defence units and many of the troops on training were using the No. 3 (Patt. 1914) rifle, .30 U.S. Model 1917, Canadian Ross and even Charger Loading Lee-Enfields. The No. 4 does not appear in photographs with the forces until the latter part of 1942, and then this was generally with specialist units, such as the Airborne Division.

Acute shortage of rifles during the 1st World War resulted in large scale acquisitions from the United States, Canada, India, Japan and Australia; during 1939-45 the situation was also to prove little different. Again, the largest supplier was the United States which provided huge quantities between 1941 and 1944 under the Lend Lease agreement as well as smaller direct purchases from U.S. arms manufacturers and trade suppliers.

Those weapons supplied by or through the U.S. Government during the war were recorded as:

	British Empire	*Canada*	*French*
.303 No. 4	1,196,706	-	-
.30 M1 Garand	38,001	8,014	58,114
.30 M1903 A1 & A3	64,003	1	11,015
.30 M1917	119,000	-	-
.30 M1 & M2 Carbine	25,362	230	96,983
.38 S & W Revolver	650,551	13,650	-
.45 M1911 ACP	39,592	1,515	19,325
.45 Submachine guns	651,086	1,321	20,856

From late in 1941, after German forces had overwhelmed the remaining resistance in Europe and the invasion of the British Isles seemed imminent, sympathisers in the United States donated arms, ammunition and other supplies to British civilians for the defence of their homes. Such aid was organised through groups like the American Committee for Defence of British Homes which operated at the highest social and political levels.

These donations were separate from the military system and additional to the government to government and Lend-Lease agreements; this letter excerpt from the Duke of Windsor to Suydam Cutting, Chairman of the aforementioned American Committee in New York, indicates the situation:

"It is not necessary for me to assure you and all those who have responded so generously to your call for arms and ammunition, binoculars, steel helmets and stop watches, how valuable is this form of assistance to Britain at the present time. It is just one more example of friendly and sympathetic co-operation it has recently been my privilege to witness in America and which, as you know, means so much to my country right now."

Cutting's comments to donors also indicates the amount of such aid at this point in time: "You'll be interested to know that to date we have shipped to our Civilian Committee for the Defence of Homes abroad the following:

4,472 guns
70 Thompson Sub-Machine Guns
6,139 revolvers
2,924 binoculars
16,604 steel helmets
375 stop watches
540,032 rounds of ammunition

Prime Minister Churchill inspects the new rifle & bayonet. *Imperial War Museum*

In addition to those firearms listed above, our Committee secured over 12,000 revolvers for civilian defense from the United States Government under the Lend-Lease Act and with the approval of the United States Treasury Procurement Division and the British Purchasing authorities 600 weapons confiscated by the United States have been made available for British home defense and to be distributed by our Civilian Committee for the Defence of Homes abroad. We are advised that we may expect further shipments from this source."

British factories were equipped with these rifles, the binoculars and stop watches used for roof spotting, and the steel helmets issued to fire wardens, doctors, nurses and other civilians. Cutting's closing comment was *"Probably nothing could bring greater satisfaction at Christmas than an increased feeling of security to those British civilians who face the possibility of invasion today without firearms. There could be no finer expression of good will at this season than to send such material for defence to those stout-hearted men and women to increase their security."* How markedly different public attitudes are today!

The Prime Minister, Sir Winston Churchill inspected the "new rifle and short bayonet" on 20th November 1942 and it was from about this time that the No. 4 started to appear with front line units, issued in quantity and on a regular basis. Sufficient quantities had been stockpiled from the three British factories and the U.S. and Canadian contracts and adequate supplies of spare parts became available through the system to service the new model.

Due to various complications in supply, the war in North Africa and to a large extent the Italian campaign, was fought with the No. 1 rifle rather than the No. 4. Production of the No. 1 had been severely limited as a result of the bombing raids on Birmingham in August 1940 and subsequent air attacks in November, three months later.

The cost of the No. 1 rifle as produced by B.S.A. in 1943 was listed at £9/10/-, which included their profit margin. The No. 4 was costed at £8/5/- per unit using imported timber and it was considered that this could be reduced to £7/15/- using home-grown timber. By comparison, the Bren light machine gun was costed at £34 and this too was produced in an expanded sub-contractor scheme by the Monotype Group as well as R.S.A.F. Enfield.

The first deliveries of the wartime No. 4 rifle after passing inspection were:

1941	June	25	Maltby's first production.
	July	363	Fazakerley 300, Maltby 63.
	August	508	Includes Shirley's first 300.
	September	2,397	
	October	6,761	
	November	11,111	
	December	12,749	

A variety of concessions or relaxations in the specifications were approved soon after production of the No. 4 commenced, in the interest of simplification of manufacturing procedures. Some of the parts were also made common to both No. 1 and No. 4 rifles. The first of the interchangeable parts produced were butt stocks, which only required a minor modification at the socket and this was approved on 25th July 1940. Later on, the screws for the No. 4 bands and swivels were declared interchangeable with those of the No. 1, although the slightly different thread type necessitated cleaning out the No. 1 rifle thread with a No. 8 tap when the No. 4 screws were fitted to No. 1 rifle bands.

Production concessions applied to the No. 4 components are better related as individual parts and items rather than a chronological sequence. It will also be found that some of these relaxations and changes were applied by different contractors and factories at different times and in other instances were not applied at all. The situation may also be confused because many of the rifles in service were subsequently refurbished by armourers and in the course of factory refits, so some interchange of these parts also occurred.

Backsights. The Mk I or "Singer" backsight as introduced on the No. 4 Mk I rifle was one of the first items to be replaced, because of the shortage of these sights due to the more complicated and precise manufacture; this shortage was delaying the final assembly of rifles. The Mk I sight has click adjustment on the elevation screw and is graduated from 200 to 1,300 yards in 50 yd. intervals.

The Mk II sight consists merely of two fixed apertures in the shape of an "L," mounted on a squared shaft similar to the Mk I base. The same plunger and spring were used in the receiver body and the two sight ranges are for 300 and 600 yards. The 300 yd. range is sighted for use with the bayonet fixed while the 600 yd. calibration required removal of the bayonet. This sight was only a stopgap measure to enable final assembly of rifles, until a

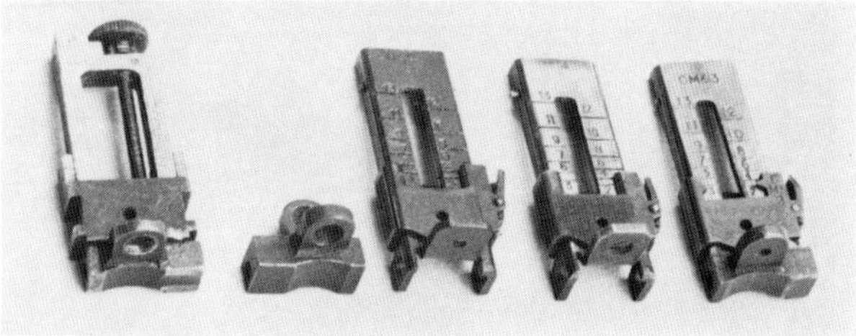

Above & below: No. 4 Backsights, from left to right in each picture:- Mk 1, Mk 2, Mk 3, Mk 4, C Mk 4. B. Labudda

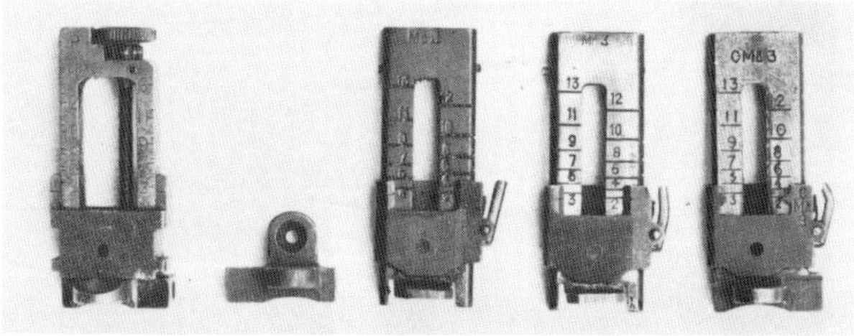

satisfactory sight with adjustable range could be developed. Many of the U.S. Stevens-Savage rifles were so fitted.

Two simplified pressed-metal sights were developed almost simultaneously, the Mk III designed by B.S.A. and the Mk IV by the Government Design Dept. These sights have pressed sheet steel leaves graduated from 200 to 1,300 yards. The Mk III sight is stamped "II" on the leaf as the Mk II sight has no such leaf, therefore in systematic progression, the Mk III sight incorporates the Mk II leaf. The slide has a spring-loaded catch on the right hand side which engages in notches along the right side of the leaf. This sight was first sent to the Small Arms School for testing in September 1942 and put into production soon afterwards. Like the Mk I sight, it is graduated from 200 to 1,300 yards and has a battle aperture for use in the folded down position.

The Mk IV sight was an improvement over the Mk III in that it has a curved release catch less liable to damage, and the spring is a stronger "mousetrap" type. The Mk III spring is a spiral type and it was claimed that it did not always lock positively. For the Mk IV sight, the designation "III" is marked on the leaf, an advance on the Mk II leaf for the Mk III sight. The Mk IV sight was adopted on 23rd April 1943 and like the Mk III sight, the battle aperture was intended for ranges to 400 yards and the leaf graduated from 200 to 1,300 yards. It was found that the battle aperture on Marks III and IV sights could be broken off if the rifle was handled carelessly.

Many rifles originally issued tithe the Mk II "L"-shaped battlesight were refitted with Mk III or Mk IV sights by unit armourers or during repair programs. The Mk II and Mk IV backsights were declared obsolete on 27th January 1947, most having been replaced with the Mk I sight which was in more plentiful supply by that time.

Three other variations of the No. 4 backsight may be encountered; the first consisting of the Mk I type with battle aperture removed for fitting on the No. 4 sniper rifle and the other two being Canadian models designated "C Mk2" and "C Mk3," which have slightly different aperture slides. The last two are further described in Chapter 8.

Barrel. This too was manufactured in a number of variations intended to simplify production. Tests were carried out in May 1941 with 2-groove barrels on the No. 1 rifle and it was found that the accuracy, wear and fouling were similar to that of the 5-groove barrel and so the 2-groove rifling was approved for the No. 1 and No. 4 rifles, designated the Mk II barrel. Many of the North American made rifles were fitted with Mk II barrels. The 2-groove rifling was declared obsolete on 11th July 1945; it had also been approved for use in the Sten Marks I, I*, II and III machine carbines and the Revolvers Nos. 1 and 2 as well.

In 1942, three-groove barrels were tested and found to be satisfactory so this was approved for manufacture should any of the contractors find that it simplified production, but very few were made and it was not awarded a Mark designation. This 3-groove rifling was also applied to an experimental "dollied" barrel.

A Mk III rifle barrel was introduced in 1943, developed by a commercial steel company, Accles & Pollock. This consisted of a solid-drawn steel tube with the knox form made as a separate sleeve, shrunk and pinned to the breech end; two pins were used to further secure both sections. The Mk III barrel did not have as bright a bore as the other barrels and so special instructions were issued for troops who had rifles with these barrels as they appeared to be dirty; such a barrel was not likely to make the soldier's lot a happy one come inspection time!

The Mk III barrel is stamped "III" on the flat at the rear of the barrel. The 3-groove dollied process comprised hammering the hot metal around a former which was shaped to the bore and grooves, so that when the former was withdrawn, the bore and rifling remained in the metal. The outside of the barrel was then turned down to the required profile. It was later found that some of these barrels developed a slight amount of play in the joint and manufacture was discontinued. Although about 100,000 were produced, this barrel was declared obsolescent soon afterwards. The method of dollying or hammer forging is the basis of some successful barrel making and rifling procedures today, but it did not prove effective at the time.

A four-groove rifling form was also acceptable in instances where the manufacturer considered that it simplified production although not many of these barrels were produced. In North America, a number of 6-groove barrels were produced, one type being designated the C Mk 4; details of this will be found in Chapter 8.

Following some experiments made with slightly bent barrels, it was found that, so long as the last 6 inches were perfectly straight, accuracy was not impaired. So the requirement for a perfectly straight barrel over its full length was relaxed and special 6-inch long gauges were introduced for testing. Thus, another bottleneck in production was overcome, that of examination by skilled barrel viewers, because the barrels could now be tested by the application of special gauges for the bore.

Cocking Piece. The early No. 4 cocking piece had a flared rim at the back end although it was found that by making both sides flat and cutting three grooves on each side, considerable time and machine milling could be saved. The early cocking piece was then designated the Mk I and the new pattern, the Mk II. The Mk II is similar to the type used on some of the No. 1 Mk III* rifles.

Top: Cocking Piece Mk II.
Bottom: Wartime alternative.

Some Mk II cocking pieces made in Birmingham have no half-bent, although this omission was shortlived because it was considered unsafe. Another production shortcut effected to some Mk II cocking pieces was the omission of the vertical slots but this was also discontinued.

Foresight Protector. Wherever possible, milling was replaced by pressings, stampings and fabricated parts. A Mk II foresight protector was introduced on 22nd April 1943 consisting of pressed steel with two inserts for the securing screw; although similar in outline to the Mk I machined pattern, it can be quite readily distinguished. Yet another type was introduced in which the sight block is an integral part of the foresight protector, the whole assembly being milled from one piece. This was approved on 10th April, 1943.

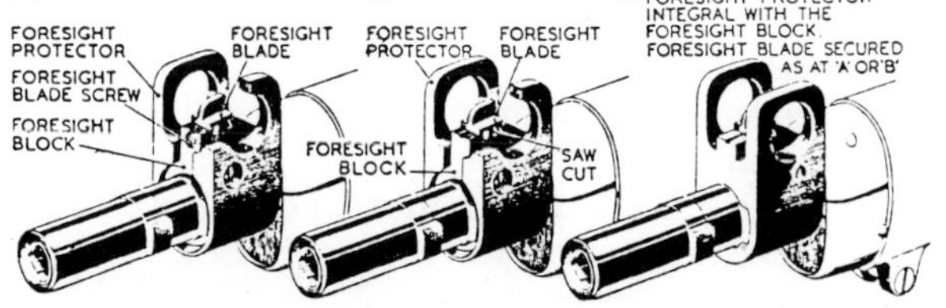

Different Foresight Types and Foresight Protectors.

Foresight. Four types of foresights were used, designated Marks I, I*, II and III, with the following differences:

Mk I	Machined solid, full length blade.
Mk I*	Similar to Mk I but with slot in base for use with solid type foresight block.
Mk II	Similar to Mk I but with longer platform at the rear of the blade.
Mk III	Similar to Mk I* with slot in base, but longer platform at the rear of the blade.

Magazine. The main source of magazines made to an alternative pattern was North America, where many were made in two pressed sections and then the halves welded together. This only occurred in later production as this relaxation in specifications was not approved until June 1943. Another variation found in some of the late manufacture magazines is in the two stop nibs at the back of the case. Many of the North American magazines were engraved or stamped with the serial number of the rifle, particularly the Canadian ones.

Buttplate. On the Enfield-made trials rifles these were made of gunmetal which has an appearance akin to brass, although some seem to have a more bronze colour when compared with the later production No. 1 rifles. In January 1940, before wartime production of the No. 4 got under way, the Ordnance Board approved the use of Mazak No. 3 alloy for the buttplate, although the use of aluminium alloy was originally recommended.

Trigger Guard. Fabricated trigger guards were also utilised instead of the milled pattern and these were made up from square rod and bent to shape, with suitable inserts, and then welded at certain points.

Locking Bolt. A Mark II locking bolt was introduced for a short time; the terms "locking bolt" and "safety catch" can be confusing because they may be one and the same, or different but associated parts of the safety catch mechanism. Such misinterpretation may also apply to the earlier Short Lee-Enfield.

With this Mark II "safety catch" variation, the quick thread was not machined and no locking bolt fitted; thus, with this safety catch applied to a cocked action, the breech bolt could be opened and drawn back the distance of the striker drop. In instances where this safety catch was used in conjunction with the Mk II cocking piece without half-bent, problems could quite easily occur with the safety mechanism.

Sling Swivel. Many were bent up from heavy gauge wire.

Oil Bottles. In April 1940, either celluloid or bakelite was recommended as an emergency substitute for brass and this was implemented when bakelite was approved on 10th July 1940. Due to the additional expense of cellulose acetate, bakelite was that most commonly used.

In many instances, these specification concessions resulted in limited interchangeability amongst some components and although many related problems were rectified after inspection at Enfield or Weedon, some rifles were still found to be fitted with non-interchangeable parts; these were marked with the letter "A" on the left side of the butt socket. The earlier upgraded No. I Mk VI and No. 4 Mk I trials rifles were similarly marked with the "A" suffix after the serial number.

It was found that quite a few magazines were defective as they could not be loaded with 10 rounds, so special loading instructions had to be issued for use with the 5-round charger clip. Consequent to this, new magazines were inspected and to indicate their acceptance and compliance with the proper standards, those which conformed were marked with a "B" on the rib at the back. A quantity of No. 1 magazines were also converted to the No. 4 pattern by B.S.A. in Birmingham.

Those British factories engaged in No. 4 rifle production during the war were R.O.F. Maltby, R.O.F. Fazakerley and B.S.A. Shirley. Inspection and certain adjustments were implemented at Enfield, where a large repair program for the No. 1 rifle was carried out early in the war; large stocks of the No. 3 rifle were brought out of storage and upgraded to W.R.S. (Weedon Repair Standard) at Enfield and by the gun trade. The major B.S.A. factory at Small Heath was mostly engaged in the manufacture of machine guns, the Boys anti-tank rifle and No. 1 rifle. Total No. 4 rifle production at Fazakerley, Maltby and Shirley was projected at 25,000 per month during 1942.

Some of the commercial trade manufacturers of component parts for the Nos. 1, 3 and 4 rifles during this period had already participated in the S.M.L.E. rifle peddled scheme during the Great War. Those more specialist No. 4 parts such as the receiver and barrel were made at Maltby, Fazakerley and Shirley, although Accles & Pollock produced some barrels. Most of the smaller components and furniture were supplied by private industry, one of the successes of British wartime production. The contractors were allocated codes which often appear marked on components; the "N" prefix indicates northern district, "M" for the Midlands and "S" for southern England.

The particular contractors and codes for the No. 4 rifle along with the components they produced are listed hereafter. A complete register of these makers' codes, registered business addresses and the small arms inventory produced in Britain during the war will be found in "British Small Arms of World War 2 – The Complete Reference Guide to Weapons, Makers' Codes & 1936 – 1946 Contracts" by Skennerton.

Some of the manufacturers were not allocated code numbers, particularly those who commenced supply towards the end of the war. No. 4 rifle contractors' names and addresses, and their applicable codes for the component parts produced during the 2nd World War period were:–

Manufacturer	*Code*	*Component/s*
Accles & Pollock Ltd., Oldbury	M1	Mark III barrel tubes.
Adams Bros. & Burnley Ltd., Harrow St., Harrow	S3	Handguard retaining rings, fore-end caps, magazines.
Addressograph & Multigraph, Oxgate La., Cricklewood	S134	Bolthead release catches.
Aero Zipp Fasteners, Southampton Row, London WC1	S312	Safety catch components.
Albion Pressed Steel Metal Co., Birmingham	M3	Safety catch springs, handguard retaining rings.
Anstey & Wilson Ltd., Kenyon St., Birmingham	M8	Handguard retaining rings.
P. Antrobus Ltd., Regent St., London W1	S308	Pins.
Associated Brassfounders, Old Oak Lane, Willesden NW10	S314	Mk II backsights.
W. & T. Avery Ltd., Birmingham	M14	Buttplates.
Baker & Finnimore Ltd., Birmingham	M17	Magazine auxiliary springs.
Bellgrett Industries, Kilburn High Rd., London NW6	S301	Backsight catch springs.
L. Benn & Sons, Denmark St., Leytonstone, London E11	S12	Triggers, Mk II backsights.
I. L. Berridge & Co. Ltd., Leicester	M601	Magazine catches.
Betterwear Products, Romford	-	Catch plungers.
Bifurcated & Tubular Rivet Co. Ltd., Aylesbury, Bucks.	S415	Rivets.
Birmingham Tempered Spring & Presswork Ltd., Birmingham	M29	Magazine springs.
Bloore & Pillar Ltd.	M33	Triggers.
J. Boss & Co., Albemarle St., London W1	S156	Extractors.
British Machine Chain Ltd., Malden	-	Handguard liners, stock tieplates.
British Oxygen Co. Ltd., Angel Rd., Edmonton, London N18	S167	Cocking pieces, pins, screws.
British Salmson Aero Engines Ltd., Raynes Pk., London SW20	S21	Foresight blocks, foresight protectors.
British Screw Co. Ltd., Leeds	N12	Buttplate & swivel mount screws.
British United Shoe Machinery Co., Leicester	M603	Stockbolts.
British Vacuum Cleaner & Engineering Co., Ermyn Way, Surrey	S22	Pins, screws.
J. B. Bruce Ltd., Birmingham	M46	Fore-end & handguard caps.
Burdon & Miles Ltd., Cockfosters, Barnett, Herts.	S166	Mk II backsights.
B.B.H. Tool & Gauge Co., High St., Teddington, Middlesex	S316	Mk II backsights.
B.S.A. Guns, Redditch, Hereford & Worcester	M47B	Some springs, small components.
B.S.A. Guns, Shirley, Birmingham	M47C	Complete rifles, most components.
B.S.A. Guns, Small Heath, Birmingham	M47A	Various assorted small components.
Carr & Smith, Birmingham	M54	Extractor springs.
Chambon Ltd., Standish Rd., London W6	S323	Mk II backsights, magazine catches.
W. C. Cheney & Son Ltd., Willenhall, West Midlands	M56	Magazine assembly.
Chiswell Wire Co., Sandown Rd., Watford, Herts.	S169	Springs.
Cogswell & Harrison, Park Rd., East Acton, London W3	S171	Sear springs.
Colostat Combustion, Victoria St., London W1	S403	Magazine springs.
Cooper & Turner	-	Rivets.
Cope & Timmins Ltd., Paul St., London EC2	S174	Springs.
George Cotton & Son, Goswell Rd., London EC1	S176	Bolthead release catches, butt trap springs.
Crown Bedding Co. Ltd., Tyseley, West Midlands	M66	Butt trap springs.
John Curtis & Son Ltd., Leeds, West Yorkshire	N22	Furniture.
Davis & Mawson Ltd.	-	Fore-end & handguard caps.
G. P. Dennis Ltd., Chester, Cheshire	N23	Backsight plungers.
Executors of James Mills	-	Safety catch pins.
E.S.S. (Signs) Ltd., Feeder Rd., Bristol	S223	Magazines, bands, trigger guards, swivels.
E.W. Engineering	-	Stockbolts.
Fenriss (1932) Ltd., Gorst Rd., Willesden	S187	Barrel pins (for Mk III barrels).
Fisheries Foils Ltd., Exhibition Grounds, Wembley, London	S189	Foresights.
Flexal Spring Co., Hanwell, London W7	S259	Magazine springs.
General Presswork & Stampings Ltd., Leicester	M607	Buttplates, brackets, swivels.
G. W. Gladhill & Sons	-	Trigger guard pads.
Gramophone Co., Hayes, Middlesex	S200	Pins.
Graylon Engineering Co., New Wharf Rd., London N1	S40	Sears.
A. C. R. Greene, Pinner Rd., Harrow, Middlesex	S381	Stockbolts, pins.
W. W. Greener Ltd., Birmingham	M94	Sling swivels, bolthead release catches.
Greville Engineering Ltd., Finchley Rd., London NW2	S329	Bolthead release catches.
S. B. Grundy & Co., Derby	M633	Stockbolts.
Guest, Keen & Nettlefold Ltd., Birmingham	M99	Screws, brackets.
T. Haddon & Stokes Ltd., Birmingham	M100	Rivets.
Hall Motors, Paddington	-	Springs.
Hardwick Engineering Co., Leicester	M639	Foresight blades.
Hildick & Hildick Ltd., Walsall	M119	Collars.
Holland & Caesar Ltd., Leighton Place, London NW5	S215	Brackets.
Holland & Holland Ltd., New Bond St., London W1	S51	Sniper rifles, extractors, swivels, cocking pieces, sniper rifle cheek rests.
G. W. Hughes, Legge Lane, Birmingham	M130	Magazine springs.
Imperial Engineering Co., Prospect Rd., London NW2	S56	Mk III backsights.
Junction Engineering Co., High St., Harlesden, London NW10	S385	Stockbolts.
K. Garage & Service Station, Watford Way, London NW7	S378	Stockbolts.
Lang Pen Co., Hope St. Works, Liverpool	N37	Buttplates, pins.
Lewford Engineering Co. Ltd.	-	Foresight block bands.
B. Lilley, Birmingham	M151	Brackets.
Lines Bros. Ltd., Morden Works, Merton, London SW19	S68	Trigger guards.
Littlewood & Sharp Ltd.	-	Stockbolts.
J. Lucas Ltd., Birmingham	M158	Trigger guards, upper bands.
Ludlow Bros. (1913) Ltd., Birmingham	M159	Magazines.
L. M. & S. Railway	-	Furniture.
Manchester Die Casting Co. Ltd., Manchester	N132	Buttplates.
G. S. Mascalls & Sons, Lawrence Rd., Tottenham, London N15	S232	Foresight band blocks, foresight protectors.
L. S. Mayer Ltd., Southey Rd., Tottenham, London N15	S132	Magazines.

Manufacturer	*Code*	*Component/s*
Metal Box Co., Barclay & Fry Branch, Southwark Rd., London	S362	Fore-end & handguard caps.
Metal Box Co., Carlisle	N94	Fore-end & handguard caps.
Metal Box Co., Gipsyville, Hull	N45	Fore-end & handguard caps.
Metal Box Co. Ltd. (Machine Section), Acton, London W3	S77	Fore-end & handguard caps.
Metal Box Co. Ltd., Newcastle-on-Tyne	N46	Fore-end & handguard caps.
Metal Box Co. Ltd., North Circular Rd., London N13	S76	Fore-end & handguard caps.
C. F. Moore & Sons Ltd., Sekforde St., London EC1	S238	Backsights, safety catch springs, collars.
Moore & Sons Ltd., Sheffield 1	N95	Backsights, safety catch springs, collars.
H. Morris & Co. Ltd., Glasgow	N49	Furniture.
Mortimer Springs	-	Magazine auxiliary springs.
M. Myers & Son Ltd., Oldbury	M175	Magazine springs.
McDonald Manufacturing Co., London	-	Butt trap pins, collars, barrel pins.
Nash Engineering (Staines) Ltd., London Rd., Staines	S87	Stockbolts, foresight protectors.
National Cash Register Co. Ltd., Marleybone Rd., London N1	S88	Cocking pieces, safety catche springs.
North Bridge Engineering Co., Leicester	M638	Stockbolts, collars, screws.
R. B. Page & Co., High St., Kingston-on-Thames, Surrey	S91	Foresight blades.
Peterborough Die Casting & Machine Co. Ltd., Peterborough	M189	Buttplates.
G. D. Peters & Co., Windsor Works, Slough, Bucks.	S95	Stockbolts.
Philco Radio & Television Corp., Greenford, Middlesex	S252	Backsights, extractors, safety catch components.
A. Poole & Son, Redditch	M194	Handguard liners, sling swivels.
Precision Engineering Co., London	-	Sears.
Prince-Smith & Stells Ltd., Keighley	N56	Foresight bands.
Pruners Ltd., Lonsdale Rd., Kilburn, London NW6	S98	Fore-end & handguard caps.
Raleigh Cycle Co.	-	Handguard caps.
A. T. Ralphs (N.C.H.) Ltd., Leicester	M612	Strikers.
J. Rawson & Sons Ltd., Mt. Pleasant, Tunbridge Wells	S99	Foresight blades, triggers, butt swivel brackets.
Repose Seating Co. Ltd., London	-	Magazine springs.
Restwell Spring Co., Darwen, Lancashire	N165	Mainsprings.
Rex Meters Ltd., Kingsbury Rd., London NW9	S261	Foresight blades.
Roneo Ltd., Roneo Works, Romford, Essex	S262	Cocking pieces.
George Salter & Co. Ltd., West Bromwich	M215	Springs.
Sanbra Ltd., Birmingham	M216	Buttplates.
Savaker Ltd., Birmingham	M217	Brackets.
Sheffield Steel Products Ltd., Sheffield	N64	Foresight protectors, cocking pieces, magazines.
Sidcup Tool & Gauge Co. Halfway St., Sidcup, Kent	S104	Mk II backsights, triggers.
Silversmiths & Jewellers War Peoduction Group, London W1	S105	Sears.
Singer Manufacturing Co. Ltd., Clydebank, Scotland	N67	Foresights, strikers, cocking pieces, pins, stockbolts, Mk I & II foresight protectors, screws, magazines, Mk I & Mk II backsights, backsight plungers, sling swivels, handguard liners & retaining rings, collars, marking discs, sears, washers, brackets, handguard & fore-end caps, brackets, collars, rivets.
Southern Railways (Chief Mechanical Engineer), Eastleigh	S360	Furniture, screws.
J. W. Spear & Sons Ltd., Green St., Enfield, Middlesex	S106	Triggerguards.
Spikins Ltd., Heath Rd., Twickenham, Middlesex	S390	Stockbolts.
Spring Washer Co. Ltd., Wolverhampton	M232	Triggerguard washers.
Streetly Tools & Stampings, Streetly, Birmingham	M241	Brackets, bolthead catch plates.
William Sykes Ltd., Horbury, Yorkshire	N74	Furniture.
Tempered Spring Co., Sheffield	N76	Fore-end caps, tie plates.
H. B. Thomas Co. Ltd., Grays Inn Rd., London WC1	S282	Stockbolts, pins.
Trafalgar Engineering Co., Croft St., Deptford, London SE8	S113	Foresight protectors.
W. Trueman & Sons, Birmingham	M249	Extractors.
Tube Fittings Manufacturing Co., Coseley, Staffs.	M250	Collars.
Thomas Turner & Co. Ltd., Sheffield	N77	Magazine cases & assembly, sear springs.
United Motors Ltd., North Rd., Islington, London N7	S117	Foresight protectors.
Univeral Stampings Ltd., Birmingham	M254	Bands, handguard liners.
Viners Ltd., Sheffield	N79	Bands, magazines.
Waddells Ltd., Millmarsh La., Brimsdown, Enfield, Middlesex	S122	Brackets, sling swivels.
Walker & Hall Ltd., Sheffield	N80	Butt trap springs, swivels.
Howard Wall Ltd., Hackney Rd., London E2	S123	Safety catches.
Walls Ltd., Birmingham	M260	Bolthead catchplates, fore-end & handguard caps.
Ward & Sons, Birmingham	M262	Cocking pieces, alteration to half-bent.
Weir Larsen Tool Co., High St., Potters Bar, Middlesex	S124	Triggers, sears, bolthead catchplates.
C. E. Welstead, Andrews Works, Tanfield Rd., Croydon	S126	Bands.
Wembley Tool Co., Bridge Works, Bridge Rd., Willesden	S128	Foresight blades.
Wickhan Precision Tool & Turning Co., Hatton Gdns, London	S352	Strikers.
W. J. Wild Ltd., Birmingham	M269	Stock washers.
Messrs. Wildt & Co., Leicester	M614	Extractors.
Wilkinson's Ltd., Bradford	N85	Furniture.
Willey & Co., Haven Rd., St. Thomas, Exeter	S129	Retaining pins.
John Wilson Ltd., Sheffield	N88	Drop forging.
John Wilson & Son Ltd.,	N89	Drop forging.
Woodhouse & Smith Co. Ltd., Nottingham	M630	Magazine catches.
S. Wright & Sons, Birmingham	M276	Sear springs.
Wright Bros., Sutcliffe Rd., London SE18	S131	Bolthead release catches, foresight protectors.
Zip Fasteners Co., Angel Rd., Edmonton, London N18	S300	Backsight caps, safety catches.

Quantities of foresights, screws and pins were manufactured by S.R.O.F. Blackburn; R.O.F. Theale produced some cocking pieces and magazines. A comprehensive array of components, except perhaps for the screws, pins and some springs, was manufactured at Fazakerley; Maltby, however, did not produce such a wide range of component parts.

Smaller components such as the pins, springs, screws and washers made by the contractors were not usually marked, however the larger components usually have various manufacturer's markings, such as the code numbers previously listed. Initials and other such markings are also often encountered, a listing of these will be found in Chapter 13. The B.S.A. production at Shirley did not rely upon many supplies from external contractors and Fazakerley likewise made most of the components. Maltby, though, relied largely upon parts from the various contractors and this factory was often referred to as "the pool" which would indicate that it was more of an assembly facility.

Some production and parts irregularities have been noted in Maltby production rifles; the markings vary from "R.O.F.M.," a stylised "M" character or just a plain "M" letter, or there may be no factory marking or letter at all. Where the receiver and/or components have an "A" letter stamped by itself, not part of any other stamp or series of markings, this usually indicates component non-interchangeability due to some minor substandard characteristics. Two 1944 Maltby rifles have even been noted stamped "NO. 4 MK I*," by the correct one-piece roller die, during the receiver body production. Yet these have the usual Mk I rifle features with bolthead release catch behind the charger guide. The Mk I* variant was not produced in Britain, only in North America.

At R.O.F. Fazakerley, near Liverpool, another section of the factory was allocated to Sten gun production and huge quantities of Sten parts were manufactured by the contractors. Total production of the various Marks of the Sten gun was about twice that of the No. 4 rifle. For the Fazakerley factory, Sten production represented nearly four times the figure of No. 4 rifles, assisted greatly by the huge quantity of parts provided by private industry, an incorporate part of the Sten design.

With both the No. 4 and Sten, the components were forwarded either to the factories for assembly or into store as spares. The Ordnance ensured that the Shirley factory was not utilised to its full capacity in No. 4 rifle production as a percentage was left in reserve in case of any urgent, additional output being required.

Various research was carried out in an attempt to increase rifle production but three main difficulties remained; a shortage of machine tools, particularly milling machines, scarcity of skilled labour, and the lack of suitable timber. Tests were carried out on a rifle with fabricated action, Mk III barrel and a skeleton stock which eliminated the requirement for timber, but this "ersatz" model proved to be unacceptable even by 1942 standards when the criteria was much less demanding.

For earlier military rifle production, the walnut used in stocking-up had to be seasoned for years prior to cutting and machining. Before the introduction of the No. 4 rifle, tests were conducted with kiln dried timber and birch and beech were used as an alternative to walnut. Walnut had to be imported whereas some quantities of beech and birch could be had from forests in Britain, although most of the wooden furniture still had to be imported from North America, especially during the war. During the period before the Great War, much of the walnut timber had been imported from the Continent, particularly France.

During World War 2, most of Britain's No. 4 rifle production was stocked in beech or birch and by the middle of the war when poorer quality timber had to be used, the bedding of rifles suffered due to warping and shrinkage. Whilst skilled labour could have cut this poorer quality timber specially so as to minimise warping, there were insufficient suitably experienced workmen available. The standards and requirements for bedding of rifles was relaxed due to this problem and the barrel "floated" by not being bedded forward of the reinforce. In late 1939 and early 1940, experiments had been carried out using scrap timber for handguards but these measures also proved to be futile.

Some other options considered and trialled were wood laminates and sheet metal. The metal-stocked version was a development of the Ministry of Supply's Production Branch and consisted of three sections; buttstock, fore-end and handguard. These are made entirely of reinforced sheet metal, covered by a fibre-treated layer with a synthetic rubber called "thiokil" which was unaffected by petrol, oil, acid or alcohol. This form of furniture eliminated a number of the small metal components normally fitted to the wood for stocking-up the rifle, such as are found on the handguards and fore-end. The first design was shown to the Ordnance Board in September 1941, after which requests were made that user trials be conducted with this type of manufacturing alternative.

The example shown here has two screws through the fore-end and the handguard extends up to the foresight protector. A pressed metal buttplate is fitted and the coloured finish is a green camouflage.

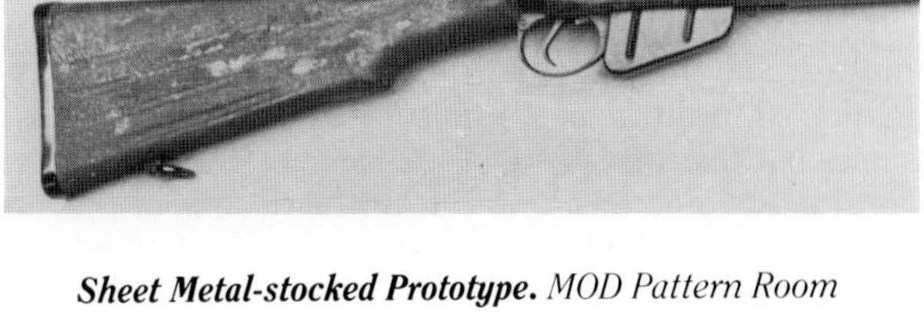

Sheet Metal-stocked Prototype. *MOD Pattern Room*

A number of tests were conducted with the metal-stocked rifle at Pendine, so as to establish:

- Its accuracy in extremes of temperature
- Handling facilities in hot and cold temperatures
- Rough handling
- Resistance to rust and the weather
- Suitability of the stocking-up
- Serviceability of the rubberised coating
- The effect caused by firing grenades (from cup discharger).

Seven of these rifles were assembled, differing slightly in the protective coatings and finish. The first rifle was tested against a service No. 4 and found to be wanting in several respects, namely stocking-up and accuracy. It must have been considered that the concept had merit as a further six rifles were fabricated by Adams Bros. & Burnley in 1942 at a cost of £234/17/-, for further trials. Unreliable accuracy was still the main problem, largely because of the difficulties in altering or adjusting the metal during stocking-up. The user reports did not favour the handling and feel of the rifle and considering other shortcomings, in April 1943 the project was abandoned.

The Germans had overcome their timber shortage problems by that same year, 1943, with the introduction of a laminated stock. Blocks of laminate were built up with strips of wood about 1mm thick, fibre, and a special substance called "tegofilm," and compressed at about 23 atmospheres in a hot press. These large blocks were then trimmed down and stock blanks for 8 rifles cut from each block. From this stage, normal machining procedures were followed. A captured German Kar 98k with such a stock was tested at Enfield in March 1943 for accuracy, after exposure to extremes of temperature, and it was found to be quite satisfactory. This test was contemporary with the metal furniture trials and any comparison between both types must have favoured the wood laminate configuration.

Some trials were also conducted using bakelite and a form of plastic, but although these compounds permitted bedding of the action and barrel more easily than the all-metal type, it was still unsatisfactory due to a liability to fracture when screws were pulled up tight and in being too heavy and not resilient enough. Another problem was that it retained the heat more than wood and after being exposed to tropical temperatures it became uncomfortable to handle whereas the wood did not. After the war, further tests were conducted with plastics and these proved to be more successful, largely as a result of further advances in the field of plastic composition development and incorporating certain North American technology.

In Australia, similar tests with the No. 1 rifle and a fibre-reinforced furniture of a resin composition which appears somewhat like that of the No. 7 bayonet grips, were also considered to be unsuccessful; further details of this are provided in Chapter 9.

A laminated stock similar to that on the German Kar 98k was tested on the No. 1 rifle in 1939 but the laminate was found to open up when exposed to heat. A No. 1 rifle was used instead of the No. 4 as the latter did not come off the production line until 1941. In Canada, different forms of laminated stocks were tested on the No. 4 rifle and a description will be found in Chapter 8.

The 2½-in. grenade discharger cup as applied to the No. 1 rifle was found to be much more suitable than with the No. 4, most probably because the attachment to the No. 1 nosecap and adjacent fore-end served to absorb recoil from the cup and barrel during firing. Tests in late 1943 confirmed the inferiority of the No. 4 in this regard and future grenade-launching development for the No. 4 and No. 5 rifles was focused on spigot-type projectors.

An improved sniper rifle had been requested on the proposed new action as early as April 1925, and when the No. 1 Mk VI was produced for trials the matter was again considered. Overhead mounts similar to those on the Pattern 1914 sniper rifle were suggested along with the Pattern 1918 telescope. The front mounts were claw type while the rear mount fixes onto the left side of the receiver, similar to the No. 3 Mk I* (T) and Australian No. 1 Mk III* H.T. Even at this early stage, other types of mounts were considered such as the illustrated pattern which bears a resemblance to the Griffin & Howe model, although all these investigations were shortlived.

Experimental No. 4 Sniper Mounts. *MOD Pattern Room*

This particular rifle was tested in 1939 and a trials No. 4 model of 1931 vintage was used as the basis for this conversion because the No. 4 rifle was not then in production. Unfortunately, the upper mount and telescope are missing; some Canadian trials sniper rifles later used a similar type of mount although the lower mounts were pinned and screwed onto the receiver instead. A description of the Griffin & Howe mount which is similar, is included in Chapter 8.

Enfield DD(E) sketches and drawing lists indicate that Watson Weaver and Patern 1918 telescopes were considered for fitting to the No. 4 rifle in 1939 and 1940.

Drawings dated April 1940 show such a Weaver scope mounted offset to the left, so that the conventional No. 4 flip-up Singer sight could still be used. The rifle mounts are very much similar to those on the No. 3 Mk I* (T) and Australian No. 1 Mk III* H.T. rifles.

There was a desperate shortage of sniping rifles at the outbreak of World War 2, some No. 3 Mk I* (T) sniper rifles were available fitted with the Patt. 1918 sight, although many of the 2,000 units assembled at the end of the Great War had been sent overseas and were in service with other Empire forces. In the early 1920's, quantities of Aldis telescopes were stripped from S.M.L.E. sniper rifles and put into store in the event of their being required at a later date. Many of these scopes were mounted onto Winchester-made No. 3 Mk I* (F) rifles by Alexander Martin in Glasgow, when over 400 of these rifles were set up with the scopes mounted offset to the left in a non-detachable mount. This rifle was designated the No. 3 Mk I* (T) A, the "A" indicating fitting of the Aldis sight.

Shortages of optical munitions is one of the most likely reasons why the Weaver telescopes were considered for fitting to the No. 4 rifle, there were no Patt. 1918 scopes available in quantity and the serviceable Aldis scopes were set up on the No. 3 rifles. Selection of the telescopic sight was one of the most critical aspects of a sniping rifle and the Weaver was reported as being liable to damage in regular service use.

The No. 32, a new telescope, was initially designed for use as a straight sighting unit with the Bren Light Machine Gun and then applied to the No. 4 rifle, although it was never ever fitted or tested on the Bren gun. The No. 32 was fitted to the No. 4 rifle early in 1940. Because of the additional weight and bulkiness of the proposed telescope, a new design of detachable mount was fabricated which added even more to the all-up weight, but the system was very rigid and robust. In the interests of durability and reliability, the additional weight was not considered to be much of a disadvantage.

The No. 4 rifle was not yet in production when this development of the No. 4 sniping unit was being carried out and even when the design of the new No. 4 telescope fittings had been decided upon, rifle supplies were still not available. So, the first batch of No. 4 sniping rifles utilised the early Enfield-made No. 4 trials rifles as the basis, of which about 2,500 were in store. Because of the shortages after Dunkirk, these were earmarked for reconditioning in order that they be put back into service. Two thousand of the No. 1 Mk VI and trials No. 4 rifles were refurbished as No. 4 rifles and about 1,400 of the trials No. 4 rifles were stripped and carefully restocked at Enfield for rebuilding as No. 4 sniper rifles. Some of the new No. 4 sniper units were in use for a time before their official introduction, which occurred on 12th February 1942.

The official nomenclature and announcement in the List of Changes was:

Para B6861:-

Equipment, Rifle, No. 4 Mk I (T)
Comprising-

Chest, S.A., No. 15 Mk I	*1*
Rifle, No. 4 Mk I (T)	*1*
Rest, cheek, telescope rifle (wood) & 2 screws	*1*
Sling, leather, U.S.A. (loop pattern)	*1*
Case, telescope. No. 8 Mk I (with carry strap)	*1*
Telescope, No. 32	*1*
Ends, cap, leather	*2*
Tool, adjusting, telescope	*1*
Polishing cloth (for lenses)	*1*

Chests, S.A. No. 15 Mk I
for Rifle, No. 4 Mk I (T)

Rifles, No. 4 Mk I (T)
Normal butt only. Complete with fine adjustment backsight; sight telescopic No. 32.

Rests, Cheek, Telescope Rifle

1. Introduction.
2. Nomenclature.

1. The above-mentioned equipment has been approved. The equipment will be issued complete, as shown, but the separate stores making up an equipment can be demanded separately for replacement purposes except that telescopes and rifles will not be separated. Both will be returned to store for repair or replacement if either needs adjustment. A brief description of each store is given below:

Equipment, Rifle, No. 4 Mk I (T): the complete equipment is as shown above and is housed in the chest.

Chest, S.A., No. 15, Mk I: is of wood, with hinged lid, hasps and turnbuckle and two hide handles. Suitable fittings are provided to house the contents of the chest when filled. A label pasted inside the lid of the chest gives a list, and notes on care of the equipment.

Rifle, No. 4 Mk I (T): is fitted with a graduated tangent backsight, not the double battle sight. Provision has been made on the body for fitting a suitable bracket which locates the telescope. The left hand side of the body has been machined at the front end slightly below the remaining surface, to facilitate fitting the front pad of the above-mentioned bracket.

Note: a number of rifles of early manufacture are not suitable for use with the standard telescope and fittings as the surface on the left hand side of the body (referred to above) has not been machined for that purpose. Each telescope is zeroed to the particular rifle to which it is fitted, it is therefore essential that both rifle and telescope are kept together. To facilitate identification the following markings have been approved and these should be checked before assembly.

No. 4 Mk I (T) Equipment, showing wooden chest and sheet steel telescope case containing Mk 3 telescope and adjusting tool. *Ian Skennerton*

(a) The rifle will be marked with the number of the telescope, stamped on the flat portion of the butt behind the cocking piece.
(b) The box housing the telescope will be stencilled with the number of the rifle on the outside of the box under the word "RIFLE."
(c) On the inside of the lid of the box for the telescope, will be marked the number of the rifle and the number of the telescope.

Rests, cheek, telescope rifle: this is made of wood and is fitted to the stock butt by two wood screws.

2. Consequent upon the foregoing, the nomenclature of the above-mentioned No. 4 rifles has been amended to read as now shown.

Weighing 2 lb. 3 oz., the No. 32 telescope was a little on the heavy side, but the telescope is extremely robust and reliable; it was first recommended for adoption with the No. 4 rifle for sniper use on 20th March, 1940. The telescope is marked "TEL STG No 32 MK I" over the name of the manufacturer, part number and telescope number, e.g. "H.B.M. Co. O.S. 466A" and "REGd No 3836." It is ranged from 100 to 1,000 yards, with click adjustment for every 50 yds on the top drum. The drum on the left side is for windage, and each click represents 2 minutes of angle, or 2 inches at 100 yards. Magnification is rated at 3 power, and a field of view of 9° obtained from the 19mm objective lens. The telescope body is a cylindrical steel tube.

The first No. 4 sniper rifle units assembled used receivers from the No. 4 trials models, marked on the left side of the butt socket as illustrated at the beginning of this chapter. The No. 4 sniper rifle first went into action in North Africa but the desert warfare was not really suited to sniper operation, and those equipments sent there could not be properly maintained due to lack of necessary information and because no one had been trained in their use. The field units which tried the rifles experienced difficulties with the sights and chose not to burden themselves with this extra equipment, so the sniping role was disregarded. The Germans also did not use snipers in this theatre of operations to any marked extent.

However, the invasion of Italy in 1943 presented a new situation. The battleground was far more stable with defensive positions and a terrain of mountains and many ruined towns and villages, but there were no trained snipers, special equipment or handbooks available. Hasty moves were made to alleviate this situation but the early handbooks and instructions contained many errors and confused information; the sniping rifles sent out by the early spring were not as accurate or reliable as was desired. So, the standard of preparation was upgraded and improvements made to the telescope and the quality of sniping improved to an effective level. No. 4 sniping equipments were set up at Long Branch, Canada from 1944, to a very high standard, and Britain placed orders for 950 units to supplement her own supplies. Details of these Canadian rifles are included in Chapter 8.

1933 Trials No. 4 Rifle converted to No. 4 Mk I (T) at R.S.A.F. Enfield, and fitted with the No. 32 Mk I telescopic sight. *J. Alley*

The reputable London gunmakers, Holland & Holland took up the conversion of Britain's No. 4 sniping rifles from 22nd September 1942 and records indicate that they contracted to set up a total of over 26,000 units costed at about £2/7/6d per unit. Nearly all the rifles for sniper conversion were especially selected from B.S.A. Shirley production although a few Maltby actions have been noted.

Hollands also made up some No. 4 rifle components, repaired No. 1 and No. 3 rifles, fitted night sights to Colt pistols, Sten machine carbines and De Lisle Commando carbines, and worked on the Welrod pistols. Other members of the gun trade took on specific jobs; Cogswell & Harrison made Sten components and 4-inch smoke dischargers; W. W. Greener and Boss & Co. assembled Lanchester machine carbines; and Westley Richards made the 4-inch smoke discharger.

The initial batch of No. 4 Mk I (T) rifles was set up at Enfield, commencing in May 1940, when the No. I Mk VI and trials No. 4 Mk I rifles underwent Factory Thorough Repair and many of these No. 4 rifles were fitted with the No. 32 Mk I telescope. In all, 1,403 of these conversions were done at Enfield. The next contract was for 12,100 units, and it was signed with Holland & Holland on 13th February 1943, for a rate of 800 per month. These wartime contracts were often not signed until after work had commenced and in many cases no contracts were issued at all as I.T.P.'s (Instructions to Proceed) were the only authority. Further orders for the No. 4 (T) conversions were placed with Hollands and this became a continuing program.

The last wartime contract for one hundred No. 4 (T) sniper rifles is recorded as having been passed on to B.S.A. Shirley, although this particular batch now appears to have been .22 RF training variants of the No. 4 (T) model.

According to the Contract orders, quantities of the British No. 4 (T) sniping equipment were:

ROF Enfield	1,403
Holland & Holland	12,101
Holland & Holland	4,824
S.A.L. Canada (British order)	950
Holland & Holland	9,517
B.S.A. Guns, Shirley	100

Of the 950 S.A.L. order, it is likely that only 500-600 were delivered and the B.S.A. contract batch of 100 rifles seems to have been .22 No. 4 rifles converted to sniper configuration for training. Such a rifle with."22 rifle, Sniper Wing Use Only" on a brass plate has been noted (although "wing" could well be "Trg.," an abbreviation for training, due to a poor illustration) and it is reported that a number of these rifles fitted with No. 32 Mk I scopes were used for Target acquisition and Fire control in indoor ranges on landscape targets, in much the same way as the .22 No. 8 with the sights set on "H." Some of these No. 9 type "sniper rifles" were in use at the School of Infantry.

The Holland & Holland wartime records have been recently studied in detail and from the individual entries and a comparison of daily, weekly and monthly totals, Peter Laidler puts the exact total at 23,177 sniping units set up by Hollands. Part of the unfinished contract of Holland's No. 4 (T) rifles are those fitted with mount pads, but not set up with the upper mount and telescope; see Skennerton's "British Sniper" pages 113, 138, 139, 246 & 247.

An exhaustive study of the No. 4 sniper rifles and telescopes, "An Armourer's Insight; the No. 4 (T) Sniper Rifle" by Laidler and Skennerton is now available. This describes many of the special features and behind-the-scenes modifications applied to the long-serving No. 4 (T) and L42A1 models.

Canada produced nearly 1,000 units for its own use in addition to the British orders and it appears that not all of the 950 units ordered by Britain were delivered by the end of the war, some of this shortfall was likely to have fallen prey to enemy action against Allied shipping across the Atlantic Ocean.

In May 1945, James Purdey & Sons fitted up twelve No. 4 Mk I (T) rifles with "special butts". These special butts appear to have been sporting models with pistol grip, roll-over cheekrest and rubber kickpad; unit cost was £16/19/6d. Some prototype Canadian No. 4 (T) sniper rifles underwent tests in England, fitted with similar sporting type butts; they were reported as showing up quite successfully.

Assembly procedure of a sniper rifle commenced with the selection of rifles which proved to be more accurate during the sighting-in tests. These rifles were then stripped down and carefully re-stocked and the tabs or pads were fitted to seat the telescope mounts, secured by screws and final finished. To ensure that the telescope, mount and rifle were not accidentally separated or exchanged with components from other sniping equipments, the telescope number was marked on the mount and stamped on the top of the butt at the socket, just behind the cocking piece. The rifle serial number should not appear on the telescope or mounts on the British rifle, although the Canadian No. 4 equipments have the rifle serial number marked on the telescopic sight mount.

It will also be noted that the North American manufacture No. 4 rifles incorporate a particular letter after the first one or two digits of the serial number, providing an indicator of its factory of manufacture. The Long Branch numbers have an "L," e.g. 30L5379, while the Stevens-Savage rifles use a "C," e.g. 12C8265. Further details on the North American serial number ranges will be found in Chapter 8.

The British-made No. 4 sniper rifles also have identification numbers stamped on the clamping rings in order that these might not be accidentally swapped around. Usually of two or three digits, these comprise of consecutive numbers on the front and back rings. The mount bracket is a steel casting, with two bearing surfaces to accept the body of the telescope tube.

Numbered clamping rings on the No. 4 Mk I (T) mount, so as to avoid accidental exchange. *I. Skennerton*

Backsights of the sniper rifles are special in that the Mk I pattern was used with the battle aperture removed, as it would foul the telescope if retained. On 23rd September 1940, cheek rests were approved for fitting to the No. 3 Mk I* (T), No. 3 Mk I* (T) A, No. 4 Mk I (T) and No. 4 Mk I* (T), secured by 2 flathead 1.5-inch, No. 8 gauge, iron screws. This approval date is of interest because the No. 4 sniping equipment as a unit was not approved until February 1942, an indication of how long the No. 4 Mk I (T) was "in the system" before the official announcement. A special sling swivel assembly was approved for fitting to the No. 4 sniper on 10th April 1943, replacing the normal front triggerguard screw.

The reference to "No. 4 Mk I* (T)" rifles in the aforementioned announcement regarding cheek rest introduction is interesting because the British production of the No. 4 rifle did not include the Mk I* model, it was only made in North America. A small number of the Stevens-Savage rifles were converted to sniper models in 1942. Readily identified by the "U.S. PROPERTY" marking on the left side of the receiver because of the Lend-Lease agreement, they also feature a different bolthead release to the Mk I model in that the gap in the ridgeway is at the front of the receiver, instead of behind the charger bridge.

The operational British sniper team consisted of two men, one with the No. 4 (T) rifle, the other with a Sten gun or No. 4 rifle. Other items usually carried were:

- 50 rounds .303 S.A.A.
- 2 grenades (No. 36 or No. 77)
- 5 rounds tracer (kept separate from the ball ammunition)
- 5 rounds Armour Piercing
- 1 telescope, scout regiment
- 1 pr. binoculars, prismatic
- 1 compass, liquid, prismatic
- 1 watch, G.S.
- 2 face veils

No. 32 Telescopic Sights. From top: Mk I, Mk II and Mk III. *J. Alley*

1 or 2 Denison smocks
1 water bottle
1 emergency ration pack
and carry cases for the telescopic sight, telescope and binoculars.

There were three marks of the No. 32 telescopic sight used on the British No. 4 sniper rifle:
Mark I, approved with the No. 4 Mk I (T) rifle on 12th February 1942,
Mark II, approved on 23rd April 1943, and
Mark III, approved on 7th October 1944.

The Mark III model (and some upgraded Mark II) sights proved to be successful enough to warrant their use on the 7.62mm L42A1 sniper rifle. The difference between the various marks of these sights are not very obvious, although the designations are engraved on the tube which makes their identification quite easy.

The **No. 32 Mk I telescopic sight** usually has a sliding brass eyeshade and the elevation and windage drums are slightly offset. A "clicker" device is incorporated in the drum mechanism so as to give positive adjustment; each click represents two minutes of angle or 2 inches at 100 yards and the lateral adjustment or deflection is up to 16 minutes, left and right. On the range drum, each click represents 50 yards, and this is graduated to 1,000 yards.

Externally the **No. 32 Mk II sight** is similar to the Mk I but for the omission of the eyeshade, the main differences are internal. On the Mk II, the click adjustments represent one minute of angle or 1 inch at 100 yards, on both the range and deflection drums. In the process of change, an alteration was made to the lens arrangement which also resulted in a slightly narrower field of view. It would also appear that some early Mk II scopes were supplied with the sliding eye shade, but these were removed along with some on the Mk I scopes, according to Army modification instructions (EMER's).

Troubles had been experienced with the retention of zero in the Marks I and II scopes; this was largely rectified by the introduction of the Mark III. Backlash problems also resulted in the fitting of anti-backlash springs in the Mk I and improvements to such a spring fitted in the Mk II sight. A special tool was needed for the drum adjustment, by rotating the inner drum; the use of a third operator's hand was also required to make such adjustments.

Some of the Mk II scopes were later upgraded during the mid-1950's to Mk 2/1; the use of Roman numerals for the mark numbers had changed by this time, as had the use of "*" suffixes which were replaced by "/1," "/2," and so on. The repair program involved fitting of the later Mk III style range and deflection drums with a slipping scale ring.

The **No. 32 Mk III** telescopic sight has a different drum arrangement as the range and deflection drums are in line instead of offset as on the previous Marks. Other improvements were incorporated as the Mk III supply proceeded; lenses were bloomed to improve light gathering (indicated by a blue "B" painted on the tube) and better sealing made the telescopes more waterproof (these were painted with a red "W" about ½-in. high). Those sights which did not pass the waterproof tests but were satisfactory in all other respects, were issued for use in temperate climates.

A simplified anti-backlash mechanism was employed and the range and deflection scales were retained on the drum by spring-loaded friction. The angle of view of the Mk III is 8.5° and the eye relief, 2.5 inches. Many of the Mk III scopes were applied to the 7.62mm L42A1 sniper rifle by merely re-calibrating the deflection drum and reading yards as metres; tubes were re-engraved with the new designation L1A1.

The sights are marked "TEL STRT STG No 32" followed by the Mark designation; this abbreviation represents "Telescope, Straight Sighting." The optical store part numbers, e.g. "O.S. 2039 A" differ between the three marks of sights and a variety of makers' names will be encountered, for example:

C.T.& S. logo	Cooke, Troughton & Simms
H.B.M. Co.	Houghton, Butcher Manufacturing Co.
A.K.& S.	A. Kershaw & Sons Ltd. (Soho Ltd.)
K.L.	Kershaw Ltd.
W logo	Watson & Sons
V.I.C.	Vickers Instrument Co.
T.H.& Co. logo	Taylor Hobson & Co. (similar to C.T. & S. logo)
R.E.L.	Research Enterprises Ltd., Canada

The cast steel mount is secured by two knurled detachable thumbscrews through the pads on the left side of the receiver; these mounts are usually marked "J G," "K D" or "N 92." The code "N 92" indicates production and/or finishing by John Dalglish & Sons, Avenue Ironworks, Cogan Street, Pollochshaws, Glasgow. The "J G" mark indicates manufacture by Rose Bros., Gainsborough, Lincs., and like those with the "K D" mark, these may well have been finally machined and finished by Hollands.

Accuracy tests for the No. 4 sniper rifle involved placing 7 out of 7 shots within a 5-inch diameter at 200 yards and 6 out of 7 shots into a 10-inch diameter at 400 yards. Weight of the No. 4 sniper rifle varies between 11 lb. 2 oz. and 11 lb. 11 oz. and it was issued in a wooden chest which contained a telescopic sight case, cleaning accessories and a U.S. type leather sling. The graticule or reticule of the No. 32 sight is a vertical, pointed post with crosshair.

A number of trials were carried out with the No. 4 Mk I (T), against sniping rifles of allied and enemy powers; the British equipment emerged as the best outfit. Other telescopes were tested as well and the Canadian No. 4 with a new 2.5 power R.E.L. sight designated the No. 32 Mk 4 was rated very highly. The Canadian made sniper rifle was produced at Long Branch, Ontario and is essentially the same as the British model, although some were set up with different telescopes and mounts, as described in Chapter 8.

The Canadian Long Branch No. 4 sniper rifles used the No. 4 Mk I* action whereas the British No. 4 sniper equipments utilised the No. 4 Mk I, except for a small number of U.S. contract Stevens-Savage rifles that were converted in Britain in 1942. The comparative tests were conducted with British and Canadian No. 4 equipments as well as a variety of R.E.L. 2.5 and 5 power, Lyman Alaskan and Weaver telescopic sights, the Australian No. 1 Mk III H.T. rifle, U.S. Springfield and Garand sniper rifles, and some captured German equipment.

Other British scopes, similar to the No. 32, but without the range and deflection drum adjustment were also tried on the No. 4 sniper rifle, set up in special mounts with vertical and lateral adjustment. The No. 53 Mk I telescope saw extended trials in 1944-45 and the No. 42, a similar type of sight, was also tested. These sights were straight-line types for use with artillery pieces, such as the 6 or 17 pdr. anti-tank or 25 pdr. field gun and they differ mainly in the eye lens and form of graticule. The No. 53 sight had a duplex or modified cross-hair while the No. 42 version had only the vertical wire.

The No. 42 sight with vertical crosswire and no lateral indicator seems to have been applied to special silenced .22 Model 74 Winchester self-loading rifles. This unusual adaption appears to have been intended for use by Home Guard style commandos after German landings in England because contract records list the preparation by Parker Hale of "660 .22 rifle silencers, fitted for use by Home Guard" in 1942 (see "British Small Arms of World War 2" by Skennerton, page 22).

Laidler's survey of the British No. 32 and L1A1 telescopic sight is suggested for in-depth detail of this series, their construction and maintenance, as well as the various contractors. Published privately in England, it is titled "Telescope Sighting No. 32." "An Armourer's Insight; The No. 4 (T) Sniper Rifle" by Laider is also highly recommended. Skennerton's "The British Sniper: British & Commonwealth Sniping and Equipments, 1915 – 1983" is suggested for further detail.

Some small details from the armourers and workshops where the No. 4 sniper rifles were maintained and tuned are worthy of mention here. Backsights had the slide phosphate grey finished or blackened, the grooved or ribbed trigger was fitted by armourers in instances where the smooth type had been originally supplied, and an unofficial mod was often effected whereby the right side of the body near the trigger guard screw was rounded off to prevent the sharp edge pressing into the trigger finger while the sniper lay in wait for long periods.

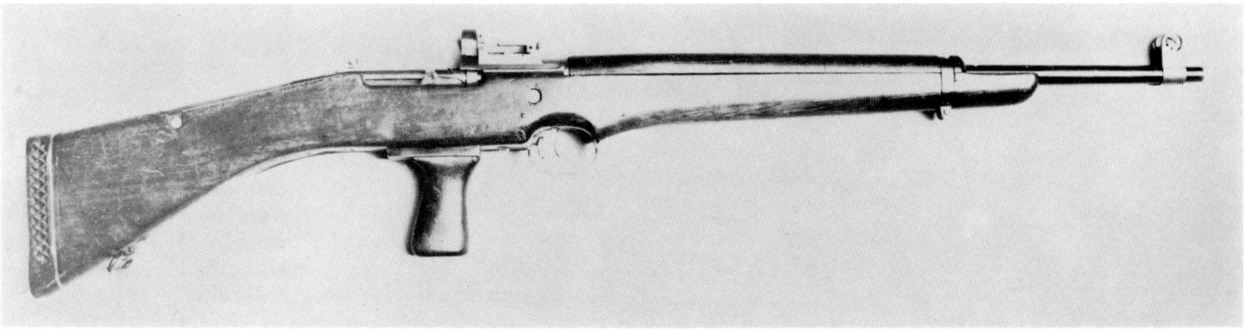

Above & below: Experimental Sniper Rifle S.R.E.M. I.
MOD Pattern Room

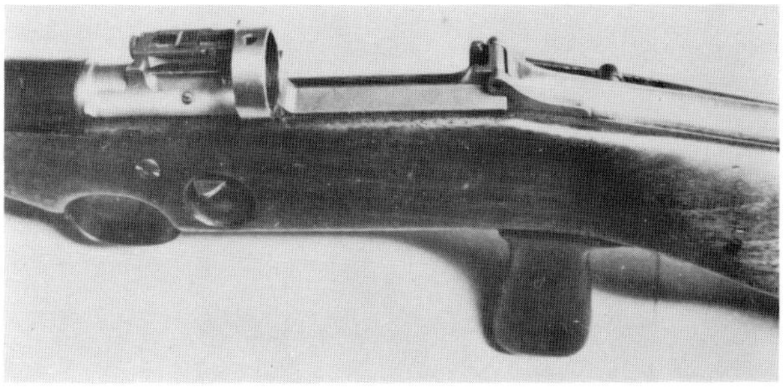

Work was commenced at Enfield in early 1944 on a new sniper rifle which did not require the arm or elbow to be moved or raised so as to manipulate the bolt. Complaints had been made that a sniper's position was often given away by arm movement in manipulating the bolt, and that a quick second shot was sometimes required which was not easy with the bolt action. By 1945, the Germans were using the G43 in considerable numbers in a sniper role, the Russians had a counterpart in the Tokarev and the U.S. snipers could use the M1 Garand unit. All of these self-loaders required no movement of the arm to reload after each shot because the action re-cycled by itself.

A completely new and unusual rifle was developed in England, called the "Sniper Rifle Experimental Model I" ; this had a rotating bolt worked by the backward and forward movement of the pistol grip. Known as the S.R.E.M. I, it is a bullpup design in which the trigger group is forward of the action and thus permits a shorter overall length and considerable weight reduction. Although the illustrated rifle has a 23.75-inch barrel, it is only 38 inches in overall length. An excellent analogy is the No. 4 rifle which has a 25.2-inch barrel and is 44.5 inches overall. The S.R.E.M. magazine is housed in the butt, behind the pistol grip.

The concept of the bullpup design is an outgrowth of Thorneycroft's ideas after the South African War but the S.R.E.M. program encountered further difficulties and problems in design which led to its suspension in 1945 because a new self-loading rifle was under development, the EM1 and EM2 series. This system was quite an advanced design, again on the bullpup principle, and it introduced a new range of cartridges which had largely been tested and developed using the No. 4 rifle as a test bed. Some details of these experiments are provided later in this chapter.

The production pressure during the war curtailed development of the No. 4 to a certain extent as there were certain modifications and improvements that might well have been incorporated in new models had not production been such a high priority. The first of these was a proposed No. 4 Mk II in October 1942 when two rifles were made up with bodies featuring "a slot in the left hand side to facilitate machining of the cam and elimination of many cuts from the right hand side of the body." Trials conducted by the Chief Inspectorate of Small Arms were satisfactory as regards the accuracy and functioning of the modified actions.

These two particular prototype "No. 4 Mk II rifles" were made at Maltby and fitted with a sear stop. The modifications incorporated in the design necessitated a new butt and fore-end and it was proposed to make up 1,000 rifles using reject bodies for trials, consisting of:

400 with 2-groove barrels
400 with 5-groove barrels
100 with Accles & Pollock Mk III barrels
100 with Canadian barrels.

Nevertheless, the new rifle was decided against on 1st March 1943, because of the interruption to manufacture and supply that would result from such changes. Instead, 100 rifles were ordered for extended trials although there is no record of these trials, or indeed of any of the 100 rifles being produced. This "No. 4 Mk II" must not be confused with the No. 4 Mk I* or the Mk 2 models which were later manufactured.

In the wartime production of the No. 4 rifle, one of the earliest bottlenecks was the late supply of the bolthead retaining catch which fits into the receiver just behind the charger bridge. Although only a small component, difficulties were encountered by the sub-contractors who first manufactured it. On 25th February 1941, trials were conducted in England with an alternative to this catch and it might well have been incorporated into the British No. 4 design under different situations.

Left: British Prototype (initially a No. 4 Trials model with magazine cut-off) of the Mk I* receiver.
J. Alley

Below: No. 4 Mk I* action. Note that the release slot is a little further forward than on the prototype model (shown at left).

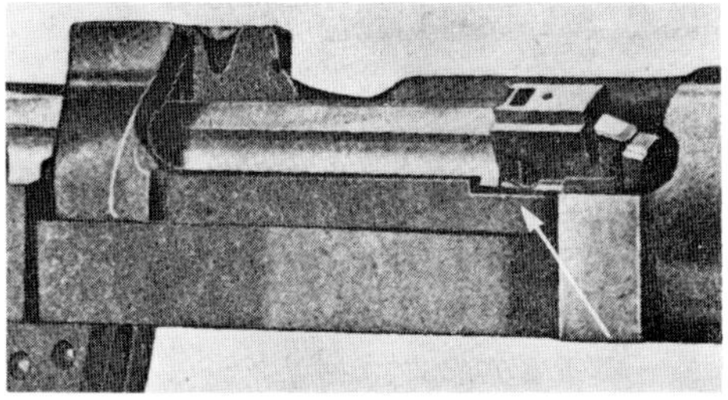

A problem which occurred at the outset of this new design related to the bolthead jumping out of the guide rib when the release slot was slightly oversize or worn but this was soon remedied. The modification was considered to be successful and the advance from Mk I to I* recommended for British production rifles but not implemented due to the advanced stage of No. 4 tooling-up and sub-contracting arrangements. It will be noted that the position of the bolthead release slot on the illustrated British prototype is a little further to the rear than that on the subsequent No. 4 Mk I* action manufactured in North America.

The particular changes in the receiver body for the advance in pattern to Mark I* were introduced into the Long Branch and Stevens-Savage production; approval for this alteration was given on 14th June 1942 although its official introduction as per the List of Changes did not occur until after the war, in 1946. It was announced as follows:

Para C2791:-
*Rifles, No. 4 Mk I**
with long butt
with normal butt
with short butt

New Pattern.

A drawing (D.D.[E] 2838) has been sealed to govern manufacture of the above-mentioned new pattern rifle which is hereby introduced for Land Service.

The new pattern is similar in design to the Mk I (§ B4737) with the following exceptions:
(a) The cutting of a slot and omission of machining for catch head breech bolt.
(b) Omission of the catch head breech bolt, also spring and plate.
(c) A new pattern bridge piece is incorporated in the body to permit removal of the breech bolt.
(d) The magazine catch screw is replaced by a pin and the sear pin has been increased in length.

This variant, the No. 4 Mk I*, was the main type produced at the Stevens-Savage and Long Branch factories and this represents nearly 90% of their total production of the No. 4 rifle. More particulars of the North American production and variations are provided in Chapter 8. Many of the early U.S.-produced rifles required rectification of certain faults, carried out at the Royal Ordnance Factories. One such faulty batch of 6,647 rifles was sent to Maltby in May 1943 for correction.

An adjustable aperture disc aperture sight for various ranges was proposed in 1941, on 20th March twelve were ordered from Enfield. These were prepared soon afterwards and tested, but not considered to be totally satisfactory. In 1944 a similar sight was proposed by Sapper E. Maidment, but differed in that the aperture was above the disc axis, whereas the earlier Enfield pattern had the aperture positioned below the axis of the sight.

Experimental Aperture Sight *MOD Pattern Room*

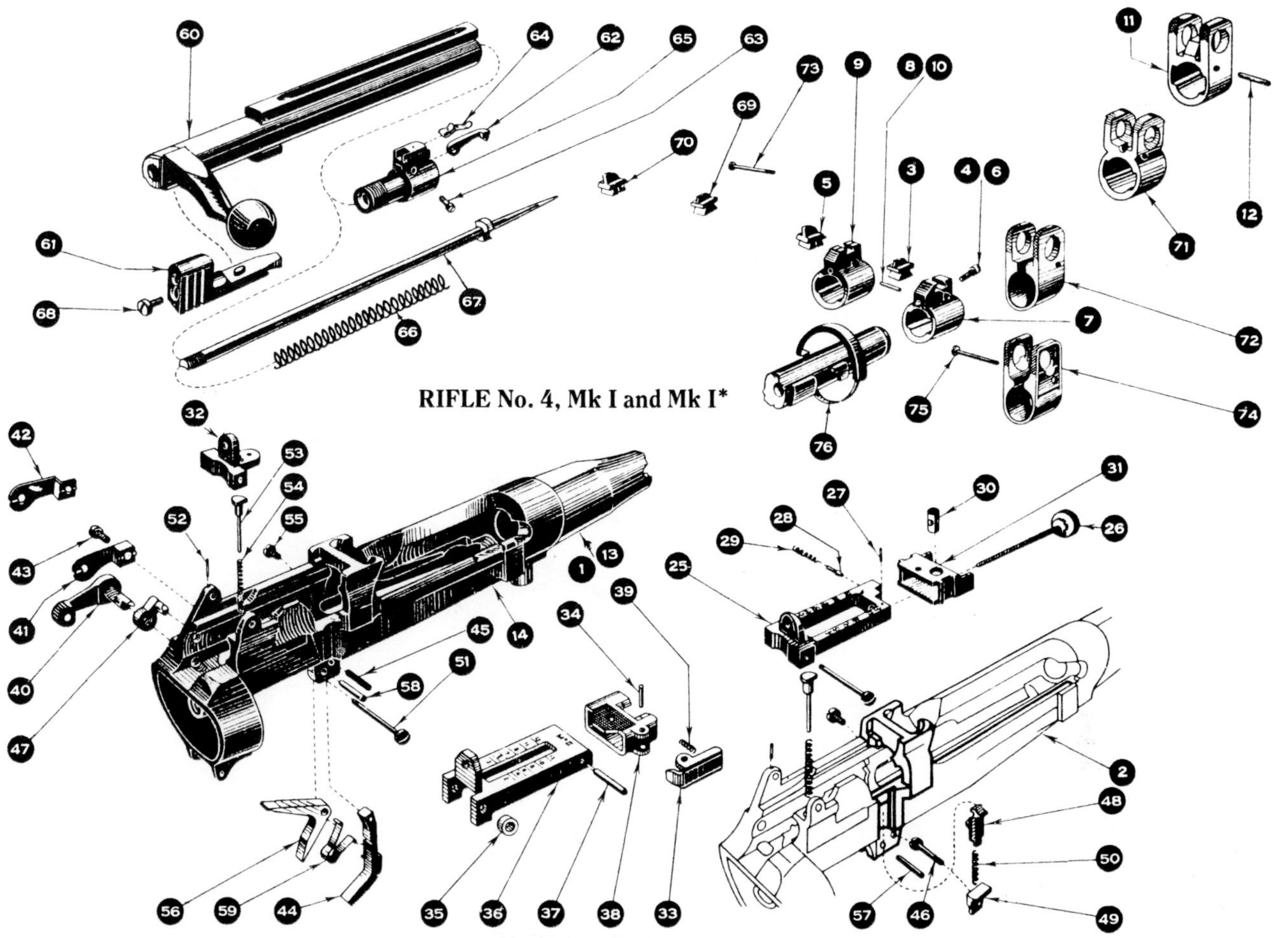

ACTIONS, SIGHTS & RELATED COMPONENT PARTS

No.	*Designation*	*No.*	*Designation*	*No.*	*Designation*
...	Barrel with Mk I body	31	Backsight, Mk I, slide	55	Screw, ejector
1	Barrel, Mk I (5-groove rifling)	32	Backsight, Mk II (300 yds & 600 yds.)	56	Sear
	Barrel, Mk II (2-groove rifling)	...	Backsight, Mk III, assembly	57	Sear, pin, Mk I
	Barrel, Mk III (fabricated)	33	Backsight, Mk III, catch	58	Sear, pin, Mk II
2	Body, Mk I	34	Backsight, Mk III, catch, pin	59	Sear, spring
3	Foresight, blade, "O", Mk I	35	Backsight, Mk III, collar	...	Bolt, breech, assembly
4	Foresight, blade, screw	36	Backsight, Mk III, leaf, Mk II	60	Bolt
5	Foresight, blade, "O", Mk I*	37	Backsight, Mk III, pin, stop	61	Cocking piece, Mk II
6	Foresight, blade, screw	38	Backsight, Mk III, slide, Mk II, assd.	62	Extractor
7	Foresight block, band, Mk I	39	Backsight, Mk III, slide, spring, catch	63	Extractor screw
8	Foresight block, band, pin	40	Bolt, locking	64	Extractor spring
9	Foresight block, band, Mk II	41	Bolt, locking, spring	65	Bolt head (Sizes 0 – 3 increase in length by increments of 0.003-in.)
10	Foresight block, band, pin	42	Bolt, locking, spring (alt. pattern)		
11	Foresight bracket	43	Bolt, locking, spring, screw	66	Spring, main
12	Foresight bracket pin	44	Catch, magazine	67	Striker
...	Barrel with Mk II body (for Mk I* rifle)	45	Catch, magazine, pin	68	Striker screw
13	Barrel (same as]1)	46	Catch, magazine, screw	69	Foresight blade Mk I
14	Body, Mk II	47	Catch, safety	70	Foresight blade Mk I*
[Nos. 15-24, same as Nos. 3-12]		48	Catch, head, breech, bolt	71	Foresight protector Mk I
25	Backsight, Mk I, leaf	49	Catch, head, breech, bolt, plate	72	Foresight protector Mk I (alt. pattern)
26	Backsight, Mk I, screw adjusting	50	Catch, head, breech, bolt, spring	73	Foresight protector screw Mk I
27	Backsight, Mk I, screw, pin, retaining	51	Pin, axis, backsight	74	Foresight protector Mk II
28	Backsight, Mk I, screw, plunger	52	Pin, axis, backsight, pin, retaining	75	Foresight protector screw Mk II (also for screw, guard, trigger, back)
29	Backsight, Mk I, screw, plunger, spring	53	Plunger, backsight		
30	Backsight, Mk I, screw, nut	54	Plunger, backsight, spring	76	Ring, retaining, rear, hand-guard

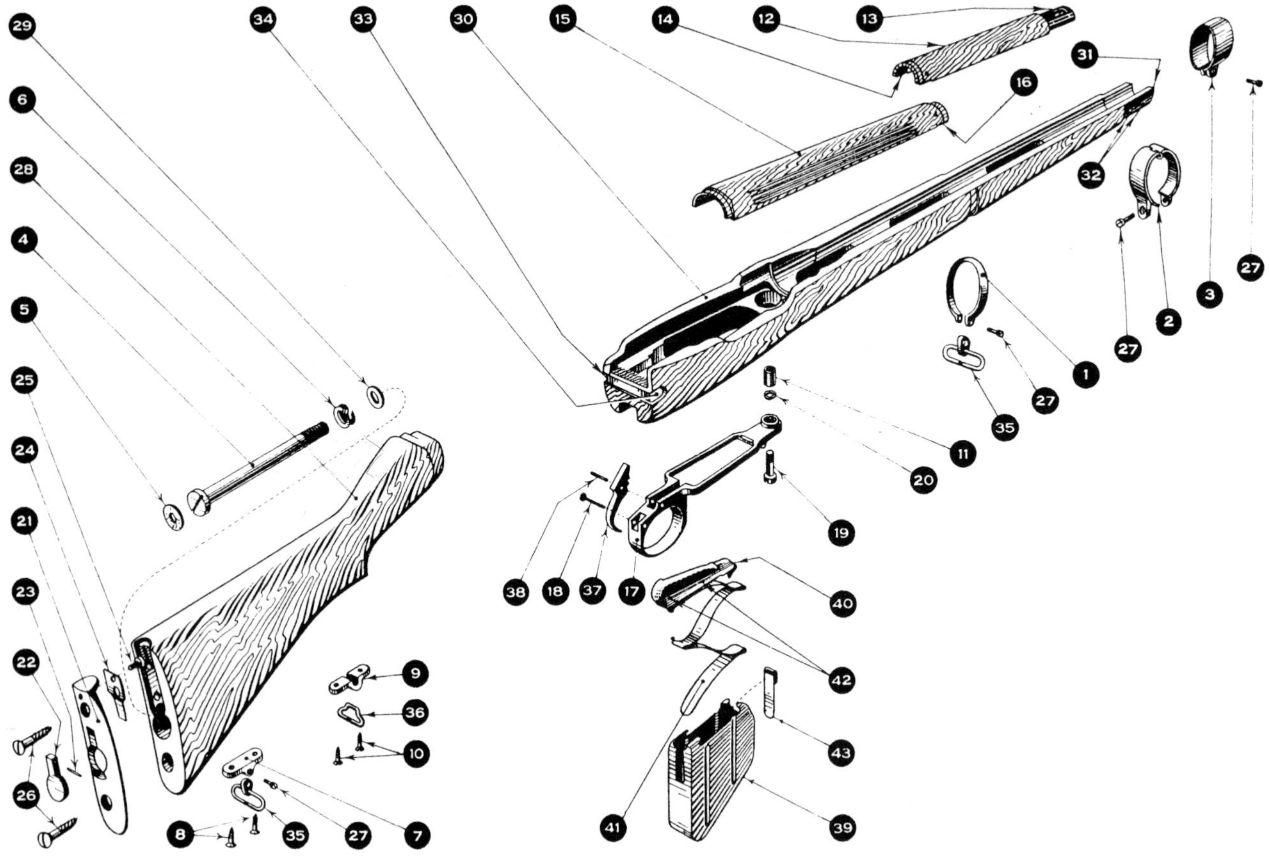

*RIFLE No. 4 Mk I & Mk I**
FURNITURE & FITTINGS

No.	*Designation*
1	Band, lower
2	Band, upper (hinged type)
3	Band, upper (solid type)
4	Bolt, stock
5	Bolt, stock, wad
6	Bolt, washer, spring
7	Bracket, Mk I
8	Bracket, Mk I, screw
9	Bracket, Mk II
10	Bracket, Mk II, screw
11	Collar
12	Guard, hand, front, assd.
13	Guard, hand, front, cap
14	Guard, hand, front, liner
15	Guard, hand, rear, assd.
16	Guard, hand, rear, liner
17	Guard, trigger
18	Guard, trigger, screw, back
19	Guard, trigger, screw, front
20	Guard, trigger, washer
...	Plate, butt, assembled
21	Plate, butt, plate
22	Plate, butt, trap
23	Plate, butt, trap, pin
24	Plate, butt, trap, spring
25	Plate, butt, trap, spring, screw
26	Screw, plate, butt
27	Screw, swivel
28	Stock, butt (long, normal & short)
29	Stock, butt, washer, bolt, stock
30	Stock, fore-end, assd.
31	Stock, fore-end, cap
32	Stock, fore-end, cap, pin
33	Stock, fore-end, tie-plate
34	Stock, fore-end, tie-plate, rivet
35	Swivel, sling, Mk I
36	Swivel, sling, Mk II
37	Trigger
38	Trigger, pin
...	Rifle No. 4, Magazine assd.
39	Magazine case
...	Magazine platform, assd.
40	Magazine platform
41	Magazine platform spring
42	Magazine platform spring rivet
43	Magazine spring, auxiliary

RIFLE, No 4, ·303

MK 1 & MK 1★

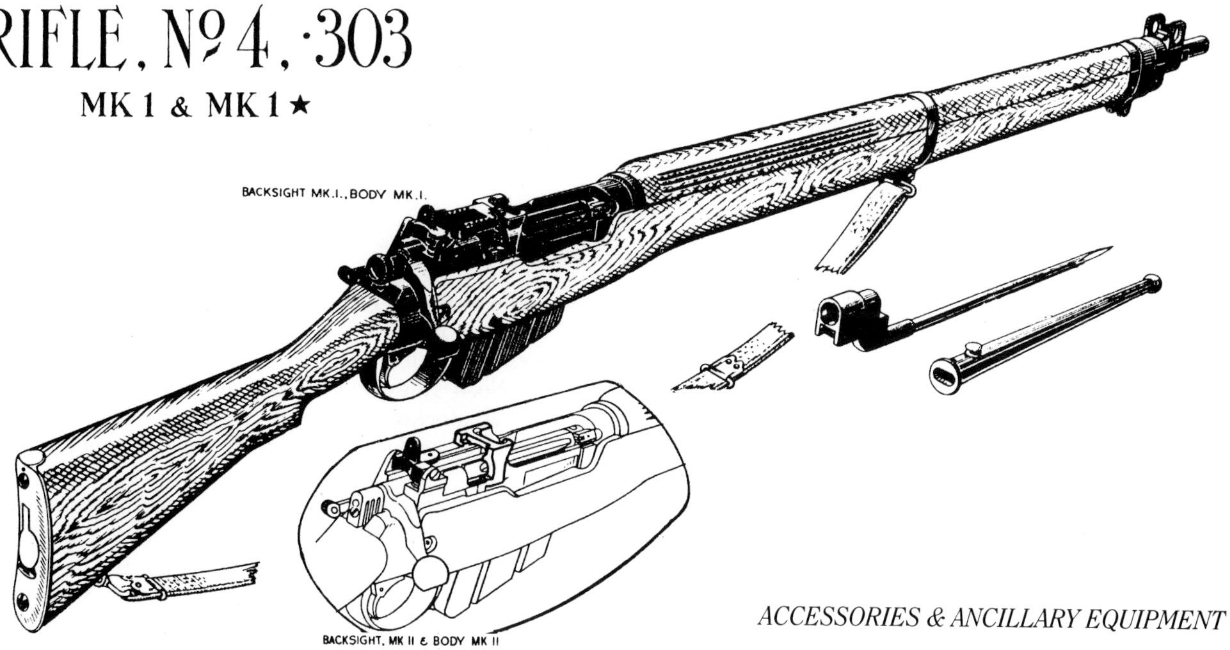

ACCESSORIES & ANCILLARY EQUIPMENT

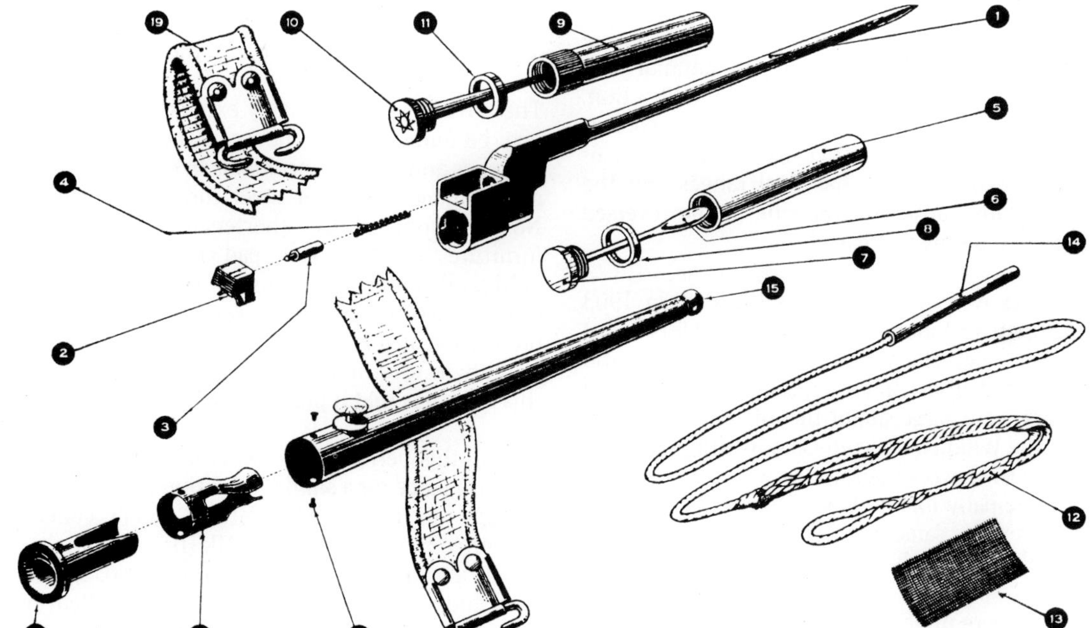

No.	*Designation*
...	Bayonet, No. 4, Mks II & II*
1	Bayonet blade
2	Bayonet catch
3	Bayonet catch plunger
4	Bayonet catch plunger spring
...	Bottle, oil, Mk IV
5	Bottle, oil, Mk IV, body
6	Bottle, oil, Mk IV, spoon
7	Bottle, oil, Mk IV, stopper
8	Bottle, oil, Mk IV, stopper, washer
...	or Bottle, oil, Mk V
9	Bottle, oil, Mk V, body

No.	*Designation*
10	Bottle, oil, Mk V, stopper, with rod
11	Bottle, oil, Mk V, stopper, washer
...	Pull-through, single, Mk IV
12	Pull-through, cord, single
13	Pull-through, gauze
14	Pull-through, weight
...	Scabbard, bayonet, No. 4
15	Scabbard, bayonet, body
16	Scabbard, bayonet, mouthpiece
17	Scabbard, bayonet, mouthpiece, screw
18	Scabbard, bayonet, mouthpiece, spring
19	Sling, rifle, web

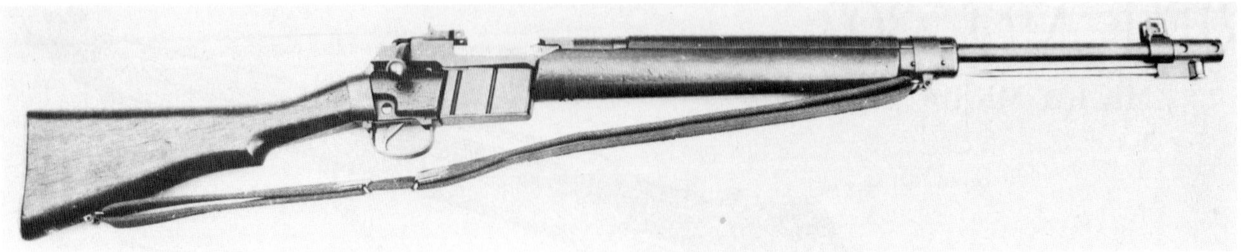

Simplified Rifles; bottom model has skeleton butt.
Top: R. Dynes. Bottom: MOD Pattern Room

The pressures upon Britain's manufacturing facilities resulted in the assembly of some "simplified rifles" in mid-1942 in an endeavour to see if such a rifle could be produced which was much easier to make. Designed by the C.S.A.D., it is .303-inch calibre with a front locking action; the body was made up from rectangular bar, somewhat reminiscent of the Pattern 1914 Enfield action. A skeleton butt is fitted on one model with a wooden cheekpiece and butt trap. The magazine capacity is 6-rounds, the pressed backsight has dual ranges and the bayonet is normally carried on the rifle in the reversed position, mounted on the rifle barrel.

Some No. 4 rifles were modified to fix the Pattern 1903 and 1907 bayonets in June 1943, due to complaints about the No. 4 spike model. The mounting for the Patt. '03 and '07 bayonets was identical and it consisted of a fabricated nosecap. There were large quantities of these bayonets available and the Wilkinson Sword Co. was still producing the No. 1 (Patt. '07) bayonet during the war, although this supply was principally for the 9mm Lanchester Machine Carbine; however this same bayonet was interchangeable on the No. 1 rifle and Lanchester.

Lightened rifles were under serious consideration as early as late 1942; the drawings DD(E) 3410 show a half-stocked No. 4 and annotations on this particular drawing also make mention of a duralium buttplate and trap. This particular example has over 10-in. of exposed barrel beyond the fore-end and the front handguard and fore-end retain the usual service sheet steel front caps.

A proposal for a lighter rifle was forwarded by the Infantry Weapons Development Committee in June 1943, mainly intended for use in the Far East where jungle fighting and the more inaccessible terrain made the No. 4 unsatisfactory because of its weight and length. Due to the shorter combat ranges for jungle fighting, the accuracy requirement was to be comparable to that of the No. 4 up to 400 yards only; a few weeks later, on 2nd July, the first samples were submitted to the Committee. These were merely No. 4 rifles shortened, and fitted with sporting style, half-stocked fore-ends made by reducing the original fore-wood length and removing the front band.

A few days later, on 13th July 1943, the firing tests of these shortened No. 4 rifles were conducted at Bisley, the sights were found to be unsatisfactory and the recoil and muzzle flash were considered excessive. Regarding accuracy, it was found to compare favourably with the No. 4 up to 500 yards so a batch of thirty lightened rifles were ordered for troop trials, to be completed by October the same year. These were prepared at Enfield and thirty-two were made in all, completed by September 1943. The No. 4 rifles were drawn from regular production and then converted; flash eliminators were fitted to this batch, although none had provision for the fixing of a bayonet.

These trials were reasonably extensive and fittings details may be found to differ between the DD(E) drawings and sketches and the few specimens in existence. The example shown here has sheet steel foresight protector wings, No. 4 bayonet lugs, shortened and modified No. 4 furniture, a rubber kick pad and a large hole in the bolthandle knob. The backsight is a fabricated sheet steel type marked to 800 yds. It was taken from B.S.A. Shirley production as it is marked "M47C" and "No 4 Mk " rather roughly in electric pencil on the left side of the receiver; this indicates that it predates the No. 5 nomenclature.

During this period of shortened rifle trials, consideration was also made for a similar rendition of the No. 1 rifle. The example in the M.O.D. Pattern Room collection was converted from a 1916 B.S.A. Mk III* rifle and fitted with a typical No. 5 butt (converted from a No. 1) and muzzle flash hider. The bolthandle has been drilled and filled, sheet steel caps are fitted to the handguard and the flip aperture backsight is ranged at 100 and 200 yards. Some of the troop trials Mk V S.M.L.E. rifles were also reduced to carbine proportions for tests.

The DD(E) drawings and sketches also indicate that a spring-retained handguard was initially under consideration; DD(E)SK No. 1564 shows such a handguard, similar in shape to the eventual No. 5 version, but with an internal spring secured by rivets (not unlike that on the No. 1 rifle rear handguard) and a short, sheet steel edge guard at the rear end. This particular drawing was dated 21st March 1944.

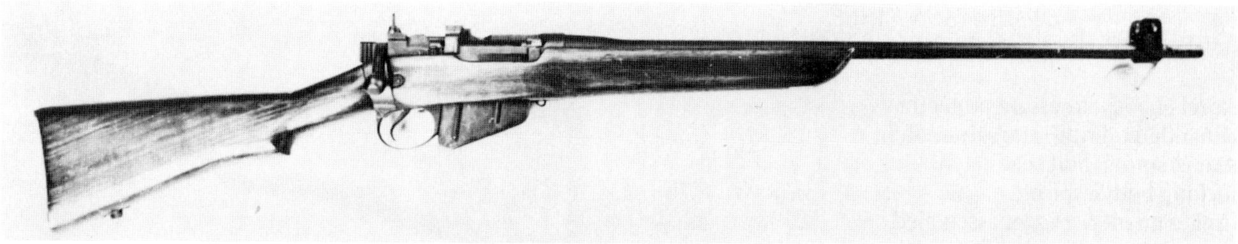

This early production No. 4 rifle is tagged "EXPERIMENTAL (woodwork cut-away)" and appears to have been a prototype lightened rifle. I. Skennerton

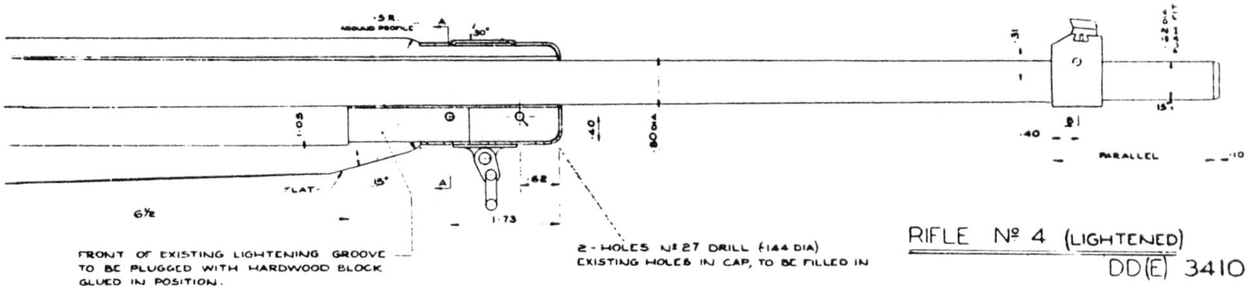

Enfield drawing of a Lightened No. 4 Rifle; similar to the example shown above, but with a handguard and sheet steel fore-end caps. Dated 25/11/42.

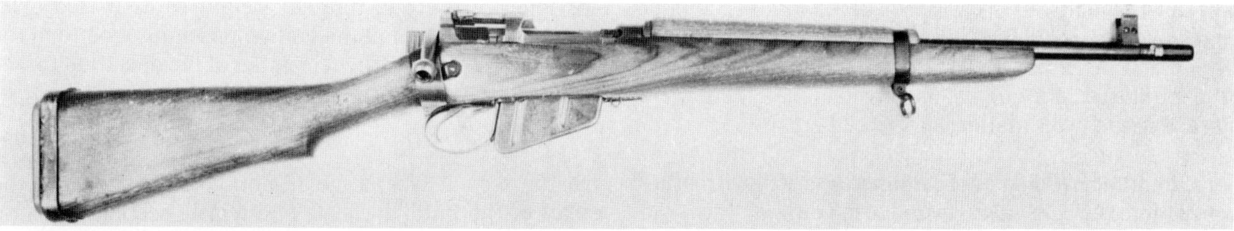

Above & below: Lightened No. 4 Rifle; note the converted wood furniture. The bayonet lugs are a brass type material, so as to enable easier shaping. MOD Pattern Room

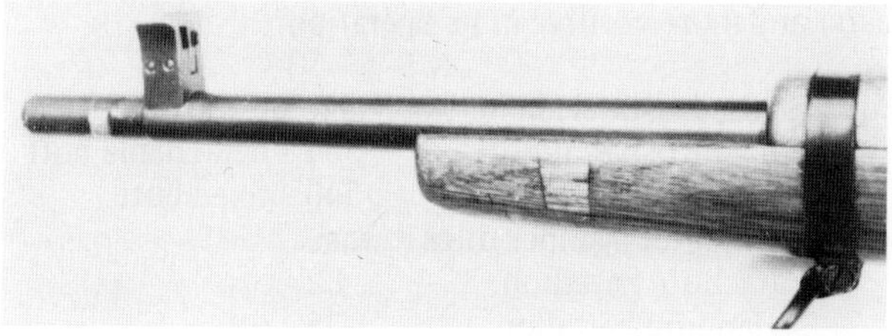

Some trials were carried out with variations of the flash eliminator to compare effectiveness, velocity trials substantiated that the loss in muzzle velocity between the No. 4 and shortened rifle was only marginal. Initial tests in 1943 with a flash eliminator were carried out on a No. 4 rifle. It was announced that a further order was likely to be placed with Fazakerley for 1,000 "lightened rifles" and these were in production by December 1943, although the new rifle design was not yet sanctioned.

This first batch of production No. 5 rifles have an "FE" prefix to their serial numbers (maybe "Fazakerley Ex.") and all those recorded examples have had 3-digit numbers. The particular manufacture dates recorded on the receivers has been 2/44 and 3/44, indicating February and March production.

Of the first batch of 1,000 rifles, six samples were requested for endurance and rough usage trials which were executed at Pendine in July 1944, using three of these rifles. Tests were also made regarding the accuracy and effectiveness of the flash eliminator. It was still regarded that the muzzle flash was likely to give away the firer's position and as jungle fighting relied much more upon concealment and close range fighting, this was a major consideration.

In the firing trials, two No. 4 rifles were used for comparison; one in part-worn condition and the other, new. The results were:

Muzzle flash not visible in daylight, no greater than No. 4.

Accuracy not as good as No. 4, especially at longer ranges, but still considered to meet requirements.

Accurate shooting was considered possible from the barrel after 6,000 rounds.

Accuracy suffered markedly more than with the No. 4 after rough usage and exposure tests.

COMPARISON TABLE.

	No. 4 *lb. ozs.*	*Lightened* *lb. ozs.*
Barrel & body, foresight protector & handguard ring, magazine catch, sear & spring, bolthead catch, locking bolt & spring	4 3.75	3 8.75
Trigger guard & trigger assembled	4.5	4
Buttplate, sling bracket & screws	6	5
Butt stock	1 3	14.5
Fore-end stock	1 2	9.25
Handguard(s)	5.5	1.5
Breech bolt, assembled	10.75	10
Stock bolt	3.75	1.5
Upper band	1	0.75
Backsight assembly & axis pin ...	1	1.75
Rifle assembled, empty magazine ...	9 1.25	7 0
Barrel length	22.22 in.	18.7 in.
Overall length	44.5 in.	39.5 in.

A requirement was stated in December 1943 for some lightened No. 4 rifles with detachable butts for use by airborne troops, so twenty were ordered additional to the 1,000 lightened rifles from the Fazakerley contract. Although the lightened No. 4 was made at Fazakerley, the take-down modifications were made elsewhere; a pilot take-down was built by the Chief Inspector of Armament Design at Cheshunt and two basic types appear to have been produced, although it is possible that there were more than the two variations described here.

Some folding butt models of carbines in service with other armed forces were also introduced during the war, principally the U.S. M1A1 carbine and Japanese Type 38 carbine. These were specifically intended for use by paratroops. Some World War 2 sub-machine guns also had folding or extending stocks, such as the German MP38 and MP40, Japanese Type 100, Russian PPS M1943, Australian Austen and the U.S. M3 "grease gun."

Take-down stocks, however, were not as popular, probably because it entailed the carriage of two separate sections and one part could become lost in the process of disembarking or landing. Despite this drawback, the Japanese produced some Type 99 Short Rifles and the Germans some Mauser short rifles in take-down configuration. The Thompson, Owen and all service marks of the Sten gun also had quick detachable butts.

The R.S.A.F. Enfield order for twenty detachable butt conversions was halved to ten and the balance of the original twenty was ordered from Parker Hale, Sterling Engineering and Chambon Ltd. The two types illustrated here disassemble on quite different principles; in the first, the stockbolt is unscrewed, while the second has a two-piece butt joined at a reinforced mortice slot. These particular examples also have the "FE"-prefix serial numbers, from the first one thousand Fazakerley production.

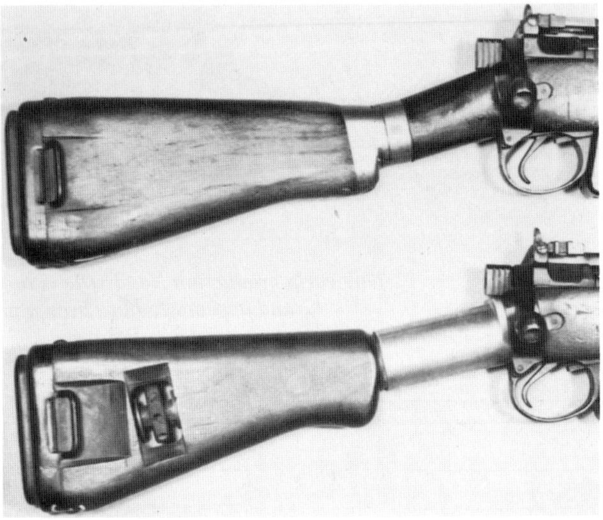

Trials Paratroop Take-down Models of the No. 5 Rifle.
Imperial War Museum

The first-mentioned pattern has a solid "wheel" attached to the stockbolt head which may be tightened or released from either side of the buttstock. This is knurled around the circumference and also has holes provided for the insertion of the pull-through weight so as to add extra leverage in tightening the bolt or loosening it off from the assembled position. The remainder of the operation can be effected without the use of any implement.

On the second type, a metal mortice slot is fixed at the small of the butt, the male and female sections forming caps on each of the corresponding ends of the butt. A catch on the underside acts as a locking device in the assembled position. This means that the overall "taken-down" length is slightly different to the first-mentioned model, although the assembly and take-down procedure is faster and more positive in its operation.

The proposed issue scale for No. 5 take-down rifles was:

Parachute battalion:	364 – detachable butt
Airlanding battalion:	546 – fixed butt

while for the No. 4 sniper rifles it was:

Parachute battalion:	32
Airlanding battalion:	38

The balance of these specialist battalion small arms comprised of machine carbines, light machine guns and side arms; no standard No. 4 rifles were proposed. Issue scale of the No. 4 sniper rifle for these groups was much higher than that of an infantry battalion, where it was only 8 rifles; the revised issue scales were due to the unique requirements of these elite forces. When the orders were first submitted for the take-down models, it was indicated that some 6,000 were required, but in September 1944 the order was cancelled and the whole program, including the trials, discontinued.

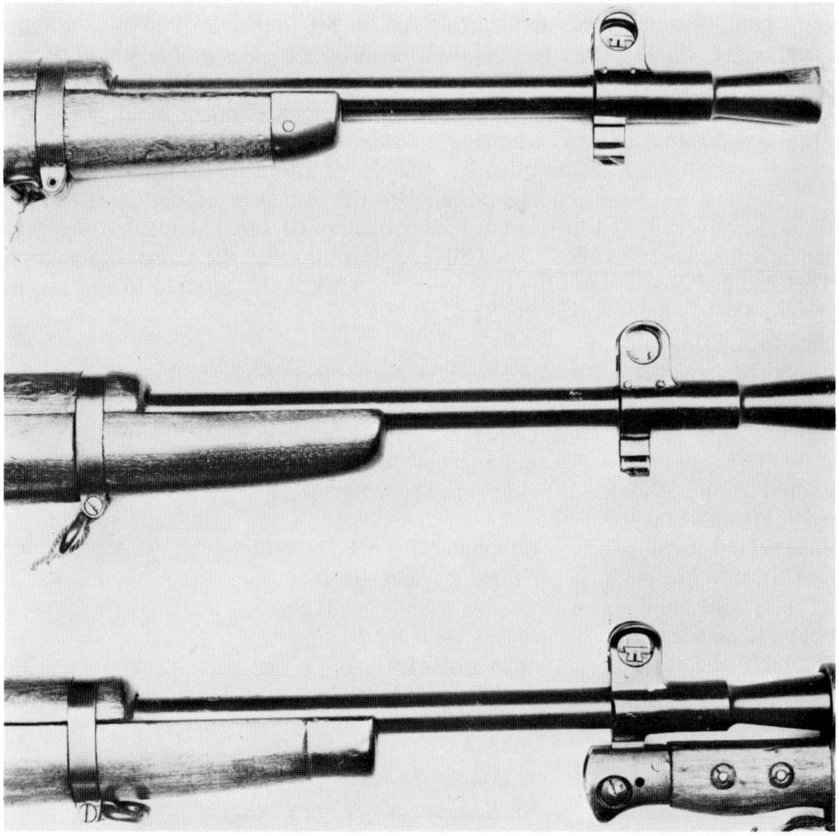

Progression of the No. 5 Rifle Fore-end Caps. From top: rounded metal type as on early trials models; rounded-off fore-end with no metal cap; and the later squared-end type.
MOD Pattern Room

The initial batches of trials lightened No. 4 rifles were not fitted with a fore-end cap and it was thought desirable to protect the end of the forewood so as to prevent moisture from entering the end grain. In early May 1944, a request was forwarded to *"arrange for the design of a lightened rifle to be modified to incorporate an end cap for the fore-end, to be introduced into production as and when convenient, without interference to out-put."* The type fabricated in answer to this request is worthy of mention as it is different to the later pattern which was adopted.

This new "jungle carbine" fore-end nosecap was first fitted in July 1944 to an experimental rifle and it is larger than the later production model. The front is also curved and follows the rounded tip of the fore-end; it is secured by a transverse pin through the fore-end. A few were fabricated and some fitted to experimental lightened No. 1 rifles as well, but the cap design was regarded as being too difficult to produce so it was changed in September of 1944.

A new, smaller and more easily manufactured model fore-end cap was approved and fitted to many of the No. 5 rifles although the rounded-off fore-end with the grain end sealed (no cap fitted at all) appears to have been the favoured type.

Some 1944 Fazakerley production No. 5 rifles, dated 7/44 and with 4-digit serial numbers, have also been noted with the rounded wood type fore-end; this early wooden furniture (butt and fore-end) appears to have been converted from unfinished No. 4 rifle production blanks. However, this was only applied to some initial production batches.

Variation in the production No. 5 rifle fore-ends without the metal nosecap, in a basic chronological sequence has been noted as:

Square-cut end after the curved taper to the tip;
Square-cut end with obvious tapering on the top edges;
Plain rounded end (usual production without metal cap).

Until later in 1944, the new short rifle was merely referred to as the "Lightened No. 4" and although it was accepted that this would in due course become the No. 5, instructions were issued to the effect that such nomenclature was not to be used until it was officially approved and introduced. The term "jungle carbine" is colloquial and not included in official references. In December 1943, a contract was issued for Fazakerley to commence production of 100,000 "No. 4 lightened rifles" even though some of the details for the new rifle had not been decided upon exactly. But as these particular details were relatively unimportant, they did not really affect the production under way. On 21st March 1944, the finalised design of the new "jungle carbine" was approved.

A rubber compound kickpad secured by a steel cap was developed to absorb the additional recoil caused by the carbine's lighter weight and shorter barrel, but minor difficulties were experienced with the buttpad composition. For satisfactory wear and longevity, a hard composition was required but this partly defeated the purpose of a kick pad. It was also found that the rifle grouped shots better with the No. 4 type butt and metal buttplate; this poorer shooting was also then attributed to the composition rubber recoil pad.

Location of the sling swivels was altered on the new rifle as well. With the No. 4, the sling could also be utilised as a shooting aid to steady the rifle but it was decided that with the lightened rifle, the sling would be used solely for carrying. So the lower sling loop was moved to the side of the butt, attached to the butt cap. The upper sling swivel remained attached to the band, as on previous rifles; there was no piling swivel requirement.

Some trials and investigations were conducted using high velocity machine gun ammunition, the Mk VIIz ball, as well as standard Mk VII ball, to ascertain expected barrel life and compare accuracy with and without the new knife bayonet fixed. It was found that the size of groups shot with the bayonet attached increased by about 20% and that the mean point of impact was considerably affected. This group size was also increased because the new bayonet was heavier than the No. 4 spike types.

So as to test flash eliminator design, a trial was conducted at R.S.A.F. Enfield on 30th September 1944 between two No. 4 rifles fitted with experimental flash eliminators and the Indian designed flash eliminator of the experimental Indian lightened No. 1 rifle. Comparative efficiency proved the new British lightened rifle type to be the most effective in various applications.

On the lightened rifles, barrels were found to wear twice as rapidly as those on the No. 4, using Mk VII ammunition; it was estimated that barrel life would be about 6,000 rounds with the Mk VIIz ball and about 12,000 rounds with the standard service Mk VII ball. Rapid fire tests showed that the accuracy of the lightened rifle suffered markedly at an early stage, although it was considered that this was most likely caused by deterioration in the bedding due to charred wood furniture from overheating in the rapid fire tests.

A new knife bayonet was introduced into service with the No. 5 rifle, a type which has influenced design until the present day. This has an 8-inch Bowie knife blade with an overall length of 11.75 ins. and wooden grips secured by two screws, although the very early No. 5 bayonets had only one such securing screw.

The introduction of the No. 5 was announced in the List of Changes, along with the Mk 3 bayonet for No. 4 rifle; it will be noted that these particular transcripts are less descriptive and more abbreviated after the 1930's, although much useful information can still be had regarding the interchangeability, components and accessories.

The use of Arabic numerals to replace Roman numerals was applied to S.A.A. in September of 1944 and extended to cover Small Arms also; *"the stars and letters at present constituting a part of some marks will follow the Arabic numeral, but when new nomenclature is allocated, this will substitute a suffix number for all stars and letters."* Subsequently, the Mark number utilised Indo-Arabic numerals rather than the Roman style; e.g. 1 instead of I, 2 instead of II, 3 instead of III, etc. Later, the Rifles No. 4 Mk 1/2 and Mk 1/3 were allocated the revised designations and the new Rifle No. 4 Mk II also referred to and marked up as the No. 4 Mk 2.

Approved on 12th September 1944, the LoC introduction for the No. 5 rifle and bayonet was:

Para C2125:-

Bayonets, No. 4 Mk 3
Fabricated, for Rifle, No. 4

Bayonet, No. 5 Mk 1
For Rifle, No. 5, Mk 1

Rifles, No. 5 Mk 1
with long butt
with normal butt
with short butt

Scabbard, Bayonet, No. 5 Mk 1
For Bayonet, No. 5, Mk 1

Tool, Foresight, Cramp, No. 4
For Rifle, No. 5, Mk 1

1. Introduction.

Rifle, No. 5, Mk 1
Backsight, Mk 2
Consisting of leaf, slide, catch and spring assembled.
Blades, foresight, Mk 2
With slot in base of dovetail; these blades are below or above the height of the "0" blade by the amount shown:
-.03 in.
-.015 in.
0
.015 in.
.03 in.
.045 in.
.06 in.
.075 in.

2. New Pattern.

Rifle, No. 5, Mk 1
Bolt, breech, unassembled. Also for Rifle, No. 4.

Scabbard, Bayonet, No. 5 Mk 1
For Bayonet, No. 5 Mk 1 and Bayonet No. 7 Mk 1.

Rifle, No. 4
Collars, backsight,
for Backsights, Mks 3 and 4, also for
Backsight, No 2, No. 5 rifle.

3. Extension of Use.

Rifle, No. 4
Bolt, breech, unassembled. To be demanded under "Rifle No. 5."

4. Detail nomenclature.

Rifle No. 4
Bolt, breech, unassembled.
Guard, trigger, fabricated pattern.

Rifle No. 5
Blades, foresight, Mk 1
With slot in base of dovetail; these blades are below or above the height of the "0" blade by the amount shown:
-.03 in.
-.015 in.
0
.015 in.
.03 in.
.045 in.
.06 in.
.075 in.

Plates, shoulder pad.

5. Obsolescent.

1. Drawings have been sealed to govern future manufacture of the above-mentioned stores, and a brief description of each is given below.

Bayonet, No. 4 Mk 3. This bayonet is of fabricated design and differs very little in appearance from the existing Mks. 2 and 2. The fabrication is mainly in the construction of the socket.*

Bayonet, No. 5 Mk 1. The design of this bayonet is similar to the No. 1 bayonet. It has a flat type of blade, and a wooden grip which incorporates a bolt for locking the bayonet in position on the rifle. The main difference is in the length of the blade, that of the new design being shorter, and is approximately 8-ins. long.

Rifle, No. 5 Mk 1. Compared with the No. 4 rifle (§§ B4737 & B6861) the new pattern is approx. 5 inches shorter with a sight radius of 23.25 ins.; other distinctive features are a flash eliminator on the barrel, one hand-guard, a short fore-end stock and a rubber shoulder pad in the butt stock in place of a metal butt-plate. The Mk 1 backsight assembly is fitted with a leaf graduated to 800 yards and the slide is adjusted by means of a finely threaded screw. The action and magazine are generally similar to those on the No. 4 rifle.

Scabbard, bayonet, No. 5. This is similar in appearance to the No. 1 Bayonet Scabbard, though it is much shorter and is made of steel throughout, whereas the former design is made mainly of leather.

Tool, foresight, cramp, No. 4. This is generally similar to the existing No. 3 but is designed to clear the flash eliminator on the No. 5 rifle.

2. Drawings have been sealed to govern the manufacture of the above-mentioned new pattern.

Backsight assembly, Mk 2. This is designed primarily to assist production; the leaf, slide and catch being manufactured from steel pressings. The slide is adjustable by means of a spring loaded catch on the right side of the assembly. The complete backsight is positioned in the body of the rifle by an axis pin and the same backsight collar that is used for locating the Mks 3 and 4 backsight assemblies to the No. 4 rifle body.

Rifle No. 5 backsights; Mk 1 (right) and Mk 2 (left), graduated to 800 yds. *MOD Pattern Room*

Blades, foresight, Mk 2. These blades differ from the Mk 1 in the same respect as the Mks 2 and 3 on the No. 4 rifle, described in § C1181.

3. The use of the above-mentioned stores has been extended as now shown.

4. The detail nomenclature has been amended to read as now shown.

5. The trigger guards referred to above were fabricated designs.

The plate, shoulder pad is a piece of flat steel plate fitted originally between the rubber pad and the butt stock, but is now considered to be no longer necessary. These items together with those shown above are now hereby declared obsolescent. Existing stocks will be used up.

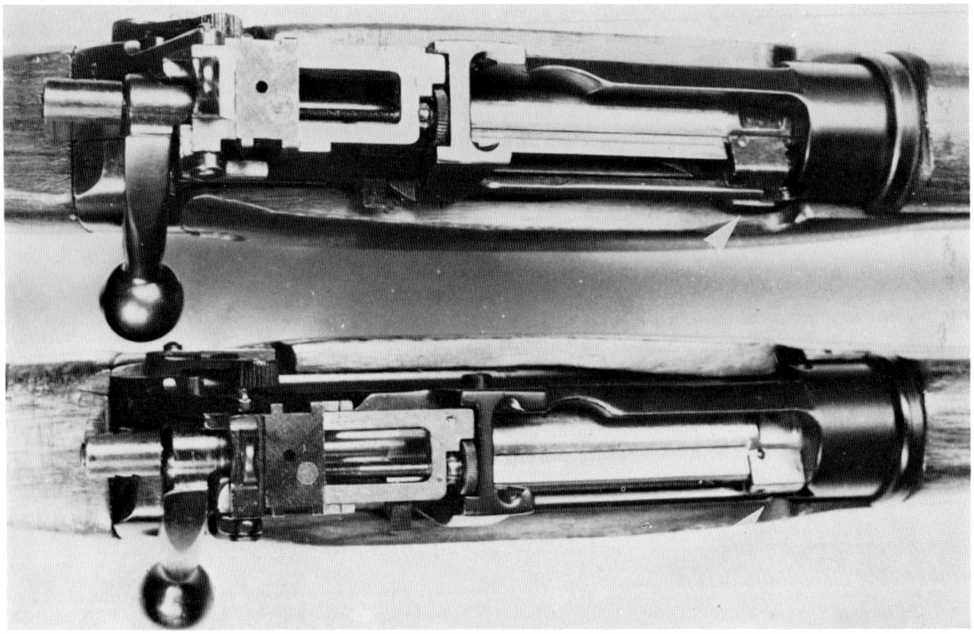

Comparison: No. 4 (top) and No. 5 rifles. Note the absence of the magazine cut-off projection on the No. 5.
MOD Pattern Room

The No. 5 was available with three lengths of butt, and two marks of foresights and backsights, evident from the descriptions in the List of Changes entry. The Mk 1 foresight is similar to the Mk I* of the No. 4 rifle, while the Mk 2 foresight of the No. 5 rifle has a longer platform at the rear of the blade.

Similar to the No. 4 rifle, the Mk 1 pattern backsight is the machined Singer type with fine adjustment on a vertical screw with clicker device, although it is graduated to only 800 instead of 1,300 yards. The Mark 2 type for the No. 5 is also graduated to 800 yards but is a pressed sheet steel type similar to the Marks III and IV of the No. 4 rifle. Soon after the end of the war, the Mk 2 backsight for No. 5 rifle was declared obsolete. References to the backsight collars, breech bolts and fabricated trigger guards indicate their being interchangeable on the No. 4 rifle and listed under the same nomenclature.

Early No. 5 rifles fitted with the steel plate underneath the kick pad would have been so supplied to prevent the possibility of the rubber pad collapsing into the large lightening slot in the butt under recoil; this was declared obsolete in the previous-mentioned List of Changes entry. Two methods were employed to fix the No. 5 buttplate assembly by screws; the conventional long male and female steel thread screws which extend from top to bottom, and two wood screws with a length of wooden dowel.

The serial number ranges are interesting; the first Fazakerley one thousand had an "FE" prefix and generally 3-digit numbers, while the following contracts used 4-digit numbers that commenced without a letter prefix, but then progressed through the alphabet with single letter prefixes. The B.S.A. Shirley rifles, code-marked "M47C," had double letter prefixes followed by up to 4-digit numbers. The only change from this series seems to have occurred with some very late Fazakerley production in 1947, with "AA" and possibly "AB" prefixes.

The following examples confirm this serial number format:

ROF(F)	FE387	2/44	M47C BB1427 1945
ROF(F)	FE478	2/44	M47C BD8056 1945
ROF(F)	8614	7/44	M47C BE7457 1945
ROF(F)	A2249	8/44	M47C BJ5346 1945
ROF(F)	P4398	7/45	M47C BJ3343 1946
ROF(F)	V887	3/46	
ROF(F)	AA3129	4/47	

Note that BJ M47C numbers are out of sequence by date (1945, 1946); this can often be found to occur.

An analysis of the serial number ranges and some of the later serial numbers, with a possible 9,999 rifles for each letter of the 4-digit series, shows some anomalies when these are compared, and also with the contract record figures. It appears that more No. 5 rifles were produced at Fazakerley after the 1946 contract production figures, but the BSA Shirley production was recorded as 81,329 and the "BJ" prefix would indicate a theoretical "90,999 plus" figure for their "B" double letter prefix run. It has been noted that serial number sequences for the No. 1, No. 4 and No. 5 rifles often display such anomalies and that attempts to establish a progression, especially relating to production dates, can often be misleading using this approach.

Changes introduced for the No. 5 rifle were the marking of the fore-end and the magazine with the rifle serial number; the former was approved on 11th November 1946 and announced in the List of Changes § C2792, the latter approved on 3rd June 1946 and also applied to the No. 4 rifles. All new production rifles were subsequently engraved or stamped with the serial number on the bottom of the magazine as well.

As with the No. 4 rifle, many components for the No. 5 were produced by private contractors and some parts are interchangeable with those on the No. 4. Maltby did not produce the No. 5; orders were reduced from 250,000 to 169,807 on the Fazakerley contracts and from 100,000 to 81,329 on the B.S.A. Shirley order. Early termination of these contracts was largely due to some problems with the design.

Those contractors and applicable codes, producing the No. 5 rifle components, were:

C.W. Cheney & Son Ltd.	M56	Magazine assembly.
Elkington & Co., Birmingham	M78	Shoulder butt pad.
Haddon & Stokes, Birmingham	M100	Fore-end cap rivets.
Imperial Engineering, London	S56	Mk 2 backsight assy.
Kigu Ltd.	–	Fore-end cap.
Singer Mfg., Scotland	N67	Backsights, pins, screws, foresight blades.
Walls Ltd., Birmingham	M260	Fore-end cap.

The first No. 5 rifles were produced in lieu of No. 4 rifles to complete original No. 4 contracts, the remaining No. 4 orders were then cancelled and new ones initiated with the Fazakerley and Shirley factories. The No. 5 contracts were also later terminated short of completion. Many of the No. 5 bodies appear to have been converted from No. 4 body forgings, as the machining marks on the lightened sections behind the charger bridge often appear to be differently cut than other body milling.

Lightening cuts in the No. 5 receiver body were made:

(a) Underneath, between the projections for the rear trigger guard screw;

(b) At the top of the receiver walls, just behind the ears for the backsight;

(c) On the left and right walls just behind the cuts listed in item 2., above;

(d) On the right side of the receiver towards the front end, near the barrel thread ring, where the magazine cut-off axis screw block was previous positioned on the early No. 4 trials rifles.

Four or five variations in fore-ends may be encountered with the No. 5 rifle; some prototypes have a rounded thin sheet metal fore-end cap while the more common parallel metal cap is smaller and another pressed steel variation of this type is of similar length to the No. 4 fore-end cap, secured by two rivets. Another type of fore-end has no cap at all but the fore-wood is rounded off in the same fashion as a sporter rifle; on this type the fore-end is thicker towards the lip and therefore has a slightly different profile. The last type mentioned was also neatly converted from the No. 4 fore-end and a sample examined was marked "No. 5 Mk I ROF(F) 9/44."

In continuation of this "jungle carbine" series, a No. 5 Mk 2 rifle was proposed and some experimental models produced, but no such pattern was eventually adopted. The first of these experimentals in July 1944 features a strengthened action for grenade firing. The first two of these particular rifles fabricated had metal blocks fitted between the sear lugs and body socket so as to reinforce the receiver.

During the early stages, two lightened No. 4 rifles were also used for this conversion and this was considered to show promise, so a further six were ordered. These also appear to have been modified from No. 4 lightened rifles at Enfield. In February, one hundred "Modified No. 5 rifles with inserted bodies" were contracted from B.S.A. Shirley and the factory was advised of this order late in 1944. However, difficulties were experienced in making the receivers although these difficulties were soon overcome.

Over this period, tests were carried out on the No. 4 rifle with the trigger mounted onto the receiver instead of the trigger guard, so this concept was incorporated in the experimental grenade-firing No. 5 rifles as the reinforcing blocks provided a ready mount for the trigger. Use of the designation "No. 5 Mk 2" was then cancelled and applied to a later model which also was cancelled; so the No. 5 Mk 2 was never approved for production, indeed this nomenclature could be applied to a number of different prototype models.

Two adjacent brackets were fitted between the body socket and sear lugs on these grenade-firing trials rifles, on the left and right sides; the space between these brackets housed and mounted the trigger and sear. This bracket made the removal of portion at the rear of the fore-end necessary, although it is not obvious on the rifle with furniture fitted. A large screw and threaded collar is attached through the back end of the fore-wood, in a similar manner to that of the No. 4 Mk 2 rifle, except that on the latter, the screw and nut are of a smaller diameter and size.

The strengthened pattern was reasonably successful and of the hundred plus rifles made, many were later used in other tests and experiments, such as stocking up, accuracy and cartridge trials. This form of trigger mounting, onto the reinforcing brackets, was applied in November of 1944 but any further development of this particular design was discontinued early in 1945.

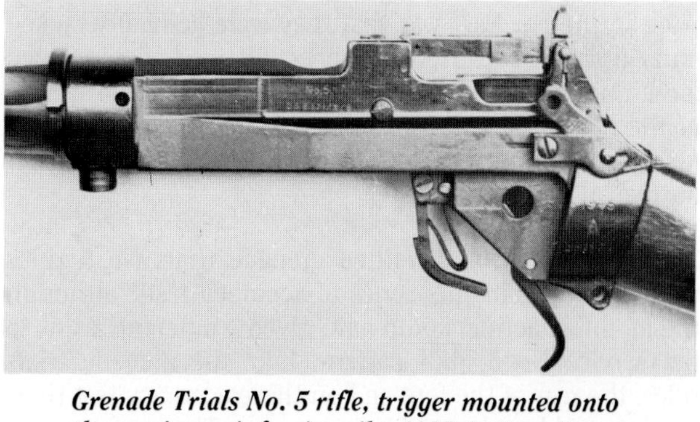

Grenade Trials No. 5 rifle, trigger mounted onto the receiver reinforcing ribs. *MOD Pattern Room*

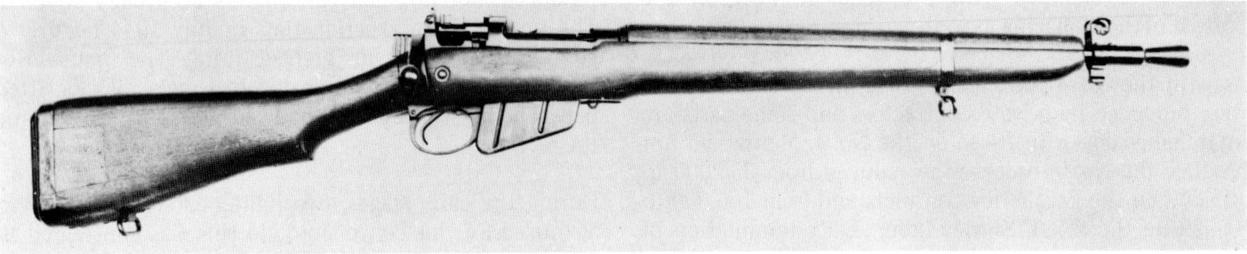

Stocking-up Trials "No. 5 Mk 2" rifle, D5E 171/X/8
MOD Pattern Room

The "jungle carbine" or No. 5 Mk I rifle was popular with the troops who liked its reduced length and weight, and its general handiness. Those who used it seem to hold it in high regard and moves were even made to have it adopted as the standard service rifle, but there were also numerous complaints about its inability to maintain zero and shoot consistently.

The reason for this "wandering zero" was something of a mystery and the shooting of the No. 5 could be found to vary from batch to batch, although some armourers claimed that it was more a case of simply "wouldn't zero." These problems were at first attributed to the stocking up and so other methods were tried in 1945 and 1946. A marginal improvement was accomplished by extending the fore-end and handguard almost to the rear of the flash eliminator and by moving the outer band forward some three inches.

A study of these accuracy problems with the No. 5 receiver would indicate that the shot pattern was similar to that of a No. 4 rifle in which the bolthead was not square with the chamber; the consensus of opinion was that the removal of steel from the rear of the receiver body near to where the locking shoulders were, allowed to body to flex at the moment of maximum stress. This was considered by some more experienced armourers to have been the main cause of the zeroing problems.

The remedy was relatively simple, replace the No. 5 action body with that from a No. 4, and this occurred in some Far Eastern theatres, particularly in Malaya. British armourers attached to some of these Commonwealth units report that such jungle carbines with No. 4 rifle bodies "behaved well." Another interesting feature of the No. 5 rifles in store in the Far East was that they were generally racked muzzle down, because the rubber butt pads deteriorated more quickly as the oil drained downwards when racked in the normal manner. Standing on the rubber buttplate for a long period also flattened and deformed the rubber buttpad at one end.

Some re-stocked and refitted grenade trials No. 5 rifles were employed in this series of accuracy tests, although the modified action would have given a different result to the service No. 5 Mk I pattern. Like the grenade trials rifles, the rear of the fore-end on these accuracy test rifles is reinforced by a large transverse screw. The configuration of the sling mounting was altered, with the lower swivel mounted at the toe of the butt as on the No. 4 rifle, and so a standard No. 4-type butt was used with a modified No. 5 buttplate and kick pad. Another rifle in this series of stocking-up trials models in the Pattern Room collection has a sling bar on the left side of the butt and a conventional upper band and swivel fitted.

Different designations and markings were used on the experimental rifles of this series, engraved on the left side of the receiver in the same position as for the No. 4 and No. 5 rifles, but the designations are unusual. In the drawings, the first models were marked "D5E 124/X/8" while the grenade launching and stocking up trials rifles are marked "No. 5" over "D5E 171/X/8"; these classifications refer to the particular trials programs.

The strengthened No. 5 rifles were also used in other investigations which involved reducing the length of the flash eliminator. It was another attempt to discover the cause for wandering zero and to see if there was any relationship with this problem to the flash hider. It was discovered that removal of the flash eliminator cone forward of the barrel muzzle improved the accuracy, even moreso than the special stocking-up procedures.

In all, fifty No. 5 rifles were ordered with the triggers mounted or "hung" from the body, converted from No. 5 Mk I models in a similar manner to the pattern eventually adopted as the No. 4 Mk 2. However the problem of wandering zero remained unsolved and so the proposal to advance the mark number was shelved when the decision was made to withdraw the No. 5 from service.

The reason for the inherent accuracy problem or "wandering zero" had been investigated from nearly all angles and the theory of incorrect bedding or stocking up or the use of improperly seasoned timber discounted. Further tests in 1948 were associated with the flash eliminator and it was found that this adversely effected the grouping and accuracy after more than 200 of the No. 5 rifles were subjected to a variety of tests. Rifles with the flash hider removed or a weight at the muzzle to replace the flash hider gave better results in the grouping and accuracy tests.

It was officially concluded that the fault was "inherent in the design." In July 1947, the No. 5 Mk I was declared obsolete although it was not until 25th September that R.O.F. Fazakerley was directed to revert to production of the No. 4 rifle. A No. 5 rifle has been noted with the manufacture date as late as 12/47, so the cessation order must not have been acted upon immediately although the production certainly ceased before the end of that year.

The concept of the lightened jungle rifle was also applied to the No. 1 rifle and in July 1943 the first models were developed, parallel with the No. 4 lightened series. Some Mark V S.M.L.E. rifles were converted as well as No. I Mk III and III* models and these early experimental models may be noted with the early trials lightened No. 4 type of rounded sheet metal fore-end cap. Such a prototype has been previously illustrated. The new No. 1 jungle carbine was awarded the designation "No. 6" and it was intended to duplicate the British No. 5 for use in the Indian and Australian services.

Initially, this shortened and lightened No. 1 was referred to as the "C.E.A.D. design" (Chief Engineer & Superintendent of Armaments Design) and while it may be considered as a pattern for conversion of No. 1 rifles in British service, it was ultimately put up for Australian and Indian adoption. Lithgow and Ishapore did not manufacture the No. 4, so it was not possible to produce the No. 5 without major changes to their machinery, tooling and equipment; the No. 6 was proposed because it used the No. 1 action. In April 1944 a further £2,000 was allocated towards "lightening of No. 1 rifles" because "India and the Dominions do not produce the No. 4." The recent events in the Far East dictated the need for such a rifle and as Australia and India were more geographically effected in the region, it was a logical conclusion.

A monopod and a bipod were considered for the No. 4 rifle; Enfield drawings DD(E)SK 1294 and 1296 were for "Monopod, Rifle No. 4" and "Bipod, Rifle No. 4" respectively. These are dated in August and September 1943 and are quite interesting; some prototypes were fabricated but they never went into full production. One type is readily detachable, with long spring clips which grip around the rifle fore-end. Another was designed to be attached to the clamping screw of the front band. The tubular steel legs were adjustable for length. The contemporary Japanese Type 99 short rifle was fitted with a folding monopod.

The monopod was to permit more accurate shooting from the prone position and further tests were carried out in April 1947 with the monopod fixed to the rifle so that it folded to the rear when not in use; it was designed for fitting to both No. 4 and No. 5 rifles. The height of the leg was adjustable by means of a release clip, but this assembly was found to be unsatisfactory and the project was again terminated soon afterwards.

In Britain, further development of the No. 4 rifle continued towards the end of the war, mainly related to the mounting of the trigger. The idea of mounting the trigger onto the receiver was investigated in North

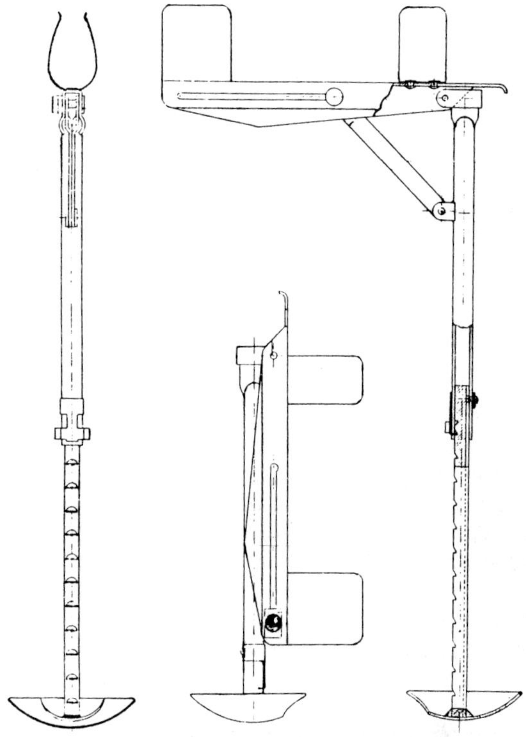

D.D.(E) Sk 1294 - 2/9/1943
Proposed Detachable Monopod for No. 4 Rifle. Front & Side Views, also View folded-up with extending leg retracted.

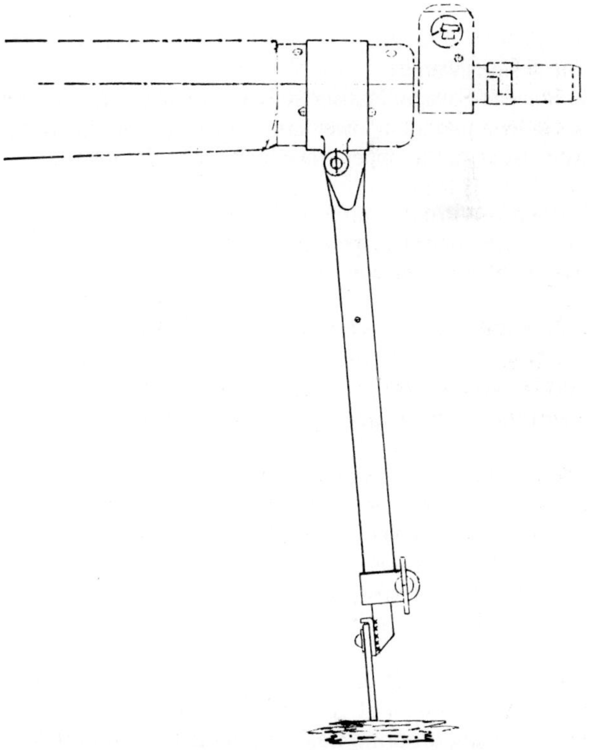

Another variation of D.D.(E) Sk 1294
Folding Monopod for No. 4 Rifle. Spade feet at the bottom of the extending leg folded out to provide better stability.

America as well as England; two patterns were under consideration in England, one devised by the Chief Superintendent of Armament Design and the other by the Chief Inspector of Armaments, Woolwich. In North America, Stevens-Savage had developed another type which was passed on to S.A.L. Canada when the U.S. Lend Lease contracts for the No. 4 were completed and Canada produced yet another model which was undergoing development and consideration.

These special trigger mechanisms were designed to overcome disadvantages incident upon fitting the trigger onto the trigger guard, mainly involving the wood stocking-up. Due to wartime shortages of timber, much of the rifle furniture was not seasoned properly when it was cut up and it often warped after being fitted onto rifles. The trigger guard is fitted onto the underside of the fore-end and this warping could therefore affect the contact of the trigger and sear. The desirability of having the trigger mechanism mounted onto the receiver becomes obvious when other bolt action designs are compared; the remedies or alternatives as applied to the No. 4 were developed quite independently of each other, in England, the United States and Canada.

This trigger development was mostly carried out during 1944, although it was commenced earlier than this in North America. In England, the C.E.A.D. pattern was considered the most promising and fitted to ten rifles for limited troop trials. The first six were completed by May 1944, three months after the order was placed; samples of the American designs were sent over for consideration.

The designs were:–

1. Stevens-Savage. Mauser style with coil spring and two major components; lower arm consisting of the trigger with ribs, and the upper arm which forms the sear.

2. Canadian Production Authorities. Similar to the Mauser rifle mechanism except that the normal sear spring was used in place of the coil spring.

3. C.E.A.D./S.A. A plate onto which the trigger was mounted, secured to the receiver by three screws underneath the trigger guard. This also required minor alterations to the triggerguard.

4. C.I.A./Woolwich. Basically the same as C.E.A.D. pattern but the block was welded to the front of the body socket and required that the fore-end be split at the rear.

Of these four patterns, No. 3 as proposed by C.E.A.D. was regarded as being the best.

Six of the C.E.A.D. modified patterns were sent to the Small Arms School at Bisley in June 1944, but returned for the fitting of spring washers after it was found that the screws worked loose. One of these rifles was then forwarded to Canada. This design also influenced the trigger mounting on a .22RF training rifle under development, the .22 No. 5, described in the following chapter. Illustrations of the Stevens-Savage and Canadian trigger mechanisms will be found in Chapter 8.

Mainly because there was no real urgency regarding the alteration of the trigger mechanism, these tests were continued over the next few years. With the success of the rifles tested at Bisley, 100 more with the "hung" trigger were ordered, comprising 50 each, No. 4 and No. 5 rifles, and in 1947 as a result of the hung trigger success, the "No. 4 Mk 2 and No. 5 Mk 2" rifles were recommended for adoption by the Board. The No. 4 Mk 2 design was approved on 4th December 1947 but because the decision was made to abandon the No. 5 at about this time, no further action was taken on the No. 5.

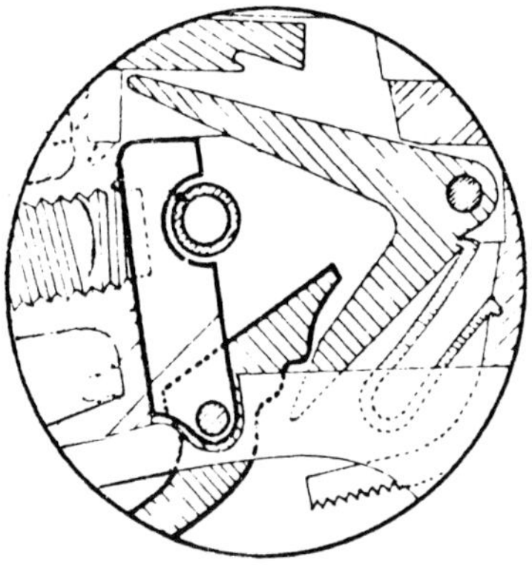

Proposed Trigger Mounting, fixed to receiver.

Another trigger modification tested on the No. 4 was the "hooked sear," this was tried in order to achieve a more regular pull-off. Developed in 1945, the sear has a forked tail and could be fitted to the standard No. 4 body, although it was assembled with a new action body which could not be used with the older (service pattern) sear. The arrangement necessitated the modification of the magazine by lengthening the back strip. However, this project was not proceeded with any further, apart from the assembly of some experimental models.

Further improvements to the service No. 4 rifle were instituted after the war, including a strengthened pattern bolt with higher grade steel in 1946, the elimination of the hole or flat in the bolthandle knob in August 1948 (it was to be spherical after this), and the introduction of a new type of spring and 3/32-in. diameter steel ball to replace the spring and plunger for the backsight; however the last-mentioned item does not appear to have been implemented.

These particular changes were announced in the List of Changes § C3727 in 1949, along with the official introduction of the Rifles No. 4 Mk 2, No. 4 Mk 1/2 and No. 4 Mk 1/3 and No. 4 Mk 1/2 (T). This new rifle and the trigger-improved conversions are described and illustrated in the following pages.

After the end of the war in 1945, production was continued at Fazakerley, Maltby and Shirley and although

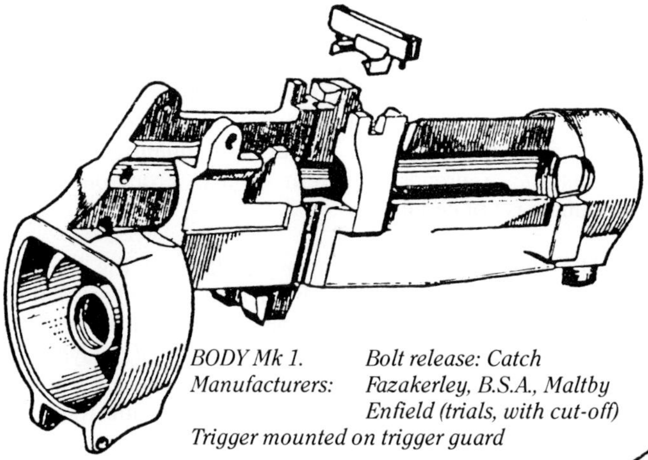

BODY Mk 1. Bolt release: Catch
Manufacturers: Fazakerley, B.S.A., Maltby
Enfield (trials, with cut-off)
Trigger mounted on trigger guard

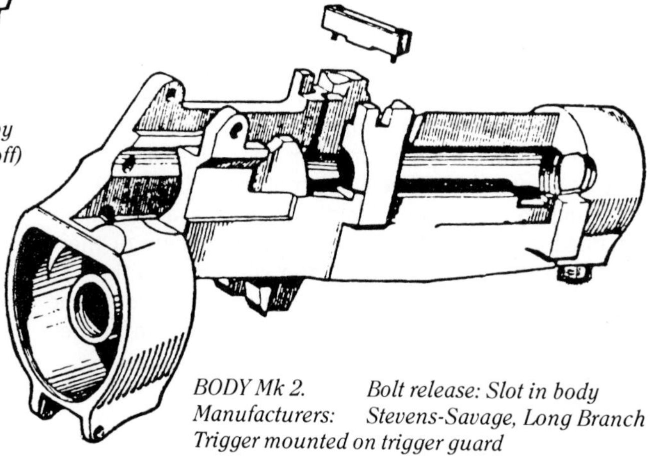

BODY Mk 2. Bolt release: Slot in body
Manufacturers: Stevens-Savage, Long Branch
Trigger mounted on trigger guard

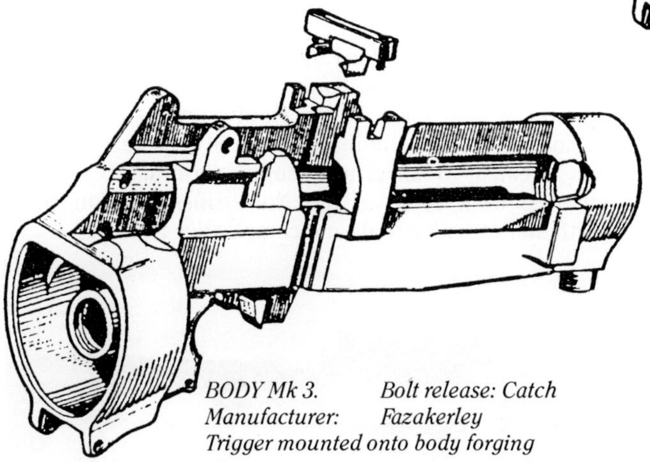

BODY Mk 3. Bolt release: Catch
Manufacturer: Fazakerley
Trigger mounted onto body forging

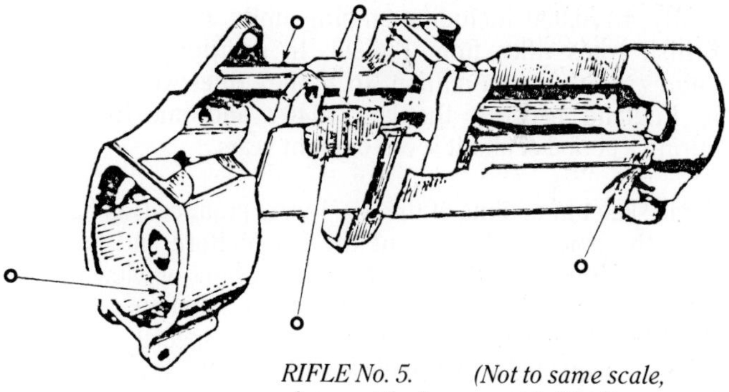

RIFLE No. 5. (Not to same scale,
shown with charger guide fitted)
Lightening cuts marked by arrows.

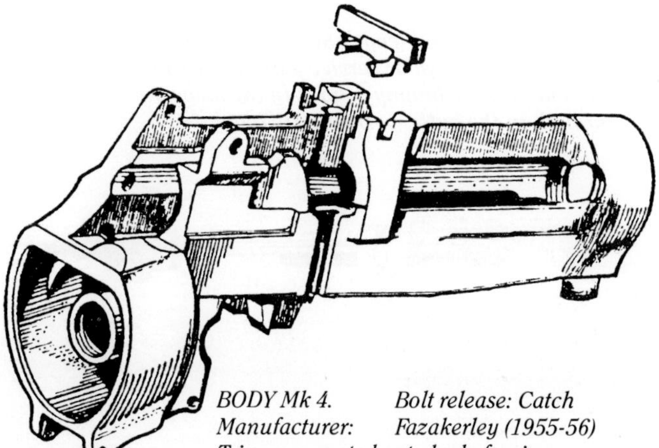

BODY Mk 4. Bolt release: Catch
Manufacturer: Fazakerley (1955-56)
Trigger mounted onto body forging

the quantity of new rifles and components was reduced drastically, a solid program of repairs was carried out on rifles so that adequate supplies should be available in the event of further outbreak of war. The repair of No. 1 rifles ceased in 1946 and the repair programs for No. 4 rifles were completed in 1947. No. 1 repair was largely undertaken at Enfield, No. 4 repair at Fazakerley, Maltby and Shirley and the No. 5 repair at Fazakerley. The Maltby plant then ceased rifle production and repair work upon conclusion of the No. 4 FTR programs.

The last No. 4 rifle action bodies produced at Fazakerley had no cut-off hinge screw protrusion or lump towards the front on the right side and they are similar in form to the No. 5 rifle body. Even after the omission of the magazine cut-off after the No. 4 trials rifles in the 1930's, this lump on the right hand side of the action body remained, although it served no purpose. Many of these new Fazakerley receivers were later converted to the L39 and this type of receiver is also often encountered on the .22 No. 8 trainer, maybe because some spare, unfinished (or "black") actions were still in store.

The total production figures for the No. 4 rifle in the British factories from 1941 until the introduction of the No. 5 in 1944 were 2,021,913, to which is added more than 1,236,000 from Stevens-Savage and 330,000 in Long Branch orders from Canada. This puts the total of No. 4 rifles made by and for Britain, at over 3.5 million, far short of the S.M.L.E. figures. In addition to the 330,000 Long Branch No. 4 rifles on British contract, Long Branch S.A.L. produced another 575,731 rifles for Canadian service and orders for New Zealand, Newfoundland and China. This puts total production of the No. 4 rifle at nearly 4,170,000 including sniping units and the early British trials rifles from Enfield. In addition to these figures, nearly half-a-million No. 4 Mk 2 rifles were made in England between the late 1940's and mid-1950's, during a period of about seven or eight years.

Some skeleton actions of the No. 4 were produced, as had been the practice throughout the Lee-Metford and Lee-Enfield series, the machining and location of viewing slots is somewhat similar. The No. 4 versions were virtually all modified at Enfield and a special serial number sequence commencing with "SKN" was used. Examples of the No. 4 Marks I, I*, 2, 1/2 and 1/3 actions have been noted.

A full-length cut-a-way No. 4 rifle was also made with viewing cuts in the barrel and locking lug recesses in the body action, and a bronze bolt and bolthead fitted. Such a rifle in the M.O.D. Pattern Room collection is marked "D.P.," but this model does not appear to have been officially introduced or produced in quantity. Whether this was a proposed pattern or just a one-off factory sample is not certain.

The appropriate, official approval dates of the skeleton No. 4 actions seem to have been 11th December 1950 for the Marks I and I*, and 3rd February 1955 for the Marks 2, 1/2 and 1/3. This is another instance of "tidying up paperwork" after the event, as most of the Mk I and I* skeleton actions were made prior to 1950 and the hung trigger models were converted at Enfield between 1951 and 1955, with a run of about 2,500 actions.

Service cartridge trials were conducted during this period and the No. 4 and No. 5 rifles were utilised as "test beds." Much of the initial work was undertaken by B.S.A. and the German 7.92mm Kurz cartridge was one of the earlier types used, chambered in a reinforced No. 5 rifle previously built up for the grenade firing tests. Another No. 5 chambered for the 7.92 x 33mm Kurz cartridge has the last 2.5-ins. of the barrel free-bored, most likely to obtain the same effective length of rifling as the STG 44. Because the flash eliminator was not required and as it was deemed to affect the bullet performance, it was removed from some of the No. 5 rifles used in these particular cartridge trials and comparative tests.

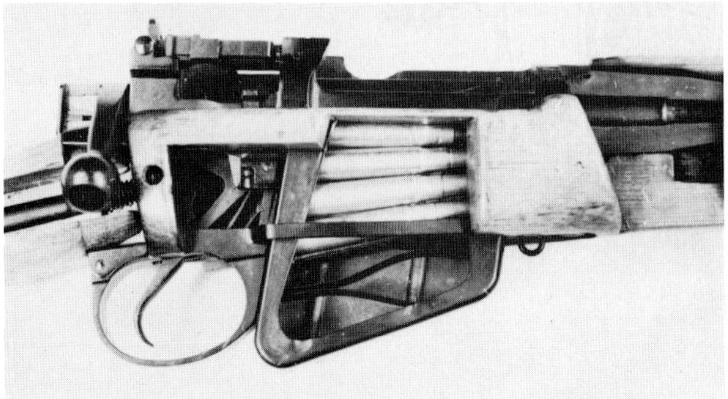

Above & below: Full-length Sectionised No. 4 Mk I Rifle. Although not a regular service pattern, this Pattern Room example reveals dummy rounds in the magazine and a pull-through in the butt. *MOD Pattern Room*

A variety of calibres was tested in the following "Ideal Calibre" trials, including .256-in., .30-in. and .330-in., and then the .270-in. and .280-in. Some No. 4 rifles used in these tests have been noted with the new calibre engraved on the left side of the receiver, e.g. "FOR 7.92 CARTRIDGE" or ".330 CARTRIDGE". B.S.A. produced a special belted cartridge in 7.92mm and .330-in. which was used with specific 4-groove R.H. twist rifling. The calibre trials eventually resulted in the 7mm or .280-in. being chosen and an "Optimum Round" was developed; the trials process extended over more than ten years until the final adoption of the U.S. T65 as the 7.62mm N.A.T.O. round. To understand the relativity of these particular cartridge and calibre trials, the EM2 and early F.N. F.A.L. rifle trials and development should be considered.

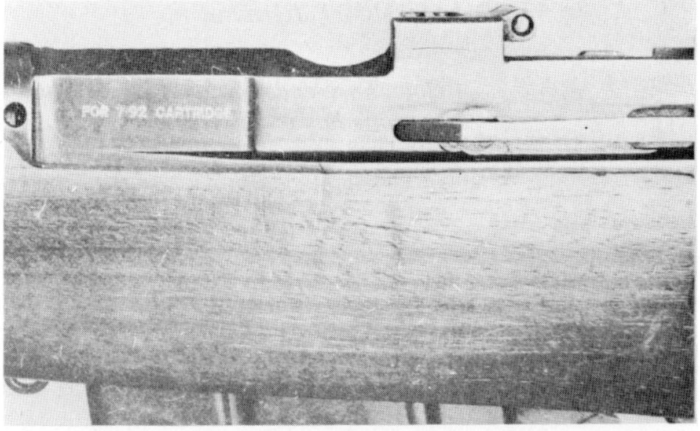

Ideal Calibre Trials rifle for 7.92mm cartridge. Note the special ejector for the longer 7.92 case. *MOD Pattern Room*

Three calibres were under consideration at the beginning of these tests; .256-in., .30-in. and .330-in. While three different cartridge cases were utilised, these in themselves were only vehicles for the calibre trials and not under direct consideration as future service cartridges. It was intended that, after the "ideal calibre" was decided upon, the cartridge could then be developed in association with the new gas operated rifles. Of course, the form of the cartridge was directly related to the rifle systems and design, and only the calibre would effectively be predetermined, hence the decision to conduct ideal calibre trials first.

The specific tests and experiments with these three calibres were generally conducted using three different types of rifles; the 6.5mm (.256-in.) Norwegian Krag, the .30-in. (or .30-06) using the Model 1903 Springfield, and the .330-in. on a modified No. 4 Lee-Enfield. The .330 No. 4 rifles were specially modified by B.S.A. and a small batch assembled, with the serial number range commencing with S0001. The particular .330 No. 4 rifle serial no. S0001 and dated 1946 was recorded as being unserviceable after 57 rounds with a bulged chamber and damaged extractor. Four such rifles were issued by the Design and Development Department in 1946 in the Ideal Calibre trials. Some other cartridges and calibres were also involved but these were mainly to provide certain standards or for comparison.

Eventually, the ideal calibre was narrowed down to .270 and .280, after which the .280 was chosen. After its selection, a number of case variations were developed, only to be further complicated by attempts to boost the cartridge's performance to meet U.S. requirements for the proposed NATO round. This resulted in the 7mm High Velocity and 1st & 2nd Optimum classifications. The .280-in. cartridge as originally designed for the new British rifle was modified slightly in that the extractor groove was widened and the angle altered, to become the .280/30. When it was considered that this particular round would be adopted, it was designated the "7mm Mk Iz."

These cartridges were specifically developed for a new rifle series and not intended for the No. 4 or No. 5. This new automatic rifle eventually became the EM2, which was chambered for the .280/30 cartridge and would have become Britain's new service rifle but for American pressure as a NATO partner. The 7.62 x 51mm round was adopted instead, along with the F.N. designed F.A.L. rifle which became the L1A1 rifle in British service.

In 1949, a No. 4 rifle was prepared at Enfield to see if the No. 4 rifles could be converted to .280-in., with 5-groove, L.H. twist rifling. The bolthead face was recessed and the extractor claw widened, while a special ejector was fitted. The magazine was modified for single loading and ".280" was stamped on the barrel knox form of the particular test rifle.

The No. 4 Mk 2 rifle was approved on 4th December 1947 but it was not introduced until 1949. Early C.E.A.D. trigger modifications had the new trigger plate screwed onto the receiver but a more permanent mode of fixing the trigger mount was desired. The two methods suggested were by brazing or by riveting, of which the former was preferred, although there was some doubt as to the effect of heat upon the receiver, so some tests were conducted on receivers for conversion. It was found that the heat required in brazing was not detrimental to the action body so this was incorporated in the conversion and it was reported that half-a-million body-forgings in stock could be modified and assembled as new rifles. Eventually, about 70% of the new No. 4 rifles were made up from fresh forgings.

Whilst the 1947 trials with fifty No. 4 rifles incorporating the trigger modifications had led to the recommendation of the design for adoption, it was considered that some aspects of the new design had not been sufficiently investigated. So, in 1948, 200 Mk 2 rifles were built up for accuracy tests; that same year, the requirement for a new bayonet for the No. 4 was stated, with a Bowie-knife blade instead of a spike. This was approved the following year when the tests with the No. 4 Mk 2 rifle reported that the new mounting of the trigger and the relevant accuracy, were "entirely beneficial" and desirable.

Introduced in 1949, the No. 4 Mk 2 therefore differs from its earlier counterpart in the trigger mechanism (when compared with the Mk I and I*). This change also required

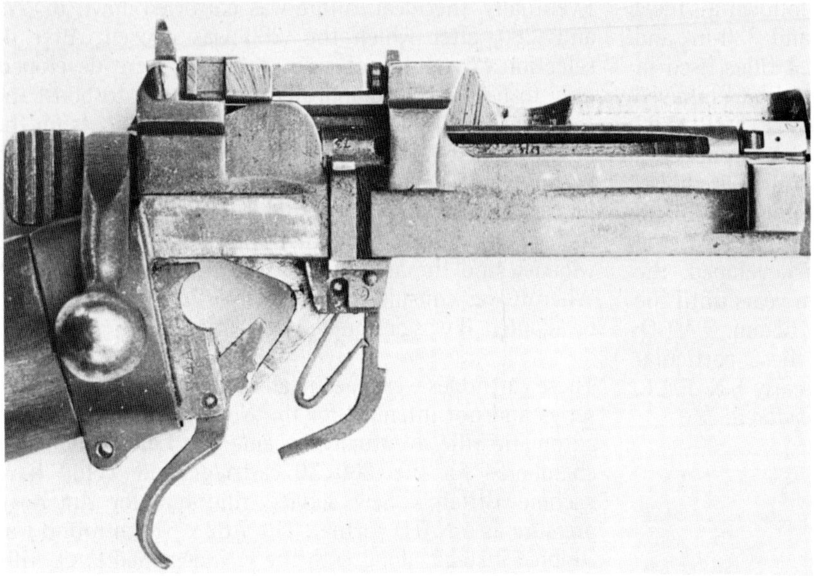

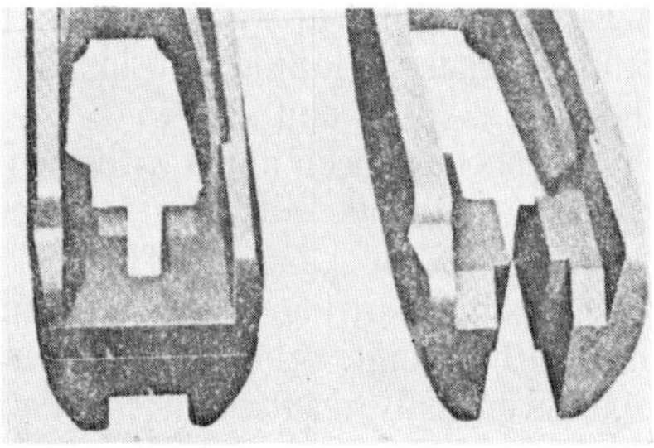

Above: Fore-end for "hung" trigger Mk 2, Mk 1/2 and 1/3 is at right.

Left: Hinged trigger of the No. 4 Mk 2 series rifles, mounted onto the action body. *MOD Pattern Room*

Below: Fore-ends for Mk 2 series; the lower one has its cross-screw fitted.

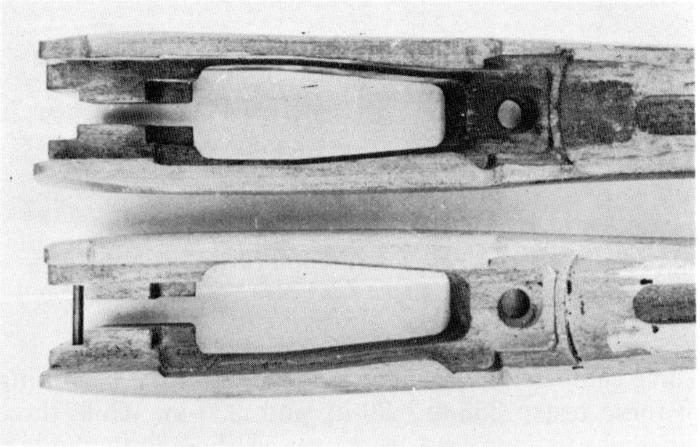

certain changes to the fore-end and trigger guard; whilst the trigger guard has the trigger and pin removed, it was also milled at the rear. The fore-end was relieved at the back for the extra fittings mounted onto the front of the butt socket. Because of the removal of this section at the back of the fore-end, there is no reinforcing tie strap across the back of the wooden fore-end.

Those Mk I and I* rifles upgraded in the 1/2 and 1/3 conversion also had the fore-ends modified by the removal of the sheet steel tie plate and the fitting of wooden plugs into the resulting recesses; a relatively thin, transverse screw or crosspin was fitted which readily identifies this series of rifle. Some of the converted rifles were fitted with newly manufactured Mk 2 fore-ends which still have the cross screw, but do not have the wooden conversion dowels.

Three different butt lengths were introduced in the announcement on the Mk 2 rifle; long, normal and short. Many of the wartime economies were also dispensed with as the better quality milled parts were reverted to and the Mk I backsight again fitted. General finish of these rifles is quite good and the wooden furniture is usually a light coloured beech. The No. 4 Mk I style of bolthead is retained with the small catch plate to the rear of the charger bridge, as opposed to the Mark I* type on which there is no catch plate, but a release slot in the bolthead ribway near the front of the action. The official approval date for the No. 4 Mk 2 was 31st March 1949, as per the List of Changes entry.

The new method of trigger mounting was then applied to many of the No. 4 rifles already in service or in store, which resulted in the following conversions:

Rifle No. 4 Mk I	became Rifle No. 4 Mk 1/2
Rifle No. 4 Mk I*	became Rifle No. 4 Mk 1/3
Rifle No. 4 Mk I (T)	became Rifle No. 4 Mk 1/2 (T).

The List of Changes entry, § C3728 has only a short description of these items; by this time the LoC's were used more for stores nomenclature and catalogue number references. Sections from this entry read:

1. Approval has been given for the introduction of the above-mentioned Mk 2 No. 4 rifle which differs from the existing Mks 1, 1, 1 (T) and 1* (T) patterns (§§ B4737, B6861, C2791) in being provided with a new design of trigger mechanism. The trigger will now be hinged about a pin located in a bracket forged integral with the Mk 3 body (Part No. CR503) instead of being hinged to the trigger guard as hereto.*

2. Rifles in current production will be modified to assimilate the new pattern in accordance with the undermentioned drawings:

D.5(E) 6154/IM/32:	*Modification of Mks I and 2 bodies to enable fitting of modified trigger mechanism.*
D.5(E) 6155/IM/32:	*Modification to trigger guard.*
D.5(E) 6156/IM/32:	*Modification to stock, fore-end.*
D.5(E) 6157/IM/32: (sheets 1 & 2)	*Modification of bodies by fitting trigger bracket.*

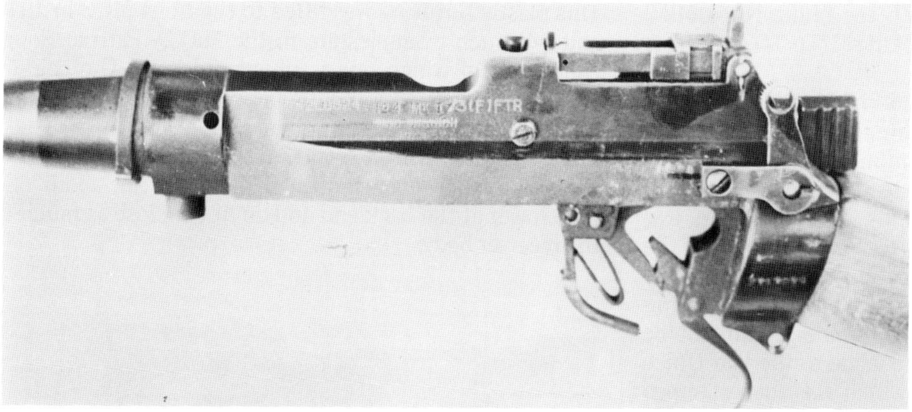

No. 4 Mk 1/3 Conversion. Long Branch marking and original "*" designation have been barred out and the new name added.
MOD Pattern Room

Rifles so modified will have their Marks advanced as follows:

Rifle, No. 4 Mk 1 to Rifle No. 4 Mk 1/2
Rifle, No. 4 Mk 1 to Rifle No. 4 Mk 1/3*
Rifle, No. 4 Mk 1(T) to Rifle No. 4 Mk 1/2(T)

3. Existing rifles will be similarly modified and the Marks advanced as indicated above, the service being carried out when rifles are returned for Factory Thorough Repair.

The "new" Mk 1/2 and 1/3 rifles have the designations engraved on the left side of the body and the original markings were left, or in most cases, added to. The Mk I* converted models were originally manufactured by Stevens-Savage and Long Branch. Typical action markings (with original and subsequent conversion markings) are:

1. N^{O}4 M^{K}1/2(F)FTR 49

2. N^{O}4 M^{K}1/3(F)FTR 49
LONG BRANCH
1944

Example (1) was a British manufactured rifle upgraded from Mk I and the markings "/2 (F) FTR 49" were added by electric pencil upon conversion in 1949; the "(F)" indicates that the conversion was effected at Fazakerley. Example (2) was made at Long Branch in 1943 and later converted to /3 configuration at Fazakerley. As R.O.F. Fazakerley was the only government rifle factory in operation by this time, the conversions and FTR work were effected there.

Fazakerley production was scaled down in the mid 1950's although massive overhaul and F.T.R. programs were undertaken there as well. In its later years the factory was plagued by industrial labour troubles and its eventual closure has even been attributed to the strikes and worker-related problems. It was Fazakerley tooling for the L1A1 rifle which was sent to Australia's Lithgow factory in the late 1950's. Some Enfield converted No. 4 rifles to the Mk 1/2 and I/3 models were produced after the closure of Fazakerley although most of these appear to have been the sniper models. One such rifle updated to Mk 1/2 specs has been noted with the 1/63 date, indicating January 1963.

The very last production No. 4 Mk 2 rifles, with the UF prefix serial numbers, have the "lump" or reinforce at the front of the action body on the right side removed (where the former cut-off axis screw was positioned) similar to the No. 5 rifles. Wooden fore-ends for these late Mk 2 rifles were also machined to accommodate this small difference on the right side of the receiver body, making it necessary to remove a small section of the fore-end with a chisel or rasp when these late fore-ends were fitted onto the earlier type of No. 4 Mk 2 rifle body.

The variety of service No. 4 sniping rifles catalogued under the new nomenclature is of interest; the new system of Indo-Arabic lettering as opposed to the earlier Roman numerals was also applied to the telescopic sights in service on the different models of rifle, compounded by the new bracket-mounted trigger models. The List of Changes entry, § C3729 records:

Rifle No. 4 (T2) Mk 1	Rifle No. 4 (T/W2) Mk 1
Rifle No. 4 (T2/1) Mk 1	Rifle No. 4 (T/W2/1) Mk 1
Rifle No. 4 (T3) Mk 1	Rifle No. 4 (T/W3) Mk 1
Rifle No. 4 (T2/1) Mk 1*	Rifle No. 4 (T/W2/1) Mk 1/2
Rifle No. 4 (T3) Mk 1*	Rifle No. 4 (T/W3) Mk 1/2
	Rifle No. 4 (T2/1) Mk 1/2
	Rifle No. 4 (T3) Mk 1/2

The bracketed figures, e.g. (T2/1) or T/W3) refer to the telescopic sight. The Marks 2, 2/1 and 3 scopes were used, in addition, the "W" indicates waterproofed. The mark number at the end of the designation was that of the rifle, i.e. Marks 1, 1* and 1/2.

From this list it appears that there were no waterproofed telescopes fitted to the Mk 1* rifles. The Mk 2/1 telescope comprises the Mk 2 upgraded to certain Mk 3 specifications. Using this nomenclature, the Mk 1 or Singer backsight with battlesight removed for use on the sniper rifles, was designated the Mk 1/1 backsight. It will be noted that there was no sniper rifle configuration of the new-made No. 4 Mk 2 rifle although some E.M.E. Regulations in

1954 have been sighted in which the "Rifle No. 4 Mk 2 (T)" was listed along with the Marks 1(T), 1*(T), 1/2(T) and 1/3(T).

These E.M.E.'s also describe re-numbering procedure for rifles; those on which the serial number was indistinct or illegible were to be re-stamped clearly on the left side of the butt socket and any previous marking there was barred out. The number was not to be re-stamped on the left side of the body as this location was suitable only when the number was rolled on or engraved during manufacture. In cases where the number could not be determined, application was to be made to the War Office for allocation of a new number which was then stamped on the butt socket, bolt, magazine and fore-end. Some inconsistencies may be noted in the serial number ranges, dates and factories of production, as some new or "black" action bodies were fitted-up and numbered in repair programs at the main Ordnance Depots and some factories.

Alternative furniture and different methods of stocking up were again considered after the war, some of these being set up on the No. 4 Mk 2. The first comprehensive stocking-up trials were conducted by the Ordnance Board in 1944 in which the Ishapore method was closely investigated and although it stood up well, no change was recommended; ten such rifles were stocked up on the Indian system. In 1949, more tests were conducted with modified bearings, using 50 rifles consisting of 10 stocked up in each of five different ways. Again, no major changes were introduced as a result of these trials and the main Committee recommendation was that, for the purpose of accuracy, a bearing at the reinforce or in front of it, was essential.

Due to timber shortages and the poor quality available for rifle furniture, the Director of Ordnance Factories (Fuzes and Small Arms) submitted nine butts of laminated wood, bonded by Phenolite resin cement. Three were impregnated with a Bakelite solution. Six of these butts, including two with the Bakelite finish, were fitted to rifles and forwarded to Pendine for trial. Tests were in progress with plastic furniture for the Bren machine gun and so it was suggested that some be made up for the No. 4 rifle also.

This plastic furniture was fitted to the No. 4 Mk 2 in 1949 and two such examples are in the M.O.D. Pattern Room collection, of which one is illustrated here. The colours and plastic composition differ between these two samples, one is a black while the other is more brownish. These fore-ends are also a little differently shaped to the conventional wooden model in that they are about one inch lower at the back end and do not enclose as much of the receiver body.

Above & below: Modified wooden fore-end on a No. 4 Mk 2 rifle, probably for stocking-up trials. *D. Booth*

Reports indicate that the plastic was generally more durable, and a big improvement over bakelite, but at extremes of temperature it was a better conductor than wood and therefore not as comfortable to hold, although the heat or cold was not sufficient to alter the grip or

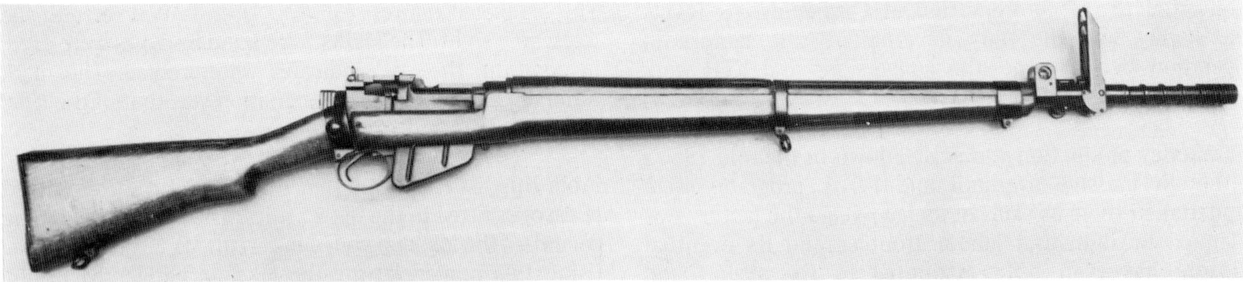

Plastic Fore-end and Handguards on a 1949 Trials Rifle, with wood butt. The Grenade launcher is not related to these trials. *MOD Pattern Room*

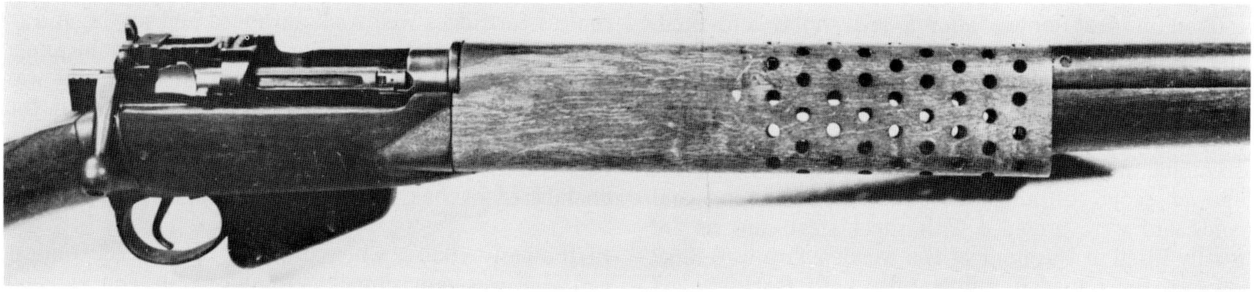

No. 4 Test Rifle with triangular cross-section fore-end, possibly associated with certain bedding and firing trials. *J. Alley*

firing techniques. The main disadvantage was that it became much harder to hold when the hands were wet or oily and this was a handicap under wet and muddy conditions.

Differences between these test models were that on some rifles the wooden butt and handguard were retained while others had all-plastic furniture. In an effort to provide a better grip on the plastic fore-end, grooves were cut along each side and the use of plastic permitted the sheet metal fore-end and handguard caps to be dispensed with. This was by far the most successful of all the synthetic furniture tried to date but the matter was not proceeded with any further. Black plastic furniture was eventually adopted for the 7.62mm L1A1 rifle.

Wooden rifle furniture was also impregnated with a plastic in an attempt to provide a more durable finish. This particular furniture was otherwise the conventional service type and the protective finish made it impervious to moisture as well as more resistant to wear and abrasion.

During the war, a protective finish had been approved for the barrel, in a form of paint. This idea was introduced in India and after tests in Australia, many No. 1 rifles were also treated in this manner. A green colour was used although the composition varied slightly between the countries due to different requirements. In Australia and India it was mainly intended for rifles in service in tropical conditions, so as to prevent rust.

Even in the mid-1950's, minor changes were still being made to some components of the No. 4, for example, the decision in 1953 to approve a cast trigger guard in place of the machined one. Because the trigger guard of the current service rifle no longer supported the trigger, it was not required to be as strong, and its less critical role permitted such a change without compromising the component efficiency and general service durability.

In the late 1950's, a new form of manufacturer code was introduced to replace that used since 1941. B.S.A. had used their code M47, followed by a "C" for Shirley, or "A" for Small Heath, while Fazakerley had used "ROF(F)" or merely "F." At Maltby, "R.O.F.M." was used on very early production, later abbreviated to "R.M." or "M." Some No. 4 receivers have been noted with no makers code or initial at all and these are also considered to be of Maltby origin.

After the change, the factory initial was used after a "U" (taken to indicate United Kingdom) and so "UE" signifies Enfield, "UF" Fazakerley and "UB" for B.S.A. Guns. Another arms manufacturer, Sterling Engineering Co. was allocated "US" although the Sterling factory produced machine carbines, not rifles. By this time, B.S.A. was no longer making any .303-in. rifles. This new factory code system was announced in the List of Changes in December, 1959.

After 1944, some research and trials were carried out with the sniping rifles but these were either "one-off's" or done in very small quantities; and the standard No. 4 (T) rifle with No. 32 sight remained unchallenged as the service sniping rifle. So, only a short summary is given here.

Some Canadian sniper rifles with a variety of telescopes were tested against the British equipment in 1944 and these investigations carried over into 1945, although they were not continuous, but rather a series of comparative trials as the equipment became available. Lyman, Weaver and experimental R.E.L. 2.5 and 5 power telescopes were fitted to Long Branch No. 4 rifles and some North American designed mounts such as the Griffin & Howe underwent quite successful tests.

Some lightened sniper rifles were trialled in 1945 and these can be considered to have influenced the later design. A No. 4 Mk I (T) rifle was modified by reducing the fore-end to a sporter configuration in Italy in October 1944, and it was considered to be successful. So a rifle was similarly altered in England for tests with the fore-end cut off level with the front of the rear handguard and the outer band moved back about 1.5-in. to a recess cut in the new position. This weight reduction amounted to about 1 lb. after the forward portion of the fore-end and the front handguard were removed.

However, this cut-down rifle was considered to be somewhat "rough" so it was suggested that a No. 5 fore-end be utilised; another No. 4 sniper rifle was thus altered and also set up with a sporting style butt with pistol grip and rubber kickpad. The two types of rifles were subjected to comparative tests and the main complaint seems to have been that the additional length of exposed barrel resulted in heat mirage after a few shots were fired, although it must be conceded that such rapid fire was not the normal modus operandi of a trained sniper in the field.

A somewhat similar "sporter style" sniper rifle was assembled in Canada, fitted with the experimental R.E.L. 5-power scope, although the shortened fore-end and handguard were provided with sheet steel caps as are usually fitted at the front of the wooden furniture. Like many of the Canadian sniper rifles with R.E.L. No. 67 telescopes, a sporting style of butt was also fitted.

As mentioned earlier in this chapter in the description of the No. 32 telescopic sight, another British service telescope was tested with the No. 4 rifle, the No. 53 Mk I, but it was not suitable because the cross wires were too thick and there was excessive parallax error. So, fifteen more No. 53 scopes were ordered with thinner crosswires and other minor improvements for further tests. These scopes were set up on three different types of mounts, including a Griffin & Howe type which had fared well in the Canadian tests.

Further telescopic sight mount trials were conducted in 1947 and some special sniping rifles constructed and tested. A variety of telescopes and mounts was used, the most unorthodox probably comprising the Pattern 1918 sight mounted on a U.S. Model 1903 Springfield. Some adjustable mounts were also tested with the range and lateral adjustment on the base mount, instead of on the telescope; these were fitted onto a U.S. 1903 Springfield.

Various models of the No. 4 rifle may be encountered with markings that would indicate production in Pakistan. These are really only re-works although in some cases, all the original manufacturer's marks have been removed and the only ones noticeable are those made by the Pakistan Ordnance Factory (P.O.F.) during their F.T.R. program. Some re-built No. 5 carbines have also been noted with similar markings. Such factory repair programs were also effected by the Indian ordnance factories.

Pakistan Ordnance Factory FTR markings. Also note the "P" over "60" on the butt socket, which signify Pakistan and 1960, the year of refurbishing. *MOD Pattern Room*

Prior to the outbreak of the 2nd World War, tests with rimless cartridges were well advanced and by the end of the war a new rimless service cartridge was again under consideration. As described previously, the No. 4 rifle was used as a test bed for many such cartridges and the Canadians also modified the No. 4 for rimless cartridges. One such rifle was forwarded to England late in 1944, chambered for the German service 7.92mm cartridge.

In the early 1950's, when the new British .280 rimless cartridge was provisionally introduced, its concept and design had to be changed considerably to meet U.S. requirements but the whole project was eventually abandoned and the 7.62 x 51mm round adopted along with the Belgian F.N. F.A.L. The F.N. design was then introduced into British service as the Rifle, L1A1. This new letter and number code replaced the previous system of numbers and marks, and was subsequently applied to the ensuing conversions of the No. 4 rifle.

The first letter, "L" is used for all the weapons series and is recorded as having indicated "Land Service," although this has been questioned because this nomenclature applies to all the services and there are no references to weapons with "A" or "N" prefix designations. The following number (e.g. the first numeral "1" in L1A1) is the class or type of weapon, while the remaining letter and number replace the earlier mark and star or oblique and number sequence; this denotes subsequent modification or variations.

Rifle L1 is the series for the F.A.L. rifle and the 7.62mm conversions of the No. 4 rifle were allocated L8 with A1, A2, A3, A4 and A5 variations. L39 was allocated to the heavy barrelled, half stocked, No. 4 based target rifle, and L42 to the telescopic sighted sniping variant. L59 was applied to the D.P. No. 4 training rifles which were further rendered innocuous for ongoing cadet issue.

These are machined away in certain places so that the body is severely weakened, then reclassified as Rifle Drill Purpose L59A1. The bolts have been copper plated and the bolthead striker holes welded up, machining cuts are also visible so as to emphasise their dewat classification.

The L59 conversion is a more recent extension of the Lee-Enfield family and it represents the "end of the line" for this series in more than one way. Notes on this modification first appear in E.M.E.R.'s during the mid-1970's and subsequent armourers updates describe the L59A1 as comprising two different methods of converting Rifles No. 4 to Rifles L59A1 for drill purposes. Additionally, it authorises the mutilation of skeleton action derivatives so as to render them innocuous.

Machining or hand fitting were the two methods employed for L59 conversion and these were effected by certain unit, field or base repair workshops. Unserviceable components were utilised as much as possible and the conversion entails:

Machining barrel and body assembly, welding plugs to D.P. configuration and machine finish welding.
Welding of barrel to action body.
Machining bolt head, fitting plug and welding in place.
Machining of bolt body.
Machining of striker (firing pin); removal of point.
Modify stock fore-end and handguard by removal of wood, effecting a viewing slot to reveal D.P. mods to barrel.
New designation "DRILL RIFLE L59 A1" engraved and the original markings barred out.
Barring out of original proof marks, stock numbers, &c.
All bright surfaces due to machining, painted black.
Copper plating of bolt body and striker.
"DP" stamped on bolt body and magazine; painted in black letters on white background, on left side of butt and underneath fore-end.

L59 drill purpose conversion, showing the machined action, barrel and furniture from the left side. *MOD Pattern Room*

The alternative "hand fitting" method of conversion was applied with a jig, thus simplifying the cutting and drilling operations. This was detailed as follows:
Drill, pin and weld barrel, forward of knox form.
Tap holes, fit plugs and weld through receiver ring into barrel.
Drill hole into left side of fore-end and handguard, so as to reveal D.P. condition of barrel.
Drill and plug bolthead.

In cases of conversion of the Action Skeleton, Rifle No. 4, the barrel was drilled, pinned and welded through the receiver ring, bolthead striker hole plugged, striker ground, breech bolt body sectioned and weakened, and the body rib drilled and pinned.

But, going back to the beginning of the 7.62mm NATO era, the strength of the No. 4 action was sufficient for the higher pressure 7.62mm cartridge, however, as the No. 1 action was not as strong, no serious moves were made in England in this direction. Tests with a reinforced No. 1 action were made in Australia, in India the 7.62mm NATO variant of the No. 1 rifle was produced as the Rifle 2A. Details of these Australian and Indian models will be found in Chapters 9 and 10.

For the refurbished 7.62mm No. 4 rifles, the first "conversion kit" was developed by Sterling Armament and patented in 1959. Such a conversion was an economic proposition for countries with quantities of the No. 4 rifle on hand, considering the high cost of total re-equipment with the new self-loading rifle. Canada and Australia had commenced production of the F.A.L. rifle, but many developing Commonwealth countries could not rationalise the expense of total re-equipment; by having a smaller quantity of the new L1A1 rifle, and reserve or training issues of converted 7.62mm NATO No. 4 rifles, such countries could financially justify the changeover.

The Sterling Armament conversion kit consisted of a new 7.62mm barrel made in Belgium by Canons Delcour, a new bolthead, extractor, charger guide adaptor, ejector and magazine made by Sterling. The new barrel fitting and conversion could be effected by a competent armourer. Patents were taken out by Sterling in U.K., Australia, Belgium, India, New Zealand, Pakistan and South Africa.

Sample Sterling 7.62mm Conversion. *Sterling Armament*

Some kits were sent out for trial to Malaya, Hong Kong, Kenya, India and other Commonwealth members and tenders were placed for 80,000 units for India, comprising of 20,000 rifles converted and 60,000 kits. With the possibility of this commercial market, R.S.A.F. Enfield also made up a conversion kit in the early 1960's and legal wrangles ensued between Enfield and Sterling over breach of patent rights.

But by the time the legal argument neared resolution, the urgency for stop-gap 7.62mm No. 4 rifles had diminished and only 1,000 Sterling conversions were eventually sent to Sierra Leone. Smaller batches were also sold off to some overseas government bodies such as New Zealand's Forestry Service and the Department of Lands, after trials had been conducted with the No. 4 Mk 2 rifle, Sterling and Parker Hale barrels, and the Sterling conversion components. The New Zealand Armed Forces adopted some of these 7.62mm conversions in 1960 and the Police Force was so equipped for a short period.

India set up the F.A.L. rifle and it was in production by 1963 and so the requirement for No. 4 conversions was no longer as urgent. Canada considered altering its sizeable stock of No. 4 rifles to the new cartridge, the outcome is described at the end of Chapter 8. A conversion kit for the No. 5 rifle was also available commercially, similar to that applied to the No. 4 but differing in the barrel form.

R.S.A.F. Enfield's 7.62mm conversion kit consisted of:
- Barrel with 4-groove, 1 in 12-in., R.H. twist rifling;
- Breeching washer, available in varying thicknesses;
- Extractor with extended claw;
- Magazine with 10-round capacity and extended left lip to form an ejection face; Pat. No. 39081 of 1961;
- Charger guide insert for the smaller-base 7.62mm round;
- Foresight block fixing pin, an oversize taper pin to allow reaming of the foresight block and barrel.

Similar to Sterling's commercial approach, Enfield listed a 7.62mm conversion for No. 5 rifle but few were made. In the 7.62mm RSAF Conversion Kit Instruction Handbook, it is pointed out that the No. 5 is not ideal for conversion as the body was lightened behind the locking shoulder, making it slightly weaker than the No. 4. Also, the No. 5 was generally inferior in accuracy to the No. 4 and there was additional recoil due to the higher pressure cartridge in a relatively lightweight rifle. No official designation appears to have been allocated to the No. 5 conversion as it was not adopted, merely offered as an alternative in the commercial application.

7.62mm Magazine (left) compared with the .303 model.

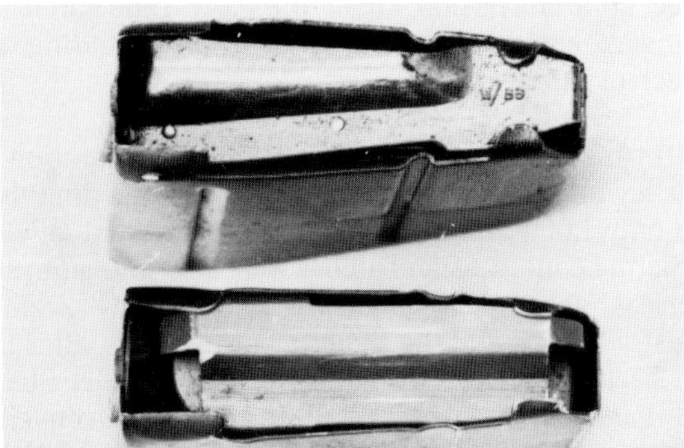

View of Magazine Floorplates, 7.62mm model at bottom.

The 7.62mm Extractor (bottom) has a longer claw.

A special barrel was developed for the Enfield L8 rifles, of chromium molybdenum alloy steel which gave a longer life; it also had a hard-chromium plated chamber. On the action body, a charger guide insert was fitted into the charger bridge and a new pattern extractor provided for the rimless cartridge. The bolthead has been reproofed and marked "19T" which indicates passing a proof pressure of 19 tons per sq. in. Some machining of the receiver body was also required for the 7.62mm L8 and L42 conversions because of the new magazine, but not for the L39 as it retained the .303 magazine and only utilises it as a loading platform.

To allow for the shorter and more parallel-sided cartridge case, the new magazine was fabricated with noticeably more square edges than its .303 forebear; this has a slightly different profile to Sterling's 7.62mm version. The service L8 magazines are marked with the part number "CR12A," whereas the identical L42A1 sniper rifle version is marked "CR141A." This "A" suffix indicates an assembly or group of parts.

Whilst the 7.62mm projectile trajectory is flatter than that of its .303-inch counterpart, the change from yards to metres as the service unit measurement meant that the old .303-in. calibrations somewhat approach those of the 7.62mm when yards are read as metres, up to 800m. At ranges greater than this, the shots strike above the point of aim, so a replacement backsight was available to match the 7.62mm cartridge at all ranges. Or so the Sterling conversion advertising leads one to believe. In reality, the original sight was generally ineffective at ranges over about 300 metres.

L8 rifle backsights to 1,000 metres are marked a little differently to their .303 forebear, the 7.62mm marked backsight slide has a lowered datum line for the NATO round. The L42 sniper rifle is usually found with a blackened face on the backsight so as to avoid any possible reflection. The L39 does not use the original backsight because competition sights are fitted.

The main differences between the commercial Sterling and Ordnance Enfield conversion is found in the style and shape of the ejector and the magazine. On the Sterling model, a new ejector is positioned through the receiver wall, secured by and part of a small bent steel piece attached to the left side of the receiver wall and retained by the original ejector screw. R.S.A.F. overcame this problem of the re-positioned ejector (caused by the shorter 7.62mm case) by incorporating it in the magazine lip and therefore no modifications were necessary to the receiver or action body.

New designations were allocated to those No. 4 rifles converted to 7.62mm for service use, in line with the new series of nomenclature:

No. 4 Mk 2	became	L8A1 (CR 60 GA)
No. 4 Mk 1/2	became	L8A2 (CR 61 GA)
No. 4 Mk 1/3	became	L8A3 (CR 62 GA)
No. 4 Mk I	became	L8A4 (CR 63 GA)
No. 4 Mk I*	became	L8A5

(The service Part Numbers after the L8A1, L8A2, L8A3 and L8A4 refer to the receivers).

Due to the slightly different lips of the new magazine, part of the receiver magazine well has been machined; this is readily identified upon examination of the underside of the receiver body. Because the new 7.62mm round is shorter than its .303-in. counterpart, the magazine lips are positioned closer to the rear and it will be found that these come into contact with the action body when fitted to an unmodified rifle, often causing loading misfeeds and jams. This was applied to the L8 and L42 conversions.

A similar removal of the receiver body metal from the magazine well, further towards the rear, as per the drawing, will remedy difficulties experienced in picking up the top round in the magazine during loading into the chamber.

Many of the conversions were carried out at Enfield and the original rifle designation on the left side of the receiver has been cancelled out and the new markings inscribed by electric pencil. Rifles of Long Branch and U.S. manufacture were converted as well, having been previously modified to Mk 1/2 and 1/3 at Fazakerley. While provision was made in the nomenclature for the conversion of No. 4 Marks I and I* to the 7.62mm form, it appears that mostly the "hung" trigger rifles (the Marks 2, 1/2 and 1/3) were utilised.

Some form of conversion was also applied to the fore-ends as these have the appropriate rifle serial number marked on the inside along with the conversion date which seems to range from 8/70 to 9/71. The stock was checked for bedding and seating of the action body and barrel; maybe the fore-end detail applied to this. The top of the fore-end adjacent to the receiver wall was usually stamped with the new rifle designation (L8A1, L8A2, &c.). Armourers applied the gauging limits as per the L42 sniper rifles in service.

Sample 7.62mm L8A3 conversion, from Long Branch Mk I* upgraded to Mk I/3 at Fazakerley. *MOD Pattern Room*

Fig. 1
Unmodified magazine well (viewed from below).

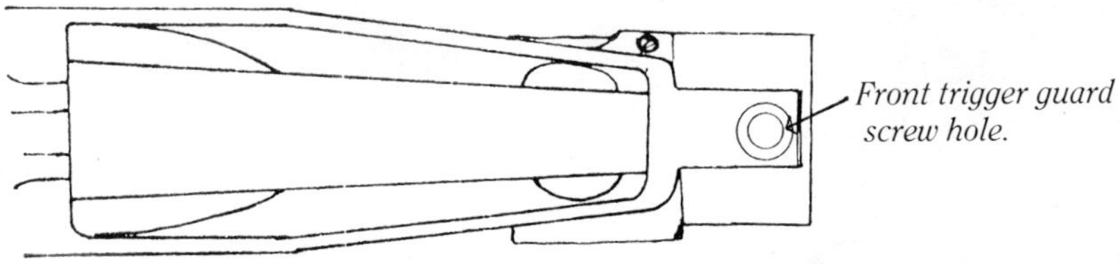

Fig. 2
Modified magazine well for 7.62mm.

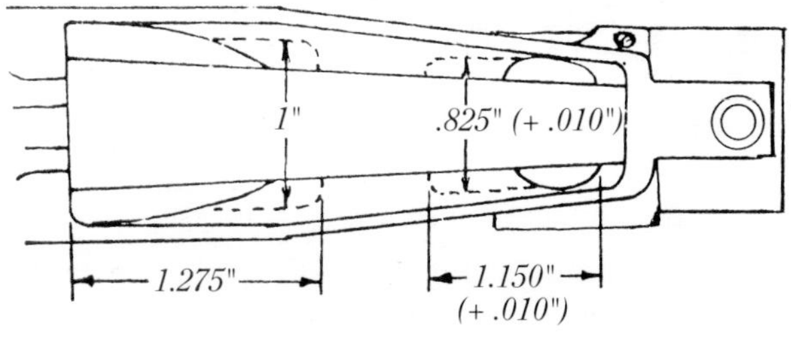

Showing recess machined .075" deep at front to allow clearance for 7.62mm magazine lips.

If difficulty is experienced with bolt not picking up top round in magazine, modify rear of well also, to dimensions shown. *Courtesy of Peter Laidler*

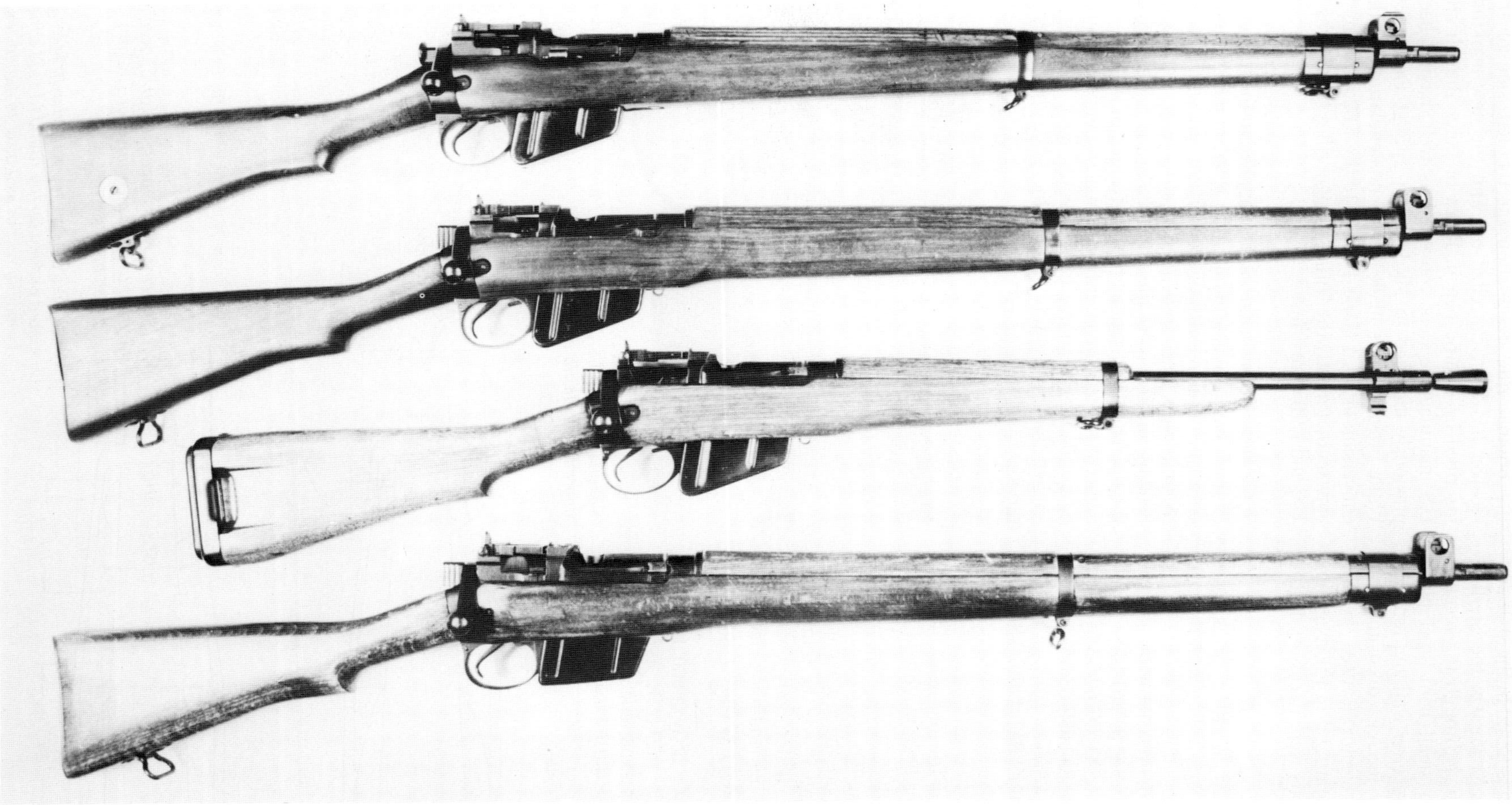

From top: ***Enfield-made Trials No. 4 Mk I Rifle of the 1930's***
British Production No. 4 Mk I of the World War 2 period
Rifle No. 5 Mk I, the "Jungle Carbine"
7.62mm L8 Series (note the different magazine)

MOD Pattern Room

Typical receiver markings are:

No.4 Mk2 RIFLE 7.62MM L8A1
UF56 A 44816 (CR 60 GA)

No4 MK1/2(F)FTR RIFLE 7.62MM L8A2
/52 Y21698 (CR 61 GA)

/51 56L5563 No4 MK I*/3(F)FTR RIFLE 7.62MM L8A3
LONG BRANCH (CR 62 GA)
1943

The new 7.62mm barrel rifling has a right hand twist compared with the previous .303-in. left hand twist and in 1961, 6-groove rifling was used experimentally for the 7.62mm barrel. A small number of 7.62mm barrels were also ordered for the No. 1 and No. 3 rifles in late 1959 and produced by mid-1960 for test purposes. Many of the 7.62mm barrels for L8 rifles are of Canadian manufacture with the Canadian Arsenals logo stamped on the barrel; some have been noted without the usual bayonet attachment lugs or studs and/or 4-groove, left hand twist rifling.

After the introduction of the 7.62mm conversion for No. 4 service rifles, some sniper models were also altered for the new cartridge. At first, these were intended for service in Ireland but they are reported as never having been issued as such. Like the L8 series, there is no external difference to its predecessor except for the magazine shape; one of these rifles has been examined, tagged "Rifle No. 4 Mk I T converted to Rifle 7.62mm L8." In some other instances these have been merely referred to as the "L8 (T)." The wooden case, telescope and accessories were those of the No. 4 sniper and the markings on the No. 32 Mk 3 scope also remained unaltered.

Six of these 7.62mm L8 (T) rifles are recorded as being converted during the 1960's but they proved to be most unsatisfactory. In shooting, it was found that the M.P.I. was inconsistent and the group sizes were larger than was acceptable. In fact, the wartime .303 No. 4 (T) model over iron sights (i.e. with telescope removed) performed better than the L8 (T) with telescopic sight fitted!

Some other No. 4 (T) rifles were similarly converted to the new 7.62mm specifications although comparable failures were recorded, generally relating to the accuracy. The example at the M.O.D. Pattern Room was engraved "RIFLE No.4.Mk.I.(T)" over "7.62.CONV.E 68" (the "E" being the Enfield logo which incorporates the E, F and D letters). Serial numbered N33975, it is not one of the six previous-mentioned conversions and the markings are also different to other examples examined.

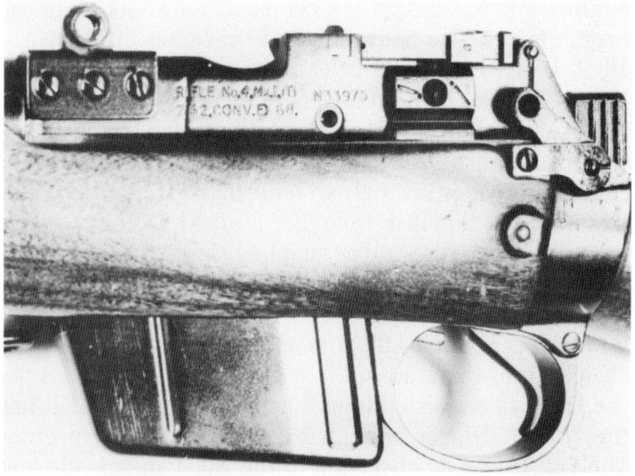

An Early Prototype 7.62mm Conversion of the No. 4 Mk I (T)
MOD Pattern Room

This particular new 7.62mm barrel has a differently cut knox form, with vertical flats on each side, probably to assist in the breeching up. A sample barrel examined also had hardness test marks on the top of the knox form. The early 7.62mm box magazine inspected is also special in the markings, a serial number "8" on the rib and the inscribed Enfield Design Department drawing number "DD(E) 20454/408"; there are hardness test marks near the top. Apart from the squared 7.62mm magazine and special barrel, this rifle is very similar to its .303 No. 4 Mk 1 (T) progenitor.

A considerable quantity of No. 4 (T) rifles must have been converted to 7.62mm in addition to the ensuing half-stocked L42A1 models because some have since filtered down through the supply and ordnance system amongst the series of L8 rifles, many without telescopic sights or brackets. Some are engraved "L8" without the usual "A" and number suffix, the action markings indicate that the L8 conversions were effected at Enfield in 1970-71. Others have been noted marked "L8A2T," presumably denoting a sniping variant of the original conversion.

Because the original L8 conversion of the No. 4 (T) was unsuitable for sniper application and as the L1A1 was not satisfactory for such a role, a program was employed to develop a suitable sniper rifle from the original No. 4 model. The end result was the L39 and L42 rifles, which were quite unlike any previous Lee-Enfield service model, except perhaps for some superficial resemblance to the .22 No. 8 trainer in the form of stocking-up.

A 7.62mm experimental half-stocked sniper rifle in the M.O.D. Nottingham Pattern Room collection is illustrated on the previous page, considered to be a prototype of the eventual L42A1 series. An original No. 4 Mk I (T), the receiver markings are "RIFLE No. 4 Mk I (T)" over "7.62mm Conv. Ð 68." A heavy profile barrel was used and the new furniture is half-stocked, reducing the revised all-up weight a little and free-floating the barrel.

Primarily, the L39A1 is an L42A1 without telescope and mounts, and was designed for competition shooting in the British forces. The body mounted trigger, as on the No. 4 Mk 2, is used; the L39 rifles were converted from No. 4 Mk 2 and Mk 1/2 receivers. There is no 7.62mm magazine, only the original .303-in. pattern which serves as a loading platform. The L39 was intended for use with competition aperture sights, so no sights were fitted onto this rifle as supplied from the Enfield factory. It was intended that these be supplied and fitted by the particular service units.

Whilst the 7.62mm L39A1 is similar to the L42A1 in its "bare" form, there are quite a few subtle differences. As the L39 was converted from the No. 4 Mk 2 or 1/2, it has the "hung" trigger, mounted directly onto the receiver. The L42A1 was a conversion of the No. 4 sniper rifle and therefore the trigger is attached to the triggerguard; both No. 4 Marks I (T) and I* (T) were converted. Because of the difference in trigger mounting, the back of the fore-end renders the L39 and L42 non-interchangeable in this respect.

The trigger pressures on the L39 are lighter than those of the L42, at a first pressure of 2 – 3 lb. and second pressure of 4 – 4.5 lb. In order to obtain the lighter trigger pressures, the angle of the face of the cocking piece has been altered slightly. Additionally, a second recess was cut into the magazine catch to allow the short arm of the sear spring to be seated in a lower position, in case the trigger pressures were still too heavy.

The Enfield "Envoy" is a commercial model of the L39 fabricated at the R.S.A.F. and fitted with a 10-round magazine similar to the L8 and L42 rifles. General finish of this rifle is better and more highly polished, as expected for a commercial rifle; the fore-end is wider underneath than that of the L39 and it is fitted with the No. 8 outer band or front band which mounts the sling swivel.

7.62mm Envoy Target Rifle with Micrometer Adjustment Backsight *MOD Pattern Room*

The "CENTAR" was another by-product of the L39A1 and it was originally intended for use by cadets. It is identical to the service L39 except for the fitting of the sporting style butt, being that of the .22RF No. 8 rifle. The Centar

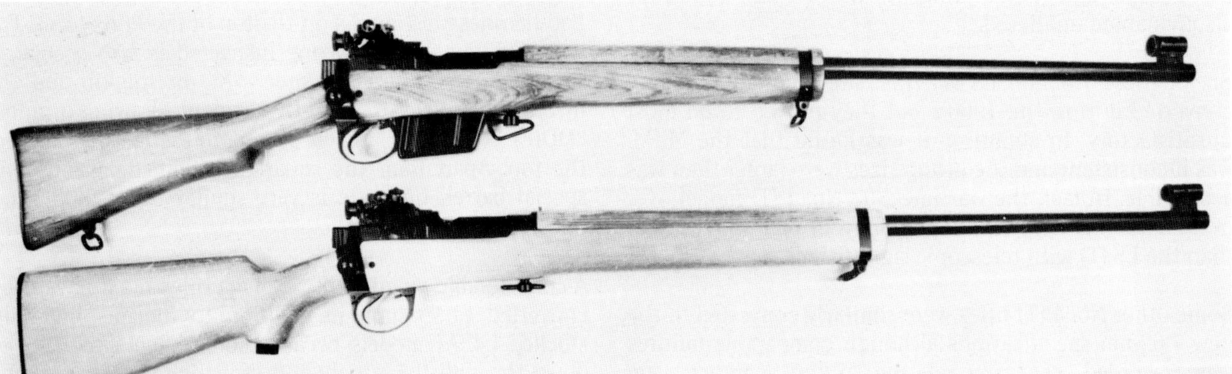

7.62mm Models in Target Rifle Configuration. These are often seen fitted with the original .303 magazine which serves as a single-shot loading platform for the cartridge. Envoy model is at bottom. *MOD Pattern Room*

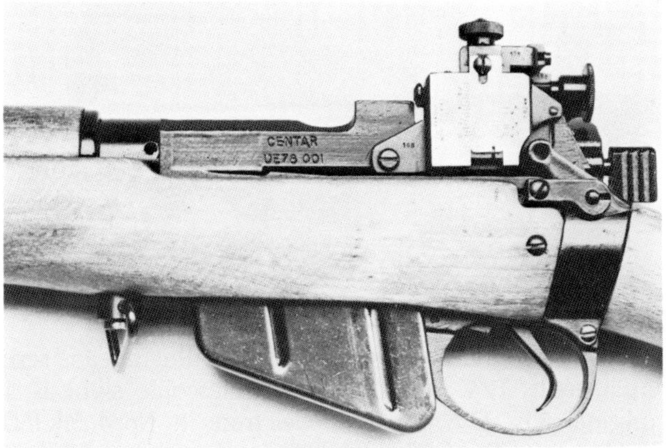

A 7.62mm Centar model with .303 magazine.
MOD Pattern Room

is fitted with an A.J. Parker backsight and a Parker "Matchmaker" tunnel foresight assembly. However, the 7.62mm Centar was not adopted and only a few prototypes were assembled at R.S.A.F. Enfield in 1978.

A prototype 7.62mm No. 4 sniper rifle was designated and marked "XL42E1," it only seems to differ from its production counterpart in the action markings on the left side of the converted No. 4 receiver. The R.S.A.F. Enfield production logo and year stamped below this XL designation indicate that the conversion/remanufacture was done in 1970, also curious because the initial L42A1 rifle conversions appear to have been commenced in 1969.

Enfield 7.62mm Sniper Model Prototype.
MOD Pattern Room

Nearly 82 years after the first .303-inch magazine Lee rifle was introduced, the 7.62mm L42A1 sniper rifle was approved on 24th August 1970, and it remains in use in a combat role with British forces today. This model incorporates the special, heavy, chrome-moly barrel with 4-groove, right-hand twist rifling, which is largely responsible for the all-up weight of nearly 12.5 lb., about 1 lb. heavier than the No. 4 sniper. It is half-stocked with a wider than normal fore-end for improved grip and steadiness, and the handguard and fore-end finish just beyond the position of the outer band, about half-way along the barrel.

This sniper rifle was set up at R.S.A.F. Enfield and the actions used were mostly conversions from existing No. 4 Mk I (T) models, although it appears that a few No. 4 Mk I* (T) models were originally included as well. This means that, unlike the L39A1, the trigger is hinged on the triggerguard, as on the original No. 4 series.

On the converted action bodies, the earlier rifle designation markings were generally machined off from the left wall of the receiver and replaced by the new nomenclature, factory and year, although the original markings can still be deciphered on some of the L42A1's. One such example is a 1931 No. 4 trials rifle converted to No. 4 Mk I (T) at Enfield in 1940 and converted again at Enfield to an L42A1 in 1971, still in service some 60 years after its original production date!

Although the original foresight assembly of the No. 4 sniper rifle is retained and the Mk I/I backsight is fitted, the datum line on the slide has been lowered by .07-in. and an "M" marked on the right side, indicating its metric calibration. The original yard graduations are retained, to be read as metres, and so the graduations are ranged to 1,300 metres. The fixed battle sight was removed from all the folding No. 4 sniper backsights in the original conversion because it came up against the telescopic sight eyepiece.

The optics comprise a slightly modified No. 32 Mk 3 telescope with the original markings cancelled out and the following code, designation and part numbers inscribed:

1240 – 99 – 963 – 2063
TEL. STRT. STG. L1A1
OS 2429 GA

To further emphasise the new scope markings, the original designations were painted over and the new ones highlighted in white, in a similar fashion to the graduations on range and windage drums. The modifications merely consist of the insertion of a small neoprene "O" ring for improved sealing of the range and windage drums, and the metric range drum. A few No. 32 Mk 2/1 telescopic sights were also converted for this role.

A small number of war reserve L8 rifles have been noted recently converted from original No. 4 sniper rifles, without telescopic sights or upper mounts, but retaining the original scope mount pads. Such a rifle is marked "RIFLE 7.62mm L8A2 (T)" over "(CR 61 GA) Đ71" and the serial number, the "Đ" being the Enfield factory logo. Some of the L8 refits have L42 type backsights with the "M" marking, others have No. 4 backsights, and some have magazine inserts. Such a situation is obviously due to shortages of some spare parts, when the 7.62mm No. 4 action rifles have been refurbished for war reserve storage.

A more recent specialist rifle in this 7.62mm series is the "Enforcer" which was set up for police use. Although it is not a military issue, it is of interest because of its lineage

Above & below: Enfield 7.62mm Enforcer has non-detachable Pecar scope, used by Police Forces. *MOD Pattern Room*

and it comprises an L39A1 rifle fitted with Pecar variable 4 – 10 power telescope and a sporting butt that incorporates a cheekrest, pistolgrip and rubber composition recoil pad.

Some cartridge tests in the late 1960's utilised L42 and L39 Lee-Enfield rifles as "test-beds" in evaluation trials. This particular 6.25 x 49mm cartridge was a necked-down version of the earlier .280/30 round which was associated with the EM2 rifle. The 20-rd box magazines are special to this group and consist of a 6-in. long box magazine welded onto the rear portion of the original .303 No. 4 magazine. The magazines are stamped "6.25mm" on the back.

These trials progressed through 1968 to 1971 and the rifles illustrated here are marked with the serial number prefix "T/V" along with the Enfield Design Department drawing number, DD(E) 25768-SK-464. Other particular features relating to the new barrels also differ from their forebears.

The 6.25mm L42A1-configuration rifle illustrated is serial numbered "T/V No. 1" and the telescopic sight is an original A.K. & S. Mk III model from its No. 4 Mk I (T) predecessor. The 6.25mm L39A1 version shown here was converted from a No. 4 Mk 2 (like the 7.62mm L39A1 rifle) and is serial number "T/V No. 3."

So, more than a century after the adoption of Lee's bolt action rifle into Her Majesty's Service, the basic design is still in commission today in a variety of forms, in Great Britain, India and some third world countries, testament to one of the best all-round military rifles of all time.

Only in April 1992 did the L42A1 become obsolescent.

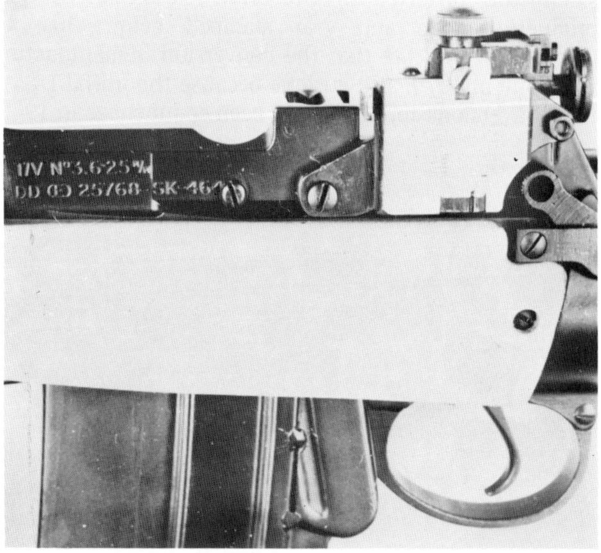

Above & below: 6.25mm Test Model with narrower magazine fitted to back of the .303 version. *MOD Pattern Room*

SMALL BORE TRAINERS

CHAPTER 7

Morris tubes in .23-in. calibre fitted inside the .450-in. Martini-Henry rifle and carbine barrels represent the first of the service small bore training rifles; initially approved on 14th March 1883. They were intended for short range practice or instruction in aiming and consist of a part-rifled tube which was placed into the bore and screwed into a threaded chamber which had been inserted into the rifle or carbine breech. With this assembly screwed together, a miniature round (initially a .297/230 cartridge) could be fired. Specific details of the first Morris tube will be found in the List of Changes reference, § 4452, (Vol. 1, 1860-1886).

An aiming tube was approved for the Lee-Metford rifle on 29th December 1891 and the announcement published in the List of Changes:

Para 6602:-
Tube, Aiming, Lee-Metford Magazine Rifle, .303-inch. Morris.

A pattern of the above-mentioned tube has been sealed to govern manufacture.

It consists of a steel tube (A) rifled throughout with eight grooves. The breech piece (B) is solid, and concentric with the tube, and is chambered to suit the "cartridge, aiming tube" (§ 6391). The sliding extractor (C) is of steel and is attached to the tube by means of a small steel stud, working in a cam slot, which gives a revolving motion to the sliding extractor. The set nut (D) and washer (E) are of gun metal, and the washer (F) is of leather, they are fitted to the muzzle end of the tube for the purpose of fixing the tube in the rifle.

As the calibre of the tube is the same as that for the M.H. rifle, the same description of "brush, cleaning" (§ 4452) will be used with it.

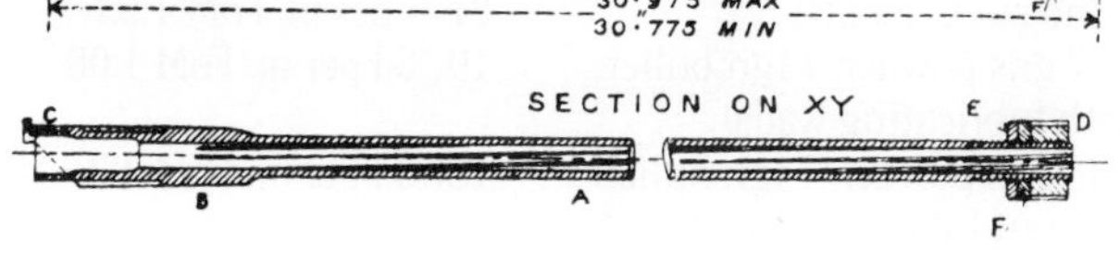

Instructions for Use

Draw the bolt fully to the rear, turn back the bolt head, and remove the bolt; unscrew nut, and remove gunmetal and leather washers from end of tube; insert the tube into the barrel at the breech end, placing it in such a position that the projection on the end of the sliding extractor coincides with the extractor way in the barrel; replace washers and nut, and screw up tightly with thumb and finger, replace bolt, and turn down bolt head. If the tube is fixed correctly in the barrel, the extractor should leave the sliding extractor after withdrawing it about 7/16th of an inch from the face of the chamber. Neglect of care in fixing the tube in the proper position would probably result in the tube being rendered unserviceable.

To remove the tube from the barrel, reverse the above operations, using the head of the cleaning rod belonging to the rifle to push the tube out of the barrel. The tube should be pushed in level with the muzzle of the barrel before the rod is used.

The tube for this .303-in. model is a little different to the Martini-Henry pattern because the tube and the breech/chamber section are all in one piece, which permits the insertion of the complete assembly from the breech end of the rifle with the bolt removed. This was not possible with the Martini due to the arrangement of the receiver and the breech block. On the Martini tubes, the tightening nut at the muzzle could not be moved whereas the tubes for the Lee-Metford and Lee-Enfield rifles and carbines require the removal of this brass end nut so as to insert or remove the Morris tube.

Not long after the Lee-Metford carbine came into service, an aiming tube was also approved, similar to the Lee-Metford rifle tube, but correspondingly shorter at 21.3-ins long. This was approved on 21st March 1895 and announced in the List of Changes § 7779.

The sub-calibre cartridge was commercially known as the .297/230 but its miliary designation was "Cartridge, Aiming Tube, Morris's" and it was advanced through a number of Marks. At first it was introduced for naval use only but later extended to the Land Service. The lead bullet weighs about 37 grains with a propellant charge of 3.25 grains of Curtis & Harvey's No. 2 Diamond or its equivalent. The cartridge is of drawn brass and it is a Berdan-primed, centrefire round.

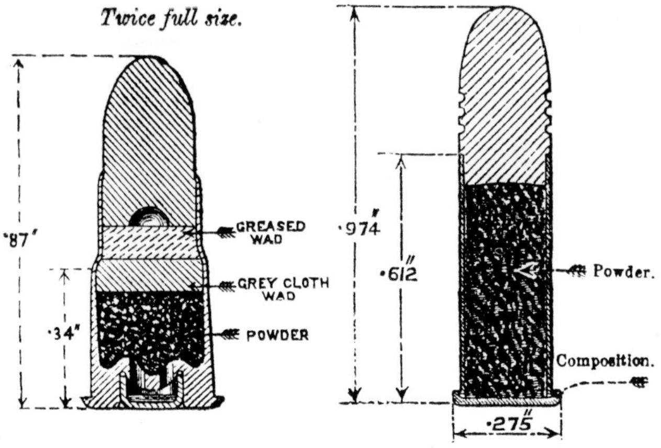

Left: .230 Aiming Tube cartridge, § 6391, 27/11/1890.
Right: .22 RF Aiming Tube cartridge Mk I, § 13642, 29/11/1904 & 6/7/1905.

No changes were required for the Morris sub-calibre tubes after the introduction of the Lee-Enfield rifle and carbine as there were no alterations to the barrel lengths or the chamber from the Lee-Metford configuration. Arms with worn rifling were usually allocated for use with the Morris tube, these were stamped "M.T.." Aiming tubes were later used with the S.M.LE. rifles, as well as some Lee-Enfield and Lee-Metford rifles and carbines; these are described later in this chapter. However, they utilised the .22 rimfire cartridge as they were not announced until 1907.

Some other sub-calibre tubes and training systems were recommended to the Small Arms Committee for consideration between 1900 and 1903; these included the Richards' system, Wilkinson Practice Cartridge Syndicate Adaptor, the Canadian Gaudet system, Ewart re-capped cartridge and adaptor, Andrews adaptor, Eley adaptor and four Kynoch systems of re-capped cartridges and adaptors. Comprehensive tests were conducted, especially with the Kynoch and Gaudet systems in M.L.E. and S.M.L.E. rifles.

Tests on these various systems were carried out in 1903 at Hythe and it was found that Kynoch's system B was the most accurate, followed by the Gaudet. The Kynoch B and Gaudet cartridges were shot in both the Mark I* Lee-Enfield and Mark I S.M.L.E. rifles and although the Kynoch ammunition was superior in accuracy, it was considered that the trajectory was too flat and had excessive range and penetration for use on miniature ranges. The Belgian system of miniature ammunition and a Swedish system invented by Capt. Nardrup were tried also. Although the Kynoch and Gaudet ammunition were both considered superior to the Morris system in a number of respects, the higher cost rendered them as unsuitable proposals.

Details of these quite extensive 1903 tests are interesting as comparisons were drawn between the various tubes, adaptors and re-capped cartridges. Some systems, such as the Morris tubes, Andrews adaptors and the Kynoch re-capped cartridges, each presented as many as five variations or combinations.

System / Design	*Comments*
Morris Aiming Tube & Ammunition Co.:-	
Early Aiming Tube	Frequent misfires & extraction failures.
	17/6d per m. FoM 1.69
	Cost of tube 18/6d.
Later Adaptors-	Ammunition extremely difficult to insert in adaptor.
System A (long bullet). FoM .65	
System A (short bullet). FoM .85	
System B (long bullet). FoM .71	Adaptor & cartridge had to be oiled before insertion.
System B (short bullet). FoM .79	
System C (long bullet). FoM .72	
System C (short bullet). FoM .88	
Two adaptors split.	
Recapped cartridges-	One long bullet stuck in the barrel after firing.
System D (long bullet). FoM 1.30	
System D (short bullet). FoM 1.81	
Adaptor-	
System E	FoM 1.4
Messrs. Kynoch Ltd.:-	
Re-capped cartridges-	
System A	65/- per m. FoM 1.00
System B (fired cases)	48/9d per m. FoM .40
System B (new cases)	64/4d per m. FoM .40
System C	75/- per m. FoM .58
Adaptor-	
System D	76/- per m. FoM .79
Mr. Ewart:-	
Adaptor No. 1	3 cases split.
	35/- per m. FoM .79
Re-capped cartridges	Difficulty in loading as cartridges were of different length.
	75/- per m. FoM .51
	Excessive penetration.
Major Richards:-	
Modified Adaptors-	Some split cases.
10 gns powder	7/6d per m. FoM 2.83
8 gns powder	7/6d per m. FoM 3.23
6 gns powder	7/6d per m. FoM 2.56
Mr. Andrews:-	
Adaptors-	Split cases & blow-backs, types A & B.
A (4 gns powder, 93gn bullet)	26/- per m. FoM 1.36
B (no such details)	27/- per m. FoM 1.53
C (7 gns powder, 74gn bullet with lubricating wads)	19/6d per m. FoM 1.00
D (8 gns powder, 74gn bullet without lubricating wads)	18/6d per m. FoM .83

Messrs. Eley:-

Adaptors-	Cases bulged, removal from adaptor hard.
Mark I	No price. FoM .64
Mark II	13/3 per m. FoM 1.36

Messrs. Wilkinson & Co.:-

Adaptors-	Cases bulged, difficult extraction. Some cases & one adaptor split. Strikers pierced two caps.
Up to 100 yds, 4 gn powder	25/6d per m. FoM 1.45
Up to 200 yds, 8 gn powder	25/6d per m. FoM 1.49

Major Gaudet:-

Re-capped cartridges-	
Using fired cases	42/3d per m.
Using new cases	58/3d per m.

[FoM = Figure of merit, grouping accuracy.
Cost per m. = cost per 1,000 cartridges.]

The introduction of the .22 rimfire cartridge into service, approved on 29th November 1904 and described and illustrated in the List of Changes § 13642, resulted in some changes to the sub-calibre training rifle situation; the Morris tube cartridge nomenclature was amended to include "C.F." for central fire (centre-fire) as opposed to rimfire. These were black powder cartridges and due to the introduction of the rimfire cartridge, a new bolthead was required to allow for the offset position of the rimfire detonating composition at the back of the case.

Implementation of the new .22 rimfire aiming tube system changed the previous concept for these sub-calibre training devices. With the Morris tubes, the "host" rifles and carbines could be used with service .303-in. ammunition by merely withdrawing and removing the tube and its fittings. The Aiming Tube however was a more permanent conversion and rifles so fitted could not be converted as easily because of the different bolt and bolthead combination. One thousand of these Aiming Tube short rifles were issued for trial in 1905-06, which resulted in the adoption of the system, although these differed slightly from the pattern that was eventually approved for adoption.

The R.S.A.F. Enfield Pattern Book lists a .23-in. Aiming Tube S.M.L.E. Mark I rifle sealed on 31st December 1904. Although no such specimen has been noted, it is considered that it was similar to the later .22-in. Aiming Tube S.M.L.E. rifle but for the different chamber and a centrefire bolthead.

The various .22-in. rimfire Aiming Tubes were first approved for the S.M.L.E., Lee-Enfield and Lee-Metford rifles in 1904, consisting of a newly-manufactured pattern as well as a conversion from the .230-in. Morris tube. Although the date of approval was 29th November 1904, patterns of these trainers were not sealed until 21st December 1906, so it is unlikely that there was any significant production until the sealed patterns were authorised. Special bolts were introduced for use with the .22-in. R.F. cartridge and a description is incorporated in the original List of Changes declaration.

Para 13650:-

Tube, Aiming, .22-inch, M.L.M. rifle, converted.
Also Rifles, M.L.E., for R.F. cartridges; converted from Tube, aiming, M.L.M. rifle, Morris; including extractor and screw, set nut and washers.

Tube, Aiming, .22-inch, M.L.M. rifle (Mark I).
Also Rifles, M.L.E., for R.F. cartridges; including extractor and screw, set nut and washers.

Tube, Aiming, .22-inch, M.L.E. rifle, short (Mark I).
For R.F. cartridges; including extractor and screw, set nut and washers.

Bolt, Breech, M.L.M. rifle, Mark I; for R.F. aiming tube cartridges.*
Assembled; converted from Bolt, breech, M.L.M. rifle, Mark I.*
Bolt, Breech, M.L.M. rifle, Mark II; for R.F. aiming tube cartridges.
Assembled; converted from Bolt, breech, M.L.M. rifle, Mark II.

Bolt, Breech, M.L.M. rifle, Mark II; for R.F. aiming rifle, Mark II*.*
Also Rifles, M.L.E.; assembled; converted from Bolt, breech, M.L.M. rifle, Mark II.*

Bolt, Breech, M.L.E. rifle, short; for R.F. aiming tube cartridges.
Assembled; converted from Bolt, breech, M.L.E. rifle, short.

Patterns of the above-mentioned aiming tubes and breech bolts have been sealed to govern manufacture and conversion.

Tubes.

The new tubes are similar to the tubes described in LoC §§ 6602 and 6723, but differ in the size of the chamber and extractor which are arranged for the .22-inch rimfire cartridge.

The conversion is made from the "Tubes, aiming, M.L.M. rifle, .303-inch, Morris" described in §§ 6602 and 6723, and consists in fitting a bush to the chamber, and fitting a new extractor, suitable for the .22-inch rimfire cartridge.

Length overall of Tubes, aiming, .22-inch M.L.M. rifle, converted, Mark I ... ... 30.94 inches.

Length overall of Tubes, aiming, .22-inch M.L.E. rifle, short, Mark I ... ... 26.00 inches.

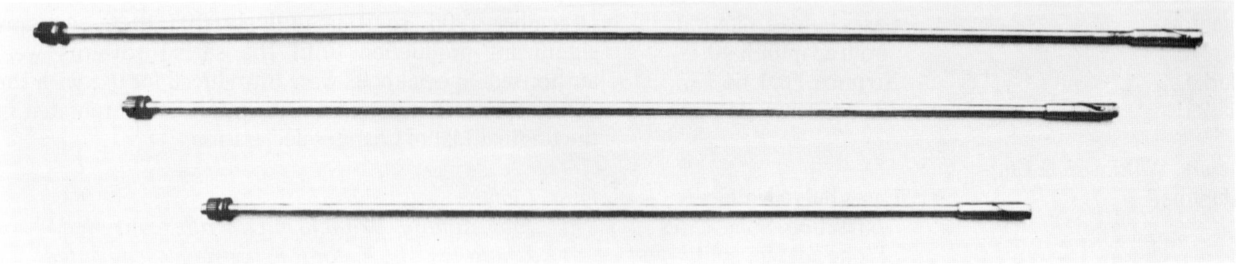

Morris Aiming Tubes
From top: for M.L.M. & M.L.E. rifles, S.M.L.E., and M.L.M. & M.L.E. carbines.

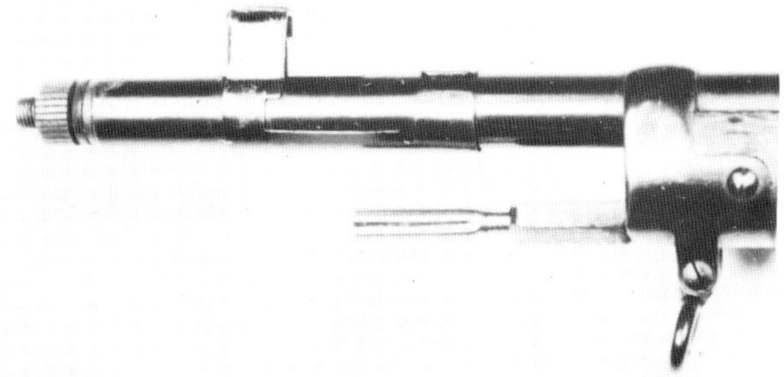

Muzzle section of Morris tube with gun-metal set nut and washer, fitted to M.L.E. rifle. MOD Pattern Room

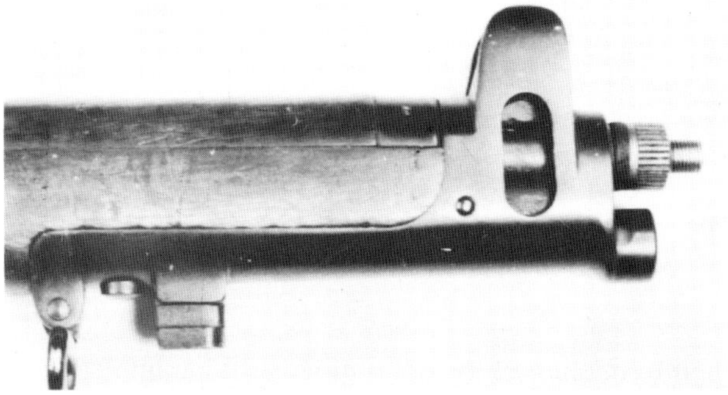

Muzzle section of Morris tube fitted to a rifle of the Mark I S.M.L.E. series. MOD Pattern Room

Modified bolthead and extractor for early model MLM or MLE .22 Rimfire trainer. MOD Pattern Room

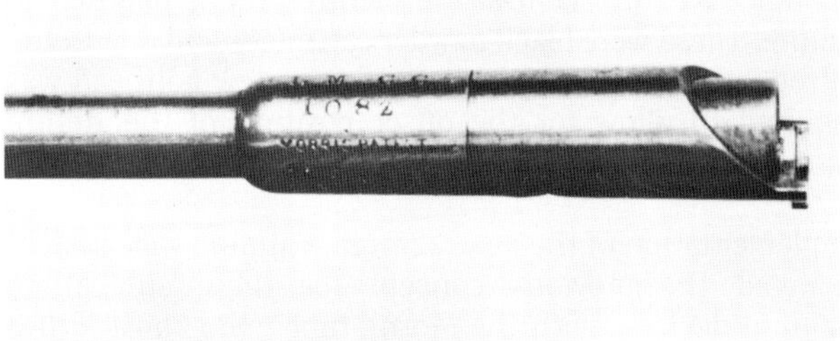

Breech section of Morris Aiming Tube stamped "L M C C 1082 Morris Patent" (for MLM Cavalry Carbine).

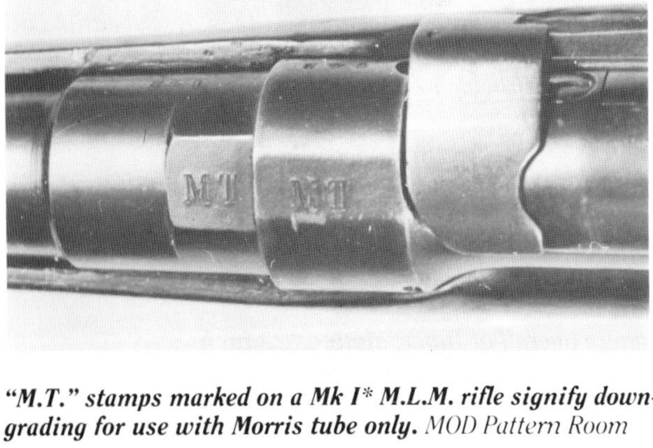

"M.T." stamps marked on a Mk I* M.L.M. rifle signify downgrading for use with Morris tube only. MOD Pattern Room

"A.T." stamp on this Mk II* M.L.M. rifle indicates a similar downgrading for use only with the Aiming Tube.

The instructions for use given in LoC § 6602 will apply to these tubes, with the exception of that part of the last paragraph referring to the cleaning rod. The cleaning rod and brush, and packing washers described in LoC §§ 7880 and 9178 are suitable for these tubes.

These tubes can only be used in rifles fitted with a special bolt adapted for rimfire ammunition. The removable components of these tubes are interchangeable.

Bolts, breech.

The Bolt, Breech, Rifle, M.L.M. Mark I, for .22-inch R.F. aiming tube cartridge is converted from Bolt, Breech, rifle, M.L.M., Mark I*. The conversion consists in removing the cover stops and a portion of the rib, plugging the bolt cover and bolt-head screw holes, and fitting and brazing in a screwed bush to receive a new bolt-head, which is arranged to receive a new striker, the point of which is placed .1-inch eccentric to the axis, suitable for .22-inch rimfire cartridges. The extractor is converted to clear the new striker.*

The Bolts, Breech, Rifles, M.L.M., Marks II and II, and Rifle, Short, M.L.E., for .22-inch R.F. aiming-tube cartridges are converted from Bolts, Breech, Rifles, M.L.M. Marks II and II*, and Rifle, Short, M.L.E., respectively. The conversion consists in altering the extractors to clear the new strikers and fitting new bolt-heads and strikers suitable for .22-inch rimfire cartridges.*

The bolts should not be stripped by means of the bolt-head. The bolts will be marked "A.T." (aiming tube) on the top of the lever.

The following table shows the interchangeability of the components of the breech bolts:

COMPONENT	DETAIL
Bolt, breech, M.L.M. rifle-	
Mark I* R.F. ...	Special
Mark II R.F. ...	Special
Mark II* R.F. ...	Interchangeable in rifles, M.L.M., Mark II*, and M.L.E., Marks I and I*.
Bolt, breech, M.L.E. rifle, short, Mark I. R.F. ...	Interchangeable in all short rifles.
Heads, breech bolt, M.L.M. rifle, Mark I*. R.F. ...	Interchangeable in all R.F. bolts, breech.
Extractor, M.L.M. rifle,	
Mark I*. R.F. ...	Special.
Mark II. R.F. ...	Interchangeable in rifles, M.L.M., Marks II and II*, and M.L.E., Marks I and I*.
Extractor, M.L.E. rifle, short, Mark I. R.F. ...	Interchangeable in all short rifles.
Strikers, M.L.M. rifle,	
Mark I*. R.F. ...	Interchangeable in rifles, M.L.M., Marks I* and II, and rifles, short, M.L.E.
Mark II*. R.F. ...	Interchangeable in rifles, M.L.M., Mark II*, and M.L.E. Marks I and I*.
Springs-	
Main, M.L.M. ...	Interchangeable in all.
Extractor, M.L.M. ...	Interchangeable in all.
Screws, extractor, M.L.M. ...	Interchangeable in all.
Cocking-piece, M.L.M. rifle,	
Mark I*. ...	Special.
Mark II. ...	Special.
Mark II*. ...	Interchangeable in rifles, M.L.M. Mark II*, and M.L.E. Marks I and I*.
Cocking piece, M.L.E. rifle, short, Mark I. ...	Interchangeable in all short rifles.
Screw, keeper, striker, M.L.M. rifle-	
Mark I*. ...	Interchangeable in all rifles, M.L.M., Marks I* and II.
Mark II. ...	Interchangeable in all rifles, M.L.M. Marks II*, and M.L.E. Marks I and I*.
Screw, nut, keeper, striker, M.L.E. rifle, short, Mark I.	Interchangeable in all short rifles.
Nut, keeper, striker, M.L.E. rifle, short, Mark I. ...	Interchangeable in all short rifles.
Nut, spring, keeper, striker, M.L.E. rifle, short, Mark I.	Interchangeable in all short rifles.
Cover, breech, bolt, M.L.M. rifle, Mark II.	Interchangeable in rifles, M.L.M., Marks II and II*, and M.L.E., Marks I and I*.
Catch, safety, M.L.M. rifle, Mark II*.	Interchangeable in rifles, M.L.M., Mark II and M.L.E., Marks I and I*.
Spring, pin, catch, safety, M.L.M. rifle, Mark II*. ...	Interchangeable in rifles, M.L.M., Mark II, and M.L.E., Marks I and I*.
Pin, catch, safety, M.L.M. rifle, Mark II*. ...	Interchangeable in rifles, M.L.M., Mark II, and M.L.E., Marks I and I*.

The .22-in. Aiming Tube S.M.L.E. rifle was approved on 13th November 1906, while some of the 1,000 trials short rifles were also adopted for conversion to the same pattern on 1st February 1907, differentiated from the earlier model by the designation "N.I." for non-interchangeable. The announcement in the List of Changes was:

Para 13649:–

Rifles, Short, M.L.E., A.T. (Mark I).
For .22-inch rimfire cartridge aiming tube and for drill purposes.

Rifles, Short, M.L.E., A.T./N.I. (Mark I).
For .22-inch rimfire cartridge aiming tube; non-interchangeable; part of 1,000 experimental Rifles, Short, M.L.E., that were issued for trial.

1. In future, in the Land Service, only one class of Rifle, Short, M.L.E. will be issued for aiming tube practice and drill purposes.

These rifles will be fitted with boltheads and strikers suitable for .22-inch rimfire aiming-tube cartridges; the striker hole in the face of the bolthead being bored .1-inch eccentric to the axis of the bolt, the point of the striker being arranged suitably, and the rear end of the extractor being altered to clear the special striker.

The barrels of these rifles will be so chambered that Service ball or blank ammunition cannot be fired in them. These rifles will be marked A.T. (aiming tube) on the body, barrel, bolt and stocks, butt and fore-end, and classified as such. The components of these rifles (with the exception of the barrel, bolthead, striker and extractor) will be interchangeable according to the "Mark" of the rifle from which they are prepared.

2. Certain of the rifles are not suitable for instruction in the parts of the rifle, as the components (particularly the sights) differ from the Service pattern. Two patterns of windgauge-backsights are fitted to these rifles, and are placed in two positions.

With the exception of a few small components, such as screws, &c., and the components mentioned below, the components of these rifles are non-interchangeable, and will not be replaced as they become unserviceable. The rifles will, consequently, not be repaired (except for such repairs as can be effected locally without the supply of new components), except in the case of the following, which are interchangeable:

> *Heads, breech bolt, M.L.M.R., Mark I*, R.F.*
> *Strikers, M.L.M.R., Mark I*, R.F.*
> *Extractors, M.L.E.R.S., Mark I, R.F.*

These rifles will be marked A.T./N.I. (Aiming Tube, non-interchangeable) on the body, barrel, bolt, and stocks, butt and fore-end, and classified as such. The boltheads, strikers and extractors of the above rifles (A.T. and A.T./ N.I.) are interchangeable.

N.B.: The strikers of these rifles should not be stripped by means of the bolthead.

These Non-Interchangeable Aiming Tube rifles were converted from the 1902 Short Lee-Enfield trials rifles, hence the reason for their scarcity in original condition today. This was also the reason for their non-interchangeability, and why components could not be replaced after they became worn or broken, because they were unique to these 1902 trials rifles and differed in various respects.

A total of 937 S.M.L.E. rifles were adapted for the Wake Aiming Tube in 1907; converted and issued on an experimental basis. These are appear to have been the non-interchangeable pattern made up from the 1902 S.M.L.E. trials rifles.

A modification to the extractor on 13th November 1906 resulted in the advancement of the Mark I .22-inch tube for S.M.L.E., to Mark I*, as indicated in the List of Changes:

Para 13651:-

> *Tube, Aiming, .22-inch, M.L.E. rifle, short (Mark I*).*
> *For rimfire cartridges; including extractor and screw, set nut and washers.*

A pattern of the above-mentioned tube has been sealed to govern future manufacture.

It differs from the Mark I tube described in § 13650 in the following particulars:

The cartridge-head recess of the extractor is continued towards the centre, so that it grips nearly one-half the circumference of the cartridge head, and thus improves the extraction of the empty cartridge case.

The rim of the extractor has been extended to the left, so that when the right side of the rim of the extractor becomes thin by wear, the worn portion of the rim can be filed away by an armourer, and the unworn portion brought into position when the tube is adjusted in the rifle.

But for the sealed pattern, no Mark I .22-in. Aiming Tubes were made for the Lee-Metford rifle as per LoC § 13650, so the pattern was resealed including the altered extractor as above. Approved 23rd May 1907, the converted Morris tube was resealed as the Mark II aiming tube, incorporating the modified extractor.

Para 13856:-

> *Tube, Aiming, .22-inch, M.L.M. rifle (Mark I).*
> *Also Rifles, M.L.E.; for R.F. cartridges, including extractor and screw, set nut and washers.*
>
> *Tube, Aiming, .22-inch, M.L.M. rifle (Mark II).*
> *Also Rifles, M.L.E.; for R.F. cartridges; converted from Tube, Aiming, M.L.M. rifle, Morris; including extractor and screw, set nut and washers.*

Patterns of the above-mentioned tubes have been approved to govern future manufacture and conversion. None of the Mark I tubes described in § 13560 have been made; the Mark I tube described herein has been sealed in substitution, and, consequently, no advance in numeral has been made.

The tubes differ from the converted and Mark I tubes described in § 13650 in the following particulars:

The cartridge head recess of the extractor is continued towards the centre, so that it grips nearly one-half of the circumference of the cartridge head, and thus improves the extraction of the empty cartridge case.

The rim of the extractor has also been extended to the left, so that when the right side of the rim of the extractor becomes thin by wear, the worn portion of the rim can be filed away by the armourer, and the unworn portion brought into position when the tube is adjusted on the rifle.

Mark II tubes will be marked with the numeral "II," on the breechpiece to facilitate identification.

It becomes obvious that the concept of the tube was not satisfactory; care had to be taken with cleaning as rust often appeared in the bore of the rifle, between the outside of the tube and the original barrel, and the accuracy was not really satisfactory. A test rifle was built up chambered for the .23-in. round, but using a solid barrel made especially for the smaller cartridge; as the rifle was assembled from parts, the barrel was the only special component, and the results were considered to be quite successful.

A direct consequence was the construction of a similar type of rifle chambered for the .22-in. R.F. cartridge, it was approved on 13th December 1907 and designated the .22-in. R.F. Short Rifle, Mark I. This rifle was quite different to any previously in service and it looks like a hybrid, using components from a variety of earlier rifles. It has a modified Mark I* Lee-Metford action and fore-end, a solid .22 barrel, Charger Loading Lee-Enfield type foresight protector, modified Mark I S.M.L.E. rear handguard and Mark III S.M.L.E. backsight. Obviously, the Mark I* Lee-Metford rifle was the basis for the conversion, and these particular markings are found on the right side of the butt socket.

The conversion markings, which comprise of the factory and year of conversion along with the designation "SHT. .22" over "I," are stamped on the left side of the butt socket. Particular modifications special to this trainer are the enlarged gas vents in the receiver, and the bayonet lug having been removed from the nosecap.

This training rifle was promulgated in the List of Changes, § 14139, with an official approval date of 13th December 1907.

Para 14139:–
Rifle, Short, .22-inch, R.F. (Mark I).
Converted from Rifle, M.L.M., Mark I.

A pattern of the above-mentioned rifle has been approved to guide conversion as may be ordered.

The conversion is made from Rifle, M.L.M., Mark I, and consists generally in altering the rifles to approximately as nearly as possible in weight, length, &c., to the short rifle, and in fitting a new barrel with back and front sights similar to those of the Rifle, Short, M.L.E., Mark III described in § 12853, but bored and chambered to suit the Cartridge, Aiming Tube, R.F., Mark I (§ 13642). The bolt action is converted, a suitable bolt-head and striker being fitted.*

Body: The gas vents left and right of the body are enlarged.

Bolt, breech: The bolt has the cover stops and a portion of the top of the rib removed, and the cover screw holes filled in. The front end is bored to receive a threaded bush, which is fixed by a portion of the original bolt-head screw, the bush and screw being brazed in. The bush is tapped to receive the screwed tenon of the bolt-head.

Bolt-head: The striker hole in the face of the bolt-head is bored .0885-inch eccentric to the axis of the bolt-head, and the hole for the extractor axis screw is bored nearer the axis of the bolt-head than in bolt-heads of Rifles, short, M.L.E., A.T., Mark I, described in LoC § 13649. For the purpose of identification, bolt-heads of this pattern are marked."22" on the top of the wing.

Striker: The striker is arranged suitably for the bolt-head, having a flat on the front portion to clear the rear end of the extractor.

Stock, fore-end: The stock, fore-end, is cut shorter, and a liner is fitted, glued in, and grooved suitably for the smaller barrel. The dial-sight seating and the clearing-rod groove are filled up.

Extractor: An extractor of suitable form, having a long and narrow hook, is fitted. It is so arranged that the outward movement is limited by the body to prevent injury to the extractor spring in the event of a burst cartridge case.

Bands (upper and lower): The upper and lower bands have liners, suitable for the smaller barrel, fitted and brazed in.

Nosecap: The bar for the sword-bayonet is removed, and the barrel groove has a liner, suitable for the smaller barrel, fitted and brazed in.

Protector, foresight: A foresight protector is fitted on the barrel, over the foresight block, and is fixed by a screw.

Particulars relating to rifling, sighting, weight, &c.

Length of barrel	*25.2 inches.*
Calibre	*.214-inch.*
Rifling	*Segmental.*
Number of grooves	*8.*

***See* § 14139, "Rifle, short, ·22-inch, R.F. (Mark I.)."**

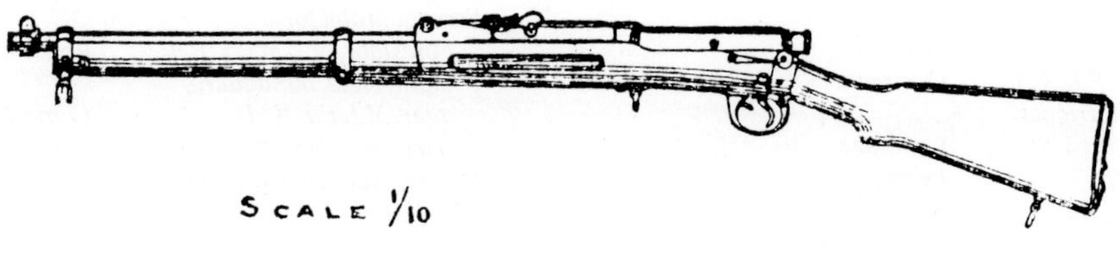

Mean depth of grooves	*.0052-inch.*
Width of grooves	*.05-inch.*
Width of lands	*.03-inch.*
Twist of rifling ...	*R.H., 1 turn in 16 inches.*
Sighting system ...	*Adjustable blade foresight, radial backsight with fine adjustment and windgauge, as on Rifle, short, M.L.E., Mark III.*
Distance between blade, foresight and backsight, U	*19.47 inches.*
Length of rifle	*3 ft. 8.5 in.*
Weight of rifle	*8 lb. 4 oz.*
Ammunition	*Cartridge, aiming tube, R.F. Mk I.*

The following are the approximate elevations required at different ranges:

Range.	*Elevation of backsight.*
25 yds.	*300 yds.*
50 yds.	*350 yds.*
100 yds.	*450 yds.*
150 yds.	*550 yds.*
200 yds.	*650 yds.*

The rod and brush, cleaning, described in §§ 7880 and 13855, are suitable for the rifle, and should be inserted from the breech end. Under no circumstances should the rod and brush be inserted from the muzzle for cleaning the barrel, as the friction of the rod is liable to enlarge the bore and make the barrel bell-mouthed.

The following is a list of components showing which are special to this rifle, and which are common to other arms. Those which are common to other arms are denoted by the word "Common" and their interchangeability is as stated in the Priced Vocabulary of Stores.

COMPONENT	*SPECIAL or COMMON*
Bands-	
Lower, R.F.R.S., I. ...	*Special. Converted from M.L.M.R., I*.*
Upper, R.F.R.S., I. ...	*Special. Converted from M.L.M.R., I*.*
Barrels, R.F.R.S., I. ...	*Special.*
Beds, backsight, M.L.E.R.S., I.	*Common.*
Blades, foresight, M.L.E.R.S., III. ...	*Common.*
Blocks, band, foresight, M.L.E.R.S., I. ...	*Common.*
Bodies, R.F.R.S., I. ...	*Special. Converted from M.L.M.R., I*.*
Bolts-	
Breech. M.L.M.R. I, R.F.*	*Common. Converted from M.L.M.R., I*.*
Stock, M.L.E., No. 1 ...	*Common.*
Caps, nose, R.F.R.S., I. ...	*Special. Converted from M.L.M.R., I*.*
Catches-	
Magazine, M.L.M.R., I.*	*Common.*
Slide, backsight, M.L.E.R.S., III. ...	*Common.*
Cocking pieces, M.L.E.R.S., I.	*Common.*
Collars, screw, front, trigger guard, M.L.M. ...	*Common.*
Discs, marking, butt. ...	*Common.*
Extractors, R.F.R.S., I. ...	*Special.*

COMPONENT	*SPECIAL or COMMON*
Guards-	
Hand, R.F.R.S., I. ...	*Special. Converted from Guard, hand, rear, M.L.E.R.S., I.*
*Trigger, M.L.M.R., I** ...	*Common.*
Heads-	
Breech bolt, R.F.R.S., I.	*Special.*
Screw, wind-gauge, M.L.E.R.S., III. ...	*Common.*
Keys, block, band, foresight, M.L.E.R.S., I. ...	*Common.*
Leaves, backsight, M.L.E.R.S., III. ...	*Common.*
Nuts, screw, band, upper, M.L.M.R.	*Common.*
Pins, axis-	
Backsight, M.L.E.R.S., III.	*Common.*
Trap, butt-plate, M.L.M.	*Common.*
Worm, fine-adjustment, catch, slide, backsight, M.L.E.R.S., III. ...	*Common.*
Catch, magazine, M.L.M.	*Common.*
Pins, fixing-	
Bed, backsight, M.L.E.R.S., III. ...	*Common.*
Block, band, foresight, M.L.E.R.S., I. ...	*Common.*
Head, screw, wind-gauge, M.L.E.R.S., III. ...	*Common.*
Washer, pin, axis, back-sight, M.L.E.R.S., I.	*Common.*
Stop band, lower, M.L.M.R.	*Common.*
Stop band, upper, M.L.M.R.	*Common.*
Trigger, M.L.M. ...	*Common.*
Plates, butt, M.L.M.R., I.*	*Common.*
Protectors, foresight, R.F.R.S., I.	*Special.*
Protectors, sight, handguard, rear, M.L.E.R.S., I. ...	*Common.*
Rivets-	
Fore-end, M.L.M. ...	*Common.*
Spring, handguard, M.L.E.R.S., I. (2) ...	*Common.*
Screws, band-	
Lower, M.L.M.R., I.* ...	*Common.*
Upper, M.L.M.R. ...	*Common.*
Screws-	
Cap, nose, M.L.M.R. ...	*Common.*
Catch, slide, backsight, M.L.E.R.S., III. ...	*Common.*
Disc, marking, butt. ...	*Common.*
Ejector, M.L.M. ...	*Common.*
Extractor, M.L.M. ...	*Common.*
Screws, guard, trigger-	
M.L.M., back. ...	*Common.*
M.L.M., front. ...	*Common.*
Screws, keeper-	
Fine-adjustment; handguard cap; and sight protector top; M.L.E.R.S., I. ...	*Common. (For sight protector top only).*
Striker, M.L.M.R. ...	*Common.*
Screws-	
Plate, butt, iron. (2).	*Common.*
Screws, protector-	
Foresight, M.L.M.C.L.R. II.	*Common.*
Sight, slide, handguard, rear, M.L.E.R.S., I.	*Common.*
Screws, sear, M.L.M. ...	*Common.*
Screws, slides, backsight, M.L.E.R.S., III. ...	*Common.*

COMPONENT	*SPECIAL or COMMON*
Springs-	
Catch, slide, backsight, M.L.E.R.S., III. ...	*Common.*
Extractor, M.L.M. ...	*Common.*
Guard, hand, rear, double, M.L.E.R.S., I. ...	*Common.*
Main, M.L.M. ...	*Common.*
Retaining, breech bolt, M.L.M.	*Common.*
Screw, wind-gauge, back-sight, M.L.E.R.S., III.	*Common.*
Sear, M.L.M. ...	*Common.*
Sight, back, M.L.E.R.S., III. ...	*Common.*
Trap, butt-plate, M.L.M.	*Common.*
Wind-gauge, backsight, M.L.E.R.S., III. ...	*Common.*
Stocks-	
Butt, M.L.M.R.S., I. ...	*Common.*
Fore-end, R.F.R.S., I.	*Special. Converted from M.L.M.R., I*.*
Strikers, R.F.R.S., I. ...	*Special. Suitable to replace Strikers, M.L.M.R., I*. R.F.*
Swivels-	
Band, M.L.M.R., I.* ...	*Common.*
Butt, M.L.M.R. ...	*Common.*
Guard, trigger, M.L.M.R.	*Common.*
Piling, M.L.M.R., I.* ...	*Common.*
Traps, butt-plate, M.L.M.R., I.*	*Common.*
Triggers, M.L.M.R., I.* ...	*Common.*
Washers-	
Bolt, stock. ...	*Common.*
Guard, hand, or rivet, fore-end-	
M.L.M. (2) ...	*Common. For Rivet, fore-end.*
M.L.E.R.S., I. (2) ...	*Common. For Guard, hand.*
Pin, axis, backsight, M.L.E.R.S., I. ...	*Common.*
Wind-gauges, backsight, M.L.E.R.S., III. ...	*Common.*
Worms, fine-adjustment, back-sight, M.L.E.R.S., III. ...	*Common.*

The Magazine Lee-Metford Aiming Tube rifle and the Magazine Lee-Enfield Aiming Tube rifle were approved on 19th July 1907 and introduced in the List of Changes for Land Service. Similar to the Short, Magazine Lee-Enfield Aiming Tube rifle announced in § 13649 and described a few pages before, they were marked "A.T." on the body, barrel, bolt, butt and fore-end. A .22-in. R.F. bolt was fitted as described LoC § 13650 and the chambers were bushed so that the service .303-in. ball or blank ammunition could not be chambered.

A .230-in. (.297/230) experimental rifle was produced at about this time, converted from the Lee-Enfield rifle and with backsight and foresight assemblies from the .303 Mk III S.M.L.E. The Mk I S.M.L.E. rear handguard with attached sight protectors and a foresight protector like that of the C.L.L.E. are fitted. The particular example referred to here is in the Pattern Room collection, RB 374; it has a newly manufactured, solid barrel chambered for the .230 aiming tube cartridge. No bolthead or striker modifications are required as this cartridge is a central-fire type.

The furniture and fittings have been modified from the M.L.E. so as to resemble the .22 Short Rifle Mark I, although the fitting of Mk III S.M.L.E. sights and a double-pull trigger would indicate that it is a little later in vintage. The .22 Short Rifle Mark I* introduced in October 1911 has a similar sighting configuration and the .22 Short Rifle Mark II approved in January 1912 has a double-pull trigger.

For Naval Service, a .23-inch Aiming Tube Mk I was introduced for the S.M.L.E. rifle, being approved on 4th February 1910; its overall length is 26 ins. because it was designed for the 25.2-in. S.M.L.E. barrel. Although this might seem to be out of date because the .22-in. R.F. tube for the S.M.L.E. had been approved in 1906, the subsequent .22-in. tubes and rifles, including the .22-inch Short Rifle Mark I, were introduced for Land Service; it appears that the Navy was still using the .230-in. C.F. Morris Tube cartridge and had adequate supplies of this ammunition, so the introduction of the .22-in. R.F. cartridge or rifle would present some problems regarding supply and maintenance.

As well as providing practice on the miniature ranges, the small bore rifles were intended to duplicate the sighting of the service issue .303 rifle and while in some instances the external appearance of these small bore rifles may not look like the current service pattern (an excellent example is the .22-in. Short Rifle Mark I), the form of sighting follows that of the service full-bore pattern closely.

In keeping with this concept, after the introduction of the Charger Loading Lee-Enfield, the sights of the Aiming Tube Lee-Metford and Lee-Enfield rifles were upgraded. This was approved on 16th August 1910 and declared in the List of Changes:

Para 15303:–

Rifles: M.L.M. A.T. & M.L.E. A.T.

Alteration of sights to approximate to those of Rifles, Charger-loading, M.L.E. Mark I.*

Approval has been given for the sights of the above-mentioned rifles (§ 14168) to be altered so as to approximate to the sights on Rifles, Charger-loading, M.L.E. Mark I (§ 13992).*

The work of altering existing rifles is to be carried out in central armourer's shops (except in the Eastern Command and London District, where the rifles will be called in to the Ordnance Factory under arrangements made by the Chief Inspector of Small Arms) in conformity with the sample rifle supplied to each shop, in the following manner:

Foresight.

The foresight is to be filed to approximate to the shape of the blade foresight of charger-loading rifles, and is to be blacked in accordance with the instructions for the temporary blacking of sights given on page 52 of

"Instructions for Armourers, 1904."

Backsight.

The backsight leaf will be fitted with a Slide, sight, windgauge, M.L.M. or M.L.E., A.T. rifles, Mark I (§ 15305), and the sight cap will be reduced in height by filing so that it does not interfere with the line of sight, and to produce a stop to prevent the windgauge slide being easily removed. The sight cap will then be blacked as before described for the foresight.

Future supplies of these rifles will be made to conform to these alteration when undergoing conversion in O.F. [Ordnance Factory].

A specific backsight slide with a windgauge was approved for the M.L.M. and M.L.E. Aiming Tube rifles on 16th August 1910, similar to that introduced for the Lee-Enfield rifle, LoC § 13314 and illustrated in Chapter 4 of this study, except that it has a "U" sighting notch instead of a "V," in order to make this the same as on the Charger Loading Lee-Enfield.

On 15th December 1910, an improved bolthead was also approved for the .22-in. Short Rifle Mk I, and it was then published in the List of Changes:

Para 15498:–

Rifle, Short, .22-inch, R.F., Mark I.
Converted from Rifle, M.L.M., Mark I.*

Fitting of strengthened bolt-head, &c.

In future conversions of rifles to the above-mentioned pattern (§ 14139), or when spare bolt-heads are required, the rifles will be fitted with a stronger bolt-head, and with a striker in two parts.

The following is a description of components:

Bolt-head.

The bolt-head has a firing-pin hole bored parallel with, but eccentric to, the axis of the bolt-head. In other respects it is similar to the existing pattern. To facilitate identity it is marked ".22 No. 2" on the top of the wing.

Firing pin.

The firing pin is a compound pin in the form illustrated.

Striker.

The striker is of the same form as the central-fire striker for Rifles, M.L.M., Mark I, with the point removed down to the collar, so that it acts as a hammer on the firing pin. The key grooves are so arranged that the blow on the firing pin is given by the solid portion of the collar.*

In fitting these bolt-heads and firing pins as spare, the following instructions will be observed:

1. Assemble the bolt-head to the rifle without the firing pin, and adjust the front face, where necessary, to enable the bolt to be closed with the bolt-head bearing on the face of the barrel.

2. Remove the bolt-head from the rifle, insert the firing pin and press home to the small shoulder (A). If the rear end (B) of the pin is found to project beyond the rear face (C) of the bolt-head, reduce the pin at (B) until it is just below the face (C). This ensures that the bolt-head, and not the small shoulder (D) of the firing pin, acts as stop to the forward movement of the striker.

3. Assemble the bolt complete; place the cocking-piece in the long cam of the bolt; test the protrusion of the point of the firing pin, which should be the same as in the Service rifle (i.e. .04-inch to .042-inch), and adjust the point where necessary, maintaining the same radial form at the point as illustrated.

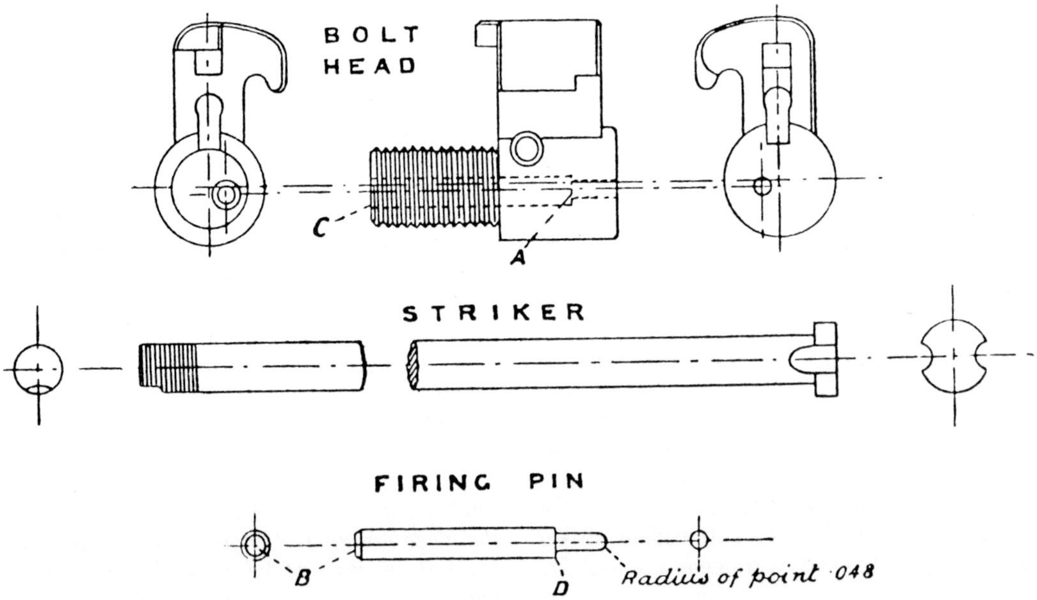

Note: The internal shoulder (A) in the bolt-head controls the firing pin in its most forward position, and when the firing pin is properly adjusted, prevents it from striking the edge of the chamber when snapping. Therefore, when adjusting the rear end (B) of the firing pin, care should be taken to see that the pin is not more than .003-inch below the rear face (C) of the bolt-head, as whatever distance it is below, so there will exist a similar clearance between the shoulder (D) of the pin and the shoulder (A) of the bolt-head when the pin is gauged for protrusion, and this added to the amount of protrusion as gauged gives the amount of protrusion of the striker point in its most forward position.

When No. 2 bolt-heads are required to replace No. 1, they must be accompanied by firing pins and No. 2 strikers.

The following table shows the interchangeability of the new components:

COMPONENT	*INTERCHANGEABILITY*
Head, breech bolt, R.F. rifle, short, No. 2.	*Special.*
Pin, firing, R.F. rifle, short.	*And for M.L.M. and M.L.E. (including short), A.T. rifles, fitted with No. 2 bolt-head.*
Striker, R.F. rifle, short, No. 2.	*And for M.L.M. Marks I* and II, and M.L.E., short, A.T. rifles, fitted with No. 2 bolt-head.*

This new bolthead, designated the Pattern No. 2, was also incorporated in the S.M.L.E., M.L.E. and M.L.M. Aiming Tube rifles, with the approval date being the same as that for its fitting to the .22 Short Rifle Mark I, 15th December 1910. The announcement in the List of Changes also abolished the bushing of the chamber.

Para 15499:–

Rifles: *M.L.M. A.T.*
M.L.E. A.T.
Short, M.L.E., A.T.
Short, M.L.E., A.T./N.I.

For .22-inch rim-fire aiming-tube.

1. Fitting of strengthened bolt-head, &c.
2. Bushing of chamber discontinued.

1. In future conversion of rifles to the above-mentioned patterns (§§ 13649 & 14168), or when spare parts are required, they will have their bolts (§ 13650) fitted with a bolt-head, &c., similar in design to those fitted to Rifles, Short, .22-inch, R.F., Mark I.

The components for these rifles are the same as for the Rifles, Short, .22-inch, R.F., Mark I, with the following exceptions:

Bolt-head.

The position of the extractor axis screw hole is further from the axis of the bolt-head. To facilitate identity it is marked "A.T. No. 2" on the top of the wing.

Striker.

The striker for Rifles, M.L.M., Mark II and M.L.E., A.T., is longer than that for Rifles, Short, .22-inch, R.F., Mark I, which is suitable for all other A.T. rifles.*

In fitting the bolt-head and firing-pin as spare, the instructions given in § 15498 will be followed, except that the bolt-head should be adjusted, where necessary, to enable the bolt to close with the bolt-head bearing on the .064-in. gauge, distance of bolt from end of chamber.

When A.T. rifles are fitted with the new pattern components, the extractors used in central-fire bolt-heads for .303-inch service cartridges may be used without conversion.

To render the firing pin readily accessible for cleaning, bolt covers, where fitted to A.T. rifles, will be removed and returned to store.

The interchangeability of the components is as shown in § 15498 and hereunder:

COMPONENT	*INTERCHANGEABILITY*
Head, breech bolt, M.L.M. rifle, Mark I, R.F., No. 2.*	*For M.L.M. and M.L.E. (including short) A.T. rifles.*
Strikers, M.L.M. rifle, Mark II, R.F., No. 2. ...*	*For M.L.M. Mark II*, and M.L.E., (excepting short), A.T. rifles, when fitted with No. 2 bolt-heads.*

2. In consequence of the introduction of the strengthened bolt-head, the bushing of the chambers of these rifles will be discontinued.

Another .22-in. Short Rifle was introduced in 1912, the Mark I*, which incorporates the improved pattern bolthead and the newer pattern sights, those of the Mark III and Converted Mark IV S.M.L.E., upgraded for Mk VII ammunition. The approval dates were 13th October 1911 and 12th January 1912. The changes are not obvious at a glance, but comprise:

Modified backsight ramp
Lower foresight blade
Improved (Pattern No. 2) bolthead
Straightened-out foresight protector wings.

The last-mentioned modification is the most obvious; the rifle designation is marked on the left of the receiver butt socket, "SHT .22" over "I*." The sights were obviously intended to conform with those of the current S.M.L.E. Mark III rifle for Mk VII ammunition.

Official details of this rifle as advised in the List of Changes after its approval on 13th October 1911 were:

Para 15895:–

Rifle, Short, .22-inch, R.F., Mark I.*
Converted from Rifle, M.L.M., Mark I.*

A pattern of the above-mentioned rifle has been approved to guide conversion as may be ordered.

It differs from Rifle, Short, .22-inch, R.F., Mark I (§§ 14139, 15077 & 15498) in the following particulars:

Backsight, and blades, foresight: These components are of the same pattern as those described in § 15638 for future manufacture of Rifles, Short, M.L.E., Mark III and Converted Mark IV.

Protector, foresight: The protecting wings are straight vertically instead of incurved. It is suitable for all patterns of Rifles, Short, .22-inch, R.F.

The following is a list of components whose interchangeability is as shown in the Priced Vocabulary of Stores, and above:

Bands, lower, R.F.R.S., I.
Bands, upper, R.F.R.S., I.
Barrels, R.F.R.S., I.*
Beds, backsight, M.L.E.R.S., III.
Blades, foresight, M.L.E.R.S., III.
Blocks, band, foresight, M.L.E.R.S. I.
Bodies, R.F.R.S., I.
Bolts, breech, R.F.R.S., I.
Bolts, stock, M.L.E. No. 1.
Caps, nose, R.F.R.S., I.
Catches, magazine, M.L.M.R. I.*
Catches, slide, backsight, M.L.E.R.S., III.
Cocking-pieces, R.F.R.S., I.
Collars, screw, front, trigger guard, M.L.M.
Discs, marking, butt.
Extractors, R.F.R.S., I.
Guards, hand, R.F.R.S., I.
Guards, trigger, M.L.M.R., I.*
Heads, breech bolt, R.F.R.S., I., No. 2.
Heads, screw, windgauge, M.L.E.R.S., III, No. 2.
Keys, block, band, foresight, M.L.E.R.S., I.
Leaves, backsight, M.L.E.R.S., III.
Nuts, screw, band, upper, M.L.M.R.
Pins, axis, backsight, M.L.E.R.S., III.
Pins, axis, trap, butt plate, M.L.M.
Pins, axis, worm, fine adjustment, catch slide, backsight, M.L.E.R.S., III.
Pins, catch, magazine, M.L.M.
Pins, firing, R.F.R.S., I.
Pins, fixing, bed, backsight, M.L.E.R.S., III.
Pins, fixing, block, band, foresight, M.L.E.R.S., I.
Pins, fixing, head, screw, windgauge, M.L.E.R.S., III, No. 2
Pins, fixing, washer, pin, axis, backsight, M.L.E.R.S., I.
Pins, stop band, lower, M.L.M.R.
Pins, stop band, upper, M.L.M.R
Pins, trigger, M.L.M.
Plates, butt, M.L.M.R., I.*
Protectors, foresight, R.F.R.S., I.*
Protectors, sight, hand-guard, rear, M.L.E.R.S., I.
Rivets, fore-end, M.L.M.
Rivets, spring, hand-guard, rear, M.L.E.R.S., I.
Screws, band, lower, M.L.M.R. I.*
Screws, band, upper, M.L.M.R.
Screws, cap, nose, M.L.M.R.
Screws, catch, slide, backsight, M.L.E.R.S., III.
Screws, disc, marking, butt.
Screws, ejector, M.L.M.
Screws, extractor, M.L.M.
Screws, guard, trigger, M.L.M., back.
Screws, guard, trigger, M.L.M., front.
Screws, keeper, fine-adjustment, M.L.E.R.S., I. [for protector, handguard, rear].
Screws, keeper, striker, M.L.M.R.
Screws, plate, butt, iron (2).
Screws, protector, foresight, M.L.E.C.L.R., I.*
Screws, protector, sight, side, hand-guard, rear, M.L.E.R.S., I.
Screws, sear, M.L.M.
Screws, spring, sight, back, M.L.E.R.S., III.
Screws, spring, trap, buttplate, M.L.M.
Screws, swivel, trigger guard, M.L.M.R.
Screws, windgauge, backsight, M.L.E.R.S., III.
Sears, M.L.E.R.S., I.
Slides, backsight, M.L.E.R.S., III, No. 2.
Springs, catch, slide, backsight, M.L.E.R.S., III.
Springs, extractor, M.L.M.
Springs, guard, hand, rear, double, M.L.E.R.S., I.
Springs, main, M.L.M.
Springs, retaining, breech bolt, M.L.M.
Springs, screw, windgauge, backsight, M.L.E.R.S., III.
Springs, sear, M.L.M.
Springs, sight, back, M.L.E.R.S. III.
Springs, trap, butt plate, M.L.M.
Springs, windgauge, backsight, M.L.E.R.S., III.
Stocks, butt, M.L.M.R., I.*
Stocks, fore-end, R.F.R.S., I.
Strikers, R.F.R.S., I, No. 2.
Swivels, band, M.L.M.R. I.*
Swivels, butt, M.L.M.R.
Swivels, guard, trigger, M.L.M.R.
Swivels, piling, M.L.M.R., I.*
Traps, butt-plate, M.L.M.R., I.*
Triggers, M.L.E.R.S., I.
Washers, bolt, stock.
Washers, guard, hand, or rivet, fore-end, M.L.M. (2).
Washers, guard, hand, or rivet, fore-end, M.L.E.R.S., I. (2)
Washers, pin, axis, backsight, M.L.E.R.S., I.
Windgauges, backsight, M.L.E.R.S., III.
Worms, fine adjustment, backsight, M.L.E.R.S., III.

The following are the approximate elevations required at different ranges:

Range.	*Elevation of Backsight.*
25 yds.	*200 yds.*
50 yds.	*300 yds.*
100 yds.	*450 yds.*
150 yds.	*600 yds.*
200 yds.	*750 yds.*

Abbreviations used in these listings may be found to vary in the official Ordnance listings and List of Changes references at different times and due to some changes in nomenclature. During this period and for comparable references to the .22 Short Rifles, Marks I, I* and II, and the following .22 Long Rifles, .22 Short Rifle Mk III & Patt. '14 series, these abbreviations are:

M.L.E.	Magazine Lee-Enfield
M.L.E.C.	Magazine Lee-Enfield Carbine
M.L.E.C.L.R.	Magazine Lee-Enfield Charger Loading Rifle
M.L.E.R.	Magazine Lee-Enfield Rifle
M.L.E.R.S.	Magazine Lee-Enfield Rifle, Short (SMLE)
M.L.M.	Magazine Lee-Metford
M.L.M.C.	Magazine Lee-Metford Carbine
M.L.M.R.	Magazine Lee-Metford Rifle
R.F.	Rim Fire
R.F.R.L.	Rim Fire Rifle, Long
R.F.R.S.	Rim Fire Rifle, Short

In many cases, the particular rifle Mark number follows these abbreviations, so "M.L.M.R., I*" indicates the Mark I* Magazine Lee-Metford Rifle.

Certain inconsistencies will be found and some misunderstandings may result from these particular parts listings, relating to the rifle type abbreviations; for example, one particular part may be specified for the M.L.M.R. I* and another part shown to be common to all Lee-Metford rifles (presumably including the Marks II and II*) by the use of the abbreviation M.L.M. (which would indicate rifles and carbines also). Such an inconsistency pointed out by Charles Woodhouse, a contemporary student of the subject, is that the "Screws, band, lower" are shown to be special to the M.L.M. I* rifle, whereas the "Screws, band, upper" are shown for M.L.M. rifles generally; both these screws were interchangeable on the Marks I and I* M.L.M. rifles but not on Marks II and II* M.L.M. rifles.

It is assumed that these parts lists, and other such contemporary lists as are found in the List of Changes, generally indicate that the M.L.M. Marks I and I* parts (the Mk I was obsolete and most were upgraded to Mk I* anyway) are special to that rifle only, all interchangeable Metford rifle parts are shown as M.L.M.R., and all interchangeable Metford rifle and carbine parts shown as M.L.M.

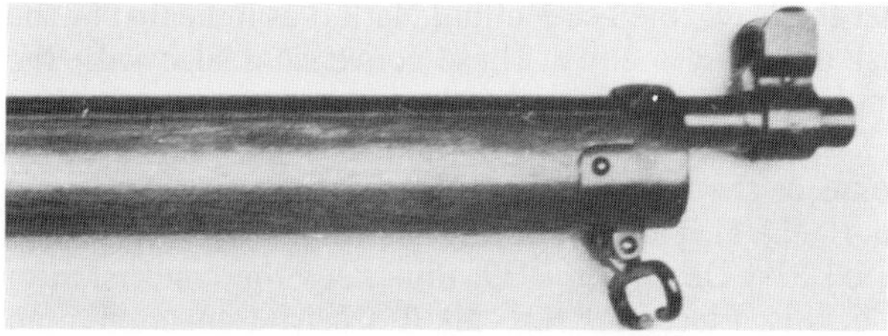

Typical muzzle and foresight protector configuration of the earlier .22 Short Rifles. *MOD Pattern Room*

The subsequent conversion in this numerical sequence was the .22 Short Rifle Mark II, approved on 5th January 1912 for Naval Service. This therefore introduced the .22 RF cartridge into Naval use as the .230-in. C.F. Morris cartridge had been used until then. A detailed description of this rifle was included in the List of Changes:

Para 15969:–
Rifle, Short, .22-inch, R.F., Mark II.
Converted from Rifle, M.L.M., Mark II.

A pattern of the above-mentioned rifle has been approved to guide conversion as may be ordered. This rifle is converted from the Rifle, M.L.M., Mark II, to approximate as nearly as possible to the Rifle, Short, .22-in., R.F., Mark I (§ 15895).*

The following is a description of its principal converted and new components:

Body: Gas vents, left and right, enlarged.

Bolt, breech: Flat ground on radius of stud to clear higher position of sear for double-pull trigger action. Bolt cover lugs retained, although the converted bolt is not fitted with a bolt cover.

Stock, fore-end: The stock, fore-end, is cut shorter and a liner is fitted, glued in and grooved suitably for the smaller barrel.

Nosecap: The sword-bayonet bar and the front end of the nosecap are removed; the corners of the barrel hole are removed to clear the barrel, and a groove is cut in the top of the barrel hole to clear the blade foresight in stripping and assembling.

Cocking-piece: The radius of the bent of the cocking-piece is ground flat to angle, to suit the double pull trigger action.

Barrel (with sights), foresight protector, bolt-head, firing pin, striker and extractor: These components are the same as in the Mark I rifle.*

The following is a list of components showing which are special to this rifle and which are common to other arms. Those which are common to other arms are denoted by the word "Common," and their interchangeability is as stated in the Priced Vocabulary of Stores, &c.

COMPONENT	*REMARK*
Band, lower, M.L.M.R., II.	*Common.*
Barrels, R.F.R.S., I.*	*Common.*
Beads, pointer, dial sight, M.L.M.R., II. ...	*Common.*
Beds, backsight, M.L.E.R.S., III.	*Common.*
Blades, foresight, M.L.E.R.S., III.	*Common.*
Blocks, band, foresight, M.L.E.R.S., I.	*Common.*
Bodies, R.F.R.S., II.	*Special.*
Bolts, breech, R.F.R.S., II.	*Special.*
Bolts, stock, M.L.E., No. 1.	*Common.*
Bolts, stock, M.L.E., No. 2.	*Common.*
Caps, nose, R.F.R.S., II.	*Special.*
Catches, magazine, M.L.M.R., II.	*Common.*
Catches, slide, backsight, M.L.E.R.S., III. ...	*Common.*
Cocking-piece, R.F.R.S., II.	*Special.*
Collars, screw, front, trigger guard, M.L.M. ...	*Common.*
Extractors, R.F.R.S., II.	*Common.*
Guards, hand, R.F.R.S., II.	*Common.*
Guards, trigger, R.F.R.S., II.	*Common.*
Heads, breech bolt, R.F.R.S., I, No. 2.	*Common.*
Heads, screw, windgauge, M.L.E.R.S., III, No. 2	*Common.*
Keys, block, band, foresight, M.L.E.R.S., I. ...	*Common.*
Leaves, backsight, M.L.E.R.S., III.	*Common.*
Pins, axis, backsight, M.L.E.R.S., III.	*Common.*
Pins, axis, trap, butt plate, M.L.M.	*Common.*
Pins, axis, worm, fine-adjustment, catch, slide, backsight, M.L.E.R.S., III.	*Common.*
Pins, catch, magazine, M.L.M.	*Common.*
Pins, firing, R.F.R.S., I.	*Common.*
Pins, fixing, bed, backsight, M.L.E.R.S., III. ...	*Common.*
Pins, fixing, block, band, foresight, M.L.E.R.S., I.	*Common.*
Pins, fixing, head, screw, windgauge, M.L.E.R.S., III, No. 2.	*Common.*
Pins, fixing, washer, pin, axis, backsight, M.L.E.R.S., I.	*Common.*
Pins, stop, band, lower, M.L.M.R.	*Common.*
Pins, trigger, M.L.M.	*Common.*
Plates, butt, M.L.M.R., II.	*Common.*

COMPONENT	*REMARK*
Plates, keeper, stockbolt, M.L.M. [b]	*Common.*
Plates, sight, dial, M.L.M.R., II.	*Common.*
Pointers, sight, dial, M.L.M.R., II.	*Common.*
Protectors, foresight, R.F.R.S., I.*	*Common.*
Protectors, sight, hand-guard, rear, M.L.E.R.S., I.	*Common.*
Rivets, fore-end, M.L.M.	*Common.*
Rivets, spring, hand-guard, rear, M.L.E.R.S., I. (2)	*Common.*
Screws, band, lower, M.L.M.R., II.	*Common.*
Screws, cap, nose, M.L.M.R.	*Common.*
Screws, catch, slide, backsight, M.L.E.R.S., III.	*Common.*
Screws, ejector, M.L.M.	*Common.*
Screws, extractor, M.L.M.	*Common.*
Screws, guard, trigger, M.L.M., back.	*Common.*
Screws, guard, trigger, M.L.M., front.	*Common.*
Screws, keeper, fine-adjustment, M.L.E.R.S., I.	*Common, for Hand-guard sight protector, top.*
Screws, keeper, striker, M.L.M.R.	*Common.*
Screws, plate, butt, iron (or brass). (2)...	*Common.*
Screws, protector, foresight, M.L.E.C.L.R., I.*	*Common.*
Screws, protector, sight, side, hand-guard, rear, M.L.E.R.S., I.	*Common.*
Screws, sear, M.L.M.	*Common.*
Screws, sight, aperture, M.L.M.R.	*Common.*
Screws, sight, dial, M.L.M., fixing.	*Common.*
Screws, sight, dial, M.L.M., pivot.	*Common.*
Screws, spring, sight, aperture, M.L.M.	*Common.*
Screws, spring, sight, back, M.L.E.R.S., III.	*Common.*
Screws, spring, trap, buttplate, M.L.M.	*Common.*
Screws, strap, butt plate, iron (or brass).	*Common.*
Screws, swivel, piling, M.L.M.R.	*Common.*
Screws, windgauge, backsight, M.L.E.R.S., III.	*Common.*
Sears, M.L.E.R.S., I.	*Common.*
Sights, aperture, M.L.M.R.	*Common.*
Slides, backsight, M.L.E.R.S., III., No. 2.	*Common.*
Springs, catch, slide, backsight, M.L.E.R.S., III.	*Common.*
Springs, extractor, M.L.M.	*Common.*
Springs, guard, hand, rear, double, M.L.E.R.S., I.	*Common.*
Springs, main, M.L.M.	*Common.*
Springs, retaining, breech bolt, M.L.M.	*Common.*
Springs, screw, windgauge, backsight, M.L.E.R.S., III.	*Common.*
Springs, sear, M.L.M.	*Common.*
Springs, sight, aperture, M.L.M.	*Common.*
Springs, sight, back, M.L.E.R.S., III.	*Common.*
Springs, sight, dial, M.L.M.	*Common.*
Springs, trap, butt plate, M.L.M.	*Common.*
Springs, windgauge, backsight, M.L.E.R.S., III.	*Common.*
Stocks, butt, M.L.M.R., II., No. 1. (a)	*Common.*
Stocks, butt, M.L.M.R., No. 2. (a)	*Common.*
Stocks, fore-end, R.F.R.S., II.	*Common.*
Strikers, R.F.R.S., I., No. 2	*Common.*
Swivels, band, M.L.M.R., II.	*Common.*
Swivels, butt, M.L.M.R.	*Common.*
Swivels, piling, M.L.M.R., II.	*Common.*
Traps, butt plate, M.L.M.R., II.	*Common.*
Triggers, M.L.E.R.S., I.	*Common.*
Washers, bolt, stock.	*Common.*
Washers, guard, hand, or rivet, fore-end, M.L.M. (2).	*Common.*
Washers, guard, hand, or rivet, fore-end, M.L.E.R.S., I. (2)	*Common.*
Washers, pin, axis, backsight, M.L.E.R.S., I.	*Common.*
Washers, screw, fixing, dial sight, M.L.M.R.	*Common.*
Windgauges, backsight, M.L.E.R.S., III.	*Common.*
Worms, fine-adjustment, backsight, M.L.E.R.S., III.	*Common.*

(a) Not in rifles with No. 1 stock bolt.
(b) According to whether fitted for stock-bolt keeper plate or not.

The following are the approximate elevations required at different ranges:

Range.	*Elevation of Backsight.*
25 yds.	*200 yds.*
50 yds.	*300 yds.*
100 yds.	*450 yds.*
150 yds.	*600 yds.*
200 yds.	*750 yds.*

The foregoing description indicates that the .22-in. Short Rifle Mark II was a conversion of the Mark II Lee-Metford rifle, intended to resemble the .22-in. Short Rifle Mark I* as closely as possible, but due to the different rifle from which the conversion was made, it has no finger grooves in the fore-end and no two-piece nosecap/upper band assembly. Unlike the previous .22 Short Rifle, the long range dial sight and rear aperture were retained, although other components such as the barrel, sights, foresight protector, bolthead, extractor and firing pin are interchangeable. Similar to the previous .22 Short Rifles, the original rifle markings are located on the right side of the butt socket, with the .22-in. conversion markings on the left side. The magazine and magazine cut-off were also removed.

Some .22 Short Rifle Mk II conversions have also been noted, made from the Magazine Lee-Enfield rifle, and not just the Magazine Lee-Metford Mark II as indicated by the List of Changes entry. These conversions also with the single-pull trigger instead of double-pull.

Some of the wartime Naval issue trainers utilised later Mk II M.L.M. parts; a factory drawing, DD(E) 2398, and dated 21st December 1939, illustrates the conversion of "Cocking Piece, R.F.R.S. Mk II (Navy) Converted from M.L.M.." In the conversion, the knurled and rounded button at the back end was ground flat, so that the cocking piece appeared flat-sided, similar to the Mk III* S.M.L.E. cocking piece.

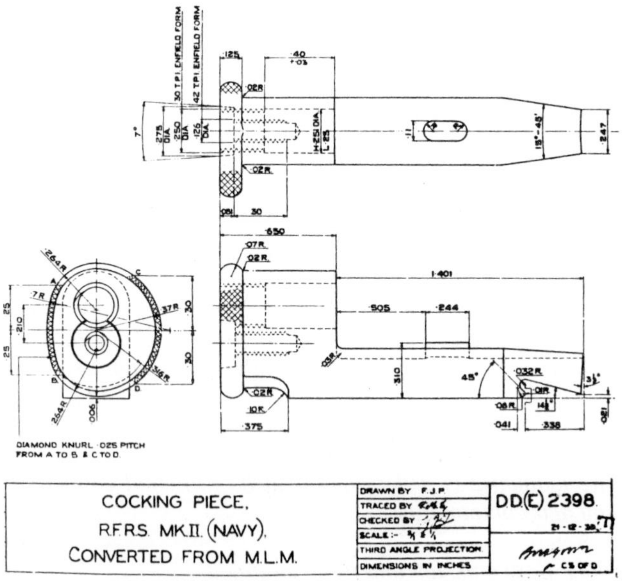

Introduction of the .22-in. RF as a training cartridge into the Navy resulted in the .230-in. C.F. Morris tubes and ammunition being phased out and other new .22-in. calibre rifles introduced. In order that Naval personnel armed with the Magazine Lee-Enfield rifle could be provided with short-range practice, a long .22-in. rifle was approved on 7th February 1912, the .22-inch Long Rifle Mark I, converted from the Mark II Lee-Metford rifle. A Long Rifle Mark I* was approved on 11th March and introduced in the same paragraph of the List of Changes, being a conversion of the Mark II* Lee-Metford and Marks I and I* Lee-Enfield rifles.

Like the previous .22-in. training short rifles, a solid barrel was fitted and the cut-off and magazine were dispensed with. The original wood stock and overall configuration was retained, the difference between the Mark I and I* .22-in. Long Rifles being in the rifles from which the conversions were made; the Mark I .22-in. Long Rifle was from the Mark II Magazine Lee-Metford and therefore has no safety catch, while the Mark I* .22-in. Long Rifle has a safety catch because it was converted from the Mark II* Lee-Metford and Marks I and I* Lee-Enfield rifles. The original markings on the right side of the butt socket will also confirm the rifle from which the conversion was made.

The description in the List of Changes reads:

Para 16044:–

Rifle, Long, .22-inch, R.F., Mark I.
Converted from Rifle, M.L.M., Mark II.

Rifle, Long, .22-inch, R.F., Mark I.*
Converted from Rifles, M.L.M., Mark II, and M.L.E. Marks I and I*.*

Patterns of the above-mentioned rifles have been approved to guide conversions as may be ordered.

The conversions are made from Rifles, M.L.M., Mark II, or M.L.M., Mark II and M.L.E. Marks I and I*, and consist in fitting new barrels with back and front sights similar to those of the rifles from which they are converted, the sights being central, the barrels being bored and chambered to suit the Cartridges, .22-inch, R.F., No. 1, Mark I (§§ 13642 & 15628) and converting the bolt action by fitting a suitable bolt-head, striker and firing pin similar to those described in § 15498; the striker stem in the Mark I* rifle being longer, to suit the longer breech-bolt as described in § 15499.*

An extractor, similar to that for Rifle, Short, .22-inch, R.F., Mark I is fitted.

Body: The gas vents left and right are enlarged.

The Mark I rifle differs from the Mark I in the patterns of the following components:*

Bolt, breech
Cocking-piece
Striker
Screw, keeper, striker;

and has the following additional components:

Catch, safety
Pin, catch, safety
Spring, pin, catch, safety.

The following is a list of components whose interchangeability is as shown in the Priced Vocabulary of Stores, and herein:

	Bands, lower, M.L.M.R., II.
(a)	*Barrels, R.F.R.L., I*
	Beads, pointer, dial-sight, M.L.M.R., II.
	Beds, backsight, M.L.M.R., II.
	Bodies, R.F.R.S., II.
(b)	*Bolts, breech, M.L.M.R., II.*
(c)	*Bolts, breech, M.L.M.R., II*.*
(d)	*Bolts, stock, M.L.E., No. 1.*
(e)	*Bolts, stock, M.L.E., No. 2.*
	Caps, nose, M.L.E.R. [M.L.M.R., II or M.L.E.R., I.]*
	Catches, magazine, M.L.M.R., II.
(c)	*Catches, safety, M.L.M.C.*
(c)	*Cocking pieces, M.L.M.C.*
(b)	*Cocking pieces, M.L.M.R., II.*
	Collars, screw, front, trigger guard, M.L.M.
	Extractors, R.F.R.S., I.
	Guards, hand, M.L.M.R., II.
	Guards, trigger, M.L.M.R., II.
	Heads, breech-bolt, R.F.R.S., I, No. 2.
	Leaves, backsight, M.L.M.R., II.
	Pins, axis, backsight, M.L.M.R.
	Pins, axis, trap, buttplate, M.L.M.
	Pins, catch, magazine, M.L.M.
	Pins, catch, safety, M.L.M.C.
	Pins, firing, R.F.R.S., I.
	Pins, stop, band, lower, M.L.M.R.
	Pins, trigger, M.L.M.
	Plates, butt, M.L.M.R., II.
(e)	*Plates, keeper, stock-bolt, M.L.M.*
	Plates, sight, dial, M.L.M.R., II.
	Pointers, sight, dial, M.L.M.R., II.
	Rivets, fore-end, M.L.M.
	Rivets, spring, handguard, M.L.M. (4)
	Screws, band, lower, M.L.M.R., II.
	Screws, bed, backsight, M.L.M.R.
	Screws, cap, backsight, M.L.M.R.
	Screws, cap, nose, M.L.M.R.
	Screws, ejector, M.L.M.
	Screws, extractor, M.L.M.
	Screws, guard, trigger, M.L.M. (back)
	Screws, guard, trigger, M.L.M. (front)
(c)	*Screws, keeper, striker, M.L.M.C.*
(b)	*Screws, keeper, striker, M.L.M.R.*
	Screws, plate, butt, iron (or brass) (2)
	Screws, sear, M.L.M.
	Screws, sight, aperture, M.L.M.R.
	Screws, sight, dial, M.L.M., fixing.
	Screws, sight, dial, M.L.M., pivot.
	Screws, spring, sight, aperture, M.L.M.
	Screws, spring, sight, back, M.L.M.
	Screws, spring, trap, butt plate, M.L.M.
	Screws, strap, butt plate, iron (or brass).
	Screws, swivel, piling, M.L.M.R., II.
	Sears, M.L.M.R., II.
	Sights, aperture, M.L.M.R.
	Slides, backsight, M.L.M.R., II.
	Springs, extractor, M.L.M.
	Springs, guard, hand, M.L.M.R., II, lower.
	Springs, guard, hand, M.L.M.R., II, upper.
	Springs, main, M.L.M.

(c) *Springs, pin, catch, safety, M.L.M.C.*
Springs, retaining, breech-bolt, M.L.M.
Springs, sear, M.L.M.
Springs, sight, aperture, M.L.M.
Springs, sight, back, M.L.M.R.
Springs, sight, dial, M.L.M.
Springs, trap, butt plate, M.L.M.
(d) *Stocks, butt, M.L.M.R., II, No. 1.*
(e) *Stocks, butt, M.L.M.R., II, No. 2.*
(d) *Stocks, fore-end, M.L.M.R., II, No. 1.*
(e) *Stocks, fore-end, M.L.M.R., II, No. 2.*
(c) *Strikers, M.L.M.R., II*, R.F., No. 2. or R.F.R.L., I*.*
(b) *Strikers, R.F.R.S., I., No. 2.*
Swivels, band, M.L.M.R., II.
Swivels, butt, M.L.M.R.
Swivels, piling, M.L.M.R., II.
Traps, butt plate, M.L.M.R., II.
Triggers, M.L.M.R., II.
Washers, bolt, stock.
Washers, guard, hand, or rivet, fore-end, M.L.M. (6)
Washers, screw, fixing, dial sight, M.L.M.R.

(a) *Special to Mark I and I* long .22-inch rifles.*
(b) *For Mark I rifles.*
(c) *For Mark I* rifles.*
(d) *For rifles not fitted for Plate, keeper, stock-bolt.*
(e) *For rifles fitted for Plate, keeper, stock-bolt.*

The following are the approximate elevations required at different ranges:

Range.	*Elevation of backsight.*
25 yds.	*350 yds.*
50 yds.	*425 yds.*
100 yds.	*500 yds.*
150 yds.	*650 yds.*
200 yds.	*750 yds.*

The rod and brush, cleaning (§§ 7880 & 13855) are suitable for these rifles and should be inserted from the breech end.

Protectors, frontsight, pattern C and No. 1 (§§ 9358 & 11714) are suitable for these rifles.

Both these rifles were introduced for Naval service.

The .22-inch Long Rifle Mark II was approved on 11th November 1911 for the Land (Army) service, and was somewhat similar to the .22-inch Long Rifle Mark I* except that it has adjustable sights and the foresight protector of the Mark I* Charger Loading Lee-Enfield. The following paragraph from the List of Changes also announces the obsolescence of the M.L.M. and M.L.E. Aiming Tube rifles.

Para 16042:–

Rifle, Long, .22-inch, R.F., Mark II.
Converted from Rifles, M.L.E., Marks I and I.*

1. New Pattern.

Rifles, M.L.M. and M.L.E., A.T.

2. Obsolescent.

Tubes aiming, .22-inch, M.L.M.R., Marks I and II.

3. Obsolescent. To be used up, but without repair in Ordnance Factory.

1. A pattern of the above-mentioned rifle has been approved to guide conversion as may be ordered.

The conversion is made from Rifle, M.L.E., Marks I and I and is the same as the Rifle, long, .22-inch, R.F., Mark I*, § 16044, except that the new barrel is fitted with back and front sights and a foresight protector similar to the Rifle, charger-loading, M.L.E., Mark I* (§§ 13992, 14758 & 14835).*

The barrel is special to this rifle. It is bored and chambered to suit the Cartridge, .22-inch, R.F., No. 1 Mark I (§§ 13642 & 15628). The reference in §16044, to rod and brush cleaning, and the instructions with regard thereto, apply to this rifle also.

The following are the approximate elevations required at different ranges:

Range.	*Elevation of backsight.*
25 yds.	*400 yds.*
50 yds.	*450 yds.*
100 yds.	*525 yds.*
150 yds.	*600 yds.*
200 yds.	*725 yds.*

2. & 3. Consequent on the adoption of the rifle at 1., Rifles M.L.M. and M.L.E., A.T. (§§ 14168 & 15499) and Tubes, aiming, .22-inch, M.L.M.R., Marks I and II (§ 13856) become obsolescent, but the tubes will be used up without repair in Ordnance Factory.

The last of this particular series of conversions was the .22-inch R.F. Short Rifle Mark III, approved on 9th August 1912. Altered from the S.M.L.E. Converted Marks II and II*, this presented a minor problem as regards the markings; the right side of the butt socket had the original Lee-Enfield or Lee-Metford markings, and the left side was marked with the .303 Short Rifle Mk II or II* conversion stampings. So, to stamp the new .22 rifle markings it was positioned either below the prior rifle designation on the right hand side or else some of the original markings were removed, making a combination of old and new markings. In this case, the Royal cypher, factory and year of manufacture apply to the original rifle, while the designation, mark number and factory abbreviation belong to the .22-in. conversion.

In the illustrated example, the original rifle was a Mark I* Magazine Lee-Enfield manufactured in 1900 and the subsequent markings were applied during or after 1912, as the Short .22-in. Mark III was not approved until mid-1912. A close examination will usually reveal traces of the original marks below, or else machining or grinding marks where they were removed.

V. R.
ENFIELD
1900
SHT .22
III
EFD

.22 Short Rifle Mk III markings, probably on an Enfield sample arm. Original M.L.E. designation markings have been removed.

The markings above are from a pattern or sample arm at the Pattern Room and may well differ from those models converted by contractors. The markings underneath are from a rifle converted by B.S.A.

V. R.
ENFIELD
1897
L.E.
I
SHT.22
III
B.S.A. Co.

.22 Short Rifle Mk III markings on a B.S.A. Co. converted model. This example has all of the original M.L.E. marks remaining, with the new conversion marks below.

It was intended that the .22-in. Short Rifle Mark III duplicate the sighting of the Mark III S.M.L.E. as closely as possible and so the .303 Mark I S.M.L.E. type rearsight was replaced by a Mark III model which required some change to the stock fore-end, and the Mark I S.M.L.E. nosecap protecting ears were drawn out and straightened similar to those of the .303 Mark I*** and Converted Marks II** and II*** S.M.L.E. rifles. A solid (rather than sleeved) .22-inch barrel was fitted, along with the .22-in. Pattern No. 2 bolthead. Further details are given in the List of Changes entry:

Para 16291:–

Rifle, Short, .22-inch, R.F., Mark III.
Converted from Rifles, Short, M.L.E., Converted Marks II and II.*

A pattern of the above-mentioned rifle has been approved to guide conversion as may be ordered.

The conversion consists in fitting Rifles, Short, M.L.E., Converted, Marks II and II, with .22-inch barrels (§ 14139) with sights as in § 15895 and inner band, replacing the central fire bolt-heads, strikers and extractors by rim-fire bolt-heads, firing pins, strikers (§ 15498) and extractors (§ 14139), making the wings of the nosecaps vertical, similar to those on Rifle, Short, M.L.E., Mark I**, § 16091, making clearance for backsight bed and leaf in stocks (fore-end), handguards and sight protectors, and in enlarging the gas vents, right and left of the bodies.*

The following is a list of components whose interchangeability is as shown in the Priced Vocabulary of Stores, and hereunder. Components special to this rifle are preceded by an asterisk (*).

Bands, inner, M.L.E.R.S., I.
Bands, outer, M.L.E.R.S., III.
Barrels, R.F.R.S., III.
Beads, pointer, dial sight, M.L.M.R., II.
Beds. backsight, M.L.E.R.S., III, No. 2.
Blades, foresight, M.L.E.R.S., III.
Blocks, band, foresight, M.L.E.R.S., I.
* *Bodies, R.F.R.S., III.*
Bolts, breech, M.L.E.R.S., I.
Bolts, locking, M.L.E.R.S., I.
(a) Bolts, stock, M.L.E., No. 2.
(b) Bolts, stock, M.L.E.R.S., I.
(c) Brackets, swivel, butt, M.L.E.R.S., I.*
Caps, hand-guard, front, M.L.E.R.S., I.
(d) Caps, nose, R.F.R.S., III.
Catches, magazine, M.L.M.R., II.
Catches, safety, bolt, locking, M.L.E.R.S., I.
Catches, slide, backsight, M.L.E.R.S., III.
Cocking-pieces, M.L.E.R.S., I.*
Collars, screw, front, trigger guard, M.L.M.
(b) Discs, marking, butt, brass.
Extractors, R.F.R.S., I.
* *Guards, hand, front, R.F.R.S., III.*
* *Guards, hand, rear, R.F.R.S., III.*
Guards, trigger, M.L.E.R.S., I.
Heads, breech bolt, R.F.R.S., I., No. 2.
Heads, screw, windgauge, M.L.E.R.S., III, No. 2.
Keys, block, band, foresight, M.L.E.R.S., I.
Leaves, backsight, M.L.E.R.S., III.
Nuts, screw, back, nosecap, M.L.E.R.S., I.
Pins, axis, backsight, M.L.E.R.S., III.
(a) Pins, axis, trap, butt-plate, M.L.M.
(e) Pins, axis, trap, butt-plate, M.L.E.R.S., I, No. 2.*
Pins, axis, worm, fine-adjustment, M.L.E.R.S., III.
Pins, catch, magazine, M.L.M.
Pins, firing, R.F.R.S., I.
Pins, fixing, bed, backsight, M.L.E.R.S., III.
Pins, fixing, block, band, foresight, M.L.E.R.S., I.
Pins, fixing, head, screw, windgauge, M.L.E.R.S., III, No. 2.
Pins, fixing, washer, pin, axis, backsight, M.L.E.R.S., I.
Pins, stop, band, lower, M.L.M.R. (2)
Pins, stop, bolt, locking, M.L.E.R.S., I. (2)
Pins, trigger, M.L.M.
(e) Plates, butt, M.L.E.R.S., I, No. 2.*
(a) Plates, butt, M.L.M.R., II.
Plates, keeper, stock-bolt, M.L.E.R.S., I.
Plates, sight, dial, M.L.E.R.S., I.
Pointers, sight, dial, M.L.M.R., II.
Protectors, sight, hand-guard, rear, M.L.E.R.S., I.
Rivets, fore-end, M.L.M.
Rivets, spring, hand-guard, M.L.E.R.S., I.
Screws, band, inner, M.L.E.R.S., I.
Screws, cap, nose, M.L.E.R.S., back.
Screws, cap, nose, M.L.E.R.S., front.
Screws, catch, slide, backsight, M.L.E.R.S., III.
(b) Screws, disc, marking, butt.
Screws, ejector, M.L.M.
Screws, extractor, M.L.M.
Screws, guard, trigger, M.L.M., back.
Screws, guard, trigger, M.L.M., front.
(f) Screws, keeper, fine-adjustment, backsight, M.L.E.R.S., I. (3)
Screws, keeper, striker, M.L.E.R.S., I.*
Screws, plate, butt, iron. (2)
Screws, protector, sight, slide, hand-guard, rear, M.L.E.R.S., I.
Screws, sear, M.L.M.
Screws, sight, dial, M.L.M., pivot.
Screws, sight, dial, M.L.E.R.S., I., fixing.

	Screws, spring, sight, aperture, M.L.M.
	Screws, spring, sight, back, M.L.E.R.S., III.
(g)	*Screws, spring, trap, butt plate, M.L.E.R.S., I*.*
(a)	*Screws, spring, trap, butt plate, M.L.M.*
(h)	*Screws, strap, butt plate, iron. (3)*
	Screws, swivel, M.L.E.R.S., I. (3)*
	Screws, windgauge, backsight, M.L.E.R.S., III.
	Sears, M.L.E.R.S., I.
	Sights, aperture, M.L.M.R.
	Slides, backsight, M.L.E.R.S., III., No. 2.
	Springs, catch, slide, backsight, M.L.E.R.S., III.
	Springs, extractor, M.L.M.
	Springs, guard, hand, rear, double, M.L.E.R.S., I.
	Springs, main, M.L.M.
	Springs, retaining, breech bolt, M.L.M.
	Springs, screw, band, inner, M.L.E.R.S., I.
	Springs, screw, windgauge, M.L.E.R.S., III.
	Springs, sear, M.L.M.
	Springs, sight, aperture, M.L.M.
	Springs, sight, back, M.L.E.R.S., III.
	Springs, sight, dial, M.L.M.
(i)	*Springs, stud, fore-end, M.L.E.R.S., I.*
(a)	*Springs, trap, butt plate, M.L.M.*
(g)	*Springs, trap, butt plate, M.L.E.R.S., I.*
	Springs, windgauge, backsight, M.L.E.R.S., III.
(j)	*Stems, swivel, butt, M.L.E.R.S., I.*
(j) & (b)	*Stocks, butt, M.L.E.R.S., Cond. II.*
(a)	*Stocks, butt, M.L.E.R.S., Cond. II*.*
*	*Stocks, fore-end, M.L.E.R.S., I.*
	Swivels, band, M.L.E.R.S., I. (2) [Also I]*
	Swivels, piling, M.L.M.R., II.
(a)	*Traps, butt plate, M.L.M.R., II.*
(k)	*Traps, butt plate, M.L.E.R.S., I*, No. 2.*
	Triggers, M.L.E.R.S., I.
	Washers, bolt, stock.
	Washers, guard, hand, or rivet, fore-end, M.L.M. (2)
	Washers, guard, hand, or rivet, fore-end, M.L.E.R.S., I. (2)
	Washers, pin, axis, backsight, M.L.E.R.S., I.
	Washers, screw, fixing, dial sight, M.L.M.
	Washers, spring, band, inner, M.L.E.R.S., I.
	Windgauge, backsight, M.L.E.R.S., III.
	Worms, fine-adjustment, backsight, M.L.E.R.S., III.

(a) *In rifles converted from Mark II* only.*
(b) *In rifles converted from Mark II only.*
(c) *In rifles converted from Mark II* and in rifles converted from Mark II if they have been fitted with Mark I* butts as spare.*
(d) *Common to M.L.E.R.S., I** in Naval Service only.*
(e) *In such rifles converted from Mark II, where the steel butt plates have been replaced by the M.L.E.R.S., I*., No. 2.*
(f) *Two for hand-guard cap and one for sight protector top.*
(g) *In such rifles converted from Mark II, where the steel butt plates have been replaced by the M.L.E.R.S., I., No. 2.*
(h) *Not in rifles converted from Mark II, except where they have been fitted with Mark I* butts as spare, when two will be found.*
(i) *Only on rifles fitted with these components before conversion.*
(j) *Not in rifles converted from Mark II when they have been fitted with Mark I* butts as spare.*
(k) *On brass buttplates M.L.E.R.S., I*., No. 2.*

An alteration to the ranges of .22-in. Short Rifles Marks I*, II and III and the .22-in. Long Rifles Marks I, I* and II was approved on 26th June 1914 and tabled as:

Range		Approx. elevation required: Long rifle	Short rifle
25 yds.	...	... 400 yds.	... 300 yds.
50 yds.	...	... 450 yds.	... 450 yds.
100 yds.	...	... 525 yds.	... 550 yds.
150 yds.	...	... 650 yds.	... 725 yds.
200 yds.	...	... 750 yds.	... 850 yds.

Production facilities were hard pressed after the outbreak of the Great War in 1914 and though there was a greater demand for the .22-inch training rifles, wartime requirements placed more importance upon .303-in. rifle production. A system of sleeving barrels to .22-in. calibre had been developed and commercially marketed by A.G. Parker & Co. Ltd. of Birmingham and the next three .22-in. conversions applied this system. Effected in Birmingham by A.G. Parker, these conversions permitted the Ordnance factories to allocate hard-pressed machinery and facilities to more important .303 production.

The firm of A.G. Parker, Birmingham, goes back to 1880 when Mr. A.G. Parker, a well-known rifleman of the Royal Warwickshire Regiment started to supply fellow rifle shooters with certain shooting aids. The business developed slowly and by the turn of the century he invited his nephew Mr. A.T.C. Hale to become a partner. Hale was also a competitive shooter and joined the Volunteers, he took a half share in the company. Foundation of a small-bore movement increased the demands, and problems which were being encountered with the Morris Tube paved the way for development of the Parkerifling by Hale.

By 1914, the firm was well established with a small manufacturing plant and orders followed progressively for the Parkerifling and Hiscock Parker magazines during the Great War. A.G. Parker died in 1915 and Hale's second son, Arthur, joined the business in 1926 along with three other brothers by 1932. A.T.C. Hale died in 1938. The firm of A.G. Parker and Co. became Parker-Hale in 1936 and P-H Arms Co. in 1940 and contributed to military rifle production during the war years and converted the .22 No. 9 trainers in the late 1950's.

This firm is not to be confused with A.J. Parker which was founded early in the 20th century by the son of A.G. Parker and continues business today in Birmingham. Alfred J. Parker was also a designer and manufacturer of shooting accessories and he was involved in the manufacture of ordnance components during both World Wars.

The new trainer with Parkerifled insert was called the .22-in. R.F. Pattern 1914 Short Rifle, approved on 4th May 1914, it was a conversion of the .303-in. S.M.L.E. Converted Marks II and II*. Like the .22-inch Short Rifle Mark III, both sides of the butt socket were already marked and so the new conversion markings were stamped below the original ones on the right side of the butt socket.

The official List of Changes announcement reads:

Para 17320:-
Rifle, Short, .22-inch, R.F., Pattern 1914.
Converted from Rifles, Short, M.L.E., Converted, Marks II and II.*

A pattern of the above-mentioned rifle has been approved to guide conversion as may be ordered.

Tubed barrel of a .22 Short Rifle Pattern 1914 No. 1. Note the silver solder ring at muzzle. *MOD Pattern Room*

Above & below: Other features of the .22 Short Rifle Pattern 1914 No. 1 trainer; tubed chamber and offset firing pin in the bolthead. *MOD Pattern Room*

The conversion consists, generally, in boring out the barrels of Rifles, Short, M.L.E., Converted, Marks II and II, and fitting and soldering a tube in the barrel; the tube being bored, rifled and chambered to suit the cartridge, .22-inch, R.F., No. 1 Mark I (§ 13642), and in converting and supplying the following components:*

Converted Components.

Bolt-head: Plug fitted and brazed in striker hole; firing pin hole bored and firing pin fitted; and extractor seating lowered.

Extractor: Hook made narrower and radius decreased.

Striker: Conical point removed.

Windgauge: Shape of V-notch altered to U, and sighting edge repaired.

New Components fitted.

Pin, firing, R.F. Rifle, Short, Mark I.
Blade, foresight, M.L.E. Rifle, Short, Mark III, .955-in.

The following components of Rifles, Short, M.L.E., Converted Marks II and II are not used in this conversion:*

Magazine, assembled.
Barleycorn, foresight.
Guide, charger, head, breech bolt.
Screw, stop, charger guide.
Cut-off.
Screw, cut-off.

All other components of the converted rifles are common to Rifles, Short, M.L.E., Converted, Marks II and II.*

Until 1915, all the .22-inch Lee-Enfield training rifles were single loaders and had the .303 magazines removed, which allowed the fired cases to fall out through the magazine recess upon extraction from the chamber. In order that training could be had in magazine fire, a special .22 magazine called the Parker Hiscock (or Hiscock-Parker) was approved on 21st December 1915 for use with the .22 Short Rifle Mark III and Pattern 1914 conversions for Land Service. It was promulgated in the List of Changes along with modifications required for the bolthead and extractor.

Para 17621:–

Magazine, .22-inch rifle, Mark I.
For Rifles, Short, .22-inch, R.F., Mark III and Pattern 1914.

Introduction.

A pattern of the above-mentioned magazine has been approved to govern supplies as may be ordered.

The following alterations are required to Rifles, Short, .22-inch, R.F., Mark III and Pattern 1914, when fitting the magazine locally:

The front bottom corner of the bolt-head to be ground off for about 1/8th-inch to an angle of 45 degrees, and the sharp edges of the bevel at the face and underside of the bolthead to be rounded off to a radius of about .03 inch.

The front bottom corner of the hook of the extractor of Rifles, Short, .22-inch, R.F. Pattern 1914, will be bevelled.

Instructions for Use.

Draw back the bolt of the rifle, turn the handle of the cut-off axis pin of the magazine to the 6 o'clock position, and insert the magazine in the rifle.

Turn the handle of the magazine cut-off axis pin clockwise as far as the fore-end will permit; insert, one at a time and rim first, five No. 1 .22-inch R.F. cartridges in the aperture on the right side of the magazine, and see that they lie horizontally in it; then press the cut-off axis handle down to the 6 o'clock position. The rifle can then be loaded by operating the bolt in the usual manner.

N.B.: In operating, the bolt must be fully closed, as partial closing and then withdrawing and again attempting to close will cause "double loading."

The magazine should always be removed before commencing to clean the bore of the rifle.

Instructions for stripping and assembling.

To strip: Remove the magazine from the rifle. Holding the magazine in the left hand, depress the auxiliary magazine with the thumb of that hand; then, with the finger and thumb of the right hand, turn the handle of the cut-off axis pin, clockwise, to the 12 o'clock position, and pull outwards. The auxiliary magazine, with the cut-off compression pin and magazine spring can then be removed from the case.

To assemble: Insert the magazine spring, long arm at the bottom, and bend to the front. Insert the cut-off compression pin in the auxiliary magazine, flat end first, and insert the auxiliary magazine in the case, fully compressing the magazine spring. Then, turning the magazine upside down to keep the cut-off compression pin in position, insert the cut-off axis pin with the handle at 6 o'clock, and sharply press home.

The external form of this case is like the standard .303 magazine and so it fits into the rifle in an identical way, but here the similarity ends. A machined block is fitted inside the magazine case, with an integral 5-round .22 magazine and a lever on the right side which depresses the .22 loading platform to enable the cartridges to be dropped in through the loading port.

The .22-inch Magazine Mk I fitted into the rifle, bolt open to reload from the magazine. *MOD Pattern Room*

The top of the machined steel magazine block insert is stamped "HISCOCK PARKER" along with the patent acknowledgments 189/12 and 25533/13. Enfield inspection stamps are also evident on the right side of this steel block insert, positioned just behind the .22 cartridge loading port, where it is marked with the magazine serial number. All the Parker Hiscock magazines examined have had a "G" prefix series, 4-digit serial number.

The Birmingham firm of A.G. Parker contracted to supply 1,000 Parker Hiscock magazines in June 1915, delivered by 26th July. The unit price was £1/7/-. Subsequent orders were for 4,000 in January 1916, 1,000 in January 1917 and a further 1,000 in April 1917. Thus the total quantity of these special magazines employed in British service does not appear to have been more than 7,000.

In September 1920, the Parker-Hiscock magazine was abolished; it was reported to have had little practical use as originally intended for practice in rapid fire and it was found to be unreliable unless properly looked after and well maintained. As the principal use of this item was with untrained users, it was reported to have malfunctioned quite regularly.

In 1915, the .22-inch R.F. No. 1 Smokeless Cartridge Mark I was approved, for use with the various miniature training rifles in service. This replaced the black powder rimfire loads which had been in service since about 1905.

Two more .22 conversions were approved on 28th April 1916, the .22-in. Long Rifle Pattern 1914 and .22-in. Short Rifle Pattern 1914 No. 2. Introduction of the latter also caused a change in the nomenclature of the .22-in. Short Rifle Pattern 1914 approved the previous year.

Their announcement in the List of Changes was:

Para 17755:-

Rifles, Short, .22-inch, R.F., Pattern 1914, No. 2. Converted from Rifles, Short, M.L.E., Mark III and Converted Mark IV.

Rifles, Long, .22-inch, R.F., Pattern 1914. Converted from Rifles, C.L., M.L.E., Mark I.*

1. Introduction.

Rifles, Short, .22-inch, R.F., Pattern 1914, No. 1. Converted from Rifles, Short, M.L.E., Converted Marks II and II.*

2. Change in nomenclature.

1. Patterns of the above-mentioned rifles have been approved to guide conversion as may be ordered.

The conversion consists generally in boring out the barrels and fitting and soldering a tube in the barrels; the tube being bored, rifled, and chambered to suit the cartridges .22-inch R.F., No. 1, Mark I, § 13642, and in converting and supplying the following components:

Converted Components.

Bolt-head: Plug fitted and brazed in striker hole; firing pin hole bored and firing pin fitted; extractor seating lowered. [These bolt heads were marked."22 No. 2 C" on the wing.]

Extractor: Hook made narrower and radius decreased.

Striker: Conical point removed.

Sight (short rifle): Leaf lined for 25 yards range, on top face.

Sight (long rifle): Sight bed lined for 25 yards range, on right hand side.

New Components fitted.

Pin, firing, R.F. rifle, Short, Mark I.

The following components of rifles, short, M.L.E., Mark III and Converted Mark IV, and Rifles, C.L., M.L.E., Mark I, are not used in this conversion:*

Magazine, assembled.
Cut-off.
Screw, cut-off.

The reference in § 16044 to rod and brush cleaning, and the instructions with regard thereto, apply to these rifles also.

2. In consequence of the introduction of the Rifle, Short, .22-inch, R.F., Pattern 1914 No. 2, the rifle described in § 17320 will be as now shown. [Rifle, Short, .22-inch, R.F., Pattern 1914, No. 1.]

Both of these trainers were converted from patterns in current use, with the obvious intention of providing education in the sighting, shooting and general handling. Again, the Parker rifled tube was used, the conversion (of the short rifle, at least) carried out by A.G. Parker in Birmingham; the receivers were stamped in a similar fashion to the .22-inch Pattern 1914 Short Rifle, as illustrated previously. The .22-inch Pattern 1914 Short Rifle No. 1 outwardly resembles the Mark I S.M.L.E. while the No. 2 described above looks like the Mark III S.M.L.E. Like the other .22-inch training rifles, the cut-off and magazine were removed. The 25-yard mark on the backsight is in the approximate position of the .303-inch, 400 yard mark.

Approval for the .22-inch Long Rifle Mark II and the newly-introduced .22-inch Pattern 1914 Long Rifle to have the trigger action converted from single to double pull was given on 5th August 1916, so as to be similar to the current pattern of S.M.L.E. rifle.

Para 17881:-

Rifles, Long, .22-inch, R.F., Mark II, and Pattern 1914.

Conversion of trigger action from single to double pull-off.

Conversion of the trigger action of rifles of the above-mentioned patterns, §§ 16042 & 17755, from single to double pull-off has been approved.

The rifles will be despatched to the Royal Small Arms Factory, Enfield Lock, for this service to be carried out, after it has been ascertained from the Chief Inspector of Small Arms that it is convenient to receive the rifles.

The conversion will be carried out by altering, replacing, or removing certain existing components, as follows:

Components to be altered.

Body: Alter the sear seating to allow the sear nose to bear higher up on the full cock bent of the cocking-piece.

Bolt, breech: Grind a flat on the bottom of the recoil stud top to clear the sear nose.

Cocking-piece: Alter the angle of the full cock bent by grinding.

Components to be replaced.

Sear: To be replaced by Sear, M.L.E. Rifle, Short.
Trigger: To be replaced by Trigger, M.L.E. Rifle, Short.

Components to be removed and not replaced.

Catch, safety.
Pin, catch, safety.
Spring, pin, catch, safety.

Great War Contract Records show orders for 1,702 .22RF training rifle conversions placed with R.S.A.F. Enfield in 1917; these would probably have been the .22 Short Rifle Mk III rather than, say, the .22 Patt. 1914 models, as the latter appear to have been relatively straight-forward conversions done by the trade as required. The earlier .22RF models were usually more involved conversions because of the use of older actions and the alteration of a variety of component parts and furniture.

The wartime .22RF conversion contracts were generally allocated to the trade. The ".22 Short Rifles converted from Rifles M.L.E." list 3,000 to Westley Richards and 3,329 to A.G. Parker & Co. Ltd., although the specific pattern to which the Lee-Enfield rifles were converted is not indicated. It is considered that they may have been the .22 Short Rifle Mk III pattern. A total of 1,500 .22 Patt. 1914 Short Rifles were also converted from S.M.L.E. Cond. Marks II and II* by A.G. Parker & Co. during this period, in late 1914 and 1915.

A new aperture rearsight which fits onto the long range aperture was approved on 24th April 1917, for use with any .22-inch Short Rifles, so as to duplicate the sighting of the .303 Pattern 1914 rifle. A new, long range aperture rear sight was fitted which has vertical calibrations, while the plate has graduations for horizontal adjustment. The description from the List of Changes reads:

Para 18981:-

Rifles, Short, .22-inch, R.F.:
Sight, auxiliary, aperture, Mark I.
Consisting of aperture sight, aperture plate, bracket, spring, 3 screws, and washer.

Introduction.

A pattern of the above-mentioned sight has been approved to govern supplies as may be ordered.

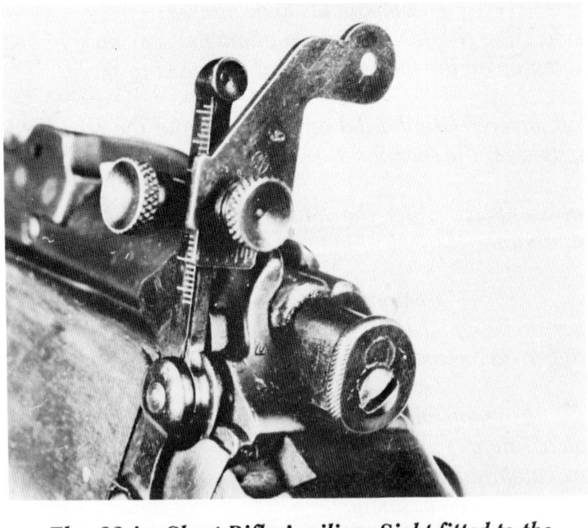

The .22-in. Short Rifle Auxiliary Sight fitted to the modified long range aperture. *MOD Pattern Room*

The sight consists of a M.L.M. rifle aperture sight with a scale marked on the left side for vertical adjustment. On this is fitted a bracket, held in position on the pillar of the aperture sight by a fixing screw; an aperture plate with fixing screw and washer, and a tension spring with fixing screw.

The rear face of the bracket is marked with a scale for the lateral adjustment of the aperture plate, each division of the scales corresponding to 1 inch on the target at 25 yards.

The auxiliary aperture sight, when fitted to Rifles, Short, .22-inch, R.F., in place of the existing aperture sight, approximates the system of sighting to that of Rifles, Magazine, .303-inch, Pattern 1914, for training purposes.

Parker-rifled tubed barrels were products of Great War economics and the next .22-inch trainer was the last of the S.M.L.E. conversions to use this system. The .22-inch Short Rifle, Pattern 1918, was approved on 10th July 1918 and is unique in that the bore is .22-inch but the chamber is for the .303-inch cartridge. This is because a conveyor and dummy cartridge like the .303 is used, chambered and bored for the .22 rimfire round. This enables the charger-loader and magazine to be used in an identical manner to the service .303 rifle. The bolt was converted in a similar fashion to the .22-inch Pattern 1914 trainers, so the Pattern 1918 externally resembles the Mark III or Converted Mark IV from which the rifle was altered, except for a special ejector on the left wall of the receiver.

On the M.O.D. Pattern Room sample, the right side of the butt socket, below the original markings, has been stamped ".22 PATT '18," although the production line examples may differ regarding some of these markings. The barrel is marked "PARKERIFLED" and the top of the receiver is stamped ".303 CUM .22 SYSTEM." Thirty adaptors and a tool for removing the spent .22 RF cartridges were issued with each rifle. The official announcement in the List of Changes best describes the technicalities:

Para 21675:-

Rifle, Short, .22-inch, R.F., Pattern 1918.
Without conveyors; converted from Rifles, Short, M.L.E., Mark III, and Converted Mark IV.

Conveyor- 30 per rifle.
Tool, removing, cartridge case- 1 per rifle; for removing cartridge case from conveyor.

Patterns of the above-mentioned rifle, conveyor and tool have been approved to guide conversion and manufacture as may be ordered.

The conversion consists generally in boring out the barrels of Rifles, Short, M.L.E., Mark III, and Converted Mark IV, and fitting and soldering a tube in the barrel; the tube being bored and rifled to suit the cartridge, .22-inch R.F., No. 1, Mark I, § 13642, and the barrel chambered, and tube cut to length to suit the conveyor. The following components also are converted and supplied:

Converted Components.

Sight leaf: Marked for 25 yards range.

25-yard mark on the rearsight of the .22-in. Patt. 1918 rifle. *MOD Pattern Room*

Bolt-head: Plug fitted and brazed in striker hole; firing pin hole bored and firing pin fitted; and ejector clearance cut.

Body: Bullet lead altered; slotted, drilled and tapped for ejector and screw; and ejector recess patch fitted.

Case, magazine: Left front rib and right front lip adjusted.

Platform, magazine: Right front edge adjusted.

Spring, extractor: Set adjusted.

Striker: Conical point removed.

New Components fitted.

Ejector.
Pin, firing.
Screw, ejector.

Appurtenances.

Conveyor: Chambered and bored to receive the .22-inch R.F. cartridge. (Conveyors may be charged into the magazine from service Chargers, .303-inch cartridges, and loaded into the rifle from the magazine similarly to .303-inch cartridges.)

Tool, removing, cartridge case: The wire standard of this tool is passed into the hole in the conveyor and the .22-inch R.F. cartridge case driven out.

All other components of the converted rifle are common to Rifles, Short, M.L.E., Mark III, and Converted Mark IV.

For this "new" pattern, an initial, trials-type batch of 1,000 rifles was ordered, assembled from Drill Purpose rifles, following the general practice of using downgraded rifles for the purpose of conversion to .22-inch trainers. The contract for the supply of what was eventually 975 rifles and 5,000 conveyors was signed by W.W. Greener in February 1919. Complaints were made about the difficulty

Above: New ejector fitted to left side of receiver.
Below: Proprietry marks on the .22-in. Patt. 1918.
MOD Pattern Room

of loading the conveyors designed by Messrs. Parker and it was determined that there were problems with the design so it was declared obsolete in 1920. After the finish of the war, the shortages of training rifles no longer applied as there was no longer such a high demand.

Research through correspondence with the War Office during this Great War period indicates that the Parker system had been turned down a number of times, only to be put up again as a suitable system. Despite what appear to have been flat refusals, it was adopted and, of course, it was not long afterwards that Parker advertisements appeared in sporting and military equipment catalogues, with the tag "approved by the W.O." or "in service with H.M. Government." Most of these advertisements were placed in the 1920's, after the system was declared obsolete. Slightly modified, it was adapted for use with the Lewis and Vickers machine guns for "Non-Automatic Firing Practice on Miniature Ranges" and sold through the trade.

After the war, Parker tubing was no longer used for new .22-inch training rifles and the next model, approved on 19th November 1921, was the .22-inch Short Rifle Mark IV. Again, in some instances, part of the original

Photographs reproduced from a contemporary British publication on the Miniature Rifle Associations and similar types of rifles in use in Europe. The uppermost rifle is merely a .303 M.L.M. Mk II converted to smallbore by tubing, fitting a rimfire bolthead, and minor alterations to the sights. The War Office Pattern at bottom is a magazine type, rather than the more usual single-shot models chambered for the .297/230 and .22 RF cartridges.

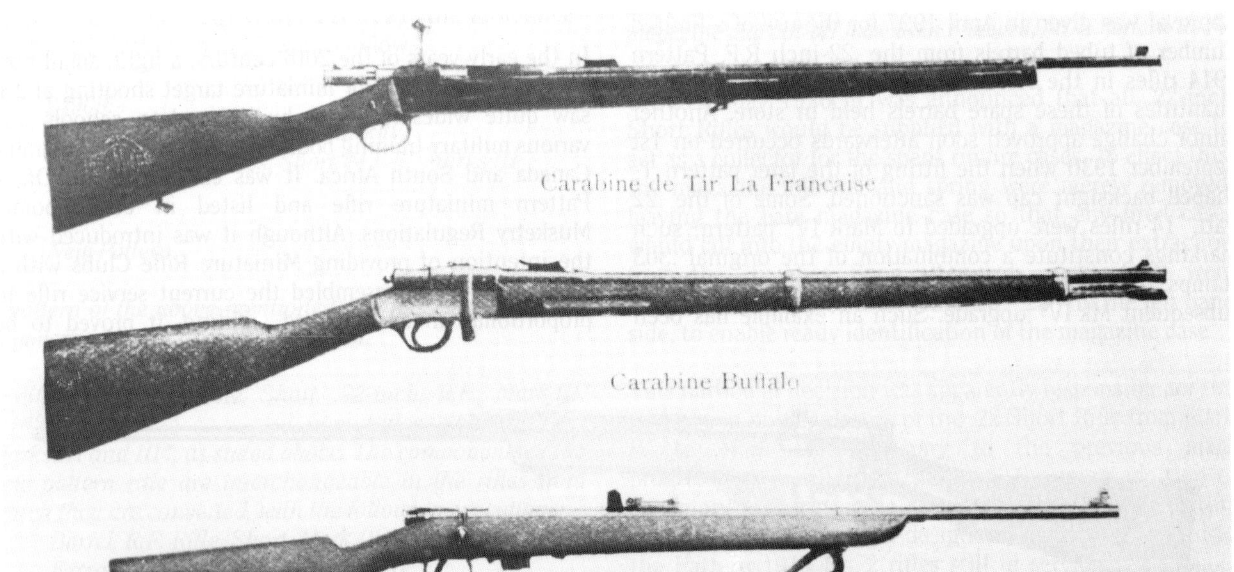

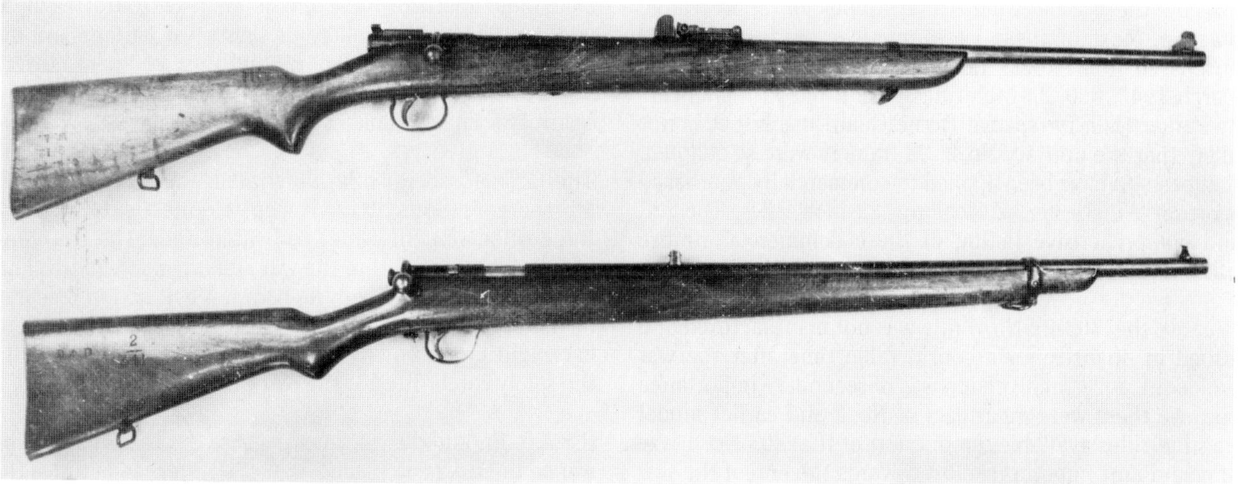

War Office Training Rifles with two different stock lengths. Other variations will be noted in bolt handle positions, cocking pieces and sights. *B. Labudda*

totally outclassed in competitive shooting by the Martini action club rifles then in general use.

The War Office Training Rifles were manufactured by B.S.A. and L.S.A. and quite a few variations may be encountered, mostly differing in the lengths of fore-end and bolt form (of the cocking piece, safety catch and position of the bolt handle). Most were half-stocked or three-quarter stocked and quite light in construction. Sighting graduations were generally from 25 to 200 yards and some were chambered for the .297/230 centre-fire cartridge as used with the Morris tube, although most seem to have been made for the .22 rimfire cartridge.

The illustrated rifles are typical, with an overall length of about 41 inches and a 23 to 24-inch barrel. Most were single loaders although a few were fitted with 5-round detachable magazines. The action resembles a Mauser perhaps more than a Lee-Enfield, and patent numbers are usually stamped on the bolt. Sighting was intended to be similar to that of the Mark III S.M.L.E. and even the backsight protectors bear some resemblance to this. Although their use was semi-official, they were purchased privately or by training units and so were never officially introduced or declared in the List of Changes.

Some .22 Patt. '14 No. 2, .22 Mark IV and subsequent No. 2 Mark IV* rifles in Canadian service were fitted with the No. 10A aperture rearsight. Whilst the flip-up sight with its screw elevation adjustment has obviously adapted from the Ross rifle sight, the base mount is special. This is marked "PATENTED 1925" and bears an inspection mark in addition to the usual Canadian stores mark, "C" and a broad arrow. The sight unit is mounted onto the receiver and secured by a screw which replaces the original ejector screw through the left wall of the receiver.

The base of this sight is marked "HWC" for Harold W. Cooey, the manufacturer, and the sight was "introduced into service to duplicate the aperture sighting of the service rifle sighting as much as possible." This would tend to indicate that the use of this sight in Canadian service was post-1940, but it could have been in use earlier because the No. 3 (Pattern 1914) rifle has aperture sights also. It is quite likely that it may have been used for service small-bore competition shooting also.

Canadian .22 Backsight. Note the "C↑" mark. *J. Cameron*

During the 2nd World War, many No. 1 rifles were converted to .22 No. 2 models, but in relatively small batches and as such requirements occurred. Quite a few of these were altered at Enfield where a few hundred were also made up as new rifles from components in store. The B.S.A. factory at Small Heath carried out some of the conversions while the remainder were effected by the trade. Parker rifled tubing was also reverted to in some cases, the barrels were tubed by W. Darlow of Bedford. These .22 No. 2 rifle wartime conversions were also undertaken by Alexander Martin, Parker-Hale and W.W. Greener.

The serious scarcity of training rifles during the early years of the 2nd World War resulted in the purchase of many .22 calibre commercial sporting and target rifles for use by the Home Guard and indoor or short range practice. Most of these were American made and many purchased from stocks held by the trade in Britain. In March 1942, 660 .22-inch rifles were fitted with silencers by Parker-Hale for special Home Guard use but it is not likely that the military No. 2 .22 trainers were so adapted. It appears to have been applied to commercially purchased sporting Winchester self-loading .22-inch rifles. The full and comprehensive listing of these commercial rimfire rifles will be found in "British Small Arms of World War 2."

Because the .303-in. No. 4 rifle was not manufactured and issued in quantity for a considerable time after the war broke out, a .22-inch version was of secondary importance because there were quantities of No. 2 and earlier model training rifles available; production of the .303 No. 4 was of paramount importance. But towards the end of the war as facilities became available and requests were made for a No. 4 based .22-in. training rifle, the B.S.A. Shirley factory looked towards developing such a rifle and produced some experimental and prototype models.

During the war, Canada also had a similar requirement and the demand there was made more acute by the non-availability of No. 2 rifles from England and because Long Branch manufactured the No. 4 rather than the No. 1 action. Some Cooey and Mossberg .22-calibre rifles were purchased, but a rifle was required which incorporated sighting and handling similarities to those of the No. 4. So a new small bore rifle was developed in Canada which became the No. 7 Mk I and its production commenced in 1944. Further details on this particular model and the special "ammunition-less" Long Branch Training rifle will be found in Chapter 8.

Canada had produced a miniature version of the .303 Ross military rifle and this was adopted by the Canadian government in 1912. Manufactured by the Ross Rifle Co., it is single shot model with straight-pull action, although the breech bolt is mechanically much more simple than its full-bore counterpart. The foresight is hooded and the adjustable aperture backsight is mounted onto the body end cap. These trainers are of Great War vintage and did not duplicate the action or sighting of then current service rifles; also they were also not available in sufficient quantity to meet Canada's requirements.

The .22 Mossberg rifle illustrated was made by O.F. Mossberg & Sons, New Haven, Connecticut. Many of these rifles were forwarded to Britain under the U.S. Lend Lease agreement during the war and are marked "UNITED STATES PROPERTY" on the barrel. A bolt action repeater, it has a 7-round box magazine and fine adjustment backsight mounted onto the left side of the receiver, with the safety catch fitted at the rear of the action.

R.S.A.F. Enfield did not manufacture the No. 4 during the war so there was no machinery set up there to produce the .22-in. No. 4 model or experimental series, even though an Inspectorate was maintained at Enfield which exerted some controls and standards on the wartime No. 4 rifles. For this reason, most of the early development of the .22-in. R.F. version of the No. 4 was undertaken by B.S.A. Shirley although it was still co-ordinated by the Ordnance Board and the normal design and experimental establishments.

Some attempts were made to modify the .22 No. 2 rifle sighting system to better resemble that of the No. 4, because one of the main purposes for these small bore trainers was to duplicate the sights of the full-bore rifle. To this end, some aperture sights were fitted to the .22 No. 2 rifles; in Canada, a Ross type backsight was utilised, mounted at the rear of the receiver. A similar British attempt is illustrated here which comprises a double-hinged type with No. 4 Mk I sight which can be swung over to the right so as to remove the breechbolt.

There were some high-level requests for a small bore rifle in July 1944, built on No. 5 lines, so a design was produced by C.E.A.D. and some more by B.S.A., one of which had a heavy barrel. Instructions were given for a

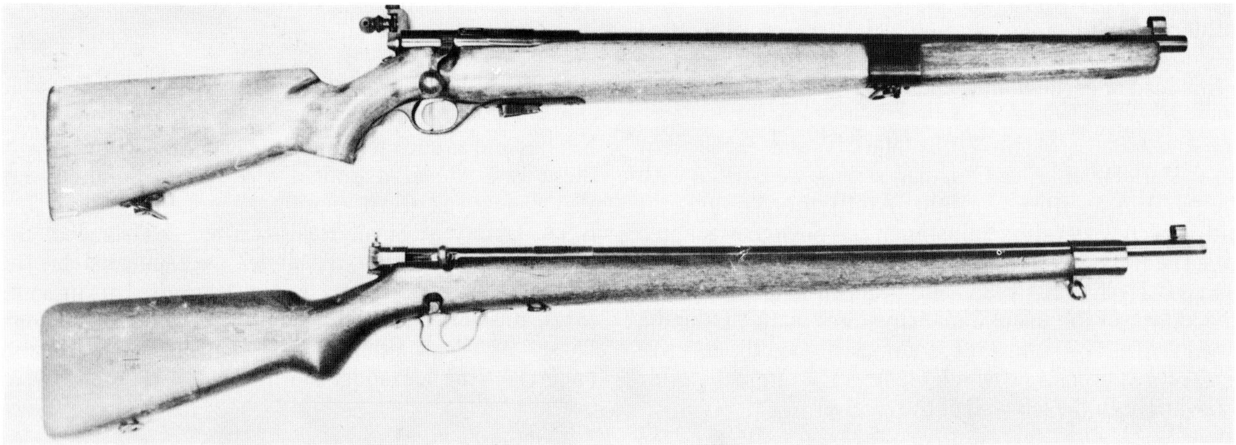

Trainers of North American origin; the Mossberg rifle (top) and Ross Cadet model (bottom). Some Mossbergs were Lend Lease during WW2. *MOD Pattern Room*

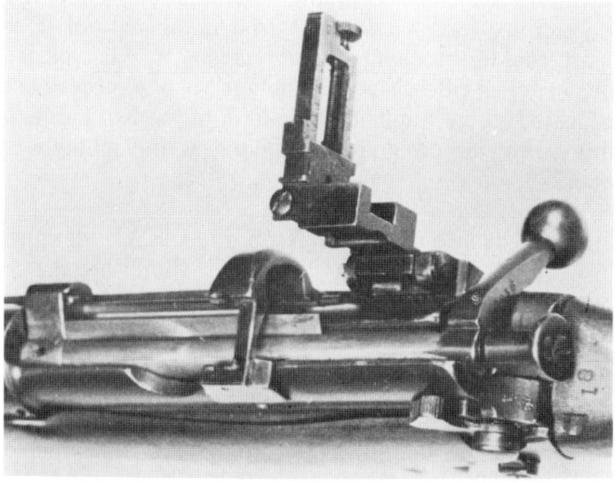

Hinged rearsight base modification on a .22 S.M.L.E. to duplicate the No. 4 sighting configuration. *J. Alley*

No. 4 style butt to be fitted as there was no requirement for a recoil pad (as fitted on the No. 5 rifle) with the .22-inch rimfire cartridge. As a result of other requests, a No. 4 was also set up in .22 calibre.

On 19th and 20th September 1944, three small bore rifles were tested; the .22 No. 5, the .22 No. 4 (set up at the request of the Society of Miniature Rifle Clubs) and the new Canadian .22-calibre No. 4, which became the C No. 7. One of the requirements for a .22-in. training rifle was to group 9 out of 10 shots into a 1-inch diameter circle at 100 feet and these three rifles proved to be very satisfactory. In view of the successful performance of the .22 No. 5, an order was placed with B.S.A. Shirley for a batch of twenty-five.

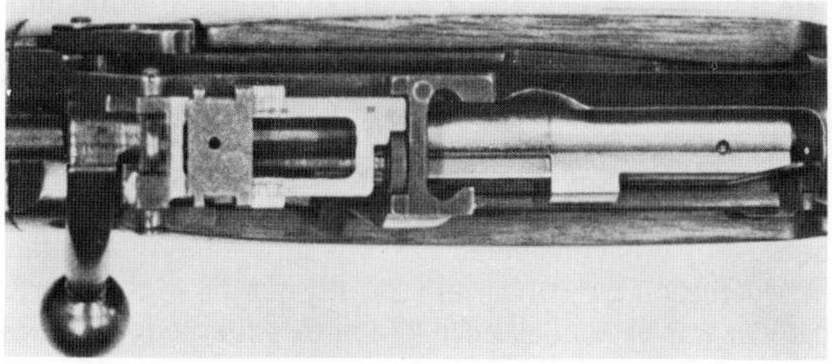

Top action view of .22RF No. 5 Prototype.
MOD Pattern Room

Although the S.M.R.C. preferred a No. 4 configuration for the proposed training rifle, the War Office requirement was for one built on the No. 5 style, mainly because the No. 5 was to be standardised as the service rifle and it was preferred that the service .22-inch trainer resemble the full-bore model. Experimental work being carried out with the .303-inch rifle at this time involved mounting the trigger onto the receiver instead of the trigger guard, making it possible to obtain a more regular trigger pull. This was obviously a desirable feature for the new .22-inch rifle, and such a design was recommended for incorporation into the next models produced. A plate was attached to the action body by three screws and the trigger mounted onto this plate. This necessitated modifications to the trigger guard and rear portion of the fore-end.

The prototype No. 5 .22-inch rifles, externally, were similar to the No. 5 rifle but the following batches of trials rifles were a little different. One hundred were ordered late in 1944 for extended trials and these rifles have chequering at the wrist and competition type sights. At first, the 6-hole Parker-Hale aperture sights were rejected in favour of a special sniper aperture backsight developed for the No. 4 but eventually the Parker-Hale sight was added to the sniper sight to form an excellent backsight. The bolt and bolthead system of the .22 No. 5 and the subsequent No. 8 and British No. 7 is unique, and one of the main features of this series of rifles.

Whilst a two-piece firing pin is used, as in the earlier .22 No. 2 rifles, here the bolt similarities end. The new bolt bodies are shortened and the boltheads lengthened, with a tapering rib towards the bolt face. To cock the bolt requires only an upward and downward movement of the bolthandle due to a new arrangement of the cocking piece and trigger. Although the bolts of the trials rifle and following Nos. 8 and 7 appear to be the same, they are not interchangeable and differ in length. The No. 8 model is the shortest in this respect and therefore has the smallest amount of bolt travel required in the operation of the action.

A distinction has been made between the No. 4 and No. 5 rifle bodies, because the latter is lightened. This is noticeable on the boltway to the rear of the charger bridge, and on the right side near the front, in the original location of the magazine cut-off axis screw, a carry-over

Overall view of the .22RF No. 5 Prototype. *MOD Pattern Room*

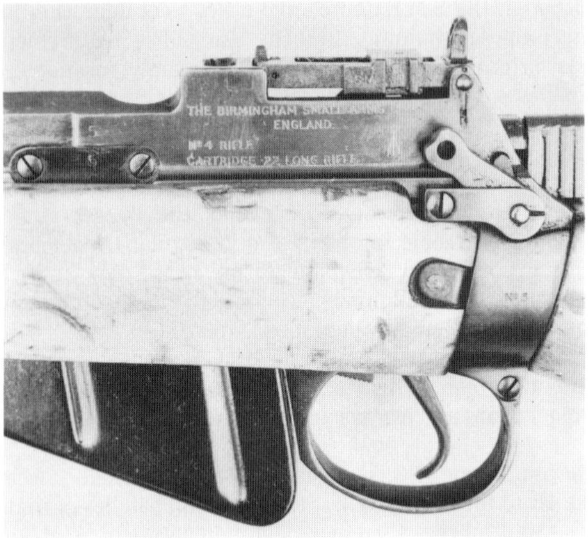

Prototype No. 7 Rifle by B.S.A. Co. *MOD Pattern Room*

from the trials No. 4 rifle which had the magazine cut-off; even though the No. 4 rifles did not have the cut-off, they still had the swelling or reinforce in this position. On the No. 5, this was machined flat. Some differentiation between the early .22-inch No. 5 prototype, experimental and trials rifles can be made in the style of receiver used, No. 4 or No. 5. Both may be encountered, depending upon the particular model of .22-inch trainer.

Even though the .22 No. 5 design was fairly well received, further work on this model was terminated when it was decided to withdraw the No. 5 from service at the end of 1947. A completely new design with a Mauser type trigger action was commenced, with a longer barrel and sighting radius to better resemble the No. 4 sighting. It was intended that the new model should be suitable for service training as well as for competitive shooting by the service teams and smallbore rifle associations.

Competitive club shooting was probably the main ideal of the new rifle and the development to date involved a certain amount of co-ordination with the Society of Miniature Rifle Clubs. When the .22 No. 5 trials rifle was assembled, civilian shooters and club members played quite an important part and they were involved in the testing of thirty of the rifles; the clubs had participated in a questionnaire and were invited to submit lists of desired features for the proposed small bore rifle. The previous attempt by the War Office in this direction had been the development of the War Office Pattern rifles at the beginning of the century, although this had been a dismal failure from the point of view of the Clubs because the accuracy fell far short of the Martini rifles then in common usage.

One particularly interesting experimental model of this series is held in the M.O.D. Pattern Room collection, it was made by B.S.A. in 1946 and it features a wide, target style fore-end and chequering. Ostentatiously enough, it is inscribed ".22 No. 6" and serial numbered SH 0008; apart from the stocking up and finer sights, it is typical of this series of rifles. The designation No. 6 was previously allocated to the jungle carbine form of the No. 1 rifle, so this particular designation marking is indisputably an error.

After the demise of the .22 No. 5, the C.E.A.D. designed two more models, with long and short barrels; in September 1947 after successful tests, a further ten of each were ordered from B.S.A. Shirley. The short barrelled rifles had Service sights and the long barrelled version was fitted with tunnel foresights and Parker Hale No. 4 aperture backsight. Two of the long and five short models were forwarded to the School of Infantry at Hythe for trials in 1948 and various samples had been sent to the Royal Air Force, Royal Navy and National Small Bore Rifle Association for tests and decisions regarding their possible adoption by the services and approval of the smallbore fraternity.

The C No. 7 was a Canadian development of 1944; the Air Ministry had ordered some .22 No. 4 rifles from B.S.A. prior to the adoption of the No. 8, so these particular .22 No. 4 rifles were allocated the designation No. 7, making the next available numeral, No. 8. Two rifles were ultimately put up for adoption which have been referred to as "Match" and "Infantry" models and for the want of better designations, this has been applied here also.

The Infantry (or shorter) model became the .22 No. 8 Mark I rifle and the longer Match model would appear to have been more target shooting orientated. The Match model has a 5.5-in. longer barrel which results in an overall length of 46.5 inches and it also has provision for a fine adjustment tangent rearsight. A choice of frontsights may be fitted in the tunnel assembly, but its action and stocking up is identical to the consequent service No. 8 model.

It was decided to adopt the shorter barrel for service use because it was considered that the difference in accuracy between each type of barrel was only marginal, the shorter pattern was better balanced and more suitable for training recruits. Just prior to its adoption, two small changes were made; a swivel was fitted to the butt and the Canadian style backsight was replaced by a pattern similar to the No. 4 Mk I, only graduated for 25, 50 and 100 yards and harmonisation calibration (H) so as to give 27-inch rise at 25 yards. On 9th September 1948, the Director of Artillery instructed that the new nomenclature be "Rifle, 0.22 inch, No. 8 Mark I" be applied.

Formal approval of the No. 8 Mark I for Land Service was announced by the Director on 24th September 1948; the Air Ministry was satisfied as to the suitability of the No. 8 for introduction into the Royal Air Force, but unfortunately they had already contracted for the supply of .22-in. No. 7 rifles with B.S.A. and thus were unable to take up any new rifles at that time. However, the Air Ministry did announce

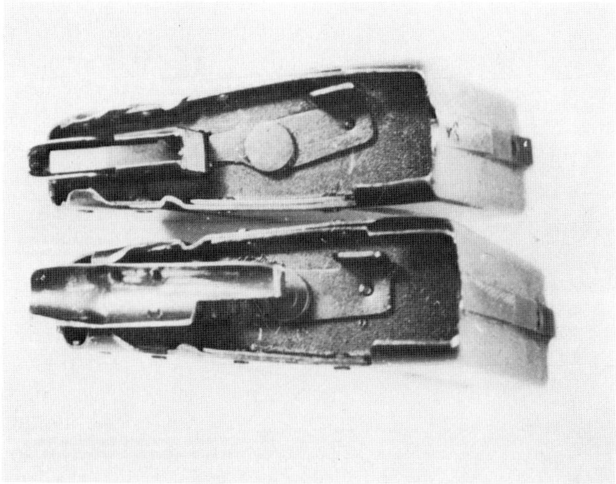

.22 No. 5 Trials model magazine options; repeating model (top) and single-loading model (bottom). MOD Pattern Room

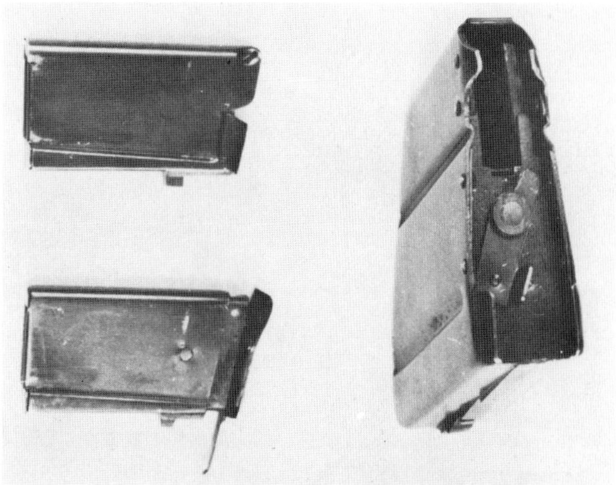

Both these .22 magazines are interchangeable in the outer .303 modified case, secured by the latch. MOD Pattern Room

Left action view of the .22 No. 5 Trials Rifle. Note chequered wrist, competition backsight and body markings, #T10027. MOD Pattern Room

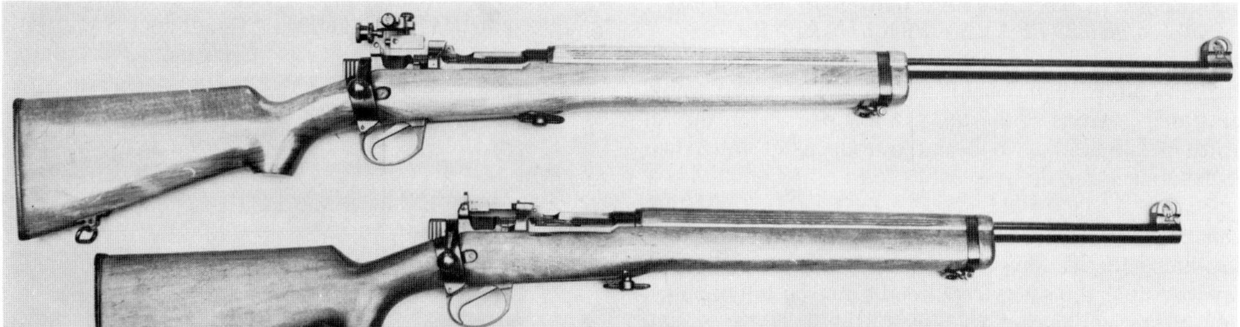

Match (top) and Infantry (bottom) Models of the No. 8 Rifle; pre-production options, although the former did not go into production. MOD Pattern Room

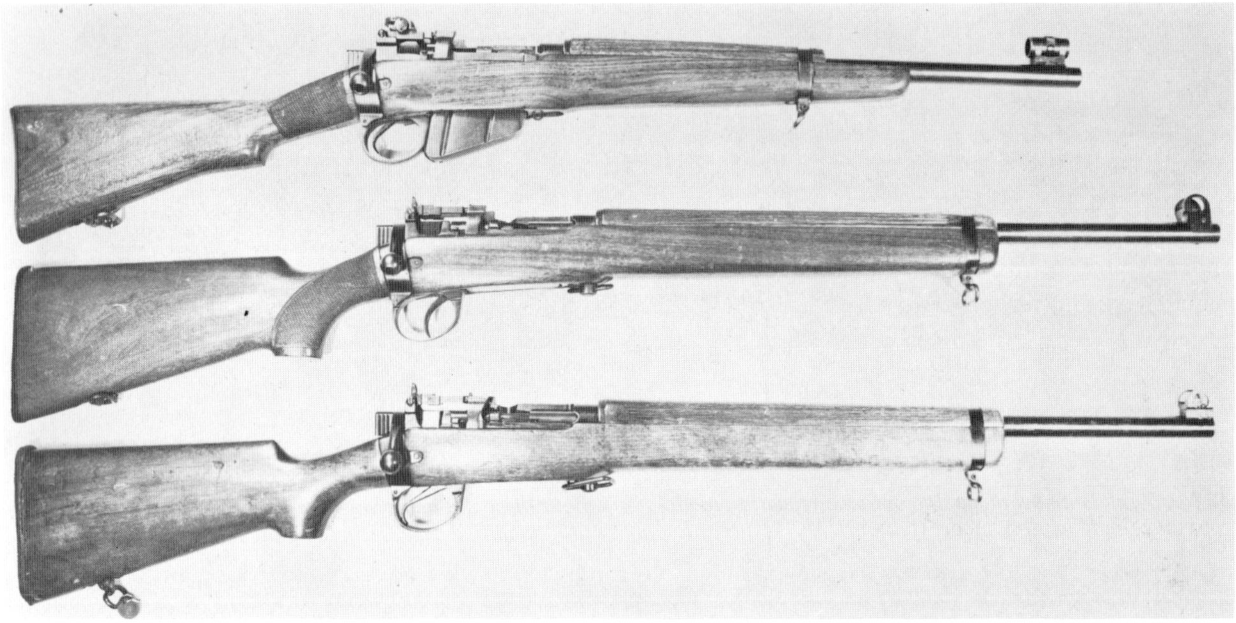

From top: .22 No. 5 Trials, Prototype No. 8, and Service No. 8 Models.
MOD Pattern Room

Backsights in raised positions; Match (left) and Infantry models of the .22 RF No. 8 rifle. MOD Pattern Room

The Parker Hale P.H. 4 backsight was designed for the No. 4 and it was therefore adaptable on the No. 8 but the vertical and lateral adjustment was in ½-minute steps only. The Navy was advised that a universal backsight for the .22 calibre rifle was being developed by the National Small Bore Rifle Association and was likely to become available in the future.

The No. 8 Mk I was listed in Vocabulary on 7th September 1950 and also promulgated in the List of Changes in relation to the introduction of long and short butts; some mention had already been made in 1949, although there was no full description or announcement in the List of Changes. This is another instance of a rifle being introduced some time before the official announcement; when first introduced it was for Land Service only and a requirement stated for 76,000 rifles; this first batch was produced at R.O.F. Fazakerley. It would appear that early that any future requirements for a .22-inch trainer would be stated in terms of the No. 8. A few years later, the Air Force acquired the No. 8 and introduced it into their instruction schedule.

One example each of the Match and Infantry models were forwarded to the Navy for trial and early in 1949 they reported that the safety catch could not be fully applied due to the omission of a groove in the body, the trigger pull was unsatisfactory, and the No. 4 Mk I type sight was inadequate for match or general precision shooting. Although the rifles were generally well thought of, it was pointed out that match shooting required a sight capable of vertical and lateral movement in ¼-minute steps and that there were no civilian type sights available commercially which might be readily attached to the rifle.

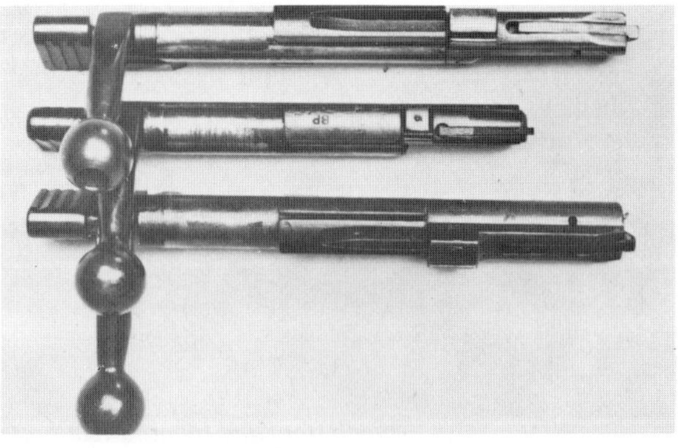

.22 RF Bolt comparison, top to bottom: Trials No. 5, No. 8 and No. 7 rifles. MOD Pattern Room

.22 No. 8 Rifle action. These were modified from No. 4 receivers, a considerable portion of the bottom was machined away and an adjustable trigger fitted. *MOD Pattern Room*

No. 8 rifles used the No. 5 body whereas some later production utilised the heavier No. 4 bodies, produced at Fazakerley. An F.T.R. program was effected at Enfield as the "U E" mark may be noticed on some of the No. 8 rifles.

Another batch produced in the early 1950's on a New Zealand contract may be found to have the B.S.A. manufacturer code mark M47C; these have "DA" serial number prefixes and utilise the No. 5 receiver. There is a strong possibility that the M47C markings on these particular rifles belong to the original No. 5 bodies used for conversion to the hung trigger system of the No. 8 trainer.

The new No. 8 trigger group was a significant advance in design, especially for a military rifle. Like the No. 4 Mark 2, it was no longer attached to the trigger guard, but here any similarity to the No. 4 Mark 2 ends. There is provision for full adjustment, which could be set up for single or double pull, a first in the military Lee-Enfield series. In 1950 there was a change in pattern of the body and foresight protector, although these changes were only minor. Fine adjustment sights for competition shooting were available through the commercial gun and related accessories trade.

Official details of the No. 8 Mk I rifle are:

Calibre:	*.22 in.*
Length of rifle, overall:	*41.75 ins.*
Length of barrel:	*23.9 ins.*
Weight of rifle:	*8.25 lbs.*
Number of rifling grooves:	*6.*
Pitch of rifling:	*1 turn in 16 in.*
Twist of rifling:	*R.H.*
Type of sights:	*Tangent.*
Range of sights:	*100 yds. max. + H.*
Stores Ref. of rifle:	*7B/1404.*

The Rifle, .22-in. No. 8 Mk I is a single shot, bolt action weapon designed for the dual purpose of the training of recruits and for competition shooting under the rules of the National Small-bore Rifle Association.

The essentials of the weapon are a heavy barrel which is capable of shooting very close groups, a trigger mechanism which can be set in the armoury for either double or single pull, and provision for the fitting of competition sights in place of the Service sights.

Used in its training role, with the trigger set for a double pull and with the Service sights fitted, the weapon closely resembles the No. 4 rifle in weight and handling characteristics.

Description.

Body: The body of the No. 8 rifle is the same as that of the No. 4 rifle. The ejector screw of the No. 4 rifle is present on the No. 8 rifle, but it serves only to plug the hole against the ingress of dirt near the bolt head. There are two additional tapped holes in the left-hand side of the body, one a 4 B.A. in the charger guide, and the other a 3 B.A. near the Service backsight mounting. The one near the backsight mounting is spot-faced and is plugged with a cheese head screw. These two holes and the 3 B.A. hole formerly accommodating the No. 4 rifle ejector screw, are provided to permit of mounting of a suitably-designed competition backsight.

Barrel: The barrel is machined from a solid forging, and is of a rigidity such that the vibration of firing is well within the maximum tolerance for close-group shooting. After the short tapered length at the rear end of the barrel, which constitutes the chamber, the bore is parallel. The rifling, however, is tapered throughout the length of the barrel; the grooves are deep at the chamber end, and run out to a smooth, or practically smooth, bore at the muzzle end. There are six rifling grooves with a constant pitch of 1 turn in 16 in., and the direction of twist is right-handed.

Bolt: The bolt is similar to that of the No. 4 rifle, but the striker is shorter. The firing pin is a separate component which is housed offset in the bolt-head; it is driven forward by the striker to strike the rim of the cartridge. The cocking piece is forced to the rear by a cam on the underside of the bolt. To cock the weapon it is sufficient merely to raise the bolt lever.

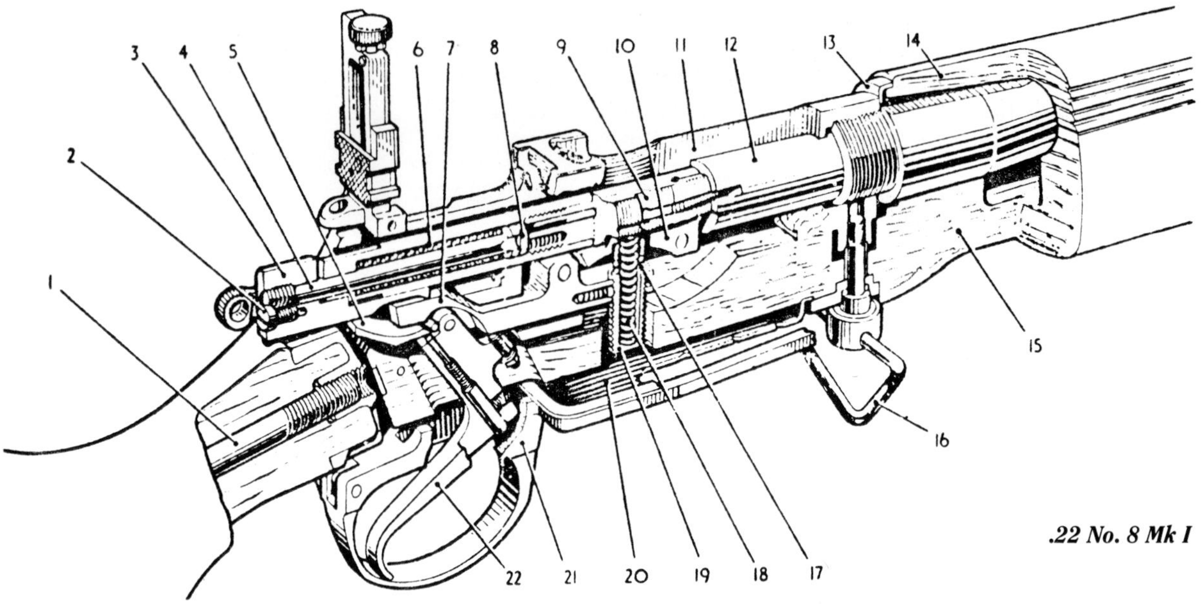

.22 No. 8 Mk I

1 STOCK BOLT
2 STRIKER SCREW
3 COCKING PIECE
4 STRIKER
5 SEAR CRADLE
6 STRIKER SPRING
7 SEAR
8 FIRING PIN
9 BREECH BOLT HEAD
10 CARTRIDGE PLATFORM
11 BODY
12 BARREL
13 REAR HANDGUARD RETAINING RING
14 HANDGUARD
15 FORE-END
16 TRIGGER GUARD SLING SWIVEL
17 OUTER (SERVICE) SEAR SPRING
18 INNER SEAR SPRING
19 SEAR SPRING CUP
20 TRIGGER GUARD FILLER
21 TRIGGER GUARD
22 TRIGGER

There is a single extractor in the bolt head which engages the rim of the cartridge on the right, a nick being machined in the end of the barrel to accommodate the extractor when the bolt is locked forward. When the bolt is drawn to the rear the cartridge is slid out of the chamber and along the cartridge platform. At the rear of the platform is a small projection which constitutes the ejector. The left side of the rim of the cartridge strikes this when the bolt nears the end of its rearward travel, and the cartridge is knocked off the face of the bolt head to be ejected to the right-hand side of the weapon.

Trigger mechanism: The trigger mechanism is specially designed to give a smooth, crisp pull-off. A very long sear is pivoted at one point well forward in the body of the rifle, on a level with the bottom of the cocking piece bent. Working from the same point as the sear is a sear cradle which extends to the rear under the cocking piece. The trigger is pivoted on the sear itself, and acts against fulcrum points on the sear cradle. The pressing of the trigger brings about a "scissor" action between the sear and the sear cradle; the sear is drawn down, while an equal upward pressure is applied through the sear cradle to the cocking piece. This counteracts the downward drag of the sear, and there is thus no axial movement of the bolt when the trigger is pressed.

The functioning of the trigger mechanism is obvious from the illustration. In referring to this illustration, it must be borne in mind that for single-pull operation the outer sear spring would be removed. This at once gives a considerable reduction in the weight of the trigger pull, and final adjustments to the poundage are made by screwing in or out of the sear cup to alter the initial compression of the inner sear spring. Adjustments to the trigger mechanism must be done by an armourer.

Furniture: The butt is of the pistol-hand pattern, with a fairly high comb, and has a black rubber butt plate. The rifle is provisioned with one of three alternative lengths of butt: long, normal, and short, there being ½-inch difference between each. The fore-end and handguard are secured at their forward end by a single band which also carries the front sling swivel. The fore-end is broader than is usual with Service weapons, so as to provide a firm and comfortable grip for the firer's forward hand.

There is an additional sling swivel which is carried on the screw securing the front of the trigger guard. The swivel works in a collar which is loosely spun on the head of the screw, and is thus capable of being turned through 360 degrees coaxially with the screw. For competitive shooting the sling is used in this centre swivel and the front swivel.

Sights: The rifle is issued with Service sights. These are similar to those of the No. 4 rifle and give the same aiming picture to the firer. The backsight graduations are for 25-yard, 50-yard and 100-yard firing, with an addition "H" graduation for firing on a landscape target. The foresight blade is mounted in a dovetail in the foresight protector, which is itself secured in a further and larger dovetail in the barrel band. Thus the foresight blade and protector can be removed complete from the barrel to enable a competition foresight of proprietary manufacture to be fitted in its place.

Safety devices: The safety catch and locking bolt are the same as those on the No. 4 rifle, except that the safety catch engages a blind hole instead of a cam groove. Also, the locking bolt recess in the cocking piece of the No. 8 rifle is semi-circular and when the safety catch is applied with the rifle uncocked, the cocking piece is not forced rearwards. When the safety catch is applied with the rifle cocked, however, the flattened end of the cocking bolt shank engages the half bent of the cocking piece, and forces the cocking piece slightly to the rear.

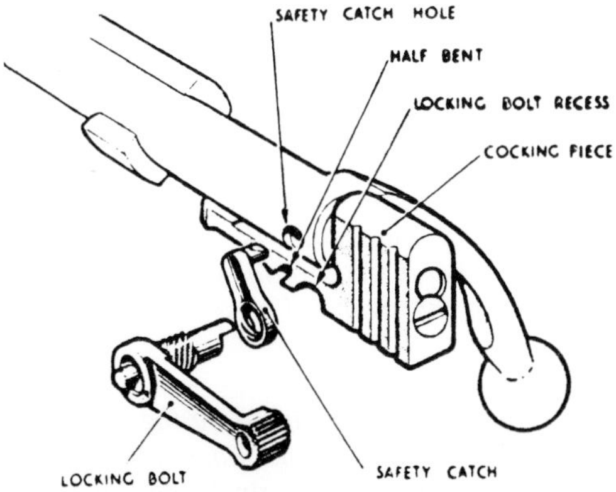

Removing the bolt from the rifle: To remove the bolt from the rifle, proceed as follows:

1. *Raise the backsight.*
2. *Press down the bolt head catch, and keep it down.*
3. *Raise the bolt lever and slide the bolt as far to the rear as it will go.*
4. *Release the bolt head catch and rotate the bolt head anti-clockwise.*
5. *Slide the bolt rearwards out of the bolt-way.*

During the course of its service life, some changes were made to the No. 8. These included a new pattern of firing pin in 1951 (.085-in. diameter with the pin riveted instead of brazed) and a new cocking piece and sear in 1952 (made from a higher grade of steel) which gave a longer life and more regular pull-off even after considerable wear.

The No. 8 was initially introduced for the Land service, the No. 7 being used by the Air Force and the No. 9 for the Navy, although the use of the No. 8 was extended in 1950. Both the No. 7 and No. 9 models externally resemble the No. 4, and are basically No. 4 rifles converted for .22-inch rimfire training functions.

The Canadian C No. 7 differs from the British B.S.A.-made No. 7 in a number of respects, including the form of the bolt and bolt-head, sights, magazine, body and ejector. The Canadian rifle is designated the "C No. 7 .22 Mk I" and marked as such on the receiver along with the factory name "LONG BRANCH" which makes it readily distinguishable from its British "name and number-sake."

Characteristic markings on the left side of the receiver of the British No. 7 are: No7 Mk1

Details on the British No. 7 from an Armourer's Instruction pam are:

Note: The physical weights and dimensions of this rifle are almost identical with those of the Rifle, .303-in., No. 4, Mk I.

Type of rifling: ...	... *Concentric.*
Number of grooves:	... *Six.*
Twist of rifling: ...	... *R.H., one turn in 16 in.*
Type of sights: ...	... *Tangent backsight, blade foresight.*
Range:	... *Rifle zeroed for 25 yards only.*

The .22-in. No. 7 Mk I rifle is a modification of the .303-in. No. 4 Mk I and is used for training purposes with the object of simulating the physical and functioning characteristics of the No. 4 rifle.

The major differences include modifications to the barrel, body and bolt assembly to suit the .22-in. rim-fired cartridge, together with the provision of an auxiliary five-shot magazine that fits inside a full-sized magazine case.

All other components, except those directly associated with the parts mentioned in the previous paragraph, are common to the No. 4 rifle and are interchangeable.

The rifle is zeroed to shoot at 25-yards range with the standard Mk I backsight set at its lowest position (200 yds), the foresight blade being slightly reduced in height to compensate for the lower muzzle velocity and consequent increased gravity drop of the .22-in. bullet.

The description given here refers only to the modified components of the No. 7 rifle. A full description of the No. 4 rifle together with details of functioning and instructions for use should be read in conjunction with this section.

Barrel: The bore of the rifle is fitted with a tube which is bored, rifled and chambered to suit the .22-in. long cartridge. The rifling is concentric, having six grooves that each make one right-hand turn in 16 inches. The tube is not removable.
Note: a number of these rifles are fitted with a new .22-inch barrel machined from a solid forging.

Foresight: The range of foresights is as provisioned for the No. 4 rifle, the one fitted being adjusted for height during manufacture to zero the rifle at 25 yards range with the backsight set at 200 yards.

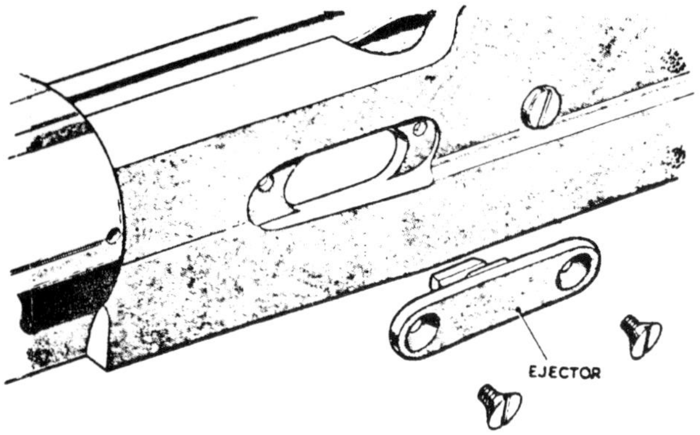

Breech assembly: The body is modified on the left side by a longitudinal recess, the centre of which is cut through into the bolt-way to accommodate a redesigned ejector nearer the breech face of the barrel. The seatings thus formed at each end of the ejector slots are drilled and tapped for the ejector retaining screws. The existing ejector screw is retained merely as a plug for the ejector screw hole and has no effect on the functioning of the rifle.

Note: a number of these rifles are fitted with a new body in which the ejector screw and ejection groove are omitted.

Ejector: The ejector consists of a locating plate which has the ejector blade machined integral with it on one side. The plate fits into the recess in the body, where it is retained by a screw at each end, while the ejector blade projects into the bolt-way to ride in the ejector slot in the bolt-head.

Bolt: The bolt is shortened from the front end to a length of 3.6 in., and is re-screwed internally to receive the bolt-head.

Housed inside the bolt is a spring-loaded striker, which in appearance, except for being slightly shorter, is similar to the normal striker with the front portion removed at the collar. The collar serves as a hammer for driving the firing pin forward in the bolt-head.

Cocking-piece: The cocking piece has a shorter arm, and consequently has no half bent. In all other respects it is identical with the normal cocking piece.

Bolt-head: The bolt head is longer than the standard one to compensate for the shortened bolt, thus making the assembled bolt and bolt head the same length as that for the No. 4 rifle. The reason for this, is the necessity for the lug on the bolt head to be further to the rear in order to shorten the bolt movement in keeping with the much shorter ammunition.

[Here the official text is incorrect as the No. 7 bolt is ½-inch shorter than its No. 4 counterpart, because the .22 barrel extends back a little into the boltway.]

The bolt head is drilled eccentrically and provided with a slot across the rear end to accommodate a spring-loaded firing pin. An annular groove, which cuts into the firing pin hole, is cut around the neck portion which screws into the bolt, and houses a circlip for retaining the firing pin.

Two extractors are fitted, the main extractor and an auxiliary one. The main extractor performs the function of withdrawing the cartridge case from the chamber while the auxiliary one helps support the case on the face of the bolt prior to ejection. [The "auxiliary extractor" is better described as a "cartridge steady" as its prime function is supporting the cartridge on the bolt face in the process of loading into the chamber, rather than in extraction.]

The firing pin is contained within the bolt head and has a flat plate at the rear which serves as an anvil for the striker. Mounted in the plate, parallel to the firing pin, is a short pin which supports the firing pin spring. The firing pin spring and its support work in a blind hole drilled in the rear end of the bolt head opposite the firing pin hole.

Magazine: The magazine holds five cartridges in a single column, and is housed inside a modified case of the standard 10-round magazine as fitted to Rifle, .303-inch No. 1 Mk 3. The magazine is removable from the modified case, either direct through the top of the body, or by first removing the modified case from the rifle in the normal way.*

The .303-in. magazine is modified by the removal of the platform and platform spring, and having the top enclosed at the rear by a flat cover plate riveted to the sides of the case.

Note: The two 3/16-in. diameter holes in this cover plate are for manufacturing purposes only.

Riveted to the inside of the case at the front end is a small rectangular housing which accommodates the .22-in. magazine in a vertical position.

The magazine, which consists of a flat rectangular case, platform, and platform spring, follows the conventional design for this type of magazine. A spring-loaded catch is for retaining the magazine in its housing, the top of the catch projecting above the cover plate of the magazine.

Functioning.

Action of firing mechanism: The action of the firing mechanism is practically identical with that of the No. 4 rifle having no half cock position. The only exceptions are the shorter bolt movement and a slightly different series of events in the withdrawal of the firing pin during primary extraction as described in the next paragraph.

When the bolt lever is withdrawn, effecting primary extraction, the striker is withdrawn about 1/8-in. in the normal way. This permits the firing pin spring to reassert itself and forces the firing pin to the rear for the same distance.

Instructions for Use.

Precautions before firing: Before firing takes place, the bolt is to be examined to ascertain that the auxiliary extractor;

1. Has a smooth, bright finish at the tip. If not, the tip is to be polished with a fine oil-stone or No. 00 emery cloth.

2. Is not deformed.

The correct form of auxiliary extractor is shown.

Examination of auxiliary extractor

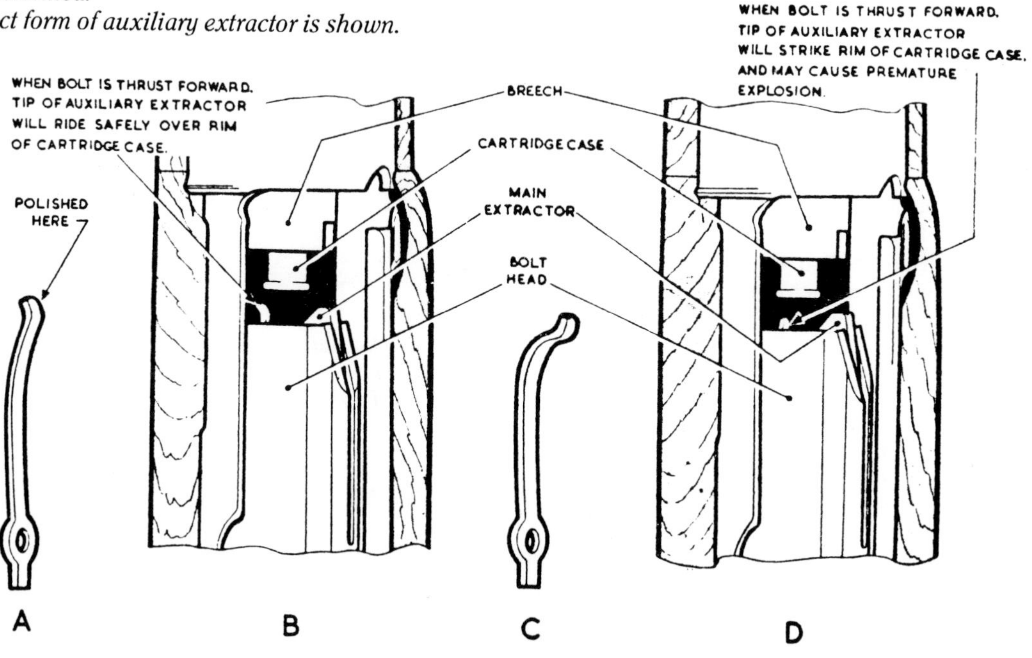

When the bolt and cartridge go forward during the normal feeding of a round from the magazine, the base of the cartridge case abuts the face of the bolt head, and the rim is engaged under both extractors. The deformed condition of the auxiliary extractor can be obtained in practice only if there is a feed fault from the magazine. Single-shot loading is prohibited with this rifle.

[The British No. 7 was also produced as a single shot rifle, in which the magazine was fitted with a spring-loaded platform or loading plate somewhat similar to that of the No. 8 rifle. In this instance the "auxiliary extractor" and screw were dispensed with; the modified No. 4 magazine was marked "7B/1061/M."]

Loading: On no account is the rifle to be loaded single shot by hand; it is to be loaded from the magazine in the following manner:-

1. *Push the locking bolt lever to the forward position with the thumb of the right hand to withdraw the safety catch form engagement with the bolt.*
2. *Grasp the knob of the bolt with the forefinger and thumb of the right hand, turn it sharply upwards and draw back the bolt to its fullest extent.*
3. *Withdraw the magazine from the rifle, complete with its housing, by first releasing the catch inside the trigger guard. Alternatively, withdraw the magazine from its housing through the top of the body, by first releasing the catch on the magazine.*
4. *Insert five cartridges by hand into the magazine by sliding them one at a time, base first, under the front lips of the magazine and pushing them as far as possible to the rear.*
5. *Replace the magazine in the rifle and ensure that the catches on both the magazine and trigger guard are correctly engaged.*
6. *Force the bolt sharply forward with the thumb and forefinger of the right hand, turn down the bolt lever as far as it will go and pull the locking bolt lever fully to the rear to engage the safety catch with the bolt.*

If it is desired to charge the magazine without loading the rifle, the top cartridge is to be pressed downwards with the thumb and held there until the bolt has passed over it.

Unloading: The rifle is to be unloaded using one of the following methods:

1st method:-

(a) Using the thumb of the right hand, push the locking bolt lever to the forward position to withdraw the safety catch from engagement with the bolt.

(b) Grasp the knob of the bolt with the thumb and forefinger of the right hand, turn it sharply upwards and draw back the bolt to its fullest extent.

(c) Work the bolt backwards and forwards without the bolt lever down until all the cartridges are removed from the magazine and chamber.

(d) Close the bolt and press the trigger.
(e) Apply the safety catch.

2nd Method:–
(a) Remove the magazine from the rifle as described in the paragraph on loading, and, using the thumb, push out the cartridges.
(b) Clear the chamber as described in 1st method.
(c) Apply the safety catch.

Firing: The procedure for firing the rifle is identical with that for the No. 4 rifle.

At a quick inspection, the bolts of the No. 8 and British No. 7 trainers seem to be identical, but there are certain dissimilarities which affect their operation considerably. The No. 7 has an auxiliary extractor or cartridge steady on the bolt head, the bolt body is longer and the cocking piece is also different.

.22 No. 7 Magazine; comprises a .22 magazine insert into a modified No. 1 magazine. Note the differences to the .22 No. 5 Trials model some pages earlier.

.22 Rifle No. 7 action with bolt open. Note the long bolt-head and shortened bolt body. *MOD Pattern Room*

The .22RF No. 9 training rifle was a special conversion of various Marks of the No. 4 rifle for the Royal Navy, effected by Parker Hale in Birmingham during the late 1950's. This conversion consists of tubing the .303 barrel to .22-inch in a similar manner to that applied to the Great War .22 Pattern 1914 rifles and by fitting a .22-in. R.F. bolthead with floating firing pin. The original striker has been shortened at the collar and so the 2-piece firing pin is of a similar design to the .22 No. 2 rifles and the Canadian No. 7. The No. 9 was only intended for use as a single loader and an empty magazine case was fitted as a container for the spent cartridge cases after ejection.

A new 25-yard mark was inscribed on the otherwise standard No. 4 backsight and the receiver re-marked with the designation ".22 R.F. No. 9 Mk 1" followed by "P-H" (Parker Hale) along with the year of conversion. It would appear that about 3,000 of these rifles were converted. Regarding the trigger mechanism, some have been noted with the hung trigger type as on the No. 4 Mk 2 rifle, while others have the earlier style mounted on the trigger guard, as on the No. 4 Marks I and I* models.

Markings and backsight of the .22 No. 9 Rifle. Assembled principally from existing No. 4 parts, the Mk I and Mk 2 series action bodies have been noted. *MOD Pattern Room*

Therefore, the No. 9 was the last of the British .22-in. conversions which spanned a period of more than 50 years. It was also the end of the Lee-Enfield numerical line after a service history which totalled more than seventy years; however the 7.62mm NATO conversions were to see the Lee-Enfield in service again, with the L42A1 sniper configuration also serving well in conflicts in the Middle East, Ireland and the Falklands.

After the application of the "L" prefix designations and introduction of the 7.62mm L1A1 self-loading rifle, some .22-in. training rifles were allocated following numbers in this new series. The No. 8 was also stripped of its wooden furniture for use as a sub-calibre trainer, mounted onto heavy, large-bore weapons and designated the L2A1 in British service.

Some .22 calibre conversion kits were fitted to the 7.62mm L1A1 rifle and some .22-in. self-loading commercially produced rifles were adopted for training. One of these was the Australian-made Sportco Model 71S which has a detachable 10-shot magazine and was designated the L29A2. The Heckler & Koch .22 RF conversion kit for the L1A1 rifle was adopted as the L12A1, enabling the service 7.62mm rifle to be readily converted for magazine-fire for use on small-bore ranges with the rimfire cartridge.

.22 No. 9 muzzle view. Note the "PARKERIFLED A.G.P." stamp on the crown, and sleeved liner. *MOD Pattern Room*

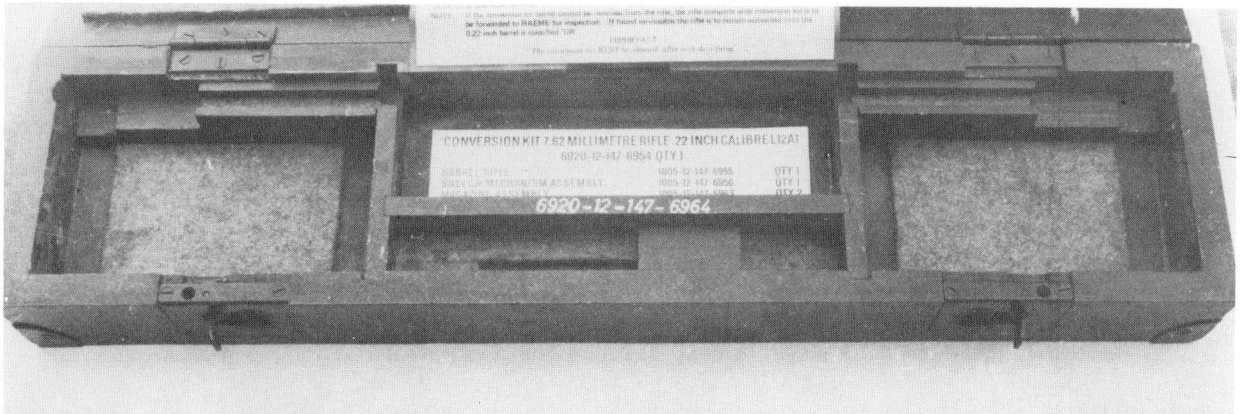

The L12A1 .22RF Heckler & Koch conversion for the 7.62mm L1A1 rifle. The Martini and early Magazine Lee rifles used Morris Aiming Tube inserts, succeeded by various conversions and then special purpose-built models after after the war. Reintroduction of the tube insert system with the L1A1 rifle is interesting as it represents the turning of a full circle. *R. Jarosz*

The Long Branch Factory.
Above: This early photograph was taken from a viewpoint which clearly shows the saw-tooth outline of the factory roof.

Below: This later aerial view is of Canadian Arsenals Ltd. Although the same site, the name was changed from Small Arms Ltd. at the end of 1945.

Both photos courtesy Canadian War Museum

NORTH AMERICAN PRODUCTION *CHAPTER 8*

The Ross Rifle Company commenced Canada's first military rifle production in 1903 and some component parts were initially brought in from the United States for assembly at the Ross Rifle factory until total self-sufficiency was achieved. But in March 1917, the factory was expropriated upon the collapse of the Ross Rifle Co. The Canadian government had signed a contract with Ross for production of the .303-in. Pattern 1914 which had been represented to the Canadian parliament as "the rifle of the future" but these plans were shelved largely because machinery was in short supply when the United States entered the Great War also.

The Ross Rifle Co. plant was taken in hand by the North American Arms Co. in 1918 and they intended to manufacture automatic pistols for the U.S. War Department but upon the signing of the Armistice, this contract was cancelled. A commission was formed to take over the premises and liquidate the business and it was proposed to have the site taken over by the National Battlefield Park. Most of the machinery was disposed of, but ironically, some of the equipment installed at Long Branch in 1940 originated from the original Ross Rifle Company in Quebec.

Britain's Pattern 1914 Enfield rifle was manufactured under contract in the United States by Remington and Winchester during the Great War and upon termination of the British orders, the Pattern 1914 was slightly modified for the U.S. .30-inch cartridge to become the U.S. Model 1917 rifle. This design was quite successful, partly because the Pattern 1914 rifle was originally set up for a rimless cartridge with certain similarities to the U.S. .30-06 round.

During the 1939-45 war, the No. 4 rifle was manufactured in both Canada and the United States; the Canadian production at Long Branch continued in a limited capacity until the 1950's. Special development was undertaken at Long Branch with some lightweight rifles and a .22-in. version of the No. 4. Wartime production of nearly one million rifles at Long Branch and a larger number at the Stevens-Savage factories was of immense assistance to Britain's struggle in Europe. The North American ordnance production took a huge amount of pressure off British manufacturing facilities as similar schemes were implemented with pistols, machine guns, machine carbines and anti-tank rifles. Similar programs were implemented in other divisions of munitions production.

Canada's Department of National Defence commenced a detailed study in 1937, into small arms production in the United Kingdom with a view to the commencement of manufacture in Canada; so when war broke out, positive steps had already been taken in this direction. On 6th June 1940, the Ordnance Branch of the Department of National Defence was authorised to commence building a small arms factory at Long Branch in Ontario. Long Branch is an outer suburb of Toronto, about 20 kilometres south-west of the city centre, on the north-western shore of Lake Ontario.

Although commenced under the auspices of the Dominion Small Arms Factory, a Crown Corporation, Small Arms Ltd., took over the operation of the plant on 7th August and the responsibility of the project was transferred from the Department of National Defence to the Department of Munitions and Supply. As a Crown Corporation, it was intended that S.A.L. operate on a cost basis, in that after allowances had been made for running and administrative expenses, and the plant depreciation, it would show no profit.

Considerable progress had been made on the new factory even by early August when S.A.L. was instituted, and arrangements regarding the site, buildings, machinery, key personnel and tools and gauges had been settled. The actual date of commencement of work on the site was 20th April 1940 and by early November 1940, some offices and 81,000 sq. ft. of factory floor space were ready, upon which machinery was delivered, a toolroom opened and the offices occupied. Soon afterwards, floor space was doubled and more machinery acquired because the proposed output was increased. The completed plant cost approximately $8,000,000.

Synonymous with its rapid wartime erection, the buildings were quite simple; the factory was single storey with saw-tooth type roof. Office accommodation was similarly unpretentious and facilities were utilitarian, directed at the earliest possible commencement production date. Most of the machinery was new, purchased in the United States, although some had originally come from the Ross Rifle Factory in Quebec, expropriated by the Canadian Government in March of 1917.

Great Britain looked towards North America's vast industrial potential at the outset of the war, when it became quite evident that it would be a protracted struggle in which industrial output and reserves would be critically relevant to the end result. Another major factor was the vulnerability of British industrial facilities to German air attacks; there was no such threat to North American production. A Joint Inspection Board for the United Kingdom and Canada was set up to control arms inspection, its inaugural meeting was held on 15th November, 1940 and from this point in time, North America's industrial commitment increased.

A Directorate of Inspection of Small Arms and Small Arms Ammunition was formed early in 1941, including officers from the British Ministry of Supply and the Ministry of Aircraft Production. The Directorate was to organise adequate proof and experimental facilities for North American production, because much of the war material was being produced under contract for British forces.

It was conceived that offices would be set up in New York and Ottawa for the Directorate, to oversee contracts in both the U.S.A. and Canada but when it became apparent that a Lend Lease program would be arranged with the United States, the Directorate Headquarters was moved to Ottawa. The Lend Lease and Mutual Aid Agreement between the governments of the United Kingdom and the United States of America was signed and sealed in Washington on 23rd February 1942. Similar circumstances also applied to an Indo-American agreement.

By the end of 1940, all the 1.5 million rifles in Britain were in use and 190,000 reservists were without arms at all; the Home Guard was largely equipped with the .303 Patt. '14 and .30 U.S. M'17 rifles, at a ratio of one rifle to two men. It was reported that nearly one-third of a million rifles were left in France by the time of the evacuation from Dunkirk and the bombing of Birmingham reduced the B.S.A. plant output to 50% by early 1941. Thus, British reserves and the capacity for production at home had reached critical levels.

In the United States, British representatives were looking for an output of around 4,000 rifles per day and they tabled a requirement for 1,747,000 rifles. Initially, an output of 1,000 a day was projected for the Savage-owned factory of J. Stevens Arms Co. in Chicopee Falls, Massachusetts, and negotiations were in progress with Remington to provide an equivalent amount. The U.S. Government had recently leased the Rock Island factory to Remington but would not permit the manufacture of .303-inch rifles for Britain because the Model 1903 Springfield rifle was required to replenish low reserves; the United States was also preparing its forces in the event of being drawn into a European conflict again.

The Remington factories were not able to manufacture the .303-in. Short Magazine Lee-Enfield and so preparations were made to produce the U.S. Model 1903 Springfield in .303 calibre. Prototypes were assembled which utilised the

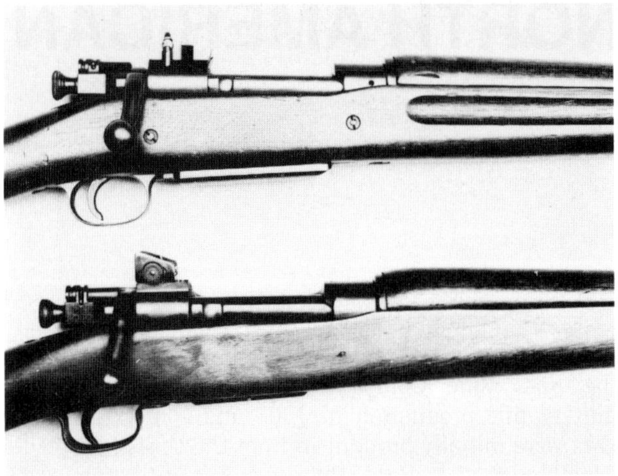

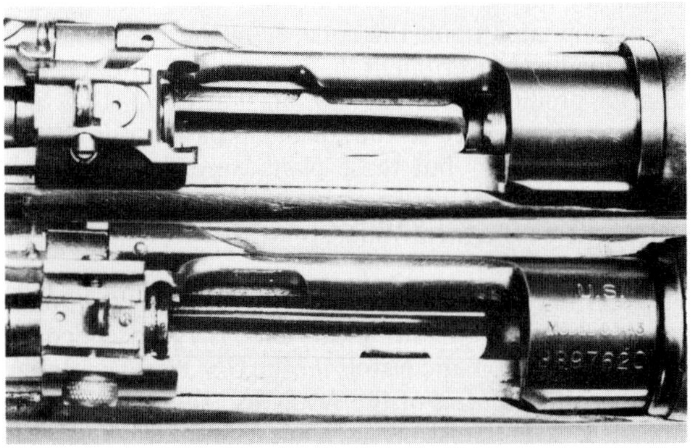

Above & below: Comparison between .303 prototype M'1903 rifle (top in each photo) and the service U.S. .30 cal. M'1903 Springfield. *MOD Pattern Room*

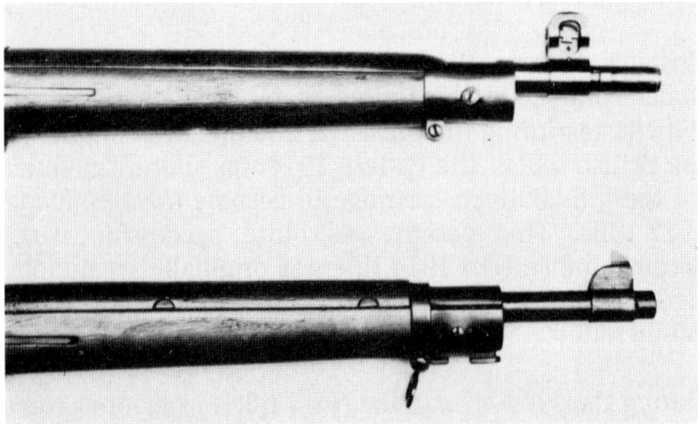

Patt. 1914 Enfield magazine platform and a modified charger guide but further development was necessary because it was prone to damage with the rimmed .303 cartridge. The No. 4 Mk I bayonet was fitted and a dual range battlesight used, similar to the No. 4 Mk 2 type.

Another U.S. arms manufacturer, the High Standard Mfg. Co. was considered as a potential manufacturer of the .303 Lee-Enfield and negotiations were entered into. Projected output at this factory was 1,000 to 2,000 per day but plans were shelved when production at the Savage plant commenced and it was indicated that its output could be increased considerably. So the .303-in. Springfield project was shelved and quantities of .30-inch rifles (probably 1903 Springfields and Model 1917's) were delivered to Britain in their place.

Savage's first contract was for 300,000 rifles at $75 each and of the 95 required components for the No. 4 rifle, some 86 were provided by about 30 sub-contractors which meant that Savage only had to produce the receiver, barrel, bolt, trigger guard, bolthead and stock; assembly was then carried out by Savage. Projected production was for 24,200 rifles and bayonets by the end of 1941 and 425,800 more in 1942. The contract had been increased from 300,000 to 450,000 units and there was no longer any requirement for another U.S. manufacturer as the Canadian S.A.L. plant also came into production late in 1941.

The first Stevens Savage .303 No. 4 Mark I rifle was test fired on Friday, July 25th 1941 at the Chicopee Falls plant, but production fell behind schedule due to the non-delivery of important machine tools. The first rifles were unsatisfactory but these problems were gradually rectified and the quality improved so that 1,325 rifles were completed by the end of 1941. Three factories at Chicopee Falls were engaged in the No. 4 production; the River plant where production commenced in 1941, the Lamb plant which came onto stream in 1942, and the wooden furniture was turned out at the Hill plant.

J. Stevens Arms Co. division of the Savage Arms Corporation records indicate that the first No. 4 rifle was finished on 22nd July 1941 and that the last rifle (presumably, serial number 0C30228 or maybe even 103C0228) was produced on 22nd June 1944. This particular account puts total production of the No. 4 rifle, according to the factory records, at 1,030,228.

When Stevens production reached more than one million rifles, there appears to have been a change in format for the serial numbers. U.S. Lee-Enfield rifle numbers incorporate the letter "C" (for the factory, Chicopee Falls) ahead of the last four digits, e.g. 0C1268, 10C3687 or 91C3646; after the number 99C9999 was attained, the presumed succeeding numbers would be 100C0000, then 100C0001, etcetera. But this does not appear to have been the case because some late production Stevens Savage Mk I* models have been noted with rifle numbers such as 0C12911 and 0C27889, with five digits following the "C," instead of the usual four.

It can thus be concluded that after the rifle number 99C9999, a five digit suffix to the "C" letter was applied, thus the millionth rifle would technically be numbered 0C00000, followed by 0C00001, 0C00002, and so on. This would then explain the serial numbers noted on some late production Mk I* models, like the 0C12911 and 0C27889 referred to in the previous paragraph.

Stevens-Savage Lend-Lease rifle. Note the US "flaming bomb" ordnance mark, "C" in the serial number and "S" logo. The New Zealand mark and rack number was a later addition. *MOD Pattern Room*

But other accounts put the total Stevens-Savage production at 1,196,706 rifles and 40,000 on Lend Lease to China, although it is possible that the Chinese quantity may have been included in the figure of 1,196,706. It is recorded that the one millionth rifle was presented at a special ceremony by Mr. P. Hassett, President Federal Labour Union to Brigadier General Guy Drewry, Springfield Ordnance District, on 18th May 1944.

Canada's progress in No. 4 rifle manufacture was similar, but of course it was an entirely new factory. By 30th June 1941, the Long Branch factory had a staff of 735 and the first five rifles had been completed and submitted for tests and trials. Some difficulties were encountered with the operation of certain machinery and the shortage of suitably experienced operators, all of which was complicated by further increases in orders and additional expansion of the factory. Browning machine guns in .303 and .30 calibre, Boys anti-tank rifles and Sten guns were scheduled for production at Long Branch in 1941 too, but only the No. 4 rifles were produced during that first year.

Increased production was required to reach the 25,000 rifles per month and although substantial quantities of most components were available for assembly by September 1941, some delays were being experienced in the manufacture of barrels, cocking pieces and the Mark I backsight. By 30th September, a total of 200 rifles had been assembled and shipped; the number of employees at Long Branch had increased to 1,220. Notwithstanding, by the end of 1941, production was moving into full swing with a total of 7,589 rifles delivered out of the factory.

The first production from S.A.L. was remarkably fault-free considering the short time in which it had been set up. Two rifles, serial nos. 1L8577 and 2L3286 were forwarded to the Ordnance Board in England for comparative trials with some of the British-made No. 4 rifles. The high quality of finish on the Canadian rifles was remarked upon and the only real complaints related to minor dimensions, which did not really affect the interchangeability anyway.

The first Mark II Sten guns were assembled at Long Branch in February 1942 and the Canadian Stens were destined for use by Canadian, British, Chinese and various European forces. A sizeable percentage of these Long Branch Sten machine carbines, Bren light machine guns and Browning pistols were sent to China under a Mutual Aid Program, the Bren guns and Browning pistols being manufactured by John Inglis Co. Ltd.

Some of the Canadian No. 4 rifle components were contracted by companies such as Canadian Cycle & Motor Co., Weston. Other Canadian firms from as far north as New Liskeard and as far east as Longueuil participated in ordnance production; these included Border City Industries at Walkerville, the Kelvinator Company of Canada in London, and Regina Industries in Regina.

Britain placed a further order with Savage in the United States in June 1942, for 720,000 No. 4 Mark I* rifles and bayonets, to run concurrent with the previous order. Added to this figure were some "foreign orders" for the U.S. government, such as 2,600 No. 4 rifles for China. Eventually, 40,000 of these rifles were forwarded to China under Lend-Lease agreement along with other armaments.

Chinese issue Lend-Lease example. The script reads "for education", indicating training use only. *N. Miers*

A total of 206,389 No. 4 rifles had been produced at Long Branch by the end of 1942, which puts the 1942 production figure at 198,800. The number of workers at Long Branch was up to 5,500 by the end of 1942, of these about two-thirds were women. In the last quarter of 1942, rifle production targets were raised to 32,500 per month. Some minor changes were implemented in 1942, the most notable of which was the modification from Mark I to I* due to the simplified bolthead release arrangement. The No. 4 Mk I produced in North America was the same as that made in England as components were designed to be interchangeable. Only a few parts are slightly different in profile, but this does not affect their interchangeability. Some such minor differences are further described in this chapter in the reports on Canadian production alternatives.

The No. 4 Mark I* model was manufactured in the United States and Canada and the changeover from Mk I occurred in 1942. The Mark I* embodies certain production shortcuts which saved machining time and deleted some minor parts. On the Mark I version, the bolt head release consists of a spring-loaded catch behind the charger guide while on the Mark I* it consists merely of a cut-away in the rib near the front of the receiver.

Three components were thus dispensed with, the catch, spring and catchplate. Machining for the catch slot in the body was not required and the screw for the catchplate was replaced by a pin for the magazine retaining catch. A larger clearance was also required on the underside of the charger guide on the Mark I* because the bolthead was disengaged from the rib for withdrawal before passing underneath the bridge, instead of afterwards as on the Mark I.

The particular serial number ranges in the Long Branch and Stevens-Savage No. 4 rifles when the changeover occurred in 1942 are not known precisely and these do not appear to have been officially recorded. In the Stevens rifles, which contain the letter "C" in the serial number, this seems to have occurred in the 78,000 – 120,000 range, somewhere between 7C8xxx and 12C3110. For the Canadian rifles this change seems to have been quite early in production, before 2Lxxxx.

Some Long Branch No. 4 Mk I bodies have been noted with high 7,000 serial numbers (0L7xxx) and 1941 dates, but early 1942 production barrels, indicating that barrel production lagged behind that of the body. All Mk I Long Branch receivers seem to be 1941 dated; the changeover to Mk I* appears to have occurred early in 1942, in the very late 1Lxxxx series.

Certain production shortcuts were implemented in British production as the war progressed and many of these concessions were incorporated into the North American manufacture as well; the American rifles maintained quite a high standard of finish, particularly the Long Branch models. Major parts for which simplified manufacture was approved were the body, barrel, backsight, trigger guard and magazine, in addition to some less important components such as the foresight protector and outer bands.

The Mark 1 and 2 backsights were made by Savage and Long Branch and many of the Savage Mk I sights were later fitted onto Canadian rifles. In Canada, special fabricated backsights were introduced as an economy, the C Mk 3 and C Mk 4 which are comparable to the British Marks 3 and 4 backsights; descriptions and illustrations of the British sights are contained in Chapter 6. These Canadian sights consist of a sheet metal leaf welded to a forged and machined base whereas their British counterparts were all sheet metal.

Differences between the Canadian backsights are:

Mk 1. Folding leaf backsight machined from solid stock with aperture slide adjustment by a worm screw, the clicker adjustment is in increments of 25 yds. Ranged from 200 to 1,300 yds and similar to the British Mk 1 pattern.

Mk 2. Dual fixed range, "L"-shaped aperture, no leaf; ranged at 300 yds with bayonet fixed and 600 yds without bayonet. Similar to the British Mk 2.

Mk 3. Leaf backsight manufactured from sheet steel pressings, ranged from 200 to 1,300 yds in increments of 100 yds. Range adjustment is made by moving a slide with a spring-loaded catch. As the previous Mark of sight does not have a leaf, the leaf of this sight is designated and marked "C Mk 2"; comparable to the British Mk 3.

Mk 4. Leaf backsight from steel pressings, like the Mk 3, but with a redesigned slide retaining catch and other minor differences; the catch is curved inward towards the leaf, rendering it less liable to damage. The slide, catch and pin are designated Mk 3 and differ slightly from those of the previous Mark; the leaf is marked "C Mk 3."

The leaf, backsight, Mk 3 differs from the Mk 2 in that the holes bored for the pin, stop, backsight are drilled in the opposite sides of the leaf. The pin, assembled on the Mk 2 protrudes on the right side whereas the pin assembled on the Mk 3 projects from the left side; the leaf Mk 3 was introduced to prevent the pin from interfering with the operator's fingers or the catch backsight when the slide was pushed to the top of the leaf. Some of the leaves Mk 3 were erroneously stamped Mk 2; the Mk 2 leaf became obsolete soon after the introduction of the Mk 3.

British and Canadian production Mk 4 sights, with the Mk 3 leaf. The "C" prefix indicates that the Canadian model differs from the British pattern in small detail.

The slide, backsight, Mk 3 differs from the Mk 2 in that a cut was made in the right side of the Mk 3 to allow clearance to depress the catch Mk 3 to a point where it could disengage from the slot in the leaf. Again, some Mk 3 slides were erroneously marked Mk 2 and these may be determined as the Mk 3 slide has a portion cut away on the right side at the top to provide a clearance for the curved Mk 3 catch; on the Mk 2 the right side was not cut other than for a slot through which the catch operates.

The Canadian patterns were developed late in 1943, and the following year samples were forwarded to England; the alternatives were approved and incorporated into Canadian production. A pressed steel buttplate was also approved as an alternative for Canadian production and changes were made to fittings such as the foresight protectors, trigger guard, safety catch and the upper and lower bands.

British approval was required for such changes and manufacturing concessions to components because a sizeable percentage of the Long Branch rifles and virtually all the Stevens-Savage rifles were to fill British orders, and were therefore subject to acceptance by the British Inspectorate. In the initial Savage production, many rifles were rejected by the British inspectors; this was not properly resolved until a Lend-Lease agreement was formulated between the British and U.S. governments.

Lend-Lease mark stamped on the left receiver wall.

Under the new agreement the British had less control in the initial inspection and refusal of rifles; conditions for payment were changed also which further reduced some of the controls Britain previously had. The Lend-Lease rifles are marked "U.S. PROPERTY" while those produced earlier under the direct purchase system were not so marked as they were for a direct British contract.

Successful tests with 2-groove rifling were carried out in England in July 1941 and resulted in such a concession being permitted for future manufacture. Virtually all of the production 2-groove barrels for No. 4 rifles were made in North America. Savage inquired whether a twist of one in 10.5 inches would be acceptable for some barrels instead of the specified one in 10 inches, because some machinery was still set up from an earlier French contract which used the slower twist; the British advised that such barrels would be acceptable as it was considered that this slight alteration would have a negligible effect.

Another concession permitted for Stevens-Savage also concerned the barrel; because their plant was already set up for 6-groove rifling, the management inquired if this too would be acceptable. Such rifling was used in early production although it was changed to 2-groove not very long afterwards. Long Branch also made some 6-groove barrels but this was later, about 1950. The various barrels were allocated different Mark numbers, in conjunction with the British series:

Mk 1	5 groove
Mk 2	2 groove (Stevens-Savage and Long Branch)
Mk 3	5 groove, fabricated (Accles & Pollock)
C Mk 4	6 groove (Canadian)

As the fabricated Mark 3 barrel was not produced in North America, it was therefore not incorporated in the Canadian nomenclature. The C Mk 4 barrels are not to be confused with the Stevens-made 6-groove wartime barrels as the Canadian 6-groove C.A.L. barrels were made in 1950 and have right hand instead of the usual left hand twist. This was because they were produced on Bren gun barrel tooling (the Bren LMG barrel has 6-groove, R.H. twist rifling).

On the No. 4 rifle, all screw threads regardless of the place of manufacture are B.S.W., B.S.F., B.S.P. or B.A. to British standard tolerances. A design relaxation or concession was allowed to Stevens concerning the permissible thread crest flattening although the relative interchangeability was not affected.

Like its British counterpart, the North American No. 4 rifle was fitted with stamped and fabricated foresight protectors, bands and trigger guards. Canadian Armourers' Instructions describe a number of different foresight protectors in service:

Protector, Foresight, Mk I: Original pattern with waisted sides and sideways extensions at the top of protector ears (from British No. 4 Trials models, not Canadian production).

Protector, Foresight, Mk I: Alternative pattern with more squared profile, machined from solid (as on early British and North American production models, and screw hole threaded right through on the right hand side of the protector screw hole).

Protector, Foresight, Mk II: Steel stamping with two steel inserts welded or brazed into position (all the Mk II protectors have a retaining screw hole which is threaded only half-way through in the right hand side, the tops of protector ears are radiused on the inside, instead of the outside like Mk I).

Protector, Foresight, Mk II: Alternative pattern, steel casting (with similar profile to Canadian-made Mark I with tops of sight protector ears radiused on the outside as on Mk I).

Protector, Foresight, Mk II: Alternative pattern, steel stamping with two steel inserts welded or brazed into position (of similar profile to first-listed Mk II pattern, tops of sight ears radiused on the inside, like the first-listed Mk II).

Components of Long Branch manufacture may be readily distinguished by the Long Branch logo mark; after the change to Canadian Arsenals Ltd. in 1946, the C.A.L. logo was stamped instead. The U.S. Stevens-Savage mark consists of a square profile "S", sometimes enclosed in a square box.

Lend-Lease No. 4 rifle fitted with Mk 2 sight.

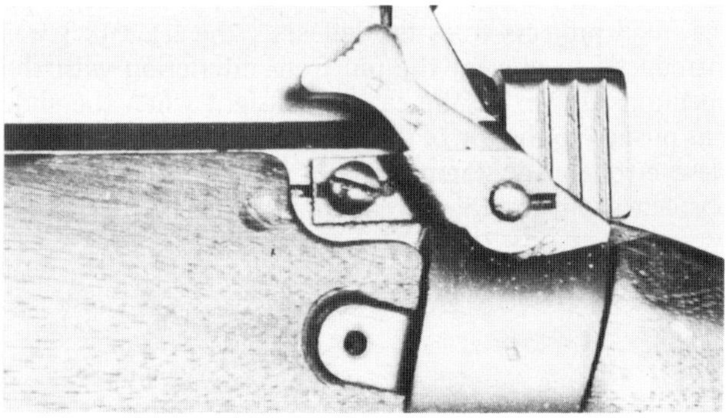

Characteristic Long Branch safety catch.

A quite distinctive safety catch was used on most Long Branch rifles; another ready means of identification of the North American No. 4 rifle is by the serial number. The letter "L" is positioned ahead of the last four digits on Long Branch rifles (e.g. 8L7909 or 56L5563) while those made at the Stevens Chicopee Falls plant incorporate a "C", placed similarly (e.g. 30C6472). Serial numbers on many of the earlier Long Branch rifles were marked with the smaller individual stamps which are not aligned perfectly.

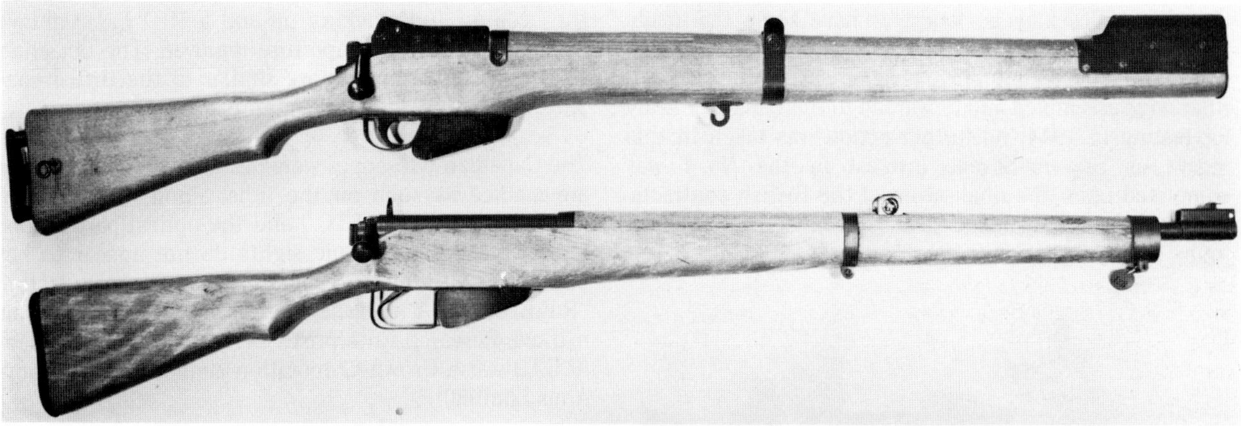

***Above:* British Swift Trainer (top) compared with the Long Branch training rifle (bottom).** *MOD Pattern Room*

***Right:* Swift training rifle panoramic target and frame.** *MOD Pattern Room*

In 1943, Long Branch produced 357,724 No. 4 rifles, the largest figure for any year and the unit price had fallen to $32.00 compared with $62.30 for the first rifles produced in 1941. Introduction of the Mark I* pattern had some bearing upon this price reduction along with the incorporation of more stamped and fabricated components instead of machined ones. Undoubtedly, more efficient manufacturing procedures and the associated savings in labour, as well as the purchase of materials in larger quantities, contributed to the lower unit costing also.

In late 1943, a new training rifle was developed and produced at the Long Branch factory and a total of 250 were completed by December 1943, with some more the following year. This is a Canadian version of the British Swift trainer which uses no ammunition, making it ideal for beginners for aiming, holding, sighting and manipulation of the action. Like the Swift, it uses a needle at the "muzzle" which is projected upon squeezing the trigger, piercing a paper target directly in front of the rifle. Not only was this a safe and convenient way to instruct new recruits, but it also saved expenditure of ammunition and did not require the use of a rifle range.

With an overall length of 44.6 inches, the weight of the training rifle is 8.5 lb., comparable with its No. 4 rifle counterpart. This trainer was used in conjunction with a special frame and stand and an aperture sight is mounted half-way along the "barrel". Two pointed needles are housed at the "muzzle" and after the rifle has been cocked with the bolt in conventional manner, the trigger is squeezed and these needles jump forward to puncture the target. To avoid accidental discharge, a buttplate pressure safety is incorporated and this has to be depressed against the shoulder before the trigger can be operated.

Its official designation was the "1943 Pattern Training Rifle" and the metal body is marked "LONG BRANCH TRAINING RIFLE" along with the year of manufacture, usually 1943 or 1944. A dummy wooden magazine is attached by wood screws but no sling swivels are fitted. In all, 3,788 units were fabricated by the end of the war and the special target stand is similar to that used with the Swift.

In mid-1944, upon completion of the Stevens contracts, instructions were issued that all spare components be forwarded to S.A.L. Long Branch. For this reason, some Canadian rifles may be found to have some U.S.-made parts, in particular the Mk I backsight. The Stevens factory developed an improved trigger design and six such rifles were assembled and forwarded to the Canadian Army for testing in 1944; no further action was taken in this regard as Stevens-Savage interest in the No. 4 was completed upon the finalisation of the British contracts. In the relevant correspondence, the total production figure is put at 1,196,706 rifles.

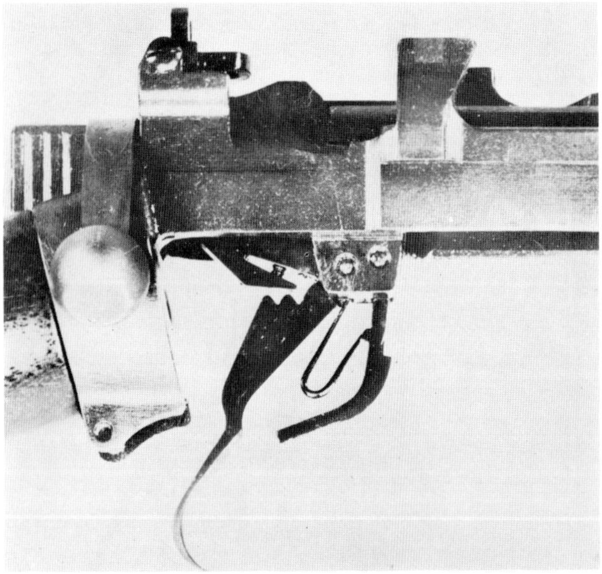

Stevens-Savage Prototype Trigger Improvement.

The U.S. designed trigger improvement incorporates a slightly re-designed body and it was considered along with the Canadian system as used on the lightweight rifle. The British C.E.A.D. design and C.I.A.(W) patterns, described in Chapter 6, were also compared against the U.S. and Canadian improved trigger systems.

British contracts and orders for the Canadian rifles were reduced late in 1943 and so the total S.A.L. production of 250,716 for 1944 was considerably lower than that of the previous year; these rifles were costed at $26.35 each. By December 1944, a total of 112,932 Sten carbines had been produced at a unit price of less than $12.50 by the end of production. In 1944, two new rifles came into production at Long Branch, the .22-in. No. 7 trainer and the No. 4 sniper model.

Long Branch manufactured Canadian No. 4 sniper rifles are similar to their British made counterparts and Britain also placed orders for 950 Canadian units, although the eventual deliveries fell considerably short of this figure. Rifles were chosen from those which exhibited superior accuracy during the sighting-in tests and then these were stripped completely. Tabs were attached to the left side of the receiver and drilled and tapped for the securing screws of the telescope mount. Special attention was then paid to the bedding and stocking up and a Mk I backsight was fitted with the battle aperture removed. The detachable telescopic sight mounts are similar to the British-made pattern as are the rifle base mounts.

The Canadian telescopes were designated "C. No. 32" and are marked as such on the tube, along with the mark number, then "TEL. STG." and the optical part number. Canadian Mk I telescopic sights do not appear to have been fitted with the moveable sunshade. The manufacturer, "R.E.L. CANADA" and the telescope serial number are also marked. Research Enterprises Ltd. of Sherbrooke, Quebec (R.E.L.) was a Crown Corporation similar to S.A.L. (Small Arms Limited).

R.E.L. produced military optics, chiefly gun sights for naval and artillery pieces, but the rate of conversion of the Long Branch No. 4 sniper rifles was limited by the supply of telescopic sights. By the end of 1944, a total of 644 sniper units had been assembled at Long Branch with a total of 1,141 by the end of the war. As production was continued in 1946, about 1,200 sniper units were eventually assembled here.

A few different models of the No. 32 telescope were manufactured by R.E.L., designated the Marks I, 1A, 2, 3 and 4, of which the Mk 3 is most commonly encountered. Like the British telescopes, when bloomed lenses and waterproofing were introduced, the steel scope tubes were marked with a red painted "W" indicating waterproofing and a blue "CB" for blooming. Whereas the bloomed lens telescopes made in Britain are marked with a blue "B", the Canadian telescopes are usually found painted "CB", which would indicate that a slightly different process was used in the blooming. The letter "C" is often prefixed to Canadian pattern designations, especially where the Canadian model differs slightly from its British counterpart, this also provides a ready means of identifying the Canadian product.

Note the "C" prefixes for Telescope designation (C NO. 32 MK 3) and bloomed lenses marking.

Catalogue part numbers and those often marked on the Canadian telescopes will generally be found to commence with "COS" whereas those of British manufacture use "OS". The following parts list of telescopic sights tabulates the full range and their relevant parts numbers.

Telescope Mount comparison. British product (top) and Canadian example (bottom). The latter has rifle serial number engraved on mount.

British:

No. 32 Mk 1 (V-5)	OS 466A
No. 32 Mk 2	OS 1650A
No. 32 Mk 3	OS 2039A

Canadian:

C No. 32 Mk 1	COS 466A
C No. 32 Mk 1 A	CGB 76GA
C No. 32 Mk 2	COS 1650A
C No. 32 Mk 3	COS 2039A
C No. 32 Mk 4	CGB 91GA (changed to No. 67 Mk 1)
C No. 67 Mk 1	CGB 91GA (previously No. 32 Mk 4)
C No. 32 T.P.	CGB 88GA (Lyman Alaskan)

Canadian No. 4 sniper rifles are marked on the left side of the receiver, "No. 4 Mk I* T" and "LONG BRANCH" followed by the manufacture year of the original rifle. Whereas the British sniper rifles usually have the telescope serial number marked on the mount and on the butt of the rifle at the wrist, the Canadian made rifles use the rifle serial number on the mount. Minor differences may be noticed in the casting of the S.A.L. mount compared with the British pattern and the mount caps are not usually numbered on the right hand side like their British counterparts.

Many Long Branch sniper rifles and R.E.L. scopes were purchased by Britain during the war and so a "mix-n-match" situation may be encountered with some of the Canadian rifles, scopes and parts from the British ordnance system, due to the various armourer and workshop repairs and replacements and F.T.R. programs. Canadian issue sniper rifles, though, were generally not as widely issued and the chief "foreign parts" encountered are usually only some Stevens-Savage backsights.

Designed in Britain specifically to assist the Canadians increase production, another model of telescopic sight was fitted to a small number of Long Branch rifles in mid-1944, the C No. 32 Mk 4. This was also made by R.E.L. and the differences in this telescope and mounts are quite obvious from the illustrations; it was later designated "Telescope, Sighting, C No. 67 Mk 1". The first examples produced and sent out for trials were marked "C No. 32 Mk 4" and the typical markings are:

TEL. SIGHTING C No 32 Mk 4
C.G.B. 91 G.A. 3.5 X 24
R.E.L. CANADA
1944 10 - C

The lower mount is a Griffin & Howe type secured by four screws and two pins; the dovetail along the top has two transverse, semi-circular cut-outs. The other part of the mount contains the telescope rings and is affixed to the telescope. This has a female dovetail which is slid onto the lower rail mount and locked into position by two cams with levers on the left side. These latches are somewhat similar in design to those used on the rear mount of the No. 3 Mk I* (T) and Australian No. 1 Mk III* H.T. sniper models.

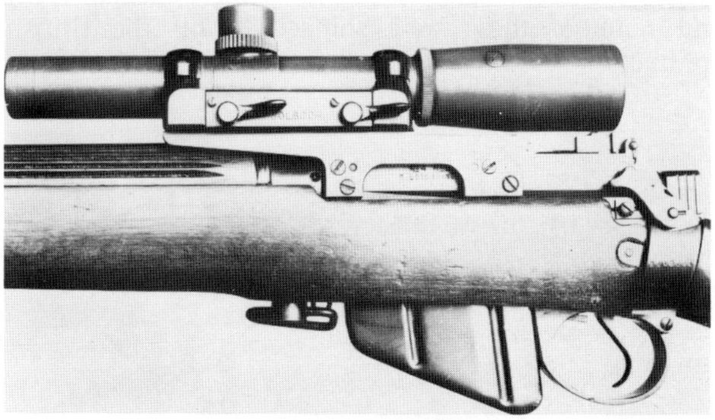

C No. 32 Mk 4 Telescope with Griffin & Howe mount. Trialled in England in 1944. *MOD Pattern Room*

The C No. 32 Mk 4 or C No. 67 Mk I sight has a 24mm objective lens and it is rated at 3.5 power with range calibrations from 0 – 1,000 yards in 50-yd. clicks. Its advantage for Canadian production was that some of its lenses were being manufactured in Canada for other optical sights. The range drum is on the top of the tube with the lateral (windage) adjustment on the right side. The illustrated specimen from the Pattern Room collection has a light green painted finish and this particular trials rifle is also fitted with a sporting style butt with rubber recoil pad, roll-over cheekpiece and pistol grip.

A North American commercial model, the Lyman Alaskan sight was also fitted onto an R.E.L. designed mount which looks like the Griffin & Howe type, except that it has a different type of dovetail attachment between the upper and lower mounts. This was designated "Telescope, Sighting, C No. 32 T.P. Mk I". and it is a trade model which fared well in the British tests and was approved by Canada when there was a shortage of R.E.L. telescopes. It was also proposed because it was weatherproof; troubles were being encountered with waterproofing of the No. 32 scopes at this time.

Lyman Alaskan Telescope with Griffin & Howe mount.
J. Alley

Canadian orders were placed for "450 sniper rifles with the U.S. telescope", apparently the Lyman Alaskan, and the records indicate that 99 of the No. 4 (T) rifles were set up with this scope in 1945, comprising 49 in February, 31 in March and 19 in April. Issue was to the Canadian Army and none of these were shipped to fill the British contracts.

The first production Stevens-Savage No. 4, serial number OC1, later converted to No. 4 Mk I (T). *P. Breakey*

The development of Long Branch's No. 4 sniping rifle was closely monitored by the British Ordnance Board and comparative trials were conducted in England during 1944 and 1945 against British, U.S., Australian and captured German sniping equipment. R.E.L. also produced some 5 power scopes for trial and the Weaver Model 330 scope was tested alongside the Lyman and German ZF 41 and Hensoldt models.

British trials records list the serial numbers and details of the following No. 4 Mk I* (T) rifles with specific telescope combinations in April 1944:

38L1957, G&H mounts, 3.5x REL scope #0144, with special butt;
4L3937, G&H mounts, 5x REL scope #0144 (possibly a Mk I action);
ASE 40-3, REL mounts, 3.5x scope #0244;
ASE 40-8, REL mounts, 5x scope #0244;
ASE 88-1, REL mounts, sent without telescope;
ASE 40-10, REL mounts, without telescope, with special butt.

During these endurance trials, the Canadian rifles performed quite well, although troubles were encountered with internal fogging of some scopes.

In July 1945, No. 4 Mk I* (T) rifles were trialled in the United Kingdom along with two Australian No. 1 Mk III* H.T., serial nos. 22859 and 60660. Details of the Long Branch rifles were:

80L8004 with 3.5x REL telescopic sight, #10-C, and
80L8007 with 3.5x REL telescopic sight, #9-C.

These sniper rifles were all recorded as having operated and trialled quite well.

The No. 32 telescopes were nearly twice the weight of most other models although they were more robust and reliable under rough conditions. Both the 3.5 and 5 power R.E.L. telescopic sights fared very well and even though the 5 power model was favoured by many who used it, the Board considered its extra magnification and smaller field of view "unsuitable for average sniper use".

This 5 power R.E.L. telescopic sight was set up on some experimental and trials snipers rifles in Canada. The illustrated example was fabricated at Long Branch and it incorporates a half-stock fore-end, sporting style butt with kickpad, pistol grip and cheekrest, the Griffin & Howe type mounts as used on the C No. 67 scope (illustrated previously) and a lightened action body. The charger guide and backsight assembly have been removed and replaced by a special rotating rearsight with 200, 300, 400 and 600 yard ranges.

Two such experimental Canadian sniper rifles were sent to Europe for battlefield evaluation and one was captured by the Germans. A special .303 round with blackened case was made for use with this rifle, the matt-finish case presumably being to avoid reflection from loose live

Above & below: 1944 Experimental Long Branch No. 4 Mk I* Sniper rifle with 5X Prototype Telescopic Sight and Griffin & Howe mount. *D. Fabok*

rounds or expended cases in the field. Further details of the specialist Canadian sniper rifle variations, sights and mounts will be found in "The British Sniper", Chapter 7, which describes the special Canadian issues and development.

The .22-inch No. 7 rifle was the other new Canadian model introduced in 1944, developed as a result of requests by the Department of National Defence for a .22 R.F. practice rifle based on a conversion of the No. 4. Although these were not actually converted from the No. 4 rifle, the No. 7 mostly utilises No. 4 components. Six such rifles were fabricated for trials and the design approved by the Department of National Defence, so an order was placed for 7,000 with production to commence in September at a rate of 1,000 rifles per month. Some 4,237 were completed by the end of 1944 and a further 2,983 in early 1945, making a total of 7,230 on the original S.A.L. contract. Rifle production was re-commenced in 1946 for one year, and re-instituted in 1951-52 during the Korean crisis.

In North America, some other .22-in. rifles were manufactured on military contracts, by Cooey and Mossberg. The Cooey Machine and Arms Co. Ltd. produced 34,819 .22-in. training rifles but these were not the No. 7 or full-size military configuration rifles. Other contracts for .22-in. rifles were filled by Mossberg in the United States. Cooey also produced 77,000 aiming tube conveyors which enabled .22-in. R.F. ammunition to be used in the Boys anti-tank rifle, for training.

The official description of Canada's No. 7 .22-in. Mark I rifle, as issued in January 1945, was:-

Weight	... ... *9 lb. 5 oz. (approx).*
Length	... ... *3 ft. 8.3 in. (normal butt).*
Barrel	... ... *25.29 ins.*
Rifling	... ... *22 cal., 6 groove, 1 turn in 16 ins. R.H. twist.* *Diameter: .2238 ins.* *Groove width: .086 ins.*
Sights	... ... *Aperture backsight, 20 & 100 yds.* *Lateral & vertical adjustment, 1 min. clicks; .2 in. at 20 yds, 1 in. at 100 yds* *Sight radius at 20 yds. = 28.75 ins.* *Aperture size: .0595 ins.* *Blade foresight, sizes: 8 of, ranging between -.030 in. and +.090 in.*
Butts	... ... *4 sizes, ½-in. difference each between the long, normal, short and bantam.*
Ammunition	... *.22-in. L.R. RF, smokeless, greased; 1100 fps M.V., 40 gr. projectile.* *.22-in. L.R. RF, tracer (.22-in. G, Mk Iz); 34 gr. projectile.*
Magazine	... *Single shot, grooved platform to facilitate single loading.*
Trigger pull	... *1st pull, 3 - 4 lb.* *2nd pull, 5 - 6.5 lb.*

This trainer utilises the No. 4 Mk I* receiver and it was first known as "Rifle, .22-in., Long Branch", although this was soon altered to "Rifle, No. 7 .22-inch., Mk I" and then "Rifle, C No. 7 .22-in. Mk I"; the "C" indicates Canada. As a result of these changes in nomenclature, three different inscriptions may be encountered on the Canadian version.

Type 1.	**.22 IN.** **LONG BRANCH** **1944**
Type 2.	**No 7 .22 IN. MK I** **LONG BRANCH** **1945**
Type 3.	**C No 7 .22 IN. MK I** **LONG BRANCH** **1946**

Some of the manufacture years may be found to overlap and a batch of about 1,000 was also assembled in 1951-52 using the Type 3 inscription.

The C No. 7 has a special disc backsight fitted; the lateral adjustment screw is graduated on the left end with six equally spaced white lines, the knurled lateral adjustment knob is on the right hand side. Elevation is effected by movement of a screw on the top of the sight, with click-stop calibrations.

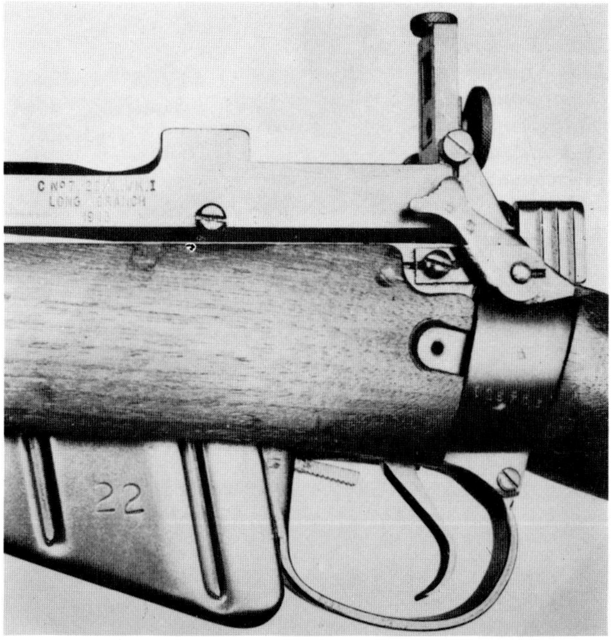

Type 3 Markings on Long Branch No. 7 .22. Also shows typical LB safety catch and fabricated trigger guard.

As the rifle is a single-loader, the magazine platform has a concave shape inclined towards the chamber to facilitate loading. The rear of the platform has been cut away to allow empty cartridge cases to fall into the magazine casing. The magazine spring is identical to that of the No. 4 except that the top leaf of the "W" is shortened and the holes for the platform rivets have been relocated.

In order to properly adjust the headspace, boltheads were available in a number of sizes. Because of the rimfire cartridges, the firing pin hole in the bolthead is off-centre and the firing pin in the bolthead is a floating type. The extractor is larger than that of the No. 4 and the claw extends further across the face of the bolt so as to engage the smaller .22-in. rimfire cartridge rim.

Those components special to the No. 7 are the barrel, foresight block band, body (with additional hole for backsight knob spring screw), magazine case (with different forward retaining lips), magazine platform and spring, and the bolthead, striker, backsight and aperture. The ejector screw in the receiver is not necessary and two different backsight leaves and aperture discs were used.

The Mk 2 disc has a "2" stamped on the rear face of the disc and the difference between Marks 1 and 2 is that one has an added lug at the lower right hand corner and is .1-inch longer. The whole unit was supplied in a "Chest, Mk I, .22-in. C No. 7 Rifle".

During the 1950's, some minor improvements were made to the C No. 7; in November 1950, the Mk I backsight disc and slide were ordered to be replaced by the Mk 2 pattern and a modification was approved in September 1953 for the backsight which permitted removal of the breech bolt without raising the backsight. This was effected by grinding and machining the leaf. In July 1954 it was announced that a modification had been made to the action body to prevent loss of the extractor spring and that certain sharp edges had been removed on the left side of the receiver and forward of the charger bridge. The particular months recorded here for these changes are the bulletin announcement dates; it is likely that the official approval dates may have been earlier.

The first Long Branch contract for .22-in. No. 7 rifles was for 7,000, upon which 7,230 were produced by early 1945. During the 2nd quarter of 1945, advice was received of an order for a further 10,450 .22-in. calibre rifles and so manufacture of components was re-commenced and 1,100 of this new contract produced by the beginning of 1946. The remainder of this contract appears to have been filled as 3,000 were completed in the first quarter of 1946 and production was continued through the remainder of the year.

The manufacture of Canada's wartime .303-in. No. 4 rifle was terminated on 17th August 1945 at Long Branch, with only 90,902 rifles being manufactured in 1945 due to the cancellation of the British contracts. All the "black action" assemblies were finished and packed in store. On 10th September 1945, cancellation instructions were received for the Sten machine carbines, a total of 126,703 having been produced since February 1942. Only 497 of the No. 4 Mk I* (T) sniper rifles were set up in 1945, bringing the total run by the end of the war to 1,141 units. The total number of Long Branch .303 No. 4 Mk I and I* rifles by the end of production in 1945 is recorded as being 905,731, less than the Stevens-Savage numbers.

These production figures have been extracted from an official history of the Department of Munitions and Supply although they do not correspond exactly with figures from the Inspectorate. According to the Inspection Board, accepted production for 1941-1945 was:

Bren Machine Guns, .303-in.	178,075
Bren Machine Guns, 7.92mm.	43,000
Boys Anti-Tank Rifles, .55-in.	50,611
Browning Pistols, 9mm.	151,818
Rifles, No. 4, .303-in.	910,368
Rifles, No. 4 (T), .303-in.	992
Rifles, No. 7, .22-in.	5,204
Rifles, Cooey, .22-in.	92,135
Rifles, Long Branch Training	3,788
Carbines, Machine, Sten, 9mm	128,238
Browning Machine Guns, .303-in.	32,653
Browning Machine Guns, .30-cal.	26,009
Polsten Machine Guns, 20mm.	8,018

Reasons for variation between these Munitions & Supply Dept. production figures and other official totals quoted earlier in this chapter is not clear, but it does provide a good indication of the overall S.A.L. production. Long Branch manufactured the No. 4 rifles, No. 7 .22, Sten machine carbine and the special training rifle, while John Inglis produced the Bren LMG, Browning pistol, and Boys A.-T. rifle. Other manufacturers involved in arms and large scale component production were Border City Industries, Kelvinator, Regina Industries Ltd. and Canada Cycle & Motor.

Instructions were received to proceed with the engineering, planning and tool changes necessary to produce the No. 5 rifle early in 1945, at a rate of 8,000 per month. This did not involve major production changes as many of the components of the No. 5 were interchangeable with those of the No. 4, but later that year the company was advised that the No. 5 orders were cancelled and all tooling should be placed in store.

At least one prototype No. 5 was produced at the Long Branch factory, the illustrated specimen is from the Canadian War Museum collection. This example utilised a 1943 vintage No. 4 Mk I* receiver with No. 5 type barrel, flash eliminator assembly and fore-end. The backsight is a No. 4 Mk 3, re-graduated to 800 yds, which makes it similar to the British-made No. 5 Mk 2 sight. The carbine is stocked within 6.25 ins. of the flash hider, similar to the British-produced No. 5 model.

Long Branch prototype No. 5. Note lightening cuts on the left side of the receiver. *Canadian War Museum*

On 31st December 1945, Small Arms Limited was wound up as wartime production had finished and no further orders were being placed by Britain. The uncompleted contracts for the No. 7 trainers and No. 4 sniper rifles were taken over by the succeeding Crown Company, C.A.L. On 1st January 1946, the Small Arms Division of Canadian Arsenals Limited took over the operation at Long Branch for peace-time production.

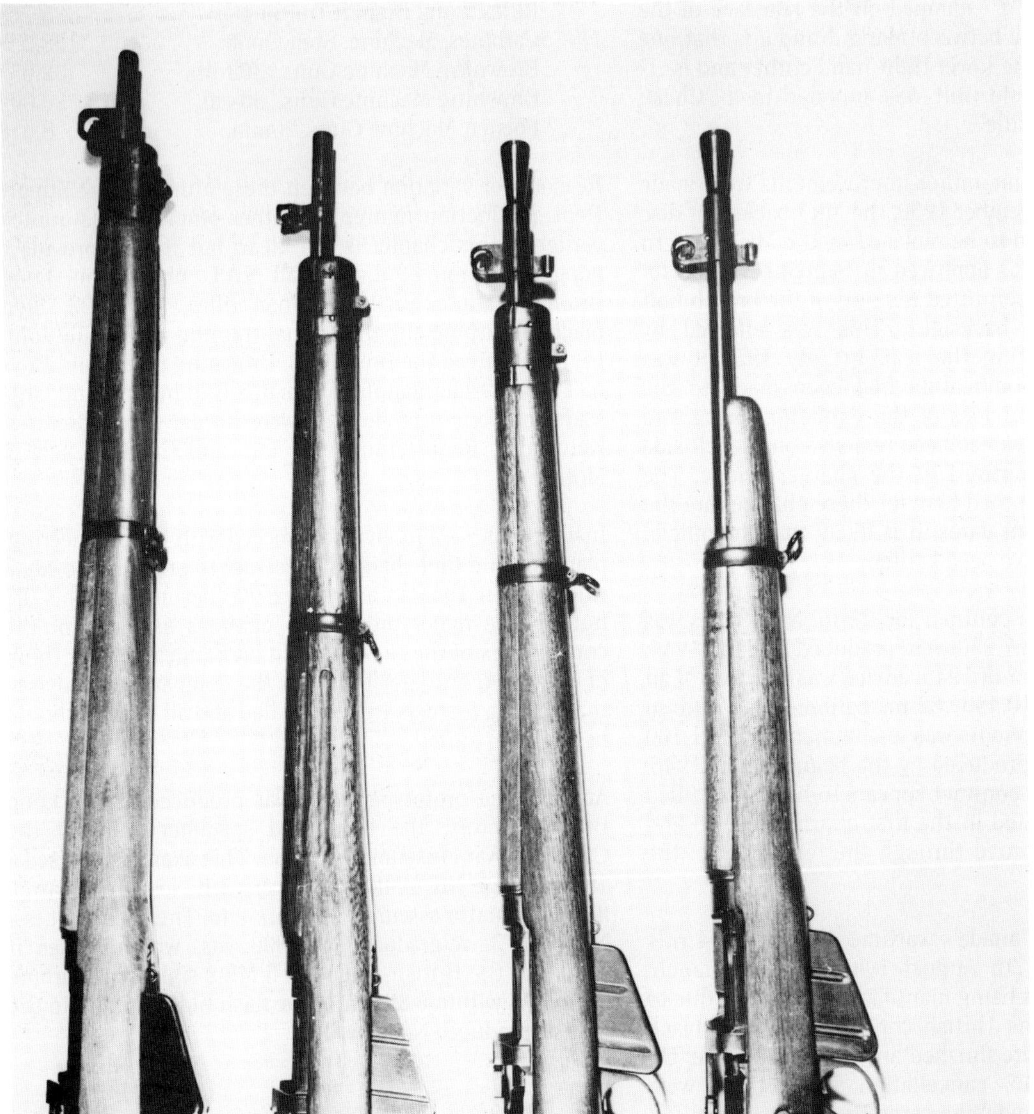

CANADIAN PROTOTYPE & TRIALS LIGHTENED RIFLES

From top: No. 4 Mk I for comparative purposes*
Canadian Lightweight No. 4 Rifle (rubber buttpadd missing)
Canadian Lightweight in basic "No. 5" configuration
Prototype No. 5 Rifle

Note:- No. 4 Mk I* actions have been utilised. *Canadian War Museum*

Over a period of about five and a half years, S.A.L. had employed more than 14,000 workers of which a very high percentage were women; the total sales amounted to $43,907,176.87¢. Capital invested had been $8,500,000 and considering that construction of the factory was not commenced until late in 1940, this was a tremendous achievement with a wartime production approaching one million rifles, in addition to the Sten sub-machine guns.

S.A.L. engineered and developed a number of prototype models of self-loading rifles and some machine carbines during 1944 and 1945 and this formed a nucleus for post-war development. Notable evolution of the wartime Canadian No. 4 includes a lightweight jungle rifle version during 1943 which resulted in production of a small batch for limited trials. These lightweight rifles incorporated many new features and were produced in a few different configurations. The weight saving of about 2 lb. represents about 20%, and adequately answers one of the main criticisms levelled against the No. 4 rifle for Far East service.

The Canadian lightweight rifle series was a complete re-development of the No. 4, not just a shortened, lightened model, and it incorporated some innovative and commendable features. A bulletin describing the main design was issued on 25th November 1943 and some major points are reproduced here.

GENERAL DESCRIPTION.

The Long Branch Light Rifle is designed as a service rifle converted from the No. 4 Rifle in such a manner as to provide a light, accurate weapon, particularly for use in the Far Eastern Field and also for general service issue.

The weight of the Rifle would not exceed 6 lbs. 12 ozs. Its overall length is 42 inches. It would be manufactured to meet acceptance tests at present establishment for the No. 4 rifle as applied to accuracy, trigger pull, firing mechanism, etc. The Rifle is also arranged to take the No. 4 bayonet.

The major items of change are as follows:

(a) A lightened barrel, 22 inches in length, designed to give accurate shooting when properly controlled by stocking up.

(b) A full length stock with body converted to provide lightness and also to provide an improved assembly condition in that the body is held to the stock by means of two screws as is the case of the No. 4.

(c) A greatly improved trigger mechanism which provides a robust sear and trigger, mounted directly to the body in such a manner that trigger pull can be established by manufacture of components to reasonable tolerances, and once established cannot without damage to the parts, change in any way. This one item is considered to be a vast improvement over the original Enfield No. 4 action where the trigger is held in the trigger guard in such a manner that variations in the fore-end and/or damage to the trigger guard is a continual source of trouble in maintaining trigger pull both in manufacture and in service.

(d) An adjustable backsight will be provided based on the Mk 3 design, graduated at 100 yard intervals from 200 to 800 yards. The battlesight will be arranged to provide accurate shooting with bayonet fixed at 200 yds. (N.B. The first samples submitted are arranged with Mk I backsight in order to save time, as samples of the special backsight for the Light Rifle are not available.)

(e) In order to offset the increased recoil resulting from the Light Rifle, a rubber recoil pad is attached to form a combined recoil pad and buttplate. This pad would be manufactured of synthetic rubber especially selected for the purpose to be resistant to oil, attacks by insects, etc.

(f) In order to arrange for a satisfactory recoil pad, it has not been possible to provide accommodation in the butt for oil bottle and pull through. It is thought that these items can be carried in the equipment of the user.

(g) The Rifle is arranged to take either a standard web rifle sling, or a lightened sling somewhat of the same weight as that provided for the Sten carbine.

(h) Dimensions and tolerances affecting the functioning points of the Rifle would be maintained to the same standards as for the No. 4 rifle.

By using an aluminium alloy for the front handguard cap, upper band and trigger guard, a saving of approximately between 3 and 4 ozs. could be effected thereby reducing the weight of the Rifle to 6 lb. 9 oz.

Shortening of the barrel below 22 inches would result in a substantial saving in weight but would offset accuracy attainable with the 22-inch barrel. The same principal of design could be used for a shorter barrel if it is felt that some accuracy could be sacrificed in favour of a reduction of weight below approximately 6 lbs., and a saving in overall length. Shortening of the barrel below 22 inches is not recommended by the designer of the Light Rifle.

The following standard No. 4 Mk I components are used on the Long Branch Jungle Rifle:*

Enfield No.	*Component Name*	*No. per Rifle*
450/111	*Bolt, Locking*	*1*
450/17	*Catch, Safety*	*1*
450/53	*Screw, Ejector*	*1*
450/77	*Spring, Bolt, Locking*	*1*
450/64	*Screw, Spring, Bolt, Locking*	*1*
450/A	*Backsight*	*1*
450/32A	*Pin, Axis, Backsight*	*1*
450/27A	*Pin, Retaining, Pin, Axis, Backsight*	*1*

450/50R	*Plunger, Backsight*	*1*
450/69R	*Spring, Plunger, Backsight*	*1*
450/C	*Bolt, Breech, Assembly*	*1*
450/26	*Head, Bolt, Breech*	*1*
450/22	*Extractor*	*1*
450/54	*Screw, Extractor*	*1*
450/72A	*Spring, Extractor*	*1*
450/15	*Catch, Magazine*	*1*
2838/3	*Pin, Sear*	*2*
450/76	*Spring, Sear*	*1*
450/25A	*Loop, Guard, Trigger*	*1*
450/59	*Screw, Guard, Trigger (Front)*	*1*
450/19	*Collar (Front)*	*1*
450/89	*Washer, Screw, Guard, Trigger (Front)*	*1*
B2/977	*Bracket*	*1*
450/48	*Screw, Bracket*	*2*
B1/710	*Screw, Swivel*	*2*
450/82A	*Swivel, Sling*	*1*
450/2	*Band, Lower*	*1*
450/F	*Magazine, Assembly*	*1*
450/31A	*Pin, Cap, Stock, Fore-end*	*2*
450/29	*Liner, Handguard*	*2*
450/33	*Rivet, Liner, Handguard*	*4*
450/44	*Rivet, Cap, Guard, Hand, Front*	*4*
450/101	*Band, Upper*	*1*
450/13C	*Cap, Stock, Fore-End*	*1*
450/12B	*Cap, Guard, Hand, Front*	*1*

The following components were converted from No. 4 Mk I parts:*

Body	*1*
Bolt, Breech (Minor Change)	*1*
Guard, Trigger	*1*
Blade, Foresight	*1*
Guard, Hand, Front	*1*
Guard, Hand, Rear	*1*
Ring, Retaining, Rear, Handguard	*1*

The following are parts of new design special to the Light Rifle:

Barrel	*1*
Sear	*1*
Trigger	*1*
Trigger Pin	*1*
Screw, Guard, Trigger, Rear	*1*
Lock, Washer, Screw, Guard, Trigger, Rear	*1*
Collar (Rear)	*1*
Butt Plate	*1*
Screw, Butt Plate	*2*
Protector, Foresight	*1*
Screw, Protector, Foresight	*2*
Pin, Protector, Foresight	*1*
Key, Protector, Foresight	*1*
Stock	*1*
Tie Bolt and Nut	*1*
Sling	*1*
Cap, Stock, Fore-end	*1*
Cap, Guard, Hand, Front	*1*
Band, Upper	*1*

Barrel: The Barrel would be 5-groove, made to the same bore and diameter dimensions and limits as the No. 4 rifle. The barrel is 22 inches in length, weighing 1 lb. 15 ozs. The outside diameter is constant for the first 12 inches from the muzzle, this diameter being established by the existing No. 4 bayonet dimensions.

A straight taper blending the parallel section into the radius of the breech end completes the barrel which has a knox form of similar dimensions to the No. 1 Mk III barrel.

Two lugs machined from the barrel forging are provided at the front end of the barrel for the standard assembly of the No. 4 bayonet. A keyway slot is cut in the muzzle of the barrel to locate and hold the foresight holder.

The barrel is designed for accurate shooting by a simple stocking up method in which the barrel is completely clear from the breech end to within approximately five inches of the muzzle, at which point it is held down on the fore-end at a weight not less than 6 lbs., or more than 10 lbs.

Body: The body is a conversion of the No. 4 body and aside from lightening cuts where surplus ineffective material is removed, involves the following component changes:

(a) The socket end is removed in such a manner as to leave a resistance shoulder which also provides an opening for the rear guard screw.

(b) A tongue projecting behind the sear lugs is provided, and hardened and polished as a bearing pad for the re-designed trigger mechanism.

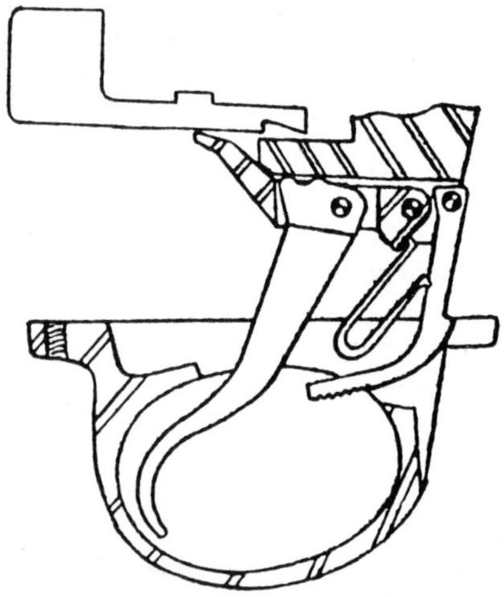

Trigger Mechanism used in Long Branch Lightened Rifle # J5550 Nos. 1 and 2 (and probably subsequent nos.)

(c) The left-hand side of the body is cut down substantially in such a manner as to equalise the amount of material between the left and right shoulders. A large and improved thumb clearance is arranged to permit of easy loading from the charger. Tests of this lightened body

Right: Different milled recesses in the stock for action body. Note trigger recess, absence of butt socket provision in the one-piece stock.
Canadian War Museum

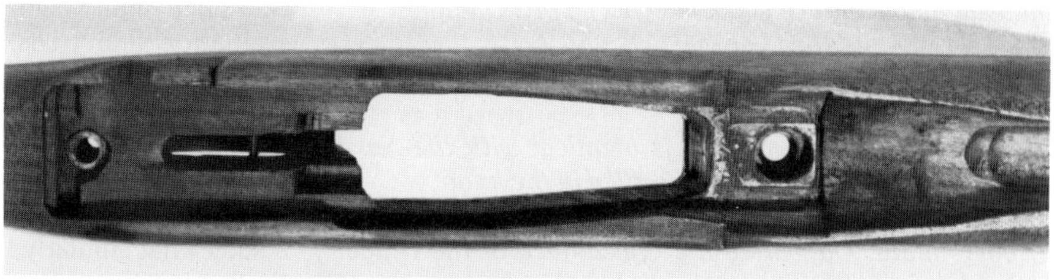

Right: Lightened fore-end and barrel muzzle.
E.D.E. Maribyrnong

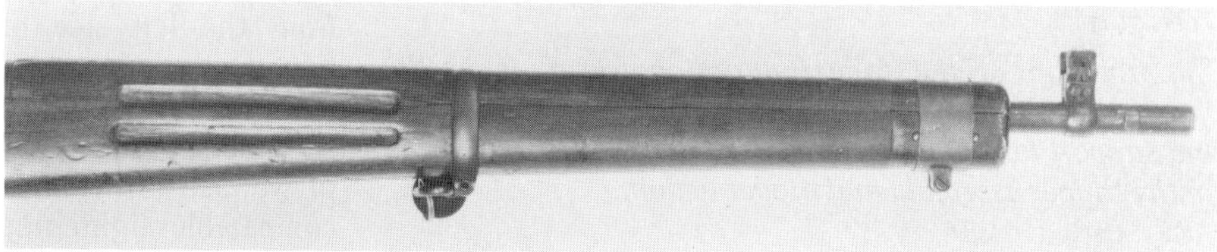

CANADIAN LIGHTWEIGHT

Right: Lightened butt, fitted with rubber kick-pad. Note fitting of the fabricated triggerguard.
E.D.E. Maribyrnong

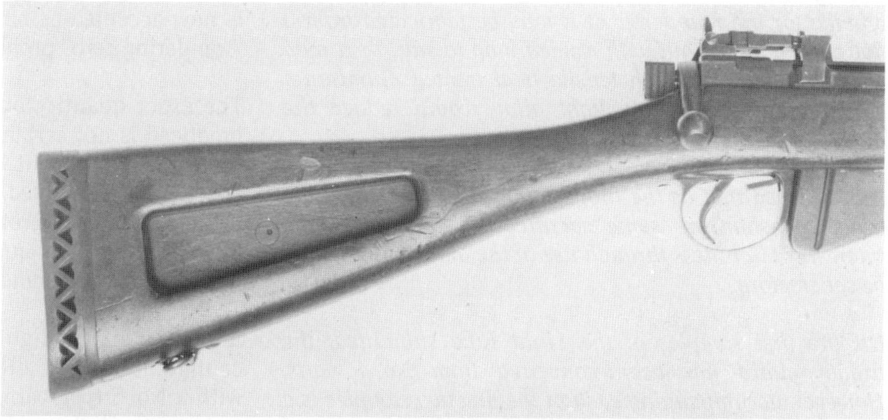

Illustrations below, left & right, credit School of Infantry, Warminster

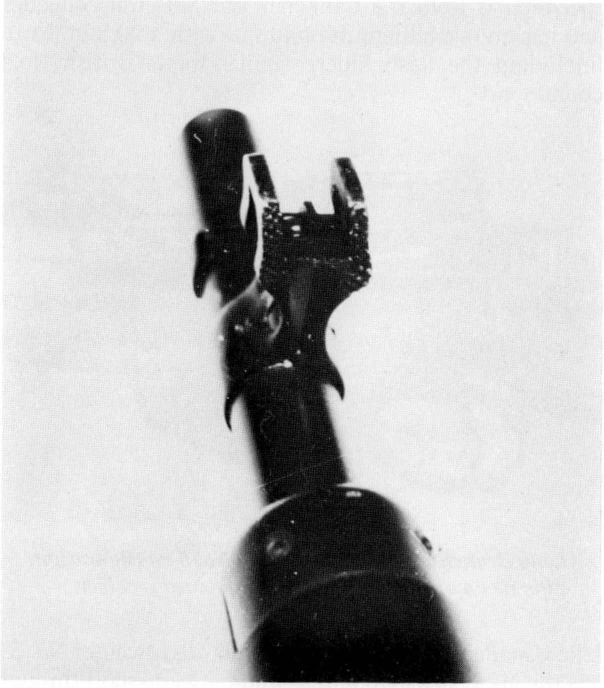

Foresight protectors and muzzle profile. The ears are bent on this sample, probably during endurance trials.

Mk I type backsight was intended to be replaced by a lightweight version. Note serial numbner on bolt.

indicate that accuracy is not affected by this change.

The weight of the lightened body is 1 lb.

Breech Bolt: The breech bolt is identical with the No. 4 except that a lightening hole is drilled in the knob.

Trigger Guard: Outwardly the appearance of the trigger guard has been changed only slightly. Such change is to remove surplus material in the interests of lightening.

The function of the trigger guard, however, is vastly changed, in that it now becomes simply a trigger guard and support for the magazine, and has no effect whatever on the trigger assembly and trigger action.

The trigger guard may be manufactured either from a forging as for the No. 4 rifle with altered dimensions to provide for the rear screw or it may be fabricated from a flat plate blanked out with welded loop to suit; or it may be die-cast from a high tensile heat treated aluminium alloy. Manufacture from light alloy would reduce the weight of the trigger guard by 2 oz.

It is believed that as the trigger guard does not carry the trigger mechanism, some sacrifice of strength in the interests of lightness through use of the aluminium would be satisfactory.

On the first samples of the Light Rifle submitted, the trigger guard has been converted from No. 4 parts. However, accompanying each of the first two samples is a trigger guard made of straight duralium. This aluminium trigger-guard is supplied for sample purposes but would be provided for manufacture if approved.

Trigger Guard Screw: This is a simple screw inserted from the top of the Rifle and held in position by a No. 8 shake-proof washer.

Butt Plate & Recoil Pad: The commercial pattern of recoil pad is attached to the sample rifles but for production purposes, the recoil pad would be made from a synthetic rubber especially selected for the purpose to provide against deterioration under service conditions.

Backsight: In order to save time in submitting sample rifles for consideration, existing Mk I backsights have been used. These will not give the proper range readings for the samples but are suitable for test purposes.

A backsight identical to the Mark 3 would be provided and would be calibrated in such a manner as to give ranges from 200 to 800 yds at 100 yd. intervals without bayonet, and would give required accuracy at 200 yds.

N.B. This suggested sighting arrangement is subject to change to suit the user's requirements.

One such lightweight rifle was delivered to Britain for evaluation and trials, initially conducted on the 29th December, 1943. Reports indicate that it shot and handled well but was considered to be weak at the "butt socket". Accuracy later fell off markedly but this was found to be due to a badly nicked bore.

A somewhat similarly lightened receiver had been tried in England in the "LT" series in 1937 but it had been found that erratic shooting resulted when metal was removed from the left side of the receiver body. One of the Canadian claims of the lightweight rifle was that it featured an even resistance bearing on the bolt lug shoulders, on which a British report commented, "a piece of eyewash". In view of this opinion, one may wonder why comparable lightening of the No. 5 receiver was not similarly regarded and why its poor shooting was attributed to so many other possibilities, for so long. This is now accepted as being the reason for the No. 5 rifle's "wandering zero" problems.

The exact quantity of Canadian lightweight No. 4 rifles produced is not certain although serial numbers ranging between 1 and about 32 would indicate that some three dozen were assembled. In keeping with Canada's practice to field test such prototypes under battle conditions in forward areas, it is quite likely that some of these rifles were sent to the Pacific and the Far-Eastern theatre.

Another Canadian lightweight was assembled on the No. 5 configuration, and although it is half-stocked and fitted with a No. 5 type barrel and flash eliminator assembly, it features the lightened receiver and one-piece stock, with the No. 5 buttplate and rubber recoil pad. The illustrated specimen is from the Canadian War Museum collection and has an overall length of 40 ins. with a 20.5-in. barrel (including the flash hider), similar to its British No. 5 counterpart.

Long Branch Lightweight rifle in No. 5 configuration. This has a one-piece stock and lightened receiver.

Like Canada's No. 4 lightweight, this "lightweight No. 5", has the Mk I backsight and re-inforcing crossbolt through the stock. The handguard has been modified as well as the fore-end and the No. 5 rubber buttplate has been drilled with six transverse holes. The rifle illustrated is stamped "J 4067-1" and it appears likely to have been modified from lightened No. 4 components rather than built up originally as a carbine. The J 4067 is reference to the particular experimental project, just as J 5550 was applied to the No. 4 lightweight rifle series.

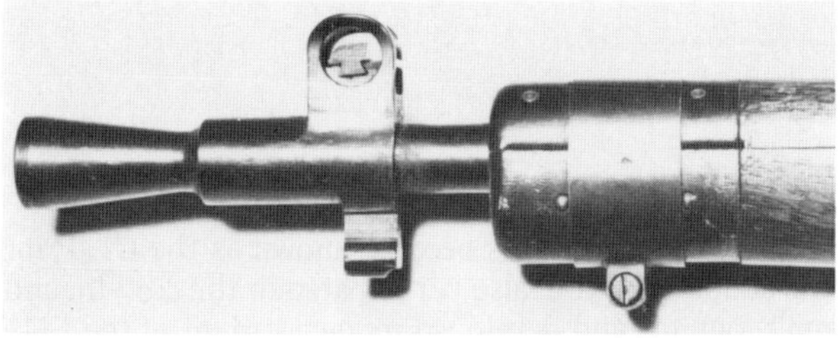

Above & below: Special Lightweight "No. 5" features.

Some other specially lightened and modified models also appear to have been fabricated at Long Branch, although no specific documentation has been sighted to confirm the particular projects or to apply specific nomenclature. Some of the later Canadian conversions may also be viewed with certain suspicion as competent gunsmiths have modified many surplus military rifles for the sporting or target shooting marketplace.

A sporterised conversion is reputed to have been produced for the Royal Canadian Air Force as a survival rifle, with the left side of the receiver milled off level from the front of the receiver to the backsight mount, removing the charger guide bridge and forward portion of the backsight rest. The fore-end has been shortened by about 11 ins. to just behind the position of the lower band, so there are no bands. A rubber kickpad replaces the buttplate. The barrel length remains standard although the foresight has been altered to a sporting configuration. The all-up weight is about 7 lbs. One such specimen is stamped "B 42" on the top of the bolt handle.

Another rifle with a similarly modified action body and fore-end configuration has had the original Long Branch markings removed and was restamped "EAL" followed by a number, presumably the serial number. It has been suggested that this was for commercial sale by Eatons Arms Ltd. and that the modifications were carried out at S.A.L. after the war, hence the marking of a "C A" logo on the top of the barrel knox form.

Some of the Canadian No. 4 rifles were converted for line throwing, using a blank cartridge and firing a brass rod to carry a thin line; the rod was inserted down the barrel from the muzzle. Prior to this, Canada had mainly used obsolete M.L.E. rifles strengthened with copper binding around the forward section of the fore-end. The No. 4 line thrower was usually marked "LINE THROWER" or "L T" on the left side of the butt or fore-end, and the front and rear sights usually removed.

Some No. 7 prototypes appear to have been assembled in 1946 as specimens have been noted with unmarked action bodies, except for "C.A.L." followed by a number on the left side of the receiver socket. Such an example at the Canadian War Museum is marked "CAL 7" and as Canadian Arsenals Ltd. was not formed until 1946, this specimen must be regarded as some form of prototype. Contracts for the production of the No. 7 were continued by C.A.L. after S.A.L. was wound up in December 1945.

Above & below: Possibly a conversion for R.C.A.F. issue as a survival rifle. *Canadian War Museum*

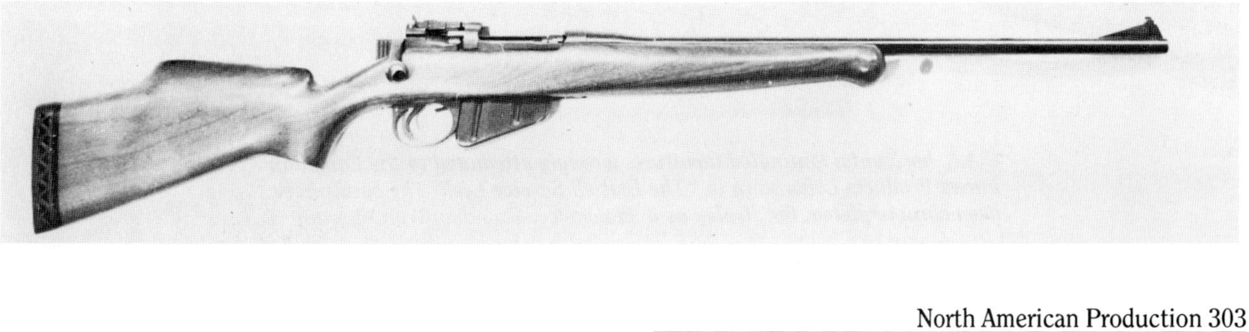

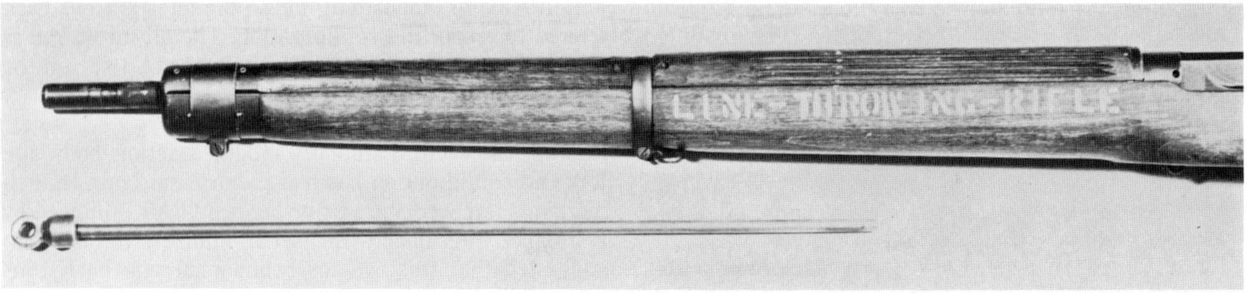

Above & below: Canadian Line-Throwing conversion of the No. 4 with line carrying rod. *Canadian War Museum*

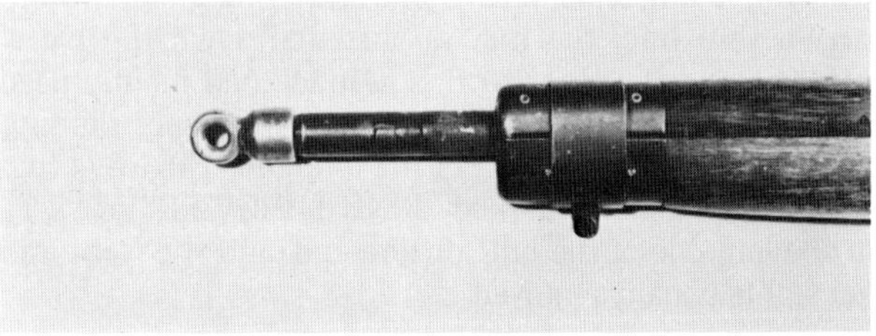

Some experiments were carried out in Canada with laminated stocks by the Forest Products Laboratory, but these were quite different to the German and British styles and also belong to a later period. The laminates are much thicker and laid horizontally instead of vertically in the stock. The purpose behind these tests was not the same as in wartime Germany and England where there were acute timber shortages, as Canada had plentiful supplies. The illustrated rifle and this series of F.P.L. tests were of a more recent vintage.

Rifle production at Long Branch from 1946 involved only the completion of those No. 4 (T) and No. 7 .22-inch contracts, with some additional experimental work; it was not until the advent of the 7.62mm cartridge and F.N. self-loading rifle in the 1950's that large scale production was recommenced. As a key power in the N.A.T.O. alliance, Canada played an important role in the automatic rifle projects which eventually saw the Belgian F.A.L. adopted. In Canadian service, this became known as the C1. Prior to its adoption, C.A.L. also participated in the .280-in. and 7mm programs and developed its own variant of the EM1 "bullpup".

Canada had realised by 1944, the advantages of a rimless cartridge and so a 7.92mm conversion of the No. 4 rifle was forwarded to England during that same year for tests and evaluation. This particular rifle was assembled in March 1944 and required a new magazine because of the longer cartridge, a recessed breech bolt and new ejector and modified extractor. The action body was opened up for the larger magazine which also resulted in a modified fore-end and trigger guard. A new barrel chambered for the 7.92mm cartridge was fitted. Even though tests were conducted and reported upon in October 1944, the project was not seriously considered at that point in time. Some Bren guns chambered for the 7.92mm round were produced in Canada to fill Chinese contracts.

In the Canadian War Museum collection, a .30 calibre No. 4 experimental rifle is marked ".300 CAL" and "W. MARSHALL" on the box magazine and receiver; these markings were made with an electric pencil. Sighting has been modified to compensate for the flatter trajectory and the foresight is marked "I". This rifle was built for the trials cartridge at that time, which eventually became the 7.62mm NATO round. Canada was a major participant in the series of trials that decided upon this cartridge and the Belgian F.N. rifle.

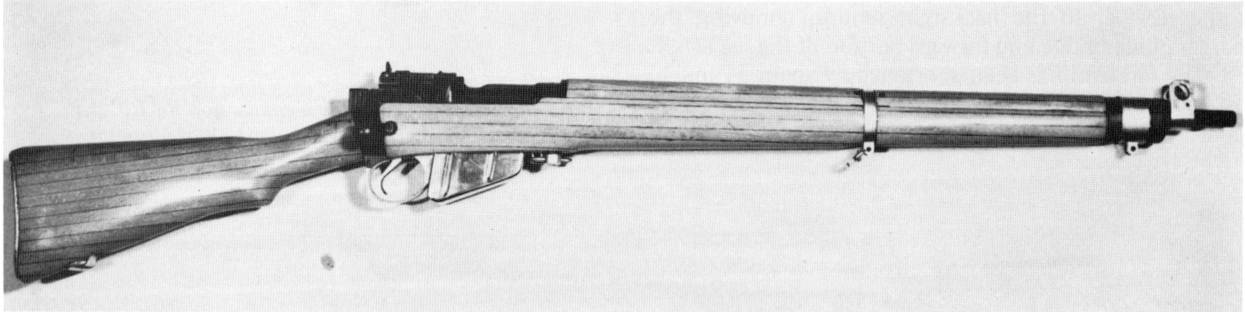

C.A.L. horizontal laminated furniture, wrongly attributed to the Canadian Forest Products Laboratory in "The British Service Lee". The bands have been chrome-plated, for display as a showpiece. *Canadian War Museum*

The F.N. F.A.L. was adopted by Canada, like Britain and Australia who were also members of the "Rifle Steering Committee" but they designated it the "L1A1" whereas Canada introduced the F.N. rifle as the "C1". Although some components of the Canadian C1 differ slightly from those made in Britain and Australia, parts are interchangeable. The No. 4 rifles produced at Maltby, Fazakerley and Shirley, along with those from Long Branch and Stevens-Savage in North America were similarly designed to be interchangeable, as were the earlier No. 1 (S.M.L.E.) rifles manufactured in Britain, India and Australia.

With the escalation of war in Korea and the possibility of large scale conflict, the Long Branch No. 4 rifle machinery, much of which had remained idle since the end of World War 2, was put into action again. In Australia also, a short run of rifles was commenced so that there would be no delays should large-scale production again be required. About 5,000 No. 4 rifles were made at Long Branch in the early 1950's and in Australia, 1,000 No. 1 rifles were produced. The Fazakerley factory in England was manufacturing the No. 4 Mk 2, as well as undertaking an extensive program of No. 4 Mk 1/2 and 1/3 conversions; the Ishapore factory in India was still producing the No. 1 rifle.

The Long Branch factory produced No. 4 .303 barrels with 6-groove right hand twist during this period, rather than with the conventional .303-in. left hand twist. Designated the "C Mk 4" barrel, this change is reported to have occurred because Bren L.M.G. rifling machinery was used. The 1950's production also saw the use of some new proof marks and stamps.

After N.A.T.O.'s introduction of the 7.62mm cartridge, the No. 4 rifle was not to play the same role in Canadian service as it did in Britain because there were no general conversions to the new rimless NATO cartridge for reserve stores or police use. However, as was the case in Britain, the new F.N. rifle was not entirely suitable for rifle club competition, so some No. 4 rifles were modified for this function.

This project was designated 3026 XP 225-28 by C.A.L. and the rifle labelled "Rifle 7.62mm No. 4 Mod (Competition shooting)"; an order was placed for 1,000 rifles. Sources indicate that only about 200 were made, the major problem encountered being the asymmetrical stress distribution in the No. 4 action and it was considered that the service 7.62mm cartridge stressed the action body to its full capacity.

No particular model number was allocated to this rifle since it was not adopted for general use; the program was shelved when the competition modifications to the C1 rifle trigger were introduced in 1964. In respect of the conversion of earlier service .303-in. rifles to 7.62mm version, Britain and Canada took different courses and the lineage of their post-war rifle production and fittings is dissimilar in a number of ways.

The new Canadian 7.62mm Lee-Enfield barrels were fitted to a variety of test and experimental rifles, even the later laminated stock models. Some were also purchased by Britain and fitted to certain rifles in the L8 series. Such barrels may be readily identified by the C.A.L. logo and because they have no bayonet lugs for the attachment of the No. 4 spike bayonet. Most of the competition components are indicated by a stamp consisting of a "C" inside an oval. The Long Branch factory, by then operated as Canadian Arsenals, closed down on June 30th, 1976.

Certain discrepancies may be noticed relating to the year of manufacture and serial number ranges on some Long Branch rifles because it was common practice to allocate the serial number from a condemned rifle to a new one, when it was assembled as a replacement. Such an example is the case of a No. 4 Mk I* Long Branch rifle, serial number 94L8407, and production dated 1955, although the 94L series was manufactured in 1950. A Mk I* rifle has also been noted with serial numbers from the usual No. 4 Mk 1 range, below 2Lxxxx. Similar circumstances may also be found to apply to the C No. 7 .22-in. training rifles.

Canadian "C" and oval marking, also found stamped on 7.62mm barrels made by C.A.L.

The Canadian marking of a "C" inside an oval, obvious in the illustration above, indicates a Competition classification. Such rifles were fine-tuned and especially fitted during assembly.

LONG BRANCH BODY PRODUCTION DATES

1941, 1942, 1943, 1944, 1945	...	*OL01 – approx. 91L*
1946 (C No. 7 Mk I rifles only)		*92L – 96L (estimate)*
1949, 1950	... ...	
1955, 1956 (replacement body)	...	*any serial number*

Canadian No. 7 Mk I rifles have body dates of 1944, 1945 and 1946, with serial numbers ranging between OL01 and 1L9999, allowing for 20,000 to have been produced.

For Long Branch production, regarding the year stamp, note that the "9" in 1949 and "6" in 1956 is an individual hand stamp, indicating that the last number was removed from the die when the year changed.

Early photograph of S.A.F. Lithgow just prior to opening, with military guard tents in front. *I. Skennerton*

Main gates of S.A.F. Lithgow today. The buildings at front have been more recent extensions. *I. Skennerton*

AUSTRALIAN PRODUCTION

CHAPTER 9

Manufacture of the Lee-Enfield rifle in Australia occurred at Lithgow which is about 140 kilometres west of Sydney in the state of New South Wales. Lithgow's coal mines, steel works and copper smelter were prime reasons for its selection; Australia's first steel had been manufactured there in the mid-19th century. The decision to establish a government factory for small arms production was made in 1907, a Lithgow site was confirmed the following year upon the purchase of 123 acres of land. This location also met with the approval of Lord Kitchener during a visit there in 1909.

Tenders were called for a complete plant with a production capability of 300 Mark III Short Magazine Lee-Enfield rifles per week and by March 1909 the four bids which had been received were from the British firms of Birmingham Small Arms Co. Ltd., Archdale Machine Tool Co. Ltd. and Greenwood and Batley Ltd., as well as Pratt & Whitney Co., of Hartford, Connecticut, U.S.A. These four tenders were:-

Archdale Machine Tool Co. Ltd.	£100,000/-/-
Birmingham Small Arms Co. Ltd.	£150,000/-/-
Greenwood & Batley Ltd.	£ 69,000/-/-
Pratt & Whitney Co., Hartford	£ 68,144/8/3d.

The £68,144/8/3d tender from the United States was lower than those of its British rivals and it also featured "repetition manufacture" (better known as mass production, which also facilitated a total interchangeability of components, closer tolerances and a production line setup); it was awarded the contract. Thus, large scale precision arms engineering was introduced to Australia. Another point raised is how close the Pratt & Whitney quote was, just below that of Greenwood & Batley; some of the British supporters claimed that there was a likelihood of "inside information" being passed on to their U.S. competition.

Pratt & Whitney supplied foremanship training for six Australian tradesmen and they also provided the first factory manager. The machinery was set up and proved in the United States at the Colt plant, so a few pre-production rifles and bayonets were turned out in North America; these are marked with a special stamp which comprises of a "C" underneath a broad arrow, inside an oval: 

The new Small Arms Factory was officially opened by the Governor-General Lord Denman on June 8th 1912, with a factory floor area of nearly 2 acres and a payroll of 250 workers. The most modern factory in Australia at this time, it had its own power house, tool room and forge, 340 individual machines, 11 forging hammers and 22 oil-fired furnaces. At the opening ceremony it was claimed that 22 hours and 36.5 minutes were taken to build a rifle under demonstration compared with 72 man hours at the Enfield factory and 48 man hours at the B.S.A. plant in England. Pratt & Whitney had contracted for 28 man hours per rifle with an annual output of 15,000 units.

Manufacture was commenced in February 1912 and the first batch of forty S.M.L.E. rifles was completed and sent to Melbourne in May 1913. Production was accelerated to 1,600 rifles per week after the outbreak of the Great War when factory employment rose to 1,400 persons. In August 1912, the Army placed an order for 20,000 rifles at a cost of £81,400 but this requirement was later reduced to 7,500 of which the first 1,500 rifles were completed by December 1913.

The unit cost per rifle in this initial Army order was therefore just below £4/1/6d; Lithgow's first commercial sale occurred in October 1912 and comprised of 476,262 .303 5-round cartridge clips to the Colonial Ammunition Company, at the price of £952/10/6d.

Some other Commonwealth Government munitions factory were established quite early:

Ammunition Factory, Footscray, Melbourne, Vic. (Colonial Ammunition Co. until 1921)	1888
Explosives Factory, Maribyrnong, Melbourne, Vic.	1911
Ordnance Factory, Maribyrnong, Melbourne, Vic.	1925
Munitions Laboratories, Maribyrnong, Melbourne, Vic. (preceded by Defence Laboratory 1907-1920)	1921

Rifle production was well under way at Lithgow by March 1914 and a few months afterwards, the British Government placed urgent orders to alleviate small arms shortages upon the outbreak of the Great War. Britain requested the supply of any surplus rifles and so all of the Australian S.M.L.E.'s except for about 10,000 were forwarded to England. Australia's shortages caused by this exportation was soon remedied as Lithgow more than doubled its original capacity and some 30,460 Mk III rifles were produced during 1915-16 financial year, the highest during the Great War period.

After the Great War, rifle production was reduced dramatically and commercial manufacture was implemented in an effort to retain many of the skilled workers. This change in direction saw the production of more than a million combs and cutters for hand shears, Western Electric talkie machines, millions of gold club heads, handcuffs, and sporting rifles and guns. These commercial firearms comprised some 250,000 .22 calibre single shot rifles, nearly 100,000 .22 calibre repeating rifles, and many .22 Hornets and .410 shotguns. The .22 rifles were made under contract for Slazengers (who later produced rifle furniture for Lithgow) while the Hornet and .410 shotguns utilised modified S.M.L.E. actions.

However, the depression resulted in large staff reductions which left only 250 workers at one stage. Figures of personnel employed at the S.A.F. were:–

Fiscal year	*Lithgow*				
1912	200				
1913	285				
1914	373				
1915	526				
1916	1,238				
1917	1,378				
1918	1,161				
1919	1,080				
1920	789				
1921	829				
1922	302				
1923	327				
1924	345				
1925	370				
1926	371				
1927	348				
1928	350				
1929	359				
1930	257				
1931	245				
1932	278				
1933	271				
1934	268				
1935	285				
1936	283				
1937	328				
1938	385				
1939	536				
1940	1,991				
1941	4,098	*B'thrst*	*Orange*	*Forbes*	*Wel'gtn*
1942	5,058	1,600	1,339	231	-
1943	5,480	1,576	2,356	256	368
1944	2,750	701	1,011	148	172
1945	1,669	496	763	–	–
1946	1,285	51	48	–	–

A Vickers machine gun section was added in the late-1920's and the first gun test-fired on 3rd May 1929. Preparation was also under way to commence production of the Bren light machine gun at about the same time as the outbreak of World War 2; the first Lithgow Mk I Bren gun was fired on 8th January 1941. Just before the beginning of World War 2, factory extensions were also made to facilitate increased production of the No. 1 rifle (the new nomenclature for the S.M.L.E.).

After the abandonment of much of Britain's war material at Dunkirk in 1940, Australia was again asked to forward any arms that could be spared; about 30,000 rifles were sent and urgent moves made to increase Australian production. A subsidiary factory was then set up at Bathurst, about 40 miles west of Lithgow, commencing production in 1941; barrel manufacture and the rifle assembly was still carried out at Lithgow, along with the machine gun production.

Barrel making facilities were then established at Orange, another 40 miles west of Bathurst, and it also took over bayonet production and assembly of the No. 1 rifle which enabled Lithgow to concentrate on the Vickers and Bren gun manufacture. With the two main annexes at Bathurst and Orange taking over the assembly and major component production of the rifles, smaller feeder factories were established at Wellington and Forbes in 1942 to supply finished parts for assembly at Orange. In Sydney, Slazengers Pty. Ltd. took over wood furniture production for the rifles and bayonet grips from 1941.

Similar feeder factories for Lithgow, supplying parts for the Bren and Vickers programs, were established at Mudgee, Cowra, Young, Parkes, Dubbo and Portland; all within a 150-mile radius of the main Lithgow factory. These factories, however, do not appear to have provided No. 1 rifle components. At peak production in 1943, Lithgow and the feeder factories were supplying 4,000 rifles, 150 Bren guns and 70 Vickers guns plus spares each week although small arms production was cut back from 1944 and the subsidiary factories were closed during 1945 and 1946.

Another proposed feeder factory at Main St. in Katoomba, in the Blue Mountains west of Sydney and not far from Lithgow, was established but never went into production. Its proposed staff strength was 123 male and 238 female workers, about the same size as the Forbes establishment.

Details of these annexes and feeder factories committed to the No. 1 and .22 No. 2 rifle production and accessories were:

Lithgow Small Arms Factory: The principal factory. The only rifle, bayonet and parts manufacturer until 1942 when the rifle barrel manufacture, rifle assembly, bayonet and accessories production were passed over to Orange.
Size: 148 acres with 25 buildings. Code marking:– MA.

Bathurst Rifle Factory No. 2: Commenced rifle component production in late 1941 as a feeder factory, until the opening of Orange Factory No. 3 in mid-1942, after which the parts were forwarded there, instead of to Bathurst.
Size: 30 acres with 9 buildings. Code marking:– BA.

Bathurst Rifle Factory No. 2
Olympic Way & Stuart St.
Current photograph; this site now owned by Timber Industries.

Orange Rifle Factory No. 3
Edward St.
Current photograph; this site now owned by Email.

Forbes Feeder Factory
in Lachlan St.
Original wartime photograph.

Wellington Feeder Factory
at 43 Maughan St.
Original wartime photograph.

Orange Rifle Factory No. 3: The 2nd largest factory, with parts and rifle barrel production commencing in early 1942 and No. 1 bayonet production from July 1942. After this, rifle assembly was changed from Lithgow, and Orange became the main rifle factory.
Size: 29 acres with 19 buildings. Code marking:- OA.

Forbes Feeder Factory: Commenced in June 1942 and supplied rifle components to Orange; the bed backsight, bolt locking, and bolt bayonet.
Size: 1 acre (nominal) with 2 buildings.
Code marking:- FA.

Wellington Feeder Factory: Commenced last week in June 1942 and supplied backsight leaves and four remaining small components to Orange.
Size: 1 acre (nominal) with 3 buildings.
Code marking:- WA.

Slazengers (Australia) Pty. Ltd. Annexe, Sydney: Commenced wood furniture production for rifle and bayonet in 1941, supplying Lithgow. After the transfer of rifle assembly to Orange, the furniture was forwarded there instead. No factory size listed as Slazengers was not a government establishment.
Code marking:- SLAZ.

J. Ludowici Ltd., Sydney: Took over the leather bayonet scabbard production (less metal locket and chape) circa late 1941. No size listed as it was another commercial concern.
Code marking:- MANGROVITE.

Some other feeder factories for S.A.F. Lithgow were established at Mudgee, Portland, Dubbo, Cowra, Young and Parkes, but these supplied parts for the Vickers and Bren machine guns, not rifle parts.

During the Great War, with the exception of rifles and .303-in. ammunition, Australia was dependent almost entirely on the resources of the United Kingdom for the technical equipment of her armed forces, both locally and in the war overseas. In 1939-45, not only was Australia able to supply her own Forces with a substantial proportion of warlike stores, but she also made shipments on account of orders for overseas Governments; Australia had become an exporter of munitions.

Substantial quantities of these munitions were exported to Britain, something in the vicinity of £20 million. In a reciprocal Lend Lease agreement with the United States, orders for £30 million were filled for U.S. forces in the Pacific region, although this figure also included radio and signals equipment, vehicles and small craft. India imported more than £3 million worth and New Zealand nearly £8 million.

The No. 4 rifle was not produced in Australia or India, only the No. 1 series. The Australian No. 1 rifle may be readily distinguished from its British and Indian counterparts by the factory name "LITHGOW" on the right side of the butt socket and in many cases by the previous-listed annex factory codes on certain components. Rifles which were assembled at Orange are still marked "LITHGOW" on the butt socket, although the Orange mark "OA" will be found on barrels and other components produced there.

The "MA" mark is a code for the Lithgow factory and its origin is most likely to have merely been a compression of "Made in Australia." When the supplementary factories were set up, the first letter of the particular town was combined with the "A" to produce a two-letter code, so Orange became "OA," Bathurst became "BA," Forbes became "FA" and Wellington was "WA." In some inspectors marks the first letter was also applied, so "L" over a broad arrow and number was used at Lithgow and "O" in a similar configuration was used by inspectors at Orange.

In early production, a seven-pointed star with an "A" inside, was used in various inspectors markings; by the 2nd World War period the "MA" marking is much more common. The Australian inspection marks and proofs are illustrated in Chapter 13.

This changeover from the "A" and star inspection mark to "MA" seems to have occurred in 1926; the elaborate star and suspended shield marking on the receiver butt socket also appears to have been changed to the more mundane "MA" over "LITHGOW" and rifle designation and mark number during that same year.

MA
LITHGOW
SMLE
III*
1944

Australian rifles were mostly stocked up in Australian Coachwood; this differs from British Walnut in that it has a lighter colour and the grain is not as dense and therefore more prone to splitting. For this reason, the Australian made rifle was often reinforced with threaded brass rod screwed into the fore-end and handguards. These brass rods were also used by armourers to repair cracks or splits and to pin breaks in the woodwork. Earliest use of Coachwood appears to have been during 1917.

The earlier rifles produced at Lithgow from 1913 until as late as 1916 were stocked in imported seasoned Walnut, most likely supplied in unfinished form from Britain. Queensland Maple was also used on many earlier Lithgow rifles, concurrent with the Walnut. This is an attractive timber, but even softer than Coachwood and more easily damaged, dented and soiled; it was particularly used during the 1920's and 1930's, up until 1940.

Early Lithgow Mk III S.M.L.E. rifles were sighted for the Mark VI .303 ammunition and had the small No. 1 pattern of "Head, Screw, Wind-gauge" which had been changed in British service between 1909 and 1911. The change in new Australian rifle production and conversion of earlier Mk III rifles for the new Mk VII ammunition occurred during the earlier years of the Great War, long before Australian production of the Mark VI .303 round ceased in late 1917. Details of these changes will be found in Chapter 5 as well as the appropriate British List of Changes reference, § 15638; the most obvious indicators of the conversion are the reduced backsight bed ramp and "2" marking on the underside, and the regraduated dial sight plate marked "L.E.S. 2." Magazine case lips were also changed due to the differently shaped projectile of the Mk VII round.

Some of the following special features of the Australian manufactured S.M.L.E. rifles apply to all the production there, while others were implemented during the 2nd World War period.

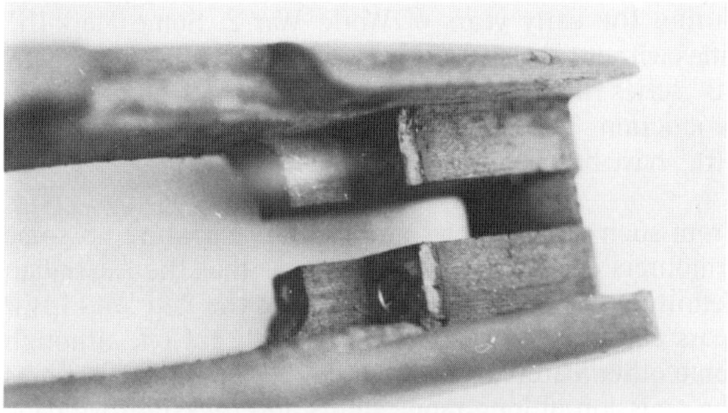

Copper blocks fitted inside Australian fore-end, each secured by a screw. *B. Labudda*

Copper blocks are fitted inside the fore-end and secured by screws, to act as recoil shoulders for the action body. These are positioned behind the body lugs which mount the sear and magazine catch because it was found that the softer timber used in Australian rifle furniture crushed quite easily under recoil. New, unfitted stocks do not have these copper blocks as they were fitted by armourers or during the F.T.R. programs as required.

This problem with the softer timber fore-end requiring recoil blocks at the rear of the action also resulted in the use of walnut blocks being fitted during 1926 and 1927, instead of the copper blocks. Another method was tried in 1935 with short steel horizontal rods incorporated in the stock bolt plate casting. But this application was shortlived also.

Australian nosecaps may be noted without provision for a piling swivel as a solid lump replaces this slot; the change was implemented in late 1941. On some 1930's production, a smaller pattern piling swivel was fitted, being that approved in the British List of Changes § 24014 and illustrated in Chapter 5.

Butt marking discs were omitted during 1918 until 1922, then re-introduced between 1922 to 1925; after 1925 they were dispensed with altogether.

Rectangular "windows" were cut into nosecap foresight protectors by field and base R.A.E.M.E. workshop units so as to permit foresight replacement or adjustment without removing the nosecap. This modification was introduced for production nosecaps in 1944 and was probably due to favourable experience with the foresight protector of the No. 4 rifle.

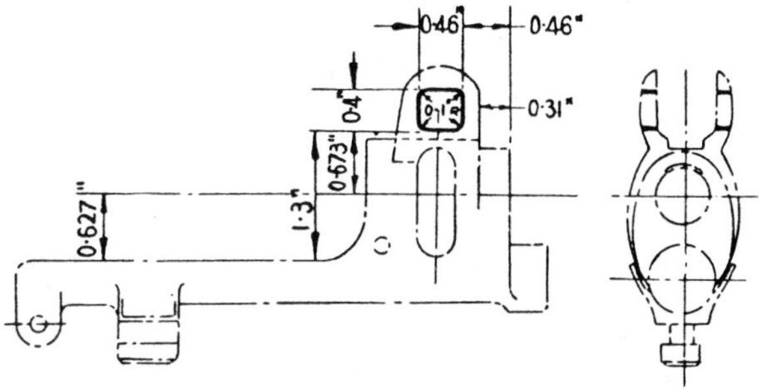

Original drawings of Armourer modifications to nosecaps, rear and side views, from an R.A.E.M.E. manual.

In the 1956 Armourers' Instruction for this modification, an alternative was advised for convenience in machining, in which the rectangular aperture could be cut on the right side only, the aperture on the left side being replaced with a 3/8-in. diameter hole, conforming with the centre of the rectangular aperture on the opposite foresight protector.

Rearsight protectors had no lightening recesses milled into the outside surfaces of the projecting ears from the end the Great War, in 1918; similar concessions were applied to the British and Indian production rifles.

Pressure die castings were substituted for forgings of the buttplate, inner band, outer band and swivel bracket; and the stem of the stockbolt was turned parallel for its full length, instead of being waisted.

Production and serial numbers for the Australian S.M.L.E. rifle commenced in 1913 with the number 1, and ran to F 40580 dated 1953, although the last group was not completed until 1956. The first series had no prefix letter and were made between 1913 and 1918, going to serial number 99,999. After this, the prefix letters A through to F were used, which technically provides for 99,999 to be manufactured in each letter range, similar to the British and Indian production.

During the 1930's, full-scale production of the No. 1 rifle was suspended although some parts production and assembly took place during most of these years. Some of the finest examples of Lithgow workmanship may be found in rifles of this period, many being stocked in Queensland Maple instead of Coachwood.

Some official figures for .303 rifle production at Lithgow list the following by fiscal years:–

1913-14	(uncertain)	1927-28	3,200
1914-15	13,786	1928-29	3,200
1915-16	30,460	1929-30	340
1916-17	23,960	1934-35	500
1917-18	23,251	1935-36	1,141
1918-19	42,129	1936-37	900
1919-20	25,570	1937-38	421
1920-21	20,420	1938-39	860
1921-22	14,440	1939-40	3,480
1922-23	2,700	1940-41	35,040
1923-24	3,320	1941-42	82,098
1924-25	3,140	1942-43	136,262
1925-26	2,760	1943-44	113,460
1926-27	3,000	1944-45	38,310

The figures just prior to W.W.2 are interesting; a few hundred rifles were produced annually, just to keep the plant moving, but sufficient few so that a considerable number of machines could be released for Bren Gun production. There was even a general belief held by influential members of the hierarchy that future wars would call for very few rifles due to the quantities already held in store and the increasing influence of light and medium machine guns. During the first year of the war (not the fiscal year), only 1,460 rifles were produced.

Such nonchalant attitudes were rudely changed after Dunkirk and new German tactics and their application of armour and paratroops; demands were received for a production of at least 2,000 per week. The British government also requested 10,000 rifles annually, which raised these estimates to 4,000 a week. Japan's entry into the war in June 1942 brought the reality of war onto Australia's very doorstep.

The 2nd World War years saw Australia's most intense rifle production, with assembly being transferred to Orange in 1942; the last rifle produced at Orange was number F 39580 in October of 1945. In 1940, Britain ordered 15,000 No. 1 rifles and an additional 100 for Borneo; in 1941, Britain ordered a further 100,000 at £7/10/- each. The last batch of 1,000 No. 1 rifles was produced at Lithgow between 1953 and 1956 so as to prove the machinery and tooling, finishing with serial number F 40580; these particular receivers are dated 1953.

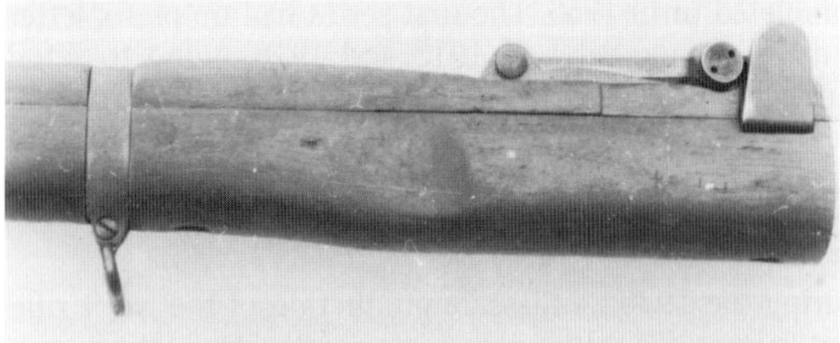

Fore-end profile soon after omission of long range dial sights. Note the recess in its original position, probably because original fore-end blanks were used. *B. Labudda*

The batches of Australian No. 1 rifles and serial number prefixes were:–

No prefix-	1913-18.
A prefix-	commenced in 1918.
B prefix-	commenced in the early 1920's.
C prefix-	during 1941; by then, the No. 1 Mk III* pattern.
D prefix-	later in 1941 to 1942; many Orange production.
E prefix-	from 1942 until 1944; Orange production.
F prefix-	from 1944 to 1945, to F39580; Orange production.
F39581 – F40580	1953; Lithgow machinery trials batch of 1,000.

The year of manufacture marked on the receiver is not necessarily when the rifle was assembled, but rather when that particular receiver was made.

Occasional discrepancies may be found in these sequences as replacement receivers or action bodies were sometimes allocated to earlier numbers (even including British rifles in Australian service). Some unfinished Mk III receivers were set-up in basic Mark III* configuration during the early years of World War 2. Some Mark III* rifles without the cut-off slot were made by 1920 yet many "B" series rifles were produced with the cut-off provision; production "MA '41" dated magazine cut-offs, marked as "III," have been noted.

From such discrepancies and the nomenclature on some armourers drawings, it would appear that the Australian definition of the Mark III rifle during the 2nd World War years came to mean a No. 1 rifle with cut-off, although some other features were from the Mk III* model, such as the omission of long range sights and backsight windage adjustment. Some Mk III* rifles have been noted with the magazine cut-off slot but no cut-off plate.

For a general outline of the application of the cut-off and cut-off slot, and the long range sights, the successive order seems to have applied for the following production years:

1913-late 1916	Mk III with long range sights (true Mk III).
1916-1918	Mk III without long range sights; the windage adjustment was also gradually omitted.
1918	Mk III* with a cut-off slot.
1918-1922	Mk III* without cut-off slot.
1922-1923	Mk III* with cut-off slot.
1924-late 1941	Mk III with cut-off.
1941-1945	Mk III*, no cut-off provision. Also 1953.

Many 1918 manufactured Mk III* rifles with cut-off slot have been noted with the "*" designation cancelled on the butt and receiver, and a cut-off plate added, possibly after 1924. Some changes in requirements for the cut-off are likely to have occurred because certain units and forces (such as the Navy) retained the cut-off long after it was not required for general service use, so if the cut-off slot was present in the receiver, the plate could be inserted by an armourer.

Another feature of many Australian service rifles is the "F.T.R." mark often seen on the receiver and butt along with the "MA" and date, and additional broad arrow mark on the bolt handle adjacent to the serial number and the butt socket above the original factory markings, marked again by the Army Inspectorate at the factory.

Particular variations of the Australian No. 1 rifle include some shortened models and jungle carbines, sniper rifles and the fitting of heavy barrels; all unique to Lithgow production.

Larger diameter or "heavy" barrels were produced initially for the Rifle Association from the early 1930's (probably 1933) until as late as 1960. The heavy barrel weighs nearly 11 ozs. more than the standard barrel and it is larger in diameter from the knox form to the foresight band keyway. The difference here is quite noticeable as the heavy barrel has a step-down whereas the standard barrel does not. In cross-section profile it is similar to the 30.2-in. Lee-Enfield barrel, cut off at 25.2 inches. The knox form of the heavy barrel is marked with an "H" on the flat, which makes it readily identified. Lithgow also manufactured spare 30.2-in. barrels for the Magazine Lee-Enfield, although these were fundamentally intended for replacements in National Rifle Association rifles, used in club competition.

Lithgow heavy barrel; note diameter reduction. *B. Labudda*

In competitive Rifle Club shooting, some Lee-Enfield rifles were shortened to S.M.L.E. length by fitting 25.2-in. barrels, removing the last 5-inches of the fore-end, and then refitting the nosecap. Three types of barrels were thus utilised:

1. Original long barrel cut to S.M.L.E. length, re-crowned.
2. Newly manufactured Lee-Enfield barrel shortened.
3. S.M.L.E. heavy barrel with the Lee-Enfield foresight band fitted over the appropriately turned-down muzzle.

The original Lee-Enfield ladder-type sights were refitted and many rifles had these barrel modifications effected at Lithgow. Some such upgraded rifles even saw military service during World War 2 when the public was requested to hand in serviceable rifles and many ex-Rifle Club .303 rifles were used in 2nd line and V.D.C. roles. These shortened Lee-Enfield rifles would therefore still have the original British-manufacture M.L.E. receiver markings and the general rifle would be similar to its predecessor but for the shorter length. For the purpose of this study, these have been designated the "Shortened Rifle Club pattern M.L.E.."

Australia's wartime dispersal policy of the ordnance manufacturing industry also saw the incorporation of commercial industry in a relatively short period of time. This process was especially accelerated after Japan's entry into the war when Australia itself became threatened. Rifle and machine gun production generally remained within the expanded Government establishments and the Bren light machine gun manufacture was commenced at Lithgow in 1941. Vickers gun production had started at Lithgow in 1929 in its Mk I (ground) configuration, then the Mark V (aircraft) variation was made for one year only in 1941, and the Mark XXI (AFV) model was produced from 1942 until 1944. Lewis gun parts and barrels were also manufactured by Lithgow and some commercial contractors for reconditioning programs.

The new 2nd World War production requirements for sub-machine guns were allocated outside the usual avenues, to private industry, with an almost official "sink or swim" attitude; Owen guns were made by Lysaghts (Aust.) Pty. Ltd. at their Port Kembla works and the Austen was produced by Diecasters Ltd. in Melbourne and W.T. Carmichael Ltd. in Sydney. For further details on the Australian machine and sub-machine gun production and models, refer to "Australian Service Machineguns" (Skennerton).

At the height of production, more than 12,000 workers were employed in small arms manufacture at the government centres; 6,000 at Lithgow and the remainder at Bathurst, Orange and the other feeder factories. During

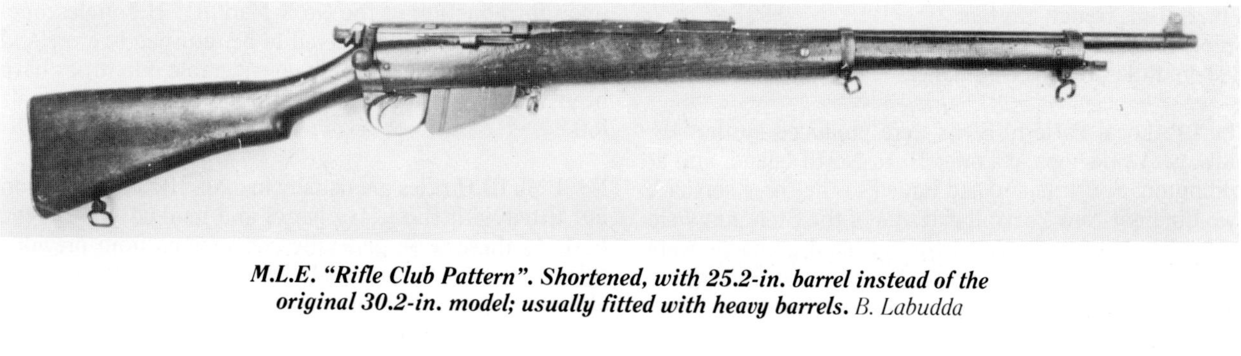

M.L.E. "Rifle Club Pattern". Shortened, with 25.2-in. barrel instead of the original 30.2-in. model; usually fitted with heavy barrels. *B. Labudda*

the period of peak production, some 4,000 rifles, 150 Bren LMG's and 70 Vickers MG's were produced each week.

Australia's wartime summary of Munitions Production from the Department in late 1945 lists the output between 1st June 1940 and 31st August 1945 as:-

No. 1 Rifles	SAF Lithgow & Orange	411,050
Bren Guns & Mountings	SAF Lithgow	17,435
Vickers Mk I Guns	SAF Lithgow	10,170
Vickers Mk V Guns	SAF Lithgow	506
Vickers Mk XXI Guns	SAF Lithgow	284
Owen Machine Carbines	Lysaghts Pty. Ltd.	45,433
Austen Machine Carbines	Diecasters Ltd. & W.T. Carmichael Ltd.	19,914
.38 No. 2 Revolvers	Howard Auto Cultivators	355
No. 1 Signal Pistols	Colonial Sugar Refinery	25,392
Mortars, 2-in.	Kelvinator Ltd. (S.A.)	2,000
Mortars, 3-in.	Gaston Bros. (Vic.)	3,006
Mortars, 4.2-in.	Gaston Bros. (Vic.)	300
Hispano 20mm Cannon	Harland Engineering Co.	214
Polsten 20mm Cannon	G.M.H. Ltd. (S.A.) & G.M.H. Ltd. (N.S.W.)	1,651
40mm Bofors Guns	Ordnance Factory Maribyrnong	290

The 2½-in. Rifle Grenade Launcher was also produced in Australia, by Metro Gas Co. in Melbourne, who manufactured about 2,000 units for Australian service. This was the No. 1 Mk I pattern of discharger cup, as used in British and Indian service.

Different Sighting Telescopes were produced in Australia for machine guns, artillery and the sniper rifle. These include a "Telescope Aust. Pattern for Telescope Layer for use with Enfield Shooting Rests fitted in Ranges Testing Rifles" made by the Commonwealth Solar Observatory in the A.C.T. and a "Layer Telescopic Rifles No. 1 – Bars Sight" by Automatic Totalisators Ltd., N.S.W. and Shailer Precision Instrument Co., also in N.S.W. The records show the supply of 48 of the first type and 49 of the second. Pattern 1918 Aust. telescopic sights was also made for the No. 1 sniper rifles near the end of the war, by the Australian Optical Company, in Victoria.

New Zealand also manufactured particular small arms during this period, in addition to the ammunition for which it is perhaps better known:-

9mm Sten Mk II LP Machine Carbines	over 10,000
.303-in. Charlton Automatic Rifles	about 1,500
2-in. & 3-in. Trench Mortars	9,527
Grenade Mortars	over 1,000
.303-in. Rifle Grenade Dischargers	3,750

The LP (Local Pattern) Stens were produced by the N.Z. Railways Workshops at Lower Hutt, North Island, and at Addington, South Island and issued to the three services. Two different New Zealand patterns of the Sten gun were made. Quantities of LP HE No. 36 hand grenades were manufactured by J.J. Niven Ltd. Founders, of Wellington, Palmerston North and Christchurch, and other contractors.

An Australian wartime National Securities Act was invoked which made provision for the surrender and appropriation of various rifles and shotguns for military service. Such firearms were marked with specific stamps, as illustrated, and the rifles included .303, .32-20, .44-40 and .22 rimfire; the Australian .310 Cadet cartridge could also be used in the .32-30 barrels. Rifles noted with the "N.S." and "D↑D" marks include various configuration Model 1892 Winchester rifles and carbines, .303 Model 1895 Winchester rifles, Remington and Colt pump action rifles and bolt action Savage .32-20 rifles.

The Australian "N.S." mark on the butt or fore-wood is taken to represent "National Security" and such arms were relegated to guard duty, prison service and the like; they were also allocated to the R.A.A.F. for security and perimeter defence. Some of the Air Force issues were carried in aircraft for use as survival rifles in the event of the crew being shot down or making a forced landing. [The "N S" mark found on some British-made boltheads however, may well indicate Nickel Steel.]

The shortages of rifles for training resulted in wooden dummy rifles or blanks being ordered, 500 of these "Rifles, Wood, Dummy" were supplied by F.K. Cox & Co. Ltd., of Racecourse Rd., Newmarket, Melbourne. The cost was £2/10/- each and these were merely 44-inch long (approximate) wooden blanks cut so as to resemble a rough outline of an S.M.L.E. rifle. They had no metal fittings.

Australia also pressed captured Great War armaments into service; the German 7.92 x 57mm MG08 and MG08/15 machine guns were officially converted and modified for the .303 cartridge at Broken Hill, N.S.W. to become known as the M.M.G. (G) and L.M.G. (G). Many captured 9mm German Luger were stamped with the "D↑D" mark (and even some "Broomhandle" 9mm M96 Mauser pistols) and subsequently issued.

Lithgow S.M.L.E. sniper rifles, produced towards the end of the war, utilised the heavy barrel as did an earlier model designated the No. 1 Mk III H (or more properly, No. 1 Mk III* H where such rifles were upgraded to the later configuration). This model comprised of a Mark III or III* rifle fitted with a heavy barrel and the appropriate furniture, it can be reliably identified by the "H" stamped on the top of the wooden butt socket and the fitting of a heavy barrel. These particular rifles were declared obsolete upon introduction of the No. 1 Mark III* H.T. (telescope-fitted sniper rifle) and ordered to be returned to store and stripped for spares, although considerable quantities have been found in Government surplus sales and disposals lots.

No. 1 Mk III H rifles are usually in a Mk III* configuration but fitted with the heavy barrel and marked "H" on the butt. As these were generally conversions from previous rifles, they have the usual Mk III markings on the receiver although some have an additional "*" after the Mark number, stamped during conversion or upgrading; some

Australian wood blank dummy rifle compared with Mk III*. For drill only. *R. Chapman*

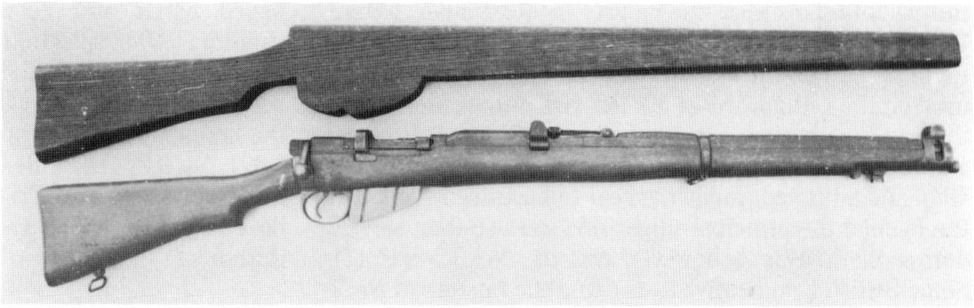

No. 1 Mk III* H rifle. Note the "H" mark on barrel knox form and at top of the butt wrist. *L. Addison*

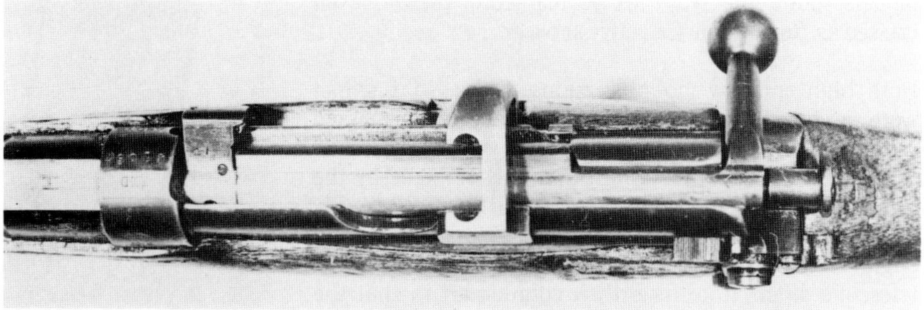

British rifles in Australian service were included as well. Fitting of the heavy barrels for Rifle Club use seems to have commenced at the factory in 1935. On the service Mk III* H rifles, the long range dial sight or just its pointer, and the rear aperture peep sight were usually removed along with the windage adjustment on the leaf backsight.

It is not uncommon to find commercially-marked heavy barrels fitted to some Mk III* H rifles, obviously ex-Rifle Club examples passed over to military service during the war and retained in the system afterwards. Such markings on the knox form as "MOTTY" and "RAWSON" and various proprietry stamps may be seen.

For the competition, marksman or sniper role, they were fitted with the typical rifle club aperture sight and plate, the removal of a small amount of wood on the left side of the fore-end will usually confirm this. Removal of this relatively thin shaving or circular shape was to enable the aperture mount plate to sit flat against the butt socket, in the vertical plane. Some Lithgow-made plates with the "MA" and ↑ service acceptance marks may be encountered, some Central plates have also been noted with the broad arrow engraved on them.

The Lithgow plate is an Argyll type; other popular plate and competition aperture sights used by competitive full-bore rifle shooters of the day included the Motty, Mues, Kings-Mues, Rawson, Ed-Mott (a trade name for Edwards and Motty), Lane, Taylor, Austral and The Don. The most commonly used models were the Central, Mues and Motty, these three as well as the Lithgow-made Argyll type have been noted on the military-marked issues and some of these sights and sideplates have been stamped or engraved with the broad arrow denoting Australian service.

To fit such a sideplate required removal of the safety catch dumbbell spring; the Central plate dispensed with the original dumbbell spring screw whereas the Lithgow type utilised this screw again. A new coil spring was usually required to provide tension onto the original safety catch. The increased thickness with the plate required a longer transverse trigger guard screw which also has a larger head.

Motty's sight. Another sideplate shown alongside with the standard trigger guard screw at bottom. *B. Labudda*

But back to the sniper rifle model. The Australian Pattern 1918 telescopic sights with detachable mounts are similar to those used on the British No. 3 Mk I* (T) rifles and were used on the Lithgow No. 1 Mk III* H.T. sniper, along with the heavy barrel and relevant furniture. The first batch of an order for 2,500 units went to store on 10th November 1944 and a limited number appear to have seen service in the Pacific; the principal sniper rifle in Australian service during World War 2, however, was the No. 3 Mk I* (T). Some British comparative tests using the Australian No. 1 sniper rifle alongside British and Canadian No. 4 (T) models found the No. 1 to be markedly inferior, even classed as "unsuitable for active service."

Both high and low mount versions of the No. 1 sniper were produced, the high mount telescope being marked "Pattern 1918 (Aust)" and the low mounted model "Pattern 1918 (Aust)/1." These sights were made by the Australian Optical Company, 246 Little Collins St., Melbourne, Victoria and the "A.O.C." mark is found engraved on the scope tube. The Australian-made telescope differs from its British counterpart in that the reticule is photo-etched on a Crystalex glass disc. "Rifle, No. 1, Mark III* H.T. (Aust)" is the official nomenclature of this sniper model.

Some of the markings on Australian Patt. 1918 telescopic sight. Rifle numbers were engraved on later, by hand.

Wooden cheek rests and two brass fixing screws were issued in the rifle chests for optional fitting from July 1945, nearly three-quarters of the way through production, so most of the sniper rifles were not thus fitted. Short and long butts were fitted to some of these rifles and a Rifle Association type sling swivel attached to the front trigger guard screw was also listed, supplied from S.A.F. Lithgow.

Ten of the rifles are recorded as having been treated with corrosion proof grease and some were coated with a khaki green paint; it was found that paint and a thin coating of the grease was best, but still inferior to sand-blasting, Parkerizing or stove-enamelling. These particular corrosion trials were conducted at Maribyrnong in April 1945 using salt water and a humidity chamber.

The Lithgow sniper rifles were conversions from older Mark III receivers as it was found that the earlier actions flexed better and generally produced more accurate shooting. After selection of suitable receivers, the rifles were stripped and in most cases only the original receiver bodies were retained for the rebuild. Approximately 10% of the actions used were of British origin, so some of these No. 1 Mk III* H.T. rifles are encountered with British-made action bodies. Original receiver manufacture years have been noted between 1907 (British) and 1921 (Lithgow).

Receiver logo, No. 1 Mk III* H.T. Original 1917 Mk III action body used, upgraded to Mk III*. *Ian Skennerton*

A total of 1,612 rifles of the order for 2,500 sniper rifles were completed by the time production ceased on 15th February 1946. The exact breakdown of telescope mount types was 1,131 high mounts (officially termed "medium brackets") and 481 with low mounts (described as "short brackets") and the change to short brackets occurred on 3rd July 1945, shortly before the cheekpieces were packed into the chests for optional fitting.

The serial numbers of these rifles converted to No. 1 Mk III* H.T. are recorded along with the designations "M" for medium and "S" for short brackets, and the dates. The first rifles were sent to store on 10th November 1944. In some cases, where British-made or post 1918 Lithgow receivers were used, the letter prefix to the serial number was not recorded in the original lists; however, such letters were recorded in the 1954 F.T.R. program. One hundred of these rifles underwent Factory Thorough Repair at Lithgow in 1954.

Competition type sling swivel. Note Lithgow inspection mark and broad arrow. Also the Slazenger inspection mark on the stock. *I. Skennerton*

Rifle Serial No.	Mount Type	Date to Store
11	M	09.04.45
72	M	17.01.45
80	S	31.08.45
103	S	04.09.45
219	S	10.09.45
220	S	27.07.45
222	M	05.04.45
235	S	18.07.45
301	S	18.07.45
314	M	17.01.45
358	S	31.01.46
419	S	27.08.45
463	M	28.05.45
594	M	19.02.45
641	M	16.05.45
657	S	11.10.45
717	M	01.05.45
	FTR	26.04.55
900	M	13.03.45
935	M	23.04.45
1082	S	02.10.45
1128	M	18.05.45
1194 U	M	26.02.45
1195	S	15.02.46
1218	M	07.02.45
1272 A	M	09.02.45
1279	M	09.02.45
1332	M	29.03.45
1352	M	02.05.45
1398	M	06.03.45
1524	M	08.05.45
1558	M	04.05.45
1562	S	27.08.45
1569	M	27.03.45
1594	M	23.04.45
1820	M	27.06.45
1822	M	07.05.45
2011	S	20.08.45
2093	S	06.09.45
2126	M	16.04.45
2137	S	22.08.45
2194	S	23.07.45
2326	M	18.05.45
2501	S	24.09.45
2507	M	15.02.45
2585	M	16.05.45
2589	S	02.10.45
2629	M	03.04.45
2660	S	15.02.46
2812	M	04.04.45
2849	S	30.08.45
3005	M	07.03.45
3046	M	28.05.45
3112	M	28.03.45
3190	S	23.10.45
3241	M	13.03.45
3242	M	04.04.45
3259	M	21.05.45
3307	M	07.05.45
3360	M	04.12.44
3409	S	29.10.45
3412	M	22.11.44
3444	M	22.11.44
3455	M	04.12.44
3498	M	29.03.45
3664	S	29.10.45
3684	S	05.09.45
3837	S	30.07.45
3862	S	15.10.45
3874	S	24.10.45
3892	M	23.11.44
3995 J	M	06.04.45
	FTR	03.08.54
4004	S	27.08.45
4014 Z	M	17.11.44
	FTR	17.08.54
4091	S	04.10.45
4152 C	M	06.04.45
4181	M	13.03.45
4214	M	24.05.45
4285	S	22.08.45
4341	M	27.04.45
4416	M	01.12.44
4425	S	06.09.45
4455	M	01.05.45
4466	M	23.05.45
4597	S	26.09.45
4764	M	07.05.45
4886	S	07.09.45
5127	S	15.02.46
5254	M	16.05.45
5322 A	S	13.08.45
5778	S	07.09.45
5823 A	M	22.11.44
	FTR	05.08.54
5833	M	18.05.45
5840	M	07.03.45
	FTR	11.08.54
5970	S	16.11.45
6040	M	15.12.44
6085	S	04.09.45
6107	M	23.05.45
6137	S	15.02.46
6143	S	07.09.45
6190	M	04.12.44
6193	M	07.02.45
6195	M	07.05.45
6200	M	27.11.44
6240	S	05.02.46
6241	M	21.02.45
6243	M	22.03.45
6264	M	10.04.45
6323	M	16.05.45
6443	S	15.10.45
6524	S	30.08.45
6549	M	23.04.45
6603	S	05.09.45
6620	S	24.09.45
6624	M	30.11.44
6926	S	06.09.45
7070	M	19.02.45
7203	M	02.05.45
7327	S	15.02.46
7348	M	02.05.45
7436	M	13.02.45
7444	S	24.09.45
7467	S	15.02.46
7534	S	16.11.45
7615	S	05.02.46
7644	M	27.03.45
7782	M	14.11.44
7794	M	29.03.45
7887	S	04.10.45
7899	S	05.10.45
7931	S	05.09.45
7965	M	12.04.45
8133	M	26.04.45
8186	M	07.05.45
8197	M	23.05.45
8267	M	07.05.45
8267	S	03.07.45
8372	M	18.01.45
8494	M	23.04.45
8614	M	20.03.45
8710	S	05.09.45
8978	S	13.07.45
9016	M	07.03.45
	FTR	17.08.54
9080	S	24.08.45
9111	S	30.08.45
9168	M	24.04.45
9206	M	02.03.45
9216	S	15.02.46
9225	M	04.05.45
9260	M	21.02.45
9292	M	27.04.45
9506	M	13.03.45
9563	M	23.05.45
9622	M	23.02.45
9663	M	13.02.45
9792	M	19.12.44
9796	M	18.04.45
9934	S	02.10.45
9984	S	22.08.45
10029	M	18.01.45
10035	M	02.05.45
10111	M	05.04.45
10372	S	16.11.45
10426	M	30.04.45
10460	M	27.04.45
11048	M	09.03.45
11106 C	M	07.03.45
	FTR	03.08.54
11788	S	30.07.45
12008	S	15.10.45
12153	S	18.09.45
12352	M	23.04.45
12423	S	26.09.45
12701	S	25.09.45
13518	M	12.12.44
13833	S	07.09.45
13835	M	24.05.45
13921	M	15.02.45
14185	M	16.01.45
14395	S	23.07.45
14811	S	03.09.45
14831	M	01.02.45
15136	M	23.11.44
15470	M	29.03.45
15579	M	18.01.45
	FTR	30.07.54
15789	M	23.05.45
15935	M	27.11.44
16020	M	15.01.45
16350	S	17.07.45
16395	S	04.10.45
16589	M	13.03.45
16686	S	15.02.46
16772	M	30.11.44
17036	S	05.09.45
17129	M	25.05.45
17270	S	31.03.45
17335	S	28.08.45
17730	S	04.10.45
17801	M	15.03.45
18222	S	10.10.45
18428	M	18.01.45
	FTR	03.08.54
18534	S	10.09.45
18590	M	08.02.45
19047	M	10.04.45
	FTR	28.07.54
19089	M	07.05.45
19435	S	20.08.45
19665	M	03.04.45
19771	M	06.04.45
20063	S	20.08.45
20719	S	20.08.45
20832	S	28.08.45
21127	M	09.04.45
21133	S	18.09.45
21135	M	29.11.44
21139	M	29.03.45
21170	M	14.12.44
21171	M	08.02.45
21181	M	11.04.45
21182	M	28.03.45
21190	M	13.02.45
21201	M	09.03.45
21205	M	21.02.45
	FTR	05.08.54
21210	M	13.03.45
21221	M	14.02.45
21236	M	14.03.45
21239	M	12.12.44
21251	M	19.02.45
21281	S	19.10.45
21290	M	07.05.45
21323	M	19.02.45
21367	M	25.05.45
21437	M	30.11.44
21510	S	29.10.45
22154	S	27.08.45
22676	S	15.02.46
22735	S	10.10.45
22796	M	15.03.45
22847	M	04.05.45
22859	M	24.01.45
22910	M	06.02.45
22948	M	23.04.45
22973	M	13.12.44
22979	M	21.05.45
22981	S	18.09.45
22984	S	02.08.45
22993	S	24.10.45
22996	S	26.09.45
23018	M	23.05.45
23062	M	30.04.45
23067	S	30.08.45
23079	M	28.11.44
23209	M	29.11.44
23369	M	16.03.45
23418	M	02.03.45
	FTR	05.08.54
23510	S	03.09.45
23615	M	16.05.45
23864	M	01.12.44
23963	M	04.04.45
24117	M	08.03.45
24145	M	23.02.45
	FTR	28.07.54
24420	S	31.01.46
24459	M	30.05.45
24514	S	12.07.45
24568	S	11.10.45
24617	M	16.01.45
24638	M	27.03.45
24643	M	01.03.45
24793	M	13.12.44
24904	S	31.01.46
25000	S	05.09.45
25019	S	15.02.46
25023	M	04.04.45
25083	M	29.11.44

Rifle Serial No.	Mount Type	Date to Store
25105	S	29.08.45
25115	S	28.08.45
25335	S	24.08.45
25339	S	30.08.45
25534	S	20.08.45
25649 L	M	06.04.45
	FTR	02.07.54
25724	M	15.01.45
25913	M	04.05.45
25943	M	19.02.45
25946	S	12.07.45
25960	S	23.07.45
26034	S	25.01.46
26040	M	27.03.45
26066	S	23.07.45
26080	M	04.12.44
26110	M	18.01.45
26159	M	08.05.45
26562	S	13.08.45
26585	M	28.02.45
26760	S	01.08.45
26824	S	13.08.45
26828	S	31.08.45
26858	M	01.05.45
26939	M	16.05.45
27012	M	19.12.44
27074	S	25.09.45
27185	M	10.11.44
27341	S	11.07.45
27431	M	27.02.45
27508	S	30.10.45
27728	S	24.08.45
27729	S	23.10.45
27766	M	25.01.45
27834	M	28.03.45
27843	M	05.03.45
27844	M	07.03.45
	FTR	30.07.54
27898	S	04.09.45
28046	M	05.04.45
28093	M	26.02.45
28118	M	15.03.45
28442	M	23.11.44
28476	M	15.03.45
28526	S	27.08.45
28528	S	31.08.45
28670	M	27.04.45
28691	M	13.03.45
28725	S	14.09.45
28748	M	13.12.44
28758	M	30.11.44
28816	M	23.03.45
28906	S	24.09.45
28916	M	19.02.45
28962	M	15.02.45
28997	M	01.05.45
29019	M	26.03.45
29036	M	08.02.45
29050	S	03.09.45
29055	M	04.04.45
29064	M	25.01.45
29090	S	15.02.46
29172	S	14.09.45
29210	M	15.11.44
29254	M	23.11.44
29337	S	16.07.45
29366	S	30.10.45
29413	M	28.11.44
29727	S	19.10.45
29740	M	23.04.45
29753	M	20.02.45

Rifle Serial No.	Mount Type	Date to Store
29778	M	01.03.45
29782	S	27.08.45
29786	S	02.10.45
29802	M	13.12.44
29878	M	26.01.45
30172	M	01.12.44
30212	M	04.04.45
30256	M	28.03.45
30281	S	16.07.45
30286	M	26.06.45
30355	S	18.09.45
30369	M	09.03.45
30389	M	22.11.44
30396	S	18.07.45
30420	S	24.08.45
30563	S	19.09.45
30567	S	23.10.45
30608	M	10.04.45
	FTR	06.08.54
30617	S	30.07.45
30694	M	22.01.45
30749	M	17.01.45
30763	M	09.03.45
30850	S	20.08.45
30855	M	25.05.45
30871	M	06.12.44
30874	M	04.04.45
30994	S	29.10.45
31090	M	01.03.45
31116	M	12.03.45
31264	M	26.03.45
31273	S	03.09.45
31572	M	10.11.44
31753	M	28.03.45
31774	M	18.01.45
31781	M	18.05.45
31788	M	25.01.45
31801	M	23.05.45
31823	S	15.02.46
31916	M	14.11.44
31933	S	29.10.45
32098	M	11.04.45
32111	M	25.01.45
32121	S	24.09.45
32174	S	16.11.45
32197	M	04.05.45
32338	M	18.04.45
32433	S	27.07.45
32458	M	19.12.44
32496	M	16.02.45
32530	M	15.02.45
32595	M	15.11.44
32601	M	14.12.44
32623	M	18.01.45
32634	M	16.03.45
32653	M	15.02.45
32810	S	15.10.45
32824	S	27.08.45
32981	S	08.08.45
33099	S	11.07.45
33118	M	30.05.45
33264	S	11.07.45
33370	M	31.01.45
33418	M	19.02.45
33435	M	04.05.45
33466	M	07.03.45
33470	M	01.05.45
33683	M	01.02.45
33694	M	01.02.45
33713	S	29.10.45
33715	S	15.10.45

Rifle Serial No.	Mount Type	Date to Store
33728	M	12.12.44
33731	S	31.01.46
33738	S	11.07.45
33769	S	28.08.45
33773	S	29.08.45
33808	S	10.07.45
33811	M	15.01.45
33825	M	28.05.45
33870	S	14.09.45
33912	M	16.02.45
34014	M	21.03.45
34066	M	16.01.45
34106	M	23.04.45
34255	M	10.04.45
	FTR	11.08.54
34293	M	06.02.45
34340	M	01.05.45
34408	M	23.01.45
34627	M	18.04.45
34681	M	28.05.45
34710	M	15.02.45
34748	M	28.11.44
34778	M	21.03.45
34852	M	13.02.45
34875	M	07.12.44
34899	S	24.10.45
34903	M	13.03.45
34953	S	24.08.45
35012	M	16.02.45
35093	M	18.01.45
35151	M	30.11.44
35167	S	18.07.45
35188	S	15.02.46
35200	M	26.04.45
35208	S	01.08.45
	FTR	06.08.54
35339	M	28.11.44
35419	M	15.01.45
35436	M	21.02.45
35447	M	13.03.45
35462	M	21.03.45
35495	M	29.11.44
	FTR	17.08.54
35566	M	23.05.45
35623	S	31.01.46
35631	S	31.01.46
35656	M	26.01.45
35697	M	01.02.45
	FTR	06.08.54
35838	M	22.02.45
35844	M	23.02.45
35952	M	19.12.44
35962	M	09.02.45
35971	M	30.11.44
36016	M	21.02.45
36230	S	15.02.46
36271	M	05.12.44
36308	M	10.11.44
36314	M	13.12.44
36315	M	15.03.45
36373	M	05.03.45
36468	M	07.03.45
36607	M	28.11.44
36628	S	30.08.45
36629	M	01.02.45
	FTR	23.07.54
36667	M	22.02.45
36685	S	03.09.45
36693	S	27.08.45
36747	S	24.08.45
36791	M	07.03.45

Rifle Serial No.	Mount Type	Date to Store
36839	S	30.10.45
36863	S	30.10.45
36872	S	11.07.45
36904	M	21.05.45
36923	S	22.08.45
36959	M	15.02.45
36967	M	14.02.45
36985	S	10.07.45
37040	M	14.03.45
37044	M	26.06.45
37084	M	09.02.45
37112	M	26.04.45
37125	S	15.02.46
37139	M	19.03.45
37146	S	19.09.45
37181	M	03.04.45
	FTR	11.08.54
37197	M	27.03.45
37208	S	27.07.45
37226	M	13.04.45
37265	M	19.03.45
37267	M	06.04.45
37276	M	29.03.45
37309	S	30.10.45
37311	M	19.12.44
37349	S	17.07.45
	FTR	02.07.54
37383	M	06.12.44
37536	S	07.08.45
37540	M	16.01.45
	FTR	30.07.54
37678	M	02.05.45
37731	M	28.11.44
37737	M	23.01.45
37740	S	25.01.46
37779	M	17.01.45
37870	M	12.04.45
37921	M	07.02.45
38085	M	20.02.45
38320	M	20.03.45
38327	M	09.02.45
38405	M	22.02.45
38609	M	18.05.45
38670	M	25.01.45
38702	S	01.08.45
38747	M	07.02.45
39115	S	24.08.45
39153	M	12.02.45
39167	M	27.11.44
39177	M	04.05.45
	FTR	05.08.54
39210	M	18.12.44
39237	M	17.11.44
	FTR	28.07.54
39238	M	27.02.45
39281	M	15.11.44
	FTR	28.07.54
39292	M	19.01.45
39345	M	26.06.45
39346	M	06.04.45
39471	S	07.08.45
39504	M	04.12.44
39603	M	16.05.45
39663	M	13.12.44
	FTR	17.08.54
39688	M	30.05.45
39702	M	26.06.45
39712	S	14.09.45
39762	S	18.09.45
39879	M	27.04.45
39902	M	27.02.45

No. 1 Mk III H.T. Rifles. Top model has medium bracket, bottom example has the short bracket. Cheekrests could be applied to both models.*

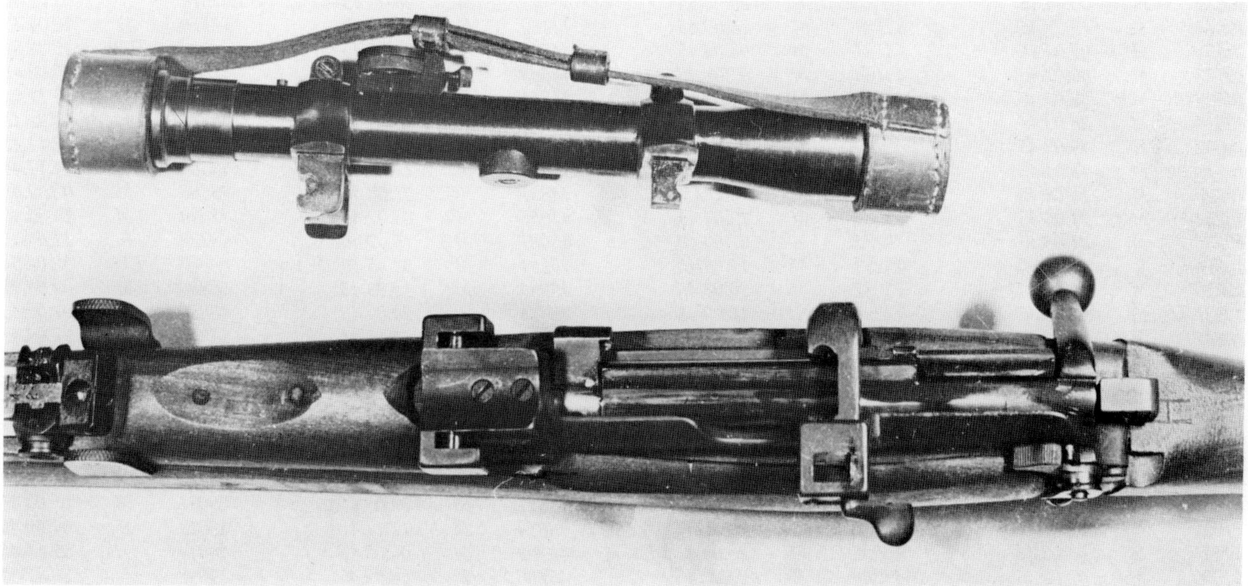

Telescopic sight detached; short bracket model with scoop out of rear handguard to allow for reduced height. I. Skennerton

Wooden plug in No. 1 Mk III H.T. fore-end, in the inner band screw hole; not required due to heavy barrel.*

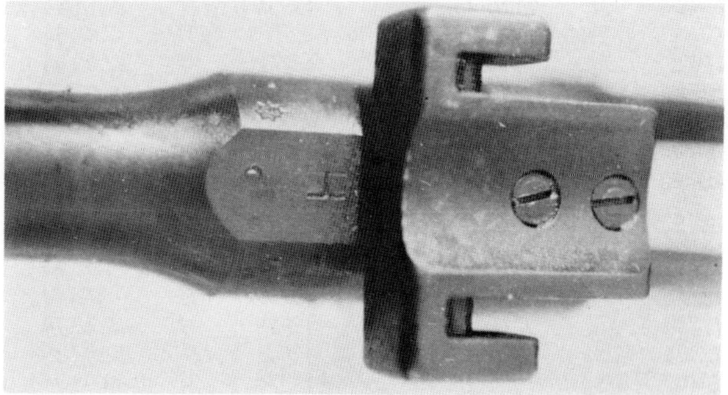

Front sight mount secured onto receiver with two screws, obscuring rifle serial number on the barrel knox form.

Rifle Serial No.	Mount Type	Date to Store
39921	M	19.02.45
39926	S	08.08.45
40002	M	26.01.45
40017	M	05.04.45
40018	S	19.09.45
40034	M	11.04.45
40068	M	15.03.45
40070	M	03.04.45
40084	S	24.09.45
40085	M	28.06.45
40097	M	13.03.45
40107	M	21.03.45
40118	M	28.03.45
40124	M	16.04.45
40134	M	28.03.45
40145	M	19.03.45
40201	M	07.12.44
40209	M	12.04.45
40248	M	14.03.45
40259	M	05.04.45
	FTR	11.08.54
40261	S	19.10.45
40279	M	06.04.45
40345	S	18.07.45
40453	M	04.04.45
40477	M	09.03.45
40496	M	02.02.45
40514	M	28.03.45
40607	M	19.03.45
40718	M	18.04.45
40723	M	16.04.45
40754	M	21.05.45
40756	M	01.03.45
40774	S	16.11.45
40820	M	29.11.44
40827	M	12.02.45
40847	M	15.03.45
40866	M	22.01.45
40869	S	16.07.45
40983	M	22.01.45
40985	S	20.08.45
41010	M	02.05.45
41013	M	13.12.44
41119	M	26.06.45
41202	M	26.01.45
41441	S	05.02.46
41650	M	07.02.45
41704	M	23.03.45
41784	M	28.11.44
41824	M	27.04.45
41830	S	18.09.45
41853	M	29.03.45
41888	M	04.04.45
	FTR	28.07.54
41925	M	04.12.44
42018	S	23.07.45
42031	S	13.08.45
42038	S	07.08.45
42065	M	26.06.45
42100	M	28.06.45
42142	M	28.05.45
42151	M	28.06.45
42344	M	12.12.44
42380	S	07.08.45
42394	S	16.11.45
42400	S	10.10.45
42403	S	07.08.45
42411	S	07.08.45
42524	M	30.04.45
42558	S	16.07.45
42637	M	08.02.45

Rifle Serial No.	Mount Type	Date to Store
42662	M	28.11.44
42670	S	13.08.45
42693	M	27.03.45
	FTR	26.04.55
42758	M	15.03.45
42765	S	07.08.45
42786	M	23.04.45
42787	S	11.10.45
42806	M	24.05.45
42858	M	01.02.45
42920	M	06.03.45
42951	M	23.03.45
42955	M	28.05.45
42971	S	17.07.45
43029	M	05.04.45
43044	S	17.07.45
43048	S	23.07.45
43049	M	30.05.45
43051	S	13.07.45
43061	M	01.03.45
43073	M	13.03.45
43101	M	27.03.45
	FTR	26.04.55
43133	M	16.05.45
43151	M	04.04.45
43168	M	01.03.45
43188	M	04.04.45
43189	M	05.03.45
43190	M	09.02.45
43203	M	28.02.45
43227	M	28.03.45
43231	M	12.04.45
43275	M	07.03.45
43283	M	19.03.45
43287	M	29.06.45
43294	S	18.09.45
43301	M	16.05.45
43306	S	02.08.45
43307	M	22.03.45
43316	M	17.11.44
43348	M	29.03.45
43354	M	05.03.45
	FTR	28.07.54
43357	M	05.03.45
	FTR	30.07.54
43358	M	20.03.45
43366	M	22.03.45
43473	M	20.03.45
43529	M	16.05.45
43649	M	13.12.44
43658	M	12.02.45
43754	M	28.03.45
43755 U	M	29.11.44
	FTR	17.08.54
43762	M	27.03.45
	FTR	26.04.55
43764	M	28.02.45
43774	M	15.11.44
	FTR	11.08.54
43788	M	16.02.45
43879	S	29.08.45
43887	M	14.02.45
43919	M	07.12.44
43939	M	13.12.44
43941	S	12.07.45
43970	M	09.02.45
44007	M	13.02.45
44037	M	23.05.45
44040	M	25.01.45
44047	M	15.02.45
44051	M	18.01.45

Rifle Serial No.	Mount Type	Date to Store
44067	M	27.04.45
44116	M	23.01.45
44121	M	05.03.45
44160	M	02.05.45
44171	M	06.02.45
44181	M	28.05.45
44206	M	15.02.45
44209	M	07.12.44
44215	M	18.04.45
44221	S	01.08.45
44226	M	18.05.45
44262	M	05.12.44
44263	M	08.02.45
44266	S	05.02.46
44271	M	02.02.45
44279	M	02.05.45
44344	M	10.11.44
44394	M	16.02.45
44403	M	13.02.45
44404	M	17.11.44
44415	S	25.01.46
44442	M	13.02.45
44445	M	26.04.45
44462	M	28.03.45
44511	S	22.08.45
44603	M	19.02.45
44641	S	28.08.45
44652	M	06.03.45
44710	M	26.06.45
44836	M	12.03.45
44837	S	01.08.45
45055	M	08.02.45
45081	M	15.02.45
45221	M	02.05.45
45234	M	14.03.45
45318	S	24.09.45
45342	M	14.11.44
45363	M	01.02.45
	FTR	30.07.54
45426	M	08.05.45
45432	M	23.04.45
45508	M	01.12.44
	FTR	26.04.55
45611	S	08.08.45
45613	M	17.11.44
45702	M	06.03.45
45736	M	06.02.45
45792	M	22.01.45
45801	M	23.11.44
45828	S	13.08.45
45831	M	08.02.45
45889	M	26.01.45
45905	M	25.05.45
45973	M	28.11.44
46017	S	05.10.45
46036	M	18.01.45
46127	M	28.06.45
46180	M	30.05.45
46214	S	30.07.45
46239	M	27.06.45
46247	S	08.08.45
46305	M	12.04.45
46313	M	12.03.45
46320	M	18.01.45
46341	M	11.04.45
46361	S	02.08.45
46398	M	04.05.45
46423	M	20.02.45
46440	M	29.03.45
46456	M	01.03.45
46469	M	29.03.45

Rifle Serial No.	Mount Type	Date to Store
46473	M	24.04.45
46485	M	10.11.44
46512	S	19.10.45
46535	M	27.04.45
46555	M	31.01.45
46564	S	01.08.45
46571	M	02.03.45
46600	M	14.03.45
46704	M	04.04.45
46724	S	18.09.45
46738	M	04.04.45
46809	M	08.02.45
46852	M	16.01.45
	FTR	17.08.54
46934	S	30.07.45
47047	M	09.04.45
47058	M	02.02.45
47136	M	18.04.45
47240	M	21.03.45
47303	S	31.08.45
47438	M	14.12.44
47517	M	28.02.45
47574	S	05.02.46
47779	M	10.04.45
	FTR	17.08.54
47857	M	16.02.45
48103	M	15.11.44
48270	M	12.02.45
48334	S	16.11.45
48500	M	14.11.44
48586	M	02.02.45
48620	M	15.03.45
48728	S	11.07.45
	FTR	03.08.54
48853	M	02.05.45
48880	M	01.02.45
48907	M	27.04.45
48962	M	08.02.45
48970	M	12.03.45
48989	M	19.03.45
48997	S	11.07.45
	FTR	26.04.55
49000	M	13.11.44
49031	M	05.04.45
49043	S	04.09.45
49044	M	07.05.45
49098	S	27.07.45
49117	S	07.08.45
49134	S	30.10.45
49137	M	31.01.45
49147	M	19.03.45
49238	M	04.04.45
49260	S	05.10.45
49290	M	20.04.45
49310	M	18.12.44
49312	S	01.08.45
49316	S	14.09.45
49321	S	26.09.45
49347	M	16.05.45
49353	M	01.05.45
49364	M	28.02.45
49382	M	06.12.44
49394	M	09.03.45
49425	M	26.04.45
49517	S	03.07.45
49520	M	14.03.45
49579	M	01.03.45
49599	S	30.10.45
49660	S	01.08.45
49661	S	29.08.45
49693	M	05.03.45
	FTR	02.07.54

Rifle Serial No.	Mount Type	Date to Store
49777	M	13.12.44
49780	M	14.03.45
49787	S	28.08.45
49794	M	29.06.45
49942	M	23.03.45
49964	M	28.03.45
49983	M	15.11.44
50004	M	15.02.45
50057	M	17.01.45
50184	M	13.12.4
	FTR	17.08.54
50197	M	06.02.45
50214	S	26.09.45
50218	M	27.03.45
	FTR	05.08.54
50247	M	18.01.45
50262	M	14.12.44
50269	M	03.04.45
	FTR	26.04.55
50287	M	27.11.44
50432	M	23.04.45
50442	M	09.02.45
50445	M	20.02.45
50500	M	17.11.44
50563	M	18.01.45
50650	M	07.12.44
50690	M	21.02.45
50704	S	27.08.45
50857	M	23.02.45
50869	S	02.10.45
50925	M	26.03.45
50971	S	25.01.46
51030	M	29.11.44
51088	M	22.01.45
51261	M	19.01.45
51264	M	07.02.45
51334	M	02.02.45
51464	M	03.05.45
51503	M	18.12.44
51506	S	03.09.45
51596	M	22.11.44
51638	M	17.01.45
51689	S	20.08.45
51768	S	02.08.45
51772	M	01.12.44
51800	S	11.07.45
	FTR	11.08.54
51821	M	02.03.45
51882	M	07.02.45
	FTR	06.08.54
51884	M	19.01.45
51982	M	28.03.45
52249	M	28.05.45
52278	M	06.02.45
52280	M	16.05.45
52353	M	25.01.45
52397	M	14.03.45
52429	M	01.03.45
52453	M	20.03.45
52460	M	07.03.45
52475	M	12.02.45
52489	M	26.06.45
52518	M	19.01.45
52520	S	27.07.45
52527	M	29.03.45
52546	M	26.06.45
52548	M	06.04.45
52571	M	10.11.44
52617	M	20.02.45
52634	S	10.10.45
52662	M	01.12.44
	FTR	23.07.54

Rifle Serial No.	Mount Type	Date to Store
52670	S	12.07.45
52680	M	21.03.45
52685	S	18.07.45
52700	M	06.04.45
52714	S	24.09.45
52728	M	29.06.45
52736	M	04.04.45
52742	M	19.03.45
52792	M	29.03.45
52854	M	09.03.45
52860	M	13.02.45
52916	M	27.04.45
52974	M	01.05.45
52987	M	07.12.44
53012	M	02.05.45
53026	M	01.03.45
53059	M	30.11.44
53098	M	18.12.44
53104	M	14.11.44
53116	M	20.04.45
53151	M	02.02.45
53154	M	17.01.45
53154	S	14.09.45
53181	M	15.12.44
53208	M	13.02.45
53224	M	16.02.45
53234	S	11.07.45
53240	M	08.02.45
	FTR	02.07.54
53284	S	08.08.45
53295	M	17.01.45
53381	S	13.07.45
53447	S	30.08.45
53481	M	16.04.45
53552	M	15.02.45
53555	S	28.08.45
53567	M	02.02.45
53667	M	06.02.45
53677	M	13.03.45
53680	M	09.02.45
53693	M	12.02.45
53816	S	05.02.46
53831	S	11.10.45
53858	M	28.05.45
53901	M	01.12.44
	FTR	17.08.54
53903	S	05.10.45
54259	M	28.11.44
54285	S	01.08.45
54338	M	29.11.44
54391 V	M	02.05.45
	FTR	23.07.54
54417	M	16.02.45
54427	S	08.08.45
54432	M	13.04.45
54640	M	28.02.45
54733	M	09.04.45
54744	S	01.08.45
54748	M	23.04.45
54751	S	05.09.45
54764	M	07.03.45
54827	M	23.04.45
54860	M	04.12.44
54920	M	07.02.45
54936	M	27.02.45
	FTR	28.07.54
54953	M	27.03.45
55024	M	14.11.44
55055	M	21.02.45
55080	M	12.04.45
55144	M	04.12.44

Rifle Serial No.	Mount Type	Date to Store
55158	S	23.10.45
55190	M	18.05.45
55208	S	18.09.45
55216	M	12.02.45
55250	M	15.11.44
55268	M	10.04.45
	FTR	23.07.54
55296	M	01.12.44
55312	S	11.10.45
55333	S	30.07.45
55560	S	10.10.45
55605	M	21.05.45
55697	M	23.02.45
55764	M	30.11.44
55773	M	12.02.45
55873	M	06.02.45
56001	S	05.10.45
56008	M	07.02.45
56021	M	27.11.44
56138	M	01.12.44
56215	S	02.08.45
56225	S	02.08.45
56382	M	10.04.45
56404	M	23.01.45
56418	M	31.01.45
56440	M	05.04.45
	FTR	06.08.54
56573	S	17.07.45
56591	S	07.09.45
56624	M	15.02.45
56692	S	19.10.45
56794	S	15.10.45
56906	M	03.05.45
56907	S	28.08.45
57026	S	19.09.45
57050	M	22.02.45
57145	M	23.05.45
57407	M	12.02.45
57430	M	13.11.44
57458	M	18.12.44
57479	M	08.02.45
57548	M	20.03.45
57667	S	26.09.45
57680	M	09.04.45
57820	S	10.10.45
57829	M	01.12.44
57845	S	08.08.45
57914	M	16.05.45
57920	M	06.02.45
57936	M	26.06.45
57946	S	10.10.45
58028	M	06.12.44
58135	S	05.09.45
58147	M	19.12.44
58150	M	05.12.44
58214	M	12.02.45
58254	M	25.01.45
58259	M	22.11.44
58305	M	26.03.45
58325	S	28.08.45
58417	M	08.05.45
58436	M	26.01.45
58443	S	10.10.45
58453	S	30.08.45
58666	S	20.08.45
58788	M	19.01.45
58798	S	20.08.45
58835	S	23.07.45
58917	S	12.07.45
58929	M	14.03.45
58955	S	07.08.45

Rifle Serial No.	Mount Type	Date to Store
58957	S	03.09.45
59017 B	S	10.09.45
59038	M	29.03.45
	FTR	03.08.54
59209	M	05.03.45
59334	M	02.03.45
59342	M	10.04.45
59362	S	13.07.45
59390	S	31.01.46
59526	M	08.03.45
59529	S	11.10.45
59587	S	11.10.45
59615	M	01.03.45
59622	M	02.02.45
59670	S	24.10.45
59713	M	01.05.45
59891	M	01.05.45
60068	M	15.02.45
60103	M	26.04.45
60117	M	23.01.45
60181	M	16.05.45
60215	M	06.12.44
60247	M	16.02.45
60323	M	18.01.45
60347	S	25.01.46
60418	M	29.11.44
	FTR	26.04.55
60606	M	15.02.45
60624	M	26.02.45
60660	M	24.01.45
60713	M	16.01.45
60815	M	06.12.44
60816	M	26.04.45
	FTR	30.07.54
61000	M	16.02.45
61105	M	27.06.45
61148	M	10.11.44
	FTR	26.04.55
61152	S	16.11.45
61164	M	26.01.45
61177	M	04.05.45
61295	M	30.04.45
61298	M	09.03.45
61338	M	14.11.44
61340	S	28.08.45
61453	M	26.02.45
61478	M	29.03.45
61563	S	30.07.45
61677	M	18.01.45
61700	M	07.03.45
61795	M	26.02.45
61804	M	15.03.45
61824	M	04.04.45
	FTR	23.07.54
61887	M	12.04.45
61890	M	18.01.45
	FTR	28.07.54
61905	M	19.03.45
61914	S	27.07.45
61992	M	02.07.45
62003	M	27.02.45
62048	M	29.03.45
62059	M	15.03.45
62082	M	02.03.45
62206	S	04.10.45
62237	M	26.03.45
62275	M	29.03.45
62375	S	24.10.45
62399	M	08.02.45
62440	M	13.02.45
62614	M	05.12.44

Rifle Serial No.	Mount Type	Date to Store
62617	M	13.11.44
62694	S	05.10.45
62699	S	29.10.45
62729	S	05.10.45
62837	M	21.02.45
63210	M	14.12.44
63434	M	29.03.45
63708	M	04.05.45
63717	M	02.02.45
63828	S	07.09.45
63853	M	05.03.45
63893	S	28.08.45
63932	M	06.03.45
63968	M	06.12.44
63987	M	28.11.44
64153	M	15.11.44
64174	S	08.08.45
64188	M	27.02.45
	FTR	05.08.54
64361	M	05.04.45
64707	M	07.02.45
64709	M	22.03.45
64742	M	17.01.45
64795	S	06.09.45
64796	M	18.05.45
64803	M	19.03.45
64835	M	27.02.45
64851	M	29.06.45
64852	S	20.08.45
65005	M	04.04.45
65027	M	13.11.44
	FTR	11.08.54
65073	M	29.03.45
65209	M	19.01.45
	FTR	28.07.54
65245	M	09.03.45
65248	M	08.05.45
65257	S	31.01.46
65280	M	22.03.45
65343	S	23.10.45
65516	S	30.10.45
65549	M	15.02.45
	FTR	17.08.54
65590	M	08.03.45
65693	S	24.09.45
65727	S	24.09.45
65766	M	02.05.45
	FTR	17.08.54
65805	M	28.11.44
66040	M	29.11.44
66646	S	07.09.45
67221 V	M	01.02.45
	FTR	23.07.54
67251	M	08.03.45
67441	S	10.09.45
67766	M	26.04.45
68000	M	24.04.45
68161	M	26.02.45
	FTR	06.08.54
68542	S	16.07.45
68555	S	30.10.45
68578	M	02.07.45
68589	M	26.04.45
68615	S	26.09.45
68674	S	18.09.45
68724	M	01.03.45
68763	M	10.11.44
68809	M	16.05.45
68848	S	17.07.45
68856	M	05.03.45
	FTR	28.07.54
68895	M	08.03.45
68900	M	13.04.45
	FTR	03.08.54
68902	M	08.02.45
68908	S	01.08.45
68954	M	04.12.44
68981	M	14.03.45
69038	M	09.03.45
69078	M	28.03.45
69090	M	19.03.45
69143	S	27.07.45
69144	S	19.09.45
69165	M	14.03.45
69189	S	16.11.45
69488	M	26.01.45
69557	S	24.09.45
69600	M	07.05.45
69911	S	22.08.45
70024	S	15.02.46
70431	S	04.09.45
70537	M	07.05.45
70653	M	19.01.45
70657	S	19.10.45
70994	M	22.02.45
71034	M	26.02.45
71142	M	26.01.45
71463	S	28.08.45
71514	M	12.03.45
71514	S	24.09.45
71517	S	07.08.45
71525	M	09.03.45
71526	S	27.08.45
71544	M	23.04.45
71563	M	12.12.44
71601	M	09.02.45
71615	M	26.02.45
	FTR	30.07.54
71619	S	15.02.46
71630	M	28.02.45
71650	M	18.12.44
71654	S	30.07.45
71671	M	10.04.45
	FTR	11.08.54
71679	S	03.07.45
71714	M	09.02.45
71721	M	15.12.44
71789	M	23.02.45
71791	S	01.08.45
71801	M	26.01.45
71829	S	30.07.45
71834	M	13.04.45
	FTR	28.07.54
71852	M	06.04.45
71864	S	15.02.46
71877	M	12.04.45
71884	M	16.03.45
71895	S	24.09.45
71901	M	06.03.45
71907	S	30.08.45
71932	M	27.03.45
71935	M	29.03.45
71939	M	29.03.45
71972	M	05.03.45
71991	S	10.07.45
71996	M	08.05.45
72005	M	19.01.45
	FTR	28.07.54
72008	S	27.07.45
72009	M	05.04.45
	FTR	06.08.54
72034	M	21.02.45
72067	M	09.02.45
72094	S	30.10.45
72098	M	13.11.44
72149	M	01.12.44
	FTR	28.07.54
72153	S	24.09.45
72233	M	15.11.44
	FTR	30.07.54
72294	M	26.04.45
72374	M	23.02.45
72450	S	15.02.46
72492	M	05.12.44
72498	M	06.12.44
72572	S	11.07.45
72598	S	03.07.45
72624	M	12.12.44
72660	S	15.02.46
72701	M	06.12.44
72756	M	25.01.45
72769	S	04.09.45
72820	M	12.02.45
72860	M	22.02.45
72933	M	06.04.45
72988	M	23.11.44
72993	M	20.04.45
73026	M	08.02.45
73100	M	30.11.44
73136	S	14.09.45
73150	M	21.05.45
73154	S	02.10.45
73269	M	02.07.45
73408	M	18.05.45
73754	S	16.11.45
73827	S	10.09.45
73837	M	31.01.45
74107	M	19.02.45
74177	S	29.10.45
74265	M	28.02.45
74300	M	05.04.45
74305	S	07.09.45
74322	M	18.05.45
74331	S	27.07.45
74414	M	06.03.45
74473	M	01.02.45
74491	S	27.08.45
74492	M	04.05.45
74499	M	28.02.45
74526	M	16.04.45
74545	S	08.08.45
74565	S	15.02.46
74568	S	26.09.45
74577	M	03.05.45
74609	M	06.02.45
74623	S	24.10.45
74629	M	02.02.45
74637	M	12.04.45
74674	S	31.08.45
74685 or 74689	M	16.03.45
	FTR	06.08.54
74691	S	25.09.45
74694	M	27.02.45
74704	M	09.03.45
74735	M	13.04.45
74738	M	20.04.45
74749	M	28.02.45
74753	M	26.02.45
	FTR	11.08.54
74760	M	31.01.45
74771	S	01.08.45
74789	M	02.02.45
74796	M	16.04.45
74801	M	30.11.44
74827	M	29.06.45
74838	M	20.04.45
74871	S	06.09.45
74875	M	29.06.45
74882	M	27.04.45
74888	M	26.06.45
74909	M	29.03.45
74967	M	07.05.45
75054	S	25.09.45
75104	M	19.01.45
75122	S	01.08.45
75132	M	26.03.45
75257	M	23.04.45
75259	S	30.10.45
75266	M	24.04.45
75275	M	28.02.45
75287	M	27.06.45
75288	M	02.05.45
75297	M	23.05.45
75310	M	26.04.45
75315	M	14.02.45
75341	M	22.11.44
	FTR	30.07.54
75355	M	02.07.45
75482	M	21.05.45
75492	M	16.05.45
75502	M	23.05.45
75527	M	21.05.45
75529	M	28.05.45
75536	S	16.07.45
75571	M	23.05.45
75580	S	07.08.45
75646	S	31.01.46
75672	M	27.11.44
75706	M	14.11.44
75775	M	19.02.45
75777	M	28.03.45
75778	M	12.04.45
75809	S	02.10.45
75828	M	01.02.45
75854	M	06.02.45
75894	M	05.12.44
75921	M	19.02.45
75955	S	30.10.45
76005	M	02.07.45
76006	S	05.02.46
76031	M	28.06.45
76032	M	06.03.45
76066	M	28.03.45
76102	S	23.07.45
76115	S	12.07.45
76174	M	27.02.45
	FTR	05.08.54
76189	M	15.11.44
76222	S	19.09.45
76268	M	14.11.44
76314	M	06.04.45
76316	M	28.11.44
76375	S	14.09.45
76382	M	03.05.45
76383	M	10.04.45
76388	M	12.03.45
76401	M	23.03.45
76406	M	26.06.45
76422	S	28.08.45
76444	S	07.09.45
76477	M	29.11.44
76538	M	26.02.45
76553	S	23.10.45

Rifle Serial No.	Mount Type	Date to Store
76589	M	24.05.45
76597	M	13.04.45
76601	S	25.01.46
76630	M	09.02.45
76674	M	13.02.45
76747	M	26.04.45
76807	M	28.11.44
76884	M	15.11.44
76900	M	09.02.45
76908	M	15.12.44
76935	M	15.02.45
76952	M	28.06.45
76976	M	04.12.44
76979	S	05.10.45
77034	M	25.01.45
77038	M	09.04.45
77082	M	21.02.45
77127	M	15.12.44
77258	M	06.02.45
77276	M	13.12.44
77333	M	06.12.44
77414	M	06.12.44
77426	M	26.06.45
77451	S	26.09.45
77456	M	27.03.45
77493	M	15.02.45
77741	M	24.05.45
	FTR	23.07.54
77746	M	26.02.45
	FTR	06.08.54
77783	M	21.05.45
77849	S	13.07.45
77885	M	23.05.45
77970	M	19.03.45
	FTR	11.08.54
77987	M	20.04.45
78054	S	11.07.45
78097	M	01.05.45
78128	S	16.11.45
78189	M	12.02.45
78194	S	30.07.45
78224	M	06.03.45
78360	M	14.02.45
78460	M	25.01.45
78493	M	29.03.45
78530	M	11.04.45
	FTR	26.04.55
78543	M	15.11.44
	FTR	05.08.54
78616	M	23.04.45
78630	S	27.08.45
78642	M	04.12.44
79041	S	27.07.45
79057	S	31.08.45
79156	S	26.09.45
79197	S	10.09.45
79301	M	09.02.45
79314	S	02.10.45
79403	M	28.03.45
79504	M	03.07.45
79601	M	29.11.44
79808	M	06.12.44
80221	M	16.01.45
80267	M	06.03.45
80287	M	22.11.44
80300	S	23.07.45
80378	M	28.02.45
80405	M	14.12.44
80474	M	16.01.45
80523	M	04.05.45
80647	M	08.02.45

Rifle Serial No.	Mount Type	Date to Store
80689	M	26.02.45
	FTR	03.08.54
80740	M	18.05.45
80765	M	28.05.45
80780	M	28.11.44
80847	M	26.06.45
80893	M	13.11.44
	FTR	28.07.54
80917	S	02.10.45
80932	S	05.10.45
81050	M	30.11.44
81059	S	31.01.46
81157	S	10.10.45
81171	M	16.05.45
81174	M	25.01.45
81257	M	22.11.44
81283	M	24.04.45
81285	M	23.04.45
81381	M	04.12.44
81457	M	15.11.44
81558	M	15.11.44
81664	M	09.02.45
81696	S	15.02.46
81776	M	23.11.44
81797	M	28.11.44
81820	S	07.08.45
81841	S	19.10.45
81907	S	31.08.45
81964	S	05.02.46
82127	S	10.07.45
82138	S	23.07.45
82261	M	16.05.45
82271	S	05.02.46
82492	M	10.04.45
	FTR	06.08.54
82505	M	16.03.45
82556	M	16.03.45
82611	M	10.04.45
	FTR	05.08.54
82613	M	23.03.45
82625	M	13.12.44
82663	S	20.08.45
82694	M	05.12.44
82747	S	06.09.45
82751	M	15.11.44
82864	M	29.03.45
82901	M	15.12.44
82957	M	06.03.45
83012	S	31.08.45
83039	M	02.05.45
83047	M	15.11.44
83177	M	14.11.44
83193 U	M	19.01.45
	FTR	03.08.54
83218	M	15.01.45
83258	M	22.11.44
83264	M	30.04.45
83291	M	10.11.44
83414	M	07.02.45
83437	M	08.02.45
83483	M	16.01.45
83530	S	30.10.45
83535	M	13.12.44
83573	M	05.12.44
83600	M	14.11.44
83676	M	18.05.45
83677	M	23.01.45
83699	S	24.08.45
83724	M	10.11.44
83728	M	10.11.44
	FTR	26.04.55

Rifle Serial No.	Mount Type	Date to Store
83729	M	15.11.44
	FTR	03.08.54
83739	M	27.02.45
83793	M	13.12.44
83798	M	16.02.45
83837	S	20.08.45
83888	M	14.11.44
83910	M	07.02.45
83925	M	23.05.45
83993	S	30.10.45
84010	M	06.03.45
	FTR	26.04.55
84015	S	02.08.45
84091	M	26.02.45
	FTR	03.08.54
84137	S	13.07.45
84141	M	16.01.45
84162	M	15.03.45
84165	M	27.06.45
84202	S	05.02.46
84204	S	14.09.45
84254	S	11.10.45
84277	M	14.02.45
84291	M	18.01.45
84294	M	20.02.45
84346	M	16.01.45
84351	S	10.09.45
84356	M	01.12.44
	FTR	23.07.54
84369	M	05.12.44
84380	M	15.11.44
84426	M	23.02.45
84440	M	14.03.45
84465	M	13.12.44
84466	M	13.02.45
84472	S	10.10.45
84486	S	20.08.45
84510	S	02.10.45
84521	M	04.04.45
84554	M	09.02.45
84558	M	28.03.45
84565	M	18.01.45
84567	S	26.09.45
84600	M	21.02.45
	FTR	05.08.54
84618	M	23.05.45
84665 C	M	24.04.45
	FTR	30.07.54
84704	M	02.07.45
84717	M	15.01.45
84755	S	19.10.45
84765	M	21.05.45
84769	M	23.02.45
84809	M	26.01.45
84854	M	05.12.44
84866	S	16.11.45
84880	M	12.04.45
84919	M	16.01.45
84927	M	08.03.45
85002	M	13.12.44
85008	S	25.09.45
85010	M	08.02.45
85024	M	22.11.44
85070	S	31.08.45
85111	M	19.01.45
85130	M	05.12.44
85146	S	25.01.46
85156	M	13.12.44
85204	M	22.01.45
85242	S	24.08.45
85287	M	17.01.45

Rifle Serial No.	Mount Type	Date to Store
85301	M	28.05.45
85332	M	26.06.45
85372	M	21.02.45
85383	M	23.02.45
85403	M	11.04.45
85420	M	22.03.45
85444	S	28.08.45
85457	M	05.03.45
85489	M	05.12.44
85537	M	23.02.45
85567	M	23.02.45
85573	S	10.07.45
85588	S	25.01.46
85615	M	27.06.45
85628	S	10.07.45
85658	S	16.11.45
85674	M	10.11.44
85703	S	07.09.45
85708	M	28.11.44
85798	S	11.07.45
85819	M	26.06.45
85904	M	03.04.45
85918	S	29.08.45
85984	S	08.08.45
85992	S	24.08.45
86069	S	03.07.45
86100	S	30.08.45
86154	M	26.06.45
86173	S	28.08.45
86359	S	03.09.45
86438	M	24.05.45
86479	M	26.06.45
86485	M	25.05.45
86674	M	21.05.45
87003	M	23.05.45
87075	M	23.04.45
87121	M	19.01.45
87423	M	01.02.45
87779	M	26.01.45
87869	M	27.02.45
87880	S	25.09.45
87925	M	25.05.45
87995	M	08.02.45
88016	S	29.10.45
88017	S	16.11.45
88067	M	30.11.44
88069	S	02.08.45
88222	M	17.01.45
88483	S	30.08.45
88490	M	22.11.44
88505	M	05.04.45
	FTR	17.08.54
88535	S	29.10.45
89220	S	26.09.45
89494	M	30.04.45
89823	S	14.09.45
90104	M	23.05.45
90180	M	27.04.45
91198	M	03.05.45
92108	M	17.01.45
92920	M	02.05.45
92956	S	04.10.45
93742	M	02.05.45
93804	M	30.05.45
94029	M	12.04.45
94047	S	10.09.45
94150	M	02.02.45
94250	M	03.05.45
94404	M	09.02.45
94534	M	02.05.45
95042	S	06.09.45

Rifle Serial No.	Mount Type	Date to Store	Rifle Serial No.	Mount Type	Date to Store
97738	M	21.05.45	99369	M	01.05.45
97940	M	29.03.45	99556	S	06.09.45
98342	S	05.10.45	99649	M	16.02.45
98671	S	30.08.45			

Note: The diligent student may well notice that three of the above H.T. rifle serial numbers appear twice, #8267, #53154 and #71514. This is because a letter prefix is missing in each instance, not having been recorded. Some other examples will be found in the above listing where the letter prefix or suffix was not noted in cases of British or post-1918 Lithgow receivers.

Another Lithgow rifle with a special barrel was the "X" series which used a barrel made of stainless steel. Only ten were assembled for trials, using new receivers, and serial numbered X1 to X10. As the stainless steel barrels could not be blued, they were painted black. Special attention was paid to the stocking-up and the receivers appear to have been 1928 or 1929 dated, with the "X" prefix serial number marked on the bolt, receiver and barrel as well.

From 1942, there had been numerous requests from forward areas in the Pacific and South-East Asia for a shorter, lighter rifle, particularly for jungle fighting. In the early years of the war, the top priority was production of sufficient quantities of the No. 1 rifle, so such requirements were not seriously considered by the Ordnance system. By 1944, however, two shortened No. 1 Mk III* rifles had been produced for Army trials, with 20.2 and 18.2-inch barrels; these rifles were referred to as the "intermediate" and "short" lengths. On the short model, it will be noticed that the outer band is positioned at the base of the nosecap, something like that of the Mark V S.M.L.E.; on the intermediate length rifle the outer band is positioned about 2-ins. back from the nosecap.

It was then decided to produce one hundred of the intermediate length model at Lithgow, designated the "Shortened, Lightened" rifle; these were completed by the end of March 1944. A special bayonet was supplied with each rifle, having a 10-inch blade with the standard Pattern 1907 hilt and grips. This bayonet was later adopted for the Owen Machine Carbine. The shortened, lightened rifles have a special pattern aperture backsight on the charger bridge, ranged at 200 and 500 yards; the serial numbers are ranged between XP 1 and XP 100.

A similar conversion had been effected in Britain as the basic design was merely a shortened, full-stocked No. 1 rifle, although the Australian variant has a simple, dual-setting aperture sight on the charger bridge whereas their British counterparts used the original bed backsight. With the development of the No. 4 lightened rifle (eventually designated the No. 5), the No. 1 rifle was similarly fitted with a shortened barrel and flash hider, and a half-stocked fore-end. This form of jungle carbine, on the No. 1 action, eventually became referred to as the Rifle No. 6.

British factory production drawings of such a jungle carbine were sent out to Australia and India, with a folding aperture backsight designated Mark V, probably in continuation of the backsights Marks 1 to 4 as fitted to the No. 4 rifle. The Australian "jungle carbines" are distinctive in that they were assembled on the No. 1 action instead of the No. 4, the No. 1 action has curved walls at the top of the receiver body for the boltway whereas the No. 4 action is slab sided.

In September 1944, four different half-stocked "jungle carbines" were tested at the Williamstown Rifle Range:

1. Detachable commercial aperture sight, as used by Rifle Clubs for competitive shooting.
2. British Mk V backsight mounted at rear of modified No. 1 receiver; adjustable aperture to 800 yds.
3. Leaf backsight, similar to the standard No. 1 rifle.

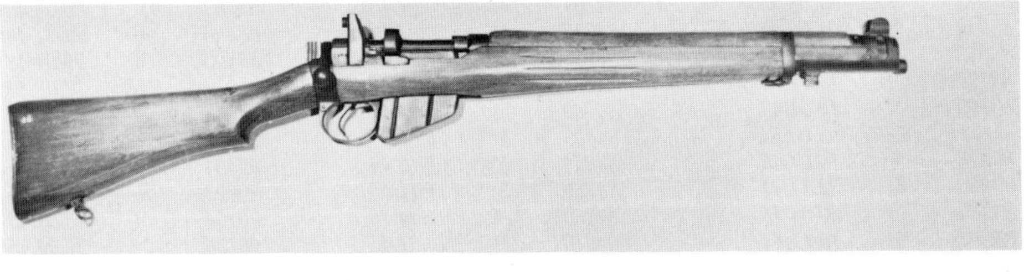

Short Length "Shortened Lightened Rifle" has an 18.2-in. long barrel.
Infantry Centre Museum

Intermediate Length "Shortened Lightened Rifle" has a 20.2-in. barrel.
B. Labudda

Backsight comparison, XP short rifles. Top model is Intermediate Length Shortened Lightened Rifle, bottom one is flip-over type mounted onto rifle with provision for the "Mk V British style" backsight. R. Jarosz

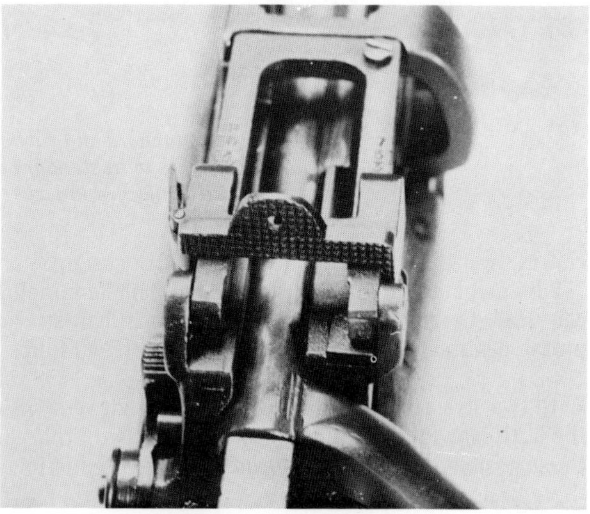

Two views of Mk V British style backsight, trialled on half-stocked "jungle carbines" at Williamstown in September 1944. Infantry Centre Museum

4. Flick-over aperture backsight with dual ranges, mounted onto the charger guide; 200 and 600 yard ranges.

These rifles were fitted with two types of buttplate, the standard brass No. 1 pattern and a rubber kickpad type mounted on a hinged, cupped sheet steel model. Flash eliminators were fitted to the barrels and the fore-ends had one groove along each side.

Regarding the sights, it was found that the aperture mounted on the charger bridge gave poor definition of sight as it was too far from the firer's eye while the standard rifle open sight gave good all-round performance but accuracy dropped away as the range increased and the light faded. The most accurate type was an aperture mounted close to the eye, on a sound base. Some prototype carbines were assembled with both the bed backsight and the aperture backsight mounted at the rear of the receiver.

As a result of these particular trials, 200 rifles were ordered for user trials; 100 fitted with an aperture backsight and 100 with the standard open sight. These trials rifles are a little different to the previously mentioned prototypes; the easiest means of differentiation is that the fore-ends on the prototypes have one wide groove along each side whereas the user trials rifles have four thin grooves along the fore-end with additional grooves in the corresponding position on the handguards.

In line with the British nomenclature, the Australian "jungle carbines" have been designated the No. 6 Mk I although these 200 trials rifles were not so labelled at the time. For the purpose of this study they have been named "Trials No. 6 rifles" even though their official nomen-

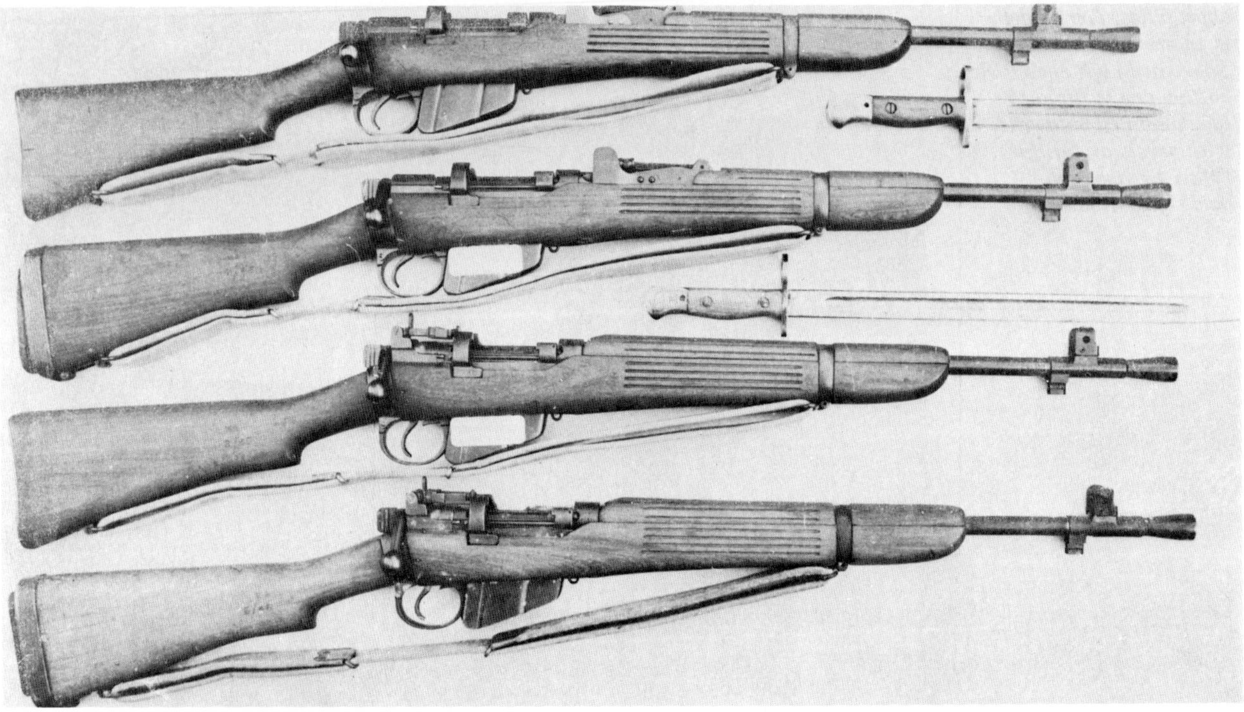

Bed backsights, standard and rubber buttplates; Aperture backsights with standard and rubber buttplates. Top bayonet has the 10-in. blade with enlarged muzzle ring, bottom one has 17-in. blade. R. Jaroz

clature was "Lightened Pattern, Bed Backsight" and "Lightened Pattern, Aperture Backsight." This terminology can lead to confusion with the previous half-stocked prototypes and the earlier Shortened & Lightened Rifles.

With two styles of backsights and two different versions of the buttplate, four variations of these Trials No. 6 rifles are thus possible. Lithgow records indicate the production of these trials rifles as being:

No. 6 Mk I type; Bed backsight, standard brass buttplate, 49 ordered and finished between 21.12.44 and 15.3.45.
Serial numbers, basically XP 101 – XP 149.

No. 6 Mk I type; Bed backsight, rubber buttplate, 49 ordered and finished between 15.2.45 and 19.4.45.
Serial numbers, basically XP 150 – XP 198.

No. 6 Mk I/I type; Aperture backsight, standard buttplate, 48 ordered and finished between 16.5.45 and 18.5.45.
Serial numbers, basically XP 200 – XP 248.

No. 6 Mk I/I type; Aperture backsight, rubber buttplate, 48 ordered and finished between 28.5.45 and 8.8.45.
Serial numbers, basically XP 249 – XP 296.

The Mk I/I nomenclature was later applied to the aperture sight variation, so it has been used here also. The serial number range with "XP" prefix carries on from the XP 1 to XP 100 of the shortened, lightened trials rifles from earlier in 1944. All these trials and prototype rifles were fabricated at S.A.F. Lithgow. The minor discrepancy in these XP serial numbers and the quantities ordered (48 and 49, although 50 of each were required) may well be accounted for by the fact that a few of the original shortened, lightened rifles (XP1 – XP100) were also converted. Such a rifle numbered "XP3," but in No. 6 Mk I configuration and with a brass buttplate, is in the Singleton Infantry Centre Museum collection.

The manufacturing order for these trials rifles was completed by August 1945 and although some endurance and accuracy tests were carried out along with bayonet trials in consideration of 8-in., 10-in. or 17-in. blade models, most of the rifles were returned to store. The Australian List of Changes announced the introduction of the No. 6 rifle but declared it obsolescent in the same issue, obviously a case of "tidying up paperwork" after the event. An order had been made for 40,000 of the No. 6 rifle and it was looked upon as a conversion of existing stocks of No. 1 Mk III* rifles rather than a pattern for new production.

The official introduction reads:

Rifles, No. 6 Mk I.
Rifles, No. 6 Mk I, with rubber butt pad.
Rifles, No. 6 Mk I/I.
Rifles, No. 6 Mk I/I, with rubber butt pad.
Bayonets, No. 6 Mk I.
Scabbards, Bayonet, No. 6 Mk I.

The Rifle No. 6 Mk I is a No. 1 Mk III rifle modified as shown below to closely resemble the No. 5 rifle:*

Prototype No. 6 Mk I rifle conversion from No. 1 Mk III*. Difference to the trials patterns is mainly in the fore-end grooves. *Australian War Memorial*

(a) The barrel is shortened by approximately 4-inches and is fitted with a flash eliminator.
(b) The stocking has been shortened, extended for approximately 13 inches from the breech, leaving the remainder of the barrel unsupported.
(c) The backsight, of the same pattern as that fitted to the No. 1 rifle, is mounted closer to the breech to obtain the maximum sight radine.
(d) Lightening cuts have been made in the body, stocking and trigger mechanism.
(e) A groove is machined on each side of the shortened stock fore-end to improve the grip.

Brief particulars are as follows:

Weight	*7 lb. 8 oz.*
Length of barrel	*19.0 ins.*
Overall Length	*39.5 ins.*
Sight Radius	*16.5 ins.*
Muzzle Velocity	*2,400 fps.*

The Rifle, No. 6 Mk I/I differs from the Mk I in that it has an aperture backsight fitted at the rear of the body, providing an increased sight radius. Rifles No. 6 Mks I and I/I may be issued with rubber butt pads.

The Bayonet, No. 6 Mk I/I is a shortened Bayonet No. 1 Mk I with the crosspiece modified so as to engage the muzzle of the flash eliminator. The blade is 8-in. in length.

The Scabbard, Bayonet, No. 6 Mk I is a Scabbard No. 1 Mk 2 shortened so as to accommodate the No. 6 bayonet.

In this Aust. List of Changes announcement it is later noted that "Consequent upon the Rifle No. 1 Mk III* remaining the standard rifle in the Australian Service, the above-mentioned items are hereby declared obsolescent" and that "existing stores are to be used up." However, most of these rifles do not appear to have been broken up, stripped or scrapped, as they were later sold off through various New South Wales Rifle Association clubs.

Experiments were conducted with synthetic furniture towards the end of the war, consisting of a pressed fibre for the stock of the No. 1 Mk III*; this is a similar looking material to that used in the British No. 7 bayonet grips. Although it is often called Bakelite, this resin reinforced fibre is, strictly speaking, a different composition again. Tests with the synthetic material were unsuccessful as it was found to be particularly prone to warping, which markedly affected the bedding of the action and barrel.

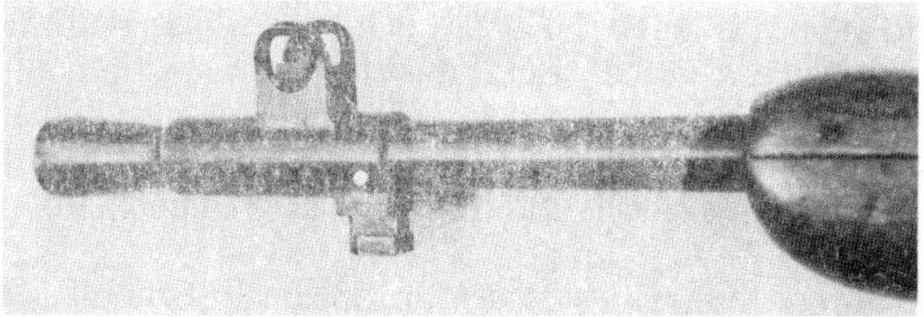

Note squared bayonet lug, compared with rounded type for the British No. 5 model. *Australian War Memorial*

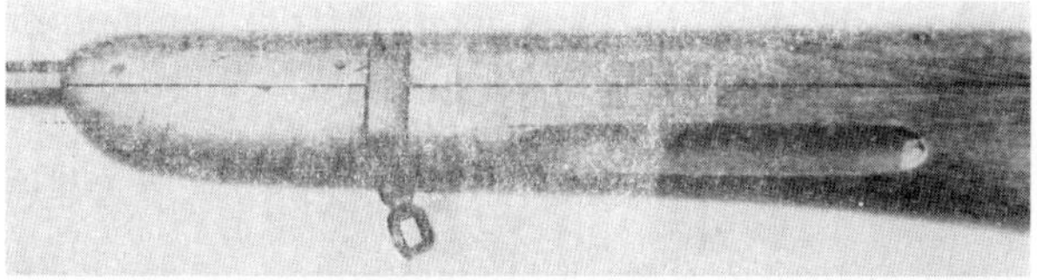

Note wood-filled hole where dial sight screw was fitted to the original Mk III rifle. *Australian War Memorial*

The "Charlton Device" was another product of the wartime years, invented by a New Zealander of the same name. This was a conversion which made the Lee-Enfield rifle capable of fully automatic fire but it was not implemented in Australia; about 1,500 Lee-Metford and Lee-Enfield rifles were so converted in New Zealand and details will be found at the end of Chapter 5.

An Australian converted version of the No. 1 rifle on Charlton's principle was to be produced by the Electrolux Company, Alexandra Ave., South Yarra in Melbourne, possibly for a New Zealand contract or as pre-production prototypes. Only a few seem to have been so converted and the Australian involvement in the project was terminated by June 1944.

The uneven distribution of weight, external protrusions and unsuitability for the sustained fire role made the Charlton only a "stop gap" weapon. New Zealand conversions are fitted with a bipod and assembled in a light machine gun configuration; the Australian conversions of the No. 1 rifle have no such accessories and are merely rifles with an automatic capability. Photographs and descriptions of both New Zealand and Australian produced Charlton automatics are to be found at the end of Chapter 12, in the illustrated technical descriptions.

Production of Lithgow No. 1 (S.M.L.E.) rifle parts ceased by 1960. The last production dates noted for many of the parts has been 1957, such as the wooden furniture and bayonet grips ("SLAZ" marked, indicating manufacture by Slazengers) and backsight beds. The last F.T.R. rifle batch at the factory seems to have been during 1957. Some metal components were produced until 1960 and may be found with the "MA 60" marking which indicates manufacture in that year. But even after production of the 7.62mm L1A1 was begun in 1958, some refits and repairs were being carried out on the .303 No. 1 rifle, such as the fitting of new barrels, furniture and major components.

Many of the .303 rifles were passed on to School Cadet units after the introduction of the 7.62mm L1A1, these Cadet units have more recently been the end-user source of surplus S.M.L.E. rifles. Many such rifles have been noted with painted bands around the butts, or the entire muzzle/nosecap portion. Colours have been green, yellow and red; the green indicates approval for ball ammunition firing, the yellow as a 2nd grade but still suitable for live firing, and the red as not to be fired. A white colour is an Air Force Cadet equivalent for the Army's green or yellow.

Some Australian S.M.L.E. ex-military or sporter configuration rifles may be occasionally encountered, chambered for a special Antipodean wildcat cartridge, the 7.7 x 54mm. This round originated after a State government (New South Wales) banned civilian ownership of military calibre rifles in the 1950's, so to reasonably reproduce the original service round, the .303 case was shortened by ⅛-inch. Some military rifle models held by various collectors were rechambered for this round so that the rifles could be legally retained, unhappily including some No. 1 Mk III* H.T. and No. 6 models.

Other better-known Aussie wildcat cartridges based on the .303 case were widely used with S.M.L.E. sporters and game-getting rifles in the post-war period in Australia; these included the .303-22 Falcon and Sprinter, Harrison Wasp, .303-25, .303-6mm, .303-270 and some rimless versions of the .303-22 and .303-25 models.

An announcement was made in 1954 that the Commonwealth Government would work with Britain and Canada and adopt the F.N. F.A.L. as Rifle L1A1, the production planning at Lithgow began early in 1955 with a sample F.N. rifle. This was followed by a batch of 20 F.N.-made X8E1 rifles for limited troop trial evaluations in mid-1955. Lithgow's production of the 7.62mm L1A1 self-loading rifle for the Australian services commenced in 1959.

Conversion of .303 rifles to 7.62mm was considered as a possible solution for the disposal of surplus No. 1 rifles but the No. 1 action was not sufficiently strong due to the use of plain carbon steel in lieu of nickel-chrome alloy steel. In 1958, some converted rifles were tested with proof cartridges and examined at critical points but it was found that headspace increased significantly, a gap developed between the charger bridge and the body, and the test rifle bodies fractured at the ejector/charger guide position.

Two different .303 rifles and been converted for these tests, a No. 1 Mk III* and a trials No. 6 Mk I, referred to in these tests as Types "A" and "B" respectively. Type "A" was fitted with a 7.62mm barrel with the usual No. 1 rifle profile, new extractor, new regraduated backsight and an 8-round magazine for the 7.62mm round which externally looked like its .303 counterpart. The Type "B" had a 7.62mm barrel with trials No. 6 rifle profile, 7.62mm box magazine from the X8E1 F.N. rifle, with the action body and trigger modified to suit, a new extractor and upgraded sights.

Ten further rifles were assembled to the Type B design for consideration by the various Services and the Design Establishment. Even though the No. 6 jungle carbine design was favoured over the No. 1 Mk III* rifle for such a 7.62mm conversion, this prospective program was being considered for the modification of a considerable percentage of now obsolete No. 1 Mk III* rifles held in store, so it was intended that the rifles be converted to the "jungle carbine" configuration as well.

It was intended that conversion/production could be under way at the end of 1959 with initial deliveries about July 1960. Production rates were estimated at between 100 and 200 per week. However, because of the excessive chamber pressure developed by the 7.62mm round, some form of action reinforcement was considered desirable.

Further correspondence followed with C.A.L. in Canada and the War Office and R.S.A.F. Enfield in England regarding the conversions applied to the No. 4 rifles; the Australian Rifle Clubs also inquired whether such a conversion could be applied to the No. 1 rifles used for civilian competitive shooting. The R.A.A.F. showed most interest in such a 7.62mm conversion, in jungle carbine configuration, and so further consideration was made of the possible flash hider and sights, utilising the F.N. F.A.L. (X8E1 & X8E2) parts where possible. Even though welding was used to reinforce critical parts of the action body, the tests were still unsuccessful and they were abandoned.

The .22 No. 2 Mk IV* rifle was also produced in Australia, although these were mostly built up on existing actions and therefore were generally conversions rather than new made rifles. The designation "No. 2 Mk IV*" is sometimes found on the butt along with the conversion date (e.g. "9/58"); the factory repair mark "F.T.R." is usually evident as well. In some instances, the receiver was marked "No. 2 Mk IV*" below the original .303 manufacture stamps and magazines were often marked ".22".

Original photograph from 7.62mm Trials Report in 1958. SAF Lithgow

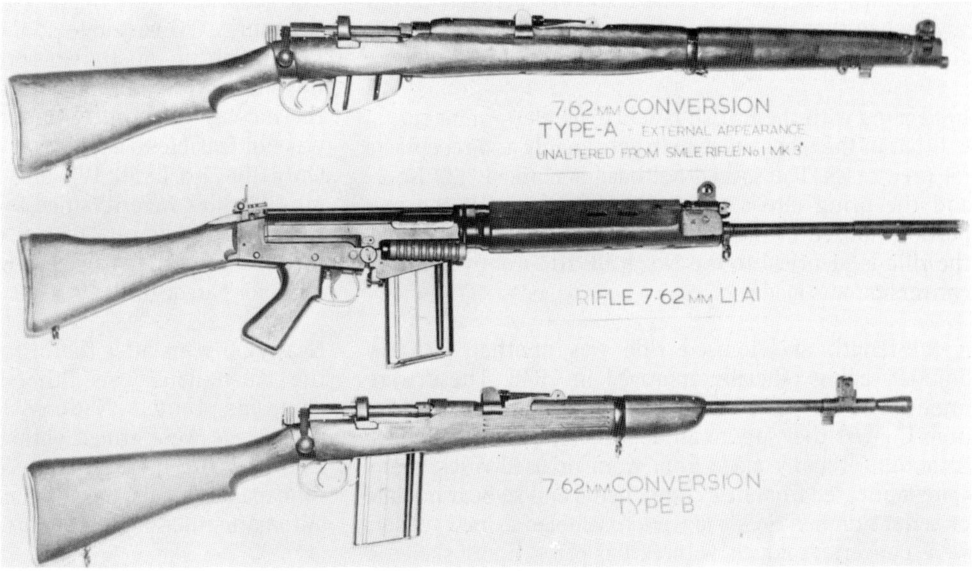

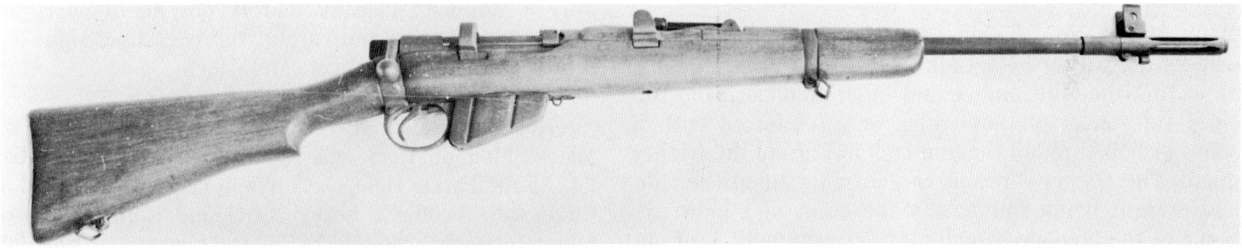

7.62mm Trials Rifle based on the No. 6 Rifle concept. This example retains the original .303 magazine, so was tested as a single-loader. *SAF Lithgow*

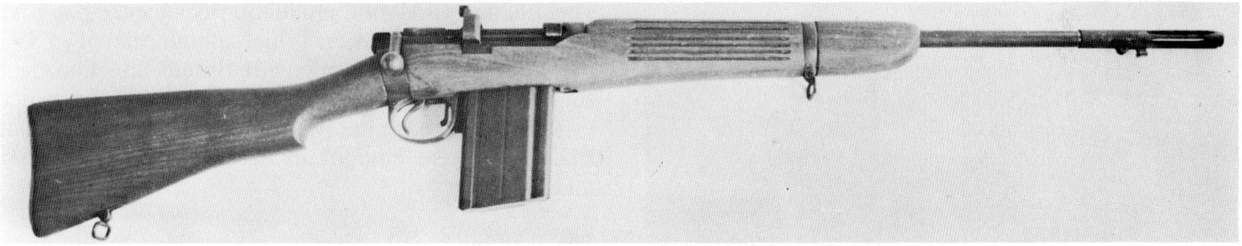

Another 7.62mm Trials model converted from a .303 No. 6 Trials rifle. Foresight not fitted, aperture backsight mounted on the charger bridge as on the earlier Shortened Lightened Rifle series in early 1944. *SAF Lithgow*

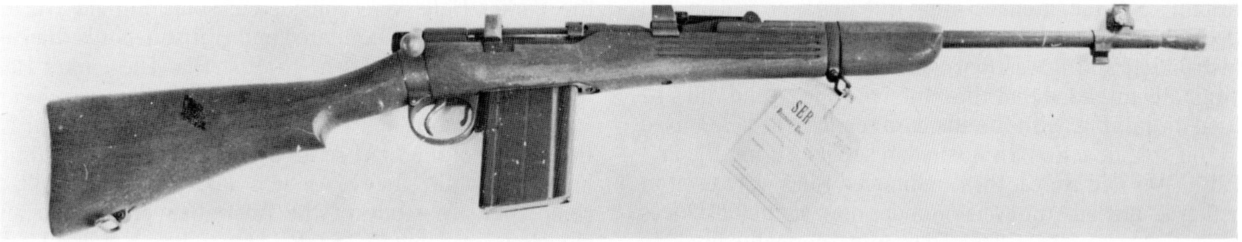

Yet another 7.62mm Trials rifle, converted from Bed Backsight No. 6 Trials model. Note original No. 6 flash hider with modified foresight wings. The magazine is a trials X8E1 Rifle model, stamped "X2E1". *SAF Lithgow*

These rifles were still on issue to School cadet forces in 1975 when they were disbanded upon the Federal Labour government taking office.

In keeping with the British practice, no platform or spring is fitted in the magazine which acts merely as a receptacle for fired cases. The special bolthead is marked ".22 No 2" and the firing pin is a two-piece design with the rear portion housed in the bolt body. In external appearance, the rifle is identical to the No. 1 Mk III* from which the conversion was made.

A full-length skeletonised rifle was another Lithgow product, being officially approved in 1948. These were intended to be viewed from the right hand side because from the left they are, to all appearances, standard rifles. Although properly made from worn or used rifles, many were fabricated from new components and so bear no date or serial number. So some were assembled as new, others were converted models with earlier parts. Short skeleton actions of the Mark III and III* models were also produced in Australia at S.A.F. Lithgow and others at Base Ordnance Depot workshops.

Many .303-in. No. 1 EY rifles were bound up with copper wire for use with the No. 1 Mk I 2½-in. discharger cup, for projecting the Mills and certain other grenades from the No. 1 rifle. Most of these rifles were reinforced with a screw or bolt through the fore-end, forward of the trigger guard. The copper wire was to generally stiffen the rifle and prevent flying splinters in the event of a blow up. Some of the grenade launching rifles were built up from new arms rather than from downgraded, reserve or E.Y. model rifles.

Soldered copper wire binding and crossbolt, used to reinforce EY Grenade Launching rifles. *I. Skennerton*

Britain's S.M.L.E. grenade-firing rifle was approved on 2nd December 1918, intended for use at the School of Instruction and it was first bound with whipcord. It was originally intended that the reinforced and bound rifles be used in training, with the standard service rifle being used in this role in action. In Australian service during the 2nd World War and Korea, these reinforced rifles were used in action as grenade throwers or under normal circumstances as ordinary service rifles. They were normally fired by placing the butt on the ground and holding the rifle at a 45° angle, the range variation was then effected by adjusting the shutter in the cup discharger.

The adoption of the Belgian designed F.N. F.A.L. rifle and 7.62mm NATO cartridge sounded the death knell for the Lee-Enfield in Australian service as the No. 1 rifle was unsuitable for conversion to the new cartridge and with the increasing shortage of supply of .303 ammunition it was not feasible to consider it even as a reserve grade arm. Only the No. 2 Mk IV* and a small number of No. 1 Mk III* rifles were retained for issue to the Cadet forces; some of these even had a wooden pistol grip fitting inserted into the magazine-way so as to duplicate the L1A1 for parade drill!

Nearly 50 years after the introduction of the Short Rifle, the Australians used the No. 1 in Korea, fighting alongside United Nations forces of which a large percentage were armed with auto-loading rifles. Yet the .303-in. Lee-Enfield still proved to be a most effective arm in capable hands, even against the Communist-block automatic rifles of their enemy.

As late as the 1970's, the No. 1 Mk III* H.T. was reintroduced into service again as a sniper rifle, pending the acquisition of a new 7.62mm sniper rifle, but this was only a stop-gap measure and it was mainly used for training in the use and care of the telescopic sight and fundamental sniper tactical employment.

After various trials in 1978 between the Austrian SSG 69, Mauser Mod. 66, Heckler & Koch G3, F.N. Mauser, British L42A1 and Parker Hale, the 7.62mm Parker Hale Model 82 fitted with 1200 TX heavy barrel and Kahles-Helia 69 telescopic sight was selected as the new sniper rifle, late in 1978. While this selection was a marked improvement on the No. 1 Mk III* H.T., the main factor in its selection was its economical price, compared with its competitors.

The 7.62mm L1A1 rifle production commenced at S.A.F. Lithgow in 1958, after an initial announcement in 1954 that Australia would work with Britain and Canada to adopt the F.N. F.A.L. rifle. Production planning commenced in early 1955 and some twenty F.N.-made X8E1 7.62mm trials rifles were brought to Australia for evaluation in mid-1955.

The first Lithgow L1A1 rifle was supplied to the Armed Forces in 1959 and a heavy-barrel automatic version, the L2A1, was manufactured from 1962. Australia conducted user trials for a new sniper rifle in early 1978, the Austrian SSG 69, Mauser 66, Heckler & Koch G3, FN Mauser and the Parker Hale were considered; some Enfield L42 A1 rifles had also been evaluated and British records indicate that 19 of these were sent to Australia. The Parker Hale was eventually chosen on economic grounds.

Production of the L1A1 series at Lithgow was discontinued in 1986, while the L2A1 was last made in 1982. A comprehensive study of the Australian production and models is available in "S.L.R. – Australia's F.N. F.A.L." by Skennerton & Balmer. The L1A1 series has since been replaced in Australian service by the 5.56mm F88, the Austrian designed Steyr AUG.

INDIAN PRODUCTION

The origins of India's Ishapore factory are not simply confirmed; a recent report by a former Proof Master at the Ishapore Rifle Factory Proof House, extracted from certain factory records, differs significantly from documentation held at the India Office Library, London. Because the historical details of the latter were recorded at about the same time as factory production commenced in 1908, these records are considered to be more reliable. Recent inquiries of the Indian government authorities, Rifle Factory Ishapore and the military have resulted in refusals to allow any further study on this subject or even a visit.

The Dutch possessed a settlement on the Hoogly River in Bengal about 1740 or 1750 in which was a fort and factory called Bankee Bazaar. This was taken by the Nawab Nizam of Murshidabad and assigned to Rajah Nubakissen Bahadur upon payment of a small annual revenue. The East India Company exchanged the old Dutch factory with the Rajah in 1794 for some lands at Setanooly near Calcutta and in their acquisition of Bankee Bazaar, laid the nucleus of a gunpowder factory which was later called Ishapur. The production of powder there commenced from that date, just before the turn of the 19th century.

The reason for this change was that black powder had previously been manufactured in Calcutta but the increasing population in near proximity to the powder mill caused the plant to be moved to Accra and thence to Bankee Bazaar in 1794. Ishapore is an anglicised version of Ishapur and it was an ideal site, about 20 miles upstream of Calcutta on east bank of the Hoogli River; it served as a gunpowder manufactory until 1902.

The site was then handed over to the Military Works Department for the construction of a rifle factory, as plant and estimates were in preparation and an output of about 25,000 rifles per annum was conceived. The Royal Small Arms Factory at Sparkbrook, Birmingham, served as a model because its output of about 500 rifles per week was similar and so details of the Sparkbrook layout and plant were supplied for comparative reference.

Rough estimates of the cost of buildings and machinery were made at £150,000; a War Office estimate amounted to £120,000 for plant and £20,000 for buildings. In 1900, Capt. H. B. Foote R.A., a British Ordnance Officer serving in India, was deputed home for instruction, to inspect the R.S.A.F. Enfield and other small arms factories in connection with the rough plan for the Indian factory. Erection of the buildings was commenced on 22nd August 1903.

On 6th January 1904, Capt. Foote, by now promoted to Major, took charge of the Ishapore factory on his return from England and it was reported that the buildings were approaching completion and that the projected date for the commencement of manufacture was 1st January 1905. The first shaft was erected on 22nd April 1904 and the power supply effected on 26th September 1905.

But the buildings were not completed by the end of March 1905, although the rolling mills were practically finished and the out turn of both steel and cartridge metal was expected to commence in the financial year 1st April 1905 – 31st March 1906. The rolling mills were erected on adjacent land to the Ishapore factory and were originally an offshoot of the Cossipore factory which produced guns and shells.

Rifle work was commenced in 1905 with the conversion of Magazine Lee-Enfield rifles to charger-loading, a special pattern which was not introduced into the British Service; this has the charger loading guide on the bolthead, similar to the Mk I S.M.L.E. New rifle production did not commence until 1907, mainly because delays were experienced in training local unskilled labour. The first rifle is recorded as having been proved on 22nd September 1907.

In the British factories there was a much higher proportion of skilled workmen and so appointments there such as foremen and leading hands were readily filled; those workmen who were not skilled could be properly advised and assisted as the ratio of workmen to skilled tradesmen was about five or six to one. In India however, the situation was quite different, as is shown in the following table.

	Enfield	Birmingham	Ishapore
Principal Foremen & Shop Managers:	8	2	1
Foreman:	12	11	5
Assistant Foremen:	22	3	12
Leading Hands:	126	1	3
Workmen:	1,090	642	1,778

From this comparison it can also be seen that there were more workmen at Ishapore than at Enfield and Sparkbrook combined, yet the output was comparable with that of Sparkbrook. Manufacturing gauges for the Ishapore rifles were produced at Enfield R.S.A.F.

Ishapore production for the 1907 and 1908 financial years (1st April – 31st March) was:

	1907-08	1908-09
S.M.L.E. Mk I*	20	3,000
C.L.L.E. (I.P.)	10,380	10,600
Barrels, S.M.L.E.	20	1,223
Bodies, S.M.L.E.	20	3,097
Nosecaps	1,569	4,300
Magazine cases	415	17,628

Most of the production difficulties had been sorted out by 1908 and production had begun in earnest. By March 1909, ten battalions of Infantry had been rearmed with the new charger loading Lee-Enfield, described shortly. Previously, all the arms for British and native troops were imported from England, evident from the following table covering 1901-05.

British imports:–	Quantity	Shipped	Issued to
M.L.E. Rifles	41,500	1901	Indian
M.L.E. Carbines	13,500	1901	Indian
M.L.E. Rifles	62,000	1902	Indian
M.L.E. Rifles	50,000	1902-03	Indian
S.M.L.E. Rifles	71,544	1904-05	British
S.M.L.E. Rifles	21,000	1905	British

Due to the situation regarding Ishapore's higher component of unskilled labour, more automatic machines were requested. The major bottleneck in production proved to be barrel manufacture although some 300 were viewed in March 1909 and a further 400 were expected the following month. More difficulties were experienced the following year when the pattern of the Short Lee-Enfield was changed from Mark I* to Mark III, in line with the British rifle progression.

Six of the new Ishapore manufactured Mark I* S.M.L.E. rifles were despatched to England for inspection on 7th October per S.S. "Ceylon," taken from run of production work. There were two different reports made on these rifles in England, one by the War Office and the other from R.S.A.F. Enfield, dated 20th February 1909.

The War Office described the Ishapore rifles as being "in general appearance, workmanship, finish and handiness, the equal in every way to those patterns manufactured at home." Some slight differences to the British pattern were listed as:

(a) Barrel protrudes 1/16-inch further beyond the nosecap;
(b) Boss of nosecap is about 3/32-inch higher;
(c) A piling swivel is fitted;
(d) A pin and spring are located underneath the nosecap;
(e) Sling swivels are 1/8-inch shorter and 1/32-inch wider;
(f) Head of wind gauge is longer and the screw has a washer;
(g) Semi-circular clearance for backsight V-notch is smaller;
(h) Steel Mk I buttplate fitted instead of Mk I* with trap;
(i) No spring link for attaching magazine;
(j) Magazine about 3/8-inch deeper at the front;
(k) Magazine cut-off fitted;
(l) Locking bolt different in shape;
(m) No butt trap, though provision is made in butt;
(n) Swivel screw ends hollow;
(o) Weight, 3 oz. lighter.

It has been pointed out that many of these criticisms [(c), (d), (e), (f), (i), (j), (k) & (n)] had already been approved features of the British Mk I* S.M.L.E. so the War Office report can be considered ill-advised in some respects.

The Enfield Inspectorate was not so complimentary; while it was recognised that they were generally correct to pattern, it was complained that the rounding off of sharp corners as instructed in §§ 13313 and 13549 of the List of Changes had not been carried out. They also listed many defects in detail which affected parts interchangeability. Further complaints were tabulated:–

Cartridges load very high, liable to cause a jam;
Butt socket very large;
Bolts not interchangeable in the six rifles;
Bolt head threads too large and under pitched;
Bolts not interchangeable with rifles of British make;
Magazine too wide at front, resulting in misfeeding;
Sight leaf too narrow and slack in the joint of the bed.

In defence of the Ishapore production, it was pointed out that some of the recent changes in the British pattern had not yet been incorporated in Indian production, and that the Indian List of Changes was a different issue to the Britain Lists, even though the British Lists were used as the basis for the Indian Lists.

In a number of respects, the Army in India operated quite differently to the War Office and there were differences of opinion between the two fraternities; many of the criticisms of the Indian Army and Indian production were obviously biased. Some of the minor modifications introduced in the India patterns were later incorporated into British production, but there were continual, ongoing complaints as to the differences between British and Indian manufactured arms of the same pattern with the same nomenclature. This was then easily rectified by India; patterns special to Indian production were designated "I.P." for India Pattern.

It would appear that only about 3,000 of the Ishapore Mark I* S.M.L.E. rifles were made, according to the production figures available, and this represents Ishapore's first production. Because of relatively minor differences to the British pattern, mainly consisting of fitting the earlier

Mk I S.M.L.E. sheet steel buttplate without a trap, this rifle was designated the Mark I* I.P. Short, Magazine Lee-Enfield. Note that the "I.P." designation has been included in the nomenclature.

Because of the relative independence of the Indian Army's from Britain's War Office, there were often major differences of opinion on various military matters and the Indian Staff Officers usually stood their ground. A certain amount of autonomy was necessary because of the distance from England and the self determination required to run such a large organisation as the Indian Army.

The Indian Army had been regarded as a second rate institution for some time and there was a certain amount of class distinction within the British Army against "Indian officers." Of course, the officers and many of the regiments in India were British regulars and due to continual border disputes and campaigns in and around India, especially in the north-west, excellent training was to be had in the soldier's trade. Such experience could not be gained on a parade ground or barracks room in the British Isles.

By the time of the South African War, most of this class distinction had been forgotten in the military as such a large percentage of regular officers had served in India and many of Britain's great soldiers had held important posts there; Roberts, Kitchener and Wood probably being the most notable. A typical example of the Indian Army's commonsense approach was the adoption of smokeless powder before Britain, and Lord Roberts' refusal to send troops from India to fight in South Africa with black powder ammunition.

The first of Ishapore's rifle conversions was the "Cavalry Carbine, Magazine Lee-Enfield, India Pattern" which was approved on 8th January 1904 and comprised an altered Lee-Enfield carbine to fix the Pattern 1888 sword-bayonet. These were intended for issue to drivers in the Corps of Sappers and Miners, and the conversion comprised:

- Shortened fore-end and Lee-Enfield nosecap fitted;
- Martini-Henry upper sling swivel fitted onto the nosecap;
- Sling swivel fitted to the butt, 2-in. from the toe;
- A slight bevel filed in the barrel holder of the nosecap, under which the front end of the handguard was fitted;
- Pattern 1888 bayonet muzzle ring bushed to suit.

A specific Indian version of the M.L.E. rifle with the "India Pattern" suffix in its nomenclature, was the M.L.E. Mk I* I.P. This consisted of fitting the Mark I* S.M.L.E. butt with gunmetal buttplate and trap and the inletted lower swivel mount secured by two wood screws, to the earlier pattern of Magazine Lee-Enfield rifle.

India introduced its own Charger Loading Lee-Enfield rifles in 1905, utilising the Mark I S.M.L.E. charger guide system on the bolthead. Three types of Lee-Enfield rifle were converted to this charger-loading system:

M.L.E. Mk I	became	C.L.L.E. Mk I
M.L.E. Mk I*	became	C.L.L.E. Mk I*
M.L.E. Mk I* I.P.	became	C.L.L.E. Mk I* I.P.

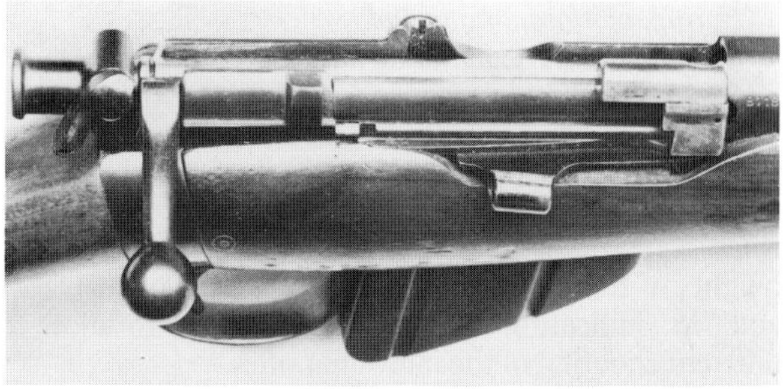

C.L.L.E. Mk I I.P., using the S.M.L.E. Mk I form of charger loading slide on the bolthead. *MOD Pattern Room*

The details of this charger-loading conversion were:

A bolt head charger guide stop will be fitted and riveted on to the right side of the body and a charger guide will be riveted to the left side. A cartridge lead will be fitted in front of the boltway, clearance cut at back for charger guide and thumbpiece, and the body will be milled for magazine stop clip clearance, and magazine seating.

Bolt cover lugs are milled off, and the barrel portion of the bolt is polished.

The bolt head is milled off at the back to increase the backward travel of the bolt, and is dovetailed to receive the charger guide slide which is brazed in and milled to size. A recess is also cut for the M.L.E. short rifle striker lug, thus rendering this component common to both rifles.

The trigger guard has a small piece brazed in and fitted in the magazine slot, a clearance is milled for the magazine stop clip, and the inside of the bow is milled out, thus rendering this component common to the M.L.E. short rifle also.

On the extractor, the angle of hook is shortened, and the back end is lightened around the screw hole.

The cut-off will be softened, altered to clear the magazine stop clip, edge turned, reset, adjusted to gauge, hardened, tempered, and oil blacked.

The fore-end stock has a clearance cut for the magazine.

The undermentioned components will be removed on conversion:
Cover, bolt.
Case, magazine.
Platform, magazine.
Spring, platform, magazine.
Pin, spring, platform, magazine.

And the following new components will be fitted:
Bolt head, guide, charger.
Case, magazine.
Clip, stop, magazine.
Stud, clip, stop, magazine.
Platform, magazine.
Spring, platform, magazine.
Spring, auxiliary, platform, magazine.
Screw, stop, guide, charger.

On 23rd September 1908, an adjustable barleycorn foresight was approved for this rifle, to be fitted while rifles were undergoing conversion at Ishapore. Three heights of the foresight were approved, marked "H," "N" and "L," indicating high, normal and low, to allow for different shooting of the individual rifles.

The barleycorn frontsight bed was made .04-inch wider than the block foresight, .02-inch at each side so as to allow for lateral adjustment when testing for accuracy. After adjustment, the projecting ends were filed off flush with the block, and centre punched at the bottom of the dovetail at each end to prevent the barleycorn from moving. It was also announced that any damaged foresights in the hands of troops were to be repaired in arsenals by the replacement of this particular sight.

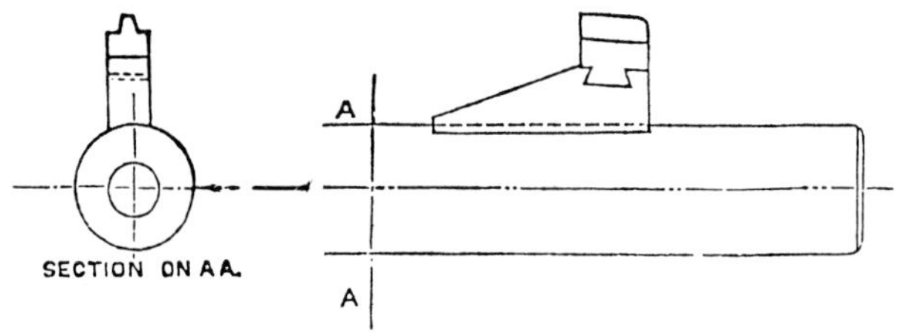

By 1907, India had arsenals at Rawalpindi, Ferozapore, Kirkee, Allahabad, Quetta, Madras, Rangoon, Fort William and Karachi, and although some of these cities are in modern-day Pakistan and Burma, such areas were once part of greater India, under British rule. These arsenal marks are often found on the butts of rifles which have seen Indian service and this especially applies to the Martini-Henry and Martini-Enfield rifles and carbines.

The British War Office announced that the fore-end and nosecap of the S.M.L.E. rifle should be numbered to the rifle so as to avoid accidental exchange and thus interfere with the bedding and possibly the shooting of the rifle. This was applied in India as well, but on 9th September 1908, approval was given for all Indian service Martini-Enfield, Martini-Metford and Lee-Enfield rifles and carbines to have the fore-ends and nosecaps numbered to the barrel. No similar numbering of the nosecap and fore-end of Lee-Metford or Lee-Enfield rifles and carbines was implemented on such a large scale in other countries of the Empire, although some Canadian issues are reported to be so numbered.

Whilst Britain introduced the Mark III version of S.M.L.E. rifle and the Pattern 1907 bayonet in 1907, Ishapore did not change production from the Mark I* S.M.L.E. rifle until 1909. Such an alteration in pattern was much more involved and complicated at the Ishapore factory due to the higher proportion of unskilled labour. The annual report for the year ending 31st March 1910 indicates that only 5,000 of the Mark III rifles had been completed, one of the main difficulties encountered being in the manufacture of the new sights. Problems were also being experienced in the bayonet manufacture.

India had refused to adopt the new Pattern 1907 bayonet and even cancelled its introduction in the Indian List of Changes. The Pattern 1888 and 1903 bayonets were preferred and even though 700 of the experimental, trials bayonets (which eventually resulted in the adoption of the Pattern 1907) were tested in India, and the Japanese-style pattern which became the Patt. 07 preferred, it was disliked because of the hooked quillon and the blade form. The 700 trials bayonets had comprised of 300 with the triangular blade, 200 of the U.S.A. pattern and 200 of the Japanese pattern. For illustration and descriptions of these trials bayonets, see "British & Commonwealth Bayonets" pages 184-186, Cat. #B232 & B233. The Patt. 07 bayonet was not accepted by India until 1910 and officially introduced there on 30th June 1911.

The Rifle Factory Ishapore converted large numbers of the Patt. 1888 bayonet to the 1903 pattern, effected from later in 1912. After India's delayed adoption of the Pattern 1907 bayonet, it was also in 1912 that its production commenced at Ishapore. Because of the large quantities of Pattern 1888 bayonets converted to Patt. 1903, it would appear that most of the Patt. 88 bayonets in India were so modified. Such quantities for the 1912 and 1913 financial years, when bayonet production was commenced at Ishapore were:

	1912-13	1913-14
Patt. 1903, Converted	2,307	23,015
Patt. 1907, Ishapore made	1,033	4,013

The Mark II India Pattern Charger Loading Lee-Enfield rifle was approved on 8th September 1909. This was converted from the Marks I and I* M.L.E. rifles, as were the previous Charger Loading Lee-Enfield rifles in India, but it differs as follows:

1. Charger guide bridge of the Mark III S.M.L.E. fitted.
2. Inside of the trigger guard bow not milled as on the Mark I* C.L.L.E.
3. Fore-end modified for the charger bridge.
4. Bolthead common to the S.M.L.E. Mark III and Converted Mark IV.

Compared with the British service patterns of Charger loaders, the Indian Mark II C.L.L.E. I.P. would resemble the Mark I C.L.L.E. more closely.

On 30th August 1910, an alteration was approved for British made Mark I S.M.L.E. rifles, consisting of fitting the Mark I* S.M.L.E. butt with butt trap and inletted sling swivel bracket; this conversion became known as the S.M.L.E. Mk I* I.P. The official announcement in Indian List of Changes reads:

The stocks, butt of rifles of the above-mentioned pattern (S.M.L.E. Mk I) will be converted, and a gunmetal buttplate and butt swivel fitted and a wad, stock bolt provided, in the Rifle Factory Ishapore. Under orders which will be issued for each regiment, the stocks, butt on charge will be supplied without discs, marking, butt, the latter being transferred from the old butts before despatch.

The only technical difference between this S.M.L.E. Mk I* I.P. and its British Mk I* counterpart is that the triggerguard has a magazine link loop and the attached No. 1 magazine, from the Mark I S.M.L.E. rifle. However, some anomalies appear in that the first sentence in the official announcement would indicate that the butts were converted, yet the last sentence intimates that the butts were replaced. Also, in the British production of the Mark I S.M.L.E. after the Mark I* introduction, some had the inletted swivel mount fitted. It has also been pointed out that early Mk I* S.M.L.E. rifles had the No. 1 magazine, without link loops.

A C.L.L.E. Mk II* I.P. was introduced in 1912, consisting of the Mark II I.P. Charger Loader rifle fitted with the Mk I* S.M.L.E. butt and buttplate, as described previously.

Because of the variety of Charger Loading Lee-Enfield rifles in Indian service by this time, a list was circulated which described the variations of the Lee-Enfield rifles that might be encountered. Although the system of nomenclature was altered soon afterwards, this list provides an indication of the conversions already applied.

Rifle	*Description*
M.L.E. Mk I	British pattern.
M.L.E. Mk I*	British pattern; no provision forcleaning rod. But some patterns in India may have the Mk I nosecap with hole for rod, but retain the nomenclature, Mk I*, as the solid fore-end is a feature of this pattern.
M.L.E. Mk I* I.P.	This rifle is the same as Mk I*, but has butt and buttplate of the M.L.E. Short Rifle.
M.L.E. C.L., Mk I	Converted from Rifle, M.L.E. Mk I.
M.L.E. C.L., Mk I* I.P.	Converted from Rifle, M.L.E. Mk I*.
M.L.E. C.L., Mk II I.P.	Converted from Rifles, M.L.E., Mk I*, I.P. & C.L., Mk I* I.P. [This appears in error, as the Indian LoC introduction specifies conversion from Marks I & I* M.L.E., not I.P.]
M.L.E. C.L., Mk II* I.P.	Converted from Rifles, M.L.E., Mks I & I*, and C.L., Mks I & I* and fitted with M.L.E. Short Rifle Mk I* butt.

In 1911, the barleycorn frontsight of the Mark I S.M.L.E. was approved for fitting onto the Charger Loading Lee-Enfield rifles, making five heights available, ranging from .93-inch to .99-inch and rendering the previous adjustable barleycorn sight used on the Charger Loader, obsolete.

Another nomenclature change was applied to the Charger Loading rifles in 1912 as the system was not readily understood, mainly due to the general application of "I.P." after the designation where the Mk I* S.M.L.E. butt was fitted; however anomalies were encountered after the introduction of the Mk II* I.P. Charger Loader as it was identical to some of the Mk II I.P. C.L.L.E. rifles. As a result, the system was altered so that all Charger Loading rifles with the Mk I S.M.L.E. type charger guides became Rifles, C.L., M.L.E. (Mark I) I.P., and those with the Mk III S.M.L.E. type charger bridge were to be known as the Rifles, C.L., M.L.E. (Mark II) I.P. This simplified the conversion listing to the following:

New Designation:	*Converted from:*
Rifle, M.L.E., C.L., Mk I I.P. (charger guide on bolthead)	Rifles, M.L.E., Mark I, Mark I*, and Mark I* I.P.
Rifle, M.L.E., C.L., Mk II I.P. (charger guide on body)	Rifles, M.L.E., Mark I, Mark I*, Mark I* I.P., and C.L. Mk I.P.

Any of the Mark I C.L.L.E. patterns which were sent in to Ordnance Depots for repair to the extent of new barrels, were forwarded to R.F. Ishapore for conversion to the Mk II pattern. Both these patterns were different to the British converted Charger Loading Lee-Enfield rifles, of which details may be found in Chapter 4.

The initial difficulties in production at Ishapore were well and truly sorted out by this time and the output figures until 1914 are tabulated here for comparison, although it must be pointed out that many of these rifles, including some of the Mark III S.M.L.E.'s, were rebuilds which utilised earlier receivers.

	SMLE Mk I*	SMLE Mk I**	SMLE Mk III	CLLE Mk I IP	CLLE Mk II IP
1906-07	–	–	–	1,000+	–
1907-08	20	–	–	10,380	–
1908-09	3,000	–	–	10,600	–
1909-10	–	–	4,980	–	7,517
1910-11	–	–	10,000	–	20,000
1911-12	–	–	6,093	–	12,370
1912-13	–	61	5,004	–	6,753
1913-14	–	4,148	4,801	–	6,057

Some remarks on this table are that the Mark I** S.M.L.E. were conversions of the Ishapore Mark I* S.M.L.E. and British Mark I S.M.L.E. and British Converted Mark II S.M.L.E. rifles. Quite a few of the C.L.L.E. Mk II I.P. rifles were Indian Mk I and Mk I* C.L.L.E. rifles upgraded, and some of the Mark III S.M.L.E. rifles used "re-manufactured" Mark I and I* S.M.L.E. receivers. In the Indian system, virtually nothing was discarded and when rifles were no longer serviceable, they were sent to Ishapore and stripped completely. Those parts which were still serviceable were put back into circulation on other rifles, while unserviceable components were smelted down and reprocessed.

Receivers were re-manufactured and there are instances of some having been used for three or four different rifles, starting life as a Mark I or I* S.M.L.E. These were not considered conversions as in the British system, because the bodies were reworked completely and even the original markings removed, so it is almost impossible to readily detect such conversions or re-manufacture.

A special single loader rifle was approved in 1912, consisting of a Mark I* M.L.E. converted for use by Frontier Levies. The conversion involves filling the magazine way in the fore-end with a block of wood, secured by wooden pegs. A thin steel sheet was inserted into the underside of the body and a special trigger guard fitted. Some 2,000 of these conversions were made at Enfield in 1911, exported for Indian service.

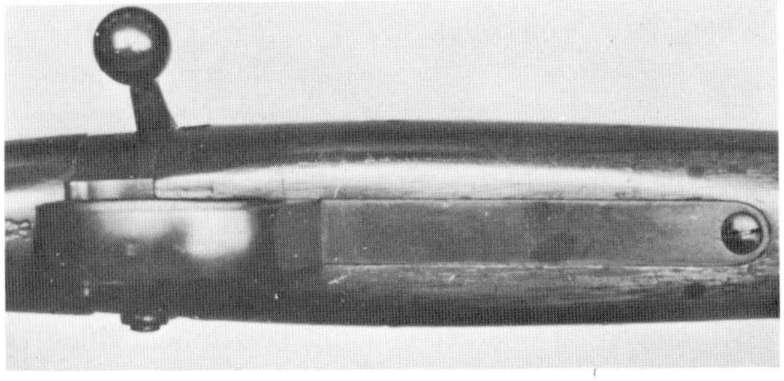

M.L.E. Single Loader. Trigger guard extension is flat with no provision for magazine. *MOD Pattern Room*

Trigger guards of these rifles were approved for slight alterations in 1913 in that they were given a radius at the front end similar to that of the Martini-Henry, and drilled so as to mount the Martini-Henry swivel fitted to the trigger guard. This was approved for rifles of new manufacture, while rifles already in use were to be fitted with this swivel locally. The butt swivels were removed as they were no longer required. It would seem that these rifles were mainly intended for use by mounted forces and that the conversion to single loader was made because of the questionable loyalties of these troops. After the experiences in the Indian Mutiny, it was policy to ensure that certain Indian forces were not supplied with current pattern or fully operational arms.

The S.M.L.E. Mark I** was introduced into Indian service as a means of converting S.M.L.E. Marks I, I* and I* I.P. rifles to the bridge charger loading system of the Mk III S.M.L.E. Approved and implemented in 1913, certain Mark I, I* and I* I.P. rifles were ordered to be returned for conversion, a pattern having been sealed to govern the conversion.

Details of these particular changes are:–
A Bridge, charger guide is riveted to the body, and the wings of the nosecap have been spread out. Where necessary, the link loop of the magazine has been removed from the trigger guard and lugs for the swivel have been fitted and brazed in. Strengthened extractor springs have been fitted. The stocks, butt and fore-end, have been similarly converted. Various other components have been suitably altered and used in the conversion.

All components of the Short M.L.E. Mark III rifle are suitable for use with this arm, and should be demanded to replace such as may become unserviceable.

Any of the S.M.L.E. rifles, Marks I, I* or I* I.P. which required repairs to the extent of new barrels, were ordered to be returned to the Rifle Factory for conversion to the new pattern. For this reason, few, if any, of the Mark I* Ishapore manufactured S.M.L.E. rifles still exist today.

Yet another alteration was approved for the Charger Loading Lee-Enfield Mark II I.P., consisting of fitting the British type C.L.L.E. Mark I* rifle foresight protector. New conversions were fitted with the protector at Ishapore while those rifles in store and in the hands of troops had the change effected locally. This consisted merely of drilling a No. 4 or .129-inch hole through the foresight base, fitting the protector and securing it by a screw through the newly drilled hole. This addition was approved in 1913.

The Government of India submitted a proposal to adopt the .22-inch R.F. rifle and ammunition as used in Home (British) service, on 1st May 1913. A quantity of 5,860 rifles was considered initially and this was costed at 17/6d per rifle and 37/- for conversion to .22-inch for each rifle. Total cost was estimated at £15,970 and this was then approved.

Short range practice had previously been conducted in Indian service with the use of a special cartridge designated "Cartridge S.A. Ball, .303-inch short range, practice, I.P., Mark I*" which had been approved on 31st August 1903 and continued in service until the Great War. This was a black powder cartridge with a relatively small rounded lead projectile, intended for use at ranges up to 100 yds. This would also indicate that the Morris and aiming tubes were not used in Indian service.

Ishapore's production was increased to meet the demands of the 1st World War and this mostly comprised of the Marks III and III* S.M.L.E. rifles and Pattern 1907 bayonet. The Mark III* S.M.L.E. was approved in India on

26th September 1916, along with an alteration to the Mark III and Mark I** S.M.L.E. rifles.

The formal announcement is a similar entry in the Indian List of Changes to the British publication, LoC § 17622, reproduced in Chapter 5, and the side by side comparison is interesting. With an approval date of 2nd January 1916 for the British entry, in India it was almost nine months later. It was intended that the Indian List of Changes be used in conjunction with their British counterparts.

The Indian announcement was:
Para 8858 (I.L.C.):-

*Rifles, Short, M.L.E., Mark I** I.P.*
Rifles, Short, M.L.E., Mark III.

1. Modifications.

Rifles, Short, M.L.E., Mark III.*
Without cut-off.

2. Introduction.

1. In future manufacture, rifles of the above-mentioned patterns may embody any of the following modifications:

(a) The omission of long range sights (dial and aperture) from stock fore-end and body. In this case the stock, fore-end, being without recesses for the dial sight, and screw and the aperture sight and a washer being employed to replace the base of the aperture sight as a seating for the active end of the aperture sight spring.

(b) The wind gauge on the back sight leaf being replaced by a fixed cap held to the leaf by a screw.

(c) The lug on the striker collar being omitted, and grooves being cut through the collar to enable the Tool, removing striker, pattern B, to be used for the removal of the striker from, and its replacement in, the bolt.

(d) The swivel lugs being omitted from the underside of the front end of the guard, and a wire loop being fitted to the left side of that position for the attachment of the breech cover, also the omission of the lightening groove in the guard.

(e) The lightening recess in the back sight protector, and the lightening holes in the sight protector and boss of the nosecap, also the lightening holes in the stock butt being omitted.

*(f) Omission of cut-off (Mark I** I.P. only).*

2. A pattern of this rifle has been approved to govern manufacture of rifles required without cut-offs.

*It differs from the Mark III rifle in that the body is not slotted and drilled to receive the cut-off and screw, but like the Mark I** I.P. and Mark III, may embody the modifications mentioned at 1. above.*

The following is a list of the new components necessitated by the changes described above:

Barrels, with body, M.L.E. rifle, short, Mk III.*	*With sights assembled (with 1.03-in. blade, foresight), inner band, and pins, stop bolt, locking; sighted for Mk VII ammunition.*
Bodies, M.L.E. rifle, short, Mk III.	*With pins, stop, bolt, locking.*
Caps, backsight, M.L.E. rifle, short, Mk III.	*And Mk I** I.P. and III; for No. 2 leaf.*
Leaves, backsight, M.L.E. rifle, short, Mk III, No.2.	*And I** I.P. and III*; assembled with cap, and screw.*
Screws, cap, backsight, M.L.E. rifle, short, Mk III.	*And I** I.P. and III* for No. 2 leaf.*
Stocks, fore-end, M.L.E. rifle, short, Mk III, No. 2.	*And I** I.P. and III*. Not recessed for dial sight.*
Strikers, M.L.E. rifle, short, Mk I, No. 2.	*All M.L.E. short rifles.*
Washers, spring, sight, aperture, M.L.E. rifle, short, Mk III.	*For I** I.P. and III and III* rifles; without aperture sights.*

When fitting the Stock, fore-end, No. 2, to rifles, with cut-off fitted, they will be slightly altered as follows:
The upward extension, on the right side of the stock, which covers the position of the cut-off slot in the body, will be cut away, and an angular recess will be cut to half the depth of the cut-off recess to clear the cut-off joint when the cut-off is opened.

When fitting these stocks to rifles fitted with aperture sights, the aperture sight will be replaced by the Washer, spring, sight aperture, but should this component not be available, the pillar of the aperture sight will be filed off at the base, and the base disc used as a washer under the spring.

Long, normal and short butts were available on the Mk I** I.P., Mk III and Mk III* rifles in the following proportions:

Long	10%
Normal	75%
Short	15%.

Typical receiver markings of a war-time Mk I** I.P. are:

R.H.

E. R.
ENFIELD
1903
SHT L.E.
I**
I.P.

L.H.

C
R ↑ F
I
1915

The sample markings shown on the previous page are of an Enfield-made Mk I S.M.L.E. of 1903 vintage, upgraded to Mk I** I.P. at Ishapore in 1915; the R.F.I. markings are located on the left side of the body, where the British conversion stampings were generally marked.

Prior to 1917, wood for the rifle stocks was imported, mostly walnut from Europe, but with supply shortages precipitated by the Great War, local supplies of walnut and maple was employed. Some other dark coloured timber similar to teak also appears to have been used during the 2nd World War production.

An India Pattern 1907 sword bayonet was approved on 9th October 1917, with an unfullered blade and special grips. The standard Pattern 1907 scabbard was used. On 30th September 1918 it was recorded that a number of these bayonets had been issued experimentally to ascertain their serviceability. The unfullered Indian blade which is a feature of Indian World War 2 production bayonets obviously resulted from the success of these trials.

Chapter 4 in "British & Commonwealth Bayonets" is devoted to the Indian special issues and local manufacture, including some special ceremonial models which appear to have been fitted to the S.M.L.E. rifles. Such general references commence on page 342 and describe R.F.I. production, markings and the series of short, unfullered blade models, through to page 355.

The Pattern 1914 rifle was on issue in India as about 100,000 rifles were forwarded direct from the United States and on 5th August 1918, approval was given for the conversion of the Mark I pattern to Mark I* at Ishapore. This conversion involved a longer locking lug recess on the bolt head and corresponding recess in the barrel. Another of the British modifications that was subsequently implemented at Ishapore was the alteration of the Marks I** I.P., III and III* S.M.L.E. rifles for the Mk VII ammunition. This approval was given in Indian service on 31st May 1920.

Somewhat similar alterations were applied to the Lee-Enfield Cavalry Carbine, for the .303 Mk VII ammunition; it was approved on 3rd April 1924 and the required changes effected at Ishapore. This conversion comprised:

1. Barrel: small cone enlarged.
2. Sight bed: graduations deleted and re-inscribed for ranges at 200, 300, 400, 500 and 600 yards, according to the sealed pattern.
3. Sight leaf: the graduations deleted and re-inscribed at 7, 8, 9 and 10, as on the pattern.

The carbines were marked "H.V." and "S.C." immediately behind the backsight.

Although the Mark III* S.M.L.E. pattern was introduced in 1916, the Mark III was still being manufactured, or just as likely, re-manufactured. On 24th October 1922, the following modifications were approved for incorporation in future Mark III manufacture:

1. Omission of long range sights (dial and aperture) and the consequent effect upon related components.
2. A cap, T-shaped in section, fixed to the backsight leaf by a screw, to replace the windgauge, and the consequent alteration to the leaf.
3. Grooves cut through the collar of the striker and lug on front of collar omitted.
4. Swivel lugs omitted from the underside of the front end of the trigger guard, and a wire loop for attachment of the breech cover.
5. Steel marking butt disc omitted in future manufacture.

A single loader (non-magazine) variant of the S.M.L.E. was approved on 21st August 1923, the conversion was effected in a similar manner to the single loader M.L.E. This alteration was made at Ishapore and consisted of filling in the magazine way in the fore-end, fitting a sheet steel loading plate in the body, and the utilisation of a special trigger guard. Again, the rifle was intended for issue to forces whose loyalty may have been questionable and for whom it was not desirable that they be armed with rifles as effective as those of regular or British forces.

Indian S.L. model. Note the filled-in magazine way and two wooden securing dowels. *MOD Pattern Room*

Typical markings for the Single Loader, which were stamped on the left side of the receiver butt socket, are:

S L
R F I
1926

However, there was still a requirement for the Mk III S.M.L.E. rifle because a conversion was approved on 6th September 1924 for the S.M.L.E. Marks I** I.P., III and III* to a pattern designated "Rifle, short, magazine, Lee-Enfield, Mark III (Converted)." Basically this re-introduced the magazine cut-off, and the milling of lightening slots in the nosecap and triggerguard, and incorporated the later improvements such as the altered cone, Mk VII ammunition sighting and omission of the long range sights. Re-introduction of the magazine cut-off is in keeping with policy as India had always advocated its use and it was largely as a result of persistent requests by India that it was re-introduced on the Mk I S.M.L.E.

The initial Indian List of Changes announcement of this new "conversion" was:

Para 12794 (I.L.C.):-

Rifle, Short, Magazine Lee-Enfield, Mk III (Converted).

*Converted from Rifles, short, magazine, Lee-Enfield, Marks I** I.P., III and III* sighted for Mark VII ammunition.*

*Conversion of Rifles, short, magazine, Lee-Enfield, Marks I** I.P., III and III*.*

A specification of the above-mentioned rifle has been sealed to govern conversion as may be ordered, at the Rifle Factory, Ishapore.

The conversion consists of the following operations being carried out:

Component	*Description of Conversion*
Barrel.	*Enlargement of small cone. Also to be lapped inside when necessary and struck off for browning.*
Body.	*Milling of cut-off slot, drilling of hole for cut-off screw and fitting cut-off and alteration of Mark I to Mark II when necessary.*
Cap, nose.	*Milling of lightening holes.*
Guard, trigger.	*Milling of lightening slot, milling off swivel lugs and fitting of wire loop.*
Stock, fore-end.	*Clearing for cut-off, filling in and pegging dial and aperture sight seatings with walnut and blending.*

The following is a list of the new components necessitated by the conversion described above:

Cap, backsight, M.L.E. Rifle, short, Mark III.
Cut-off, Mark III.
Leaf, backsight, M.L.E. Rifle, short, Mark III No. 2.
Screw, cap, backsight, M.L.E. Rifle, short, Mark III.
Screw, cap, nose, M.L.E. Rifle, short, Mark I, front.
Screw, cut-off, M.L.M.
Swivel, piling, M.L.E. Rifle, short, Mark III.
Washer, spring, sight, aperture, M.L.E. Rifle, short, Mark III.

The conversion will not advance the Mark as it merely brings rifles of early manufacture into line with current manufacture.

The low survival rate of the early Ishapore rifles is quite easy to comprehend in view of the policy to strip and refurbish any rifles that became unserviceable. This announcement of August 1924 is typical of the policy which remained throughout the Indian S.M.L.E. production:-

*When the barrel of the Rifle, short, M.L.E. Mark I** I.P. becomes unserviceable, the rifle will be returned to I.A.O.C. for break-up and recovery of serviceable components; a serviceable rifle being issued in replacement.*

As occurred in Britain, few changes were approved for the No. 1 rifle between the world wars. On 14th April 1932, India approved the No. 1 E.Y. rifle for grenade firing, but this differs from its British and Australian service counterparts in that it was fitted with rubber pads on the socket end of the fore-end and a spring washer on the stockbolt. The two pads are of hard rubber, 1/4-in. thick, cut to size and fitted as indicated in the sketch. Use of the spring washer on the stockbolt was later extended to all rifles, and an "S" 5/16-inch in size was stamped underneath the butt, near the socket.

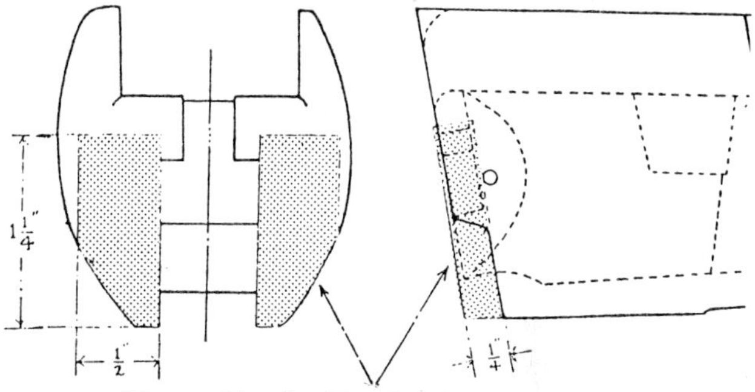

Pieces of hard rubber fitted slightly above the surface of the wood and glued in position.

Similar to their British and Canadian counterparts, the drill purpose rifles were usually stamped "D.P." but to make for more ready identification and to avoid their being mistaken for rifles suitable for ball ammunition firing, a white band, 1-inch wide, was painted on the fore-end and front handguard, midway between the nosecap and outer band. This marking was approved on 15th January 1926, in Indian service.

The quantity of rifles in store and in the hands of troops by 1926 amounted to 650,000 and manufacture of new rifles was being increased from 10,000 to 17,000 p.a. The conversion of earlier rifles was continuing, to make an additional output of 14,000 p.a. This mode of conversion resulted in remaking the rifles, to all intents and purposes, up to the new rifle standard. In this program, it was announced in 1926 that 150,000 rifles were being withdrawn from service and sent to the Rifle Factory to be stripped down to the individual components and completely rebuilt. Parts were scrapped where necessary while all suitable components were relegated as spares. Much of the repair work was also undertaken at the various arsenals or ordnance depots.

The Ishapore factory was expanded in 1927 and existing machinery had to be re-organised due to the erection of new buildings. With the impending adoption of a new rifle, the No. 1 Mk VI in England, India intended to swing into production of the new pattern as soon as possible, so as not to be "a lap or two behind as was too often the case." So, by 1929, the rifle factory was working at full pressure because the stock of serviceable rifles had been allowed to get very low, and to accumulate sufficient stocks so that the factory could shut down for a couple of years in order to change over to the new Mark VI rifle pattern. The program was for 180,000 to be produced in 3 years and this was completed by 31st March 1930.

It was anticipated that the output of the Mark VI rifle should commence on 1st April 1932 with an initial batch of 2,500 rifles out of the first order for 10,000. These 2,500 rifles were ordered for large scale troop trials because India had been advised that the Mark VI rifle would be adopted, once the difficulties had been sorted out. The introduction of the Mark VI was sanctioned in India in March 1928 and funds voted towards replacement and alteration of machinery and the purchase of drawings, components and gauges. This was proceeded with and by 1929 some components were being produced and much of the new machinery was in place.

India had been assured by the British hierarchy of the impending adoption of the Mark VI pattern, but on 15th May 1929 they were informed that there was no intention of rearming at Home, or manufacture of the Mark VI in large quantities. India was also advised that by going ahead with the Mark VI, its forces would be armed with a different rifle to the troops at Home. The War Office decision not to rearm with the Mark VI at Home resulted in the scrapping of parallel production at Ishapore.

Not long after India abandoned its plans to produce the Mark VI, the design was altered slightly, re-designated the No. 4, and manufactured at Enfield for troop trials. This of course resulted in the No. 4 Mk I rifle which was the basis for the standard British service rifle until the 1950's; India again had been misdirected by the War Office. So the forces of the Sub-Continent entered the conflicts of the 2nd World War with the No. 1 rifle when it could well have been the No. 4, and another chapter of likely Lee-Enfield rifle development and production in India was curtailed.

Most observers do not realise the magnitude of India's war effort, largely due to the poor publicity that India, and indeed most of the Empire, was given. During the period 1939 – 1945, India provided:

- 805 3.7-in. howitzers;
- 700,000 rifles;
- nearly 9 million rounds of gun ammunition;
- nearly 5 million rounds of mortar ammunition, hand grenades and mines, and
- nearly 1 billion rounds of small arms ammunition.

Until 1942 when threatened by the Japanese who actually penetrated India's eastern borders, the bulk of this production went to Indian forces fighting overseas, and into the British stores. A new rifle factory was proposed at Kanchrapara to meet the projected production of 120,000 rifles and bayonets per year, but the project was dropped when the output at Ishapore was increased due to the implementation of smaller schemes. Bayonet production was carried out by North Western Railway (N.W.R.) and Metal Industries Lahore (M.I.L.), as well as in Jhelum near Rawalpindi (J.U.); these particular bayonets comprised of the short Mk II and Mk III patterns.

These three makers of the short bayonet patterns during the war were situated in the north west, in modern-day Pakistan; before partition in 1947, this was part of India. While the NWR and MIL factories were manufacturing concerns already in commercial production, the Jhelum Arsenal, situated on the Jhelum River, was a government establishment which continued production after the war finished and after Pakistan attained nationhood.

The Indian No. 1 bayonet series proceeded through a wider variation than its British counterparts, with nominal 17, 12 and 8-inch blades, with and without fullers. These were:

Bayonet	*Approval year*	*Details*
No. 1 Mk I	1926	New nomenclature in 1926, previously known as the Patt. 1907, introduced in 1911.
No. 1 Mk I*	1941	No. 1 Mk I, blade reduced to 12.2-ins., fuller runs through point.
No. 1 Mk I**	1941	No. 1 Mk I* with top of blade sharpened for 2-ins. into a false edge.
No. 1 Mk II	1941	New-made, 12.2 in. blade, no fullers, with the standard pommel.
No. 1 Mk II*	1947	No. 1 Mk II with false edge.
No. 1 Mk III	1947	New-made, 12.2-in. blade, no fullers, square profile pommel.
No. 1 Mk IV		Uncertain, no details.
No. 1 Mk V		Uncertain, no details.
No. 1 Mk VI		No. 1 Mk I grip & crossguard with 8-in. No. 5 style, fullered blade.
No. 1 Mk VI*		No. 1 Mk VI with unfullered blade.

Illustrations and further descriptions of this series of Indian bayonets for the No. 1 rifle will be found on pages 346-351 of "British & Commonwealth Bayonets," in the chapter on India.

In case it became necessary to evacuate the Rifle Factory from Ishapore, a shadow Small Arms Factory was set up at Cawnpore, on the Ganges; Ishapore was in Bengal and also in line of the Japanese advance. This shadow factory was set up late in 1941 and ready by late 1943, but temporarily given over to the British R.A.F. Maintenance Unit in October 1943 for use as an engine and component repair depot; it appears that no rifle production or assembly took place there. A Bren gun factory was established at Hyderbad in central India but no rifles were manufactured there either.

Second World War production at Ishapore, by financial years (1st April – 31st March) is recorded as being:

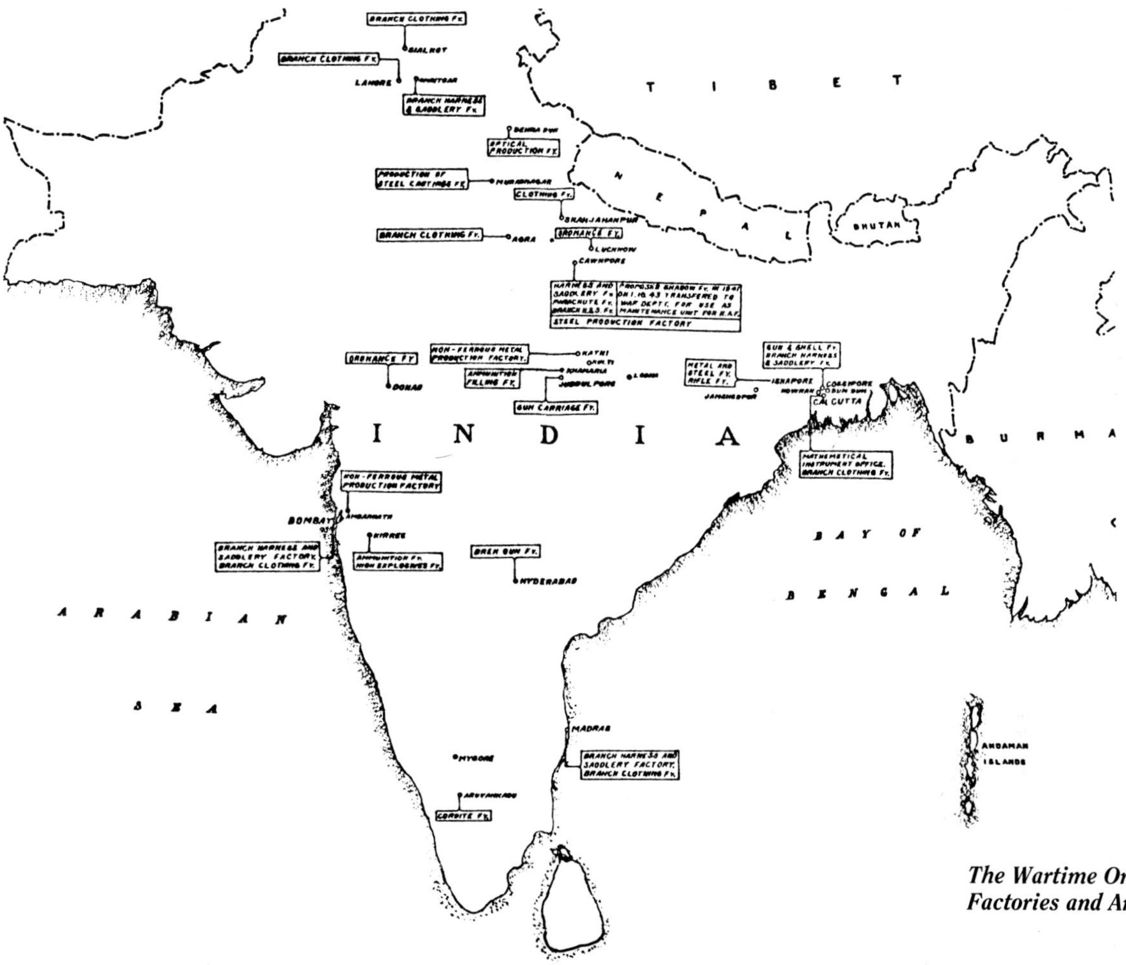

The Wartime Ordnance Factories and Arsenals.

	Rifle No 1 new & repair	*Rifle No 2*	*Bayonets*	*Dah Mk 3*
9/39-3/40	17,620	-	220	-
1940-41	73,030	1,624	15,775	-
1941-42	109,000	1,620	84,907	-
1942-43	131,999	1,200	141,605	-
1943-44	133,000	3,600	224,442	-
1944-45	160,930	3,104	213,776	84,994
4-8/45	55,840	-	70,424	149,548
Totals	681,419	11,148	751,149	234,542

And the Vickers gun, Bren tripods and grenade dischargers are listed as:

	Vickers MMG repair	*Vickers MMG new*	*Rifles, grenade discharger*	*M9A1 grenade launcher*
9/39-3/40	411	288	-	-
1940-41	402	883	-	-
1941-42	152	1,541	-	-
1942-43	123	2,414	-	-
1943-44	90	3,231	-	-
1944-45	60	-	5,108	16,318
4-8/45	40	-	1,652	9,146
Totals	1,278	8,357	7,364	25,464

The Indian ordnance factories and arsenals during the war are recorded as:

Ishapore	Rifle & Bayonet Factory.
	Metal & Steel Factory.
Hyderabad	Bren Gun Factory.
Cawnpore	Proposed Shadow Rifle Factory.
	Steel Production Factory.
Jhelum	Bayonet Factory.
Lucknow	Ordnance Factory.
Dohad	Ordnance Factory.
Cossipore	Gun & Shell Factory.
Ambarnath	Non-Ferrous Metal Production Factory.
Kathi	Non-Ferrous Metal Production Factory.
Jabulpore	Gun Carriage Factory.
Muradhabar	Production of Steel Castings Factory.
Dehra Dun	Optical Munitions Factory.
Kirkee	Ammunition & High Explosives Factory.
Dum Dum	Ammunition Factory.
Khamaria	Ammunition Filling Factory.
Aruvankadu	Cordite Factory.

Indian .22 No. 2 rifles were usually conversions from No. 1 rifles or re-manufactured receivers. R.F.I. Ishapore records that both the No. 2 Mk IV and No. 2 Mk IV* were produced, although no specific differences are described between the two patterns, as is the British case.

One particular .303-inch No. 1 rifles conversion special to India is a smoothbore model chambered for a special brass .410 gauge shot cartridge. This was designated the ".410 Musket," usually converted from downgraded rifles and intended for guard and police use. Markings on these guns are located on the left side of the receiver butt socket, e.g.:

.410
R F I
1942

India's .410 Muskets are single-loaders as a wooden plug has been fitted into the magazine way of the fore-end with a sheet steel platform on top to serve as a loading platform. The production targets were set at 10,000 muskets each for 1940, 1941 and 1942 and the total quantity converted is not certain; the figure is likely to have been included in the previous-listed annual totals of repaired No. 1 rifles. The common sporting .410 shot cartridge will not chamber in these arms.

Filled magazine way and dowels of the .410 Musket are similar to .303 S.L. (Single Loader). *MOD Pattern Room*

Shot cartridges had been used for some time in India, by police, guards and even some irregular forces. Special shot cartridges had been introduced for the percussion Enfield rifle and .577 Snider rifle and carbine; then Martini-Henry arms were smoothbored and chambered for a special .476 cartridge.

The .410 musket was first approved in September 1927 so it can be assumed that the smoothbored Lee-Enfield rifles date from this period. The .410 Indian cartridge comprises the .303-in. brass case which has not been necked or sized, with parallel sides. Two versions of the cartridge were announced in 1927, one loaded with a lead ball over which the mouth of the case was lightly coned, and the other having 12 segmental shot, with the mouth of the brass case folded over.

Until 1931, the .410 musket conversion numbers appear to have been stamped below the original year of manufacture on the right side of the receiver. But after this, the markings were changed to the left side, which is more traditional for the conversion side according to British convention.

India used many British and some Australian manufactured rifles and as it was the Indian practice to convert or refurbish old or worn rifles; some "Indian" S.M.L.E. rifles may be noted with the original British or Australian markings on the right side of the butt socket, with the Indian conversion markings on the left. An excellent example of such a conversion is the "Carbine, Smoke, for 2½-in Grenades" described here and illustrated below.

Regarding this example in the Pattern Room collection, the most easily recognised features are the discharger cup and the S.M.L.E. action. The pictured 2½-in. cup was made in England by Components Ltd., Bournebrook, Birmingham, and in this role as a projector, it was mounted onto armoured vehicles and used for firing the No. 37 smoke grenade.

The example shown here started life in 1917 as a Mk III S.M.L.E. made by L.S.A. Co. and was later sent to India. In 1942 it was converted at Ishapore to .410 Musket, ascertained by the markings on the butt socket. After this, the butt was removed, a .303-in. barrel shortened and fitted without the rearsight, and a discharger cup attached by means of a thread on the end of the barrel, with a knurled collar behind it. The handguard is secured by two strap bands. The special wooden grip at the rear and the

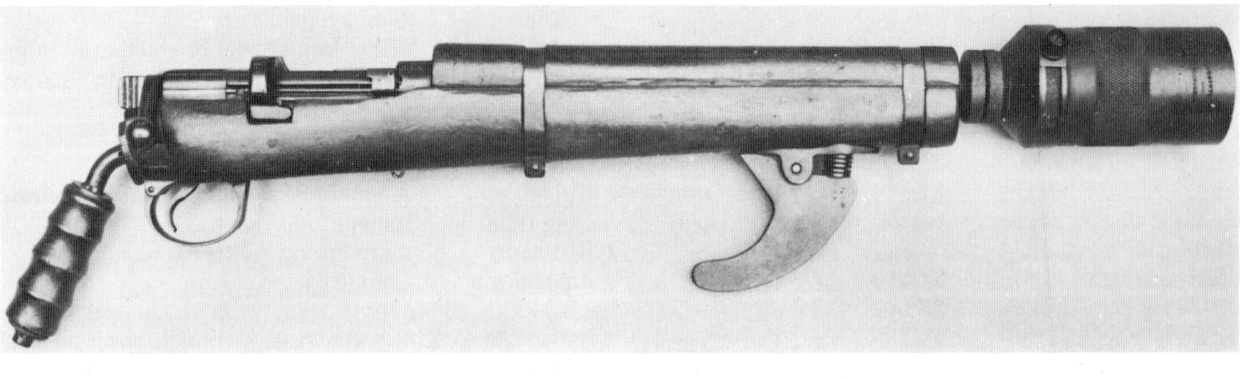

Carbine, Smoke, for 2½-in. Grenades. An Indian converted S.M.L.E. for use as a Smoke Grenade Discharger from Armoured Vehicles. *MOD Pattern Room*

spring-loaded extension below the fore-wood were intended for use in a swivel mount in the A.F.V.

Some hand-made rifles were produced on the Indian sub-continent, in the region of the North West Frontier and what is now part of Pakistan. They are often referred to as "Khyber Pass specials" and may be found to range from poor to excellent copies. Martini-Henry, Martini-Enfield, Lee-Enfield, S.M.L.E. and even No. 4 rifles have been noted. But there is always some feature evident to a sharp eye which can result in a true identification of these replicas.

Although the finish and contour of components is generally identical to the particular service patterns, these Khyber Pass models can usually be identified by certain irregularities in the markings. As they were usually made with individual letter and number stamps, these do not line up perfectly or are sometimes even mis-spelled; a prime example being "ENFEILD." A similar example on an otherwise perfect No. 4 rifle are the marks "No. 4 Mk I" over "LONG BRANCH" and a squared "S" on the receiver along with "M47C" on the butt socket. This would indicate manufacture at Long Branch, Stevens-Savage and finally B.S.A. Shirley, an absolute impossibility for the one receiver! While mistakes are not always this obvious, a keen eye can usually distinguish these copies.

Some of these Khyber Pass specials were actually authorised for issue after purchase from official funds. During World War 2, bodies of irregulars were organised in India and one such force of Pathans had locally made rifles on issue. The price paid was about 60 rupees or £4/15/- for the S.M.L.E. while Martini rifles could be purchased for about 35 rupees, about half the price. Despite a reasonable quality of finish, the safety in firing these rifles was suspect and so those arms chosen for the Pathan irregulars were subjected to a "proof test" of five rounds of MkVIIz ammunition. Some of the rifles were failed due to improperly sized chambers, excessive firing pin protrusion (piercing the cartridge cap) and excessive head space.

Some of those "Khyber Pass" replicas which passed the aforementioned tests were purchased but later found to become extremely loose in the action after firing any reasonable number of rounds; the barrels were also too soft and wore quickly. British or Ishapore rifles were not issued as it was feared that the Pathans would decamp as soon as such an arm came into their possession. It was the practice for government forces on duty in these areas to sleep with their rifles chained to their bodies, with the rifle bolts tucked inside their shirts. The irregular Pathan force was disbanded after a short time without seeing any active service; no doubt their rifles were sold off locally.

But for the manufacturer's designation and factory stamp and various proof and inspection markings, pre-2nd War Mark III and III* rifles made at Ishapore are identical to their British counterparts. A slightly darker and more dense timber (probably local walnut) was used for the furniture and most rifles have quite a deep blued finish on the metal. Some manufacturing changes were implemented during the 1939-45 period as a means of speeding up production.

The most obvious of the Indian characteristics is the nosecap, which has no piling swivel provision, the square profile sight protector ears, and the backsight protector shape which is symmetrical in end-elevation. The cocking piece is grooved like the wartime production Mark III* British and Australian models, but it is thicker and stronger. A T-shaped rearsight cap was used, something similar to later British Mark III* production and the wood furniture was thicker and not as well finished, often being proud at the butt socket, trigger guard and nosecap.

Many of these wartime Indian rifles were fitted with a steel woodscrew through the fore-end, about 1.5-ins. forward of the front trigger guard screw; whether this was intended as a reinforce for poorer quality timber in the fore-end or for the occasional use of the grenade discharger is uncertain. The 2½-in. discharger was produced in India and used rather widely. Quite a few of the 2nd World War and post-war production rifles were stocked up with a lighter and softer timber which is consequently more easily marked and damaged; it also splinters much more readily. Wide steel bands were clamped onto specialist Indian grenade launching rifles.

The manufacturer's markings on Ishapore Lee-Enfield rifles are quite distinctive. Early ones have the Royal cypher over "ISHAPORE," the year of production and the rifle designation; later examples have this abbreviated slightly, as can be seen in the accompanying drawings and photographs.

INDIAN ISHAPORE RECEIVER DESIGNATIONS

On 14th August 1947, at midnight, India became an independent nation, but it was not until after the Republic was declared on 1st January 1950 that the Royal cypher was dispensed with and the national symbol, the ashoka, inserted in its place. This consists of four tigers (or lions?) with one at each quadrant, so that only three are visible; after Independence, the ashoka was also incorporated into the proof mark. Some authorities quote

Typical receiver markings on an Ishapore S.M.L.E. rifle.

June 1953 as the month when the Indian national proof stamp of the ashoka replaced the British style George Rex marks.

Some shortened, lightened rifles were developed at Ishapore, as occurred at about the same time in Australia, and the first of these "jungle rifles" considered was merely a shortened S.M.L.E., such as those assembled in Britain and Australia for corresponding tests. These shortened rifles comprised of shortened barrels to a nominal 20-inch length with the forward wooden furniture appropriately reduced and the original nosecap refitted.

India also took part in the No. 6 program, in fact, the No. 6 was at first considered for India because of its jungle carbine configuration of the S.M.L.E. or No. 1 rifle. Because Ishapore did not manufacture the No. 4 action, the No. 5 rifle (or jungle carbine) was not suited to Indian production and so a special version of the No. 1 rifle was developed as a mode of conversion to a jungle rifle configuration.

The prototypes of this series are quite unorthodox in appearance and this particular program commenced in 1943. A few rifles were assembled for trials and some sent to the Small Arms School at Sangor, two more were forwarded to Britain in March 1944 and tested against the British-designed lightened No. 1 rifle, the C.E.A.D. pattern, which eventually became the provisional Rifle No. 6.

Comparative details of these test rifles are given as:

	India Patt.	*C.E.A.D. Patt.*
Weight:	7 lb. 8 oz.	7 lb. 0 oz.
Weight with bayonet:	8 lb. 8 oz.	7 lb. 10 oz.
Weight of bayonet:	1 lb. 0.5 oz.	10 oz.
Weight of scabbard:	7.5 oz.	4.7 oz.
Total weight with bayonet and scabbard:	9 lb. 0 oz.	7 lb. 14.7 oz.
Length:	38.4 ins.	39.6 ins.
Length with bayonet:	47.25 ins.	47.6 ins.
Length of barrel:	16.5 ins.	18.0 ins.
Overall length of bayonet:	16.6 ins.	11.9 ins.
Length of blade:	12.0 ins.	8.0 ins.

The Indian shortened rifle did not have a butt pad like the British design and it was reported that the recoil was excessive. A flash eliminator was fitted but its external dimensions and cone were different to the C.E.A.D. pattern and not as effective. The two other points of contention were the additional weight and more vulnerable sights; these sights are quite distinctive and consist of a "flag" aperture type with ranges of 200, 400 and 600 yards, mounted on a post behind the charger bridge across the receiver.

The original No. 1 rifle nosecap was retained and this probably accounted for the additional weight of the Indian model. The butt swivel is mounted on the top of the butt and the upper swivel is fitted onto the right side of the fore-end, which is also reinforced by a transverse screw like most Ishapore rifles of the period. Total barrel length is 19.3 inches which includes the flash hider.

Some No. 5 fittings were sent to India for inspection and possible incorporation in subsequent experimental models at India's request; these parts included the flash eliminator, barrel, butt and buttplate. This is probably why the ensuing Indian models of the jungle carbine more resemble the C.E.A.D. pattern, somewhat similar to the Australian No. 6 version.

The No. 6 Mark I was provisionally approved for Britain and India on 1st September 1944, but later cancelled only to be "re-approved" on 7th September 1945, most likely for Australian introduction.

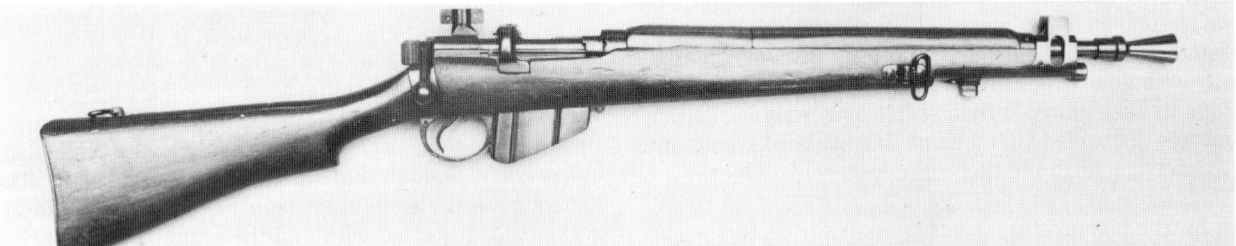

Indian Trials "Jungle Carbine" configuration S.M.L.E. conversion.
This example has an 18-in. barrel and aperture sights. MOD Pattern Room

The No. 4 and No. 5 rifles were not produced in India, although many were refurbished at Ishapore and other Ordnance Depots and factories. Such rejuvenation even included refinishing and removal of original manufacturing marks along with the stamping of Indian ordnance and repair marks, giving an appearance of Indian manufacture. Some No. 5 knife bayonets were produced at Ishapore, such an example has been noted with the marking "R.F.I. 77" on the ricasso which would indicate manufacture as late as 1977.

This No. 4 Mk 1/3 Rifle was F.T.R.'d at Ishapore; the markings are "F.R. 1980 R.F.I." *MOD Pattern Room*

Some sporting gun and rifle production at Ishapore occurred from late in 1952; the double barrel 12 bore breech-loading shotgun in 1953, a single barrel 12 bore equivalent in 1956 along with a .315-in. sporting rifle, and a .22-in. RF sporting rifle in 1971. The double barrel shotgun was copied from a 25-year old B.S.A. gun. Mass production of a .32-in. (7.65mm) ICF pistol was also commenced at the factory in 1980.

Production of the .303 No. 1 rifle at Ishapore continued after the end of the war, until the late 1950's at least. Tension in India's border areas, particularly with her new neighbour Communist China, highlighted the requirement for a self-loading rifle. During such border clashes, the Communist forces were generally armed with the 7.62mm Automat Kalashnakov and although a few Colt Armalites reached the front, it was the .303 bolt action Lee-Enfield in the hands of Gurkhas that caused most of the Communist Chinese casualties.

The introduction of a new rifle had basically been delayed by India's economic problems but when her security was threatened this was given a higher priority and progress in this direction was much more rapid. These efforts commenced early in 1960 and the introduction of the new self-loading rifle was announced in the Hindustan Times in September 1963. The new 7.62mm F.A.L. rifle, though, was quite different from its predecessor the S.M.L.E. in that it is not interchangeable with those manufactured in Britain and Australia; the Indian 1A as it became known, was a combination of both the metric Belgian F.A.L. and the British, Canadian and Australian inch-system, and is therefore not interchangeable with either type.

Introduction of the 7.62mm N.A.T.O. cartridge also saw the development of a new Lee-Enfield rifle in 7.62mm calibre. India had considered the Sterling conversion system for many of the ex-British No. 4 rifles in service, but the legal wrangles in England between Sterling and the Royal Ordnance Factory over patent and manufacturing rights halted any such ideas until other alternatives had been decided upon.

To cope with the additional stress of the new, more powerful cartridge, the Indians overcame the problem in a novel way, by using a higher quality of steel, in this case an EN grade, along with more advanced heat treatment techniques. But for the rectangular, detachable box magazine, cast alloy buttplate and trap and the much higher comb on the butt, the rifle is very similar to the No. 1 Mk III* .303-inch model.

A closer examination, though, will reveal other differences to its .303 counterpart; a regraduated and differently marked backsight, distinctive angular profile backsight bed ramp, fore-end tieplate similar to that of the No. 4 rifle, and on some examples, a slightly larger and differently contoured trigger guard.

The Rifle 7.62mm 2A conversion or "re-manufacture" commenced in 1963 and continued until 1970; production has been put at 2.5 lakhs, or 250,000 rifles.

In terms of the No. 1 rifle genealogy, the Indian 7.62mm 2A represents a finale; still in service on the sub-continent and in substantially larger quantities than its 7.62mm British No. 4 counterparts. It is perhaps ironic that the S.M.L.E. design has survived for nearly one hundred years in service and, in Indian usage at least, will probably outlive its successor, the No. 4 series.

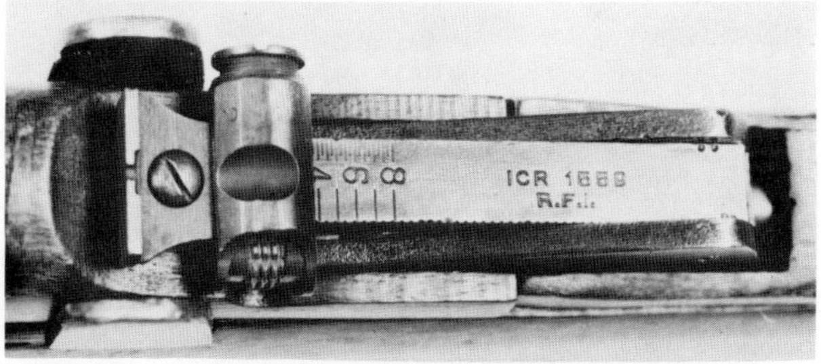

800m Backsight of a 7.62mm 2A rifle. *MOD Pattern Room*

Typical late production Ishapore nosecap; note squared shape of foresight protectors and top. *MOD Pattern Room*

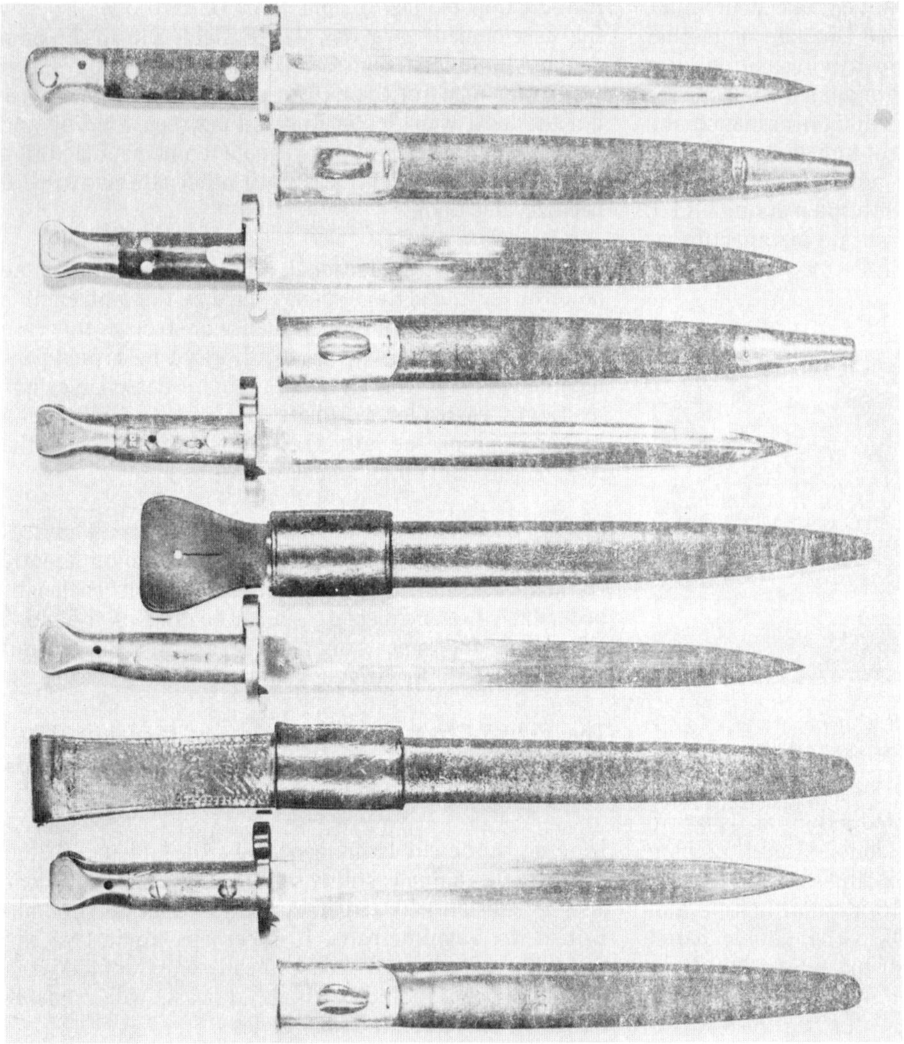

Left, from top:
1888 Troop Trials
Patt. 1888 Mk I, Early Type
Patt. 1888 Mk I, Later Type
Patt. 1888 Mk II
Patt. 1888 Mk III
J. Anthony-Carter

Below, from top:
Patt. 1903
Patt. 1907, Hooked Quillon
Patt. 1907
R. Richardson

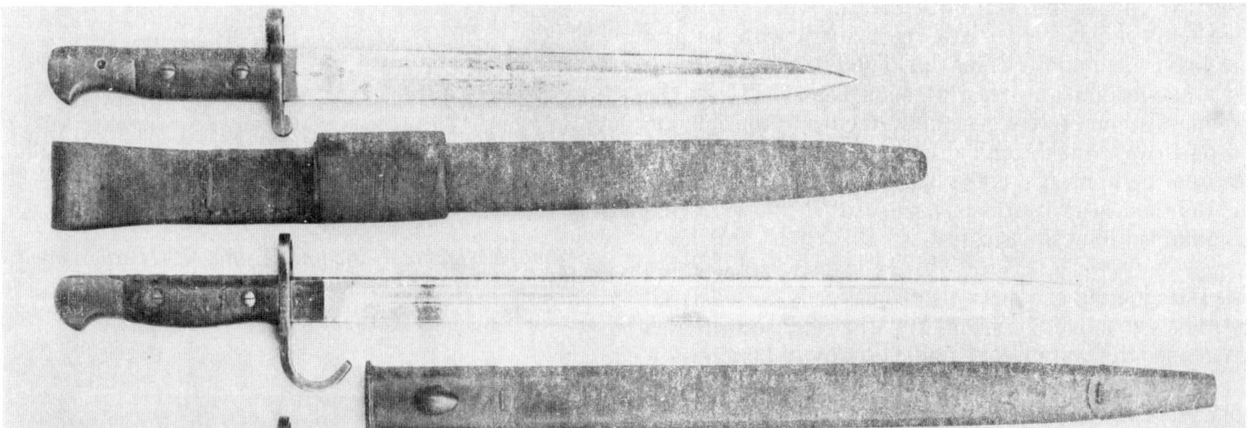

ACCESSORIES & FITTINGS

CHAPTER 11

THE BAYONETS

The topic of sword bayonets for the series of Lee-Metford and Lee-Enfield rifles is quite diverse when one considers the variations manufactured in Australia and India as well. However, Canada's production followed the British patterns much more closely and South Africa only produced a small number of bayonets during World War 2 as well as a modified No. 9 type during the 1960's.

A detailed and comprehensive study of this subject is provided in "British & Commonwealth Bayonets" by Skennerton & Richardson, with separate chapters on the British, Australian, Canadian, Indian, New Zealand and South African issues and models. The Miscellaneous section (Chapter 7) also illustrates and describes some other relevant producers such as Pakistan and the British commercial and export trade.

Catalogue reference numbers from "British & Commonwealth Bayonets" are also listed here should this particular aspect or line of interest in the British Service Lee be followed by the student or collector. Because of the complexity and wide variation in the service patterns, the production in Australia and India, experimental and prototype samples, trials models and commercial versions; the resume here only outlines the troop trials and service types.

The troop trials model sword-bayonet in 1888 represents the first bayonet for a British Lee rifle, of which about 390 were manufactured at Enfield; this is illustrated in Chapter 2 along with the particular trials rifle. The double edged, 12-inch long blade which is a feature of the Pattern 1888 and 1903 bayonets was introduced with this trials bayonet. The leather grips are secured by two rivets and the hilt and crossguard are similar to the Pattern 1887 Mk III, of which it was a contemporary.

The Mark I Pattern 1888 bayonet was approved on 22nd December and introduced along with the Mark I Magazine rifle. A subsequent variation was approved in January 1890 with two rivets securing the grips instead of three, this is referred to as the Mark I, 2nd Pattern. A Mark II version was introduced in 1899 after the clearing rod was abolished on the rifle, resulting in the complementing clearance hole in the bayonet grip being changed to the pommel. The wooden grips were secured by rivets which made their removal difficult, so two screws with special slotted nuts were used instead of the rivets and this resulted in the Mark III variation in late 1901.

These Marks of the Pattern 1888 sword bayonet were not produced in India or Australia as their manufacture was restricted to the established British makers; Enfield, Wilkinson, Sanderson and Mole, except for a small number produced in South Africa by the Railways workshops during the 2nd World War.

The Short Magazine Lee-Enfield also saw the introduction of a new sword bayonet in 1903; the blade and crossguard were carried over from the Pattern 1888 but a new pommel and slightly different grip were required because of the special nosecap on the S.M.L.E. One feature of the new bayonet is that the "muzzle ring" no longer fixes over the muzzle, but onto a stud on the nosecap. Many of the Pattern 1888 bayonets were converted to the 1903 Pattern by fitting a new pommel and grips; similar conversions were effected in India also, but the 1903 bayonet was only manufactured in England, by Enfield, Wilkinson, Sanderson and Mole. Some 1888 bayonets were converted to the 1903 Pattern in Britain also.

There were some complaints about the short reach of the pattern 1903 bayonet when fixed onto the S.M.L.E. as the new rifle was 5 inches shorter than the M.L.E., so with the same 12-in. length of bayonet blade the total reach with fixed bayonet was still 5 inches less. Trials were conducted with three different pattern bayonets with longer blades, the pattern eventually chosen was inspired by the Japanese Arisaka bayonet. Adopted as the Pattern 1907, the new bayonet features a 17-in. fullered blade which compensated for the previous 5-in. reach deficiency and also has a fighting hooked quillon on the crossguard.

During its long service life, the British Pattern 1907 bayonet dispensed with the hooked quillon in 1913 and had a clearance hole approved for the pommel in 1916. The British makers and contractors were Enfield, Wilkinson, Sanderson, Chapman, Mole and Vickers, and a batch was contracted to Remington in the United States during the Great War. The Patt. 1907 and further variants were made in Australia and India also and its manufacture was continued through World War 2. A change in nomenclature in 1926 resulted in it becoming known as the Bayonet No. 1 Mk I.

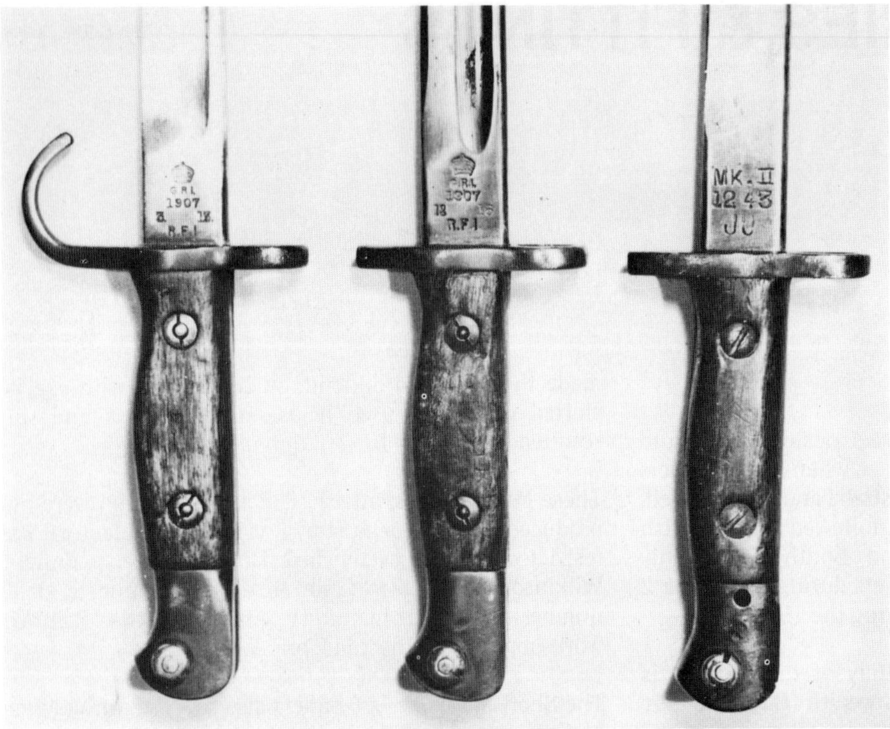

Typical Indian Patt. 1907 ricasso markings.
From left:
Patt. 1907 hook quillon made in 3 '13.
Patt. 1907 dated 12 '16.
No. 1 Mk II bayonet made at Jhelum in 12 '43.
J. Anthony Carter

The Indian pattern S.M.L.E. bayonets vary considerably in blade length and profile and in the shape of the some pommels. Even after the introduction of the Pattern 1907 in British service, India refused to adopt it as it was considered to be unwieldy and inferior to the Pattern 1903, so the Pattern 1907 bayonet was not produced there until after 1912. A Mk I.P. was approved in 1917 with an unfullered blade and special grips although the official announcement indicates that it was issued experimentally.

India's No. 1 Mk II and Mk III series have a nominal 12-inch blade which may be fullered or unfullered and some of the Indian series bayonets have a false edge or grinding on the top near the point so as to improve the penetration.

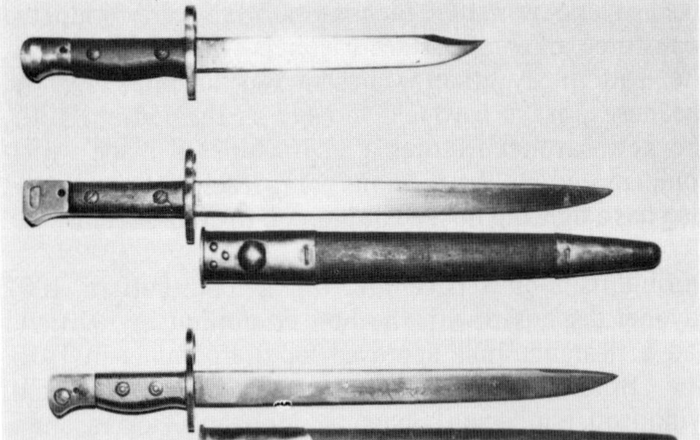

Short Indian models with unfullered blades. From top: No. 1 Mk VI*, No. 1 Mk III, and No. 5 with long blade. *D. Complin*

Some of these bayonets were also manufactured with squared pommels and parallel sided tangs and grips. This series extended to the Mk VI* model and the later bayonets have an 8-inch blade with Bowie shape.

Wartime bayonet production in India incorporated some commercial industry and other arsenals; M.I.L. (Metal Industries, Lahore), N.W.R. (North West Railways) and JU (Jhelum Arsenal). Production of the shorter 12-in. blade models in these three factories was in the north western area, in what is now Pakistan.

Australian production of the Pattern 1907 bayonet commenced at Lithgow in 1913 and the omission of the hooked quillon occurred during 1915; the bayonet production was transferred to the Orange Rifle Factory No. 2 in July 1942 and continued there until the end of the war. A shorter Patt. 1907 type was provided for some trials shortened lightened rifles and it was then manufactured in quantity for the Owen machine carbine although it could also be fixed onto the No. 1 rifle.

Some other bayonet variations were fabricated for the Australian prototype No. 6 "jungle carbines" with an enlarged muzzle ring to accommodate the flash hider.

Another distinctive model peculiar to Australia is the machette bayonet, of which two marks were approved. It was first introduced as a parachutists' bayonet and the nomenclature was changed in 1946 to Machette bayonet Mk I. The Mk II version has larger composite-material grips.

Australian production, from top:
Patt. 1907 hooked quillon
Patt. 1907
Owen Mk I
Owen Mk I/I
Machette Mk I
B. Labudda

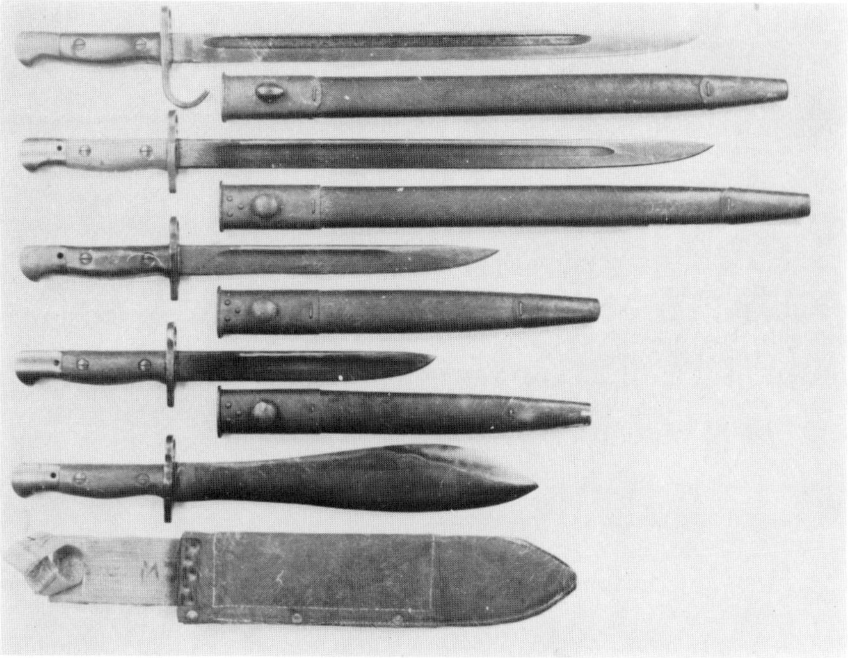

The British No. 1 Mk VI rifle also introduced a new spike bayonet in 1930 and this was eventually brought into service as the No. 4 Mark I in 1939. This series of short spike bayonets was consequently advanced to a Mark III variant, described hereunder:

No. 4 Mk I:	Cruciform blade, introduced in 1939 and produced only in the U.K.
No. 4 Mk II:	Rod spike, manufactured in England, the U.S.A. and Canada.
No. 4 Mk II*:	Rod spike or two-piece assembly, made only in the U.K.
No. 4 Mk III:	Rod spike, the bayonet being a fabricated assembly, made only in the U.K.

A variety of scabbards for the No. 4 spike bayonet may be encountered, the most common being a tapered sheet steel pattern. Plastic composite and tubular scabbards are less common, but were also produced in quantity. This subject is covered in detail in "The British Spike Bayonet" (Skennerton) as well as in "British & Commonwealth Bayonets" (Skennerton & Richardson).

Two other socket-type bayonets which also fix onto the No. 4 rifle are the No. 7 and No. 9. These feature the Bowie shaped blade like the No. 5 bayonet and the No. 7 also has a swivelling pommel. The No. 7 bayonet was originally intended for the Sten machine carbine Mk V, hence the oversize muzzle ring.

The No. 9 bayonet was also introduced for fitting onto the Sten gun. There are instances of these bayonets being used with the No. 4 rifle, particularly for ceremonial purposes, although it was not until after the end of the 2nd World War that they were introduced.

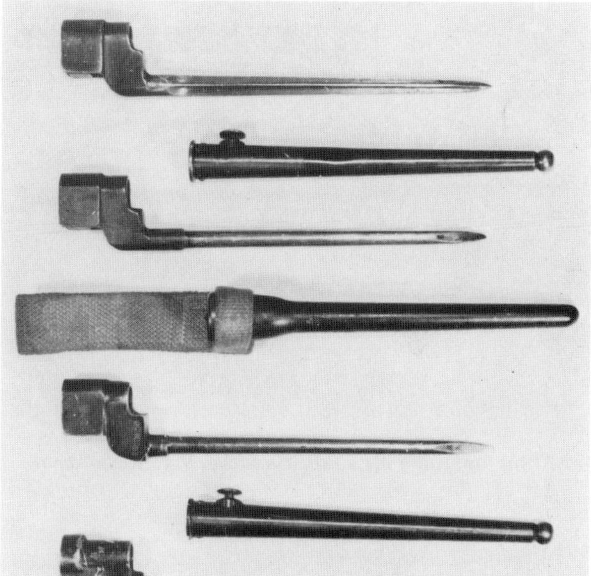

The No. 4 spike bayonet series, from top: No. 4 Mk I cruciform, No. 4 Mk II, No. 4 Mk II* and No. 4 Mk III. Shown with different scabbard variations.
R. Richardson

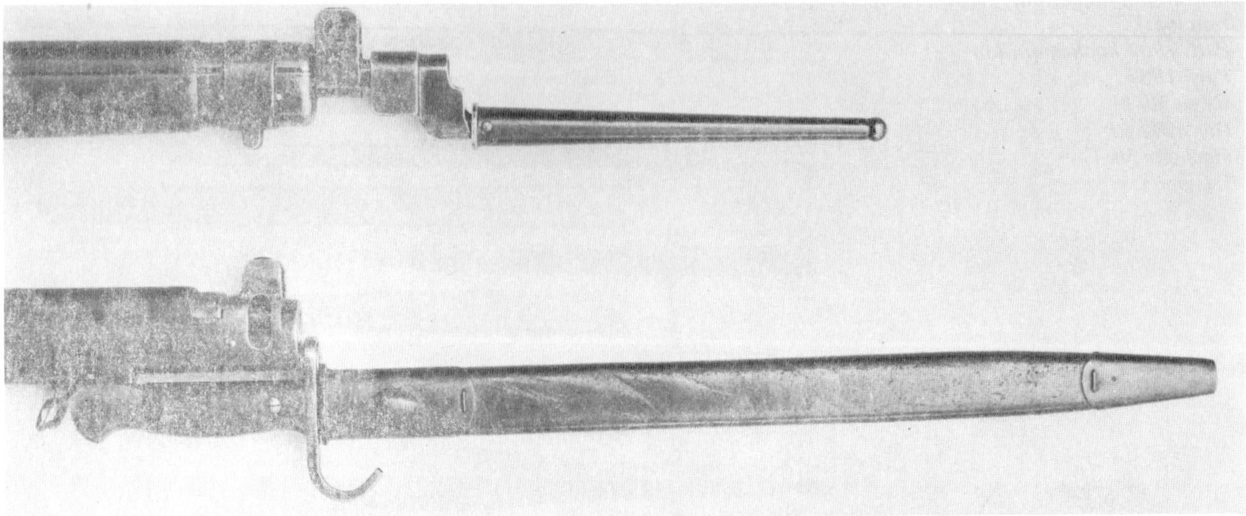

Comparison of No. 4 Rifle with Spike Bayonet & Scabbard fixed (top) and Mk III S.M.L.E. with Patt. 1907 Hooked Quillon Bayonet & Scabbard.

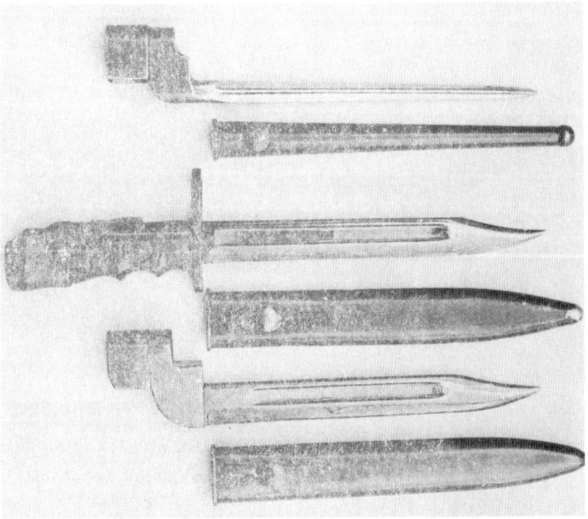

From top: No. 4 Mk I, No. 7 and No. 9. R. Richardson

A new model knife bayonet was instituted along with the No. 5 rifle and this has influenced design until the present day. A Bowie shaped blade was utilised and this form was carried over to the No. 7 and No. 9 bayonets. Introduced in 1944 with an all-steel scabbard, it proved to be much more versatile as a hand weapon and utility tool than the previous model bayonets.

An Australian version for its proposed "Jungle Carbine" was designated the No. 6 but only produced in small numbers for tests and trials, with 17-in., 10-in. and 8-in. blades. The last model is shown here, with a Bowie style clipped point.

The Australian No. 6 has a large diameter muzzle ring for the flash hider, similar to its British counterpart, although the pommel is the same as its Patt. 1907 derivative with a squared mortise slot. On the British No. 5 bayonet, the mortise slot has a rounded cross-section due to the shape of the rifle sword bar or bayonet stud.

Some Carcano bayonets were converted to fix onto the No. 1 rifle by modifying the crossguard and pommel mortise; these appear to have been in Italian service in North Africa during the 2nd World War and were converted locally.

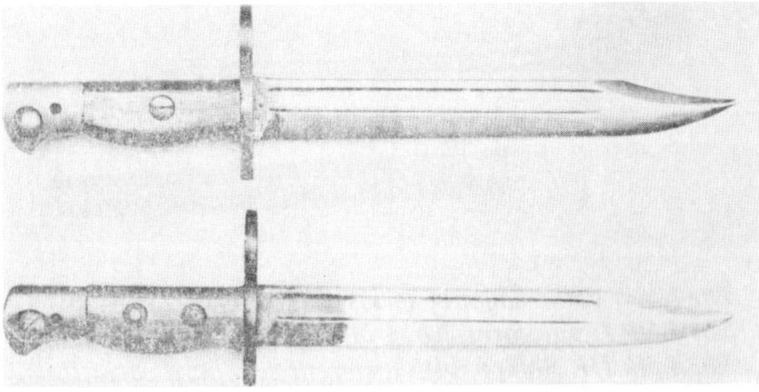

From top: No. 5 Mk I early production and No. 5 Mk I later type. MOD Pattern Room

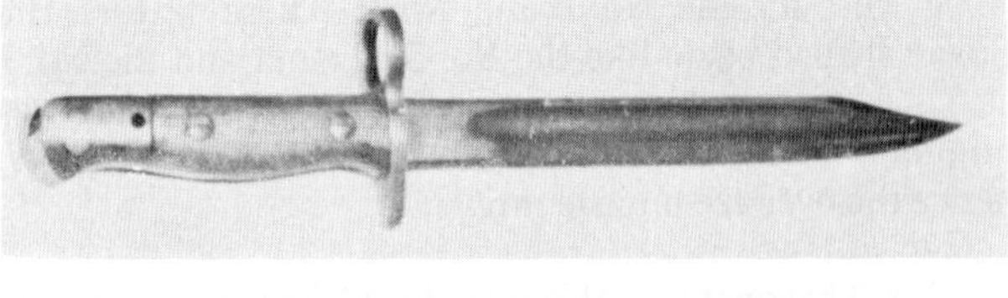

Australian trials No. 6 bayonet with 8-in. blade and Bowie point. Proposed production No. 6 model was to have a 10-in. blade like the Owen Mk I bayonet. SAF Lithgow

	Patt. 1888 Mk I early model	Patt 1888 Mk I later model	Patt 1888 Mk II	Patt 1888 Mk III
Approved:	22.12.1888	05.02.1890	19.05.1899	23.12.1901
Overall Length:	16.6 in.	16.7 in.	16.7 in.	16.7 in.
Blade Length:	12.0 in.	12.0 in.	12.0 in.	12.0 in.
Ring Diameter:	17mm	17mm	17mm	17mm
Grip Type:	Wood, 3 rivets	Wood, 2 rivets	Wood, 2 rivets	Wood, 2 screws

	Patt. 1903	Patt. 1907	I.P. No. 1 Mk I	I.P. No. 1 Mks II & III
Approved:	14.12.1902	30.01.1908	09.10.1917	?
Overall Length:	16.8 in.	21.8 in.	21.8 in.	16.8 in.
Blade Length:	12.0 in.	17.0 in.	17.0 in.	12.0 in.
Ring Diameter:	17mm	17mm	17mm	17mm
Grips:	Wood, 2 screws	Wood, 2 screws	Wood, 2 screws	Wood, 2 screws

	Sht. Lght'd XP	Machette Mk I	Machette Mk II	No. 6 XP
Approved:	made 03.1944	03.04.1944	? 1946	23.06.1946
Overall Length:	14.7 in.	16.1 in.	16.1 in.	14.9 in.
Blade Length:	10.0 in.	11.3 in.	11.3 in.	10.0 in.
Ring Diameter:	17mm	17mm	17mm	23mm
Grips:	Wood, 2 screws	Wood, 2 screws	Composite, 2 screws	Wood, 2 screws

	No. 4 Mk I	No. 4 Mk II	No. 4 Mk II*	No. 4 Mk III
Approved:	15.11.1939	13.02.1940	13.02.1940	21.03.1944
Overall Length:	10.0 in.	10.0 in.	10.0 in.	10.0 in.
Blade Length:	8.0in. cruciform	8.0 in. spike	8.0 in spike	8.0 in spike
Ring Diameter:	15mm	15mm	15mm	15mm
Grips:	None	None	None	None

	No. 5 early type	No. 5 later type	No. 7	No. 9
Approved:	21.03.1944	? 1944	02.01.1945	22.06.1948
Overall Length:	12.0 in.	12.0 in.	12.25 in.	10.1 in.
Blade Length:	8.0 in.	8.0 in.	8.0 in.	8.0 in.
Ring Diameter:	23mm	23mm	22mm	15mm
Grips:	Wood, 1 screw	Wood, 2 screws	Composite, 2 screws	None

Note: Minor varations in length may be noted due to different manufacture, sharpening and wear.

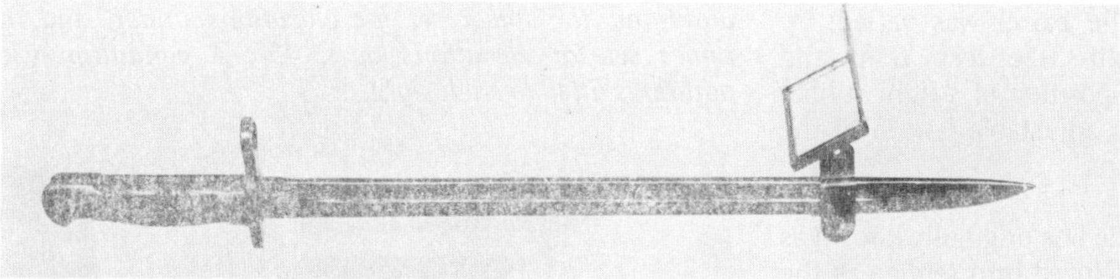

Other Great War service accessories associated with the bayonet. Periscope mirror and early grenade cup, No. 1 Mk I.
MOD Pattern Room

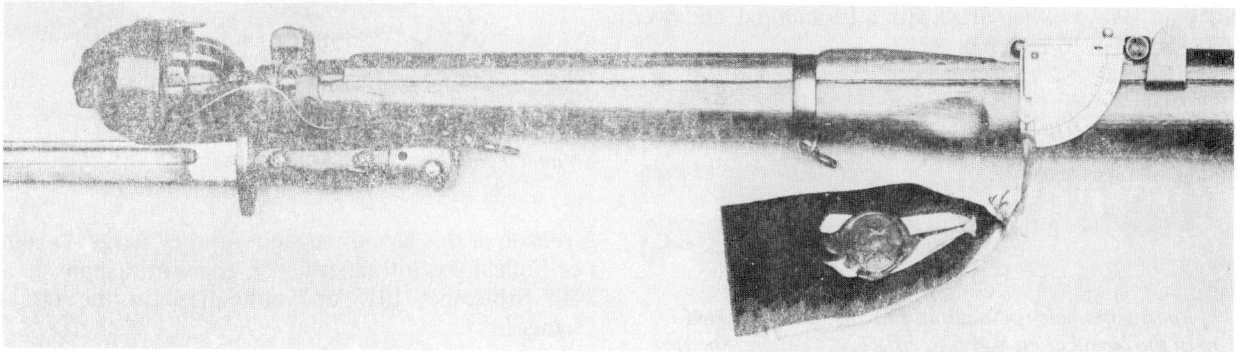

THE GRENADE DISCHARGERS & ADAPTORS

Although there was very limited use of special grenade discharging muskets in the 18th century, rifles were not commonly used as a means of discharging grenades until the early 20th century, during the Russo-Japanese war. So the stage was set for considerable advances to be made in this field by the Germans, British and French during the trench warfare of the Great War.

Changes in the mode of battle had occurred by this time; the cavalry charge and infantry square giving way to entrenched positions, barbed wire, machine guns and eventually, tanks. As a single feature, the machine gun did more to change the type of warfare than any other innovation and so the advent of the trench and machine gun post resulted in the hand grenade and its projection from the rifle assuming a new importance. Range and accuracy were the main problems with hand throwing or lobbing and it was not long before the rifle was used as a means of projection.

It was Marten Hale who designed a method for firing the grenade from a rifle, by screwing a steel rod into the base of the grenade. This was then inserted down the barrel from the muzzle, leaving only the bomb or grenade body projecting beyond the muzzle. A blank cartridge was used as a means of propulsion and this method was that most widely used by the Allies and Axis powers.

Screwed into the base of the grenade, the rod was a sliding fit down the rifle bore although a certain amount of gas escaped around it; the Germans used a form of gas check in their 1913 and 1914 model rifle grenades. To alter the projection range, three main methods were used:

1. The angle of the rifle was changed.
2. A disc was attached to the base of the grenade so as to increase the air resistance in flight.
3. Rod length inside the rifle barrel was varied; by fitting different rod lengths, between 6 in. and 17.5 in., or by altering the position of a tight rubber ring on the rod so that an adjustable length of the rod remained in the rifle barrel.

Britain's first official rifle grenade was originally known as the "J" pattern and it was introduced into service as the "Grenade .303-inch, Rifle, No. 3 Mk I", approved on 16th February 1915. Designed by Hale, its range was recorded as being 100 yds when fitted with a 10-inch rod, and over 250 yds with a 15-inch rod.

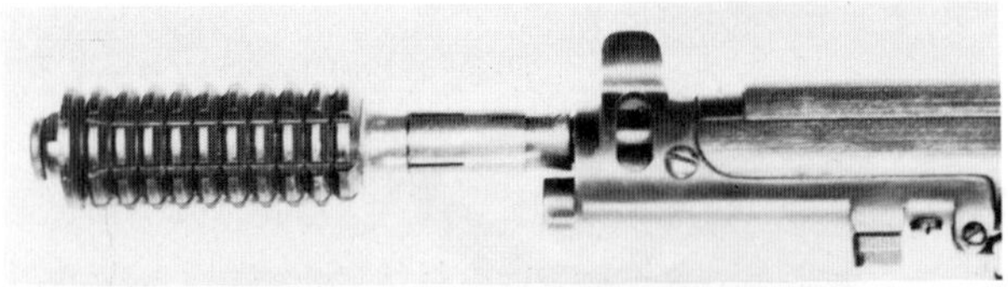

Rodded grenade (a No. 20 in this instance) inserted in the barrel of an S.M.L.E. rifle. *MOD Pattern Room*

One problem with the early systems of rifle discharged grenades was that, should the muzzle be tipped below the horizontal plane, the grenade would usually fall from the rifle. Some of the early models of British rifle grenades had a spring clip at the base to attach to the rifle nosecap so that this would not happen.

Because the range of these rifle grenades depended upon the angle of fire, a clinometer was introduced to enable more accurate lobbing. The first of these was announced in the List of Changes, having been approved on 24th July 1915. This official introduction reads:

Para 17416:–
Sights, pendulum, rifle grenade;
No. 1 Mk I
No. 1 Mk II
for all rifles, short, M.L.E.

Introduction of.

Patterns of the above-mentioned sights have been approved to govern supplies.

The sights are for use on short M.L.E. rifles when firing grenades, and are fitted over the backsight leaf with the quadrant to the left side of the rifle.

The Mark I sight is sprung on the backsight leaf of the rifle, and retained by clips pressed out of the metal. The graduations are marked on an enamelled plate.

The Mark II sight differs from the Mark I in having a hinged clip which is pressed against the sight leaf by a spiral spring, the clip having a knob for disengaging it. The graduations are marked on the metal.

Both sights have two sets of graduations marked on the quadrant; the lower set for elevations under, and the upper set for elevations over, 45°. A pendulum with pointer is fitted to each sight.

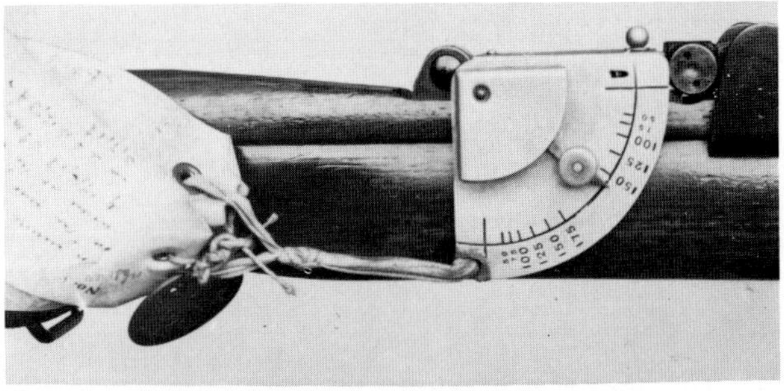

Grenade Pendulum Sight, Sealed Pattern. *MOD Pattern Room*

A version of this pendulum sight for the Charger Loading Lee-Enfield was introduced soon afterwards, approved on 14th September 1915 and announced in the List of Changes:

Para 17504:-

Sight, pendulum, rifle grenade, No. 2 Mk I.
For Rifles, charger loading M.L.E., Mk I.*

Introduction.

A pattern of the above-mentioned sight has been approved to govern supplies.

The sight is for use on Rifles, C.L., M.L.E. Mark I, when firing grenades, and is fitted on the backsight leaf with the quadrant on the left side of the rifle. It is attached to the rear sight as follows:-*

The rifle sight slide is pressed forward against the stop screw at the top end of the leaf, when down, to the 600 yards position, the tongue on the top plate at the rear end of the pendulum sight being inserted between the sides and under the bottom end of the rifle sight leaf, and the top plate pressed down on the leaf. The rifle sight slide is then pressed back, and the clamping nut tightened.

The sight has two sets of graduations marked on the quadrant; the lower set for elevations under, and the upper set for elevations over, 45°. The sight is fitted with a pendulum pointer.

Rifle and hand grenades were used in roles other than high explosive anti-personnel; such as signalling, illumination, smoke screen, tear gas and anti-tank. The rifle projected grenades were widely used for signalling and illumination.

"An Introduction to British Grenades" (Skennerton) describes the full numerical series of hand and rifle grenades from the No. 1 War Department Percussion Pattern in 1908 to the No. 95 Lachrymatory model of 1952. Many of this series were used with, or adapted for discharging from, the service rifle; this booklet illustrates and details each of these patterns and also records the special fittings and applications for rifle firing.

The Mills bomb was designated No. 5 grenade and advanced to the Nos. 23 and 36 upon subsequent design improvements. These early bombs were fired with a rod screwed into the base, but then a flat plate was attached onto the grenade base for use with the rifle discharger cup from mid-1917.

The earliest style of cup for a rifle grenade was approved on 7th July 1916, a simple sheet steel attachment secured by the bayonet and used with the rodded Mills bomb, as shown on page 351. List of Changes announcement was:

Para 17874:-

Cup, No. 23 rifle grenade, No. 1 Mk I.
Steel, for Rifles, short, M.L.E. with sword-bayonets, Pattern 1903 and 1907 fixed.

Introduction.

A pattern of the above-mentioned grenade cup has been approved to govern supplies as may be ordered.

The cup consists of a ring and a stem, and is retained in position on the rifle by the sword-bayonet when mounted.

The stem portion of the cup is formed to fit round the under side of the nose-cap, the cup portion, which retains the grenade lever in position after the pin is withdrawn, projecting over the muzzle of the rifle. A slot is cut in the stem through which the ring of the sword-bayonet crosspiece passes, and the lower end of the stem is forked to pass the base of the sword bar.

Although this was called a "cup", it is not regarded as such in terms of the cup dischargers which were to follow, because it was used with the rodded grenades and acted more as a steady and safety for the grenade striker handle. The later cup dischargers were used in conjunction with a flat baseplate or gas check on the base of the grenade to provide the method of propulsion after the blank cartridge was fired.

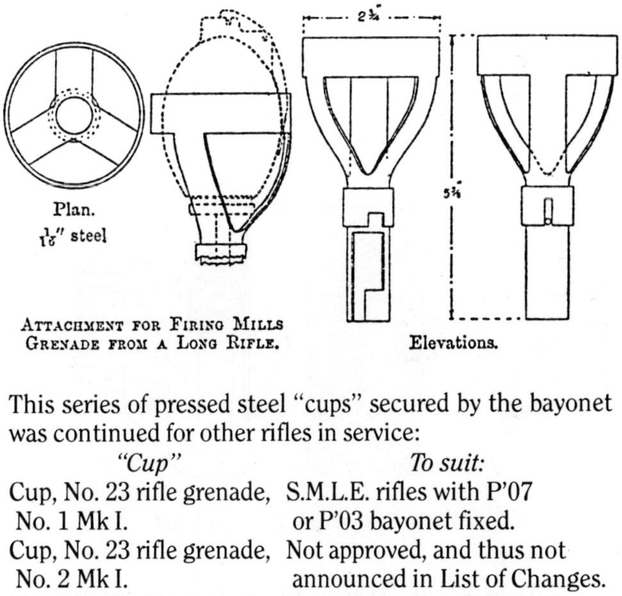

ATTACHMENT FOR FIRING MILLS GRENADE FROM A LONG RIFLE. Elevations.

This series of pressed steel "cups" secured by the bayonet was continued for other rifles in service:

"Cup"	*To suit:*
Cup, No. 23 rifle grenade, No. 1 Mk I.	S.M.L.E. rifles with P'07 or P'03 bayonet fixed.
Cup, No. 23 rifle grenade, No. 2 Mk I.	Not approved, and thus not announced in List of Changes.
Cup, No. 23 rifle grenade, No. 3 Mk I.	Not approved, and thus not announced in List of Changes.
Cup, No. 23 rifle grenade, No. 4 Mk I.	Patt. 1914 rifle, without P'13 bayonet fixed.

The succeeding machined steel discharger cups required a flat baseplate screwed into the bottom of the grenade, eliminating the need for rods attached to the rifle grenades. The first of these dischargers for the S.M.L.E. was introduced in the List of Changes in August 1917.

Para 19457:-

Discharger, Grenade, Rifle: No. 1 Mk I.
Consisting of barrel with base and adjusting screw; for attaching to Rifles, Short, M.L.E. for firing grenades with base gas checks.

2½-in. Discharger Cup fitted to an E.Y. Rifle *I. Skennerton*

No. 2 Mk I.
Consisting of barrel with base, for attachment to Rifles, Magazine, .303-in. Pattern 1914, for firing grenades with base gas checks.

This No. 1 discharger has two claws which fit into the lightening slots of the S.M.L.E. nosecap and are tightened up by screwing the cup in a clockwise direction. The outside of the cup is knurled so as to provide a hand grip for this purpose. The adjusting screw is a butterfly nut on a slide which permits stages of opening or closing of a vent in the base of the cup, and thus vary the range of projection of the grenade.

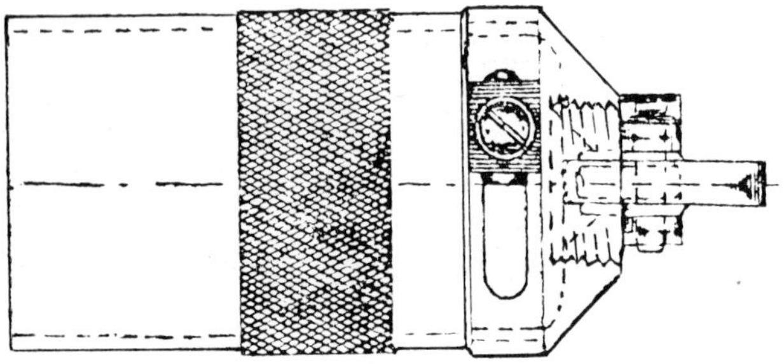

2½-in. Discharger Cup, unfitted.

Dissimilarities in the Pattern 1914 rifle fore-end, nosecap, foresight protector and muzzle arrangement resulted in the No. 2 discharger cup being quite different; there is also no adjustable vent. The Patt '14 rifle fitted with the No. 2 discharger was intended more for training purposes as this particular model of the .303-in. rifle was not used as a front line issue.

Although most infantry troops were trained in basic use of the discharger cup, a prescribed number of specially trained rifle bombers was allocated to each battalion. Accurate lobbing of rifle grenades required considerable skill and practice and there was usually co-operation between the battalion sniping teams and bombers in calling the target ranges as well as observations of the fall and pattern of projection of their shots and grenades.

By the end of the Great War, ninety-six bombers, drawn from the rifle companies, were allotted to each battalion. Each of these men carried the discharger cup and they were considered to be the battalion's assaulting power, along with the riflemen. Because of their mobility and secondary role as riflemen, they were an important factor in the trench warfare, especially in the destruction of enemy machine gun or observation posts, as well as in attack and defence.

First deliveries of rifle grenade cup projectors from the British factories were not made until the end of September 1917 and by November they were being produced by the National Projectile Factory and eight other contractors, including Components Ltd., S. & B., Bowden Wire Ltd. and W. Ward & Co. Ltd. It was also complained that the exact pattern of the discharger had not been entirely settled, even by the end of December 1917. As this discharger cup was also produced during World War 2, cups and markings of later manufacturers may also be encountered.

The German, British and French forces had largely discontinued use of the rodded grenade patterns in favour of discharger cup models by early 1918, because the rodded types were not as accurate as desired, often caused bulged rifle barrels, were inconvenient to carry, and generally had less reliable fuze systems. As range variation was obtained by altering the angle of the discharging rifle, to use this method accurately, a clinometer or pendulum sight was required and these were also liable to damage.

The discharger provided better accuracy in lobbing because of the more regular gas loss, the pendulum was soon dispensed with as the rifle was held at a regular elevation and range differences obtained by regulating the volume of propelling gas with a port or adjustable shutter on the cup. This more readily controlled range variation also permitted the standardisation of grenade launching blanks; different cartridges had been required for use with rodded and cup discharged grenades due to the variations required in gas pressure and this range of cartridges sometimes created difficulties.

This infantry employment of rifle grenades was really only a makeshift proposition but it provided the infantry with the means to barrage and bomb at short notice with increased and more accurate range than was possible with hand throwing. Reliance upon other units such as mortar crews and artillery stationed behind the front lines was not entirely satisfactory as it also involved communications, map co-ordination and synchronisation. When urgently required in a defensive role, this additional time lapse was all too often very critical.

Whilst the rifle grenade application in this role was of limited effect compared with artillery and mortars, the direct control and instant facility for front line troops had obvious advantages. Early in the war, the proportion of hand to rifle grenades was about ten to one in favour of the hand-thrown model, but by the end of the war, more of the rifle grenades were in use than their hand thrown counterparts. The static trench warfare made this application unique and unlikely to be repeated in any later conflicts.

The Great War also saw the introduction of the first anti-tank rifle grenade, the No. 44, which was issued from April 1918. Subsequent rifle projected anti-tank models were the Nos. 68, 85 and 94. These are illustrated and described along with their particular rifle application in "Introduction to British Grenades".

A new 2-inch discharger cup was approved for the S.M.L.E. rifle on 7th December 1925; so as to distinguish from the earlier pattern, the No. 1 Discharger was redesignated the 2½-in. discharger, referring to the bore and applicable rifle grenade diameter. The List of Changes introduction reads:

Para A2535:–

Discharger, Grenade, Rifle, 2-inch, No. 1 Mk I.
Consisting of barrel with bush and adjustable shutter; base, 2 bolts and 2 nuts.
For Rifles, No. 1 Mks III and III.*

1. Introduction.
Discharger, Grenade, Rifle, 2-inch, No. 1 Mk I.
Consisting of barrel with base and adjusting screw; for attachment to Rifles No. 1 Mk III and III, for firing grenades with base gas checks.*

2. Nomenclature.

1. A pattern of the above-mentioned rifle grenade discharger has been approved to govern supplies as may be ordered.

It differs from the Discharger, grenade, rifle, No. 1 Mk I, § 19457, in the following details:

(1) The internal diameter of the barrel is 2-inches compared with $2^1/_2$-inches.
(2) The base is attached to the rifle nose-cap by two bolts, the heads of which engage in the lightening holes of the nose-cap.

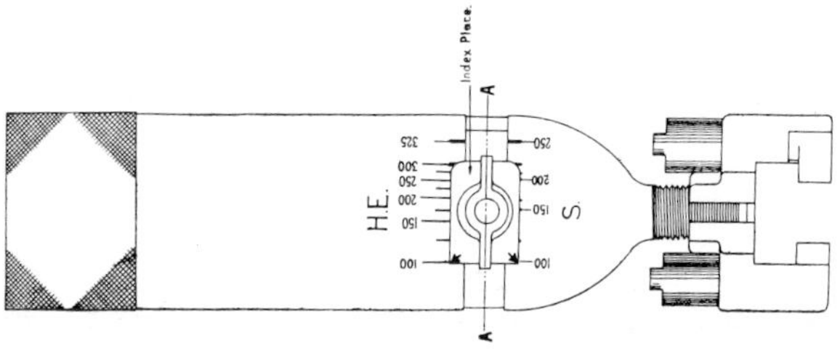

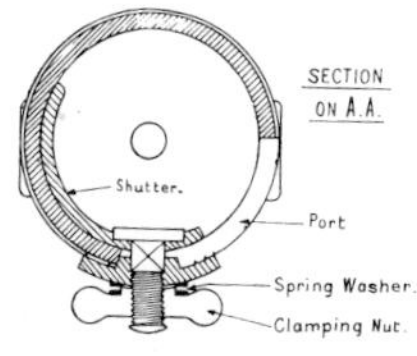

Discharger, Grenade, 2-inch (with cross-section A-A).

(3) The adjusting screw is omitted, the end of the discharger barrel being screwed down in light contact with the muzzle end of the rifle barrel.
(4) There are two range scales, the upper for "H.E." grenades above the clamping nut, being calibrated from 100 to 325 yards, and the lower for "S" grenades being calibrated from 100 to 250 yards.

Instructions to be observed in fitting dischargers to rifles.
(a) Unscrew the base from the barrel of the discharger.
(b) Place the base on the nose-cap of the rifle and turn the heads of the bolts into the lightening holes of the nose-cap. Screw down the nuts, using the pommel slot of the bayonet, No. 1 Mk I, to finally tighten. (Note: Care should be taken, during tightening, to ensure that the discharger barrel hole in the base is held concentric with the bore of the rifle barrel.)
(c) Screw the discharger barrel into the base until it is in light contact with the muzzle end of the rifle barrel. (Heavy contact will bend the rifle barrel, and must be avoided.)

General Note: When assembling spare bolts and nuts, the end of the bolt should be centre-punched at three equidistant points to prevent loss of nut.

The following is a list of the component parts:

Component	*Remarks*
Barrel, with bush.	*Bush not issued separately.*
Base.	*(none)*
Bolts & nuts.	*Two of each.*
Shutter, assembled with wing nut, &c.	*Not issued separately.*

2. Consequent upon the introduction of the 2-inch discharger, the nomenclature of the pattern introduced § 19457 will be read as now shown.

This new 2-inch discharger cup was produced at Enfield and also required an updated series of 2-inch diameter rifle grenades but as the 2½-in. pattern was reinstated in 1933, only the Nos. 54, 55 and 56 were so introduced.

A special sight was devised for fitting to the S.M.L.E. rifle (by now redesignated the Rifle No. 1) for use with the 2½-in. cup and a new anti-tank rifle grenade, the No. 68, and approved on 15th November 1940. Known as the "2½-inch Grenade Discharger Sight Mk I", it comprises a stem and steel strap. The top of the stem forms a "V" or open sight and it is mounted onto the rifle by means of a strap which incorporates two T-bolts and wing nuts for tightening. The List of Changes announcement advises that rifles should be strengthened by binding, now better known as the E.Y. rifle adaption.

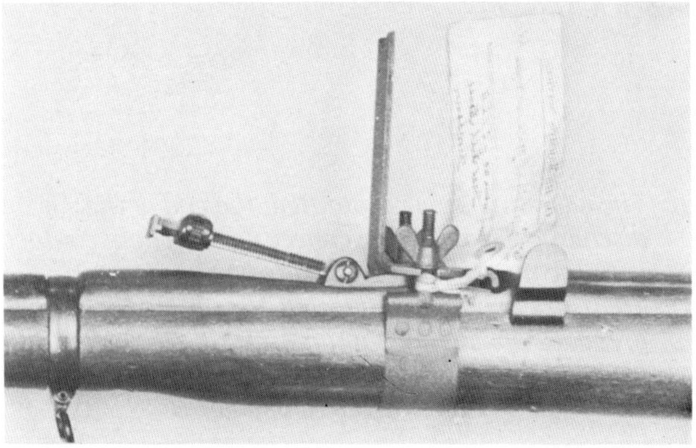

Reverse view of Mark II Grenade Discharger sights for firing the No. 68 Anti-Tank Grenade. *MOD Pattern Room*

A Mark II sight was approved on 3rd March 1942 and the Mark I pattern declared obsolete. The details are:

Para B6860:-

Sights, 2½-in. Grenade Discharger, Mk II.

1. Introduction.

Sights, 2½-in. Grenade Discharger, Mk I.

2. Obsolescent.

1. The above-mentioned Mk II sight is hereby approved. The main feature of the sight is that it can be used on Rifles, .303-in. Nos. 1 and 3 and .30-in. Patt. '17, whereas the existing Mk I pattern is for the Rifle No. 1 only. The sight is used in conjunction with the 2½-in. rifle grenade discharger for firing "No. 68 Grenade A.T."

The upper part of the sight is manufactured from flat steel strip, which, with a clamp and two wing nuts, secures the sight to the rifle.

At the top of the sight is a stepped slot; the extreme top edge is used for sighting when assembled to the .303-in. rifles, and the step is used for sighting when the sight is fixed to .30-in. Patt. '17 rifles.

The respective positions of the sight on the rifles and method of sighting are published in Army Council Instructions No. 1346 of 1942.

Rifles No. 1 should be strengthened by binding in accordance with Instructional Print D.D.(E) 2560; and Rifles No. 3 and .30-in. Patt. '17 should be strengthened as shown on D.D.(E) 3012.

2. Consequent upon 1. above, Sights, 2-in. Grenade Discharger Mk I are hereby declared obsolescent. Existing stocks will be used up.

On 7th August, only a few months afterwards, a series of dischargers, sights and adaptors, was approved for use with the 2½-in. grenades. The official announcement was:

Para B8247:-

Adaptors, Discharger, 2½-in. Grenade. No. 1.
With bolt and wing nut. For attachment to Rifles, .303-in. Patt. '14 and .30-in. Patt. '17.

Dischargers, Grenade, Rifle, 2½-in. No. 2 Mk I.
Consisting of barrel and base assembled for attachment to Rifles, .303-in. No. 1 and Patt. '14.

Dischargers, Grenade. Rifle, 2½-in. No. 3 Mk I.
Complete with cup; sleeve, locking and catch assembled.
For attachment to Rifles, No. 4.

Sights, 2½-in. Grenade Discharger, Mk III.
With clamp and wing nuts assembled.
For Rifles No. 4.

1. Introduction.

Sights, 2½-in. Grenade Discharger, Mk II.
Clamps.
Nuts, wing.

2. Nomenclature.

1. The above-mentioned Adaptors; Dischargers, Grenade; and Sights have been approved for ordering supplies as required. A brief description of each is given below.

Adaptor, Discharger, 2½-in. Grenade, No. 1: is designed to enable the No. 1 Discharger, at present fitted to Rifles No. 1 Mks III and III, to be assembled to Rifles, .303-in. Patt. '14 and .30-in. Patt. '17 U.S.A. It is made of gun metal, slides over the muzzle end of the barrel, and is held in position by a bolt and wing nut (issued with the adaptor), the bolt passing through the adaptor and behind the foresight protector. Two slots are provided in the sides of the adaptor to accommodate the levers of the discharger, this being assembled in the normal manner.*

Discharger, Grenade, Rifle, 2½-in. No. 2 Mk I: This pattern discharger is intended for assembly to Rifles, No. 1 and No. 3 (Patt. '14) and consists of two main parts; one the base, and the other the discharger cup. The base comprises three main parts; one part carries a threaded boss (into which the discharger cup is screwed), the hole adjacent to the boss is to clear the end of the nose cap on the No. 1 rifle. This hole is of no consequence when the base is assembled to the Patt. '14 rifle. The second part, channelled section with tapered sides, is situated, (when assembled to the rifle) under and parallel to the nose cap, or barrel. The stepped slot in this part is to permit anchorage over the bayonet standard. The third part is referred to as the link and retains the two parts previously mentioned in the correct position when fully assembled to the rifle.

Note: On the left side of the front part of the base, will be seen the markings "No. 1" and "P 14". The axis pin (with split pin at each end) must be fitted to the correct hole applicable to the rifle that is being fitted for use with the discharger.

When the base has been assembled to the rifle, the discharger cup is screwed into the base until the discharger cup tightens up (not too tight) on the face of the muzzle end of the barrel. The discharger cup which holds the grenade has 5 holes (early pattern, modified later to 4) at the base end, these holes being covered or uncovered by a gas regulator ring, which is adjusted to give the desired range.

Discharger, Grenade, Rifle, 2½-in. No. 3 Mk I: is designed for assembly to the No. 4 rifle. The complete discharger consists of three main parts; the Cup, the Sleeve, locking and the Catch, sleeve, locking.

The main feature of this discharger is the ease of assembly to the rifle. It is assembled to the rifle over the bayonet lugs at the muzzle end, thus preventing the rotation of the cup.

To lock the cup to the rifle, the catch is depressed and the locking sleeve rotated until the catch is engaged in the slot marked "SECURE", to remove the cup from the rifle, the catch is depressed and the sleeve rotated to the right until the catch is engaged in the slot marked "FREE".

The locking sleeve is retained to the cup by two round headed screws which extend into a cannelure machined round the boss of the cup.

Sights, 2½-in. Grenade, Discharger, Mk III: are designed for use on the No. 4 rifle when fitted with the Discharger, grenade, 2½-in. No. 3 Mk I. The clamp and wing nuts supplied with the sight are interchangeable with the Mk II sight.

The main difference between the new pattern and the existing Mk II is in the length of the blade, the new pattern being approximately 1-in. shorter.

When assembled to the rifle, the sight is positioned immediately in front of the Band, lower; the sight range is 75 yds.

2. Due to the introduction of the above-mentioned Sights, the nomenclature of the original item has been amended to read as now shown.

Above & below: 2½-in. Discharger Cup for No. 4 Rifle.
National Army Museum

British wartime Contract Records list the manufacturers of these items as having been:

Cup Discharger 2½-in. No. 1:	Adams Bros. & Burnley
	Lightfoot Refrigeration
Cup Discharger 2½-in. No. 2:	Lightfoot Refrigeration
Cup Discharger 2½-in. No. 3:	Wembley Tool Co.
	British Vacuum Cleaner Co.
Adaptors, 2½-in Discharger:	Pegles Ltd.
	Hugh Campbell
	J. Stone & Co.
	Alite Foundry
Sights, 2½-in. Discharger:	Lines Bros.
	Hygienic Wire Works

While the foregoing few pages may seem a little confusing due to the variety of dischargers, adaptors and sights for the various types of rifles in service, the following tabulation makes the situation more easily understood.

Cup Dischargers:

No. 1 Mk I	for Rifle No. 1.
No. 2 Mk I	for Rifles No. 1 & No. 3.
No. 3 Mk I	for Rifle No. 4.

Cup Discharger Adaptor:

No. 1	for Rifles No. 1 & No. 3 to take the No. 1 Mk I Discharger Cup.

Cup Discharger Sights:

Mk I	for Rifle No. 1.
Mk II	for Rifles No. 1 & No. 3.
Mk III	for Rifle No. 4.

All these items were declared obsolete by 1947 as the new spigot type discharger was introduced along with tail finned grenades which permitted greater accuracy and longer ranges.

In Australia, the 2½-in. Rifle Grenade Launcher was produced by Metro Gas Co. of Melbourne, Victoria who manufactured about 2,000 units for Australian service. This was the No. 1 Mk I pattern of discharger cup, the same as in British service. In India, patterns of the 2½-in. discharger cup were also produced, although these usually differ slightly in the machining and shape of fittings compared with their British and Australian counterparts; the edges are also more squared-off rather than being rounded. A positive identification can be made regarding the origin of the various cups by an examination of the manufacturers' markings and designations.

The .30 Model 1917 rifle was set up for grenade launching with the discharger cup, for use with the No. 68 A.-T. grenade. Special steel strengthening plates were fitted behind the receiver tang in the butt and secured by screws and copper binding. These particular modifications were effected by the Birmingham gunsmiths, Parker Hale.

A few No. 4 rifles were wire-bound for grenade firing tests but it was established that the No. 4 fitted with discharger cup was not as satisfactory as the No. 1. This is largely because the recoil shock on the No. 1 was substantially taken by the nosecap and fore-end and absorbed to a certain degree; with the No. 4, however, the shock was taken solely by the barrel and action. As a result, particularly with the No. 68 grenade, trigger guard screws were often bent or broken. This is most likely why so few of the 2½-inch cups were issued and used with the No. 4 rifle.

The 2½-in. discharger cup used with the No. 4 rifle differs in a few details from its predecessors used with the No. 1 and No. 3 rifles. Such a discharger cup was originally specified in 1923 along with the first experimental rifles which resulted in the No. 1 Mk VI and No. 4 Mk I. A new design of cup was produced in July 1926, of which one sample was manufactured. It was found that the No. 1 cup fitted onto the S.M.L.E. often worked loose, unscrewing during firing; it was also not possible to attach or remove it as quickly as was considered desirable. For this reason, the lugs on the proposed No. 4 rifle barrel were considered to be more suitable for the fixing of a discharger cup.

Because the barrel projected beyond the fore-end of the proposed new Mk VI and No. 4 rifles, a better fit could be obtained over the barrel with the cup secured to lugs on the barrel. Instructions were issued in 1931 for experiments with the grenade discharger cup to continue and it was at this time when the transition occurred from No. 1 Mk VI to No. 4 Mk I.

The No. 4 discharger cup was not officially introduced until mid-1942, designated the Discharger No. 3 Mk I. Its mode of attachment differs from the earlier cups for No. 1 and No. 3 rifles as it utilises a turning collar which locks behind the lugs on the rifle barrel, the same lugs that also fix the bayonet. The Wembley Tool Co. and British Vacuum Cleaners made 10,600 of these cups during the 2nd World War and this number is quite likely to represent the total production of the No. 4 discharger cup.

The previous 2½-in. discharger No. 1 Mk I and components were produced by Adams Bros. & Burnley Ltd., Lines Bros. Ltd., Lightfoot Refrigeration Co. Ltd., Components Ltd., and possibly some other manufacturers during both world wars. Due to the different mode of warfare, the discharger cups did not see as much use in World War 2 and the No. 4 rifle hardly seems to have been used in this role at all. These discharger cups were declared obsolete in 1946.

Some of the 2½-in. discharger cups were fitted onto shortened rifles for firing smoke grenades from armoured vehicles and many of these were converted locally upon requirement, mainly using the No. 1 action. Only a short length of barrel was retained and the cup was attached at the end of this, with some form of swivel or hinge arrangement for attachment to the vehicle. A special pattern was adopted in India for use on armoured cars, described in Chapter 10. Some Ross rifles were converted for similar application in Canadian service.

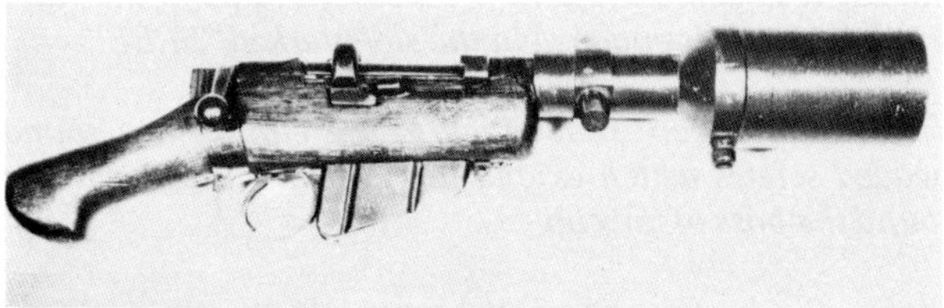

2½-in. Smoke Discharger for Armoured Vehicle.
National Army Museum

The S.M.L.E. action was also utilised on larger 4-in. smoke dischargers for tanks, as per the illustrated drawing from the Valentine IX Tank Service Instruction Book of July 1943. This mechanism was changed to an electrical system by the latter years of the war, dispensing with the requirement for the Lee-Enfield action. A 4-in. diameter smoke generator was used instead of the 2½-in. smoke grenades as described in the previous-mentioned models.

The following series of grenade launching devices were spigot types, officially called "Projectors" and these were

Discharger, Smoke Generator, 4-in., No. 2 Mk I.

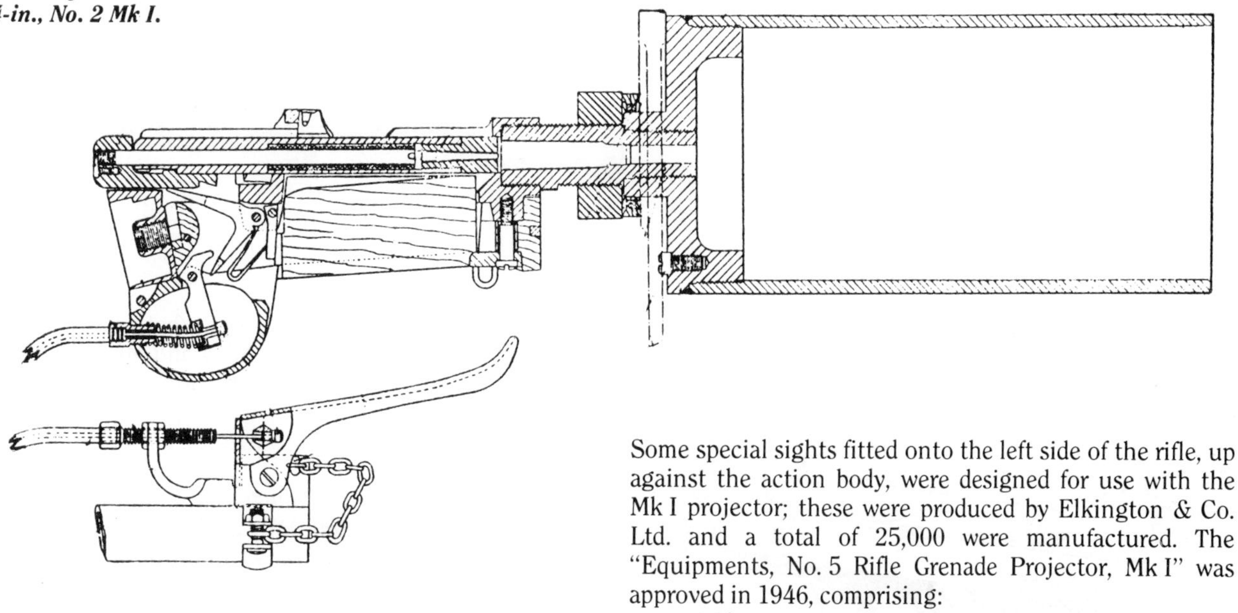

introduced late in World War 2 even though general policy decisions had been made against the use of rifle-discharged grenades. The main application of these projectors was with the anti-tank grenade, used against fortified positions or bunkers as well as armoured vehicles. These rifle spigots feature external rings turned into the body, acting as gas checks; the grenades also have tail fins for better stability in flight.

In mid-1944, grenade launcher trials were conducted with the Rifles Nos. 1, 4 and 5, in both high and low angle roles. The ranges with the No. 1 rifle were found to be much shorter than with the No. 4 and No. 5 and when a comparison was made with the U.S. rifle and grenade projector, none of the British rifles had an equivalent range. Urgent requirements for the No. 5 rifle and grenade projector were nominated for jungle fighting so these units were produced in quantity and the No. 1 and No. 4 projects were shelved.

A reinforced No. 5 action was principally fabricated for firing grenades and 100 of these strengthened rifle actions were made up at B.S.A. Shirley in 1945. Two parallel braces were positioned from the underside of the receiver to the butt socket section, enclosed by the fore-end; it was later decided to mount the trigger onto this brace for further trials, although this variation was never adopted. More details on this particular development of the No. 5 are described towards the end of Chapter 6.

Few Mk I projectors were made for the No. 4 rifle and contract records indicate that 26 were made by Sharp & Wright Ltd. in 1946. A similar type of projector for the No. 5 rifle was manufactured in 1945 by Sharp & Wright Ltd. who produced 15,006, along with Baker Perkins Ltd. who manufactured 20,000; this makes a total of more than 35,000. This was designated the Projector, Grenade, Rifle No. 5 Mk I.

Some special sights fitted onto the left side of the rifle, up against the action body, were designed for use with the Mk I projector; these were produced by Elkington & Co. Ltd. and a total of 25,000 were manufactured. The "Equipments, No. 5 Rifle Grenade Projector, Mk I" was approved in 1946, comprising:

- 4 Clips, retaining, grenade
- 1 Instruction book
- 1 Pad, shoulder, rifle grenade projector Mk I
- 1 Projector, grenade, No. 5 rifle, Mk I
- 1 Sight, Grenade projector, No. 4 and No. 5 rifle, Mk I/I.

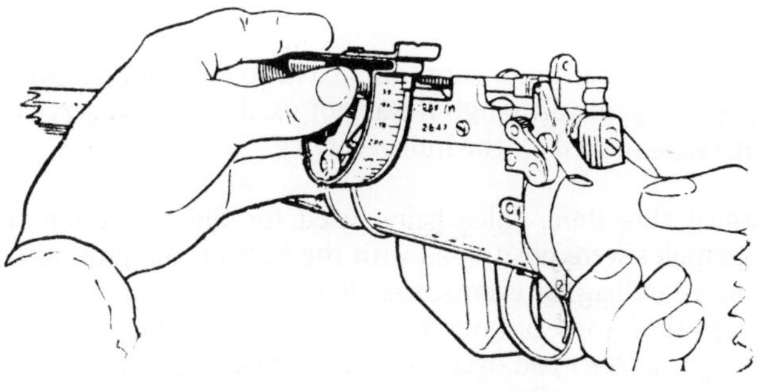

Above & below: No. 5 Rifle Grenade Projector and Sight with No. 85 Grenade. *MOD Pattern Room*

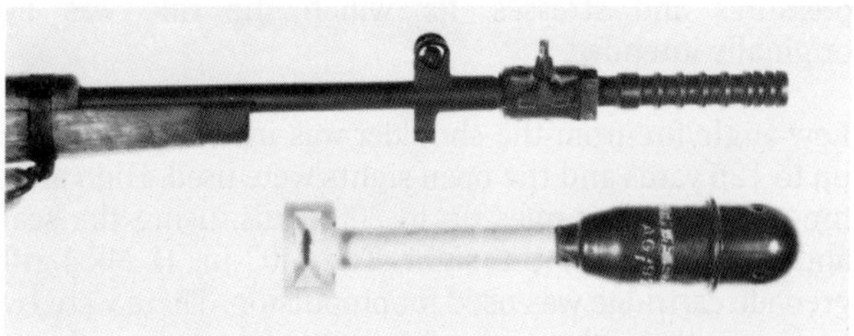

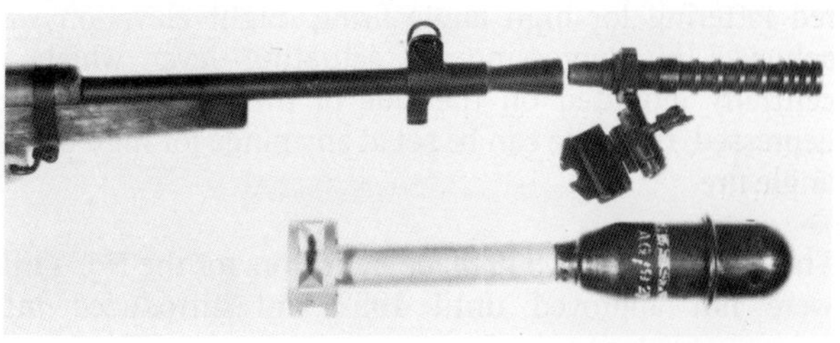

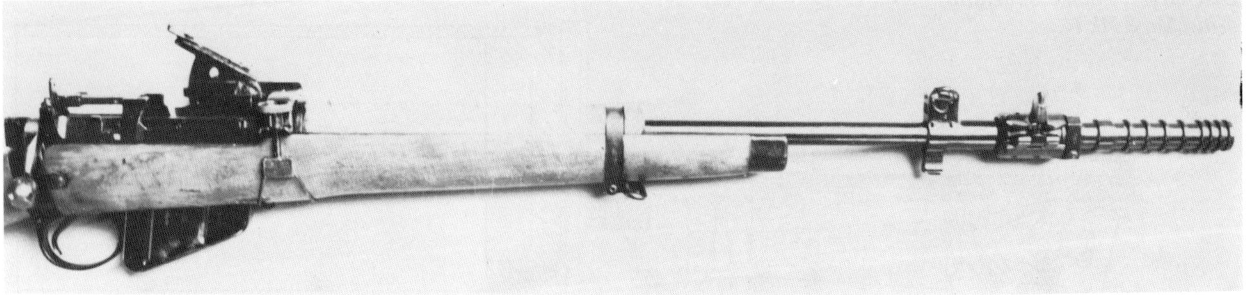

Above & below: Rifle Grenade Projector Sight fitted onto No. 5 Rifle. *MOD Pattern Room*

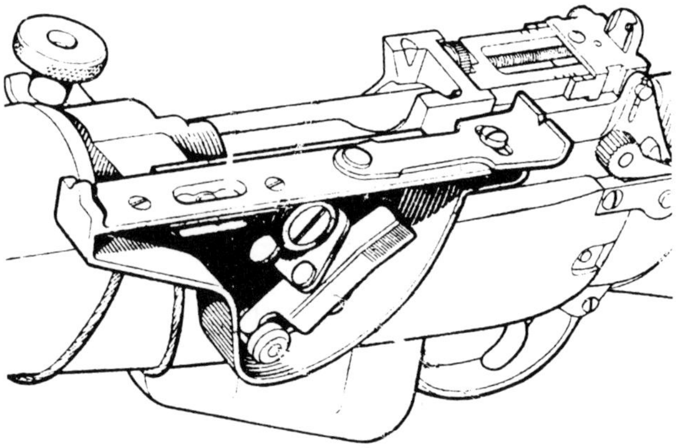

A Mk 2 grenade projector was also introduced for use with the No. 4 rifle although it was not used in British service; it was approved for the Indian service.

Until this time, rifles being used for the projection of grenades were positioned with the butt on the ground or on a sandbag, or even suspended by the sling and braced with the hands, but never fired from the shoulder. Use of a slip-on rubber pad over the original rubber kick pad of the No. 5 rifle permitted low angle fire from the shoulder and minimised the likelihood of damage to the butt when placed on the ground for high angle fire. The firing of any form of grenade from the service rifle raised certain pressures and stresses, for which the rifle was not originally intended.

Low angle fire from the shoulder was intended for ranges up to 125 yards and the open sights were used. High angle fire was used at ranges up to 200 yards, using the scale and spirit level on the sight. The .303-in. H Mk 4 rifle grenade cartridge was used for propulsion. There were two scales on the sight, with white letters for low angle and red lettering for high angle firing. Sight elevation was achieved by depressing an actuating lever which is centrally mounted on the side of the sight. With this depressed, the scale can be set at any range for high or low angle fire.

The Marks 3, 4 and 5 grenade projectors for the No. 4 rifle were not approved until 1952 and introduced into Vocabulary on 21st October 1952 for both Land and Air Services. Official details of the Mk 4 and Mk 5 projectors were circulated in Armourers Instructions:

PROJECTORS, (No. 4 RIFLE), Mk 4 & 5.

Introduction

The projectors (No. 4 rifle), Mk 4 and 5 are used to fire the Grenade, rifle, anti-tank, No. 94 from the No. 4 rifle, in conjunction with the Cartridge, .303-in. H Mk 7z. This grenade is an anti-tank weapon which has an effective range of 75 yards against moving targets and 100 yards against stationary targets. The tail tube of the grenade fits over the projector barrel, and the grenade is fired by a special cartridge which is carried in a cork plug. During transit, the cork plug fits into the tail of the grenade.

When in use, the projector is clamped to the muzzle of the rifle, the barrel of the projector permitting normal use of the rifle. When the projector is fitted, however, the rifle will fire approximately 10 inches low at 100 yards and the size of the bullet group will be approximately doubled; these factors must therefore be taken into account when firing ball ammunition.

WARNING:– If a round of ball ammunition is fired when a grenade is in position on the projector, the grenade will be detonated.

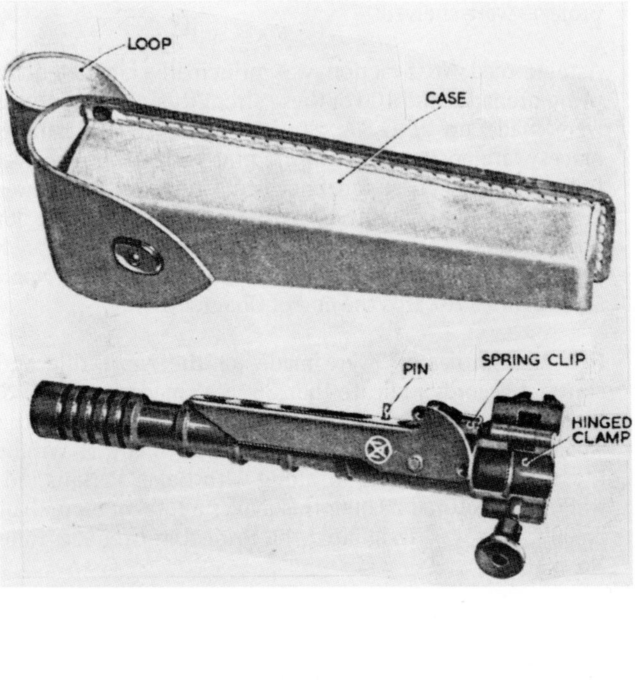

When not in use, the projector is carried in a leather case, which has a loop at its rear to enable the case to be slung from the webbing belt.

PROJECTOR (No. 4 RIFLE), Mk 5.

Projector: The projector consists of a steel barrel which is shaped into a piston at its front end, and formed into a clamp at its rear end. A sight is hinged to the barrel; when the sight is in use, it is held in the vertical position by a pin on the front face of the sight engaging a spring clip on the barrel. A circlip, seated in a groove at the rear of the barrel, acts as a stop for the tail tube of the grenade.

The projector is attached to the muzzle of the rifle by the hinged clamp, the hinged part of the clamp having a square hole in its centre. When the clamp is closed about the rifle muzzle, the hole is engaged by one of the bayonet lugs, and the clamp is secured by the tightening of the milled nut.

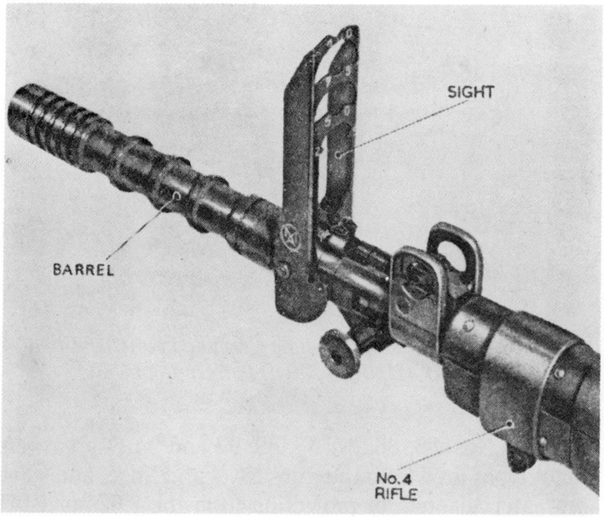

Sighting Arrangements: The sight is constructed as shown, and is graduated in ranges of 25, 50, 75 and 100 yards. If the range of the target is 25 yards, the two triangular pointers are used as aiming marks. If the range is 50, 75 or 100 yards, the semi-circular rings are used. The lower part of the sight is cut away, to allow the rifle sights to be used when ball ammunition is being fired with the projector fitted.

Sighting: After the range has been estimated, the sights are aligned so that the target will be viewed as shown. In the illustration, the estimated range is 75 yards, the upper edge of the 75 yard ring is aligned with the upper edge of the grenade body, and the selected point of the target is in the centre of the sight frame.

Firing: The projector and the rifle can be fired from several positions, standing, kneeling or sitting, or from a slit trench. Detailed descriptions of these methods and positions are given in the relevant training manuals.

WARNING: The rifle must never be fired from the shoulder, nor with any part of the butt plate in contact with any part of the body.

The standing position is an emergency one which can only be used in the open. The kneeling and sitting positions can be used in the open, in a shallow trench, or behind a bank.

Fitting and Removing the Projector.

Fitting:

1. *Ensure that the rifle is unloaded.*
2. *Raising the sight to the vertical position, then open the hinged clamp.*
3. *Position the projector on the rifle muzzle so that the sight is vertical, then push the projector rearwards until it abuts the foresight protectors. Close the clamp, ensuring that the hole in the hinged part of the clamp fits over the bayonet lug, then secure the clamp by tightening the milled nut.*

Removing:

1. *Raise the sight to its vertical position, unscrew the milled nut, open the clamp, then slide the projector off the muzzle.*
2. *Close and secure the hinged flap, then place the projector in its case.*

PROJECTOR (No. 4 RIFLE), Mk 4.

Description: The projector (No. 4 Rifle), Mk 4 is the same as the Mk 5 projector except that the sight is of a different pattern. The method of fitting, sighting, firing and removing the projector is the same as the Mk 5 projector.

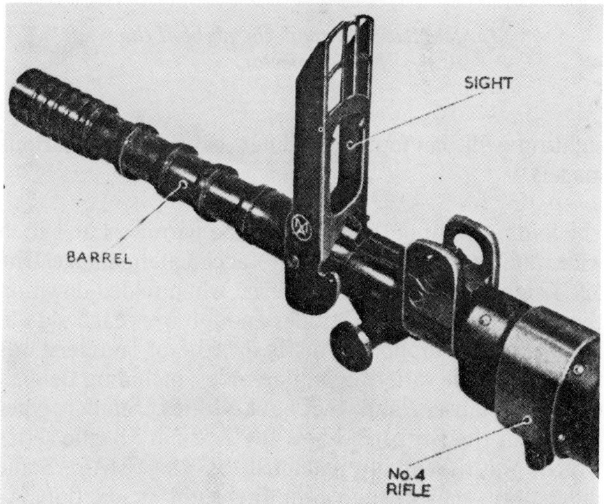

The Mark 3, 4 and 5 projectors were manufactured in Belgium by Mecar, of 4 Rue de la Chancellerie; none of these Marks seem to have been made in Britain at all. The Mecar S.A. trade mark is located on the left side of the folding sight. The major differences between the Marks 3, 4 and 5 are found in the sights. These have 25, 50, 75 and 100 yd. markings on the sights, but the form of this

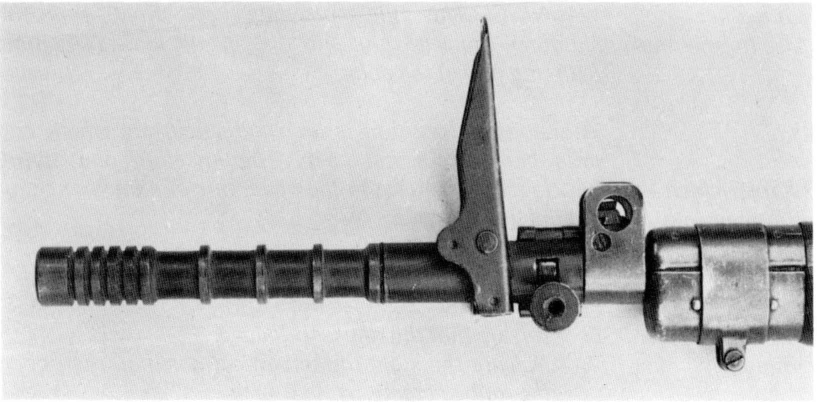

Left view of Mk 5 Projector fitted onto No. 4 Rifle, sight in raised position. *MOD Pattern Room*

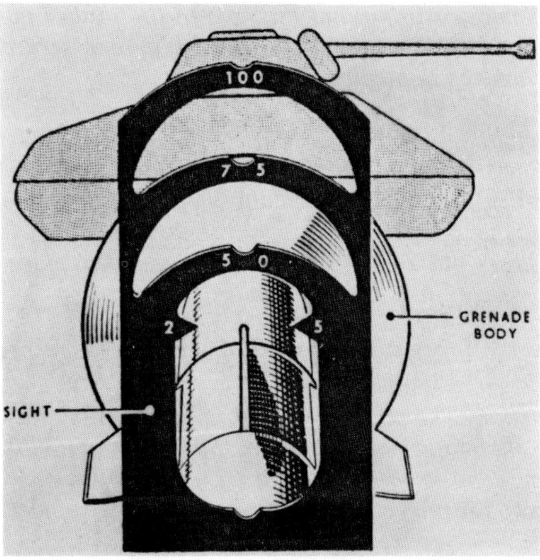

Target viewed through the sight of the Mk 5 projector.

sighting will be found to differ between the various models.

The folding sight of the Mark 5 is the narrowest at 1-inch wide, and therefore less liable to accidental damage. The Mk 4 sight is 1.5 in. wide and even when folded down on the projector tube, it protrudes enough over each side to be damaged if dropped. This basic type of launcher was made for a wide variety of military rifles, including Belgian contract Mausers and the F.A.L. series. Similar types figured in the British trials of the 7.62mm X8 rifle series which was eventually adopted as the L1A1. Some differences may be noted in the sight graduations of various types, some are in yards, others in metres.

The spigot type projectors were not easily adapted to the No. 1 rifle and as the No. 1 was readily approaching obsolescence it was only in Australia and India where moves were made in this direction. Two styles have been noted, one with a clamp fixing system through the nosecap lightening slots on a system akin to that of the discharger cup No. 1, and the other mounted onto a bayonet fitting like that illustrated. The bayonet type of mount was not found to be successful.

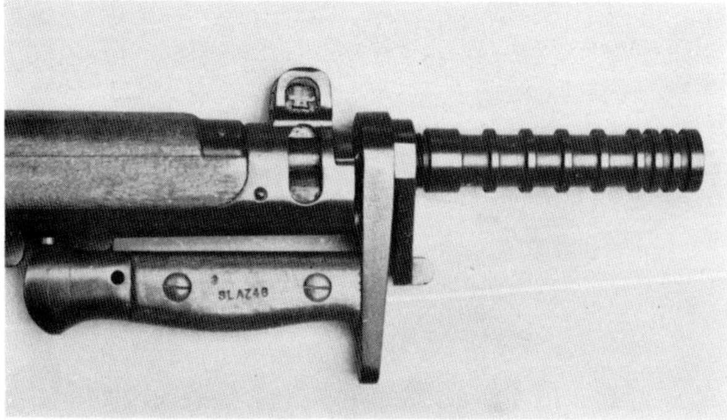

Australian Trials Spigot-type Grenade Launcher fitted onto No. 1 Mk III* Rifle. *MOD Pattern Room*

Grenades Nos. 85, 86, 87, 88, 89, 92 and 94 Energa were spigot types fired from the rifles No. 4 and No. 5, and some were later adapted for projection from the 7.62mm L1A1 rifle. A table of those grenades which were rifle-discharged has been provided because the numerical sequence of hand grenades also includes the rifle projected patterns. Some rifle grenades could be adapted for hand throwing by removal of the rod, baseplate or tail fin assembly and some of the hand thrown types could also be adapted for rifle discharge by the attachment of such fittings.

Full descriptions of the various grenade markings and colour codes for both hand and rifle projection, along with illustrations and details of the numerical series will be found in "Introduction to British Grenades".

No. 94 A.-T. Energa Rifle Grenade.

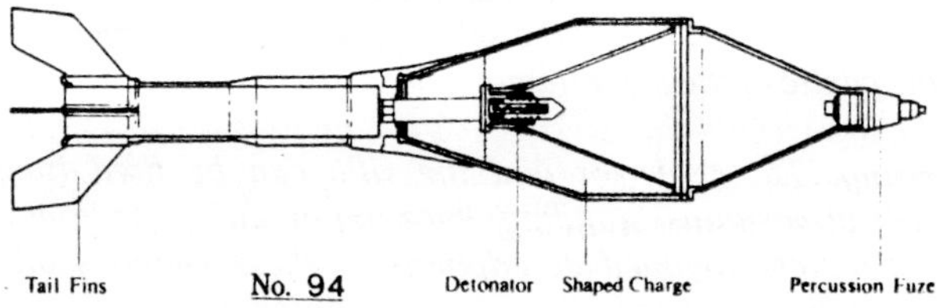

WIREBREAKERS & WIRECUTTERS

These fittings were also associated with the new type of warfare in 1914, although they were introduced shortly before as a result of the new application of barbed wire for defensive positions. The wire entanglements presented just as much an obstacle as machine gun posts and was often tactically arranged in accordance with machine-gun zones of fire. The wire cutters are more conventional types adapted for fitting onto the rifle and in most cases using the rifle as some form of leverage for operation; the wire breakers attach over the rifle muzzle so that the wire strand could be positioned accurately and severed by firing a ball cartridge.

The original military wirecutters were ordinary hand plier types, in service long before the war and mainly used as general purpose tools. Conventional wirecutters were quite regularly used for cutting wire in entanglements and leather and webbing frogs were introduced for carrying them. However, with the extended use of barbed wire, it meant carrying another item in the hand when assaulting an enemy position whilst also carrying the rifle.

British rifle-fitted wirebreakers were first approved on 9th May 1912, attaching onto the Pattern 1903 or 1907 sword bayonets fixed on the S.M.L.E. rifle. A second pattern was introduced soon afterwards which fitted directly onto the nosecap rather than the bayonet. The List of Changes announcement reads:

Para 16087:-

Breakers, Wire:
For severing wire with S.A. ammunition.

No. 1 Mk I.
For use with Patterns 1903 and 1907 sword-bayonets fixed on Rifles, Short, M.L.E.

No. 2 Mk I.
For use on Rifles, Short, M.L.E., without sword-bayonet fixed.

Patterns of the above-mentioned wire breakers have been approved to govern manufacture as may be ordered.

Instructions for fitting Wire Breakers to Sword-Bayonets & Rifles.

No. 1:- When required for use on sword-bayonets, pattern 1903, it is placed on the blade with the V-shaped wire groove over the centre of the muzzle of the barrel, and the pin end of the adjuster turned down, so that the figures "3" stamped in the left sides of the adjuster and body, respectively, are opposite each other.

When required for use on sword-bayonets, pattern 1907, the pin end of the adjuster is turned up, so that the figures "7" stamped on the right side of the adjuster and body, respectively, are opposite each other.

No. 2:- Place the spring in the neck groove of the sword-bayonet bar on the nosecap, at the same time pressing the body against the underside of the nosecap and slide home. The turned end of the spring then engages behind the sword bar, and the body of the sword-bayonet boss on the nose-cap, the V-shaped wire groove being over the centre of the muzzle of the barrel.

No. 1:- This will be carried in the pockets of the bandolier.

No. 2:- This will be carried in a leather case with strap (§ 16088) attached to the brass triangle of the bandolier.

These wirebreakers were originally intended for use by mounted troops, presumably for cutting fences or breaching wire obstacles, but their use was later extended to other forces as well, indicated by the variety of other rifles for which they were approved; Magazine Lee-Enfield, Magazine Lee-Metford, Charger Loading Lee-Enfield, Ross and Pattern 1914.

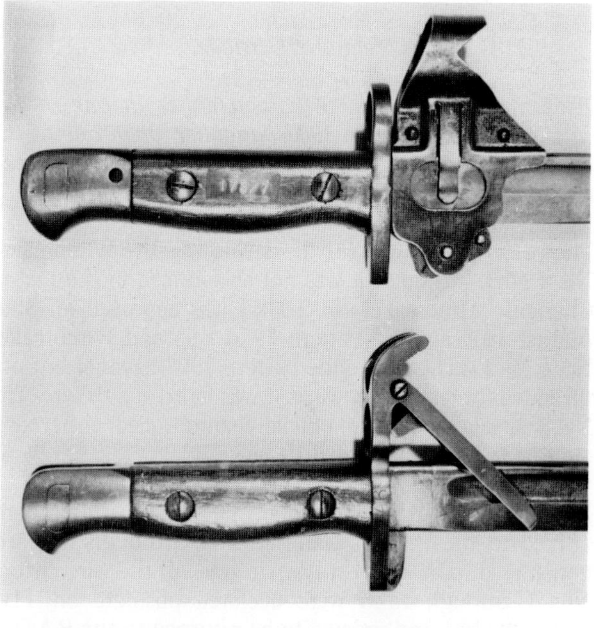

Wirebreakers fitted to Bayonets. Service pattern No. 1 Mk I (top) and folding prototype model permanently attached to the bayonet (bottom). *MOD Pattern Room*

The following No. 3 Mk I breaker was for use with the Lee-Enfield and Lee-Metford rifles, indicating an application at home as these rifles were not active service arms by this time. Approved on 16th April 1915, the List of Changes announcement also indicates their limited issue in the text of the first paragraph, "as may be ordered".

Para 17212:-

Breaker, Wire, No. 3 Mk I.
For severing wire with S.A. ammunition. For use with Pattern 1888 sword-bayonets fixed on Rifles, M.L.M., M.L.E., and charger loading, M.L.E.

Left: Prototype Wire-breaker fitted onto M.L.M. Mk I rifle.
Imperial War Museum

Below: Modified bayonets for wire cutting and breaking.
MOD Pattern Room

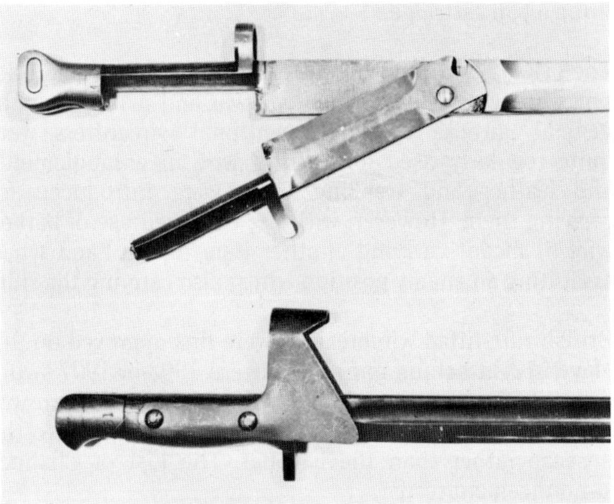

A pattern of the above-mentioned wire breaker has been approved to govern manufacture as may be ordered.

Instructions for fitting the wire breaker to sword bayonet & rifle.

Place the wire breaker on the blade of the sword bayonet with the V-shaped wire groove over the centre of the muzzle of the barrel.

Method of Carriage.

The wire breaker will be carried either in one of the cartridge pockets of the 1903 bandolier, or in one of the carriers, cartridge, 75 rounds, pattern 1908, web equipment.

The pattern for the Mk III Ross rifle was the Wirebreaker No. 4 Mk I, although it was not announced in the List of Changes. Wirebreaker No. 5 Mk I was approved on 30th August 1915 for the Pattern 1914 rifle and Wirebreaker No. 6 Mk I was fitted to the Pattern 1907 bayonet but not intended for ready removal as it was more of a fixture. The No. 6 breaker was not officially announced either and appears to have seen very little use.

Some experimental wirebreakers are also illustrated here, affixed to the Pattern 1907 bayonet, although they are considered to be of Great War vintage. All the wirecutters, Nos. 1 through to 6 were declared obsolete on 11th February 1921 and ordered to be returned to the R.A.O.C. for disposal.

Most of these breakers were made by Rudge-Whitworth Ltd., Crow Lane, Coventry or by C.H. Pugh Ltd., Whitworth Works, Tilton Rd., Birmingham. Rudge-Whitworth produced breakers Nos. 1, 2 and 3; their factory marking "R-W" is followed by the year designation (R-W 16) and stamped on their product. Pugh only produced the No. 1 breaker and their marking was usually "P" or "C.H.P.". The total production by Rudge-Whitworth and Pugh amounted to more than 150,000 breakers.

Other manufacturers of these wirebreakers according to the Contract Records were Herberts (Nottingham) Ltd. who are listed as having produced 5,000 (presumably the No. 1 model), and J. Legge & Co. Ltd., Colonial Works, Willenhall; although the Legge order was not filled by March 1917 and may well have lapsed. M. Mole & Son of Charlotte St., Birmingham, was given an order for the No. 1 and No. 3 breakers in June 1915 but as these were not forthcoming, the contract was cancelled in June 1916.

Some of the special breakers such as the Nos. 4 and 5, not produced in large quantities, have also been noted with the "EFD" Enfield production mark and may well have only been manufactured there because there is no record of the specific No. 4 or No. 5 breakers in the Contracts.

A summary of the various Wirebreakers and their appropriate rifles and bayonets and dates of official approval follows:

Wirebreaker	*Approved*	*Rifles*	*Bayonets*
No. 1 Mk I	09.05.12	SMLE	P'03 & P'07
No. 2 Mk I	08.08.12	SMLE	not needed
No. 3 Mk I	16.04.12	CLLE, MLE, MLM	P'88
No. 4 Mk I	–	Ross Mk III	Ross Mk II
No. 5 Mk I	30.08.15	Patt. 1914	P'13
No. 6 Mk I	–	SMLE	P'07

Some experimental wirebreaker fittings were attached directly onto the rifle, such as the Magazine Lee-Enfield rifle illustrated. This particular example is interesting because the slots for positioning the wire in the path of the bullet are for pushing onto the wire or pulling back, to put tension on the wire so as to locate it directly in the path of the bullet.

From left: Breakers Nos. 1, 2, 3, 4 and 5

More orthodox types of wirecutters were also fitted onto rifles for the same purpose; these were attached to the front of the rifle and operated by using the rifle as a form of lever. Three such types were announced in the List of Changes, called Wire Cutters (as opposed to Wire Breakers) and in order to avoid any confusion with the hand plier types, the abbreviation "S.A." for small arms was added to the official nomenclature. The No. 1 was approved on 11th February 1916 and the Nos. 2 and 3 on April 11th, later the same year. Cutters Nos. 2 and 3 were for attachment to the Mk III and Mk IIIB Ross rifles, which were generally only training or reserve issues.

Para 17751:-

Cutters, Wire, S.A.: No. 1 Mk I.
For attachment to Rifles, short, and charger loading M.L.E. and Magazine .303-inch, Pattern 1914.

Introduction.

Patterns of the above-mentioned wire cutters have been approved to govern supply as may be ordered.

The wire cutters are self-contained, having a clip and bolt for attachment on the right side of the rifle in the following positions:

No. 1: Rifles, short. The clip butting against, and immediately behind, the sight protecting wings on the nose-cap. Rifles, C.L. and Pattern 1914. The clip butting against, and immediately behind, the swivel lugs on the nose-caps.

Method of Using.

As the front hand-guard of the rifle touches the wire, the rifle is drawn back, and the wire entering between the jaws of the cutter rotates it, causing the rear levers of the shear blades to be opened and the jaws closed, at the same time winding up a coiled wire spring which, when the wire is severed and pressure on the cutter has ceased, expands and returns the cutter to position with jaws open.

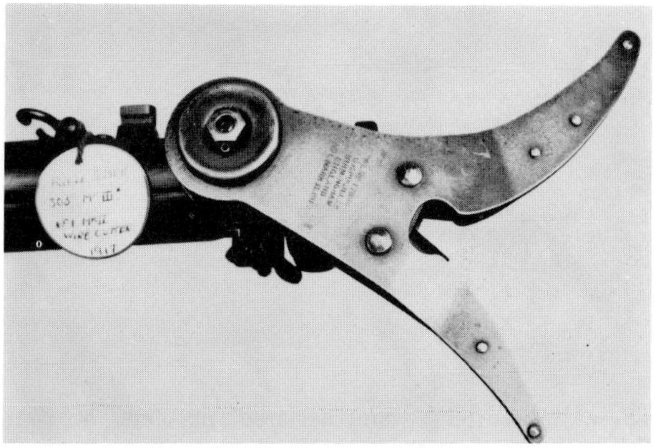

Wirecutter No. 1 Mk I fitted onto an S.M.L.E. Mk III* rifle. *MOD Pattern Room*

List of Components.

Plate, frame, right	*Not detachable.*
Plate, frame, left	*Not detachable.*
Rivets, plate, frame (4)	*Not detachable.*
Blades (2)	*Not detachable.*
Plates, blade (2)	*Not detachable.*
Links, blade (4)	*Not detachable.*
Rivets, axis, blade (2)	*Not detachable.*
Washers, blade (4) (copper)	*Not detachable.*
Bolt, axis	*Not detachable.*
Nut, bolt, axis	*Detachable.*
Pin, split, nut, bolt, axis	*Detachable.*
Studs, link, blade (2)	*Not detachable.*
Rivets, link, blade (2)	*Not detachable.*
Plate, stud, link	*Not detachable.*
Bush, bolt, axis	*Not detachable.*
Spring	*Detachable.*
Plate, clamping	*Not detachable.*
Clip	*Not detachable.*
Bolt, clip	*Detachable when burr is removed from screwed end of bolt.*
Nut, wing, bolt, clip	*As above.*
Cap, spring	*Detachable.*

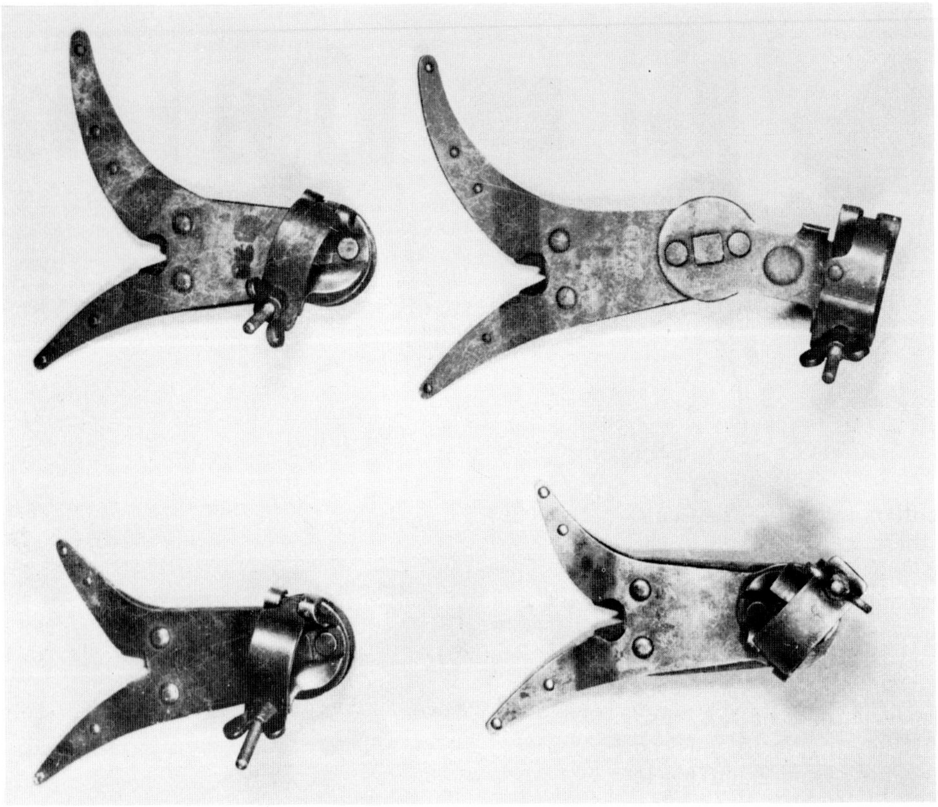

Wirecutters:-
Clockwise from top left: No. 1 Mk II, No. 1 Mk I, No. 2 and No. 4.
Imperial War Museum

These wirecutters were declared obsolete on 11th February 1921.

The next two types of wirecutters, the No. 1 Mk II and No. 4 Mk I were operated in a different manner, fitted onto the rifle so that the "horns" projected forward of the muzzle and the cutter was operated by pushing forward on the rifle instead of pulling backwards as with the previous patterns. Approved on 31st January and 13th February 1917, respectively, the No. 1 Mk II and No. 4 Mk I were announced together in the List of Changes:

Para 18516:–
Cutters, Wire, S.A.:
No. 1 Mk II.
For attachment to Rifles, Short, M.L.E.
No. 4 Mk I.
For attachment to Rifles, Magazine, .303-inch, Pattern 1914.

Introduction.

Patterns of the above-mentioned wire cutters have been approved to govern supplies as may be ordered.

They differ from those described in § 17751 in being operated by a forward push instead of a pull, when on the rifle, and in the projecting horns being longer.

They are self-contained, having a clip and bolt for attachment on the left side of the rifle in the following positions:

No. 1 Mk II. Rifles, Short, M.L.E.: The clip butting against, and immediately behind, the sight protecting wings on the nose-cap.

No. 4 Mk I. Rifles, Magazine, .303-inch, Pattern 1914.: The clip butting against, and immediately behind, the swivel lugs on the nose-cap. The cutters project diagonally upward and to the front, the long horn overlapping, and nearly touching, the sword-bayonet blade; the clip bolt being on the top, and the fly nut on the right.

Method of Using.

As the sword-bayonet blade engages the wire, the rifle being pushed forward, the wire slides along the back of the blade, enters the jaws of the cutter and rotates it, thus causing the rear levers of the blade to open and the cutting jaws to close, at the same time winding a coiled wire spring which, when the wire is severed, expands and returns the cutters to position with jaws open.

List of Components.

Plate, frame:	
Right, No. 1 Mk II	*Not detachable.* *
Left, No. 1 Mk II	*Not detachable.* *
Rivets, plate, frame (5)	*Not detachable.*
Blades (2)	*Not detachable.* +
Plates, blade (2)	*Not detachable.* +
Links, blade (4)	*Not detachable.* +
Rivets, axis, blade (2)	*Not detachable.* +
Washers, blade (4) (copper)	*Not detachable.* +

Bolts, axis	*Not detachable. +*
Nuts, bolt, axis	*Detachable. +*
Pins, split, nut, bolt, axis	*Detachable. +*
Studs, link, blade (2)	*Not detachable. +*
Rivets, link, blade (2)	*Not detachable. +*
Plate, stud, link	*Not detachable. +*
Bush, bolt, axis	*Not detachable. +*
Spring	*Detachable. +*
Plate, clamping, No. 1 Mk I	*Not detachable. ++*
Plate, clamping, No. 4 Mk I	*Not detachable. #*
Clip, No. 1 Mk I	*Detachable. +*
Clip, No. 4 Mk I	*Detachable. #*
Bolt, clip, No. 1 Mk I	*Detachable when burr is removed from the screwed end of the bolt. +*
Nut, wing, bolt, clip	*As above. +*
Cap, spring	*Detachable. +*

* *Common to No. 4 Mk I.*
\+ *Common to No. 1 Mks I & II, and Nos. 2, 3 & 4, Mk I.*
\# *Special.*
++ *Common to No. 1 Mk II.*

A listing of the various wirecutters and appropriate rifles along with the mode of cutter operation on the rifle is:

Wirecutter	Approved	Rifles	Operation
No. 1 Mk I	11.02.16	SMLE, CLLE, P'14	Backwards
No. 1 Mk II	31.01.17	SMLE	Forwards
No. 2 Mk I	11.04.16	Ross Mk III	Backwards
No. 3 Mk I	11.04.16	Ross Mk IIIB	Backwards
No. 4 Mk I	13.02.17	P'14	Forwards

The only rifle-attached wirecutters remaining in service after 1921 were the No. 1 Mk II for S.M.L.E. and No. 4 Mk I for the Pattern 1914. The other patterns were declared obsolete on 11th February 1921 and this was also announced in the monthly issues of the List of Changes.

There were two contractors who are recorded as supplying the service S.A. Wirecutters; C.H. Pugh Ltd. and Decimals Ltd. Decimals produced the No. 1 Mk I, No. 1 Mk II and No. 4 Mk I models while Pugh made the No. 1 Mk II and No. 2 Mk I. These particular cutters were marked with the maker's name, the item designation and sometimes, the year of manufacture.

Markings on the illustrated (facing page) examples are:

No. 1 Mk I:

DECIMALS LIMITED
MAKERS
SELLY OAK
ENGLAND

No. 1 Mk II:

PAT. No 17886 - 15
C.H. PUGH Ltd
BIRMINGHAM
ENGLAND
No 1 MARK II 1917

No. 2 Mk I:

C.H. PUGH Ltd
BIRMINGHAM
ENGLAND

No. 4 Mk I:

No 4 Mk I
MAKERS
DECIMALS Ltd
SELLY OAK
1917

New patterns of wire breakers to fit the No. 1 and No. 4 bayonets were fabricated in 1939 and 1940, improved models which used ball ammunition, but as there was no requirement, they were not proceeded with beyond the prototype stages. A stronger wire cutter was assembled too, but this also was shelved.

The drawings of these proposed 2nd World War wirebreakers, DD(E) 2355 and 2356, are dated 3rd November 1939 and show simplified sheet metal patterns with hinged retainers, utilising spot welding of the pressings. It is also interesting that a similar type, drawing DD(E) 2354, was proposed for the .380 Enfield No. 2 revolver. This was attached to the pistol barrel and it extended from the hinge pin, forward of the muzzle. It did not go into production either.

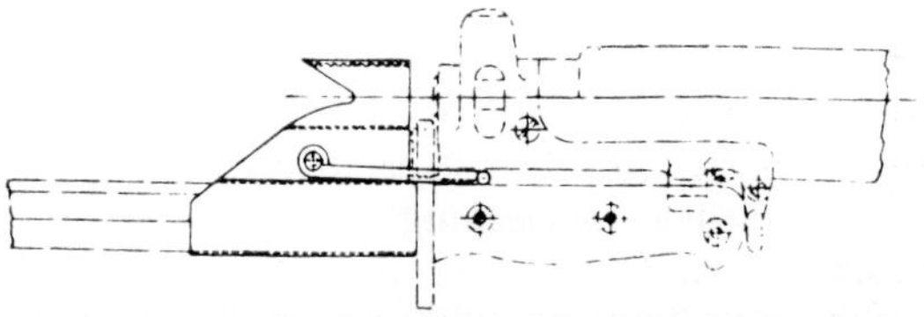

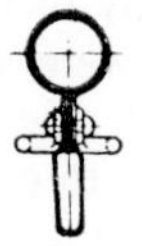

Breaker, Wire. For Rifle No. 1 (with Bayonet No. 1). D.D.(E) 2356.

PROPOSED W.W.2 WIREBREAKERS

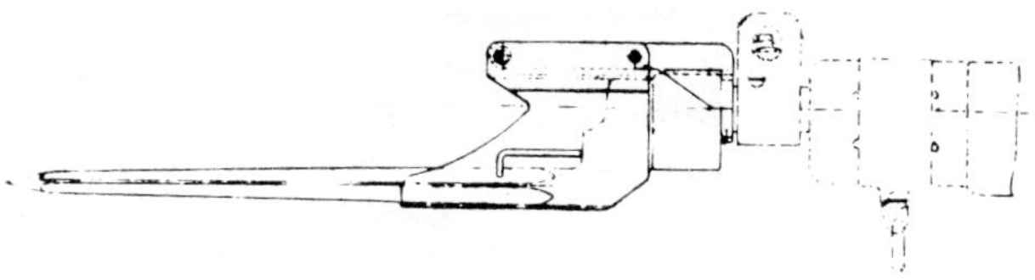

Breaker, Wire. For Rifle No. 4 (with Bayonet No. 4 Mk I). D.D.(E) 2355.

OTHER ACCESSORIES

Due to the assorted types and variations of other accessories, servicing and armourers tools, only passing mention is made here. With production, issue and use of regular service models and local patterns in Great Britain, Canada, Australia, India, Pakistan, South Africa, and virtually everywhere the Lee-Enfields have served in almost a century, any attempt to list or categorise such items would almost be futile.

Under this term we could include slings, oil bottles, rods, jags, brushes, pull-throughs, sight covers, muzzle protectors, action covers, chargers, luminous sights, optical and telescopic sights, aim correctors, aiming rests, clamps, gauges, tools, and so on, almost ad infinitum. As the appropriate oil bottles and cleaning items are included in the various LoC references, they should be examined under those particular rifle or carbine headings, or the separate entries in the List of Changes. In Chapter 5 (on the S.M.L.E. rifle), the subsequent Great War contractor lists contain some of these accessory items in the tabulation of manufacturers.

Only some of the accessories are discussed and illustrated here, largely due to the limited original source material on these particular articles. Early style muzzle protectors were generally push-on brass types with covered provision for the foresight. Some had holes in line with the muzzle so that they might be utilised as a guide for the cleaning rod and later types for the S.M.L.E. rifle are pressed steel models with spring-loaded flaps to give access to the muzzle and the bore for cleaning.

Foresight protectors are often of similar design and generally incorporate the muzzle cover. Early types tend to be of brass while later models are generally of blued steel. Some early sight protectors, though, were made from spring steel and intended to be left on the rifle at all times, only turned around or folded over so as to expose the sights for shooting the rifle.

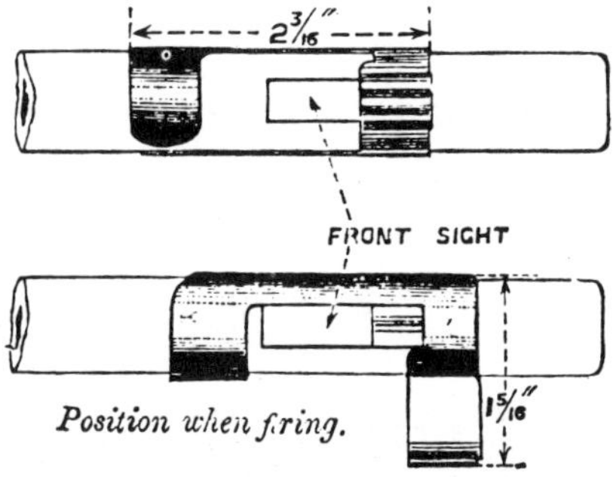

Protector, Front Sight, Magazine Rifles, Steel (Mark I). LoC § 9358, approved 19.10.1898.

A listing of Great War suppliers of oil bottles (the Mk IV pattern) designates the major contractors as:

J. & J. Bent
Thomas Bland & Sons
W.H. Briscoe & Co.
H.S. Cooke & Co.
Gabriel & Co.
S. Hall & Sons Ltd.
M. Halladay & Co.
Harcourts Ltd.
S. Heath & Sons Ltd.
Henry Jenkins & Sons Ltd.
Kings Norton Metal Co.
Lightwood & Son Ltd.
Marris's Ltd.
May & Padmore
Nicole, Nielsen & Co. Ltd.
Nobel's Explosive Co. Ltd.
Edwin Showell & Sons
Sperryn & Co.
Vickers Ltd.
Western Electric Co.

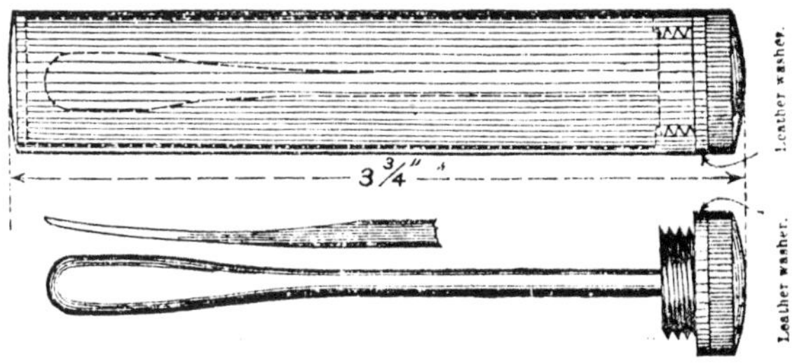

Stopper with spoon.

Oil Bottle Mk III. LoC § 10122, approved 19.12.1899.

On the metal oil bottles, R.S.A.F. Enfield, B.S.A. and L.S.A. inspection marks are often seen; such marks are illustrated in Chapter 13. Australian oil bottles are readily identified by an "A" within a star, or later "MA" and "OA" stamps. Indian markings are also typical of the Indian inspection marks in Chapter 13. Succeeding 2nd World War oil bottles were generally plastic and non-metal types, without inspection or easily-recognised makers' markings. North American oil bottles can often be distinguished by the different contractors' marks as well. See Chapter 13.

On the pull-through, makers' identifying marks will often be found on the brass weight. Oil bottle and cord pull-through were stored in the butt of most Lee-Metford, Lee-Enfield, S.M.L.E. and No. 4 rifles in service.

The khaki webbing rifle sling is 46-in. long, not to be confused with the Bren and SMG sling which is 11-in. longer at 57 inches.

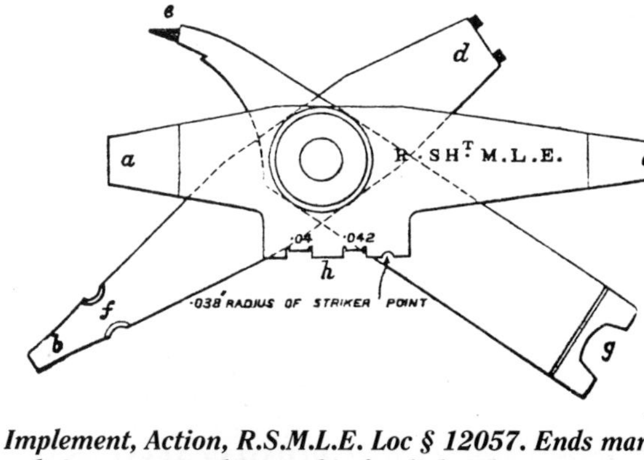

Implement, Action, R.S.M.L.E. Loc § 12057. Ends marked a, b & c are screwdrivers; d is for dial sight screw; e for extractor spring; f for expanding magazine link; g for bayonet stud nut; and h for striker point gauge.

Luminous night sights seemed to proliferate during the Great War, with the more action at dawn, dusk and under darkness. Contract records indicate that 15,000 of these items were supplied by Alex. Martin in mid-1915, some 10,000 by Addressograph Co Ltd. in 1916 and 1917, and 10,000 from Davis & Timmins Ltd., also in 1917. Orders were recorded in 1917 with Linotype & Machinery Ltd. although their delivery was not dated. Use of these sights for the S.M.L.E. rifle was later extended to South-East Asian and Pacific theatres as some Australian World War 2 examples indicate.

Above & below: Indian grenade launcher rifle sheet steel reinforcing bands. *P. Breakey*

The Aim Corrector was principally for instructional use, it enabled the coach to view the target through the rifle sights via a mirror lens which clipped over the fore-wood behind the sights. Thus, the instructor could stand at 90° the firer's position, to the left or right hand side, and observe the sight picture. These were generally supplied by contractors rather than the government factories; the Great War Mk II model was provided by T. Bland & Sons, W.W. Greener, W.H. Foster & Co., and Birmingham Small Arms Co. The Aim Teacher Mk II is also listed during the Great War period, supplied by E. Masson, Lightwood & Son, Linotype & Machinery Ltd., and G.D. Peters & Co.

11,392. Morris, R. July 4. ***1891 Patent.***

Sighting-accessories. —A spring clip A fits the barrel and forestock of the gun, and has an upright B upon which slides the frame D holding the glass plate E. The apparatus forms an "aim corrector," and enables a gun instructor to see if the marksman is aiming correctly.

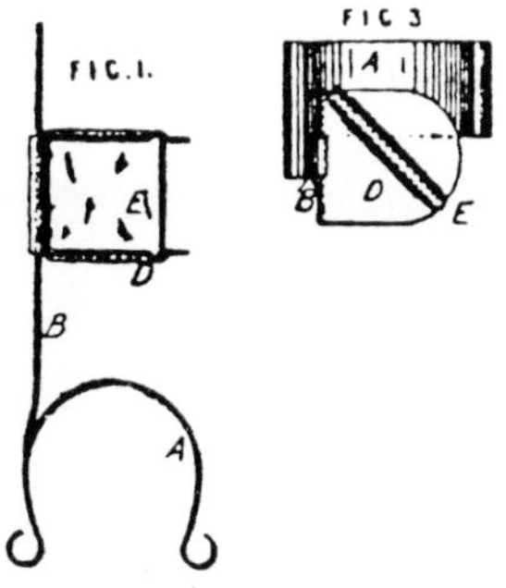

Some Indian grenade launching rifles were fitted with steel straps for reinforcement. These sheet steel bands are tightened up with screws and their finish is usually the late style, matt black, rather than the blued or parkerised type. Such straps were usually positioned around the fore-end, one forward of the outer band and one behind the band with the other behind the backsight and its protector.

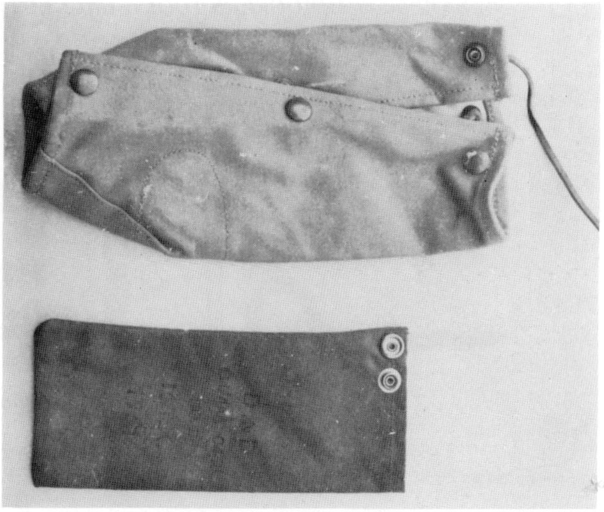

Above & below: Rifle action cover and nosecap cover.

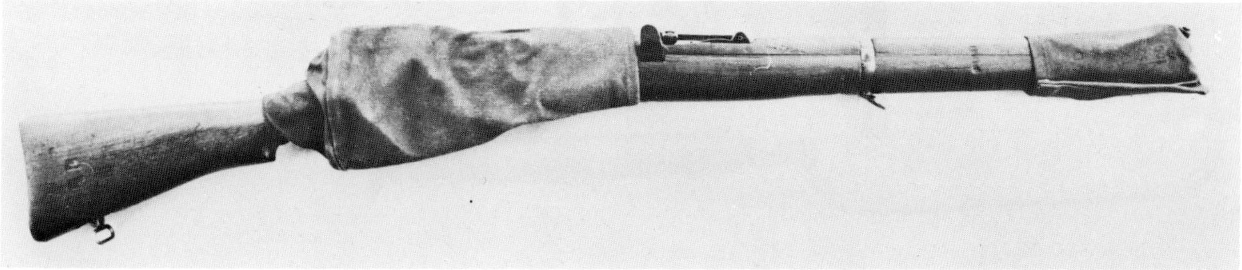

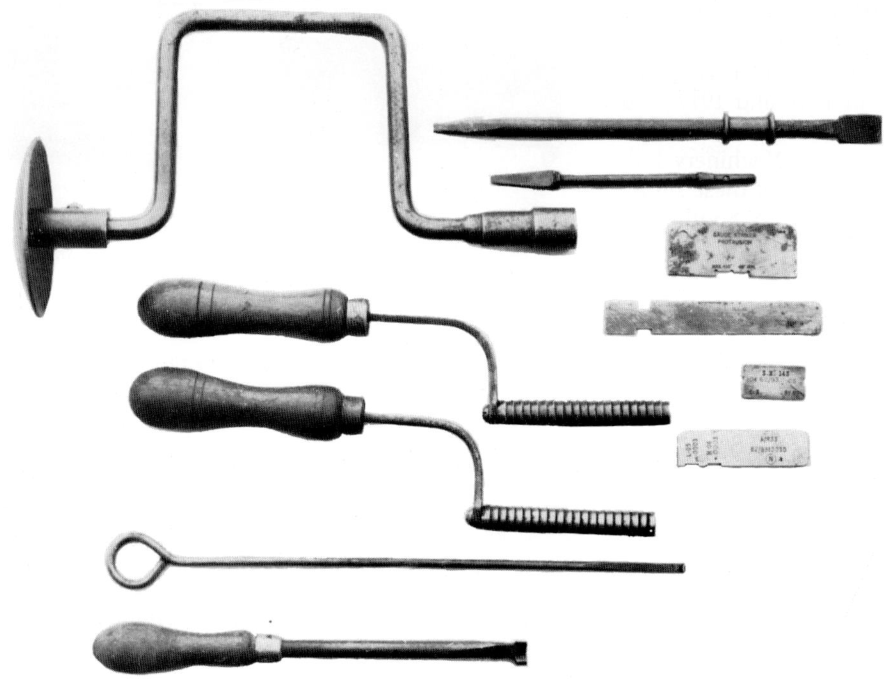

Top to bottom:
Brace and bits for stockbutt screw and cleaning chamber.
Striker protrustion and other gauges.
Fore-end routers.
Butt leather washer removing tool.
Gouge, handled, firmer.
R. Chapman

SAMPLE of ARMOURERS TOOL & GAUGES for LEE-ENFIELD.

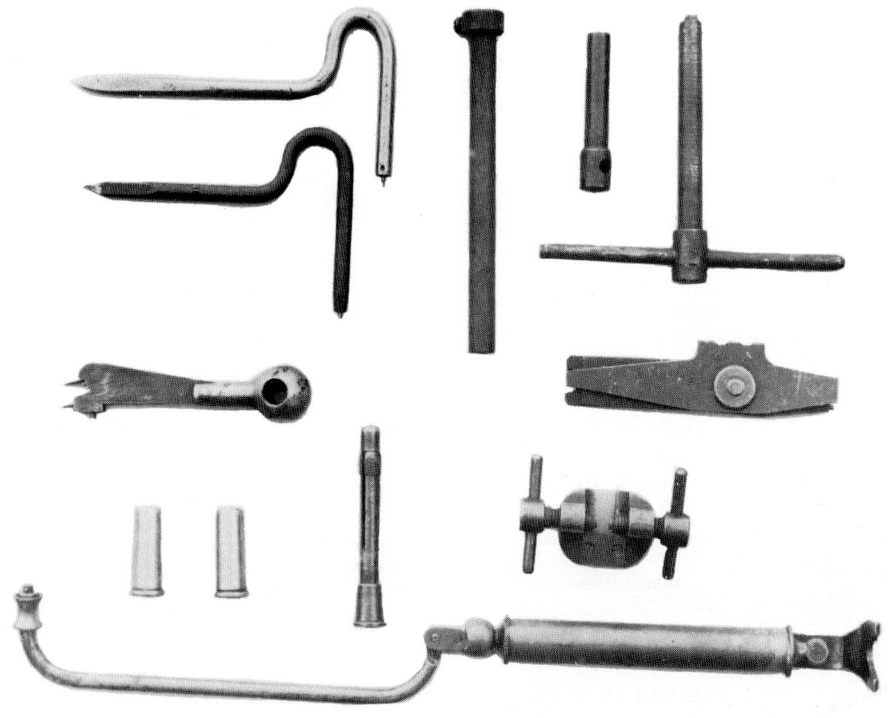

Top to bottom, from left to right:
Extractor spring tools (2 of).
Bush stop for Cleaning Rod.
Firing pin tools.
Extractor spring tool.
Combination tool, folded.
Headspace gauges.
Ruptured case extractor.
Foresight adjusting tool.
Gauge for weighing springs.
R. Chapman

RIFLE DESCRIPTIONS & ILLUSTRATIONS

CHAPTER 12

Similar to "The British Service Lee," this chapter is presented in such an arrangement to serve as an illustrated ready reference section, concentrating upon the technical details of the individual rifles and carbines rather than their background or historical relevance. So as to avoid duplication, many of the details here do not appear in previous text; to fully appreciate the origin, introduction and service implications of arms in this section, the particulars should be considered in combination with the appropriate sections of the earlier chapters. The main patterns and types of rifles and carbines are described here grouped into families; the precis at the beginning of each section provides a short background and indicates the chronological sequence. Production figures have also been provided where possible to indicate the relative scarcity of the particular model or variation.

Regarding the specifications for each rifle and carbine, some measurements may be found to vary slightly due to such factors as butt lengths and even different manufacturers. Wooden furniture is also likely to have shrunk a little in some cases, depending upon the type of original timber and country of service. The lengths given have generally been extracted from official documentation. Receiver designation markings have been provided from typical specimens, or, where these have not been available, from the pattern arms themselves.

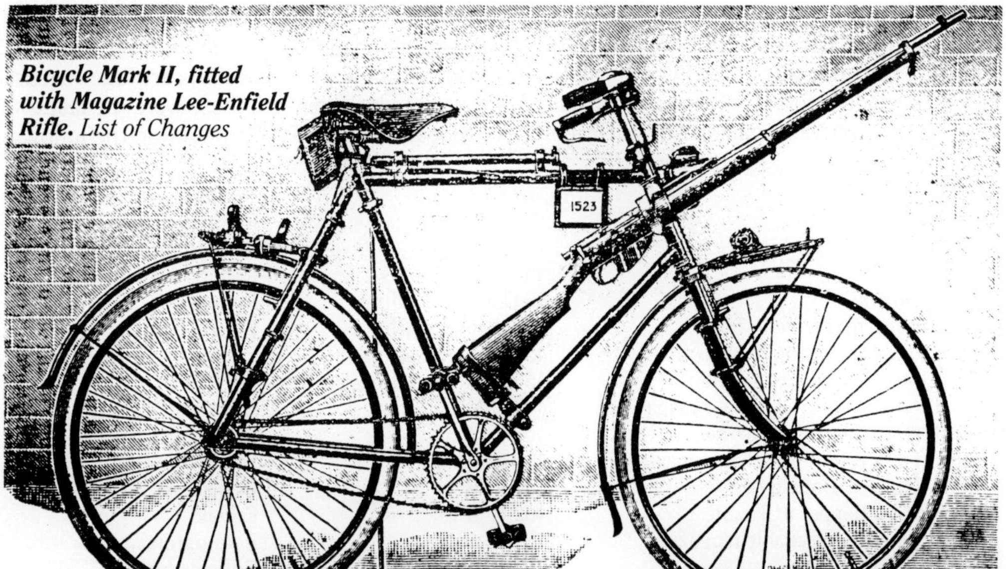

Bicycle Mark II, fitted with Magazine Lee-Enfield Rifle. *List of Changes*

EARLY TROOP TRIALS RIFLES

The question of magazine arms for British service was first raised seriously and investigated in late 1879 and this led to the initial magazine rifle trials being conducted in 1880. There were just eight contenders, two of which were the Model 1879 Lee rifle and carbine. These rifles and carbines were set up for the solid-drawn .450-inch Gatling cartridge using modified Martini-Henry pattern barrels. Three specimens each of the 1879 Lee rifle and carbine were utilised and by the end of 1880, the Lee rifle, Mauser with Lee magazine and the Green were considered to be the most successful. It was realised that the future cartridge would have a solid-drawn brass case so this was another standard applied to the trials rifles generally.

The following trials in 1883 reduced contenders to a short-list of just three systems; the Owen Jones, the Improved Lee and the Improved Lee with Bethel Burton magazine. The last-mentioned model had the sliding, side-mounted magazine applied to the Enfield produced Lee action and became known as the Lee-Burton. Again, modified Martini-Henry barrels were fitted and chambered for the .450 Gatling round, although most of the rifles in this series were manufactured at Enfield. Various improvements were incorporated in successive "Improved Lee" rifles forwarded from the United States and although some were fabricated at Enfield, the Improved Lee trials rifles were generally U.S. manufactured. The Owen Jones is an obvious adaption of the Martini with a side-mounted magazine somewhat similar to that on the Lee-Burton. The Lee-Burton and Owen Jones rifles utilised two-piece stocks, in keeping with traditional British service rifles of the period, whereas the Improved Lee had a one-piece stock.

The subsequent series of trials rifles in 1886 were chambered for the new solid-drawn .402-in. Enfield-Martini cartridge, the product of small calibre development in the quest for an improved service Martini rifle. During the ongoing trials, design improvements were incorporated in the various rifles as further samples were assembled or imported, however, the Owen Jones was eventually removed from contention and the contest limited to the Lee-Burton and Improved Lee.

A new Improved Lee rifle arrived from the Remington factory in January 1887, incorporating more recent improvements, and as 300 Lee-Burton rifles were being prepared at Enfield for trials, it was inquired whether a similar quantity of the latest pattern Lee rifle could be made at Enfield for limited troop trials against the Lee-Burton. At such short notice, this was a virtual impossibility, and so three hundred rifles were bought in the United States, chambered for the .43 Spanish cartridge. Therefore, this limited trial was conducted between the Lee-Burton in .402 calibre and the latest Improved Lee in .43 Spanish. Although the .43 Spanish cartridge was inferior in performance to the .402, this particular trial disregarded such comparison and only related the systems and their serviceability; the Improved Lee emerged as the better choice.

The testing and introduction of smaller bore rifles by some of the European powers resulted in a revision of the proposed .402-inch calibre which had been introduced with the Enfield-Martini and originally proposed as a replacement for the .450-inch Martini-Henry. Favourable reports on the small-bore Rubini in mid-1887 brought about a reconsideration; advantages of higher velocities and a flatter trajectory resulted in the Committee deciding on a .303-inch calibre, Metford rifled barrel chambered for the Swiss Rubin cartridge. This was finally decided upon in late 1887. Joseph Speed, Factory Manager at Enfield, was responsible for many improvements to these designs, and the eventual 1888 trials rifle. A nominal 350 rifles and 50 carbines were fabricated at Enfield for troop trials in 1888, but as the service .303-inch cartridge was not yet developed to a satisfactory standard, the Rubin cartridge was used. Although further changes were made before the eventual Magazine Rifle Mark I was approved, these were the last trials before the endorsement of the new service magazine rifle in December of 1888 and its official introduction the following year.

Year	*Trials Rifle*	*Cartridge*	*Manufacturer*	*Quantity*
1880	Pattern 1879 Lee Rifle	.450 Gatling	Remington/Enfield	3
1880	Pattern 1879 Lee Carbine	.450 Gatling	Remington/Enfield	3
1883	Improved Lee	.45/70	Remington	6
1883	Bethel-Burton	.45 Gardner-Gatling	Enfield	3
1883	Owen Jones, magazine in butt	.45 Gardner-Gatling	Enfield ?	?
1884	Owen Jones, magazine in butt	.45 Gardner-Gatling	Enfield	?
1885	Owen Jones, hopper magazine	.45 Gardner-Gatling	Enfield	?
1886	Owen Jones, improved hopper magazine	.402 E-M	Enfield	?
1886	Lee-Burton, hopper magazine	.402 E-M	Enfield	c.24
1886	Improved Lee, non-detachable magazine	.450 Gatling	Enfield	1
1886	Improved Lee, "Enfield Lee"	.402 E-M	Enfield	1
1887	Lee-Burton, hopper magazine	.402 E-M	Enfield	327
1887	Improved Lee, detachable magazine	.43 Spanish	Remington	300
1888	Magazine Rifle	.303 Rubin	Enfield	387
1888	Magazine Rifle, Morris Magazine	.303 Rubin	Enfield	50
1888	Magazine Carbine	.303 Rubin	Enfield	51

OWEN JONES RIFLE

Calibre:	*.402 in. Enfield-Martini.*
Rifling & Twist:	*7 groove ratchet, R.H.*
Barrel Length:	*33.2 ins. (846mm)*
Overall Length:	*51.5 ins. (1308mm)*
Weight:	*10 lb. 7 oz. (4.7 kg)*
Magazine Capacity:	*5 rounds.*
Manufactured:	*Enfield.*

Action Markings, L.H. side of the Receiver:

V. R.
ENFIELD
1886

M.O.D. Pattern Room

The Owen Jones was adapted from the Martini action and it has a tilting block mechanism with hopper magazine feed on the right side. On the last model, the magazine operates in a similar manner to that of the Lee-Burton rifle described on the following page. The breech action was manipulated by a sliding grip or handle on the underside of the butt, operating the breech block and cocking piece. Magazine feed onto the top of the block and into the chamber is via links and levers, making for a somewhat complicated mechanism.

Earlier trials Owen Jones models were chambered for the .450 Martini-Henry and .450 solid-drawn Gatling cartridges but the last rifles in this series were set up for the .402 Enfield-Martini round. Tests were undertaken by the Navy in late 1886 but the Lee bolt system was preferred, this resulted in the Owen Jones being decided against by early 1887.

Influences from the Enfield-Martini trials rifles are quite obvious in the form of furniture, fittings and sights. The folding leaf backsight is graduated to 2,000 yards and a safe and fire selector is fitted on the right side of the action. The improved magazine can be raised for magazine fire or lowered for single shot with the magazine held in reserve.

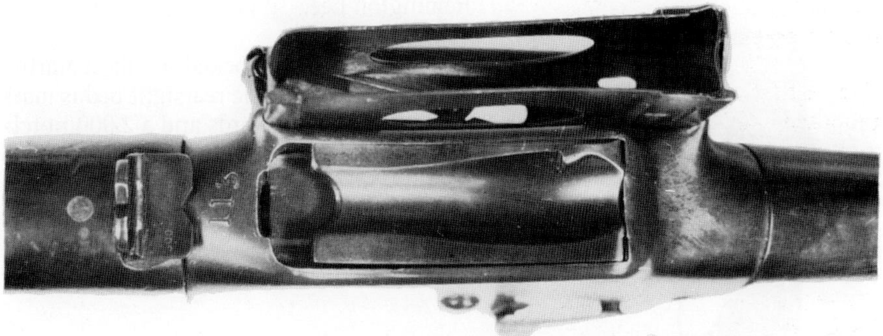

LEE-BURTON RIFLE

Calibre:	*.402 in. Enfield-Martini.*
Rifling & Twist:	*7 groove, Ratchet, R.H.*
Barrel Length:	*30.2 in. (767mm)*
Overall Length:	*50.2 in. (1275mm)*
Weight:	*10 lb. (4.5 kg)*
Magazine Capacity:	*5 rounds.*
Manufactured:	*Enfield.*

Action Markings, L.H. side of the Receiver:

ENFIELD
1886
I.
1.

The Bethel-Burton magazine was first used with the Lee action in the 1883 trials; Burton's Patent No. 4046 of October 5th 1880 was for his bolt action with hopper magazine. By 1886, only the Owen Jones and Lee-Burton systems were in contention and after the demise of the Owen Jones, the new Improved Lee became the competitor against the Lee-Burton. The Lee-Burton rifles were made at Enfield and some variations may be encountered because of the time span occupied by the trials on these rifles and the gradual incorporation of certain improvements.

In May 1886, 150 were ordered for trials but this was deferred when the Owen Jones was decided against. An order was later placed for 300 for trials against the Improved Lee in 1888. It seems that about two dozen were manufactured for the first order as well as the nominal 300 of the following batch.

The Martini style cypher, mark number and class marks were stamped on the receiver as illustrated, dated 1886; some are marked ".402". The magazine has a vertical movement of about 2 inches; in the "up" position the cartridges feed into the action while in the "down" position the rifle was used as a single loader. Although the Lee bolt system is utilised, the receiver is flat sided, unlike the Remington Lee.

The forward part is stocked-up like a Martini, some Enfield-Martini parts (marked "E-M") were used. The rearsight bed is marked at 150 and 200 yards with the leaf graduated to 1,900 yards and a 2,000 notch on the cap. Although this rifle has a bolt action, the Martini influence is quite obvious.

M.O.D. Pattern Room

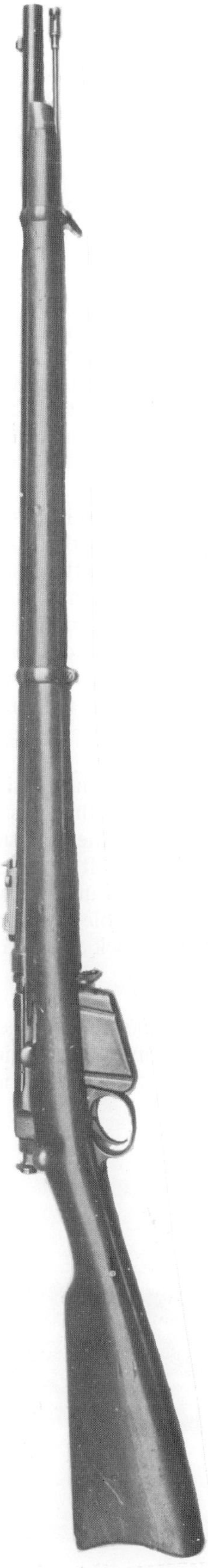

IMPROVED LEE RIFLE

Calibre:	*.43 in. Spanish.*
Rifling & Twist:	*5 groove, R.H.*
Barrel Length:	*32.1 in. (815mm)*
Overall Length:	*51.6 in. (1311mm)*
Weight:	*10 lb. 2 oz. (4.6 kg)*
Magazine Capacity:	*5 rounds.*
Manufactured:	*Remington, N.Y.*

These trials rifles were direct purchases from the Remington factory in the United States and 300 arrived in England in April 1887 for testing against a similar number of Lee-Burton rifles. As Remington had a quantity of these rifles on hand, set up for the Spanish cartridge, 120 rounds were ordered with each rifle in order that the comparative trials be conducted against the .402 calibre Lee-Burton rifles.

This particular model incorporates certain improvements over the earlier Model 1879 Lee rifles, mainly relating to the bolt, and also in that the magazine could be loaded in position on the rifle. A magazine cut-off is fitted onto the right side of the action but this was an R.S.A.F. modification to enable single loading as well as magazine fire. The rifle has two bands, stamped with the Remington "U" mark.

Due to this model being of commercial Remington configuration, the one-piece stock is fitted with a steel, American-style buttplate and fittings. The folded leaf backsight is graduated to 1,200 yards and the bed has notches for 100, 200, 300, 400 and 500 yards. The illustrated specimen is marked "W.D." on the right face of the butt and this is the only indication of its trials role apart from the magazine cut-off. The left wall of the receiver is marked with the patent date Nov. 4th 1879, and:

"E. REMINGTON & SONS, ILION N.Y. U.S.A.
SOLE MANUFACTURERS AND AGENTS"

There is a slotted jag tip at the end of the 31-inch long steel cleaning rod and this rifle only has provision for a socket type bayonet which secures around the foresight. It is not likely that the trials rifles were supplied with bayonets.

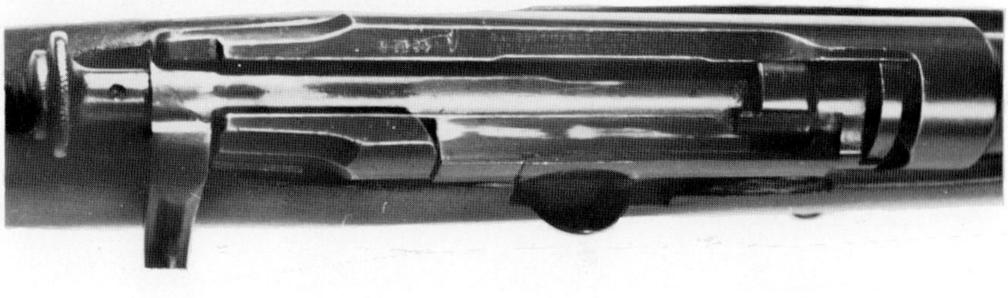

M.O.D. Pattern Room

.303-in. MAGAZINE RIFLE - 1888 TRIALS MODEL

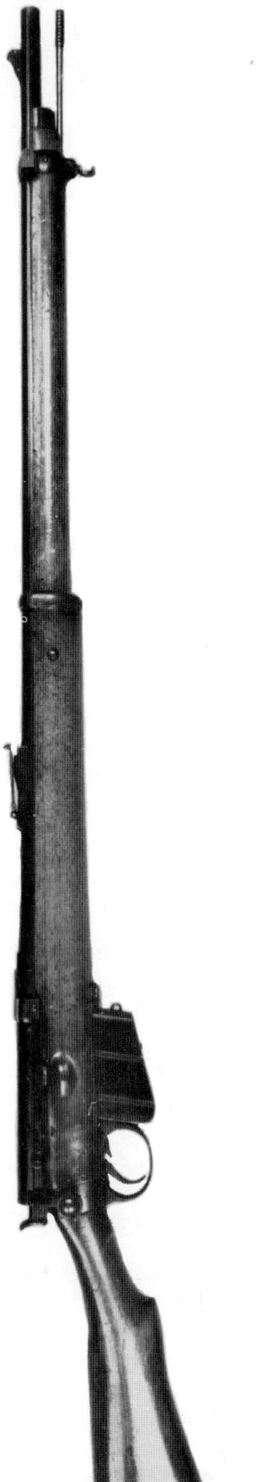

Calibre:	*.303 in. Rubin.*
Rifling & Twist:	*7 grove, Metford, L.H.*
Barrel Length:	*30.2 in. (767mm)*
Overall Length:	*49.9 in. (1267mm)*
Weight:	*9 lb. 1 oz. (4.1 kg)*
Magazine Capacity:	*7 rounds.*
Manufactured:	*Enfield.*

Receiver Markings, R.H. side of the Butt Socket:

V.R.
ENFIELD
1888

This troop trials model was developed from the Improved Lee on the previous page, largely as a result of improvements introduced by Speed and the R.S.A.F. Enfield staff. Enfield manufactured a total of 387 of these rifles but the decision to use a smaller calibre than .402 had only been recently taken and the Rubin type cartridge in .303 calibre was not properly developed by the time these rifles were made up. Some problems were caused by this particular cartridge, with its "ringed" neck.

Like other contemporary trials rifles, the stocking-up bears certain resemblance to the Enfield-Martini and the origin of some parts is proven by the "E-M" mark. The upper band, nosecap and sword bayonet fittings are from the Martini. The receiver markings on the butt socket are similar to those of the ensuing Lee-Metford rifles, but these trials rifles are dated 1888 and have no Mark number. The sighting system is similar to that of the Martini and the backsight has 200 and 300 yard marks on the bed with graduations to 1,600 yards on the right side of the leaf backsight. Long range dial sights on the left side of the fore-end are graduated to 2,800 yards.

Other distinctive features include the protrusion at the top of the cocking piece, the absence of a safety catch, spring steel ejector inletted into the left side of the receiver and secured by a screw, and the bolt release catch on the right side of the receiver, also secured by a screw. There is no handguard, although a leather protector was available. A small chain of about seven links secures the magazine to the rifle. The steel buttplate is provided with a trap to enable storage of an oil bottle in the butt and a transverse pin through the buttstock acts as a stop for the oil bottle.

The bayonet is a special trials type with pommel and grip similar to the Pattern 1887 Mk III. The 12-inch double edged bayonet was an innovation applied to the eventual Pattern 1888 sword bayonet and introduced along with the Mark I Magazine rifle.

M.O.D. Pattern Room

EARLY SERVICE RIFLES & CARBINES

This group encompasses the Magazine Lee-Metford and Magazine Lee-Enfield series of rifles and carbines which range through from the Magazine Rifle Mark I of 1888 (later re-named the Lee-Metford Magazine Rifle Mk I) to the Charger Loading Lee-Enfield which saw active service during the Great War. The Lee-Metford was the first British issue repeating rifle and soon afterwards some relatively minor changes resulted in the Mark I* variation; most of the Mark I rifles were upgraded to Mark I*. The next rifle, the Mark II, introduced a number of new features which improved the rifle considerably and was followed not long afterwards by the Mark II*. Cavalry units were issued with a special magazine carbine designed for mounted forces.

The Metford small-bore segmental rifling was specifically designed for use with black powder but after the introduction of smokeless propellant in the new ammunition it was found that this rifling deteriorated rapidly. So a new form of concentric rifling was developed at R.S.A.F. Enfield, called "Enfield" rifling, and its utilisation resulted in the first Magazine Lee-Enfield rifle; a new Magazine Lee-Enfield Cavalry Carbine was introduced as well. Both the rifle and carbine were later advanced to Mark I* upon the abolition of clearing rods in 1899. These rifles and carbines saw service during the Boer War and certain deficiencies which became quite obvious in South Africa eventually led to the introduction of a new Short Rifle soon after the turn of the century.

Two particular models of the Lee-Enfield carbine were later produced which fix the Pattern 1888 sword bayonet (as used on the Lee-Metford and Lee-Enfield rifles). The first was manufactured in relatively small numbers as a new-made arm on special order for New Zealand; the second was a conversion for the Royal Irish Constabulary. Some of these carbines were similarly converted for use by drivers in the Corps of Sappers & Miners in Indian service.

The new Short Magazine Lee-Enfield in 1903 introduced loading of the magazine by charger clips and many of the long Metford and Enfield rifles were so converted later. The first one was peculiar to India and it utilised the sliding bolthead charger guide system of the Mark I S.M.L.E. In 1907, the Mark III S.M.L.E. introduced the charger bridge across the top of the action and this was also adapted onto the Lee-Metford and Lee-Enfield rifles. The updated long rifles then became the Charger Loading Lee-Metford Mark II and Charger Loading Lee-Enfield Marks I and I* and many were later altered for high velocity Mk VII ammunition during the 2nd World War.

While the Lee-Metford and Lee-Enfield rifles were principally manufactured by the government factories at Enfield Lock and Sparkbrook, large numbers were also produced by B.S.A. and L.S.A. The service carbines were only made at Enfield, although B.S.A. produced some commercial patterns. The Indian Ishapore factory converted these long rifles to two basic patterns, the Charger Loading Lee-Enfield Mark I I.P. and Mark II I.P.; the former utilises the Mark I S.M.L.E. system of charger guide while the latter is fitted with the charger bridge of the Mark III S.M.L.E. similar to the British C.L.L.E. rifles. A special single-loading Indian model was also produced for frontier levies.

The last seven models of rifles and carbines in this list were conversions from earlier models and therefore should not be considered as new production arms.

	Enfield	*Sparkbrook*	*B.S.A.*	*L.S.A.*	*Ishapore*	*Vickers*
M.L.M. Rifle Mks I & I*:	188,930	63,072	67,500	c.40,000	-	-
M.L.M. Rifle Mk II:	84,368	47,907	73,100	c.46,700	-	-
M.L.M. Rifle Mk II*:	5,991	7,488	-	-	-	-
M.L.M. Carbine:	18,686	-	-	-	-	-
M.L.E. Rifle Mk I:	171,084	46,212	64,300	35,600	-	-
M.L.E. Rifle Mk I*:	241,547	83,294	c.192,000	c.74,000	-	-
M.L.E. Cav. Carbine Mk I:	13,926	-	-	-	-	-
M.L.E. Cav. Carbine Mk I*:	26,647	-	-	-	-	-
M.L.E. Carbine, N.Z. Model:	1,500	-	-	-	-	-
M.L.E. Carbine, R.I.C. Model:	c.11,000	-	-	-	-	-
C.L.L.E. Rifle Mk I I.P.:	-	-	-		c.22,000	-
C.L.L.E. Rifle Mk II I.P.:	-	-	-	-	c.54,000	-
C.L.L.M. Rifle Mk II:	4,454	-	}	}	-	}
C.L.L.E. Rifle Mk I:	778	-	} 120,000+	} c.40,000	-	} c.23,000
C.L.L.E. Rifle Mk I*:	144,679	-	}	}	-	}
L.E. Rifle, S.L. (I.P.):	2,000	-	-	-	-	-

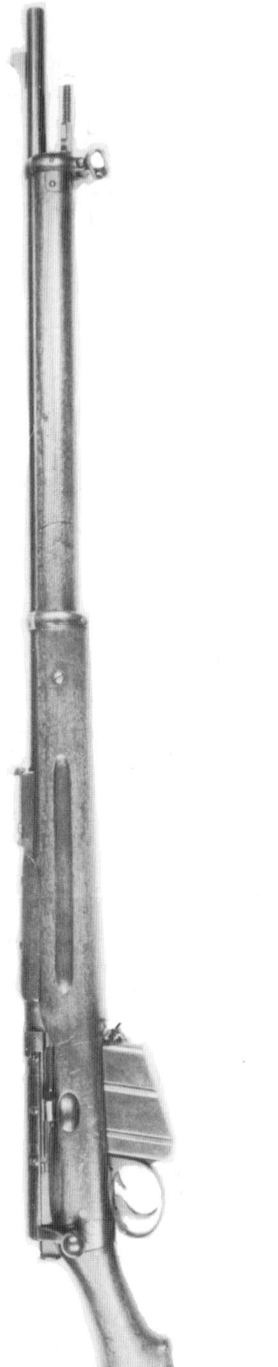

MAGAZINE LEE-METFORD RIFLE Mk I

Calibre:	*.303 in.*
Rifling & Twist:	*7 groove, Metford, L.H.*
Barrel Length:	*30.2 in. (767mm)*
Overall Length:	*49.5 in. (1257mm)*
Weight:	*9 lb. 7 oz. (4.3 kg)*
Magazine Capacity:	*8 rounds.*
Manufactured:	*Enfield, Sparkbrook, B.S.A. Co., L.S.A. Co.*

Typical Markings, R.H. side of the Butt Socket:

V.R.
ENFIELD
1889
I

Originally announced as the Magazine Rifle Mark I on 1st December 1889, this rifle was initially approved on 22nd December 1888; the nomenclature was altered to Rifle, Magazine, Lee-Metford Mark I on 8th August 1891. A number of small modifications were applied to Mark I rifle production which culminated in the introduction of the Mark I* model in January 1892. Most Mark I rifles were converted to the Mark I* pattern and factory inspection stamps are usually marked on the left side of the butt socket, representing the individual modifications.

The Royal cypher, manufacturer and year over the Mark designation are marked on the right side of the butt socket, just below the bolt handle in the closed position. Particular features of the Mark I model are the safety catch and Lewes sights. These sights have a foresight block with vertical slot cut through it while the backsight has a squared notch. The sight bed is marked at 300 and 400 yards for use with the sight leaf in the folded down position, the leaf is graduated to 1,900 yards for the raised position. Long range dial sights on the left side of the fore-end comprise a dial plate marked from 1,800 to 3,500 yards; the flip-up aperture rear sight is on the left side adjacent to the safety catch.

The detachable box magazine has a straight-line, 8 round capacity and is narrower than the later 10-round models; it is secured by a catch forward of the trigger. A magazine cut-off permits use of the rifle as a single loader and a scoop in the fore-end corresponds with the position of the tab or finger grip for the cut-off. The fore-end has finger grooves along each side, extending from the front of the receiver for about 7 ins. The nosecap/front band assembly is a two-piece arrangement which fixes the Pattern 1888 sword-bayonet as well as the double swivel set up for both sling and piling swivels. The jag head of the cleaning rod protrudes through the sword bar extension of the nosecap, the rod being 27.5 ins. long with a left hand thread.

The butt has a steel buttplate with provision for a trap, to house the brass oil bottle as well as a 7.25-in. long jag. A wooden handguard, secured by two springs, is fitted over the barrel between the breech and rear sight. Regimental markings were stamped on the buttplate or right side of the butt.

M.O.D. Pattern Room

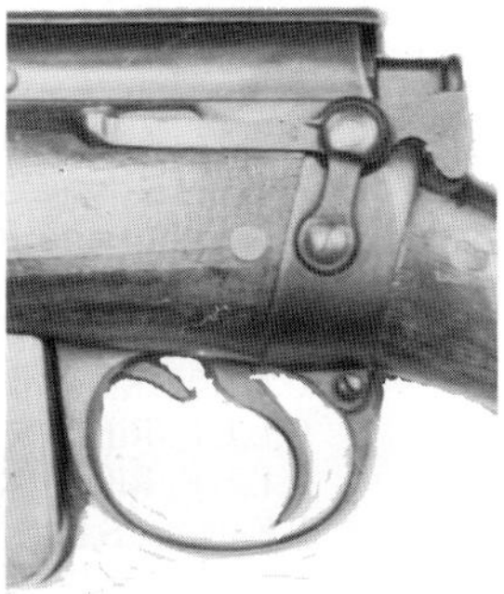

MAGAZINE LEE-METFORD RIFLE Mk I*

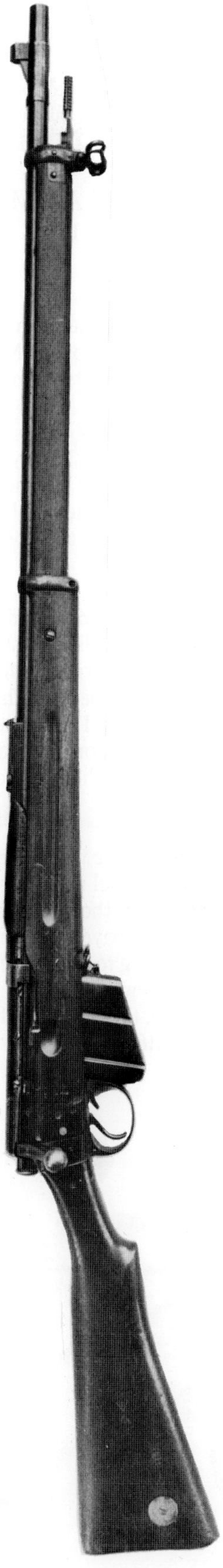

Calibre:	*.303 in.*
Rifling & Twist:	*7 groove, Metford, L.H.*
Barrel Length:	*30.2 in. (767mm)*
Overall Length:	*49.5 in. (1257mm)*
Weight:	*9 lb. 8 oz. (4.3 kg)*
Magazine Capacity:	*8 rounds.*
Manufactured:	*Enfield, Sparkbrook, B.S.A. Co., L.S.A. Co.*
Converted:	*Enfield, Sparkbrook.*

Typical Markings, R.H. side of the Butt Socket:

V. R.
SPARKBROOK
1890
1*

The Mark I* model was approved on 9th January 1892 as both a new rifle and pattern to govern alteration of Mark I rifles. The Mark II piling swivel was approved on 28th August 1890, the Mark III piling swivel on 19th December 1890 and the steel marking disc for drill arms was approved on 13th January 1892. These changes effectively increased the weight by .8 oz.

Like the Mark I model, the distinguishing features include the finger grooves in the fore-end, two-piece nosecap and front band, steel buttplate fitted with a trap, and the single row magazine. The new features include the replacement of Lewes sights by a more conventional barleycorn foresight and V-notch backsight. The sight bed is marked at 200, 300, 400 and 500 yards and the folding leaf is graduated to 1,800 yards in 100 yard increments on alternate sides of the leaf, with smaller 50-yd. lines in between. The long range dial sight plate is graduated from 1,600 to 2,900 yards compared with a maximum of 3,500 yards for the Mark I rifle.

Other obvious changes are the omission of the safety catch, rounded corners at the front of the handguard (also introduced on later Mk I models) and brass butt marking disc (or steel for D.P. arms). Less obvious modifications include clearing the body at the rear to prevent the cocking piece from catching, a leather washer over the stock bolt head inside the butt-trap, and a stronger piling swivel, firing pin spring and magazine floorplate spring.

Most of the Mark I* rifles were conversions from Mark I; this may be determined by a number of features. The pre-1892 dated receivers are obviously conversions and the asterisk or dot after the mark numeral "I" is usually not perfectly aligned. The safety catch had the operating lever, half moon cam and stop pin removed and it then was fitted back into the recess behind the long range rearsight. The converted rifle cocking piece has two semi-circular cuts, originally provision for the Mk I safety catch; the new-made Mark I* model has neither of these.

R. Richardson

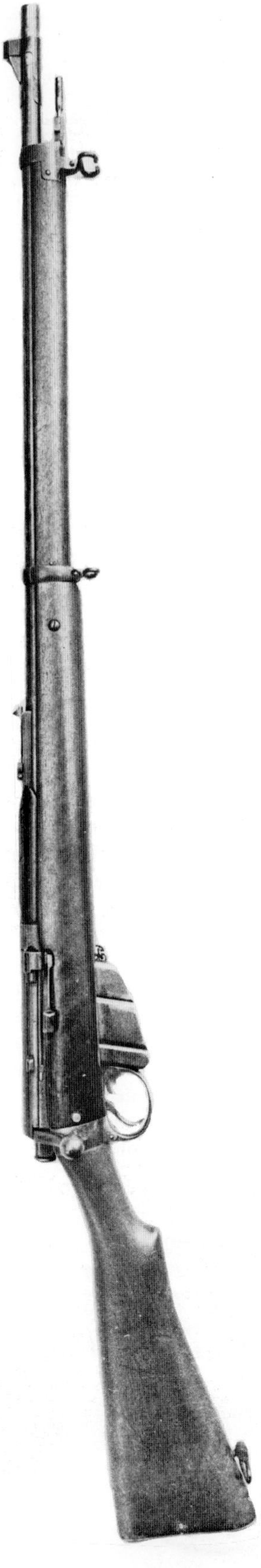

MAGAZINE LEE-METFORD RIFLE Mk II

Calibre:	*.303 in.*
Rifling & Twist:	*7 groove, Metford, L.H.*
Barrel Length:	*30.2 in. (767mm)*
Overall Length:	*49.5 in. (1257mm)*
Weight:	*9 lb. 4 oz. (4.2 kg)*
Magazine Capacity:	*10 rounds.*
Manufactured:	*Enfield, Sparkbrook, B.S.A. & M. Co., L.S.A. Co.*

Typical Markings, R.H. side of the Butt Socket:

V.R.
B.S.A.&M.Co
1894
II

Lee's improved 10-round magazine was the initial catalyst for the advance in Mark number as early as January 1890, although the final approval of the Mark II did not occur until 12th April 1893 as further improvements were also incorporated.

The obvious changes are the wider but shorter 10-round magazine, slightly wider fore-end without finger grooves, brass buttplate with long tang at the heel for regimental markings, re-designed bolt and stronger but simplified bolthead, spring steel bolt cover, simplified cut-off, and the one-piece band and fore-end cap with integral sword bar.

The rearsight is similar to that introduced with the Mark I*; the bed graduated from 200 to 500 yards and the leaf to 1,800 yards in 100-yd. intervals with shorter lines for each 50 yards. The sight slide was reported to be a little wider although inspection does not reveal this; wind gauge lines were also inscribed. On the left side of the fore-end, the long range dial sight is calibrated from 1,600 to 2,800 yards, the maximum range being 100 yards short of that marked on the Mark I* rifle. The weight reduction from the Mark I* is about 4.5 oz., which of course varies from rifle to rifle due to the varying timber density.

With the introduction of the buttplate tang, the butt marking disc was dispensed with. The sling swivel arrangement differs in that the lower swivel is mounted in the butt instead of at the front of the trigger guard. Although the nosecap and sword bar design was improved, it still fixes the Pattern 1888 bayonet. The butt trap houses an oil bottle and pull-through and the 17.2-in. long rod was intended for use in clearing cartridges, requiring that two rods be screwed together. With the change in operation of the rod from cleaning to clearing, the Mark II rifle rod is shorter than its predecessor and has a different style of fluted tip.

T. Williams

MAGAZINE LEE-METFORD CARBINE Mk I

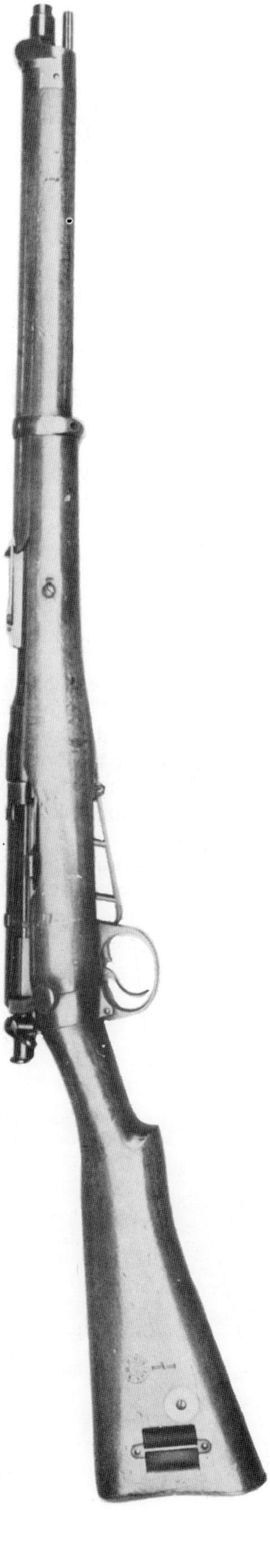

Calibre:	*.303 in.*
Rifling & Twist:	*7 groove, Metford, L.H.*
Barrel Length:	*20.75 in. (527mm)*
Overall Length:	*39.9 in. (1013mm)*
Weight:	*7 lb. 7 oz. (3.4 kg)*
Magazine Capacity:	*6 rounds.*
Manufactured:	*Enfield.*

Typical Markings, L.H. side of the Butt Socket:

V. R.
ENFIELD
1895
I

This carbine was approved on 9th June and 29th September 1894 especially for the cavalry. It is 10 inches shorter and over 2 lb. lighter than the rifle and it has a safety catch fitted on the bolt cocking piece. This safety catch was then utilised on the Mark II* Lee-Metford rifle and succeeding Lee-Enfield rifles and carbines. Virtually all the features of this carbine are special; the sling bar in the right side of the butt, lower band retained by a Mauser style spring catch inletted into the fore-end, and nosecap with integral sight protectors.

The overall shape of the carbine is smooth and the bolt handle is swept forward and down with the knob flattened. The magazine is also shortened, making for easier stowage in the saddle bucket. When originally introduced, a sling loop was fitted to the butt socket on the left side of the receiver although these "D" loops were ordered removed on 6th March 1896. Due to the position of this loop, the receiver markings were placed high on the butt socket (left side); rifle markings are stamped on the right side.

Another difference in the carbine is found in the backsight; carbine sights are graduated to 2,000 instead of 1,800 yards. The backsight slide is narrower by .025-in. and there are no long range volley sights. In 1896, a leather sight protector was approved, secured to the fore-end by two screws. There is provision for a 12.4-in. long cleaning rod and a marking disc was inletted into the right side of the butt, secured by a screw. As the butt has a marking disc, the buttplate does not require a long tang and it is identical to that of the Mark I and I* Lee-Metford rifles; it also has a trap. Due to the cavalry carbine configuration, there is no provision for the fixing of a bayonet.

M.O.D. Pattern Room

MAGAZINE LEE-METFORD RIFLE Mk II*

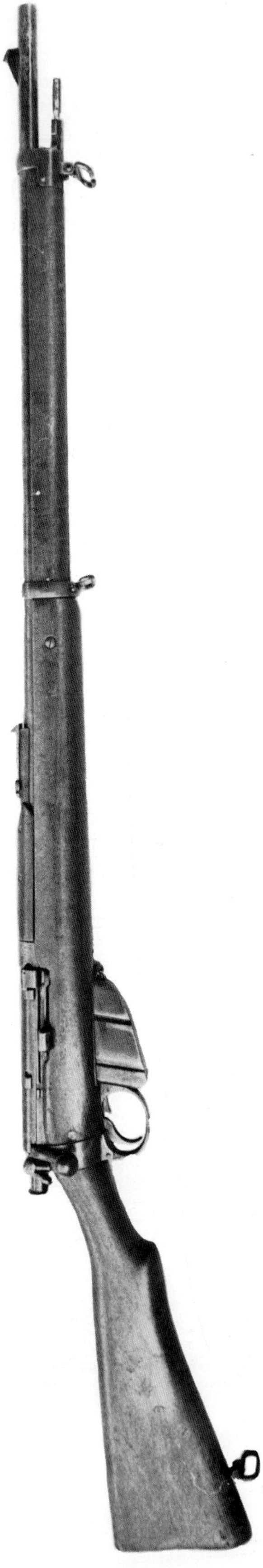

Calibre:	*.303 in.*
Rifling & Twist:	*7 groove, Metford, L.H.*
Barrel Length:	*30.2 in. (767mm)*
Overall Length:	*49.5 in. (1257mm)*
Weight:	*9 lb. 6 oz. (4.3 kg)*
Magazine Capacity:	*10 rounds.*
Manufactured:	*Enfield, Sparkbrook.*

Typical Markings, R.H. side of the Butt Socket:

The advance from Mark II to II* was due to only one change, the fitting of a safety catch to the bolt cocking piece. This was approved on 22nd April 1895 although this particular safety catch was first introduced on the cavalry carbine approved some six months earlier. All bolt components of the Mark II* rifle and cavalry carbine are interchangeable, except for the bolt body itself because the carbine has a flattened and forward-swept bolt handle.

The safety catch or locking bolt is quite different to that fitted on the Mark I rifle; there is a rearward extension of the bolt, about 1 inch long, onto which a cocking piece is fitted with the safety catch on the right hand side. Two notches in the rearward bolt extension permit the safety to be applied in either the cocked or fired position. The safety is operated by raising the catch and depressing it to disengage the safety bar. Grooves cut across the upper surface of the catch improve the thumb grip.

Designation markings are on the right side of the receiver butt socket, similar to the other Lee-Metford and Lee-Enfield rifles; the Mark number "II*" is found as illustrated. Like its predecessor the Mark II, the sight bed is marked from 200 to 500 yards for adjustment of the rearsight in the folded down position; the rearsight leaf is marked to 1,800 yards. Except for the bolt body, longer firing pin and safety catch, all parts of the Marks II and II* rifles are interchangeable.

The Mark II* had a short production life as the Lee-Enfield rifle was introduced only a few months afterwards; it was only produced at the Government factories.

T. Williams

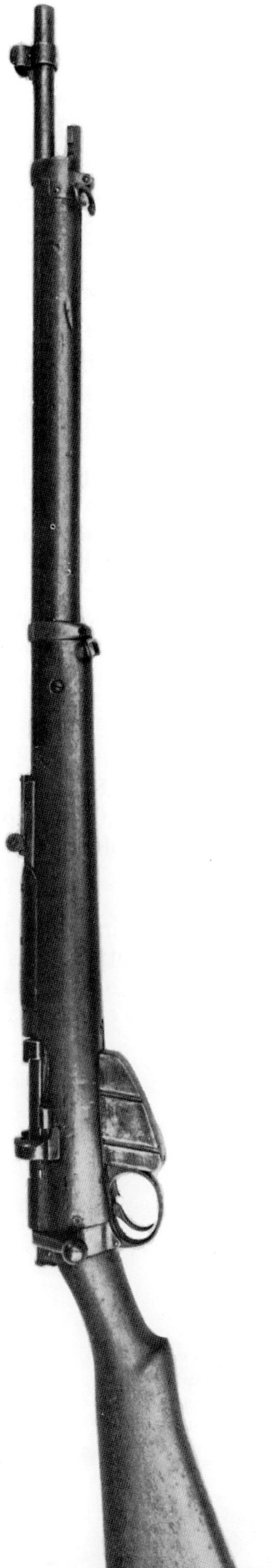

M.O.D. Pattern Room

CHARGER LOADING LEE-METFORD RIFLE Mk II

Calibre:	*.303 in.*
Rifling & Twist:	*7 groove, Metford, L.H.*
Barrel Length:	*30.2 in. (767mm)*
Overall Length:	*49.5 in. (1257mm)*
Weight:	*9 lb. 2 oz. (4.1 kg)*
Magazine Capacity:	*10 rounds.*
Converted:	*Enfield.*
	Also by the Trade ?

Typical Markings, L.H. side of the Butt Socket:

ENFIELD
1910
C.L.L.M.
II

This conversion to charger loading followed the introduction of the Mark III S.M.L.E. rifle and it also involved upgrading the rifle for use with Mark VII ammunition. Similar to the Charger Loading Lee-Enfield, a charger bridge was fitted to the receiver, the bolt dustcover and lugs removed, and a new magazine fitted which is deeper at the front end and has a lip riveted on the right hand side (requiring a relieving of the trigger guard and fore-end in the magazine way). As this conversion was made from the Mark II Lee-Metford rifle, it is not fitted with a safety catch. The bolt is similar to that of the S.M.L.E. as it was not provided with a rearward extension for safety catch. The bolt dustcover lugs were ground off in the conversion but this particular bolt may be identified because there is a groove at the back which formerly served to engage the rear lip of the dustcover.

The rearsight is quite distinctive, with an 1,800 yard leaf and no cap, only a stop screw; the slide has provision for windage and locking with a knurled screw on either side, to facilitate each of these functions. The sight bed has been modified, having been deepened at the lowest point of the ramp, near to the hinge. Markings from 200 to 500 yards on the sight bed are for the leaf in the folded down position. The long range dial sight plate is graduated to 2,800 yards. An adjustable foresight blade is fitted with protecting ears marked with an "M" to indicate Metford. (See the C.L.L.E. Mk I* description for the reason for "M" and "E" markings.)

The original rifle markings will be found on the right side of the butt socket while the conversion stamps are located on the opposite side. Apart from the charger bridge, sights and magazine, this rifle is identical to the Mark II Lee-Metford from which the conversion was made. Although it was mostly issued to reservists and some naval forces, it saw service through both World Wars.

MAGAZINE LEE-ENFIELD RIFLE Mk I

Calibre:	*.303 in.*
Rifling & Twist:	*5 groove, Enfield, L.H.*
Barrel Length:	*30.2 in. (767mm)*
Overall Length:	*49.5 in. (1257mm)*
Weight:	*9 lb. 4 oz. (4.2 kg)*
Magazine Capacity:	*10 rounds.*
Manufactured:	*Enfield, Sparkbrook, B.S.A. Co., L.S.A. Co.*

Typical Markings, R.H. side of the Butt Socket:

V. R.
ENFIELD
1896
L. E.
I

The first in a long series of Lee-Enfield rifles, this rifle is identical to its predecessor the Mark II* Lee-Metford except for a new barrel with Enfield rifling and the receiver designation markings. It was approved on 11th November 1895.

Enfield rifling has five grooves instead of the Metford's seven, and it is a little deeper at .0055-in. compared with .004-in. This rifling has a sharp 90° edge whereas the Metford style appears to be a series of smooth "hills and valleys". The foresight is set further to the left so as to compensate for the bullet drift caused by the new rifling and the barrel is marked "E" on the knox form to indicate the Enfield rifling. The Metford rifling had been found to wear rapidly with the newly introduced cordite ammunition, this was the main reason for development of the new Enfield rifling; it was found to be marginally more accurate and also to double the barrel life.

Like the Marks II and II* Lee-Metford rifles, the backsight is marked at 200, 300, 400 and 500 yards on the bed with the leaf graduated from 600 to 1,800 yds. The dial sight plate on the left side of the fore-end is calibrated from 1,600 to 2,800 yards. A detachable dustcover is fitted onto the bolt and the safety catch is mounted on the right side of the cocking piece.

The receiver markings are different to the previous Lee-Metfords in having the designation "L.E.". The butt is fitted with a swivel and the brass buttplate has an extended upper tang for regimental markings, and a trap for the oil bottle and pullthrough. Clearing rod length is 17.2 inches and it has a fluted end, the same as on the Mark II and II* Lee-Metford rifles.

M.O.D. Pattern Room

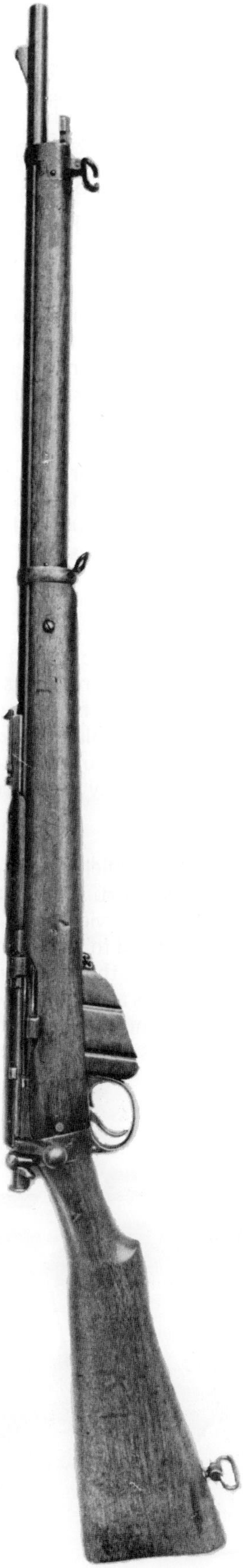

R. Richardson

MAGAZINE LEE-ENFIELD RIFLE Mk I*

Calibre:	*.303 in.*
Rifling & Twist:	*5 groove, Enfield, L.H.*
Barrel Length:	*30.2 in. (767mm)*
Overall Length:	*49.5 in. (1257mm)*
Weight:	*9 lb. 4 oz. (4.2 kg)*
Magazine Capacity:	*10 rounds.*
Manufactured:	*Enfield, Sparkbrook, B.S.A. Co., L.S.A. Co.*

Typical Markings, R.H. side of the Butt Socket:

V. R.
ENFIELD
1900
L.E.
I*

Clearing rods were abolished in British service on 19th May 1899 which meant that rifles and carbines for future manufacture were to be made without holes in fore-end and nosecap. This resulted in an advance in the designation of the Magazine Rifle to Mark I*. Clearing rods were withdrawn from arms which had been fitted with them and this is why so few of the early magazine rifles and carbines are found today complete with these rods. The official approval date for the Mark I* Lee-Enfield rifle was 7th August 1899.

The Mark I* Lee-Enfield rifle has no rod groove or nut in the fore-end, nor longitudinal hole through the nosecap sword bar. In all other respects it is identical to the Mark I rifle.

The sights are graduated at 200 to 500 yards in 100 yd. steps on the bed, and the leaf to 1,800 yards. The long range dial sights are marked from 1,600 to 2,800 yards and used in conjunction with the folding rear aperture sight. Sling swivels are attached to the lower band and butt while the piling swivel is fitted to the nosecap. Another Pattern 1888 bayonet was introduced after abolition of the clearing rods, designated the Mark II; this has no recess inside the grips for the head of the clearing rod. The oil or clearance hole of the bayonet is in the pommel instead of the grips.

In the 1903-04 financial year, R.S.A.F. Enfield produced nearly 10,000 "modified" Mark I* Lee-Enfield rifles fitted with the Mark I pattern butt, and a further 31,000 Mark I* Lee-Enfield rifles on hand were also converted by fitting this butt. This means that these "modified" rifles have a sheet steel buttplate with no trap, and a circular butt disc provided for the regimental markings.

MAGAZINE LEE-ENFIELD CAVALRY CARBINE Mk I

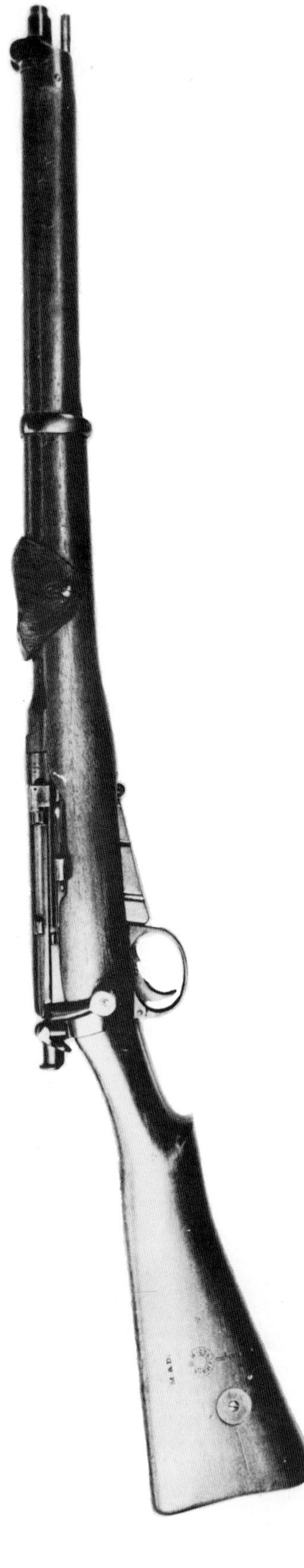

Calibre:	*.303 in.*
Rifling & Twist:	*5 groove, Enfield, L.H.*
Barrel Length:	*20.75 in. (528mm)*
Overall Length:	*39.9 in. (1013mm)*
Weight:	*7 lb. 6 oz. (3.3 kg)*
Magazine Capacity:	*6 rounds.*
Manufactured:	*Enfield.*

Typical Markings, L.H. side of the Butt Socket:

V. R.
ENFIELD
1897
L.E.C.
I

This carbine was the natural progression from the Lee-Metford carbine after the introduction of Enfield rifling and it was approved on 17th August 1896. An "E" is marked on the barrel knox form to indicate the Enfield rifling and there are also minor differences in the sights, namely in the height and position of the foresight, and the height of the backsight cap.

The rearsight leaf is still graduated to 2,000 yards although these markings are in a slightly different position, confirmed by the "EC" stamp on the bottom right hand corner of the leaf. The bed is marked at 200, 300, 400 and 500 yards and there are no long range sights as found on the Lee-Enfield rifles. The sight protecting wings on the nosecap are slightly higher than those on the Lee-Metford carbine and there is no sling bar in the butt.

A leather backsight protector was introduced and fitted to the Lee-Enfield carbine and applied to the Lee-Metford carbines in service too, secured by a screw on each side of the fore-end. Other features peculiar to the carbines are the flattened and forward-swept bolthandle, short magazine, and buttplate without a long tang. A full-length handguard is fitted from the backsight to the nosecap and a 12.4-in. long clearing rod is housed in the fore-end with only the fluted tip projecting through the nosecap.

Receiver markings include "L.E.C.", an abbreviation for Lee-Enfield Carbine, so that there can be no confusion with its Lee-Metford counterpart. Although the service carbines were only made at Enfield, some for extra-Service trade were manufactured by B.S.A. and L.S.A.; these are commonly referred to as "volunteer patterns". While some of these are similar to the service model, the Royal cypher is not included in the receiver inscription and Speed's patents are acknowledged on the appropriate components. In addition to the absence of government inspection stamps, it may also be found that certain features of the stocking up and fittings do not comply with the service patterns.

R. Richardson

MAGAZINE LEE-ENFIELD CAVALRY CARBINE Mk I*

Calibre:	*.303 in.*
Rifling & Twist:	*5 groove, Enfield, L.H.*
Barrel Length:	*20.75 in. (528mm)*
Overall Length:	*39.9 in. (1013mm)*
Weight:	*7 lb. 7 oz. (3.4 kg)*
Magazine Capacity:	*6 rounds.*
Manufactured:	*Enfield.*

Typical Markings, L.H. side of the Butt Socket:

V. R.
ENFIELD
1900
L.E.C.
I*

This Mark I* version of the Lee-Enfield cavalry carbine was brought about by the abolition of cleaning rods on 19th May 1899, in similar circumstances to the Mark I* Lee-Enfield rifle. Rifles and carbines produced after this date have no provision for rod; this involved only a minor change in the fore-end and nosecap. This new pattern was approved for production on 7th August 1899.

All other features are similar to the Mark I Lee-Enfield carbine; the backsight leaf is graduated to 2,000 yards while the bed is marked at 200, 300, 400 and 500 yards for use with the leaf in the folded down position. A leather sight protector is secured to the fore-end by two screws and a handguard is fitted between the backsight and the nosecap. The nosecap has smooth, rounded lines and also serves as a sight protector; the lower band is retained by a spring-loaded bar inletted into the fore-end. A trap is fitted in the brass buttplate for storing the oil bottle and pullthrough, while a brass marking disc is let into the right side of the butt for stamping regimental markings.

The bolt has a flattened knob and the handle is swept slightly forward, the cocking piece has a safety catch. A link loop secures the short magazine to the trigger guard and the magazine cut-off is similar to those of previous Lee-Enfield rifles and carbines. An "E" stamped on the barrel knox form denotes the Enfield rifling.

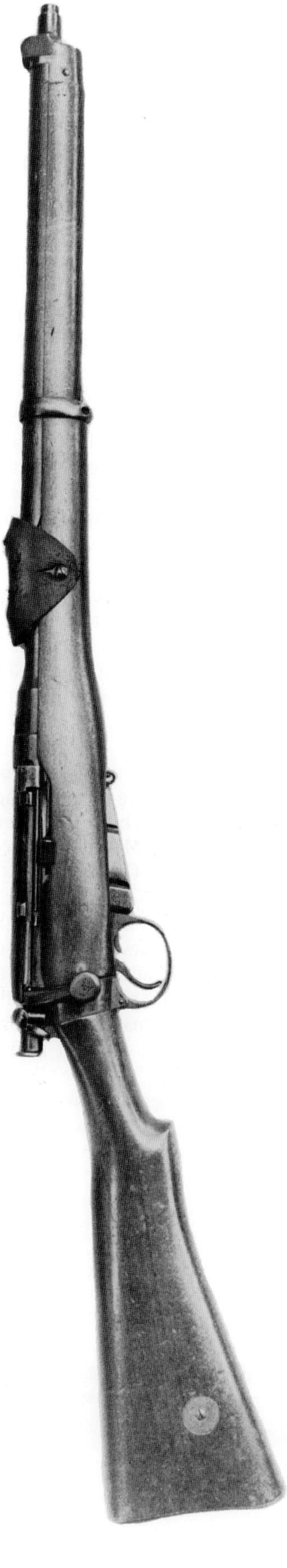

R. Richardson

MAGAZINE LEE-ENFIELD CARBINE; FITTED for PATT. 1888 BAYONET

Calibre:	*.303 in.*
Rifling & Twist:	*5 groove, Enfield, L.H.*
Barrel Length:	*21.0 in. (533mm)*
Overall Length:	*40.25 in. (1021mm)*
Weight:	*7 lb. 8 oz. (3.4 kg)*
Magazine Capacity:	*6 rounds.*
Manufactured:	*Enfield.*

Typical Markings, L.H. side of the Butt Socket:

V. R.
ENFIELD
1901
L.E.C.
I

Approved on 30th May 1900 as "Carbine, Magazine, Lee-Enfield, fitted to take 1888 pattern sword bayonet", this model is often referred to today as the New Zealand pattern, so as to avoid confusion.

This is different to the cavalry carbine in a number of ways; there is no lower band, the multi-function nosecap is that of the Martini-Enfield Artillery Carbine Mk III and it has no provision for clearing rod. The barrel has a larger diameter than other Lee carbines as it is a Martini-Enfield Artillery Carbine barrel with modified extractor slot. A barleycorn foresight is used and the handguard extends from the backsight forward to the nosecap, secured by two spring clips.

The backsight bed is marked at 200, 300, 400 and 500 yards while the folding leaf rearsight is graduated to 2,000 yards. No leather sight cover is fitted but the foresight has protecting wings as on the Martini-Enfield Artillery Carbine. In October 1902, an altered backsight was approved with lower sight cap and (supposedly) narrower slide; this was marked "E.C./88" at the lower right corner of the leaf.

Features of the Lee-Enfield cavalry carbine which are retained include the butt and buttplate, flattened bolt knob, short magazine, and absence of long range volley sights. Sling swivels are mounted on the upper band and butt. Its Pattern 1888 sword bayonet was the same as that issued with the Lee-Metford and Lee-Enfield Rifles and Martini-Enfield and Martini-Metford Artillery Carbines.

An order for 1,000 of these carbines was placed by New Zealand in 1901-02, manufactured at Enfield at a cost of £2/4/11d each. These New Zealand issue carbines are marked "N Z" on the right side of the butt socket and dated between 1901 and 1903. An additional 500 carbines were made at Enfield in 1903 so the total production of this model only amounted to 1,500. From markings on some of these receivers, it appears that they were drawn from spare stores at Enfield and assembled especially for the New Zealand contract.

N ↑ Z
1057
1903

M.O.D. Pattern Room

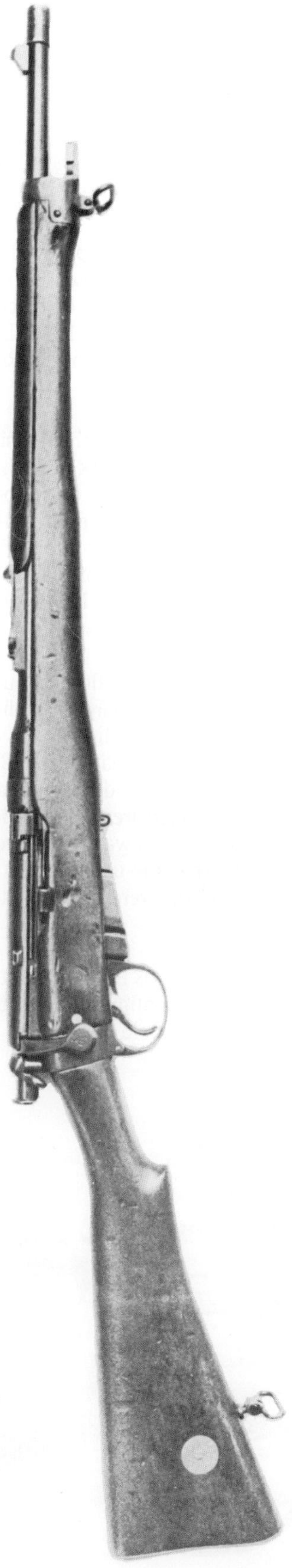

C. Woodhouse

MAGAZINE LEE-ENFIELD CARBINE; R.I.C. PATTERN

Calibre:	*.303 in.*
Rifling & Twist:	*5 groove, Enfield, L.H.*
Barrel Length:	*20.75 in. (527mm)*
Overall Length:	*40.0 in. (1021mm)*
Weight:	*7 lb. 5 oz. (3.3 kg)*
Magazine Capacity:	*6 rounds.*
Converted:	*Enfield & Birmingham.*

Typical Markings, L.H. side of the Butt Socket:

V. R.
ENFIELD
1900
L.E.C.
I*

A special conversion for the Royal Irish Constabulary, it was made from Lee-Enfield and Lee-Metford cavalry carbines so as to fix the Pattern 1888 sword-bayonet. This model was not announced in the List of Changes because it was not for military service, so no official approval date was recorded.

Because this is a conversion, the overall dimensions are similar to the carbines from which the alterations were made, the receiver markings on the left side are also those of the original models. The action is the same with a flattened bolt handle, but here the similarity ends. The lower band was removed and a wooden patch used to fill the space previously occupied by the band retaining catch. The cavalry nosecap was dispensed with and a new piece of wood grafted onto the fore-end which increases in size sufficiently enough to fit the Lee-Enfield type rifle nosecap assembly in order to fix the bayonet. A collar, about 1-in. long, has been fixed around the muzzle end of the barrel for the Pattern 1888 bayonet muzzle ring, as the carbine muzzle has a smaller diameter.

The backsight is identical to that of the cavalry model, with bed marked at 200, 300, 400 and 500 yards for use with the leaf folded down; the leaf is graduated to 2,000 yards. There is no leather backsight protector, the screw holes having been filled. The foresight has a short ramp and fixed barleycorn, being that of the original carbine (without sight protector wings).

A 1903 order for 10,000 carbines was completed by 1904; the first 2,000 conversions were costed at just over 2/- each, the remaining 8,000 were M.L.E. Mk I* Cavalry models and costed at ½d more. In 1905, 500 M.L.M. carbines were also converted to this pattern and in 1907 a further 450 carbines were altered. The last batch of 250 was converted in 1913-14 and quantities of spare receivers and barrels were supplied as well. While the carbines are recorded as having been converted at Enfield, most have the Birmingham roundel on the butt which indicates that part of the conversion was effected there. Some of these carbines have also been noticed with the forward portion of the handguard shortened by about 1 inch, most likely effected because of fractures in the front handguard. Many of these carbines converted from Lee-Metford models retain the Lee-Metford butt which has a sling bar.

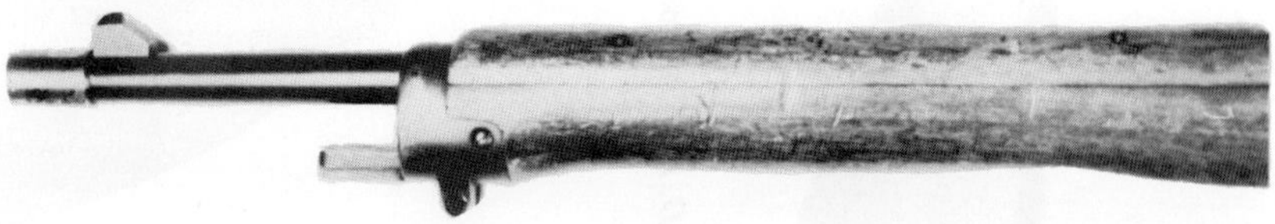

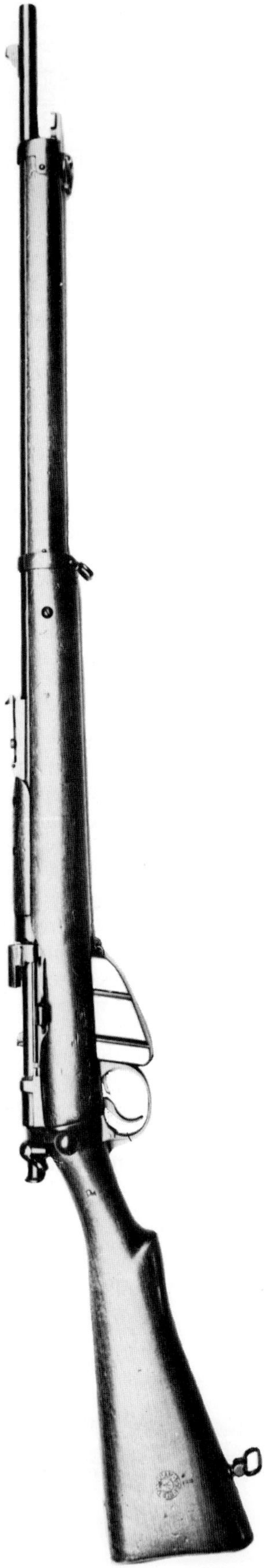

CHARGER LOADING LEE-ENFIELD RIFLE Mk I I.P.

Calibre:	*.303 in.*
Rifling & Twist:	*5 groove, Enfield, L.H.*
Barrel Length:	*30.2 in. (767mm)*
Overall Length:	*49.5 in. (1257mm)*
Weight:	*9 lb. 1 oz. (4.1 kg)*
Magazine Capacity:	*10 rounds.*
Converted:	*Ishapore.*

The first of the Charger Loading Lee-Enfield conversions, this was approved for Indian service on 28th February 1905 as the "Charger Loading Lee-Enfield Mark I, Mark I* and Mark I* I.P." but to avoid confusion, it was later instructed that all Charger Loading Lee-Enfield rifles which utilised the Mark I S.M.L.E. system of charger loading be designated the "Charger Loading Lee-Enfield Rifle Mk I I.P."

Early conversions consisted mainly of attaching the charger guide to the left side of the receiver and stop nib on the right side, and fitting of the Mark I S.M.L.E. bolthead with sliding charger guide. This also entailed the removal of the bolt dustcover and corresponding dustcover lugs on the bolt, and the fitting of a new magazine suitable for the Mark VI ammunition. A number of minor alterations were made to the bolthead, extractor, trigger guard, magazine cut-off and fore-end and most of these rifles were fitted with the S.M.L.E. Mark I* I.P. pattern butt.

Later, adjustable foresights were approved for the charger loading rifles although the foresight protector wings as fitted to the British Mark I* Charger Loading Lee-Enfield were not attached to this rifle. The backsight remains the same as on the rifle from which the conversion was made, with the bed graduated from 200 to 500 yards and the folding leaf to 1,800 yards. The long range dial sights are graduated from 1,600 to 2,800 yards and the long range aperture is on the left side of the body. Because the rifles from which the conversions were made were of British manufacture, the right side of the butt socket has the usual Mark I or I* Lee-Enfield markings; the Ishapore conversion markings were made on the left side.

There was also a Mark II I.P. Charger Loading Lee-Enfield but as this utilised the charger bridge of the Mark III S.M.L.E., it is identical in configuration to the British C.L.L.E. Mark I except for the standard pattern Lee-Enfield backsight.

M.O.D. Pattern Room

M.O.D. Pattern Room

CHARGER LOADING LEE-ENFIELD RIFLE Mk I

Calibre:	*.303 in.*
Rifling & Twist:	*5 groove, Enfield, L.H.*
Barrel Length:	*30.2 in. (767mm)*
Overall Length:	*49.5 in. (1257mm)*
Weight:	*9 lb. 2 oz. (4.1 kg)*
Magazine Capacity:	*10 rounds.*
Converted:	*Enfield, B.S.A. Co., L.S.A. Co., Vickers.*

Typical Markings, L.H. Side of the Butt Socket:

B.S.A. C°
1910
C.L.L.E.
I

No official approval date for this rifle appears to have been recorded and this pattern is not as common as the Mark I* version. The introduction of the Mark III S.M.L.E. resulted in upgrading of the earlier rifles by fitting a charger bridge and modifying the bolt to suit. The bolt alterations comprise removal of the dustcover lugs and dustcover, this is evident by the groove at the rear of the bolt body, previously used to retain the dustcover lip.

The introduction of the Mark VII .303 cartridge resulted in the first official notification of the Mark I C.L.L.E., on 1st September 1914, when a new magazine was fitted and the sights altered. The new backsight leaf is graduated to 1,900 yards and the rear of the backsight bed ramp lowered to allow for the different trajectory of the new round. The dial sight plate was altered so that graduations are from 1,700 to 2,700 yards and it is marked "C.L.". On this conversion, the barrel was stamped "H.V." just forward of the rearsight bed, indicating the new cartridge modifications.

In other respects, it is similar to the rifle from which it was altered, the original markings being on the right side of the receiver butt socket with the conversion markings on the opposite side. Most of these rifles were used in Naval Service. These rifles also had a groove cut through the top of the charger loading bridge to allow for the lower sighting line.

CHARGER LOADING LEE-ENFIELD RIFLE Mk II I.P.

Specifications:	*as for C.L.L.E. Mk I.*
Converted:	*Ishapore.*

Externally, this India Pattern charger loader resembles the British Mk I pattern, but it was introduced much earlier, approved on 8th September 1909. Conversions were made from British Lee-Enfield Mk I and I* rifles as well as the Indian Mark I I.P. Charger Loader, so there are variations in some of the smaller details. Many were fitted with the Mark I or Mark I* S.M.L.E. butts and buttplates.

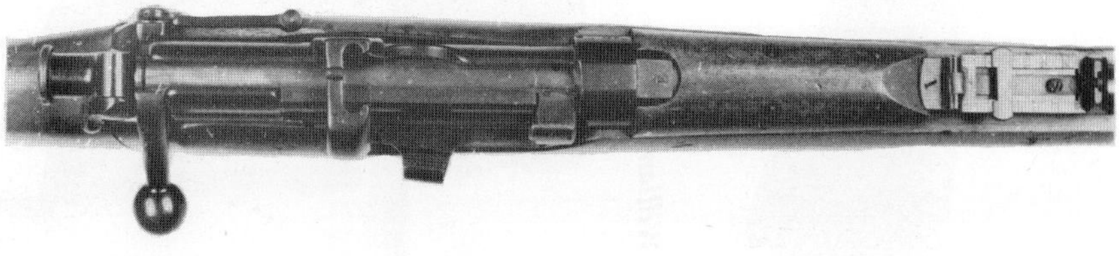

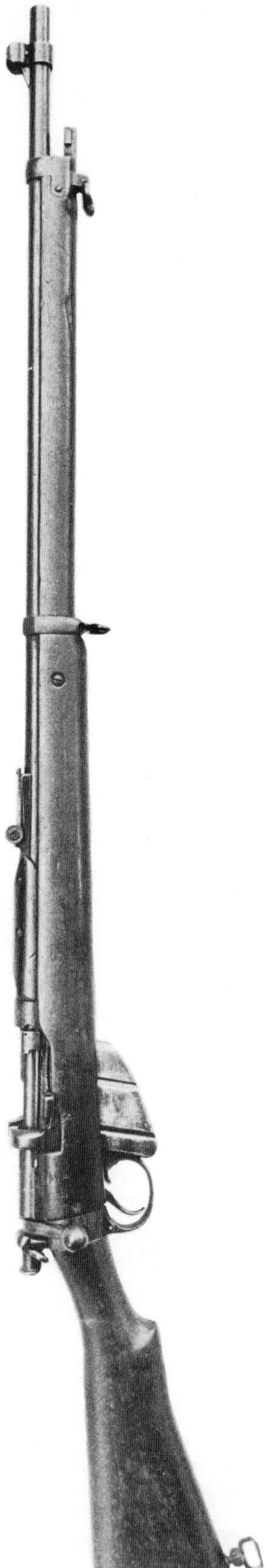

CHARGER LOADING LEE-ENFIELD RIFLE Mk I*

Calibre:	*.303 in.*
Rifling & Twist:	*5 groove, Enfield, L.H.*
Barrel Length:	*30.2 in. (767mm)*
Overall Length:	*49.5 in. (1257mm)*
Weight:	*9 lb. 4 oz. (4.2 kg)*
Magazine Capacity:	*10 rounds.*
Converted:	*Enfield, B.S.A. Co., L.S.A. Co., Vickers.*

Typical Markings, L.H. Side of the Butt Socket:

B.S.A. Co
1910
C.L.L.E.
I*

Approved on 1st July 1908 as a result of the new charger loading system of the S.M.L.E. Mark III rifle, this was a conversion from the Lee-Enfield rifles Marks I and I* and Lee-Metford Mark II* fitted with Enfield barrel. Further modifications required for the Mark VII .303 cartridge were approved on 2nd October 1914.

While the overall dimensions are similar to those of the Lee-Enfield, there are many differences to the sighting, action body, bolt and feed system. The charger bridge of the Mk III S.M.L.E. is fitted, the dustcover and its lugs removed from the bolt, a new magazine fitted for charger loading, and the action body, cut-off, stock fore-end and trigger guard altered to accommodate this. The sights were changed in that an adjustable foresight with detachable protectors and a new leaf with windage adjustment were fitted; the backsight bed is marked to 500 and the leaf to 1,800 yards. This leaf has no cap at the top, but a screw positioned at the top right corner at maximum elevation. Foresight protectors for the C.L.L.M. were marked with an "M" while those of the Lee-Enfields have an "E", because of the offset foresight required for the Lee-Enfield.

After the introduction of the Mark VII ammunition, some changes were effected to the sighting of these rifles; the backsight bed was reduced in height, "V" notch slide fitted on a new 1,900 yard leaf marked "C.L." at the bottom right hand corner and a regraduated dial sight plate from 1,700 to 2,700 yards, also stamped "C.L.".

The stocking up is identical to the rifle from which the conversion was made although no differentiation was made between the Mark I or I* Lee-Enfield fore-ends when a replacement was effected. Some were fitted with Metford rifled barrels but still referred to as the C.L.L.E. Mk I*.

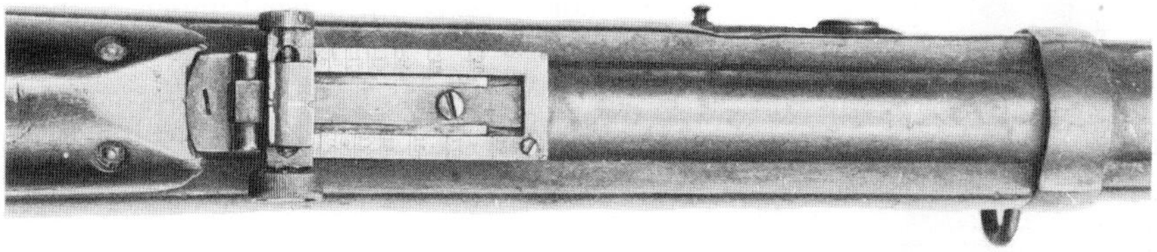

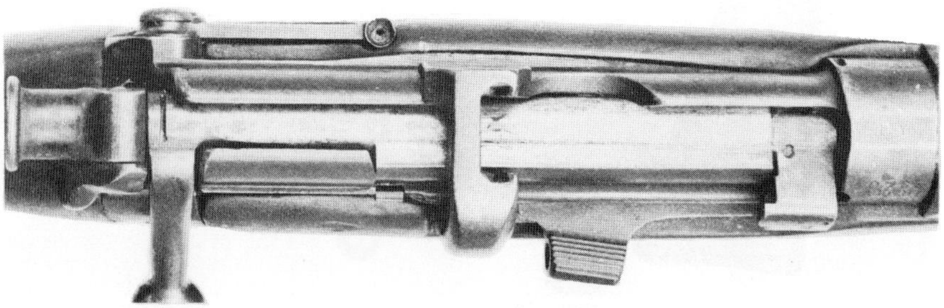

T. Williams

MAGAZINE LEE-ENFIELD, SHORTENED RIFLE CLUB PATTERN (AUST)

Calibre:	*.303 in.*
Rifling & Twist:	*5 groove, Enfield, L.H.*
Barrel Length:	*25.2 in. (640mm)*
Overall Length:	*44.6 in. (1130mm)*
Weight:	*8 lb. 6 oz. (3.8 kg)*
Magazine Capacity:	*10 rounds.*
Converted:	*Lithgow, Various Gunsmiths, Rifle Club Armourers.*

Typical Markings, R.H. side of the Butt Socket:

V. R.
ENFIELD
1896
L. E.
I

With the introduction of the S.M.L.E. into Australian Rifle Club competition, many British M.L.E. and possibly some M.L.M. rifles were reduced in length to comply with the new Rifle Association regulations on barrel length. This merely involved shortening of the fore-end and the fitting of a 25.2-in. barrel, the same length as the S.M.L.E.

Three techniques were applied to modify the barrel length, essentially the introduction of the Lithgow "heavy" barrel:

(1) The original 30.2-in. barrel was cut to S.M.L.E. length, re-crowned, and the original foresight band re-fitted after the muzzle was turned down; an adjustable blade was fitted by cutting a dovetail into the foresight block.

(2) A new 25.2-in. long barrel made at Lithgow, of M.L.E. profile, was fitted. This is marked "LITHGOW" on the knox form and an adjustable foresight was fitted.

(3) A new S.M.L.E. barrel was fitted, the muzzle part of the original M.L.E. barrel was then sleeved over the turned- down S.M.L.E. muzzle. This retained the original M.L.E. foresight assembly but an adjustable foresight was generally fitted after a new dovetail slot had been cut into the M.L.E. foresight block.

The original M.L.E. backsights were retained, with the leaf graduated to 1,800 yards. Long range dial sights were usually retained. Rifle Club type sling swivels were usually fitted and for club shooting, a sideplate and detachable, adjustable aperture sight was generally fitted onto the left side of the action body.

The fore-end was reduced in length, by about ½-inch, and the original M.L.E. nosecap re-fitted. Many of these shortened rifles were recalled or relinquished for military service during World War 2 and such rifles will have service marks for this period. The appropriate bayonet remains the Patt. 1888, as for the M.L.E. and M.L.M. rifles.

B. Labudda

A SHORT RIFLE - THE NO. 1

Often abbreviated to S.M.L.E., and colloquially called the "smelly" due to this abbreviation, the Short, Magazine Lee-Enfield was so named because it was a shortened Magazine Lee-Enfield with improved sighting and charger loading provision. This was largely the outcome of experiences in the South African war against the Boers, where the Lee-Metford and Lee-Enfield were, for the first time, pitted against proficient European type forces armed with modern firearms. The Mark I S.M.L.E. rifle was first approved in 1902 but underwent some minor changes which caused its re-introduction in 1903. As well as the newly-produced model, some Lee-Metford and Lee-Enfield rifles were converted to the new pattern and designated the Converted Mark II.

A few years afterwards, further new and updated patterns in the Mark I series, with relatively minor changes, were approved. The Mark I* and its counterpart the Converted Mark II*, with secondary refinements and subsequent modifications, resulted in the Mark I**, I*** and Converted Marks II** and II*** patterns, all of which were updated models to accommodate later ammunition, sighting systems or charger-loading features.

In 1907 a new model was approved which introduced improvements in the sighting system and charger loading, and simplified the manufacture; this was the Mark III and Converted Mark IV. Essentially the rifle that served Britain, India, South Africa, Australia and New Zealand through the two world wars, production also followed in Australia and India. Certain production economies during the Great War resulted in the Mark III*, the most common model used during the 2nd World War. In 1926, a new system of nomenclature resulted in the .303-in. S.M.L.E. being designated the Rifle No. 1 and so the Mark III became the No. 1 Mark III and the Mark III* became the No. 1 Mark III*.

Between 1921 and 1924, folding aperture sights with a longer sight radius than the earlier open sights were incorporated in a series of limited production rifles and along with some other smaller improvements, this pattern became the Mark V S.M.L.E. While this particular design was not considered to be a complete success, aperture sighting was incorporated in all future service rifles.

The last of the S.M.L.E. or No. 1 rifles was the Mark VI, a troop trials rifle designed to utilise simplified and more modern manufacturing processes, also reducing production time and the requirement for skilled and experienced tradesmen. It was also intended to cut down on the production expenses. By the early 1930's, it was not expected that the Mark VI would eventually be adopted as the service rifle because trials were in progress with new self-loading models. With some relatively minor changes, the Mk VI became the No. 4 trials rifle of 1931 and resulted in the service No. 4 model of World War 2 because the self-loading systems had not been finalised.

The S.M.L.E. Marks III and III* were also manufactured at Ishapore in India and at Lithgow in Australia. This was continued after the termination of Mk III* construction in Birmingham towards the end of the 2nd World War. Foreign production also saw some specialist S.M.L.E. models unique to these countries; sniper models, carbines and assorted conversions. Description of these variants will be found in Chapters 9 and 10, and in following sections of this chapter.

	Enfield	*Sparkbrook*	*B.S.A. Co.*	*L.S.A. Co.*	*Ishapore*	*Lithgow*
S.M.L.E. Trials:	1,055	-	-	-	-	-
S.M.L.E. Mk I:	133,457	14,640	150,000	65,000	-	-
S.M.L.E. Mk I Cond.:	9	-	-	-	-	-
S.M.L.E. Mk II Cond.:	26,532	16,763	-	-	-	-
S.M.L.E. Mk I*:	60,187	-	-	-	(IP) 3,020	-
S.M.L.E. Mk II* Cond.:	22,191	-	-	-	-	-
S.M.L.E. Mk I***:	16,208+	-	-	-	-	-
S.M.L.E. Mk III & III*:	2,235,000*	-	c.2,000,000	c.430,000	c.1,400,000	640,580
S.M.L.E. Mk IV Cond.:	98,559	-	-	-	-	-
S.M.L.E. Mk V:	c.20,000	-	-	-	-	-
No. 1 Mk VI:	1,025	-	-	-	-	-

+ Figures are likely to have been more than this.
* Approximate only, also includes more than 250,000 peddled rifles from S.S.A. & N.R.F.

SHORT, MAGAZINE LEE-ENFIELD TRIALS RIFLE

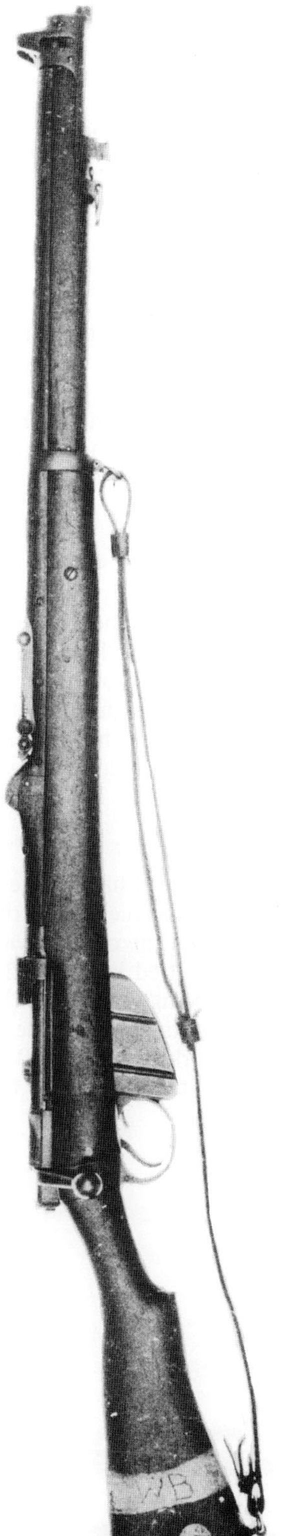

R. Faris

Calibre:	*.303 in.*
Rifling & Twist:	*5 groove, Enfield, L.H.*
Barrel Length:	*25.2 in. (640mm)*
Overall Length:	*44.5 in. (1130mm)*
Weight:	*8 lb. 2 oz. (3.7 kg)*
Magazine Capacity:	*10 rounds.*
Manufactured:	*Enfield.*

A nominal 1,000 of these rifles were assembled for troop trials in 1902 as a result of the requests for improvements to the Magazine Lee-Enfield made during and after the South African War. For the first time, a standard length rifle and bayonet was approved for trials by all arms of the services; infantry, cavalry, artillery, engineers and naval forces.

A number of innovations were incorporated in this short rifle which influenced development until World War 2. Handguards enclose the barrel for its full length and the charger loading was implemented by means of a guide on the left side of the receiver with a corresponding slide on the bolthead. A new mode of fixing the sword bayonet was introduced where the bayonet was not supported by the barrel, but by a separate nosecap instead. The nosecap is special to this particular model because it is similar to the Mark I type but has a circular hood over the foresight. In the trials, it was complained that more light would be admitted onto the foresight if the top of the hood were removed. This was effected on the eventual Mark I Short Rifle, the reason for the special shape of the Mark I S.M.L.E. nosecap protecting "ears", compared with those of the Mark III pattern. There was no provision for magazine cut-off.

Two types of backsight were used in the trials, equivalent numbers of each being produced. Type "A" was similar to the pattern eventually adopted on the Mark I S.M.L.E., graduated from 200 to 2,000 yards in 100 yd. increments with a press release slide with bone knobs and provision for windage adjustment. The "B" model was a ladder type graduated to 2,000 yards but used in the folded down position for ranges to 500 yards and raised vertically for ranges over 500 yards. Windage adjustment was effected by a wheel at the rear of the bed. In the drawings, both these sights appear to be set closer to the lower band than that on the Mark I S.M.L.E. The "B" pattern had metal protecting ears forward of the sight and a two-piece handguard arrangement; the "A" pattern had the protectors at the rear, like the Mark I S.M.L.E., but a one-piece handguard.

A sheet gunmetal buttplate is fitted with no provision for butt trap. Nearly all these rifles were converted to Aiming Tube Short Rifles in 1906-07 and so are extremely scarce today. There are no designation markings on the receiver.

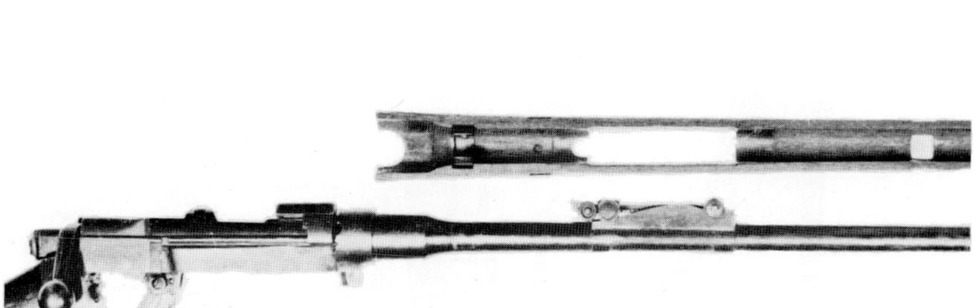

SHORT, MAGAZINE LEE-ENFIELD RIFLE Mk I

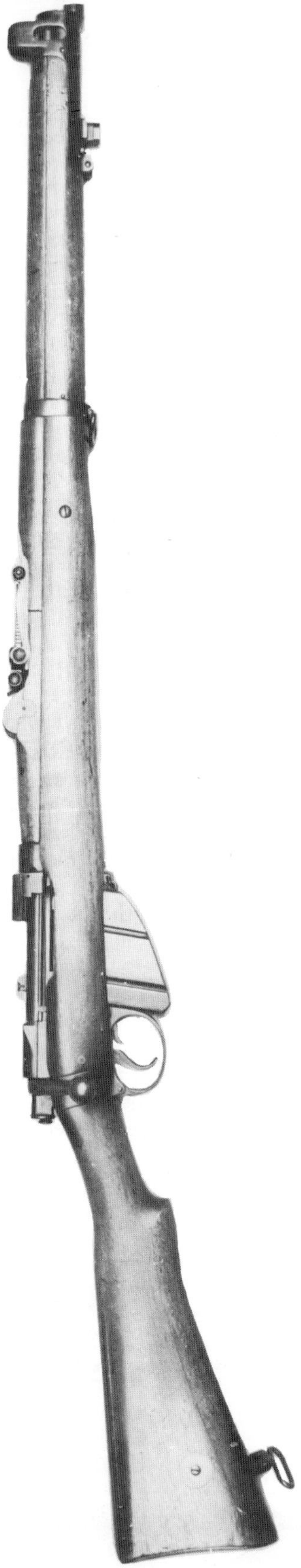

Calibre:	*.303 in.*
Rifling & Twist:	*5 groove, Enfield, L.H.*
Barrel Length:	*25.2 in. (640mm)*
Overall Length:	*44.5 in. (1130mm)*
Weight:	*8 lb. 3 oz. (3.7 kg)*
Magazine Capacity:	*10 rounds.*
Manufactured:	*Enfield, Sparkbrook, B.S.A. Co., L.S.A. Co.*

Typical Markings, R.H. Side of the Butt Socket:

E. R.
L.S.A.C°L[D]
1904
S[HT] L.E.
I

First approved on 23rd December 1902, the Mark I S.M.L.E. was re-introduced with a new approval date of 14th September 1903 after some minor changes were made, mainly the addition of a wind gauge. Some other changes were gradually introduced as production proceeded, including a double spring on the rear handguard, stud and spring under the nosecap, omission of the link loop on the trigger guard, minor changes to the backsight, and the fitting of a piling swivel and shorter sling swivels and magazine cut-off.

This rifle is full-stocked, with front and rear handguards arranged so that the barrel is enclosed for its entire length. A one-piece steel backsight protector is fitted onto the rear handguard and the front handguard differs from the following types in that there is a distinct, sharp-edged step down from the lower band forward; the later models are more rounded at this point. The nosecap has been changed from the trials model in that the top was cut out of the sight protector ring, but the pronounced incurving shape towards the top remains. The handguards are in two sections, front and rear, joining at the point mid-way along the sight bed. The sheet steel buttplate has no trap and the butt swivel screws directly into the buttstock on its woodscrew type base.

A sliding charger guide is mounted onto the bolthead while the left hand guide is fixed to the receiver. A safety catch is fitted on the left side of the receiver and the naval issue rifles were fitted with a magazine cut-off, although the other service rifles had provision for this component and many were so fitted later on. The rearsight leaf is graduated to 2,000 yards, all adjustments being made with the slide on the ramp; the slide has chequered bone knob inserts on each side. A windgauge is provided. Dial sights for volley firing are marked from 1,600 to 2,800 yards. The mounting points for the bayonet are integral parts of the nosecap.

M.O.D. Pattern Room

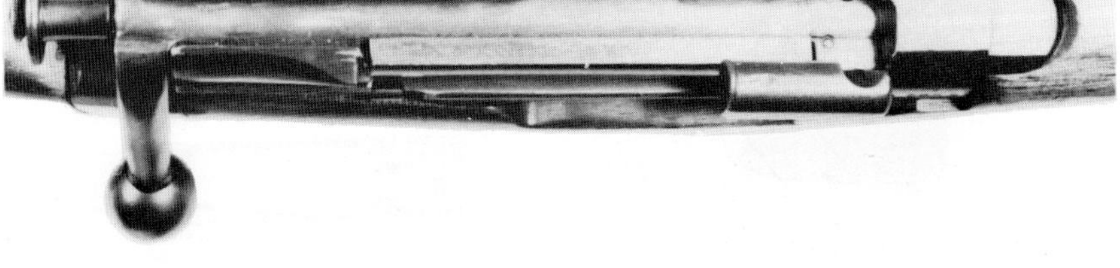

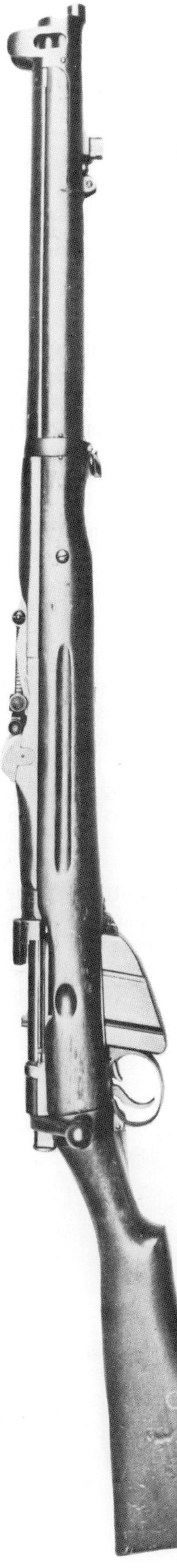

SHORT, MAGAZINE LEE-ENFIELD RIFLE Mk I Cond.

Calibre:	*.303 in.*
Rifling & Twist:	*5 groove, Enfield, L.H.*
Barrel Length:	*25.2 in. (640mm)*
Overall Length:	*44.5 in. (1130mm)*
Weight:	*8 lb. 4 oz. (3.7 kg)*
Magazine Capacity:	*10 rounds.*
Converted:	*Enfield.*

Receiver Markings, L.H. side of the Butt Socket:

ENFIELD
1904
S^HT L.E.
CON^D I

Only a small number of examples of this particular conversion were produced but it is of interest in the S.M.L.E. lineage as a conversion of the superseded early Lee-Metford model to the new short rifle pattern. The approval date is of interest too, 2nd November 1903, after the approval date for the Mark II Cond. model (which was 16th January 1903).

It is also noteworthy because of the more difficult conversion from the Mark I series of Lee-Metford rifle. While the general dimensions resemble those of the Mark I S.M.L.E., the fore-end is the most obvious pointer to its origin because of the finger grooves and recess underneath the magazine cut-off slot. The fore-end profile is also different, particularly behind the lower band; two pins have been provided for this band, behind and in front, whereas on the Mark I S.M.L.E. there is only one, at the front. New S.M.L.E. handguards have been fitted while the butt was converted from the Mark I* Lee-Metford pattern.

While a new barrel, sights, nosecap and magazine have been fitted, the action body has been considerably modified, along with the bolt, bolt head and trigger guard. The receiver was altered for the wider magazine, charger nib on the left side and the safety catch. The bolt has had the dustcover stops removed, top of the rib reduced in height, the dustcover screw holes filled and the bolthead modified so as to accept the charger guide. The Lee-Metford trigger guard was enlarged for the wider magazine, the swivel boss removed and new link loop fitted, although the original chequered magazine release catch was retained.

Sighting is identical to the Mark I S.M.L.E., graduated to 2,000 yards on the leaf and with the long range dial sight plate marked to 2,800 yards. The long range rear aperture is positioned alongside the safety catch.

As the Mark I pattern of Short, Magazine Lee-Enfield rifle was resealed with improvements, so, too, the Mark I Converted can be differentiated between an earlier and later model. The earlier type has the two handguard spring rivets in a horizontal position while the later type has them vertical, in line with the bore. The other minor changes applied to the Mark I S.M.L.E. were also incorporated in the later type of Converted Mark I.

M.O.D. Pattern Room

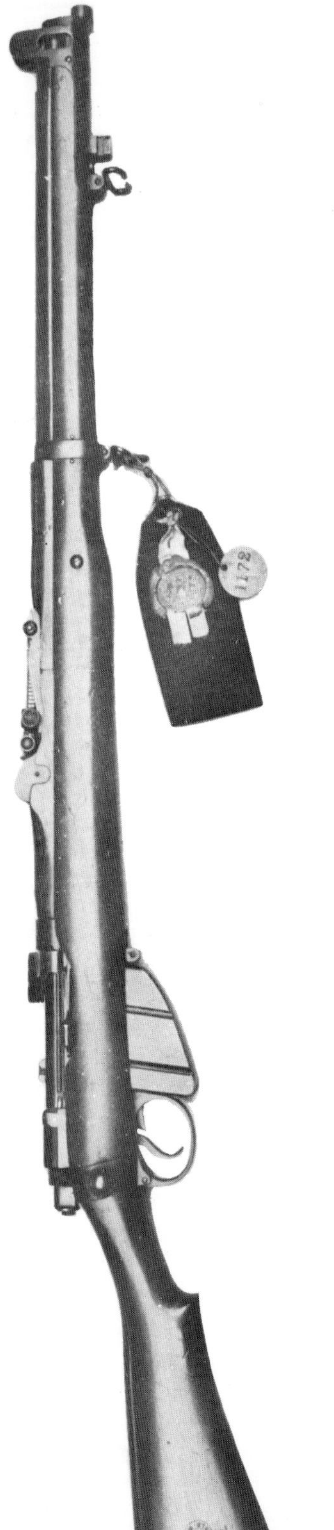

SHORT, MAGAZINE LEE-ENFIELD RIFLE Mk II Cond.

Calibre:	*.303 in.*
Rifling & Twist:	*5 groove, Enfield, L.H.*
Barrel Length:	*25.2 in. (640mm)*
Overall Length:	*44.5 in. (1130mm)*
Weight:	*8 lb. 8 oz. (3.9 kg)*
Magazine Capacity:	*10 rounds.*
Converted:	*Enfield, Sparkbrook.*

Typical Markings, L.H. side of the Butt Socket:

SPARKBROOK
1904
S^{HT} L.E.
CON^{D} II

This conversion was adopted on 16th January 1903 and it was made from the Marks II and II* Lee-Metford and Marks I and I* Lee-Enfield rifles so as to duplicate the features and characteristics of the Mark I S.M.L.E., although the conversion of the Mark I* Lee-Enfield was not utilised until later. It is of special note because it was the first mass-production model in a line of Converted rifles which updated earlier long rifles to the new short rifle patterns with their attendant new features and improvements.

Original production and designation markings of the rifle from which the conversion was made will be found on the right side of the butt socket while those relating to the conversion were stamped on the left side. The receiver was altered by attaching a charger guide tab to the left side and the S.M.L.E. safety catch fitted. The original bolts were modified by removing the rearward extension for the safety catch, the dustcover lugs were ground off and the bolt guide rib reduced in height and radiused. However, the converted models may be identified as the original dustcover locating groove at the rear is still evident. Rather than merely apply the new production bolthead, the old pattern was converted to receive the charger guide and a slot cut in the rear portion of the threaded end for stripping and assembly of the striker and cocking piece. The trigger guard was slotted at the back to lighten it, and modified so as to clear the new magazine stop clip.

Considerable work was also done to the wooden fore-end and butt to assume the short rifle configuration and accept the new nosecap, buttplate and marking disc. As the shorter and lighter Mk I S.M.L.E. barrel was used, a liner was fitted and glued into the converted fore-end. New handguards were also fitted. Like the Mark I S.M.L.E., magazine cut-offs were only fitted to those rifles supplied to the Navy.

M.O.D. Pattern Room

SHORT, MAGAZINE LEE-ENFIELD RIFLE Mk I*

Calibre:	*.303 in.*
Rifling & Twist:	*5 groove, Enfield, L.H.*
Barrel Length:	*25.2 in. (640mm)*
Overall Length:	*44.5 in. (1130mm)*
Weight:	*8 lb. 8 oz. (3.9 kg)*
Magazine Capacity:	*10 rounds.*
Manufactured:	*Enfield, B.S.A. Co., L.S.A. Co., Ishapore.*

E. R.
ENFIELD
1906
SHT L.E.
I*

Typical Markings, R.H. side of the Butt Socket:

Approved on 27th March 1906, the Mark I* S.M.L.E. applied a number of improvements to the Mark I and differs in the following particulars: the gunmetal buttplate is fitted with a trap and the butt modified to accommodate a pull through, a new type of butt swivel with mounting bracket is inletted into the butt and secured by two screws, the swivel screws are bored longitudinally at the threaded end for centre-punching, and a new magazine case has been fitted which is deeper at the front and marked with a "2". The keeper screw for the firing pin is slotted to facilitate removal with a coin.

In all other respects it is similar to the Mark I S.M.L.E., with the incurving protector wings of the nosecap, backsight protector mounted on the rear handguard, sliding charger guide on the bolthead and corresponding charger guide on the left side of the receiver. The rearsight consists of a folding leaf graduated from 200 to 2,000 yards and a slide with windage adjustment and bone finger tabs; the long range sights are graduated from 1,600 to 2,800 yards. Some Mark I rifles were upgraded to Mark I*.

SHORT, MAGAZINE LEE-ENFIELD RIFLE Mk II* Cond.

Specifications:	*As for S.M.L.E. Mk I*.*
Converted:	*Enfield.*

ENFIELD
1906
SHT L.E.
COND II*

Typical Markings, L.H. side of the Butt Socket:

Approved on 15th March 1906, this was a conversion of the Lee-Metford Marks II and II* and Lee-Enfield Marks I and I* to the Mark I* S.M.L.E. pattern. Original rifle markings are found on the right side of the butt socket with the conversion marks on the left. Modifications for the charger guides are the same as on the Converted Mark II and in external appearance it is similar to the Mark I* S.M.L.E., except where a converted fore-end is used. The original butt was modified and new butt swivel fitted. Converted bolts have a groove at the rear where the back of the dustcover was previously located. Some of the Converted Mark II rifles were upgraded to the Converted Mark II* pattern.

R. Richardson

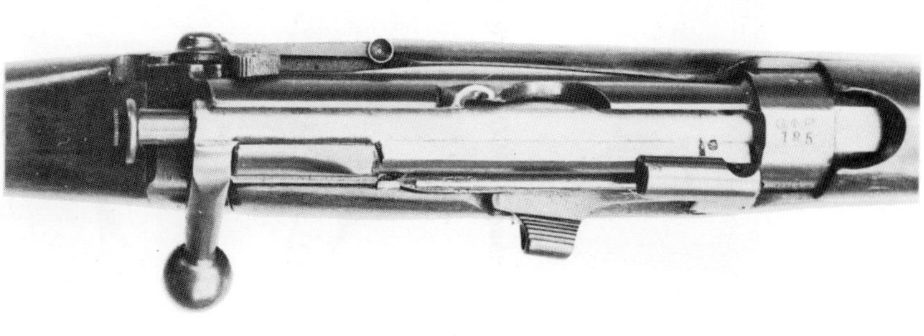

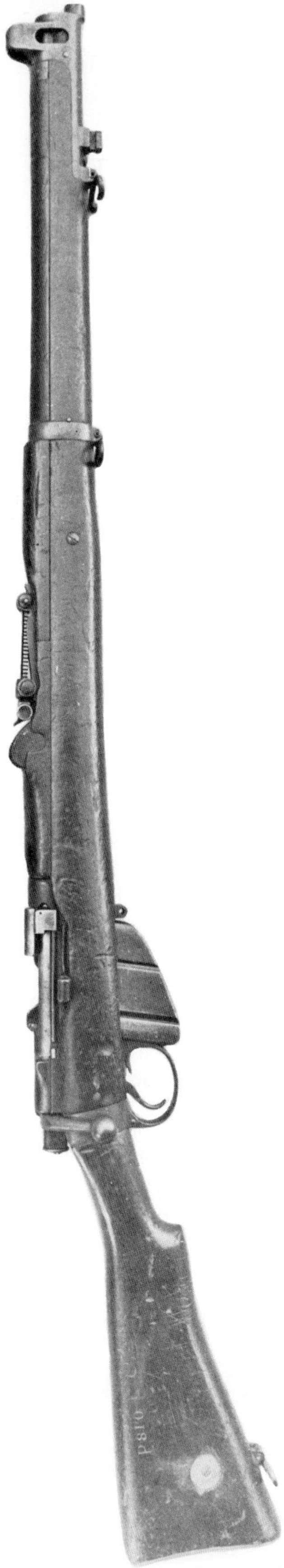

SHORT, MAGAZINE LEE-ENFIELD RIFLE Mk I***

Calibre:	*.303 in.*
Rifling & Twist:	*5 groove, Enfield, L.H.*
Barrel Length:	*25.2 in. (640mm)*
Overall Length:	*44.5 in. (1130mm)*
Weight:	*8 lb. 8 oz. (3.9 kg)*
Magazine Capacity:	*10 rounds.*
Converted:	*Enfield, Royal Naval Ordnance Depots.*

E.R.
ENFIELD
1906
S^HT L.E.
I***

Typical Markings, R.H. side of the Butt Socket:

This Land Service model was approved on 22nd April 1914, mentioned in the List of Changes §16910. It involved the alteration of sights on the Mark I* rifle to suit Mark VII ammunition; a blade foresight was fitted and for the backsight, a higher wind gauge with U-sighting notch was provided. A spring-tensioned fine adjustment screw with larger head was also incorporated, being that fitted on the Mk I** S.M.L.E.

The Navy decided to upgrade the Mark I** rifles which were not fitted with charger bridges, to the Mark I*** pattern, on August 18th 1915. They were also altered to suit the Mark VII ammunition. The additional star or asterisk was marked alongside the existing numeral designation.

Some Mark I*** rifles have been noted with the charger bridge as on the Mark III S.M.L.E.; while this does not appear to have been the originally approved pattern, it would seem that some Mark I*** rifles were fitted with such charger guides like the Charger Loading Lee-Enfield rifles, during the early part of the Great War. These may well have been those examples converted at the Royal Naval Ordnance Depots.

The backsight slide has chequered bone inserts at the ends and the sight and sight protector configuration is similar to that of the Mk I series of S.M.L.E. rifles. The backsight leaf graduations are up to 2,000 yards. Specifications are generally similar to those of the S.M.L.E. Mark I* pattern, with brass buttplate and butt trap provision, inletted butt swivel mount and long range dial sights. Naval issue rifles were usually issued with the magazine cut-off.

B. Labudda / M.O.D. Pattern Room

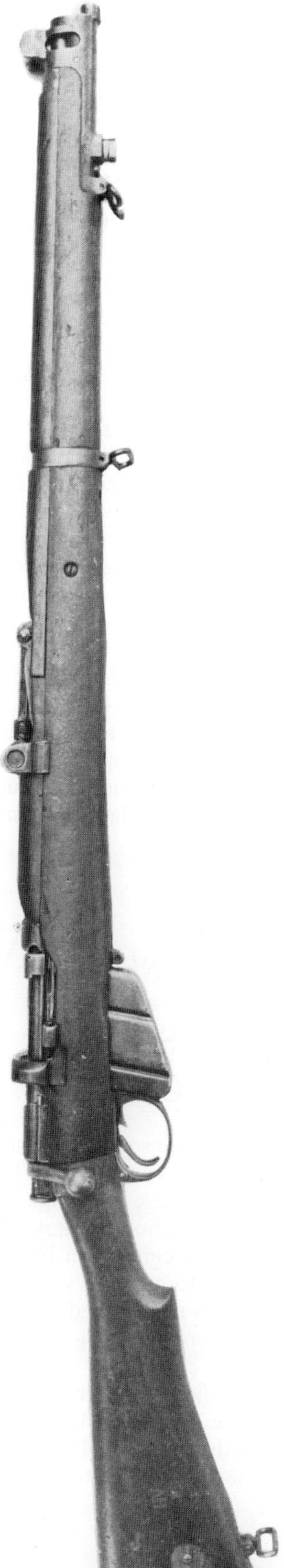

B. Labudda

SHORT, MAGAZINE LEE-ENFIELD RIFLE Mk III (RIFLE No. 1 Mk III)

Calibre:	*.303 in.*
Rifling & Twist:	*5 groove, Enfield, L.H. +*
Barrel Length:	*25.2 in. (640mm)*
Overall Length:	*44.5 in. (1130mm)*
Weight:	*8 lb. 10 oz. (3.9 kg)*
Magazine Capacity:	*10 rounds.*
Manufactured:	*Enfield, Ishapore, Lithgow, B.S.A. Co., L.S.A. Co.*

G. R.
ENFIELD
1910
S^{HT} L.E.
III

Typical Markings, R.H. side of the Butt Socket:

Approved on 26th January 1907, this short rifle features a number of improvements over the Marks I and I*. The changes were made in respect of parts which considerably improved its battle efficiency and also simplified manufacture, so this was most important in view of the role it was to play in the wartime years.

A charger bridge was fitted across the top of the receiver and the bolthead guide of the Mark I removed, simplifying production and slightly reinforcing the action. The backsight was strengthened and more finely graduated, a U-notch incorporated instead of the V-notch, the bone inserts dispensed with, and the bed made stronger, encircling the barrel. A straight-sided blade foresight replaced the barleycorn and the foresight protector wings on the nosecap were widened and shaped differently so as to admit more light.

The inner band around the barrel was moved 1 inch closer to the breech, so its securing screw is no longer hidden underneath the outer band. Due to the new charger bridge, the fore-end was changed by having a squared recess cut into the right hand side and provision made for the new sight bed. The rearsight protector was removed from the handguard and attached to the fore-end by a screw and nut. The front handguard was slightly increased in size forward of the outer band and the edges at this point were also rounded off.

Graduations on the rearsight remain from 200 to 2,000 yards and those on the dial plate from 1,600 to 2,800 yards; the long range aperture sight is located adjacent to the safety catch. The brass buttplate, fitted with a trap for storing the pullthrough and oil bottle in the butt, is similar to that of the Mark I*. This rifle was also manufactured in India at Ishapore and at Lithgow in Australia.

+ Some 4-groove barrels were manufactured by B.S.A. during World War 2.

SHORT, MAGAZINE LEE-ENFIELD RIFLE Mk IV Cond.

Specifications:	*As for S.M.L.E. Mk III.*
Converted:	*Enfield.*

ENFIELD
1911
S^{HT} L.E.
CON^{D} IV

Typical Markings, L.H. side of the Butt Socket:

Similar to earlier converted rifles, this was also a modification of Mark II and II* Lee-Metford and Mark I and I* Lee-Enfield rifles; in this instance they were upgraded to the Mark III S.M.L.E. pattern. Approved on 17th June 1907, the receiver was modified in that a charger bridge and safety catch were fitted, the bolt has had the dustcover lugs removed and in the case of the Mark II* Lee-Metford and Marks I and I* Lee-Enfield, the rear section of the bolt for safety catch removed as well. The fore-end was converted from Long Rifle pattern and the barrel, sights, sight protector and nosecap of the Mark III S.M.L.E. were fitted.

B. Labudda

SHORT, MAGAZINE LEE-ENFIELD RIFLE Mk I**

Calibre:	*.303 in.*
Rifling & Twist:	*5 groove, Enfield, L.H.*
Barrel Length:	*25.2 in. (640mm)*
Overall Length:	*44.5 in. (1130mm)*
Weight:	*8 lb. 12 oz. (4.0 kg)*
Magazine Capacity:	*10 rounds.*
Converted:	*R.N.O.D. Chatham, Portsmouth, Plymouth.*

Typical Markings, R.H. side of the Butt Socket:

E. R.
ENFIELD
1903
S^{HT} L.E.
I**

Approved in Naval Orders on 4th January 1908, this consists of the Mark I S.M.L.E. being upgraded for Naval service so as to incorporate improvements and some features of the Mark III S.M.L.E. The foresight and U-sighting notch of the Mark III were utilised and a double spring handguard and barrel centring stud and spring under the nosecap were fitted to the fore-end. The magazine link and loop were removed and sling swivel lugs were attached to the front of the trigger guard; the retaining nib has been riveted to the magazine platform spring, slotted keeper screw inserted in the cocking piece, bored swivel screws fitted, sharp edges removed from the locking bolt and action body, and the new charger guide No. 2 fitted.

From 1912, these rifles were further upgraded at Enfield to correspond with the Mark III S.M.L.E. by fitting a charger bridge. The typical Mark I S.M.L.E. foresight protector wings were straightened and the stock modified for the new charger bridge base. As this was purely a Naval conversion and (supposedly) all the Mark I** rifles were to be upgraded, there was no change in nomenclature after this update. Similar modifications were carried out on the Marks II** and II*** and so these three conversions are similar in appearance.

SHORT, MAGAZINE LEE-ENFIELD RIFLE Mk II** Cond.

Specifications:	*As for S.M.L.E. Mk I**.*
Converted:	*R.N.O.D. Chatham, Portsmouth, Plymouth.*

Typical Markings, L.H. side of the Butt Socket:

ENFIELD N
1904
S^{HT} L.E.
CON^{D} II**

This conversion was made from the Mark II Converted rifle and so the right side of the butt socket bears the original Lee-Metford or Lee-Enfield designation. The first conversion, to Mark II, was effected at Enfield and these markings are located on the left side of the butt socket. The next modification to Mark II** was done at one of the Royal Naval Ordnance Depots, this conversion having been approved in Naval Orders on 25th June 1908 for it to conform with the original Mark I** specifications. This conversion date was not marked, merely involving the extra stars or asterisks being placed after the original designation.

From 1912, further improvements were carried out at Enfield in order that the rifle resemble the Mark III pattern more closely, so the foresight wings were straightened and shortened and a charger bridge fitted to replace the Mark I S.M.L.E. style charger guides. This later conversion was not indicated in the receiver markings.

SHORT, MAGAZINE LEE-ENFIELD RIFLE Mk II* Cond.**

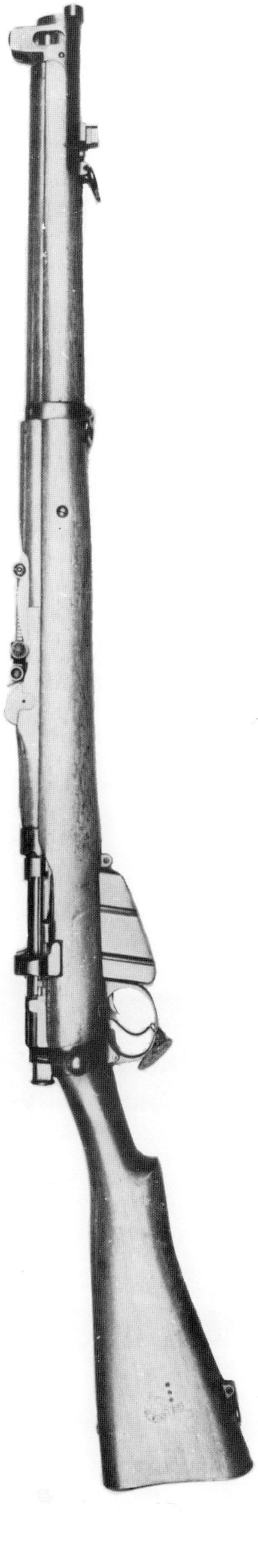

M.O.D. Pattern Room

Calibre:	*.303 in.*
Rifling & Twist:	*5 groove, Enfield, L.H.*
Barrel Length:	*25.2 in. (640mm)*
Overall Length:	*44.5 in. (1130mm)*
Weight:	*8 lb. 8 oz. (3.9 kg)*
Magazine Capacity:	*10 rounds.*
Converted:	*R.N.O.D. Chatham, Portsmouth, Plymouth.*

Typical Markings, L.H. side of the Butt Socket:

N
ENFIELD
1905
S^HT^ L.E.
CON^D^ II***

Similar in appearance to the Mark I** and Converted Mark II** rifles, this model differs only in that the Converted Mark II* was used for this conversion. A naval issue, modified at one of the Royal Naval Ordnance Depots, it was originally a Lee-Metford or Lee-Enfield rifle upgraded to the Converted Mark II* pattern at Enfield, and these conversion markings are found on the left side of the receiver while the original Lee-Metford or Lee-Enfield markings are on the right side.

Effected at the Naval establishments, the modification was quite simple; the barleycorn foresight and V-notch backsight were exchanged for a blade foresight and U-notch backsight while the bolt-head charger guide was exchanged for the No. 2 guide, similar to that effected to the Mark I**. The two additional stars were stamped on the left side of the receiver after the original conversion markings, so that the Mk II* Cond. now reads Mk II*** Cond.

After 1912 these rifles were further modified at Enfield so as to conform more closely with the Mark III S.M.L.E. pattern. A Mk III foresight blade was provided and a higher wind gauge with U-notch fitted to the rearsight. The sight protector ears on the nosecap were elongated, double spring handguard fitted, magazine link loop removed and a lug fitted at the front of the trigger guard for a sling swivel. These changes are also referred to in the previous descriptions of the S.M.L.E. Mark I** and Converted Mark II*** rifles. Most aspects of the Mk I**, Mk II** and Mk II*** were to bring them into line with the Mk I* pattern, except for the sighting.

SHORT, MAGAZINE LEE-ENFIELD MK I** I.P.

E. R.
ENFIELD
1903
S^{HT} L.E.
I**
I. P.

Calibre:	*.303 in.*
Rifling & Twist:	*5 groove, Enfield, L.H.*
Barrel Length:	*25.2 in. (640mm)*
Overall Length:	*44.5 in. (1130mm)*
Weight:	*8 lb. 10 oz. (3.9 kg)*
Magazine Capacity:	*10 rounds.*
Converted:	*Ishapore.*

Typical Markings, R.H. side of the Butt Socket:

This rifle conversion is special to India, even though it is identical in specifications to the Mark III S.M.L.E. It is most readily identified by an examination of the receiver body markings on the right and left hand sides. The markings shown here would indicate conversion from an original Enfield Mk I or Mk I upgraded to Mk I*; the additional "*" and "I.P." markings were made at Ishapore.

This pattern was a conversion from the Ishapore made S.M.L.E. Mark I* I.P. rifle, or British S.M.L.E. Mark I or Converted Mark II series, upgraded so as to resemble the Mark III S.M.L.E. The year of conversion at Ishapore was usually stamped on the left side of the butt socket along with the Ishapore factory mark, while the right side of the butt socket retains the original manufacturer's markings and designations along with the I.P. title, in most instances.

Similar to many other conversions and re-builds effected at Ishapore, most components were replaced, converted or updated and about all that remains of the original S.M.L.E. rifle is the action body, bolt and trigger guard.

Imperial War Museum

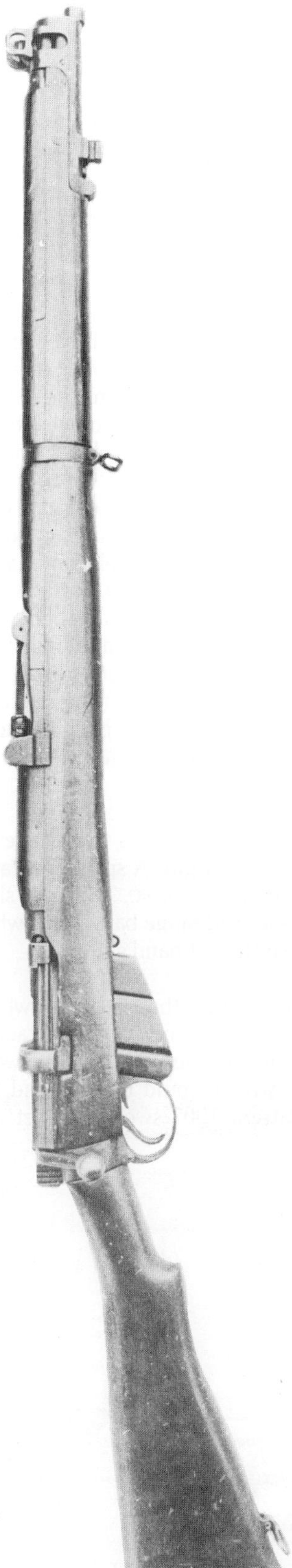

SHORT, MAGAZINE LEE-ENFIELD RIFLE Mk III* (RIFLE No.1 Mk III*)

Calibre:	*.303 in.*
Rifling & Twist:	*5 groove, Enfield, L.H. +*
Barrel Length:	*25.2 in. (640mm)*
Overall Length:	*44.5 in. (1130mm)*
Weight:	*8 lb. 15 oz. (4.1 kg)*
Magazine Capacity:	*10 rounds.*
Manufactured:	*Enfield, Ishapore, Lithgow, B.S.A. Co., L.S.A. Co., S.S.A. Co. & N.R.F. peddled.*

G. R.
B.S.A. C°
1918
S^{HT} L.E.
111*

Typical Markings, R.H. side of the Butt Socket:
For Indian & Australian marks see Chapters 9 & 10.

Approved as a wartime expediency on 2nd January 1916 (although B.S.A. and Enfield produced some in 1915), the Mark III* basically differs from the Mark III in the omission of the long range dial and aperture sights, windage adjustment on the backsight, magazine cut-off, lug on the firing pin collar, and swivel lugs at the front of the trigger guard. These swivel lugs were replaced by a small wire loop to secure a fabric action cover. Soon after the introduction of the Mark III*, the brass butt marking disc was also omitted. Some Mark III* rifles may be encountered with provision for magazine cut-off.

In other respects it is similar to the Mark III S.M.L.E. and has a backsight with hinged leaf graduated from 200 to 2,000 yards. Sling swivels are mounted to the butt and band while a piling swivel is fitted at the bottom of most nosecaps. Some Australian and Indian made rifles have differences in the metal fittings such as the nosecap, rearsight protector and backsight cap which were generally production economies; these are further detailed in the relative chapters on Indian and Australian manufacture. The buttplate is brass, or more correctly, gunmetal, and it has a trap for the storage of an oil bottle and pullthrough.

The Australian and Indian production of the Mark III* rifle continued through World War 2, the last rifles being made in the 1950's; B.S.A. continued production during most of World War 2. In 1926, the nomenclature was altered from S.M.L.E. Mark III* to Rifle No. 1 Mark III*.

Some spare barrels for the Marks III and III* were also made at the Long Branch factory, Canada.

+ Some 4 groove barrels were produced by B.S.A. during the 2nd World War.

B. Labudda

SHORT, MAGAZINE LEE-ENFIELD Mk V

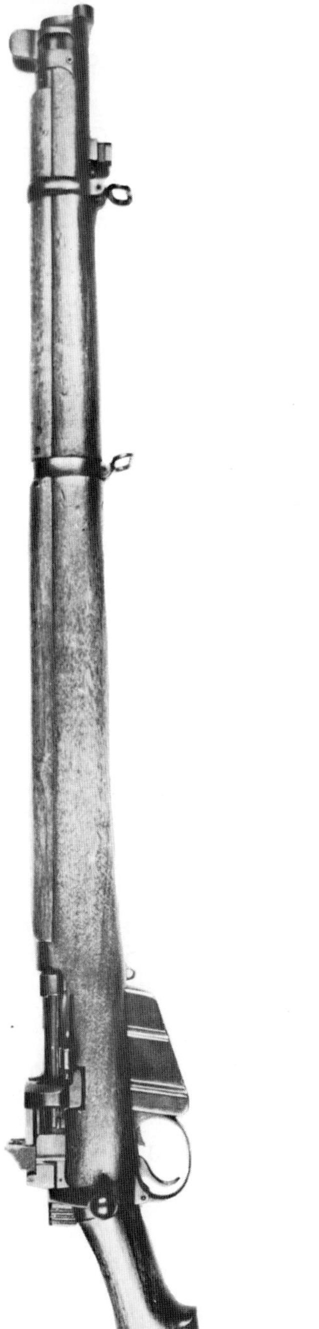

Calibre:	*.303 in.*
Rifling & Twist:	*5 groove, Enfield, L.H.*
Barrel Length:	*25.2 in. (640mm)*
Overall Length:	*44.5 in. (1130mm)*
Weight:	*8 lb. 13 oz. (4.0 kg)*
Magazine Capacity:	*10 rounds.*
Manufactured:	*Enfield.*

G. R.
ENFIELD
1924
S^HT L.E.
V

Typical Markings, R.H. side of the Butt Socket:

Although generally regarded as a troop trials rifle, the Mark V S.M.L.E., because of production figures amounting to 20,000 between 1922 and 1924, must be considered as more than just a trials rifle. Conversion of Mark III to the Mark V pattern was contemplated but not implemented due to costs and some sighting considerations.

It is a further development of the Mark III and is perhaps best compared with that rifle for the purpose of this description. The receiver body is similar except for the upper rear section which has been modified considerably for a wide aperture backsight, hinged at the rear. The safety catch is special, marked "V", with angled grooves towards the front. Because of the aperture sights, there is no backsight on the barrel and a full length handguard is fitted. An additional band is positioned at the rear of the nosecap to reinforce it for bayonet-fighting. No long range dial sights are fitted, although a magazine cut-off is provided, different in that it has no spotting hole. The transverse nosecap screw has a larger head with enlarged slot so that it might be tightened up or released with a coin, to secure a separate muzzle cover.

The folding aperture sight is graduated from 200 to 1,400 yards with the even numbers on the left side and the odd numbers on the right. A spring operated ball in the receiver permits the sight to be positioned at 45 or 90. The sight slide is L-shaped with reinforced sides and it is used as a fixed range battlesight when the leaf is folded down. The slide release catch is on the left hand side.

Like the Mark III* pattern, there is no maker's roundel on the butt. The swivels are mounted onto the butt and lower band, and a narrow piling swivel is attached at the bottom of the nosecap. Rifle serial numbers will be found to range between 1 and 9,999 after which an "A" prefix was used which would then extend the serial number range to A 9999. The standard Pattern 1907 sword-bayonet was issued with this rifle.

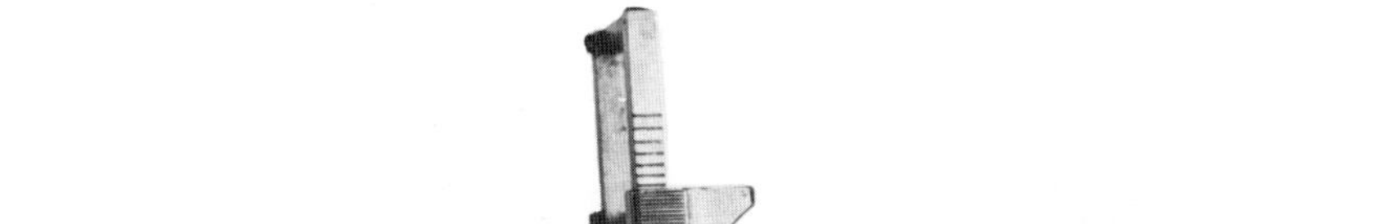

M.O.D. Pattern Room

(SHORT, MAGAZINE LEE-ENFIELD) .410 MUSKET

M.O.D. Pattern Room / J. Alley

Calibre:	*.410 in.*
Rifling & Twist:	*Smoothbore.*
Barrel Length:	*25.2 in. (640mm)*
Overall Length:	*44.5 in. (1130mm)*
Weight:	*8 lb. 10 oz. (3.9 kg)*
Magazine Capacity:	*Single shot.*
Manufactured:	*Ishapore.*

Typical Markings, L.H. side of the Butt Socket:

.410
RFI
1942

The .410 Musket cartridge was approved in India in September 1927 and is basically a brass .303-in. cartridge case which has not been necked or sized down, but remains straight-cased. Although the dimensions appear to be similar, the conventional commercial .410 shotshells will not chamber in this musket.

Apart from the absence of the magazine, there is little exterior difference between this and the Mark III or III* S.M.L.E. from which it was converted. The magazine way is filled by a wooden plug and it finishes flush with the magazine surround of the trigger guard. The top of this block is covered with a sheet steel plate which serves as a loading platform; the wood block is also secured by two transverse wooden pins through the fore-end.

The bore is noticeably larger than that of the .303 and the barrel is thinner due to the increased bore diameter. No change was required for the bolthead or action generally because the base and overall length of the .410 cartridge is the same as the .303. Conversion numbers were originally marked on the right side of the receiver below the original manufacture date, but some have been noted marked on the left or conversion side. Some were converted from the Indian S.L. (Single Loader) model, so this designation may be found on the left side of the receiver socket. The single loader model is the same as the .410 musket but for being fitted with a .303-in. barrel.

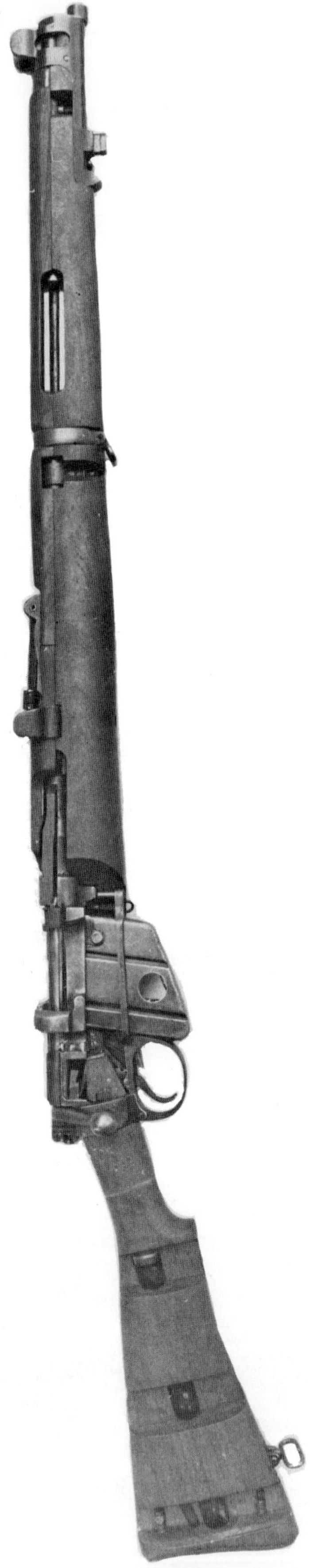

RIFLE No. 1 SKELETONISED

Calibre:	*.303 in.*
Rifling & Twist:	*5 groove, Enfield, L.H.*
Barrel Length:	*25.2 in. (640mm)*
Overall Length:	*44.5 in. (1130mm)*
Weight:	*8 lb. 4 oz. (3.7 kg)*
Magazine Capacity:	*10 rounds.*
Manufactured:	*Lithgow.*
Receiver Markings:	*None.*

This particular full-length sectionised Mark III* rifle was approved in Australia in 1948; some were produced at Lithgow while others appear to have been converted at Ordnance Depots. Other full-length rifles were sectionised at various factories in England and India as well, but most were made up by trainees and apprentices or on "one-of" orders as decorators and presentation pieces; therefore, the cuts and machining may be found to differ considerably from rifle to rifle. Some were also sectionised by unit armourers for instructional purposes but these are generally less well finished. The Lithgow pattern has been chosen to represent these full-length models as it was made to a specific pattern.

Machined so as to be viewed from the right hand side, there are no cuts at all on the left side and when observed from this side it appears to be an ordinary No. 1 Mk III*. Viewing cuts were made in the chamber and barrel, the right side of the fore-end at the receiver has been removed and certain sections of the receiver milled so as to expose the locking lugs and other functional parts. The bolt, bolthead, cocking piece and magazine are also machined and slots have been cut in the fore-end so as to bare the rearsight protector mounting, inner band, nosecap screw and stock stud and spring. The butt has been cut to display the buttplate screws, oil bottle and pull through, and the stockbolt head. Dummy cartridges were usually loaded in the magazine. In the Australian List of Changes, it is noted that worn or obsolete rifles were to be used for conversion, but some of the Lithgow modified rifles appear to have been assembled from new parts; they were not serial numbered.

It is likely that most of the other patterns of British service rifles ranging from the Lee-Metford to No. 8 have been full length sectionised in various factories, and certainly Lee-Enfield and Mark I and Mark III S.M.L.E. rifles were done at Enfield as well as some Mark III and III* rifles at Ishapore. Much larger quantities of short sectionised models were made as armourers and demonstration models, comprising of the butt, action and about 9 inches of barrel; details of the various patterns are provided later in this chapter.

Magazines from the short sectionised models were sometimes fitted to the full-length versions; the former generally have a 1-in. circle cut into the right side and are sectionised on the left side (as per the illustration). The Australian full-length sectionised rifle was usually supplied with a magazine sectionised on the right hand side and no circular cut in the other side.

T. Williams

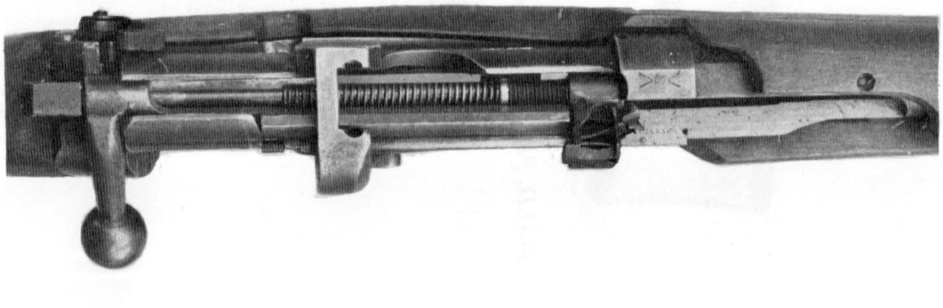

RIFLE No. 1 Mk VI

M.O.D. Pattern Room

Calibre:	*.303 in.*
Rifling & Twist:	*5 groove, Enfield, L.H.*
Barrel Length:	*25.2 in. (640mm)*
Overall Length:	*44.5 in. (1130mm)*
Weight:	*8 lb. 10 oz. (3.9 kg)*
Magazine Capacity:	*10 rounds.*
Manufactured:	*Enfield.*

Typical Markings, L.H. side of the Butt Socket:

G R
ENFIELD
1930
No1 MkVI
A 0257

This rifle was the numerical successor to the Mark V and it was basically a troop trials model which never went into full-scale production. In conjunction with the trials No. 4 Mark I, there were three patterns produced; the A pattern in 1926, B pattern between 1929 and 1931, and the C pattern in 1935. The B pattern was this troop trials model.

Perhaps the best description is made when compared with the subsequent No. 4 Mk I. The trials Mk VI has some readily noticed features such as the deep chequering of the fore-end commencing about 7 inches from the back of the fore-wood and continuing to within 2.5 inches of the lower band. This chequering is in a vertical and horizontal square pattern and the rear handguard has four 7-inch longitudinal grooves on each side. A magazine cut-off is fitted and the cocking piece has a large circular protrusion at the rear end with a transverse securing screw. The bolt knob is drilled out. The foresight protector also is special in that it bears more resemblance to the Pattern 1914 rifle style than the wartime No. 4 form.

The left wall of the receiver is lower and there is a deep scoop in the receiver behind the safety catch; these are the definitive "trade marks" of the Mark VI. The rearsight is a flip-up aperture similar to the Mk I sight of the No. 4 rifle, with a battle sight in the folded down position and screw adjusted elevation for use in the raised position. Graduations are to 1,300 yards.

A narrow piling swivel is attached to the top band and sling swivels are fitted to the butt and lower band, similar to the S.M.L.E. Mark III. The buttplate is made of gunmetal, of a more bronze colour than the previous S.M.L.E. type, although the familiar brass marking disc is fitted. Most of the Mark VI rifles were refurbished at Enfield early in World War 2 as No. 4 rifles, due to the acute shortages at that time.

EARLY .22 AND THE NO. 2 RIFLES

The Morris tube represents the first of the smallbore training rifles, initially introduced for use in the Martini rifles and carbines; this was merely a tube and breech section which was inserted into the barrel and chamber so that miniature centrefire ammunition could be used. The first of these rounds was the .297/230 and on 29th November 1908, the .22 Rim Fire cartridge was approved; many Morris tubes were bushed for the new rimfire round. Aiming Tube rifles were approved in 1906, featuring a .22-in. calibre tube and special bolthead; Lee-Metford, Lee-Enfield and Short Lee-Enfield rifles were so modified. Some of the first Aiming Tube rifles were altered from the 1902 trials S.M.L.E. rifles.

However, the system of temporarily modifying or adapting regular service rifles had certain disadvantages so a special small bore training rifle, the .22-in Short Rifle Mk I, was approved in 1907, followed by Mark I* and Mark II variants a few years later. These were factory conversions from the Lee-Metford rifle and involved many alterations, to such an extent that few of the original features are obvious. A Mark III .22-in. Short Rifle was approved in 1912 which used the S.M.L.E. as the basis for conversion. Long rifle configurations were also utilised for the .22-in. role in this series; the .22 Long Rifle Marks I and I* were converted from the Lee-Metford rifle and the .22 Long Rifle Mark II from the Lee-Enfield rifle. It was intended that these trainers duplicate the sighting and "feel" of the service rifles rather than resemble the general appearance, so some of the .22-in. trainers may appear to be obsolete types on first inspection although they feature the current form of sighting. As these rifles were only used as single loaders, the magazine and cut-off plate were removed.

The production pressures during the Great War resulted in the introduction of the .22-in. Short Rifle Pattern 1914 Nos. 1 and 2; this conversion was effected by A.G. Parker in Birmingham, using a .22 calibre liner for the barrel. The Charger Loading Lee-Enfield Mark I* was similarly modified to become the .22-in. Long Rifle Pattern 1914. Another variation of these tubed barrel models was the .22-in. Short Rifle Pattern 1918, although this model has the .303-in. chamber and uses a dummy .303 round into which was inserted the .22 rimfire cartridge to permit charger loading and magazine fire. The .22-in. Short Rifle Mark IV was approved in 1921, with a solid (as opposed to tubed) barrel and other minor changes. Converted from the Marks III and III* S.M.L.E., it outwardly resembles these rifles. This trainer was also produced in India and Australia and used in these countries until the 1960's. A nomenclature change in 1926 resulted in the .22-in. rifle becoming the No. 2, and this rifle became the No. 2 Mark IV*; some were assembled as new No. 2 rifles rather than being conversions. During World War 2, conversion was again carried out by the trade although some new production rifles were still being assembled at Enfield. The No. 2 was superseded by the Nos. 7, 8 and 9 after the war.

Rifles	*Service*	*Converted from*	*Quantity*
M.L.M. & M.L.E. Morris Tube	-	M.L.M. & M.L.E.	# 679
M.L.M. & M.L.E. Cavalry Carbine M.T.	-	M.L.M. & M.L.E. C.C.	# 136
M.L.M. & M.L.E. Aiming Tube	L	M.L.M. & M.L.E. M.T.	+ 2,521
S.M.L.E. Wake Aiming Tube	L	S.M.L.E. 1902 Trials Rifle	935
S.M.L.E. Aiming Tube	L	S.M.L.E. Mk I series	677
.22 Short Rifle Mk I	L	M.L.M. Mk I*	## 5,532
.22 Short Rifle Mk I*	L	M.L.M. Mk I*	## 22
.22 Short Rifle Mk II	N	M.L.M. Mk II & M.L.E. I*	## 1,000
.22 Short Rifle Mk III	L	S.M.L.E. Mks II Cd. & II* Cd.	over 11,000 (BSA & LSA: 8,432 Enfield: @ 2,374)
.22 Long Rifle Mk I	N	M.L.M. Mk II	?
.22 Long Rifle Mk I*	N	M.L.M. Mk II*	@ 3,178
.22 Long Rifle Mk II	L	M.L.E.	?
.22 Short Rifle Patt'14 No. 1	L	S.M.L.E. Mks II Cd. & II* Cd.	427
.22 Short Rifle Patt'14 No. 2	L	S.M.L.E. Mks III & IV Cd.	1,743
.22 Long Rifle Patt'14	L	C.L.L.E. Mk I*	?
.22 Short Rifle Patt'18	L	S.M.L.E. Mks III & IV Cd.	975
.22 Short Rifle Mk IV	L	S.M.L.E. Mks III & III*	Unknown
No. 2 Mk IV*	C	S.M.L.E. Mks III, III* & IV Cd., & new manufacture:	Britain: c.25,000 Ishapore: a few thousand Lithgow: a few thousand

#	*Converted as new production; many more downgraded .303 rifles were later converted.*		
+	*Minimum figure, possibly as high as 5,000.*		
@	*This quantity to March 1914 only. More are likely to have been made after this.*	L	*Land Service.*
##	*Possibly more; these are the only recorded figures.*	N	*Naval Service.*
?	*Uncertain, but not likely to have exceeded a few hundred.*	C	*Common to both Services.*

SHORT, MAGAZINE LEE-ENFIELD AIMING TUBE RIFLE Mk I

Calibre:	*.22 in.*
Rifling & Twist:	*8 groove, R.H.*
Barrel Length:	*25.2 in. (640mm)*
Overall Length:	*44.5 in. (1130mm)*
Weight:	*8 lb. 13 oz. (4.0 kg)*
Magazine Capacity:	*Single shot.*
Receiver Markings:	*A.T.*

The .230-in C.F. (.297/230) tube was approved for the Lee-Metford rifle in 1891, and the Lee-Metford carbine in 1895; upon the introduction of the Lee-Enfield rifle and carbine, the use of the Morris tube was extended to the new arms. After introduction of the .22-in. R.F. cartridge, some new .22-in. aiming tubes were approved for the Lee-Metford, Lee-Enfield and Short Lee-Enfield rifles, along with special boltheads.

The .22-in. A.T. S.M.L.E. rifle was approved on 13th November 1906, but it differed from the earlier Morris tube models in that it was not intended for the tube be removed so that the rifle could be used in its .303 form. Chambers of these A.T. rifles were modified so that .303-in. ball or blank ammunition could not be used and these rifles were stamped "A.T." on the action body, barrel, bolt, butt and fore-end. Boltheads were special, with an offset firing pin and new extractor.

Most of the 1,000 Short Lee-Enfield 1902 trials rifles were altered to A.T. rifles and because of the many different components to the service pattern Mark I S.M.L.E., they were designated "A.T. N.I.", the "N.I." to signify not interchangeable. This particular version of the A.T. rifle was approved on 1st February 1907.

Regarding its external appearance, the S.M.L.E. A.T. rifle is similar to the Mark I S.M.L.E., and if the aiming tube has been removed, it can be identified by the "A.T." markings and rimfire bolthead without the charger guide.

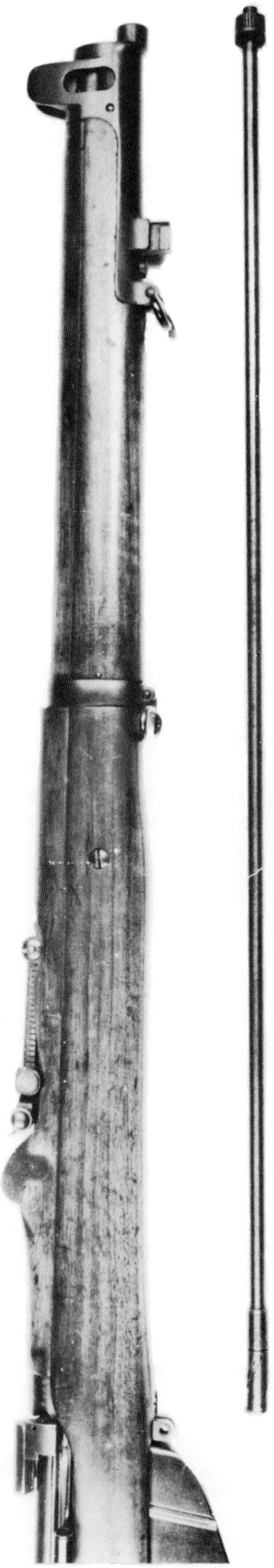

M.O.D. Pattern Room

.22 SHORT RIFLE Mk I

Calibre:	*.22 in.*
Rifling & Twist:	*8 groove, R.H.*
Barrel Length:	*25.2 in. (640mm)*
Overall Length:	*44.3 in. (1130mm)*
Weight:	*8 lb. 7 oz. (3.8 kg)*
Magazine Capacity:	*Single shot.*
Converted:	*Enfield.*

Typical Markings, L.H. side of the Butt Socket:

ENFIELD
1908
$S^{\underline{HT}}$ ·22
I

Approved on 13th December 1907, this was the first of a series of small bore trainers which replaced the Morris and Aiming Tubes. It was a conversion from the Mark I* Lee-Metford rifle and these markings will be noted on the right side of the butt socket, with the .22 conversion markings on the left side.

This particular conversion involved quite a few modifications and the result was quite a distinctive addition to the Lee-Enfield family. The magazine and cut-off plate were removed, a new .22 calibre barrel fitted, fore-end shortened by about 2.4 inches and the original Lee-Metford nosecap and front band replaced, although the sword bar of the nosecap was removed and the clearing rod and long range dial sight hole filled. Mark III S.M.L.E. sights with windage adjustment and a foresight similar to that of the Charger Loading Lee-Enfield (but marked ".22") were fitted. A modified Mark I S.M.L.E. rear handguard with sight protector was added.

The Lee-Metford bolt was altered by having the dustcover securing holes filled and cover stops and portion of the rib removed. The front of the bolt was modified to receive another bolthead and one-piece striker; in 1910 an improved bolthead with separate firing pin and striker was approved, the forward portion being contained in the bolthead itself. The bolthead is marked ".22". Gas escape vents in the receiver have been considerably enlarged. The standard Mark III S.M.L.E. backsight remains with the .303 graduations on the leaf, from 200 to 2,000 yards. Some of these rifles have been noted with locally fitted wooden "magazine" inserts to act as a loading platform.

.22 SHORT RIFLE Mk I*

Specifications:	*As for .22 Sht. Rifle Mk I.*
Converted:	*Enfield.*

Typical Markings, L.H. Side of the Butt Socket:

ENFIELD
1910
$S^{\underline{HT}}$ ·22
I*

Approved on 13th October 1911, this model differs from the Short Rifle Mk I in that the sighting was made to resemble that of the S.M.L.E. Mark III altered for Mk VII ammunition, with a lower foresight blade, straightened foresight protector wings and modified backsight ramp. The new bolthead with floating firing pin was used, designated the No. 2 pattern (approved on 15th December 1910).

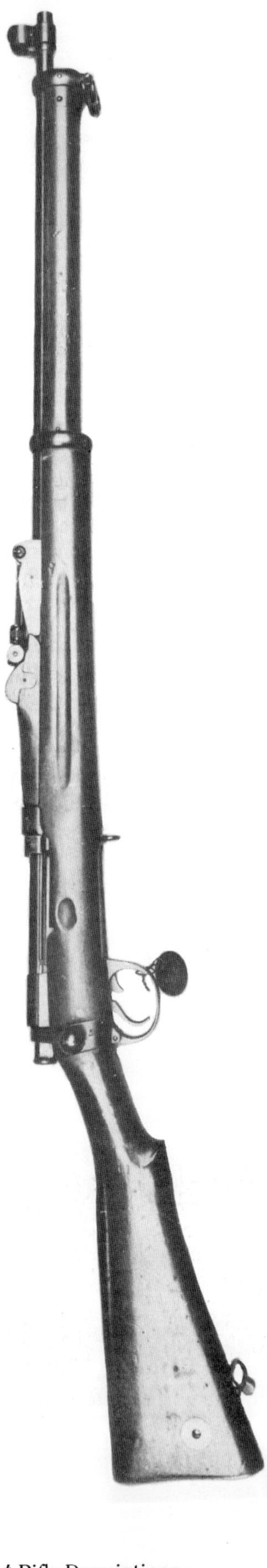

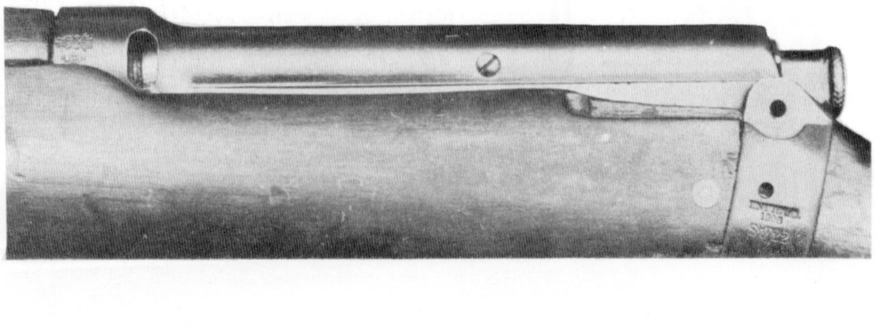

M.O.D. Pattern Room

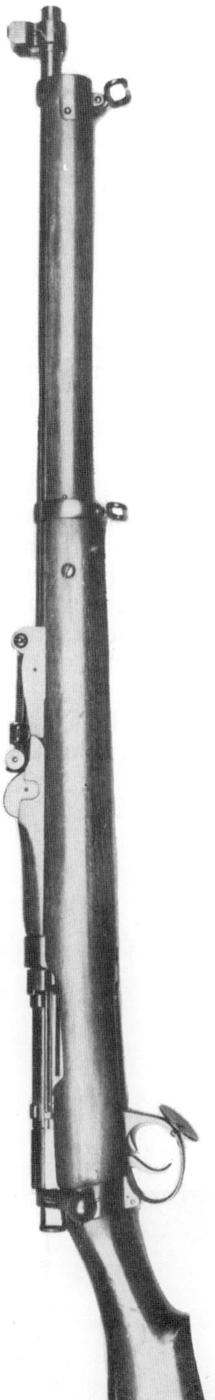

M.O.D. Pattern Room

.22 SHORT RIFLE Mk II

Calibre:	*.22 in.*
Rifling & Twist:	*8 groove, R.H.*
Barrel Length:	*25.2 in. (640mm)*
Overall Length:	*44.3 in. (1125mm)*
Weight:	*8 lb. 4 oz. (3.7 kg)*
Magazine Capacity:	*Single shot.*
Converted:	*Enfield.*

Typical Markings, L.H. side of the Butt Socket:

ENFIELD
1912
S^HT ·22
II

Approved for Naval service only, on 15th January 1912, this was intended to resemble the .22 Short Rifle Mk I as closely as possible, although the conversion was made from the Mark II Lee-Metford instead of the Mark I*, so there are no finger grooves in the fore-end and a different nosecap was fitted. Some rifles were also converted from Lee-Enfield models, with the safety catch on cocking piece. The long range dial sight and aperture were retained.

The barrel and sights, foresight protector with straightened ears, bolthead, firing pin and extractor, are interchangeable with those of the .22 Short Rifle Mark I*. The backsight is that of the Mark III S.M.L.E. and a Mark I* S.M.L.E. rear handguard was fitted, modified so that it would fit behind the Mk III S.M.L.E. backsight bed. Original .303-in. long rifle markings are stamped on the right side of the butt socket with the later .22-in. conversion markings on the left side. The original butt with its brass buttplate and trap are fitted.

Magazines and magazine cut-off plates were removed from these rifles in the conversion, as they were not intended for use with the .22-in. Short Rifle. The spring steel dustcover was removed from the bolt, although the securing lugs were not removed from all these bolts. Like the Marks I and I* Short Rifle trainers, the gas vents in the receiver have been noticeably enlarged. The backsight leaf retains the .303 calibrations and the ranges for the rimfire cartridge were worked out against these settings.

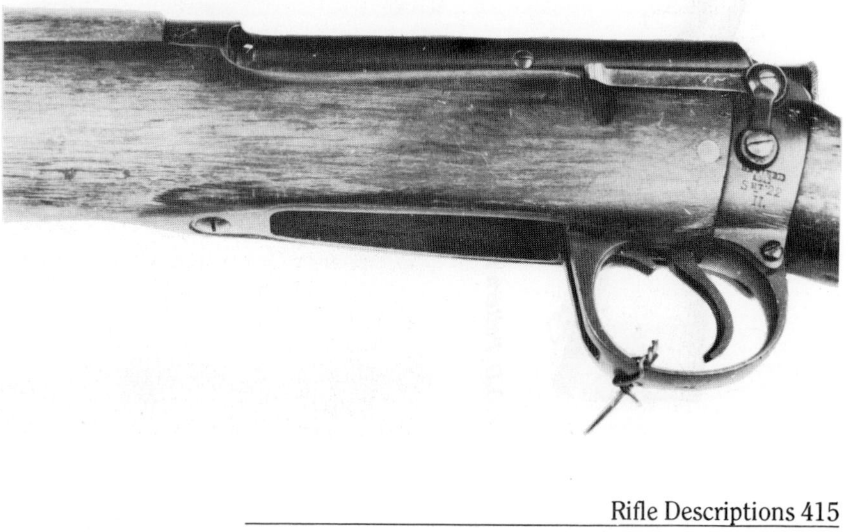

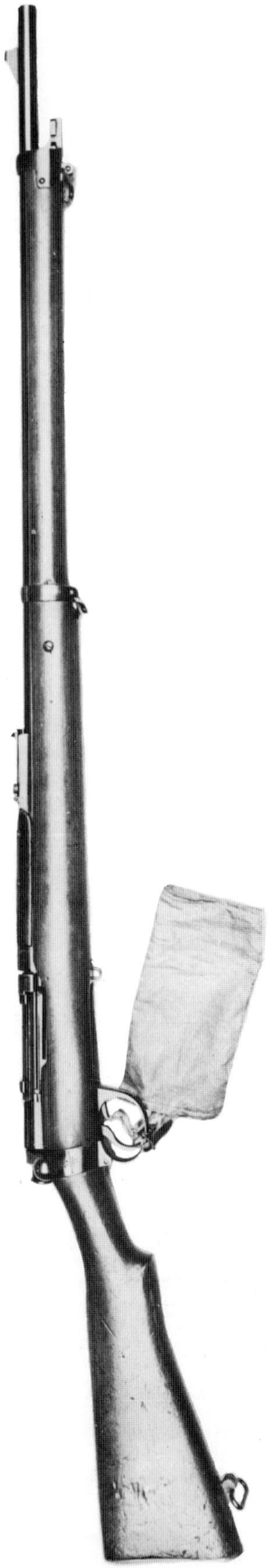

.22 LONG RIFLE Mk I

Calibre:	*.22 in.*
Rifling & Twist:	*8 groove, R.H.*
Barrel Length:	*30.2 in. (767mm)*
Overall Length:	*49.5 in. (1257mm)*
Weight:	*8 lb. 12 oz. (4.0 kg)*
Magazine Capacity:	*Single shot.*
Converted:	*Enfield.*

Typical Markings, L.H. side of the Butt Socket:

ENFIELD
1912
L^NG ·22
I

This first long rifle trainer was approved for Naval service on 7th February 1912, intended to provide practice for those armed with the Magazine Lee-Enfield rifle. Converted from the Mark II Lee-Metford rifle, the original stampings are found on the right side of the butt socket; conversion markings are stamped on the left side.

A solid .22 calibre barrel was fitted and the bolt cover, magazine and magazine cut-off were removed. The striker point was removed at the collar and a new bolthead with floating firing pin fitted, effecting a two-piece firing pin. As this conversion was made from the Mark II Lee-Metford, there is no safety catch on the bolt. All other features are identical to the Lee-Metford rifle, except for the weight; the new .22-in. Long Rifle is heavier due to the more weighty barrel.

The original sights are maintained, with their .303-in. graduations, from 200 to 500 yards on the sight bed and from 600 to 1,800 yards on the leaf. Long range dial sights were retained although obviously not intended for use.

M.O.D. Pattern Room

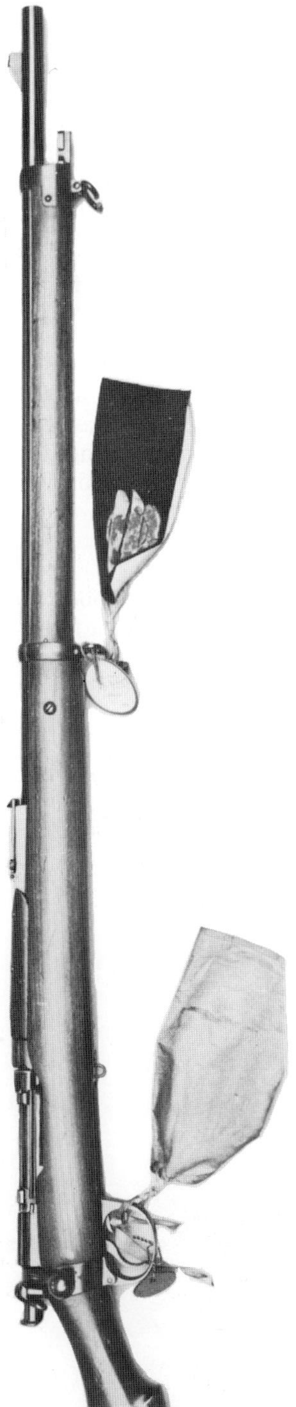

M.O.D. Pattern Room

.22 LONG RIFLE Mk I*

Calibre:	*.22 in.*
Rifling & Twist:	*8 groove, R.H.*
Barrel Length:	*30.2 in. (767mm)*
Overall Length:	*49.5 in. (1257mm)*
Weight:	*8 lb. 14 oz. (4.0 kg)*
Magazine Capacity:	*Single shot.*
Converted:	*Enfield.*

Typical Markings, L.H. side of the Butt Socket:

ENFIELD
1912
L$^{\underline{NG}}$ ·22
I*

This was another Naval service trainer and differs from its Mark I predecessor only in that the bolt was fitted with a safety catch as the conversion was made from the Mk II* Lee-Metford or Mark I or I* Lee-Enfield rifle. Therefore the bolt body, cocking piece and firing pin differ from those on the .22 Long Rifle Mk I. The markings of the rifle from which the conversion was made will be found on the right side of the butt socket and the conversion stamps were made on the left side of the action body, on the butt socket.

The .22-in. calibre barrel is a solid one of new manufacture and not a bored out .303-in. barrel fitted with a .22 calibre tube. The original sights were refitted to the new barrel. The bolt cover, magazine and magazine cut-off were dispensed with as they were no longer required. A two-piece firing pin is utilised, the front section being a floating, offset type contained within the bolt-head. This is the Pattern No. 2, also applied to other contemporary trainers.

Like the previous .22 conversions, the graduations of the original .303-in. sights remain and the different elevations for small bore shooting were taken from a table and read against the .303-in. markings. These original marked ranges are from 200 to 500 yards on the backsight bed, 600 to 1,800 yards on the sight leaf and 1,600 to 2,800 yards on the long range dial sight plate.

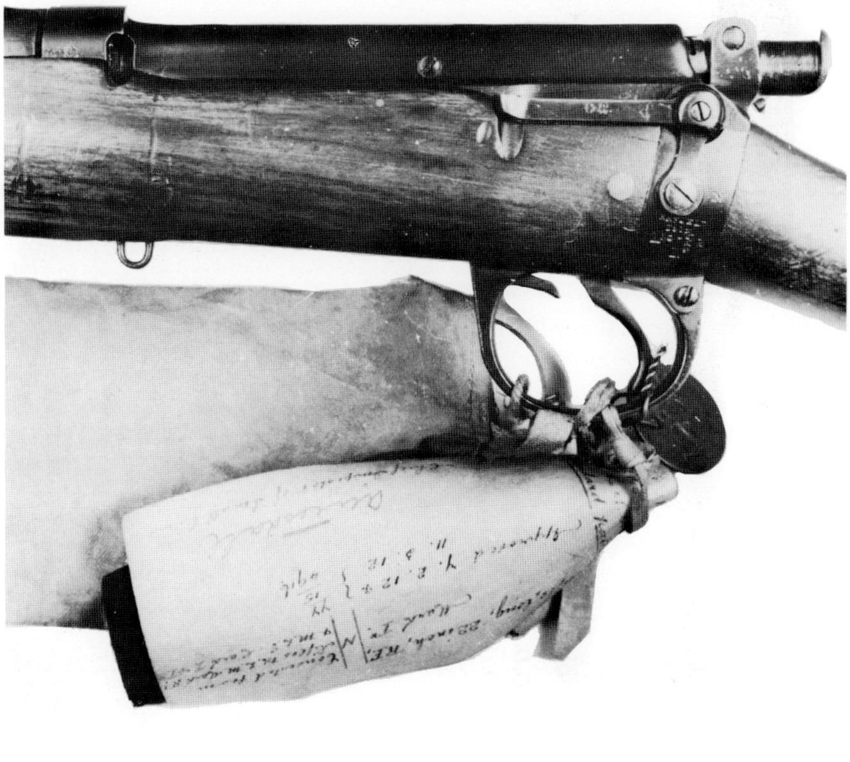

.22 LONG RIFLE Mk II

Calibre:	*.22 in.*
Rifling & Twist:	*8 groove, R.H.*
Barrel Length:	*30.2 in. (767mm)*
Overall Length:	*49.5 in. (1257mm)*
Weight:	*9 lb. 2 oz. (4.1 kg)*
Magazine Capacity:	*Single shot.*
Converted:	*Enfield.*

Typical Markings, L.H. side of the Butt Socket:

ENFIELD
1912
L^{NG} '22
II

This long rifle trainer was approved on 11th November 1911 and converted from the Mark II* Lee-Metford or Marks I or I* Lee-Enfield rifle for Army service. It was basically intended to resemble the Charger Loading Lee-Enfield as regards its sighting, although no charger bridge was fitted. The original markings of the rifle are found on the right side of the butt socket, while those of the conversion are marked on the left and comprise the factory, year of conversion, and rifle designation.

Whereas the previous .22-in. Long Rifles externally resemble the model from which the conversion was made, this Mark II version is fitted with a foresight protector, adjustable foresight, and backsight with windage adjustment, similar to the Charger Loading Lee-Enfield. The magazine, magazine cut-off and bolt cover were removed.

Like the preceding long and short rifle trainers, the original .303-in. calibrations were left on the backsight bed and leaf (up to 1,800 yards) because tables were used for conversion of the .303 reading to the .22-in. RF ranges. The long range dial sights were retained.

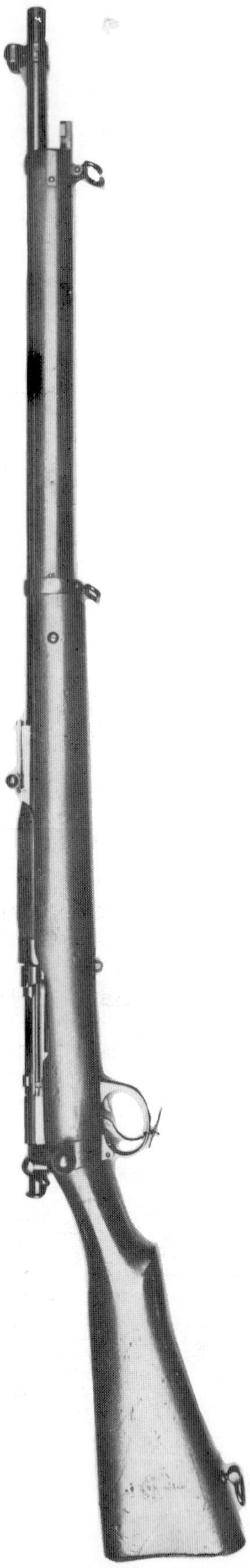

M.O.D. Pattern Room

.22 SHORT RIFLE Mk III

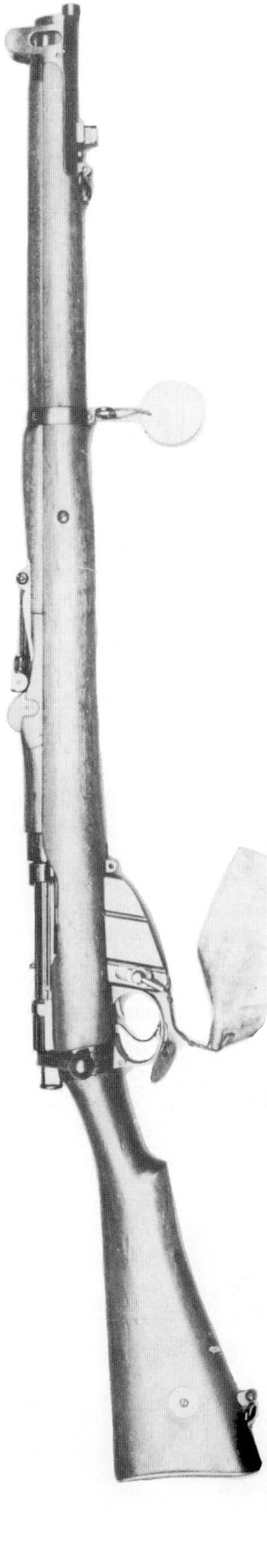

M.O.D. Pattern Room

Calibre:	*.22 in.*
Rifling & Twist:	*8 groove, R.H.*
Barrel Length:	*25.2 in. (640mm)*
Overall Length:	*44.5 in. (1130mm)*
Weight:	*9 lb. 4 oz. (4.2 kg)*
Magazine Capacity:	*5 rounds, Parker Hiscock.*
Converted:	*Enfield, B.S.A. Co., L.S.A. Co.*

Typical Markings, R.H. side of the Butt Socket:

V. R.
ENFIELD
1900
S^HT .22
III
EFD

Approved on 9th August 1912, this was made up from the Converted Mark II and II* S.M.L.E. rifles. It is an interesting model regarding its markings because the first markings were made on the right hand side of the butt socket as a Mark II or II* Lee-Metford or Mark I or I* Lee-Enfield, and the next set were stamped on the left side when it underwent .303 short rifle conversion. This left little space available for the .22 Short Rifle markings, so, in some cases, original markings on the right side were partly removed, leaving only the Royal cypher and year of manufacture. The new .22 markings were then made below these, without the conversion year; therefore the year marked above the .22 Short Rifle Mk III stamp does not usually apply to the conversion. Some of these rifles have also been noted with full inscriptions of both the original Lee-Enfield and .22-in. conversion.

A .22 calibre barrel similar to that of the .22 Short Rifle Mk I was fitted, along with the Pattern No. 2 bolthead with its offset, floating firing pin. The Mark I S.M.L.E. foresight protectors were straightened out and a Mark III backsight and bed fitted, also requiring minor modifications to the fore-end. This was done in order that the sighting resemble the form of the Mark III S.M.L.E. as much as possible. There was no regraduation of the backsight, only the standard .303-in. calibrations.

The Parker Hiscock magazine was approved on 21st December 1915, designated "Magazine .22-in., Rifle, Mark I" and fitted to the Mark III .22 Short Rifle. This magazine consists of a .22 magazine and feed mechanism within the S.M.L.E. magazine case. Because of the involved mechanism to feed the rimfire cartridges and a system which permits loading with the rifle bolt, it is considerably heavier than the standard .303-in. magazine. A lever on the right side of the case operates the spring loaded platform and permits loading the 5-round .22 magazine, then when it is released it puts tension on the cartridges so they are forced upwards into the boltway for loading into the chamber.

.22 RF PATTERN 1914 SHORT RIFLE No. 1

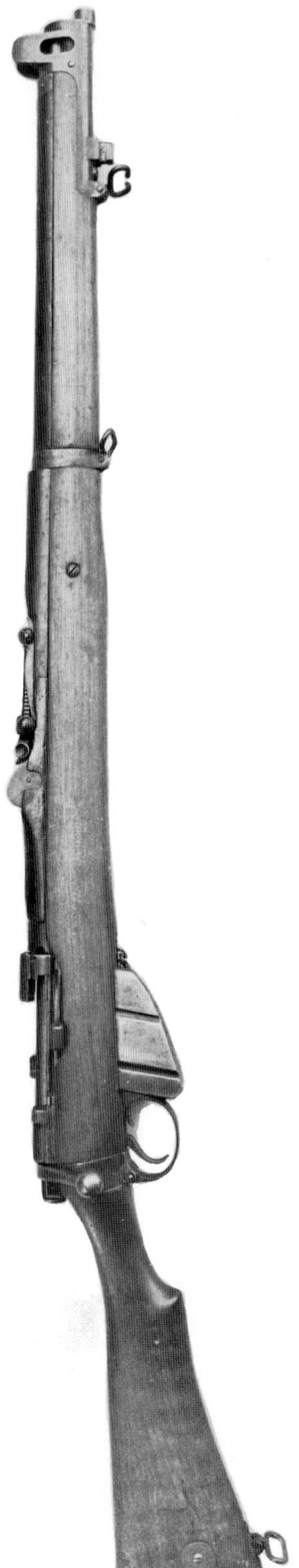

Calibre:	*.22 in.*
Rifling & Twist:	*8 groove, R.H.*
Barrel Length:	*25.2 in. (640mm)*
Overall Length:	*44.5 in. (1130mm)*
Weight:	*8 lb. 8 oz. (3.9 kg)*
Magazine Capacity:	*Single shot.*
Converted:	*A.G. Parker & Co. Ltd.* *Westley Richards & Co.*

Typical Markings, R.H. side of the Butt Socket:

This wartime trainer was approved on 24th May 1915, altered from the Converted Mark II and II* rifles. A new mode of smallbore conversion was implemented due to the wartime economies; the .303 barrel was bored out and a .22 calibre liner called a "Parker Tube" fitted inside the barrel. This was done by A.G. Parker and also by Westley Richards, both of Birmingham. The new designation and company effecting the conversion was stamped along with the original markings on the right side of the butt socket. The left side was already marked with the .303-in. Converted Mark II or II* inscription.

The original bolthead had the firing pin hole filled and a new off-centre hole drilled for the .22 RF firing pin. The extractor was altered and the original firing pin or striker cut short at the collar, where it conveys the impact to the floating firing pin in the bolthead. From 1925, the magazines were fitted again, without follower and spring; the empty magazine case serves only as a receptacle for the fired cases upon extraction. In external appearance, this rifle resembles its predecessor but for the absence of the charger guide on the bolthead, and the cut-off, and the fitting of U-notch backsight and frontsight of the Mark I** S.M.L.E. A 25-yd. graduation was marked on the rearsight.

When first introduced, the nomenclature was "22 RF Pattern 1914 Short Rifle" but upon the approval of the .22RF Pattern 1914 Short Rifle No. 2 on 28th April 1916, it became known as the ".22 RF Pattern Short Rifle No. 1".

Some commercial or "volunteer pattern" rifles were also converted to this pattern and these will be found to bear the name of the manufacturer or retailer in place of the usual government markings on the right side of the butt socket. An example of this is "WESTLEY RICHARDS & Co." over ".22 PATT '14 No. 1" and "A.G. PARKER BIRMINGHAM"; this example was a commercial Westley Richards rifle converted to .22 trainer by A.G. Parker.

T. Williams

.22 RF PATTERN 1914 SHORT RIFLE No. 2

Calibre:	*.22 in.*
Rifling & Twist:	*8 groove, R.H.*
Barrel Length:	*25.2 in. (640mm)*
Overall Length:	*44.5 in. (1130mm)*
Weight:	*9 lb. 0 oz. (4.1 kg)*
Magazine Capacity:	*Single shot.*
Converted:	*A.G. Parker & Co. Ltd.*
	Westley Richards & Co. ?

Typical Markings, R.H. side of the Butt Socket:

G. R.
ENFIELD
1914
S^HT L.E.
III
A.G. PARKER
& Co LTD
BIRMINGHAM
.22 PATT '14
No 2

This modification of the Mark III S.M.L.E. was approved on 28th April 1916 and as the alterations consisted of tubing the barrel and altering the bolt, it resembles its host rifle in external appearance. Whether a Mark III or Converted Mark IV rifle was utilised, the .22 conversion stamp was made on the right side of the butt socket, below the original markings. The .22 conversion date was not marked.

The original barrel was bored out and a .22 calibre liner inserted and soldered into place. The striker hole in the bolthead was brazed up and a new hole drilled for the floating firing pin. The extractor slot was slightly modified and a new extractor with longer claw fitted. Similar to the .22 Patt. '14 Short Rifle No. 1, the original striker was cut off at the collar, where it now comes into contact with the firing pin in the bolthead. A 25-yd. graduation is marked on the backsight and the cut-off for the magazine has been removed.

When this trainer was first introduced, the magazine was discarded, but in 1925 it was ordered that magazine cases, without the platform and spring, be fitted in order that the spent cases be collected.

T. Williams

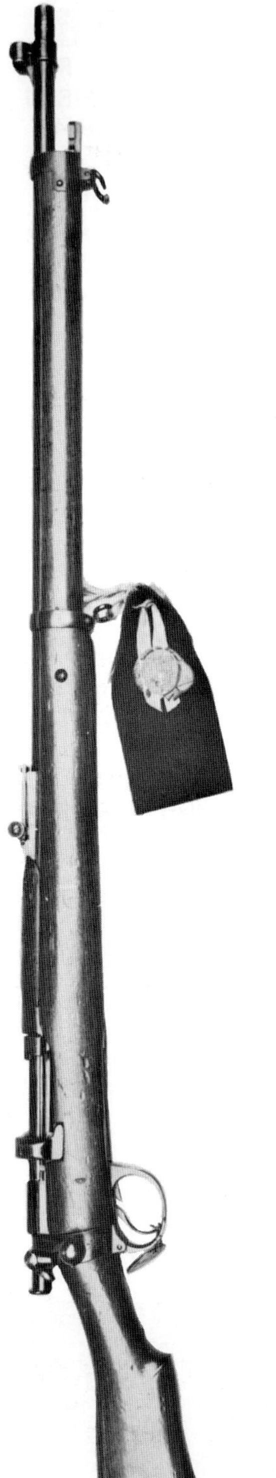

.22 RF PATTERN 1914 LONG RIFLE

Calibre:	*.22 in.*
Rifling & Twist:	*8 groove, R.H.*
Barrel Length:	*30.2 in. (767mm)*
Overall Length:	*49.5 in. (1257mm)*
Weight:	*9 lb. 0 oz. (4.1 kg)*
Magazine Capacity:	*Single shot.*
Converted:	*A.G. Parker & Co. Ltd.*
	Westley Richards & Co. ?

Typical Markings, R.H. side of the Butt Socket:

V. R.
ENFIELD
1898
L.E.
I
A.G. PARKER
& Co LTD
BIRMINGHAM
.22 PATT
'14

Approved on 28th April 1916 for Land Service only, this trainer was a conversion of the Charger Loading Lee-Enfield Mark I*. Thus, it resembles the C.L.L.E. Mk I* in external appearance as the modifications were made only to the barrel and the bolt. The original Lee-Enfield or Lee-Metford markings are found on the right side of the butt socket, with the .22 rimfire conversion marks underneath. The left side bears the C.L.L.E. conversion marks.

The barrel was bored out and a .22 calibre Parker Tube fitted, the bolthead has had the firing pin hole brazed up and a new hole drilled off centre for the small floating firing pin in the bolthead. The original striker in the bolt body was shortened back to the collar, so this bolt has a two-piece firing pin. A 25-yd. mark is stamped on the right side of the backsight bed.

In the original conversion, the magazine and cut-off were removed but in 1925, magazine cases were again fitted to the .22 trainers, with their platform and spring removed so as to act as a receptacle for the fired cases after extraction. Contract records do not indicate whether any of the Long Rifle patterns were converted by Parker or Westley Richards and it may well be that the illustrated specimen was a sample or pattern only.

M.O.D. Pattern Room

.22 RF SHORT RIFLE PATTERN 1918

Calibre:	*.22 in.*
Rifling & Twist:	*8 groove, R.H.*
Barrel Length:	*25.2 in. (640mm)*
Overall Length:	*44.5 in. (1130mm)*
Weight:	*9 lb. 0 oz. (4.1 kg)*
Magazine Capacity:	*10 rounds.*
Converted:	*W.W. Greener.*

Typical Markings, R.H. side of the Butt Socket:

G. R.
ENFIELD
1915
S$^{\underline{HT}}$ L.E.
III
.22 PATT '18

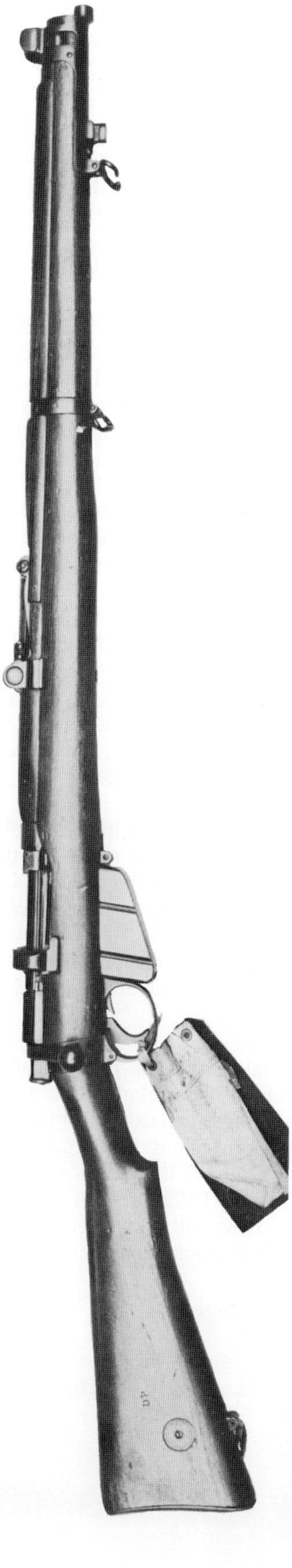

Approved on 10th July 1918, this rifle is unique in the field of British smallbore trainers in that it utilises a mock .303 cartridge or "conveyor", loaded with the .22-in rimfire cartridge. Like the .22 Patt. '14 Short Rifle No. 2, it was modified from the Mark III and Converted Mark IV S.M.L.E. rifles and has a tubed barrel.

The Parker rifling system was used and involved boring out the original barrel and soldering in a .22 calibre tube. Thus, the earlier sights remain and the barrel was marked "PARKERIFLED", the top of the receiver was stamped "PARKER-HALE .303 CUM .22 SYSTEM". The bolt and bolthead were modified in the same way as the .22 Patt. '14 in that an offset, floating firing pin if fitted to the bolthead, which has the original striker hole filled in and the S.M.L.E. firing pin cut off at the collar. The barrel is still chambered for the .303 cartridge and the .22-in. tubing commences after the chamber. A special ejector is fitted onto the left side of the action body.

Each rifle was issued with 30 conveyor cartridges and a tool to remove the spent .22RF cases from the conveyor or mock cartridge. The original magazine was retained and loaded with the conveyors, which enabled magazine fire and charger loading; a feature not available with any other .22-in. training rifles. However, this system does not seem to have been very successful as it was declared obsolete in 1920.

Some commercial variants have been noted marked "PARKER-HALE .303 CUM .22 SYSTEM", with centrefire bolts and obviously intended for the conveyor with its own off-set firing pin. Such conveyors have been examined with "EFD" markings.

M.O.D. Pattern Room

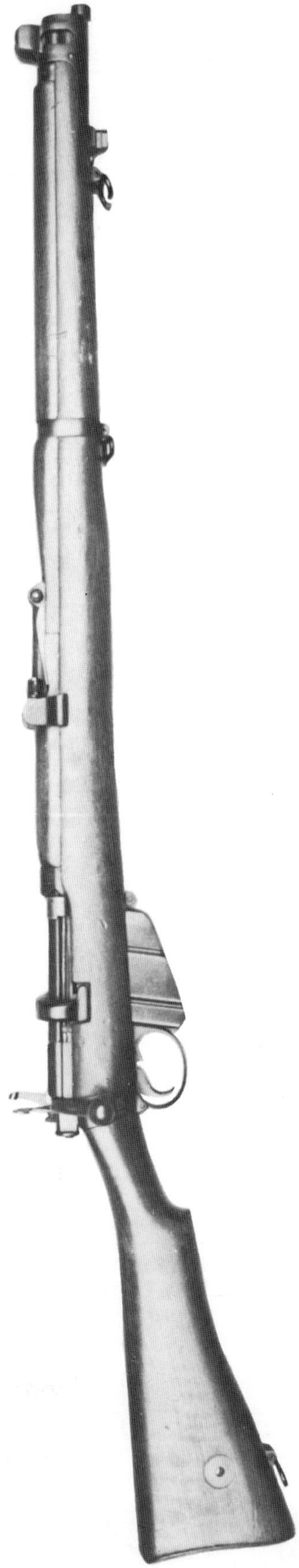

.22 SHORT RIFLE Mk IV (RIFLE No. 2 Mk IV*)

Calibre:	*.22 in.*
Rifling & Twist:	*8 groove, R.H.*
Barrel Length:	*25.2 in. (640mm)*
Overall Length:	*44.5 in. (1130mm)*
Weight:	*9 lb. 2 oz. (4.1 kg)*
Magazine Capacity:	*Single shot.*
Converted & Manufactured:	*Enfield, B.S.A., Ishapore, Lithgow.*

Typical Markings, R.H. side of the Butt Socket:

G. R.
ENFIELD
1912
SHT ·22
IV*

Approved on 19th November 1921, this was a conversion from the Marks III and III* S.M.L.E. In the 1926 nomenclature changes, it became the Rifle No. 2 Mk IV*. Because some are conversions and others assembled from new components in Australia and India as well as England, for more than 25 years, the markings will be found to vary considerably. In the case of earlier British-made rifles, the original designation markings were removed and new ones stamped below the original cypher, factory and year of manufacture, so that earlier dates than those of the .22 conversion are often found above the conversion markings. The example here is dated 1912, but the conversion was not approved until 1921. Australian and Indian converted models are usually marked in a different fashion.

The barrels are solid, not tubed, and the magazine (added after May 1925) is stamped ".22" with the follower and spring removed. The bolthead is marked ".22 No. 2" and has a floating, offset firing pin and larger extractor than that fitted to the .303-in. model. The rear part of the two-piece firing pin is housed in the bolt and attached to the usual S.M.L.E. cocking piece. Externally, this rifle is identical to the Mark III or III* from which the conversion was made.

A special aperture sight was introduced for .22-in. Short Rifles in 1917, called "Sight, Auxiliary, Aperture Mk I", and fitted in place of the long range aperture. The new long range aperture stem has graduations marked on the side and a flat aperture plate slides vertically on this, locked by a knurled screw. During World War 2, some British gunsmiths such as W. Darlow, Parker Hale, R.B. Page, W.W. Greener and Alex. Martin, were involved in conversion and the supply of parts for these rifles. Some new No. 2 Mark IV* rifles were also made up at Enfield at this time.

Australian No. 2 Mark IV* rifles were still being converted in the 1950's, many of these were effected by armourers rather than S.A.F. Lithgow. Markings vary from "No. 2 Mk IV*" on the butt and receiver, to no markings at all except for those on the bolthead and magazine case. Some Indian conversions have commercial B.S.A. .22-in. barrels and are marked on the right sight of the receiver with the year followed by the rifle designation. e.g. "1955 SHT .22 IV*". In instances where this rifle was still in service in the 1950's, it was often referred to as the No. 2 Mk 4/1, after the "*" indicator was deleted from official nomenclature.

M.O.D. Pattern Room

THE NO. 4 SERIES

The fundamental configuration of the No. 4 rifle made its debut in 1926 as the No. 1 Mark VI, an exercise in markedly improving certain aspects of the S.M.L.E. found wanting in the Great War. These were principally the application of aperture sights, a shorter bayonet and a heavier barrel for better accuracy. This new trials rifle was redesignated the No. 4 Mark I in 1931, although the No. 4 has a higher, flat-sided wall on the left side of the boltway and incorporated other notable changes to the action body, furniture and fittings. The concept of this new series was also to simplify manufacturing procedures and utilise more modern machining techniques, thus minimising the required hand fitting and finishing. The No. 4 also incorporates improvements such as aperture sights, heavier barrel and lighter bayonet with new form of mounting, as introduced on the No. 1 Mk VI. While the early test and trials rifles were made at Enfield, wartime production was carried out at Fazakerley, Maltby and B.S.A. Shirley. The No. 4 rifle was also manufactured in the United States and Canada, where the total production figures exceeded those quantities from the British factories.

World War 2 brought about the adoption of the No. 4 rifle, approved on 15th November 1939, although its production did not commence until 1941. Many of the components were supplied by contractors and the government Maltby factory acted more as an assembly point. Manufacture at Long Branch in Canada and Stevens-Savage in the United States also commenced late in 1941. Because a variety of production shortcuts were employed, quite a few differences may be encountered in many of the components. At both the U.S and Canadian factories, the Mark I* variation was produced from 1942, although it was not officially announced in the List of Changes until 1946. This modification further simplified manufacture of the receiver although this was not done in British production. Post-war design remained with the Mark I configuration and many of the wartime economies utilised in component manufacture were discontinued. A special lightweight model was trialled in Canada, but only a few prototypes were produced there.

As the war extended into South-East Asia, requests were put forward for a shorter, lighter rifle and so the No. 5 Mk I, or as it is more colloquially known, the "Jungle Carbine", was introduced in 1944. Whilst its utility was unquestionable, precise and consistent shooting of this lightened No. 4 rifle proved to be a serious problem; a Mark II version was proposed but the "wandering zero" dilemma could not be remedied; further details on the No. 5 series will be found in the subsequent section, "Jungle Carbines".

After the war was over, trials were in continued in the quest for a new cartridge and automatic rifle, but the No. 4 was not yet destined for obsolescence. The No. 4 Mark 2 was approved in December 1947, featuring an improved trigger mounted onto the receiver body instead of the trigger guard. Previous models of the No. 4 rifle were then converted to the new pattern, resulting in the No. 4 Mark 1/2 and 1/3 patterns.

A new self-loading rifle, the Belgian F.N. F.A.L., along with the rimless 7.62mm N.A.T.O. cartridge was eventually adopted, this rifle being designated the Rifle L1A1. Some No. 4 rifles were then converted for the new cartridge and an updated system of nomenclature adopted, abolishing the "No." and "Mark" which had been in use for about one hundred years. These upgraded rifles became the L8A1, L8A2, L8A3, L8A4 and L8A5 patterns, mainly allocated to reserve stores and for use by police forces. Special target and sniper versions became the L39A1 and L42A1; these are described in the later section on 7.62mm models.

The L59 is included in this category also as it is a training model of the No. 4, sectionised to a limited way. Cuts into the action and barrel chamber were done to make its D.P. status quite apparent rather than to illustrate the mechanism. These special machining cuts and modifications render the weapon innocuous and have been done in such a way as to prevent easy re-activated by replacement of certain parts. The L59A1 is a current issue to Cadet forces in the United Kingdom.

	Enfield	*Fazakerley*	*Maltby*	*Shirley*	*Long Branch*	*Savage*
Rifle No. 4 Mk I Trials:	c.2,500	-	-	-	-	-
Rifle No. 4 Mk I:	-	619,913	737,000	665,000	-	-
Rifle No. 4 Mk I & I*:	-	-	-	-	910,730	1,236,706
Rifle No. 4 Mk 2:	c.450,000	-	-	-	-	-
Rifle No. 4 Mks I/2 & I/3:	c.10,000?	c.350,000	-	-	-	-

RIFLE No. 4 Mk I; TRIALS MODEL

Calibre:	*.303 in.*
Rifling & Twist:	*5 groove, Enfield, L.H.*
Barrel Length:	*25.2 in. (640mm)*
Overall Length:	*44.5 in. (1130mm)*
Weight:	*9 lb. 0 oz. (4.1 kg)*
Magazine Capacity:	*10 rounds.*
Manufactured:	*Enfield.*

Typical Markings, L.H. side of the Butt Socket:

G R
ENFIELD
1931
N°4 MkI
A 0485

The No. 1 Mark VI rifle was changed slightly and redesignated the No. 4 Mark I in mid-1931; some test batches were manufactured for a few years afterwards to undergo a variety of trials and tests including production techniques and processes. In most respects these early No. 4 rifles are similar to the No. 1 Mark VI trials rifles except for the action body shape and designation markings, as well as the obvious absence of chequering on the fore-end.

The serial numbers have an "A" prefix and even though the No. 1 Mark VI rifles also used this numbering sequence, the numbers were re-commenced at A 0001 upon the changeover to the No. 4. Some of these rifles were made without any action markings and have their serial number prefixed by "EX". These were mostly prototypes for testing machinery and assembly. Features of the Mark VI that are apparent on these rifles are the specially shaped foresight protector, cocking piece, magazine cut-off and brass butt marking disc. The recess in the receiver behind the safety catch, which is a feature of the Mark VI rifle, only appears on some of the trials No. 4 models.

Aperture backsights are fitted, similar to the Mark VI, graduated to 1,300 yards. A battle aperture is used when in the folded down position and the slide is moved by a click stop adjustment screw for range alteration in the raised position. The foresight is a blade type, available in varying heights.

Because of the acute shortages of rifles after Dunkirk, most of these Enfield trials rifles were refurbished with wartime No. 4 parts and put into service, although many of these retained the special foresight protector and cocking piece of the original rifle. Quite a few of the trials rifles were also converted to No. 4 Mk I (T) sniper rifles.

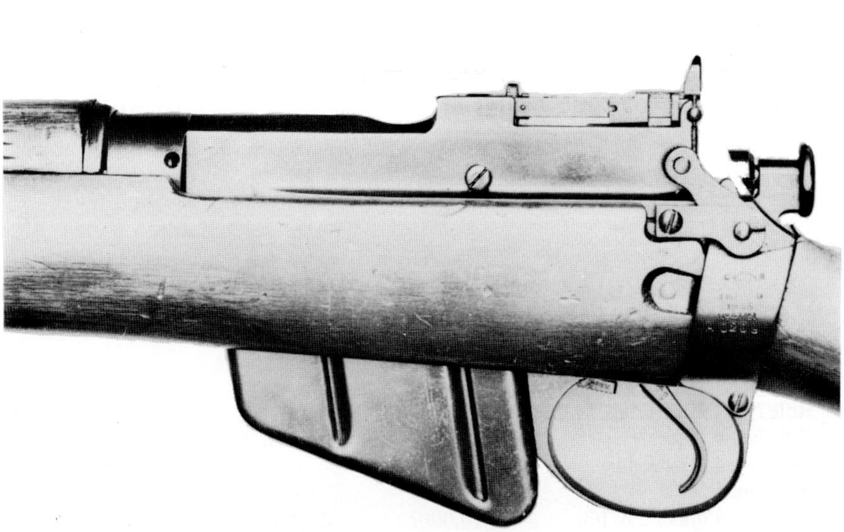

M.O.D Pattern Room

B. Labudda

RIFLE No. 4 Mk I

Calibre:	*.303 in.*
Rifling & Twist:	*5 groove, Enfield, L.H. +*
Barrel Length:	*25.2 in. (640mm)*
Overall Length:	*44.4 in. (1128mm)*
Weight:	*8 lb. 11 oz. (3.9 kg)*
Magazine Capacity:	*10 rounds.*
Manufactured:	*Fazakerley, Maltby, B.S.A. Shirley, Long Branch, Stevens-Savage.*

Typical Markings, L.H. side of Receiver:

No4 Mk1 ROF(F)
2/45 PC 26277

The service No. 4 Mark I was approved on 15th November 1939 and was a development of the No. 1 Mark VI of 1929 and the No. 4 trials rifles of the early 1930's. It was manufactured at a number of factories in England and North America; but none of these wartime rifles were produced at Enfield, the traditional home of the Lee-Enfield.

A stronger action than the S.M.L.E., it is more flat sided, easier to manufacture, and the charger guide brackets are an integral part of the receiver. The bolt-head release catch is located just behind the charger bridge and a groove in the bolt-head engages on a rib along the right inside of the receiver instead of outside as on the service No. 1 rifle. Aperture sights are mounted at the rear of the receiver and although four marks of backsight may be encountered, the earlier rifles were fitted with the Mk I pattern which was reverted to after the war.

The machined Mk I backsight has a leaf graduated from 200 to 1,300 yards in 50-yd. increments and a battle aperture for the folded down position; adjustment is by an elevation screw with click stops. The Mk II sight is L-shaped, an interim measure comprising two fixed apertures ranged at 300 and 600 yards. The Marks III and IV sights are made of pressed metal, but similar in basic design to the Mk I, with leaf graduated to 1,300 yards and battle aperture for use in the folded down position. These slides are operated by pressing a release catch and moving up or down to the desired range. Differences between the Mark III and IV are largely in the form of the release catch.

A larger diameter barrel is used in the No. 4 and unlike that of the S.M.L.E., it extends about 2.8 ins. out of the fore-end. The longer set of lugs near the muzzle are for securing the sight protector and the other unequal pair are for fixing the bayonet or grenade launcher; the muzzle is squared for the grenade attachment. The buttplate is fitted with a trap and sling swivels are secured to the butt and lower band.

As the No. 4 was of wartime production and many components were made by assorted factories and contractors, some variations will be found in components such as the bands, cocking pieces, sight protectors, trigger guards, foresights and mounts; relevant details are given in Chapter 5. The screw thread of the No. 4 rifle is different to that of the No. 1 and most parts are not interchangeable between these rifles.

+ Many North American No. 4 rifles were fitted with 2-groove barrels, some later had 6-groove rifling.

I. Skennerton

RIFLE No. 4 Mk I*

Calibre:	*.303 in.*
Rifling & Twist:	*2 groove, Enfield, L.H. +*
Barrel Length:	*25.2 in. (640mm)*
Overall Length:	*44.4 in. (1128mm)*
Weight:	*8 lb. 10 oz. (3.9 kg)*
Magazine Capacity:	*10 rounds.*
Manufactured:	*Long Branch, Stevens-Savage.*

Typical Markings, L.H. side of Receiver: S No 4 MK I*

This variant of the No. 4 was the result of economies in manufacture designed in Britain in 1941 and approved for production on 14th June 1941, but only implemented in North America and not announced in the List of Changes until after the war, another instance of "tidying up the paperwork" afterwards. The change from Mark I to I* occurred in America in 1942 and the Mk I* model represents the bulk of American production. A few Maltby No. 4 Mk I rifles have also been noted, incorrectly marked as "MK I*".

Essentially, the difference between the Mark I and I* is the mode of releasing the bolthead; on the Mark I* this is merely a slot cut into the ribway near the front of the receiver, the Mark I has a spring operated catch at the rear of the charger bridge. A new bridge piece was fitted to the Mark I* giving greater clearance so that the bolt can be withdrawn with the bolthead in the disengaged position. The magazine catch screw is replaced by a pin and the sear pin is slightly longer.

Regarding its stocking-up, barrel and general furniture, the Mark I* is similar to the Mark I, although various production shortcuts were employed in the barrels, sights, bands, trigger guards, buttplates and cocking pieces, many of which are encountered on the Mark I* model. Those rifles produced in the United States under the Lend Lease agreement are stamped "U.S. PROPERTY" alongside the usual markings.

With the closure of the North American factories, the Mark I* was no longer manufactured; the later British rifles such as the No. 8 utilise the Mark I design with bolthead release catch behind the charger bridge. The Canadian No. 7 .22-in. rifle, however, was built on the Mark I* receiver.

+ While 2-groove barrels are most common, 5 and 6-groove models may be found on the No. 4 Mk I* rifle. Some Long Branch barrels of the 1950's have 6 grooves with R.H. twist instead of L.H. twist.

RIFLE No. 4 Mk 2

Calibre:	*.303 in.*
Rifling & Twist:	*2 groove, Enfield, L.H. +*
Barrel Length:	*25.2 in. (640mm)*
Overall Length:	*44.4 in. (1128mm)*
Weight:	*9 lb. 1 oz. (4.1 kg)*
Magazine Capacity:	*10 rounds.*
Manufactured:	*Fazakerley.*

Typical Markings, L.H. side of Receiver:

NO 4 MK 2
UF 55 A 1471

Introduced on 31st March 1949, the Mk 2 model mainly remedied the problems associated with the trigger pull; although this improvement was originally approved in 1947. Some of the earlier fittings like the Mark I backsight and the machined rather than pressed or fabricated parts, were used again, and so economies and concessions permitted due to wartime pressures were generally dispensed with. The serial numbers are usually prefixed by "PF", although later rifles have an "A" prefix.

The No. 4 Mk 2 has a noticeably better quality of finish compared with the wartime British No. 4 rifles and it is usually stocked in beech which is much lighter in colour than previous timbers. The Mark 2 rifle has the trigger mounted directly onto the receiver instead of on the trigger guard; this may be readily ascertained by the presence of a long, transverse screw through the rear of the fore-end. Because of the new trigger mounting, additional wood had to be removed from inside the back of the fore-end.

The Mk I backsight is graduated from 200 to 1,300 yards with click-stop elevation and a battle aperture for use in the folded down position. Like the earlier No. 4 rifles, a protector is attached to the foresight block and barrel by a screw, metal caps are riveted to the tip of the fore-end and handguards. Sling swivels are mounted onto the butt and lower band.

RIFLE No. 4 Mk 1/2

Specifications:	*As for Rifle No. 4 Mk 2.*
Converted:	*Fazakerley, Enfield.*

Typical Markings, L.H. side of Receiver: N^O^4 M^K^1/2(F)FTR 49

Also introduced on 31st March 1949, this was a conversion of the No. 4 Mark I to the No. 4 Mk 2 pattern. The alteration of markings on the left side of the receiver is usually quite obvious and some North American made rifles were also converted. The new markings were made by an electric pencil instead of stamps; a 5-groove barrel was generally fitted.

This conversion also included the upgrading of component parts in cases where certain wartime economies had been applied, so the Mark I backsight and machined parts were usually fitted. Mounting brackets were brazed onto the action body to locate the trigger, which was removed from the trigger guard; the metal re-inforcing strap and portion at the back of the wooden fore-end was removed and a special transverse screw fitted through the fore-end at this point. As this conversion included a Factory Thorough Repair, the receiver was marked "F.T.R." and new furniture and fittings fitted where required.

J. Page

RIFLE No. 4 Mk 1/3

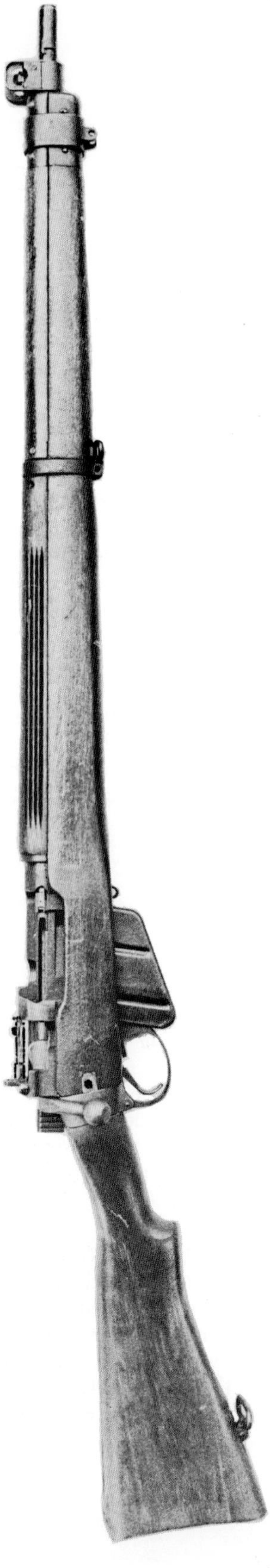

Calibre:	*.303 in.*
Rifling & Twist:	*2 groove, Enfield, L.H. +*
Barrel Length:	*25.2 in. (640mm)*
Overall Length:	*44.4 in. (1128mm)*
Weight:	*8 lb. 15 oz. (4.0 kg)*
Magazine Capacity:	*10 rounds.*
Manufactured:	*Fazakerley.*

Typical Markings, L.H. side of Receiver:

No4 Mk 1/3(F)FTR 49
LONG BRANCH
1944

Officially introduced on 31st March 1949 along with the Marks 2 and 1/2, this was a conversion of the No. 4 Mk I* up to the Mark 2 specifications, which have the receiver mounted trigger. The original Long Branch or Stevens-Savage markings are usually evident on the receiver and the alteration to the nomenclature is quite obvious, made with an electric pencil.

This conversion also included upgrading components in cases where certain wartime alternatives had been utilised and so the Mk I backsight is usually fitted along with machined parts. Some of the Mk III or IV rearsights have been noticed on certain rifles, especially reserve, Naval or Air Force issues. The Mk I, III and IV backsights are graduated from 200 to 1,300 yards and have a battle aperture for use in the folded down position.

Because this conversion involved brazing lugs onto the receiver so that the trigger could be attached directly onto the action body, it also required the removal of a portion at the rear of the wooden fore-end, along with the small metal strip reinforcing cap. A transverse screw was fitted through the rear of the fore-end and this section was often plugged with a wooden insert on each side. Therefore, the trigger guard now acts solely as a guard because the trigger and pin have been removed.

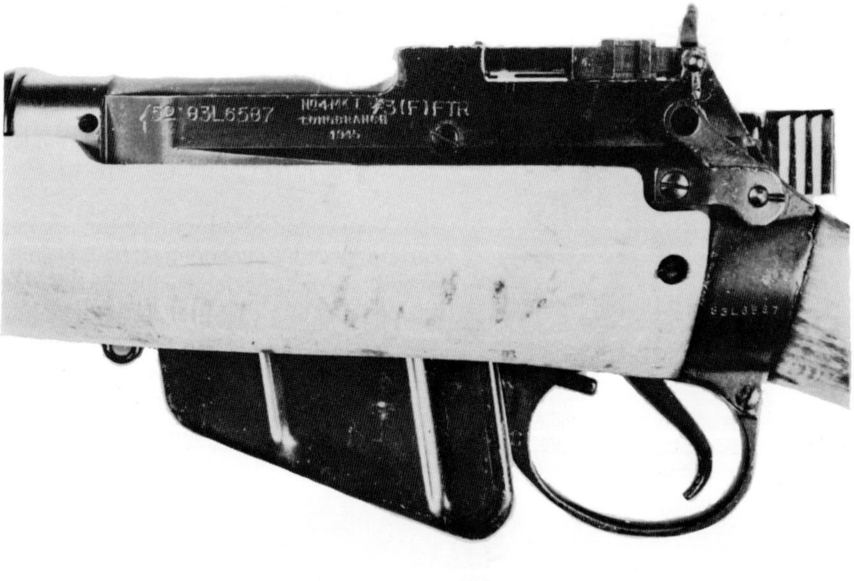

M.O.D. Pattern Room

RIFLE DRILL PURPOSE L59A1

M.O.D. Pattern Room

Calibre:	*.303 in.*
Rifling & Twist:	*5 groove, Enfield, L.H.*
Barrel Length:	*25.2 in. (640mm)*
Overall Length:	*44.4 in. (1128mm)*
Weight:	*8 lb. 8 oz. (3.8 kg)*
Magazine Capacity:	*10 rounds.*
Converted:	*Enfield, Field & Base Ordnance Depots.*

Typical Markings, L.H. side of Receiver:

DRILL RIFLE
L59 A1

Reference to the L59AI first appears in the mid-1970, a drill purpose conversion of the No. 4 rifle for cadets. Two methods of conversion were applied; some were factory machined, others effected by hand at field and base workshops mainly using a drill and jig. The No. 4 Mk I and Mk 2 variants from which the conversions were made were manufactured at the various English and North American factories, so various original markings will be encountered. Downgraded and reject components were often applied in the re-builds.

Sections of the body on both left and right hand sides are machined away so that the action is seriously weakened. The bolt and bolthead are also sectionised in critical sections and the barrel welded to the action body. The bolt is copper plated and the bolthead striker hole welded up. The breech was also machined from the left side, through the fore-end and barrel.

Some skeleton actions were also converted, the barrel drilled, pinned and welded through the receiver ring, bolthead striker hole plugged, striker ground, breech bolt body sectioned and weakened, and the body rib drilled and pinned.

The designation "DRILL RIFLE L59 A1" is engraved on the action and the original markings barred out. All bright surfaces, due to machining, were painted black. "DP" was marked on the bolt and magazine and painted in black on a white background on the left side of butt and underneath the fore-end. The Mk I No. 4 backsight is usually fitted, the leaf graduated from 200 to 1,300 yards.

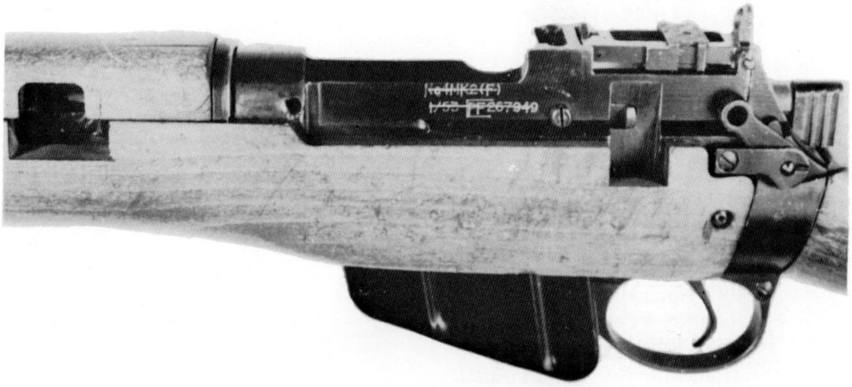

JUNGLE CARBINES

Since the Magazine Lee Cavalry Carbines and the New Zealand and Royal Irish Constabulary models of the late 19th century, there had been no service requirement for a shorter and lighter .303-in. rifle until 1942. The concept of the short rifle was to provide a standard arm for all services but the jungle warfare in South-East Asia against the Japanese brought with it some new criteria; high on the list was a lightened, shortened rifle. When these first prototypes were assembled, it was merely referred to as the "No. 4 lightened rifle".

Eventually it was designated the No. 5, following its progenitor in numerical sequence. It met the new requirements, comprising of a half-stocked No. 4 with shorter barrel, flash eliminator, new knife bayonet and special butt with rubber kickpad. The action is that of the No. 4 with relatively minor lightening modifications. Problems were encountered with its ability to retain zero so changes were later made to the trigger, similar to that applied to the No. 4. This was to have been the No. 5 Mk 2 but the idea was discarded when the No. 5 was decided against as the service rifle. A strengthened action was manufactured for grenade firing trials and further tests made with stocking-up but the consistent accuracy problems remained unsolved and so the No. 5 design was abandoned. Like many of the wartime production rifles, some components for the No. 5 were supplied by private contractors.

The No. 5 rifle production was also planned in Canada to fill British contract orders; the required tooling was set up and a few prototypes assembled at Long Branch, but as the end of the war in the Pacific drew to a close, Britain cancelled the orders and the North American production program was terminated. A special lightweight No. 4 rifle was produced at Long Branch for trials, designed for jungle warfare, but this never went into general production.

A similar "jungle carbine" configuration of the No. 1 rifle was allocated the No. 6 designation and although experiments had been carried out in England with the No. 1 Mark III, Mark III* and Mark V altered to this carbine configuration, it was in Australia that the program was implemented, although only at troop trials and production feasibility levels. Because the Lithgow and Ishapore factories produced the No. 1 action rather than the No. 4, the No. 6 was promoted as the Australian and Indian "jungle carbine".

Two models were approved in Australia, the No. 6 Mark I and Mark I/I; the Mark I has the standard bed backsight like the No. 1 Mark III* rifle and the Mark I/I has aperture sights somewhat similar to the British No. 5. This new design was contemplated as a mode of conversion for the No. 1 Mark III* rifles already in Australian service. Although an order was placed for 40,000, only trials rifles and some prototypes were produced before the requirement was rescinded soon after the end of the war.

	Enfield	*Fazakerley*	*BSA Shirley*	*Lithgow*	*Long Branch*
Rifle No. 4 Mk I Lightened:	32	-	-	-	-
Rifle No. 5 Mk I:	-	169,807	81,329	-	Prototypes
Rifle No. 5 Mk 2:	-	-	Prototypes	-	-
Rifle No. 4 Canadian Lightweight:	-	-	-	-	A few dozen
Trials Rifle No. 6 Mk I (Aust):	-	-	-	100	-
Trials Rifle No. 6 Mk I/I (Aust):	-	-	-	100	-

I. Skennerton

RIFLE No. 5 Mk 1

Calibre:	*.303 in.*
Rifling & Twist:	*5 groove, Enfield, L.H.*
Barrel Length:	*20.5 in. (521mm) including flash hider.*
Overall Length:	*39.5 in. (1003mm)*
Weight:	*7 lb. 1 oz. (3.2 kg)*
Magazine Capacity:	*10 rounds.*
Manufactured:	*Fazakerley, B.S.A. Shirley.*

Typical Markings, L.H. side of Receiver: No5 MkI ROF(F)

In early production, this jungle carbine was referred to as the "No. 4 lightened", the No. 5 Mk I was officially approved on 12th September 1944 although it was in production by that time. The reduction in length and weight make this a most handy weapon, along with its new knife bayonet, but serious problems were encountered with its accuracy, leading to its eventual demise.

Most obvious features of the No. 5 are the shortened fore-end and the sling loop in the butt, flash hider with foresight protector wings and sword bar, and the shortened handguard. It represents a saving of 1 lb. 14 oz. over the No. 4 rifle and it is nearly 5 ins. shorter. About 7 ins. of the barrel and flash eliminator assembly protrude beyond the fore-end, of which two main variations are encountered; some have a metal fore-end cap while others merely have the wooden tip of the fore-end rounded off. The short handguard is secured at the rear by a lipped ring around the barrel (like the No. 4) and at the front by the band. The butt has a rubber kickpad and sling bar and a corresponding scoop removed from the right side of the butt. Early models had a sheet steel plate between the rubber pad and the wood butt.

Aperture sights are similar to the No. 4 rifle, but ranged from 200 to 800 yards instead of 1,400 yds. There are two forms of backsight, the Mk I being a machined type with click-stop elevation like that of the Mk I sight for No. 4 rifle, and the Mk II being a pressed sheet steel model with catch operated slide like the Mk IV sight for No. 4 rifle. Quite a few lightening cuts were made in the metal components; the barrel is about the same diameter as the No. 1 barrel and has scoops cut out of the knox form, the bolt knob is lightened and so is the trigger guard. The body is lightened behind the locking lugs and at the right side near the extractor cut-out, and in the butt socket near the trigger guard attachment point. Some stockbolts were lightened by boring from end to end.

A new knife bayonet was introduced with the No. 5 and the bayonet bar on the flash hider was shaped especially to fit, being circular in cross section; previous sword bayonets and fixing bars had squared mortice slots.

EXPERIMENTAL No. 1 SHORTENED & LIGHTENED RIFLE

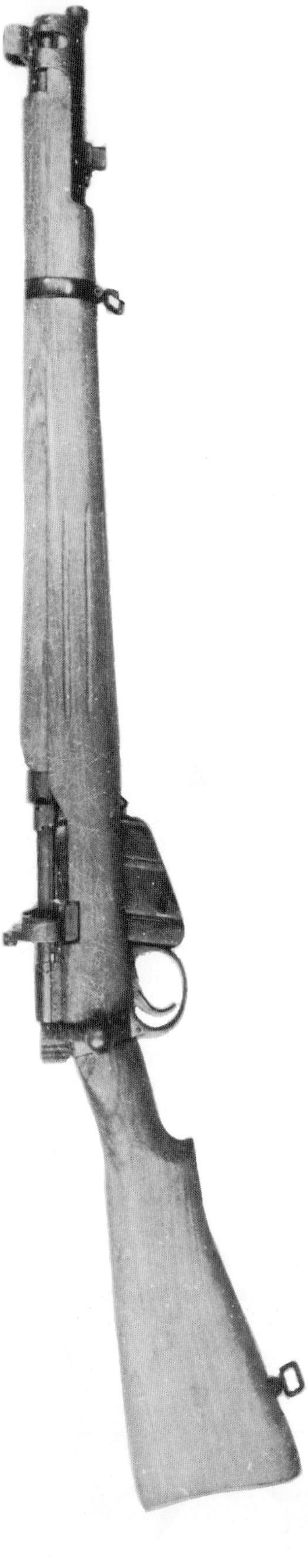

Calibre:	*.303 in.*
Rifling & Twist:	*5 groove, Enfield, L.H.*
Barrel Length:	*20.2 in. (513mm)*
Overall Length:	*39.3 in. (998mm)*
Weight:	*8 lb. 11 oz. (3.9 kg)*
Magazine Capacity:	*10 rounds.*
Manufactured:	*Lithgow.*

Typical Markings, R.H. side of the Butt Socket:

MA
LITHGOW
SMLE
III*
1944

While a few sample rifles of this basic configuration were fabricated in England, a batch of about 100 was produced in Australia in 1944 for Army trials. The serial numbers are preceded by "XP" and range between 1 and about 100.

The barrel and fore-end were shortened by 5-ins. and the nosecap re-fitted. Except for the one-piece top handguard with springs in the usual rear handguard position, and two longitudinal grooves on each side of the shortened fore-end, the furniture and fittings are generally similar to the service No. 1 Mk III* Lithgow production model, but for the backsight. The aperture sight is mounted onto the charger bridge and has only two push-pull range settings, for 200 and 500 yards; a screw on the left side of the charger bridge regulates the tension on this sight.

The first 18 rifles were completed by 6th March 1944 and the remaining 82 by the end of the month. Complete army trials did not eventuate and most of the rifles were left in store at S.A.F. Lithgow. A similar model, two inches shorter, was also initially considered, so the sample illustrated here was referred to as the "Intermediate" length. Sling swivels are of the standard type and placement, mounted onto the butt and outer band.

A different bayonet was proposed for this shortened rifle, with a 10-inch blade. This particular design was later introduced for the Owen Machine Carbine, although the markings are a little different because the trials bayonets had no service acceptance, date or pattern marking; only the "X" bending mark and "OA" factory code.

B. Labudda

RIFLE No. 6 Mk I (AUST)

Calibre:	*.303 in.*
Rifling & Twist:	*5 groove, Enfield, L.H.*
Barrel Length:	*20.5 in. (521mm) including flash hider.*
Overall Length:	*39.1 in. (993mm) with brass buttplate.*
Weight:	*7 lb. 8 oz. (3.4 kg)*
Magazine Capacity:	*10 rounds.*
Manufactured:	*Lithgow.*

Typical Markings, R.H. side of the Butt Socket:

MA
LITHGOW
SMLE
III*
1944

This particular variation was produced in Australia and about 100 of each No. 6 type were set up for trials; some other similar prototypes were fabricated also. Assembled on the No. 1 action instead of the No. 4, such a project was investigated in England and India as well, in 1944 and 1945. The No. 6 was never put into full production although it was announced in the Australian List of Changes and an order placed for 40,000 conversions from No. 1 Mk III* rifles.

Its configuration and overall dimensions are similar to those of the No. 5 but it utilises the No. 1 receiver which has a curved left wall instead of the flat, slab side of the No. 4 and No. 5 actions. Considerable lightening of metal components occurred; the bolt handle and trigger guard have been milled and the shortened barrel has also been lightened at the knox form. At 7 lb. 8 ozs., the No. 6 is slightly heavier than the No. 5. The flash eliminator, sight protector and bayonet standard assembly differ from that of the No. 5 assembly in that the bayonet lug is squared for the Patt. 1907 hilt whereas the No. 5 version has a rounded lug for the No. 5 bayonet.

The Mk I differs from its Mk I/I counterpart in the form of the backsight; this has a conventional bed backsight mounted on the barrel, with sheet steel protector wings attached by two screws on each side. Graduations are the same as for the Mk III* rifle, from 200 to 2,000 yards. The 8.3-in. long handguard extends from the front of the backsight to the fore-end tip. The fore-end has four longitudinal grooves complemented by matching grooves on the handguard. On the early trials rifles, foresight protectors were of sheet steel screwed to the block, later versions have protectors more like the British No. 5 rifle. Two variations of buttplate were available, the standard brass rifle model and a composition padded type hinged at the bottom for access to the butt trap.

Serial numbers are preceded by "XP" and range between XP 101 and XP 149 for the standard butt, and XP 150 to XP 198 for the rubber butt buttplate. Some 40,000 of these conversions were ordered but only the prototype trials batches produced. The bayonets have the conventional Patt. 1907 hilt but an enlarged muzzle ring; three different lengths of blade were tested, 8-in., 10-in. and 17-in.

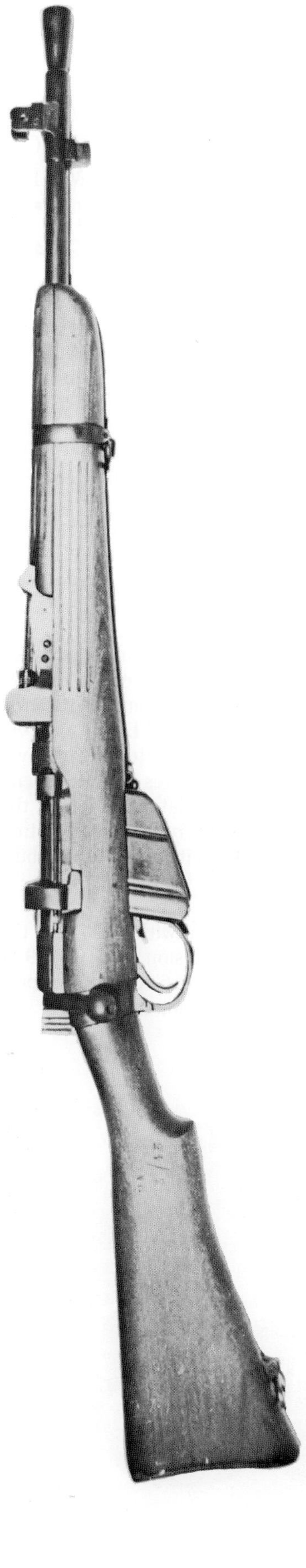

L. Addison

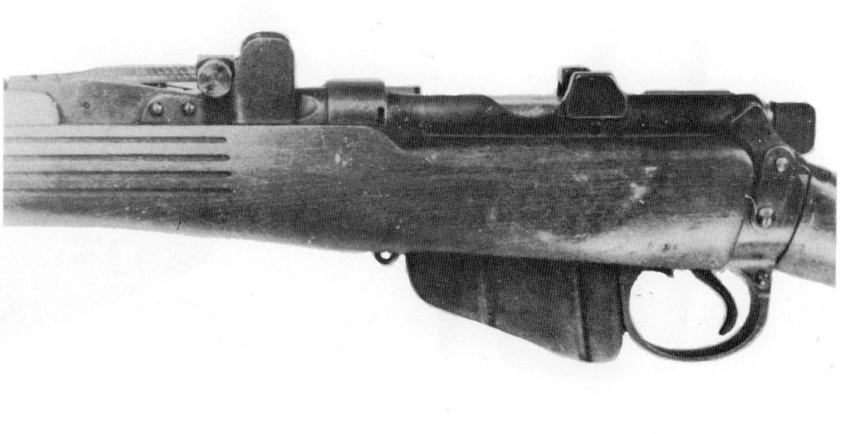

RIFLE No. 6 Mk I/I (AUST)

Calibre:	*.303 in.*
Rifling & Twist:	*5 groove, Enfield, L.H.*
Barrel Length:	*20.5 in. (521mm) including flash hider.*
Overall Length:	*39.7 in. (1008mm) with rubber kickpad.*
Weight:	*7 lb. 7 oz. (3.4 kg)*
Magazine Capacity:	*10 rounds.*
Manufactured:	*Lithgow.*

Typical Markings, R.H. side of the Butt Socket:

MA
LITHGOW
SMLE
III*
1944

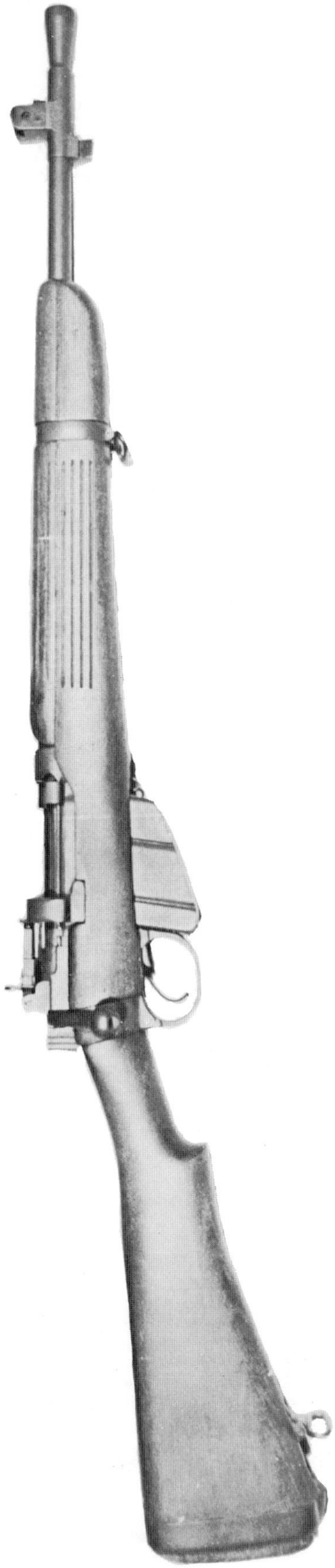

The No. 6 Marks I and I/I were developed simultaneously and similar quantities of each type produced; a few experimentals at first, then about 100 of each model for trials, and then a few more prototype conversions of the No. 1 Mark III*. It was regarded as a mode of conversion for the Mark III* rifle but this never eventuated.

The overall dimensions, profile and balance are similar to the No. 5 and this particular model also has aperture sights. The barrel has been lightened by machining and the trigger guard and bolt handle milled. As the No. 6 was a conversion from the No. 1 action and parts, none of these components are interchangeable on the No. 4 or No. 5. The flash eliminator assembly is also different as the bayonet standard has a squared lug for the Patt. 1907 bayonet hilt instead of the rounded No. 5 type.

The aperture sights are similar to those of the No. 5 jungle carbine, graduated from 200 to 800 yards and mounted on the receiver at the rear of the charger bridge. A wider backsight variation has also been noted on some of the first Australian prototypes, similar to that of the British Mark V S.M.L.E. The 12-in. long handguard is different to the Mk I model and the longitudinal grooves extend back almost to the receiver ring. The longitudinal grooves on the fore-end are similar to those on the Mark I model. Two types of buttplate were set up, the standard brass S.M.L.E. model and a hinged type with rubber composition buttplate, hinged at the lower end on an assembly that also incorporates the butt swivel mount.

Like its No. 6 Mk I counterpart, serial numbers commence with "XP" and the range was from XP 200 to XP 248 for the standard butt version, and XP 249 to XP 296 for the rubber buttplate model. The designation "No. 6" does not appear on the rifle, only the standard No. 1 Mk III* Lithgow production markings.

L. Addison

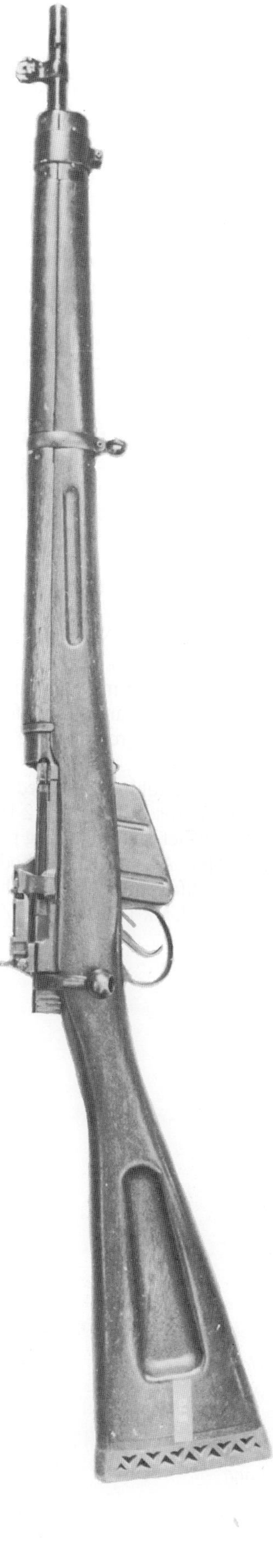

Infantry Museum, Warminster / E.D.E. Maribyrnong

No. 4 CANADIAN LIGHTWEIGHT

Calibre:	*.303 in.*
Rifling & Twist:	*5 groove, Enfield, L.H.*
Barrel Length:	*22.8 in. (579mm)*
Overall Length:	*42.1 in. (1077mm)*
Weight:	*6 lb. 9 oz. (3.0 kg)*
Magazine Capacity:	*10 rounds.*
Manufactured:	*Long Branch.*

Typical Markings, on Receiver: J 5550 - 13

Peculiar to Canada, this development incorporates many desirable improvements to the No. 4 in addition to the obvious reduction in weight and length. Designed as a conversion of the No. 4, it was intended to provide a light, accurate arm, particularly for jungle warfare in the Far East.

Few parts were unaltered in this conversion; the barrel reduced in weight and length and the action body lightened by extensive milling. The left side of the receiver has been cut down so that the amount of material is equivalent to the right side, a scoop cut out of the thumb for charger loading and the butt socket removed so that a one-piece stock could be fitted. The trigger was removed from the guard and mounted onto the action body and a slot was milled behind the safety catch, similar to the No. 1 Mark VI. The trigger guard and bolt were lightened and on some rifles, the bolt handle bent forward slightly.

The one-piece stock is most unusual for a Lee-Enfield and it has a finger groove each side of the fore-end and a milled butt. A bolt through the original butt socket also secures the action to the stock, along with vertical front screw and new rear trigger guard screws. While the fore-end and handguards appear to be similar to the usual No. 4 pattern, the lower band has been set back about 3.5 ins. and the front band about 4.5 ins. back. A commercial Hawkins rubber kickpad was fitted to those specimens examined.

A Mk I backsight graduated to 1,300 yards was fitted to the illustrated sample but it was intended that a pressed steel version like the Mk III be supplied, graduated to 800 yards, and somewhat similar to the Mk II sight for No. 5 rifle. The foresight protector has been trimmed and more resembles the Pattern 1914 type.

Some prototypes retain the original Long Branch No. 4 Mk I* receiver markings while more have been noted with no such stamps. Other variations such as one-groove or two-groove fore-ends have also been observed. The lightweight rifles are marked "J 5550", followed by the serial number which ranges from 1 to perhaps 40 or 50.

THE LATER .22 TRAINERS

This grouping consists of those more recent .22 rifles based on the No. 4 action, the Nos. 7, 8 and 9. There were two service versions of the No. 7 in the Commonwealth; the first was a Canadian production assembled as a new rifle rather than being converted, while the later British model was generally a conversion from the No. 4 rifle. Whilst the Canadian and British .22 No. 7, and the No. 9, generally resemble the .303-in. No. 4 in external appearance, dimensions and weight, the No. 8 was developed especially for competitive miniature rifle club shooting and was therefore a first for the British service small-bore and training rifle.

The B.S.A. Shirley factory developed a .22-in. No. 5 for extended trials; this features a much modified trigger mechanism, bolt and competition type sights. Although considerable development had been made on this smallbore No. 5 style trainer, after it was decided that the No. 5 jungle carbine was not to be adopted as the universal rifle for the British forces, further designs for a .22 trainer were called for. This new trainer was required to more closely resemble the No. 4 rifle in its sighting and overall length, and as the .22-in. No. 5 trials rifle was well received, the resulting No. 8 bears quite a few similarities to the preceding .22 trials rifle also. The .22 No. 5 trials rifle utilises a No. 4 action body whereas the No. 8 rifle uses a No. 5 body.

Canada's Long Branch No. 7 was the first general production model of these .22 trainers, probably because there was no local production of service smallbore rifles to meet the Canadian requirements apart from some commercial models by Cooey and the U.S. Mossberg. The C No. 7 is quite straight-forward and conventional compared with its British counterparts and it varies from the No. 4 Mk I* only in being fitted with a .22 calibre barrel, new bolthead with offset firing pin, new magazine and more precise, adjustable sights.

Britain's No. 8 however, was quite an unconventional matter; more or less developed from the .22 No. 5 trials rifle, it was not designed to duplicate the handling or sighting characteristics of the full bore model. The receiver is special, with an adjustable trigger and a unique bolt and bolthead which was also the basis for the British No. 7 trainer. Britain's No. 7 was a modification of the No. 4 by B.S.A. Shirley who had developed the first .22 trainers on the No. 4 and No. 5 actions. The No. 7 is the only service small-bore trainer in this series which was adapted for magazine fire, made possible by fitting a modified B.S.A. sporting .22 magazine inside the .303 magazine case. There are quite a few differences between the Canadian and British No. 7 rifles which render most of the components non-interchangeable.

The No. 9 was the last of the series, a conversion by Parker Hale for the Royal Navy. Parkerifled tubing was used to sleeve the bore down from .303-in. on most of these rifles and this conversion is somewhat reminiscent of the .22 RF Pattern 1914 Short Rifles converted by A.G. Parker during the Great War. The bolthead has an offset, floating firing pin and a special extractor for the .22-in. cartridge, while the original striker is shortened to the collar. A standard bolt body is utilised, unlike the previous Nos. 7 and 8. The No. 4 rifle configuration is retained, the only differences being in the barrel, bolthead, striker and magazine, although many of these rifles were converted from the Mk 2, Mk 1/2 and Mk 1/3 models with hung triggers.

	BSA Shirley	*Fazakerley*	*Long Branch*	*Parker-Hale*	*Enfield*
.22 No. 5 Trials:	103	-	-	-	-
Rifle C No. 7 Mk I:	-	-	c.20,000?	-	-
Rifle No. 7 Mk I:	2,500	-	-	-	-
Rifle No. 8 Mk I:	c.2,000?	c.15,000?	-	-	?
Rifle No. 9 Mk I:	-	-	-	c.3,000	-

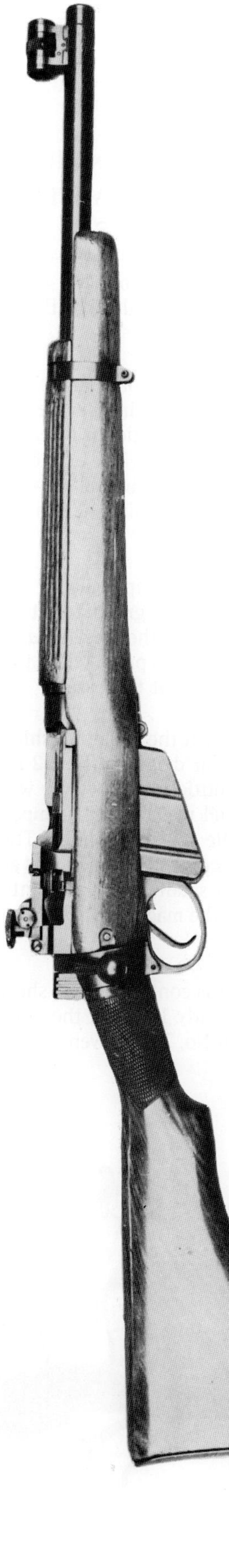

.22 No. 5 TRIALS RIFLE

Calibre:	*.22 in.*
Rifling & Twist:	*6 groove, R.H.*
Barrel Length:	*19.0 in. (482mm)*
Overall Length:	*37.5 in. (952mm)*
Weight:	*8 lb. 10 oz. (3.9 kg)*
Magazine Capacity:	*Single shot, or 5 rounds.*
Manufactured:	*B.S.A. Shirley.*

Receiver Markings, L.H. side of the Butt Socket:

M47C
1945

Effectively a prototype or trials model which resulted in the .22 No. 8, there are a number of important differences even though the action design and the configuration are somewhat similar. It was designed as a competition small bore rifle and so has special sights and more sporting style furniture.

A No. 4 butt has been fitted rather than the No. 5, with chequering at the wrist and the fore-end has been shortened to 18.5 ins., slightly shorter than that of the No. 5. The handguard is No. 5 length, but grooved on each side as on many of the earlier No. 4 rifles. The trigger is mounted onto the action body rather than the trigger guard, although in a different manner to the ensuing No. 8 rifle as a plate is secured behind the trigger by three screws, underneath the trigger guard.

A No. 1 magazine has been converted with a new floorplate riveted inside and a recess and catch provided for a small box magazine insert. However, in these trials rifles, two types of magazine inserts were available, one with a 5-round capacity and the other with a spring-loaded, oversize platform on top for single loading. Although the bolt and bolthead are similar to the No. 8, they are considerably longer because the barrel does not extend back into the boltway like the No. 8. The bolthead has a separate 2.7-in. long firing pin and a recessed bolt face. At 19 inches, the barrel is considerably shorter than that of the No. 8.

Micrometer adjustment is provided on the backsight, which is graduated at 25, 50 and 100 yards and there is also a micrometer click adjustment for windage. Six aperture sizes are available in the rotating disc backsight. The foresight is a competition tunnel type with the facility of interchangeable foresights. Serial numbers are marked on the left side of the butt socket, prefixed by "T1" and they range between 0001 and 0102. Some have been noted with very shallow engraving on the left side of the receiver, "B.S.A. .22 TRAINER".

M.O.D. Pattern Room

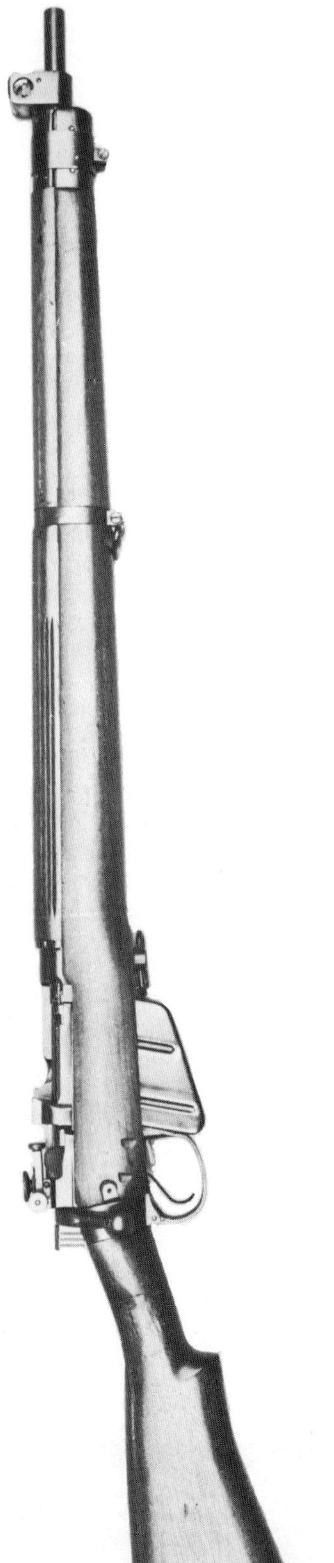

RIFLE C No. 7 Mk I

Calibre:	*.22 in.*
Rifling & Twist:	*6 groove, R.H.*
Barrel Length:	*25.2 in. (640mm)*
Overall Length:	*44.5 in. (1130mm)*
Weight:	*8 lb. 15 oz. (4.0 kg)*
Magazine Capacity:	*Single shot.*
Manufactured:	*Long Branch.*

Typical Markings, L.H. side of the Receiver:

C N° 7 .22 IN Mk 1
LONG BRANCH
1946

Canadian production commenced in 1944 due to the requirement for a No. 4 .22-in. training rifle there. Some No. 7 rifles were later produced in England by B.S.A. but these differ in a number of respects.

A solid, rather than tubed, barrel was used and there are no bayonet lugs. The bolt has a two-piece firing pin, the forward portion is carried in the special bolthead and it is offset for the rimfire cartridge while the rear portion is fitted to the bolt body in the normal manner but differs in being shortened. The extractor has a longer claw than the No. 4 .303 type. A No. 4 Mark I* receiver blank was utilised and a new serial number range used with the Canadian .22 rifles commencing again with 0L0001.

There is a slight difference in the rear part of the No. 7 when compared with the No. 4, where the backsight folds down. The backsight is special, graduated at 20 and 100 yards on the right side of the folding "ladder" and the knurled disc aperture sight is available in differing sizes. Click adjustment is provided on the elevation and a windage screw is mounted on the right side at the base.

The bolt consists of a No. 4 body with firing pin shortened at the collar, to which a special rimfire bolthead is fitted on a somewhat similar deign to the .22-in. S.M.L.E. system. Like the No. 4 sniper rifle, a competition type swivel was mounted onto the front trigger guard screw. The special magazine is stamped ".22", this serves as a loading platform and receptacle for spent cases. The follower or floorplate forms a groove and it is spring loaded so that it only acts as a loading platform. There is a space at the rear to permit spent cases to fall into the lower portion of the magazine case. The front lips of the magazine have been modified slightly from the No. 4 types.

The finish on C No. 7 rifles is excellent and production was continued for a short time after the war. Another batch was assembled in the early 1950's at the same time as some No. 4 .303-in. rifles. Some of the Canadian No. 7 rifles were issued in wooden chests similar to the No. 4 sniper unit.

M.O.D. Pattern Room

RIFLE No. 7 Mk I

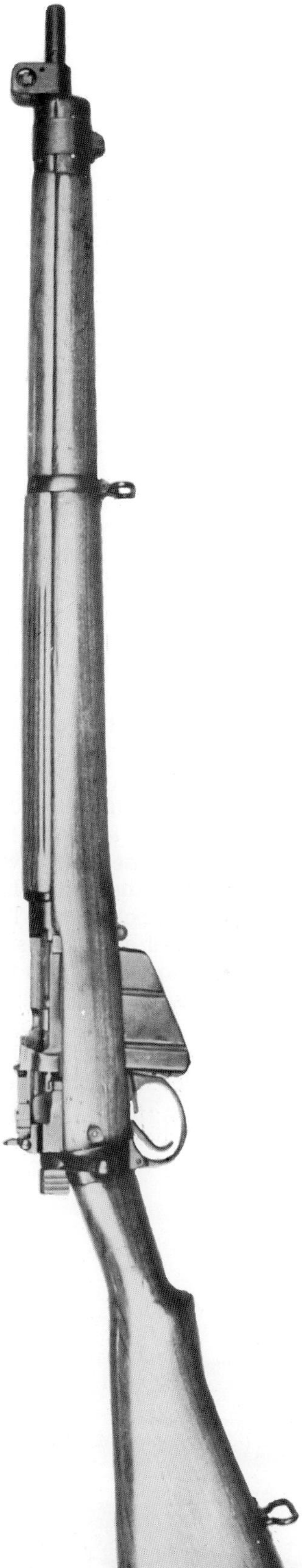

Calibre:	*.22 in.*
Rifling & Twist:	*6 groove, R.H.*
Barrel Length:	*25.2 in. (640mm)*
Overall Length:	*44.5 in. (1130mm)*
Weight:	*9 lb. 0 oz. (4.1 kg)*
Magazine Capacity:	*5 rounds.*
Converted:	*B.S.A. Guns.*

Typical Markings, L.H. side of the Receiver: No7 MkI

Different to its Canadian counterpart in a number of features, the British No. 7 was a conversion rather than new-made rifle. The contract for 2,500 rifles was set up for the R.A.A.F. during 1948 and the action design was influenced by the .22 No. 5. Serial numbers are usually prefixed by "BS".

Some early barrels were sleeved with a .22-calibre liner on the Parker system, though most were solid barrels of new manufacture. The bolt is like that of the .22 No. 5 and therefore about 1-in. longer than the No. 8 pattern. The British patterns have a shorter bolt body and longer, more tapered bolthead than the Canadian No. 7 and the No. 4 rifles. The British No. 7 has an auxiliary extractor on the left side of the bolthead and a special ejector is attached to the left side of the body by two screws, with an ejector blade projecting through the receiver wall into the boltway. A small number of rifles which were assembled with new receivers have the .303 ejection screw and groove omitted.

A No. 1 Mk III* magazine was used, different to the No. 4 rifle model mainly in the rib length on the back. A plate was spot-welded across the top of the case and a B.S.A. 5-round commercial .22 magazine inserted through the top. The .22 magazine was converted by mounting the release latch upside down and inserting the small-bore magazine bottom first through the top of the .303 case. A few No. 7 rifles were assembled as single-shot models.

Intended for use at the 25-yd. range, the standard Mk I (Singer) backsight is set at the lowest position of 200 yards; to compensate for this, the foresight has been slightly reduced in height. Only the No. 4 Mk I bodies were used in No. 7 conversions as it would not be possible to disengage the bolthead from the ribway of a No. 4 Mk I* action due to the special, longer bolthead. The British No. 7 bolt body has been shortened to 3.6 in. so as to compensate for the longer bolthead, the cocking piece has a shorter arm and consequently no half bent. The main extractor is on the right side of the bolthead and the auxiliary one on the left. Bolts of the .22 No. 5, No. 7 and No. 8 rifles are not interchangeable.

M.O.D. Pattern Room

RIFLE No. 8 Mk I

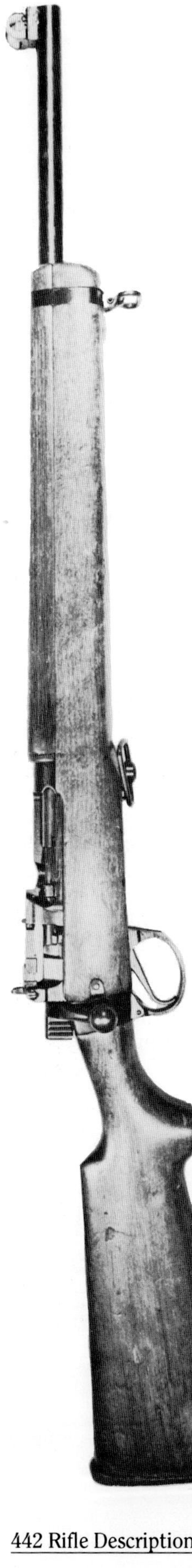

Calibre:	*.22 in.*
Rifling & Twist:	*6 groove, R.H.*
Barrel Length:	*23.9 in. (607mm)*
Overall Length:	*41.7 in. (1059mm)*
Weight:	*9 lb. 2 oz. (4.1 kg)*
Magazine Capacity:	*Single Shot.*
Manufactured:	*B.S.A. Shirley, Fazakerley.*

Typical Markings, L.H. side of the Receiver:

.22 N° 8 M^K 1
A 13386

Produced to meet a special requirement for a .22 calibre trainer which could be used in miniature rifle competition, the No. 8 was first introduced for Land Service only. Therefore its configuration is quite different to any of the other small bore rifles as previous trainers were intended to duplicate the handling and feel of the .303 service rifle as closely as possible. The No. 8 weighs 8.25 lb. and was introduced in the late 1940's.

It is half-stocked with a wide fore-end, slightly squared off at the bottom, while the sling swivels are fitted to the front of the trigger guard and the outer band. The 13.4-in. long handguard is grooved and the heavy barrel is about 2-in. shorter than the .303 No. 4 barrel, with a special foresight block and protector which allows interchange of match and service type foresights. The fully-adjustable trigger is attached to the receiver body, quite different to any of the No. 4 designs. There is no magazine as this is a single loader and the cartridge is laid onto a grooved platform for loading. A sheet metal plate is fitted into the magazine recess of the trigger guard.

The bolt is another innovation; the body is 3.6-in. shorter, the long bolthead is tapered and a two-piece firing pin is fitted. All bolt components are special to the No. 8 and not interchangeable with those of any other No. 4 series rifles, even though the design is similar to that of the .22 No. 7. Unlike any other British Lee rifle, to cock the bolt only requires a raising and lowering of the bolt handle. The No. 4 Mk I action body was used and there are 25, 50 and 100 yd. graduations on the folding Singer-type aperture backsight. An "H" is marked on the sight at the top of the slide, indicating "Harmonisation" for use on a landscape target.

A special batch produced at B.S.A. Shirley in the early 1950's for New Zealand has the serial number prefix "DA". Quite a few of the No. 8 rifles also bear Enfield receiver markings "UE" followed by the year and F.T.R. mark. These underwent factory thorough repair at Enfield in the late 1960's and some new parts of Enfield manufacture fitted, most commonly the outer band.

J. Alley / M.O.D. Pattern Room

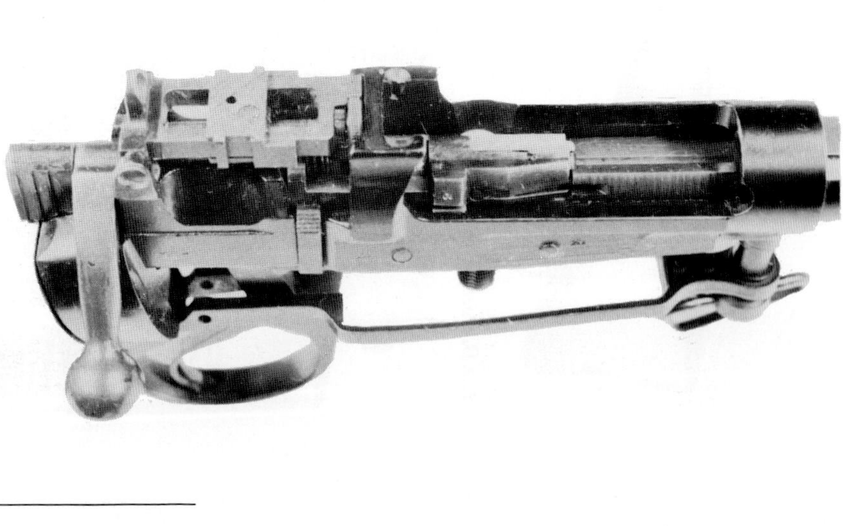

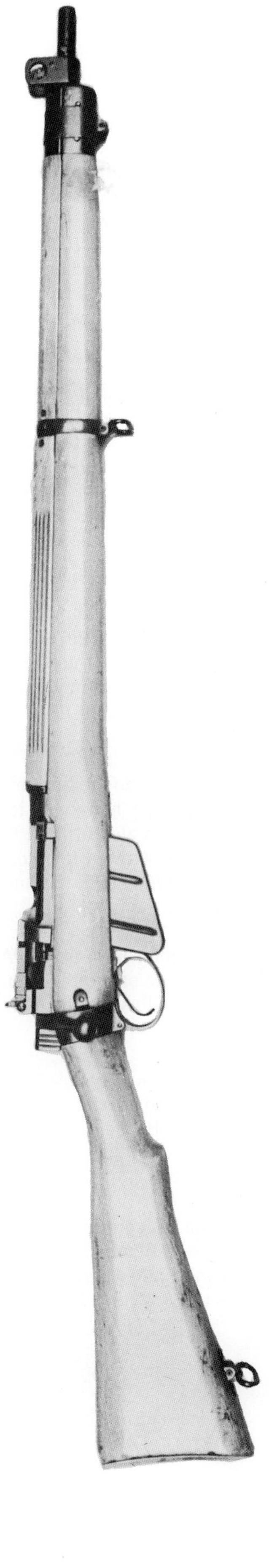

M.O.D. Pattern Room

RIFLE No. 9 Mk I

Calibre:	*.22 in.*
Rifling & Twist:	*6 groove, R.H.*
Barrel Length:	*25.2 in. (640mm)*
Overall Length:	*44.5 in. (1130mm)*
Weight:	*9 lb. 4 oz. (4.2 kg)*
Magazine Capacity:	*Single Shot.*
Converted:	*Parker Hale.*

Typical Markings, L.H. side of the Receiver:

.22 R.F. N°9 MK1
P-H '58

A special requirement for the Royal Navy, about 3,000 No. 9 rifles were converted by Parker Hale of Birmingham between 1956 and 1960. It is identical to the No. 4 rifle from which the conversion was made, the only differences being in the rimfire bolt, barrel, magazine and backsight; most appear to have been converted from the No. 4 Mk 2 type rifle with hung trigger.

The last of the .22 calibre rifles, this model utilises the same system of barrel sleeving as the Great War .22RF Patt. '14 Short Rifles converted by A.G. Parker. The barrel was bored out from .303 calibre and a .22-in. liner sleeved into the barrel and trimmed to length; a converted bolthead with offset, floating firing pin was fitted. Due to the smaller base .22 cartridge, the extractor claw is longer and groove in the bolthead deeper. The firing pin is a two-piece type with forward section retained in the bolthead, the rear part in the bolt body attached to the cocking piece. This striker is actually only a standard No. 4 type shortened by removal of the tip forward of the collar. The muzzle crown is stamped "PARKERIFLED A.G.P."

As a single loader, the magazine consists only of a bare case with spring and platform removed, because the case acts as a collector for spent cases after extraction. The sight has the original No. 4 Mk I leaf graduated to 1,300 yards, but a 25-yd. mark is inscribed on the left hand side indicated by "25 YDS". Conversions were made from the No 4 Mks I and I* and Mks 2 and I/2 and possibly some Mk I/3 rifles (British, Canadian and U.S. manufacture), so some action and fore-end tie differences may be encountered.

THE 7.62mm ROLE

As early as 1913, Britain had attempted to introduce a new service rimless cartridge but the Great War intervened and the .276 Pattern 1913 rifle was altered to .303 calibre and produced in the United States under contract as the Pattern 1914. Prior to the 2nd World War, automatic rifles were undergoing trials and a new cartridge was likely to be introduced but war again interposed and the No. 4 rifle was manufactured as the new service rifle. After World War 2, the .280/30 round and the EM2 were recommended for adoption as the 7mm Mk Iz and Rifle 7mm No. 9 Mk I, but at U.S. insistence as a N.A.T.O. partner, the T65 (7.62 x 51mm) cartridge was agreed upon in December 1953.

Effectively a shortened, updated variant of the .30-06 cartridge, the 7.62mm round has a trajectory roughly 10% flatter than that of the .303 Mk VIIz cartridge. This permits the original .303 yard graduations to be read as metres when using the new cartridge. Along with the N.A.T.O.-standard round, a new rifle was adopted, the Belgian F.N. F.A.L., which became the L1A1 in British service. This new nomenclature sequence was also applied to the 7.62mm conversions of the No. 4 rifle.

For the 7.62mm series of No. 4 rifles, the new designation was Rifle L8 and so the No. 4 Marks 2, 1/2 and 1/3 were designated the A1, A2 and A3 variations; provision was also made for the No. 4 Marks I and I* to be allocated the L8A4 and L8A5 titles. The series was commenced with the conversion of No. 4 Mark 2 rifles to L8A1 and then some of the 1/2 and 1/3 Marks to L8A2 and L8A3; few of the L8A4 and L8A5 versions seem to have been modified. Conversions for the new cartridge were also offered by Sterling Armament, effected at their Dagenham factory or supplied in kit form. R.S.A.F. Enfield joined the commercial marketplace by also offering a kit conversion but legal wrangles occurred between Sterling and the government factory over patent rights which lost orders for both parties. A conversion kit was also offered for the No. 5 rifle although this conversion was not considered entirely desirable.

As the modifications only applied to the barrel, magazine, bolthead and charger guide, the overall configuration of the No. 4 models remained unchanged. The external appearance differs merely in the shape of the magazine. The 7.62mm conversions were effected at Fazakerley and the rifle designation, factory and date marked on the left side of the receiver was altered to include the new designation. Rifles of North American origin were also converted in this program as some had previously been upgraded to the No. 4 Mk 1/2 and 1/3 patterns. These particular conversions were generally allocated to war reserve stores although some were issued to police forces.

Some of the No. 4 sniper rifles were also converted to 7.62mm by using the same procedure but these do not appear to have been satisfactory because special target and sniper rifles were developed. The L39A1 has a heavy barrel and it is half stocked, intended for competitive rifle shooting. Its sniping counterpart was the L42A1 which also utilised the No. 32 telescopic sight. Some of the L8 conversions were also later made from the No. 4 sniper rifle, without telescopic sights and they were designated according to their original nomenclature; e.g. a No. 4 Mk I (T) modified with the hung trigger became the No. 4 Mk I/2 (T) and then converted to 7.62mm without telescopic sight became the L8A2.

An Enfield produced commercial variant of the L39A1 was marketed as the Enfield Envoy and it differs from the L39A1 in having a longer and wider fore-end, higher quality finish and 10-round 7.62mm magazine. Some of these 7.62mm conversions were also set up with telescopic sights for specialist police use. Further details and illustrations of the sniping rifles will be found in the following section.

Effectively a shortened, updated variant of the .30-06 cartridge, the 7.62mm round has a trajectory roughly 10% flatter than that of the .303 Mk VIIz cartridge. This generally permits the original .303-in. yard graduations to be read as metres when using the new cartridge. Along with the N.A.T.O.-standard round, a new rifle was adopted, the Belgian F.N. F.A.L., which became the L1A1 in British service. This new nomenclature sequence with "L" prefix and "A" and number suffix was also applied to the 7.62mm conversions of the No. 4 rifle.

No production figures are available for these Lee-Enfield rifles in the 7.62mm series, but the L8 conversions were generally carried out at Fazakerley and the L39 and L42 models altered at Enfield. Many of the 7.62mm Lee-Enfield No. 4 conversions are still in service today. At Ishapore, the No. 1 action was made again using higher grade steel to better suit the NATO round; with magazine and sighting modifications, this was introduced as the 7.62mm 2A model.

THE L8 SERIES: A1, A2, A3, A4 & A5

Calibre:	*.308 in. (7.62mm)*
Rifling & Twist:	*4 groove, Enfield, R.H.*
Barrel Length:	*25.2 in. (640mm)*
Overall Length:	*44.5 in. (1130mm)*
Weight:	*9 lb. 1 oz. (4.1 kg)*
Magazine Capacity:	*10 rounds.*
Converted:	*Enfield.*

Typical Markings, L.H. side of the Receiver:

RIFLE 7.62MM L8A3
(CR 62 GA)

After adoption of the 7.62 NATO round and L1A1 self-loading rifle, it was decided to convert some No. 4 rifles for the new cartridge, as reserve arms; in earlier tests, the No. 4 showed that it was quite capable of withstanding the added pressures of the new cartridge. Conversion kits were also available, produced by R.S.A.F. Enfield and commercially by Sterling.

The conversion entails the fitting of a new barrel of special chrome molybolenum alloy steel, with new rifling. An insert was placed in the charger guide so that the new cartridges could be charger-loaded in clips of 5. The new magazine is stamped with the part number "CR12A" and year of manufacture, incorporating an ejector which obviates the need for such a fitting to the rifle action. For the new magazine lips, some action body grinding was required. The new bolthead has a wider extractor. With the appropriate parts and a few special tools, this conversion could be effected by armourers. Factory converted rifles have the No. 4 receiver markings cancelled out and the new designation made by electric pencil. A special barrel breeching washer was utilised in fitting of the barrel.

The 7.62mm cartridge has a flatter trajectory than the .303-in. round, though it was claimed that the change from yards to metres permitted original backsight graduations to be read in metres as the slightly longer distances rather matched the higher performance of the new round. Backsights may be encountered with the original .303 markings graduated to 1,300 and read as metres; others have new 1,000m Singer-style sights with a lowered datum line.

Provision was made for the alteration of the different types of No. 4 rifles, to these designations:

- L8A1, converted No. 4 Mk 2
- L8A2, converted No. 4 Mk I/2
- L8A3, converted No. 4 Mk I/3
- L8A4, converted No. 4 Mk I
- L8A5, converted No. 4 Mk I*

With the inclusion of some No. 4 Marks I* and 1/3 rifles, originally Canadian and U.S.-made models, some provision was made for their conversion with the L8A3 and L8A5 designations.

M.O.D. Pattern Room

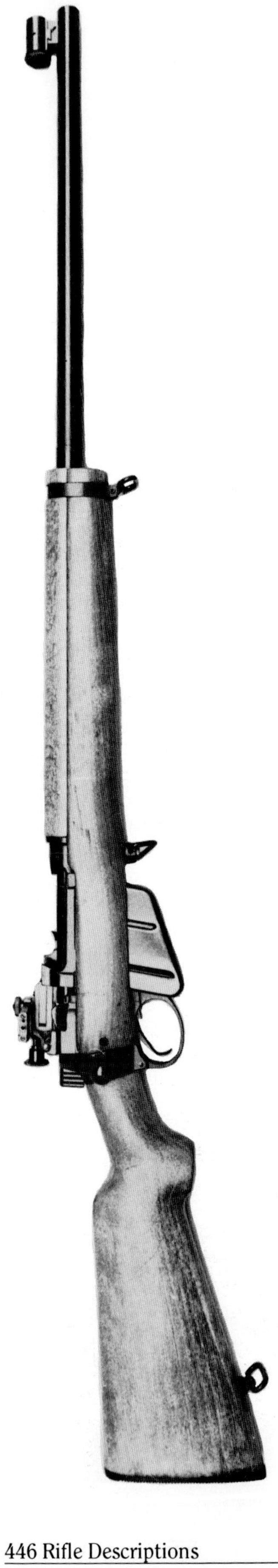

M.O.D. Pattern Room

RIFLE 7.62mm L39A1

Calibre:	*.308 in. (7.62mm)*
Rifling & Twist:	*4 groove, Enfield, R.H.*
Barrel Length:	*27.6 in. (700mm)*
Overall Length:	*46.5 in. (1181mm)*
Weight:	*10 lb. 1 oz. (4.5 kg)*
Magazine Capacity:	*Single Shot.*
Converted:	*Enfield.*

Typical Markings, L.H. side of the Receiver:

7.62 M/M L39A1
UE. 72 A1033

Another service 7.62mm variant of the No. 4 rifle, this is in essence an L42A1 sniper rifle without the telescopic sight or 7.62mm magazine, because it was designed for competitive rifle shooting in the British forces. As supplied from the factory, it has no front or rear sights because it was intended that they be supplied by the unit. The foresight is a tunnel type with interchangeable inserts, the usual competition No. 4 aperture rearsights could be fitted.

As the L39A1 utilised the No. 4 Mk 2 action, the trigger is mounted directly onto the receiver body instead of the trigger guard, permitting a more crisp trigger pull and preventing the condition of the fore-end from affecting the trigger. Some No. 4 Mark I/2 rifles were also used in this conversion. Due to the retention of the .303 magazine, there is no positive ejection. No internal action body mods were required because the original magazine was retained.

The L39A1 is half-stocked, with the fore-end and handguard terminating just forward of the outer band as the heavy profile barrel is intended to be a "floater", with the fore-end barrel channel relieved. A magazine is fitted even though the rifle is used in a single shot mode; the 7.62mm rounds do not feed through the .303 magazine.

A circular hole has been machined in the butt to mount a small screw container which houses the interchangeable foresights. The previous markings on the left wall of the receiver were skimmed off and the new inscription applied; the earlier markings are rarely visible.

A few rifles of similar configuration to the L39A1 were made up as the "7.62mm CENTAR" and this differs only in the sporting style of butt. Intended for use by cadets, this particular project was not continued with.

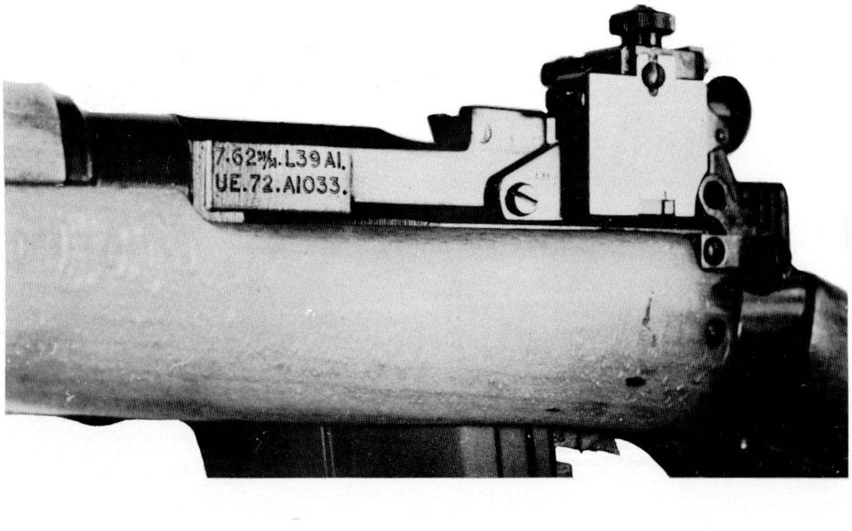

7.62mm ENFIELD ENVOY

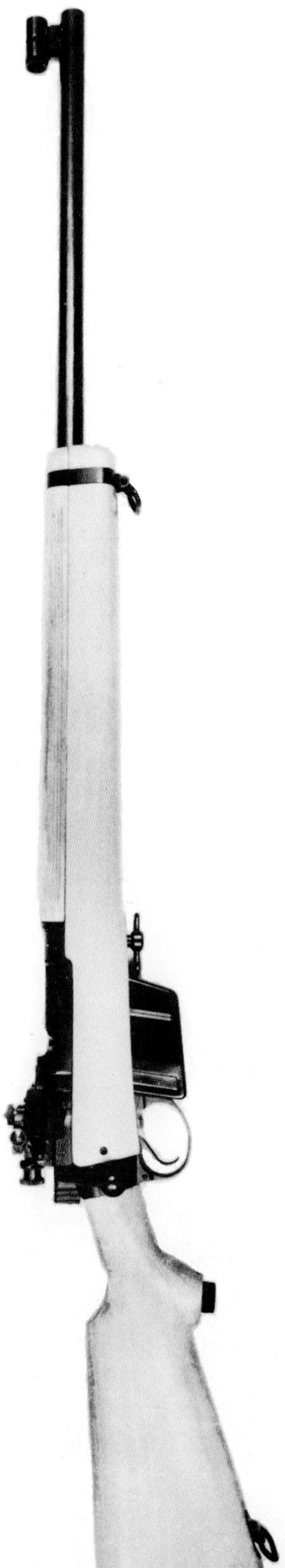

M.O.D. Pattern Room

Calibre:	*.308 in. (7.62mm)*
Rifling & Twist:	*4 groove, Enfield, R.H.*
Barrel Length:	*27.6 in. (700mm)*
Overall Length:	*46.7 in. (1185mm)*
Weight:	*10 lb. 1 oz. (4.5 kg)*
Magazine Capacity:	*10 rounds.*
Converted:	*Enfield.*

Typical Markings, L.H. side of the Receiver:

ENVOY 7.62M/M
Đ69 E. 515

A commercial variant of the L39A1 rifle, this differs from its service counterpart in only a few details. The same configuration of heavy target barrel and half-stocked furniture, with provision for the fitting of various competition front and rear aperture sights, applies. A No. 4 Mk 2 or I/2 receiver was utilised and where a conversion was made from an earlier service rifle, the original markings were machined off and the new designation and manufacture markings engraved by electric pencil.

As the No. 4 Marks 2 and I/2 actions feature the receiver-mounted trigger, this has been incorporated in the Envoy model. A special handguard and fore-end are fitted, about 2.3 in. longer than those on the L39 and L42 rifles; the .22 No. 8 rifle fore-end was modified for the receiver-mounted trigger and this has a wider fore-end with flatter underside, more suitable for competition shooting. The outer band is marked "CR 320", the same as that on the No. 8.

With a price tag of about £100, the Envoy was quite popular with the Rifle Club fraternity and it was intended for use as a single loader in Club competition, without a magazine fitted. With the magazine fitted, however, it could be used for magazine fire. For the 7.62mm magazine to be fitted, some grinding or machining of the magazine way was required because of the different-shaped lips compared with the .303 magazine.

RIFLE 7.62mm 2A (INDIA)

Calibre:	*.308 in. (7.62mm)*
Rifling & Twist:	*4 groove, Enfield, R.H.*
Barrel Length:	*25.2 in. (640mm)*
Overall Length:	*44.5 in. (1130mm)*
Weight:	*9 lb. 5 oz. (4.2 kg)*
Magazine Capacity:	*10 rounds.*
Converted:	*Ishapore.*

Typical Markings, L.H. side of the Receiver:

RIFLE
7·62mm 2A
R.F.I.
1965

This model, while essentially a conversion of the of the No. 1 Mk III* rifle, represents an interesting development of a new Lee-Enfield rifle in 7.62mm calibre. These are new-made receivers of higher quality, EN steel. The new charger guide has a smaller guide channel for the new cartridge and clip. The ejector screw is positioned in the action body similar to its .303 counterpart, but about ½-in. forward to allow for the shorter 7.62mm case. On the bolthead, the extractor is a little longer and it moves the fired case base to the left receiver wall which is grooved with a taper, deeper at the ejector screw.

The new backsight, although similar in construction to the original .303-in. version, is graduated in 200 metre steps, with the numbers 2, 4, 6 and 8. Thus, the maximum range is restricted to 800 yards, rather than 2,000 yards as on its .303-in. No. 1 rifle forebear, conforming with modern infantry small arms trends.

A new, detachable box magazine is fitted. This is more square in shape and different to the British 7.62mm L8 and L42 magazine. The buttplate is cast alloy and is noticeably different in shape to the No. 1 and No. 4 rifle versions. The rear of the fore-end has a metal tie-strap similar to that on the No. 4 rifle, rather than the through rivet as on the No. 1. Most of the nosecaps have the square foresight protector ears typical of late-war Indian production.

B. Labudda

SNIPING RIFLES

The term "sniper" was used freely in both World Wars, perhaps too widely applied as it generally referred to any marksman operating close to or behind enemy lines. "Sniper rifle" really pertains to specially set-up rifles with optical or telescopic sights. Early advances in this field were restricted by optical development and though telescopic sights were mounted onto rifles many years earlier, such sights were quite fragile and lacked many later refinements.

It was the Germans who first employed snipers widely on the Western Front, armed with scope-mounted rifles; for some time the Allies had to make do with whatever telescope-fitted rifles could be obtained. The first British formally-approved sights were a little unorthodox in the classification of accepted sniping equipment, comprising of two lenses, one placed at the front of the rifle near the foresight and the other nearer to the firer's eye, on the backsight or in place of the long range aperture. Therefore, no tube or conventional telescope was required and the aiming point was observed through the front lens.

Assorted telescopic sights were fitted to the S.M.L.E. by renowned British gunmakers and these served the Great War snipers, although the S.M.L.E. did not prove to be entirely satisfactory. The Pattern 1914 rifle was introduced late in the war as the regulation sniping unit but it did not appear in sufficient quantity to be thoroughly tested. Many of the telescopes used on the S.M.L.E., in particular the Aldis patterns, were retained and some later fitted to the Pattern 1914 for use earlier in World War 2.

Australia developed an S.M.L.E. sniper rifle designated the No. 1 Mk III* H.T. late in World War 2, using the Pattern 1918 telescope, a heavy barrel, and special stocking-up with modified furniture. An earlier pattern, the No. 1 Mk III (and Mk III*) H, was fitted with the heavy barrel and aperture rifle club sights, until the approval of the scope-mounted H.T. model.

The first No. 4 sniping rifle units appeared about 1941, set up with the No. 32 telescope. This became the standard unit and the telescope was advanced through three Marks. Most of the British rifles were set up by Holland & Holland, although Enfield and B.S.A. converted some as well. No. 4 sniper rifles were also made by S.A.L. in Canada, where some commercial and improved No. 32 sights and mounts were fitted in addition to the usual British designed types.

After development of the 7.62mm N.A.T.O. round and L1A1, the new self-loading rifle proved to be inadequate for a 'scoped sniping role, so some No. 4 (T) sniper rifles were converted for the rimless cartridge. These had problems with accuracy, so a special model was developed with a heavy barrel, stocked up like the new service competition L39A1 rifle. Designated the L42A1, the No. 32 scope was retained as the Telescope L1A1. This unit is still in service and some other 7.62mm conversions of the No. 4 have been set up with telescopic sights for police and extra-service use, such as the "Enforcer".

Optical (Galilean) Sights:	Lattey	c. 8,500	[Additional sights of most types were purchased by front-line army units.]
	Neill	4,000	
	Martin	575	
	B.S.A.	80	

Rifle	*Telescope*	*Quantity*	*Converted by*
S.M.L.E. Sniper:	Periscopic Prism Co.	4,830	Periscopic Prism Co.
	Aldis Nos. 1, 2, 3 & 4	3,196	Aldis, Purdey, Holland
	Winchester A5 & B4	907	Whitehead
	Goerz/Evans	130 ?	Evans
	Miscellaneous	725 ?	various gunsmiths
No. 1 Mk III* H.T. (Aust):	Pattern 1918, high mounts	1,131	Lithgow
	Pattern 1918, low mounts	481	Lithgow
No. 4 Mks I & I* (T):	No. 32 Mk I	1,403	Enfield
No. 4 Mk I (T):	No. 32 Mks I, 2 & 3	23,177	Holland & Holland
No. 9 Mk I:	No. 32 Mk 3	100	B.S.A. Shirley
No. 4 Mk I* (T):	C No. 32 Mk I, IA, 2, 3 & C No. 67 Mk I	c.1,000	Long Branch
	Lyman Alaskan (No. 32 TP Mk I)	99	Long Branch
L42A1	L1A1 (ex No. 32)	10,000	Enfield

SHORT, MAGAZINE LEE-ENFIELD Mk III – LATTEY OPTICAL SIGHTS

Calibre:	*.303 in.*
Rifling & Twist:	*5 groove, Enfield, L.H.*
Barrel Length:	*25.2 in. (640mm)*
Overall Length:	*44.5 in. (1130mm)*
Weight, approx.:	*8 lb. 14 oz. (4.0 kg)*
Magazine Capacity:	*10 rounds.*
Converted:	*Unit Armourer.*

Typical Markings, R.H. side of the Butt Socket:

G. R.
ENFIELD
1915
$S^{\underline{HT}}$ L.E.
III

This represents the first issue "sniper rifle" with optical sights and merely comprises of the service Mark III or III* S.M.L.E. rifle fitted with special front and rearsight which function on the Galilean principle to form a telescope. Three different designs, Lattey, Neill and Martin, were introduced and described in the List of Changes in 1915. Lattey's sight was the first introduced and it was issued in larger quantities than the other types, so it has been chosen here to represent these optical sights. Two other models that were not officially introduced, but purchased in limited quantities, were a B.S.A. pattern and the Gibbs.

The illustrated rifle is fitted with Lattey's sights, comprising of a horseshoe-shaped frame containing the front lens and it is positioned in front of the foresight so that the original foresight is still used as the aiming mark. The rearsight lens is mounted in a fixture which clamps onto the rifle backsight cap, and the original rearsight "U"-notch is used for aiming.

These optical sights utilise two lenses; the front one mounted near or to the left of the foresight, with the rear one either on the backsight cap or fitted in place of the long range aperture peepsight. Of course, these fittings are quite vulnerable to damage, had a narrow field of view and were very much affected by dust and stray light. The magnification is about 2 or 3 power, with a field of view of only 1° to 1.25°, which presented difficulties when moving quickly onto a target.

L. Walsh

SHORT, MAGAZINE LEE-ENFIELD SNIPER RIFLE with PERISCOPIC PRISM CO. TELESCOPE

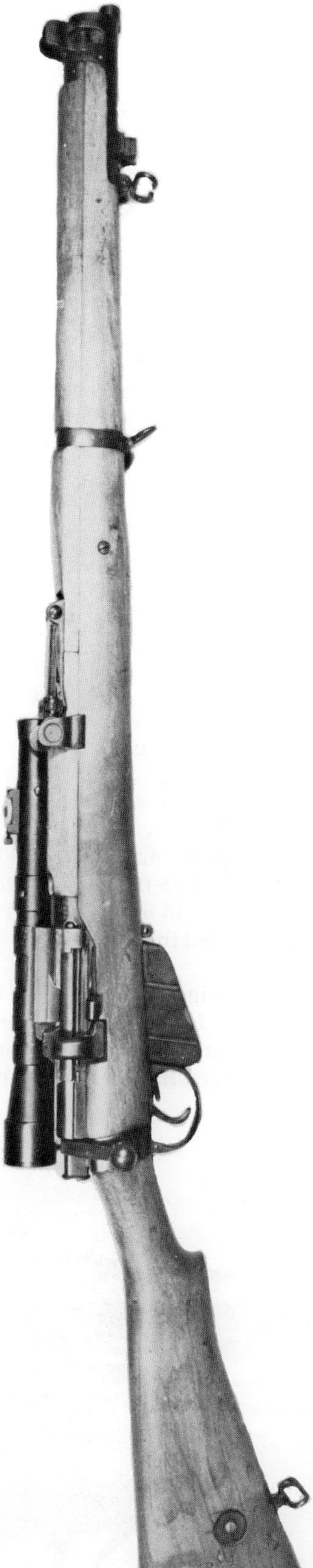

Calibre:	*.303 in.*
Rifling & Twist:	*5 groove, Enfield, L.H.*
Barrel Length:	*25.2 in. (640mm)*
Overall Length:	*44.5 in. (1130mm)*
Weight:	*10 lb. 1 oz. (4.6 kg)*
Magazine Capacity:	*10 rounds.*
Converted:	*Periscopic Prism Co.*

Typical Markings, R.H. side of the Butt Socket:

G. R.
ENFIELD
1915
SHT L.E.
III

This model was the most widely used of the British Great War sniper rifles, built on the Mark III and III* actions. Quite a variety of telescopes and mounts were used, these are described in more detail in Chapter 5. No official approval date is recorded and most of the rifles were stripped down after replacement by the Pattern 1914 sniper rifle.

Standard Mark III and III* rifles were taken from store and fitted with mounts; in the case of the P.P. Co. conversion, a lower mount was soldered and screwed onto the left side of the body. A leaf spring is positioned along the bottom of the groove with a chequered thumbpiece at the end to act as a securing and release latch. The male dovetail is part of the upper mount which also has two rings for the telescope. This mode of attachment was patented by P.P. Co. who also manufactured the telescopic sight.

The rather slim telescope is mounted offset of the boreline and has a range drum on the top of the tube. The range drum dial is brass, graduated to 600 yards; focus is adjustable via a knob in a helical slot in the tube. The scope is a little more than 12 inches long with magnification rated at 2 power and a weight of 18 ozs. The objective lens is about the same diameter as the tube, although the eyepiece is larger. The reticule housing and range drum is about 2.5 ins. from the objective lens and the range drum has a securing screw.

Britain's S.M.L.E. sniper rifles were declared obsolete on 30th September 1921 and ordered to be returned to Weedon for stripping as the Pattern 1914 rifles were issued instead. This model is the most commonly encountered of the Great War sniper rifles, although few remain today as most were broken down in the 1920's and the telescopes sold off through the gun trade.

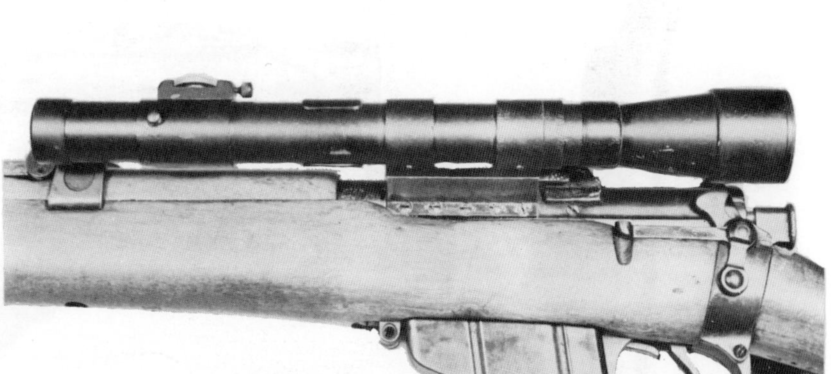

Imperial War Museum

SHORT, MAGAZINE LEE-ENFIELD SNIPER RIFLE with ALDIS BROS. TELESCOPE

Calibre:	*.303 in.*
Rifling & Twist:	*5 groove, Enfield, L.H.*
Barrel Length:	*25.2 in. (640mm)*
Overall Length:	*44.5 in. (1130mm)*
Weight:	*10 lb. 2 oz. (4.6 kg)*
Magazine Capacity:	*10 rounds.*
Converted:	*Aldis Bros., J. Purdey, Holland & Holland.*

Typical Markings, R.H. side of the Butt Socket:

G. R.
ENFIELD
1915
SHT L.E.
III

The Aldis was quite widely used and considered to have had better optics than the P.P. Co. model. Aldis telescopes were fitted by various businesses in the gun trade, using a variety of mount systems; some of these were offset to the bore, while others were overhead, in line with the bore.

Both the Mark III and III* rifles were used as the basis for conversion and Aldis manufactured four models of the telescope, designated the Nos. 1, 2, 3 and 4. The first three models are externally similar in appearance as the differences are found in the internal focusing sleeves and range elevation mechanisms. The objective lens has a similar diameter to that of the tube, like the P.P. Co. model, but the diameter is increased towards the eyepiece in a series of "steps", instead of bring coned like most other sights. The No. 4 pattern has a larger diameter objective lens and the increase in diameter from the tube to the eyepiece is coned instead of stepped as on the earlier types. The No. 3 pattern is illustrated.

The Aldis scopes were regarded as being the best available, especially the No. 4 model, some were fitted offset onto the Pattern 1914 rifles early in World War 2. When the S.M.L.E. sniper rifles were scrapped in the 1920's, many of the Aldis scopes were stored for possible later use. These sights are 11.5 ins. long and the range drum is usually graduated in 100 yd. increments up to 600 yds. Focus is possible by moving the knob in a helical slot situated behind the range drum.

Purdey's mounts are fitted to the illustrated rifle and this indicates setting up by J. Purdey in London; the rear telescope ring is engraved "J. Purdey & Sons" and the rifle number is also engraved on the scope tube. In many cases more than one such number will be noted, where the telescope has been fitted to successive rifles.

M.O.D. Pattern Room

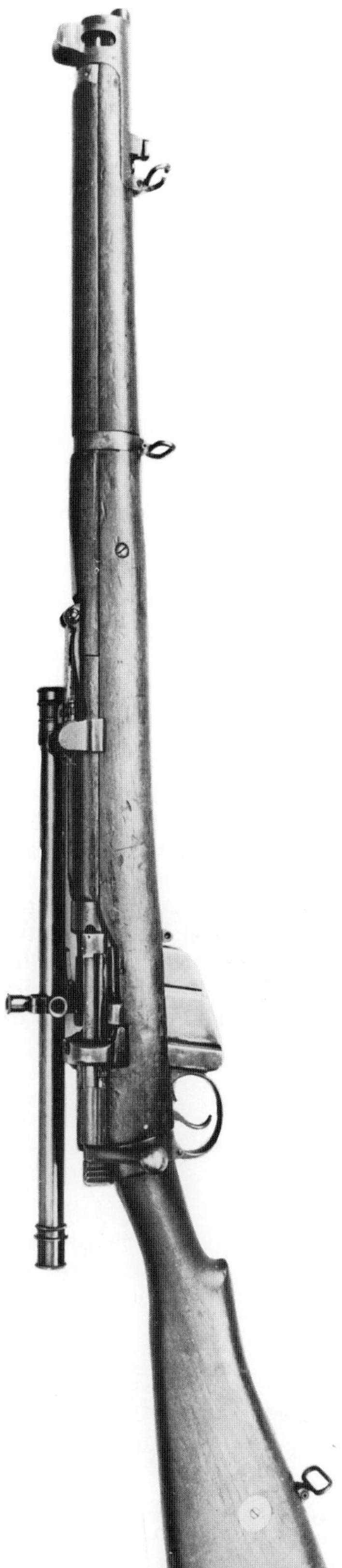

SHORT, MAGAZINE LEE-ENFIELD SNIPER RIFLE with WINCHESTER TELESCOPE

Calibre:	*.303 in.*
Rifling & Twist:	*5 groove, Enfield, L.H.*
Barrel Length:	*25.2 in. (640mm)*
Overall Length:	*44.5 in. (1130mm)*
Weight:	*9 lb. 5 oz. (4.2 kg)*
Magazine Capacity:	*10 rounds.*
Converted:	*Whitehead Bros.*

Typical Markings, R.H. side of the Butt Socket:

G. R.
ENFIELD
1915
S^{HT} L.E.
III

Two variations of the U.S. made Winchester telescopic sights were set up on the Mark III (and maybe a few Mark III*) rifles by the gunsmith Whitehead Bros.; these were the A5 and B4 models. The former scope is rated at 5 power magnification and the latter at 4 power. Supplied with screw-detachable dovetail mounts, those used in British service were marked with the broad arrow and Enfield inspection marks. Windage and elevation alignment is possible on the rear mount with its micrometer click adjustment. The illustrated scope is an A5 model.

Another feature of this 16-in. long telescopic sight is its forward movement under recoil which required that it be pulled back after each shot. The reticule is reported as being a cross-hair type although some "pin head" or post types are recorded, possibly in the B4 model. There are no yard graduations on the rear mount, only the numbers 0 – 5 – 10 – 15 – 20, similar to those on the windage drum.

The double mounts were set up so as to position the sight offset or overbore; earlier units were offset to the left and later equipments were positioned over the bore. Because of the different application of these front and rear dovetail bases and mounts, some variation occurred in the position of mounting points. Typical Winchester patent numbers and proprietary markings of the period were stamped on the telescope tube and mount bases.

Few of these sniper rifles have survived as they were broken down during the 1920's and the telescopes, upper mounts and leather carrying cases disposed of through the trade. Most of these particular British service Winchester telescopes were the A5 model.

R. Etherington

RIFLE No. 1 Mk III (or III*) H. (AUST.)

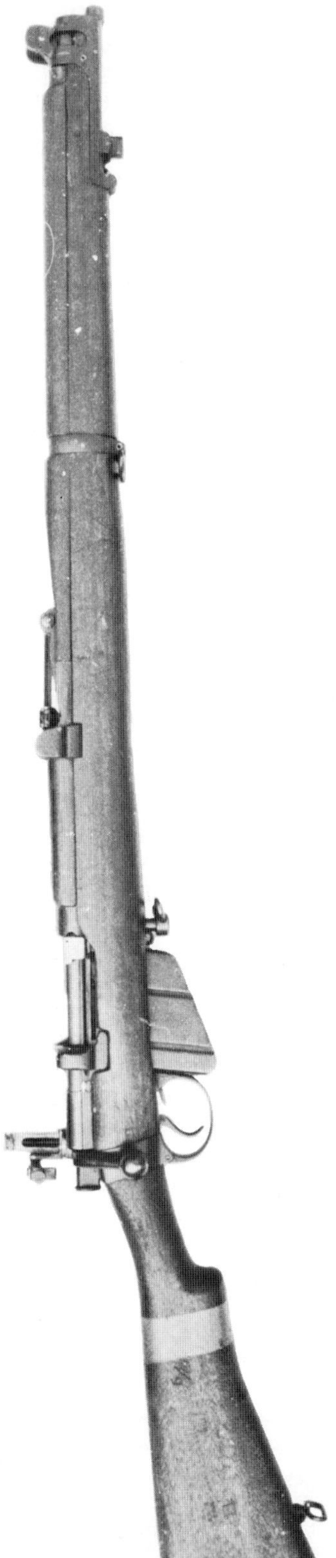

Calibre:	*.303 in.*
Rifling & Twist:	*5 groove, Enfield, L.H.*
Barrel Length:	*25.2 in. (640mm)*
Overall Length:	*44.5 in. (1130mm)*
Weight:	*10 lb. (4.5 kg)*
Magazine Capacity:	*10 rounds.*
Converted:	*Lithgow.*

Typical Markings, R.H. side of the Butt Socket:

Although no formal date of introduction has been found, this model is considered to have originated during the mid-1930's, a few years after the .303 heavy barrel production commenced at Lithgow. It was probably a natural sequence, commencing with some Lithgow-fitted heavy barrel Rifle Club Mk III rifles and then some factory set-up specimens used for Service competition shoots. Many similar Rifle Club models were also handed in for service during the early years of the 2nd World War. Some British-made Mk III actions have also been noted converted for this role.

In external appearance, without the aperture backsight and plate fitted, these rifles are very similar to their Rifle Club forbears. Probably the only means of certification is to check for a matching serial-numbered heavy barrel, "heavy" fore-end, "H" mark on the top of the buttstock, matching numbers on the leaf backsight and fore-end, and a small amount of wood removed from the left side of the fore-end so as to fit the Lithgow or Central aperture sight sideplate.

The sideplate requires a longer rear trigger-guard screw; some rifles have been noted with such screws shortened so as to conform with the standard pattern, after conversion back to the conventional model (but still retaining the heavy barrel and woodwork). The "heavy" fore-end had the barrel channel opened out to suit the larger diameter barrel, sometimes stamped with an "H" in the channel. Backsight bed and inner band have enlarged internal diameters for the heavy barrel. This model was not originally produced for snipers, it is better classified as a competition target or marksman's rifle. It was declared obsolescent upon the introduction of the No. 1 Mk III* H.T. in 1944 and ordered to be returned to Depot for conversion to, or parts utilisation for, the H.T. rifle.

B. Labudda

RIFLE No. 1 Mk III* H.T. (AUST.)

Calibre:	*.303 in.*
Rifling & Twist:	*5 groove, Enfield, L.H.*
Barrel Length:	*25.2 in. (640mm)*
Overall Length:	*44.5 in. (1130mm)*
Weight:	*10 lb. 10 oz. (4.8 kg)*
Magazine Capacity:	*10 rounds.*
Converted:	*Lithgow.*

Typical Markings, R.H. side of the Butt Socket:

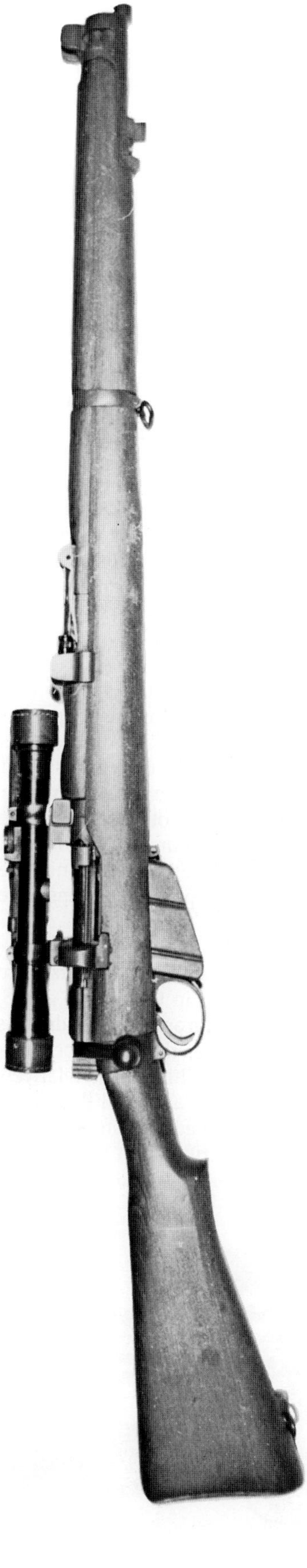

L. Addison

This model was introduced into Australian service towards the end of the 2nd World War with a variety of options; medium or short brackets, short, medium or long butts, wooden cheekpiece, and rifle club type front trigger guard screw with attached swivel. The difference between medium and short (often referred to as high and low) mounts is about ½-in. and the short bracket model has a scoop removed from the rear handguard to permit dismounting of the scope assembly.

The standard bracket telescope tube is marked "SIGHT TELESCOPIC PATT 1918 (AUST)" while the short bracket model has the suffix "/1" added to this inscription; the quick detachable mounts are similar to those used on the No. 3 Mk I* (Pattern 1914) sniper, with the release latch on the left side of the rear mount. The front male mount is a clawfoot design, incorporated in the lower portion of the front telescope ring. Leather scope caps and a canvas carry case were provided, the rifle number was engraved or scratched on the underside of the scope tube. The range drum is graduated from 100 to 1,000 yards and the sight was produced by the Australian Optical Co., Victoria, whose mark "A.O.C." is found on the steel tube.

Mostly rebuilt Mk III actions with magazine cut-off provision, the cut-offs were not refitted. Some British Mark III receivers were also converted, though mostly Lithgow bodies dated between 1915 and 1918 were utilised. These rifles were set up to Mk III* specifications and some had the additional "*" stamped after the "III" designation during conversion, in which a heavy barrel was fitted and numbered to the rifle, scope base mounts screwed and soldered onto the receiver, and the appropriate furniture fitted; this has an "H" stamped on the butt and inside the fore-end channel. The inner barrel band, screw and fittings were deleted, replaced by a wooden plug in the screw hole on the underside of the fore-end. A standard Mk III* backsight was fitted, without windage adjustment; the backsight bed had to be reamed out to accommodate the larger profile barrel.

Of the contract for 2,500 units, 1,131 medium and 481 short bracket models were produced as the order was terminated at 1,612 units; the first batch went to store in November 1944 and the last in February 1946. At Lithgow, 100 were F.T.R.'d between July 1954 and April 1955.

RIFLE No. 4 Mks I & I* (T)

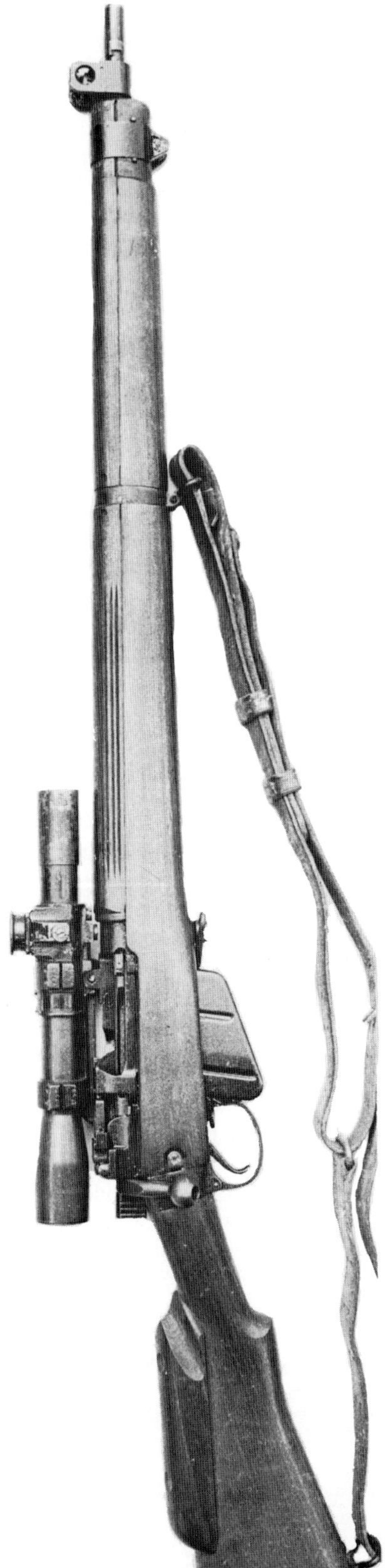

Calibre:	*.303 in.*
Rifling & Twist:	*5 groove, Enfield, L.H.*
Barrel Length:	*25.2 in. (640mm)*
Overall Length:	*44.5 in. (1130mm)*
Weight:	*11 lb. 10 oz. (5.3kg)*
Magazine Capacity:	*10 rounds.*
Converted:	*Enfield, Holland & Holland, Long Branch.*

Typical Markings, L.H. side of Receiver & Butt Socket:

No4 MkI T
M47C

The new sniping version of the No. 4 rifle was introduced early in World War 2, recommended for adoption with the No. 32 telescope which was originally developed for use on the Bren L.M.G. While the recommendation for the No. 4 rifle and No. 32 scope was officially made on 20th March 1940, the No. 4 Mk I (T) was not adopted until 12th February 1942, even though it was already in service by this time.

Early No. 4 (T) rifles were set up at Enfield utilising early No. 4 trials rifles, but from September 1942, the conversion was undertaken by Holland & Holland whose code "S-51" is marked on the butt. A number of U.S. produced No. 4 rifles were also converted in England and many Long Branch conversions of the No. 4 Mk I* were also acquired for British service. Later on, some .22 No. 4 (No. 9 type) trainers were similarly converted at B.S.A. Shirley.

The rifles were stripped after selection and re-stocked carefully, metal tabs attached to the left side of the receiver and holes drilled and threads tapped for the telescope mount clamping screws. A Mk I aperture sight was fitted with its battle aperture removed, so graduations are to 1,300 yds. with click adjustment elevation. The telescopes were matched to the rifles and the scope serial number stamped on the top of the butt at the socket. For the purpose of assembly, scope rings were stamped on the right hand side with consecutive numbers at front and rear, with matching numbers on upper and lower rings. The Canadian units are different in that the rifle number is marked on the telescope and the caps were not numbered like the British ones.

Three Marks of telescopic sight were used and in Canada some others were applied also. Appropriate details and differences are listed in Chapters 6 and 8. No. 32 sights were graduated to 1,000 yards on the top drum and the side drum is for deflection or windage, marked to 16 at each extreme; the telescope mark designations are engraved on the top of the tubes. While this unit is relatively heavy at more than 2 lb., the all-up sniper rifle weight is about 11 lb. 8 oz. A wooden cheekpiece, attached by brass screws, was attached and an adjustable U.S. type leather sling, steel telescope case and wooden rifle chest were part of the equipment.

I. Skennerton

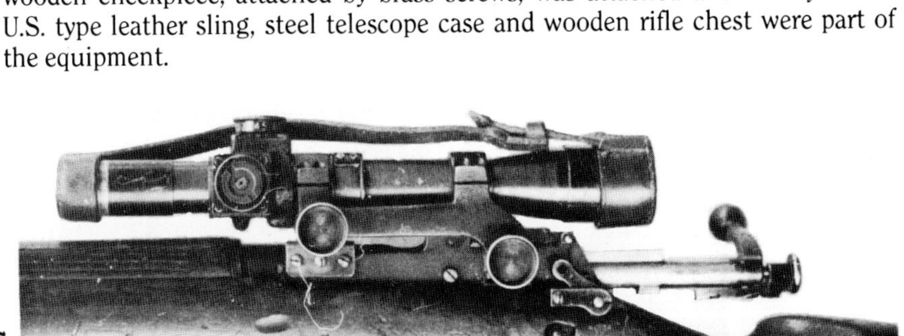

RIFLE No. 4 Mk I* (T) with C No. 32 TP Mk I TELESCOPE

Calibre:	*.303 in.*
Rifling & Twist:	*5 groove, Enfield, L.H.*
Barrel Length:	*25.2 in. (640mm)*
Overall Length:	*44.5 in. (1130mm)*
Weight:	*10 lb. 5 oz. (4.7 kg)*
Magazine Capacity:	*10 rounds.*
Manufactured & Converted:	*Long Branch.*

Typical Markings, L.H. side of Receiver:

No 4 Mk I*
LONG BRANCH
1944 T

One hundred of these Lyman Alaskan telescopic sights were purchased for setting up on the Canadian No. 4 sniper and 99 are recorded as being set up. Apart from the mounts and telescope, the rifle is otherwise identical to the standard No. 4 Mk I* (T); these particular conversions were effected during a three month period in early 1945.

While the Griffin & Howe type mount appears similar to that fitted on the C No. 67 scope (described on the following page), the upper section is different; it is an R.E.L. type with the dovetail base at 45° and a slightly different clamping arrangement to the subsequent G. & H. type. The Griffin & Howe and R.E.L. clamping arrangements appear similar, but they function a little differently and the latter is on an inclined plane. The "TP" designation indicates waterproofing.

The Lyman Alaskan is a commercial model telescope, marked:

"THE LYMAN GUN SIGHT CORP.
MADE IN U.S.A. MIDDLEFIELD, CONN.
U.S. PATENT No 2078858"

and further along the tube is the model identification "ALASKAN". Typical of the contemporary commercial telescopic rifle sights, it has a cross hair reticule, it is smaller in diameter and lighter than the military scopes and there are no graduations for set ranges on the elevation drums. The elevation adjustment drum is on the top, while the lateral adjustment drum is on the right side of the tube.

Due to serious delivery delays at the time with R.E.L. No. 32 telescopic sights, orders were placed for a total of 450 of these sights; the Alaskan was also favoured because it was waterproof.

J. Alley

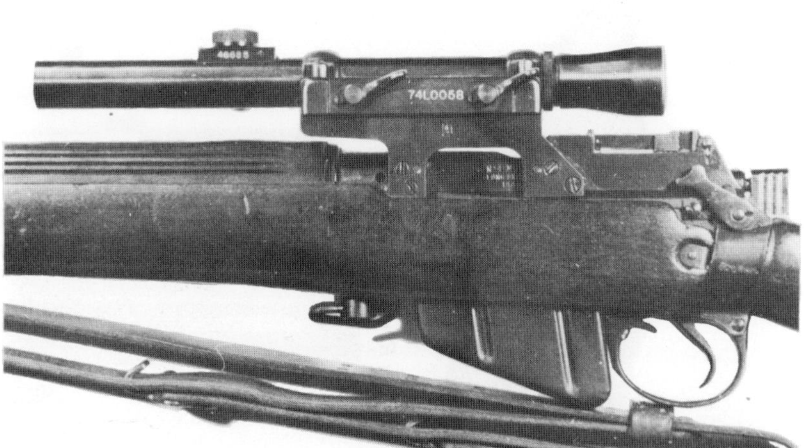

RIFLE C No. 4 Mk I* (T) with C No. 67 Mk I TELESCOPE

Calibre:	*.303 in.*
Rifling & Twist:	*5 groove, Enfield, L.H.*
Barrel Length:	*25.2 in. (640mm)*
Overall Length:	*44.5 in. (1130mm)*
Weight:	*11 lb. 8 oz. (5.2 kg)*
Magazine Capacity:	*10 rounds.*
Manufactured & Converted:	*Long Branch.*

Typical Markings, L.H. side of Receiver:

No 4 Mk I*
LONG BRANCH
1944 T

About one hundred of these units were set up at the Canadian S.A.L. factory and it differs in the telescopic sight, mount and butt configuration. While the conventional British and Canadian scope mount is a one-piece unit secured to the rifle and pads by two thumb screws, this Griffin & Howe design features roll-over quick release catches which clamp across the top of the lower mount. The lower mount is secured to the receiver by four screws and two pins. To detach the upper mount and telescope, the roll-over catches are lifted up and a cam on each disengages from a transverse slot on the lower mount, permitting the unit to be slid forward and removed.

This telescope is unlike any in service with the other No. 4 sniper rifles; early models are marked as C No. 32 Mk 4 but the designation was changed to C No. 67 Mk I soon afterwards. The sample shown here is marked:

"TEL. STG. C No 32 Mk 4
CGB 91 G.A. 3.5 X 24
R.E.L. CANADA
1944 10-C"

The magnification is rated at 3.5 power and the objective lens at 24mm. Made by Research Enterprises Ltd. of Montreal, Quebec, the serial number "10-C" is also marked on the top of the rifle butt at the wrist. The range drum at the top is calibrated to 1,000 yards and the lateral deflection drum is marked to 10 in each direction. Screw caps protect these adjustment drums.

The specimen shown here was used in British comparative trials and the sporting type butt has a rollover cheekpiece, plastic cap at the bottom of the pistol grip, and a rubber kickpad. Like many of the Canadian sniper units, a Stevens aperture backsight is fitted, graduated to 1,300 yards, with the battle aperture removed.

M.O.D. Pattern Room

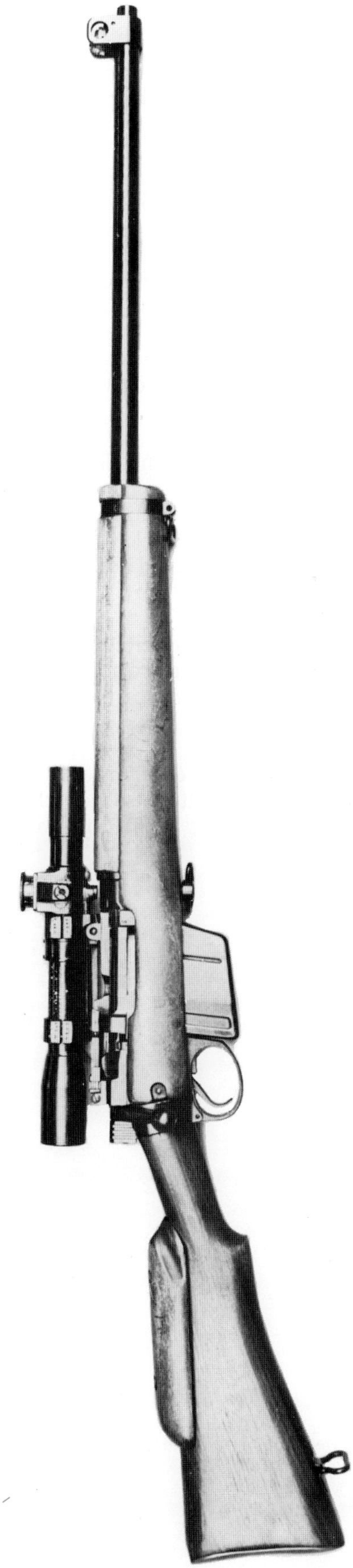

RIFLE L42A1

Calibre:	*7.62mm (.308 in.)*
Rifling & Twist:	*4 groove, Enfield, R.H.*
Barrel Length:	*27.6 in. (700mm)*
Overall Length:	*45.5 in. (1181mm)*
Weight:	*12 lb. 5 oz. (5.6 kg)*
Magazine Capacity:	*10 rounds.*
Converted:	*Enfield.*

Typical Markings, L.H. side of Receiver:

L42A1
T

Currently in service with British forces, the L42A1 was approved on 24th August 1970. Converted from the No. 4 (T) rifle, it weighs 12 lb. 5.5 oz., nearly 1 lb. more, due to the large diameter 7.62mm barrel which contributes significantly towards its excellent accuracy. Finally declared obsolescent in April 1992.

Its half-stocked fore-end serves to reduce weight; a full length fore-end is not required due to the different bedding method with the heavy barrel. A modified .22 No. 8 handguard was utilised along with its lower band but more recently, newly-made handguards with some external differences have been introduced. A new, square-shaped magazine is required for the 7.62mm NATO round, marked "CR 141 A", the year of manufacture, and the Enfield logo. A new bolthead and wide extractor are fitted; the bolthead is marked "19 T" which indicates a pressure of 19 tons per sq. in. for the 7.62mm re-proofing.

In other respects, the L42A1 is similar to its derivative, the No. 4 sniper rifle, high commendation indeed for the No. 4 action and sniping outfit of 2nd World War vintage. The standard 1,300 yd. No. 4 Mk I/1 aperture backsight leaf was usually retained, with battle aperture removed; the yards being read as metres. This recalibrates sufficiently for the new cartridge, although the backsight slide has a lower datum line and is marked "M" on the right hand side. The face was usually blackened.

The re-marked "TEL. STRT. STG. L1A1" telescope is converted from the No. 32 Mk 3 of the No. 4 sniper, with the original designations cancelled; this conversion also incorporates an improved sealing. The range drum is calibrated to 1,000 metres and the lateral adjustment drum is numbered to 16 at each end of the scale. A competition-type front trigger guard screw with swivel combination is fitted to permit use of the sling. While the L42A1 appears similar to its L39 counterpart, the trigger on the sniper rifle is mounted on the trigger guard, whereas the latter has it hung from the action body.

M.O.D. Pattern Room

SKELETON ACTIONS

These actions first emerged as "Action, Skeleton"; armourers tools for testing suspect components by substituting the doubtful part in the action and observing its operation through the viewing slots or cut-outs. With the percussion Pattern 1853 series rifles and carbines, removal of the lock from the wooden stock permitted an examination of its function should the action or any component be suspect, but with the introduction of the Martini in 1871, this was no longer possible as the mechanism was enclosed. By the time the Lee-Enfield came into service, such a role had been largely discarded and the skeleton actions utilised more for instructional purposes, and so reject or condemned components and rifles were often selected for sectionising.

Until the Great War, nearly all the skeleton actions were made at Enfield and then after the Ishapore and Lithgow factories started operation some were produced there also. The skeleton actions were mostly Marks I, III and III* S.M.L.E. models which used condemned British rifles and sometimes other components were used in the assembly which were not necessarily true to pattern. Some Lee-Enfield and Mark I S.M.L.E. skeleton actions have been noted with Metford barrels and occasionally even incorrect pattern butts, but as far as the action itself is concerned, the correct pattern was adhered to.

Mark I/I* and Mark II Lee-Metford rifles, the Lee-Enfield rifle and cavalry carbines were so modified; the Mk II* Lee-Metford was not used in this form as the action is identical to the Lee-Enfield model. There is no differentiation between Marks I and I* Lee-Enfield skeleton actions as the difference only applies to the provision for clearing rod or its omission, the actions are identical. No real distinction was made between the Lee-Metford and Lee-Enfield carbines in this role either as the actions are the same, even though there are some differences between the butts and the rifling in their original forms.

The differences between later Marks I and I* S.M.L.E. will only be found in the butt, while the converted Marks of S.M.L.E. will have to be studied carefully for signs of the alteration in order to distinguish them, though sometimes the conversion stamps remain on the left side of the butt socket. The Marks III and III* S.M.L.E. represent the largest production of the skeleton action and quite a few were made up during both World Wars by the trade.

Skeleton No. 4 actions were modified at the Royal Ordnance Factories, usually from condemned or downgraded rifles, although it would seem that some were also produced by the trade in Birmingham during the war. It is also likely that some Mark I and I* models were sectionised at S.A.L. Long Branch, Canada. All action variations of the No. 4 were skeletonised, the Marks I, I*, 2, 1/2 and 1/3.

Figures to March 1914:-

	Enfield
M.L.M. Mks I & I* Rifles:	508
M.L.M. Mk II Rifle:	751
M.L.M./M.L.E. Cavalry Carbines:	183
M.L.E. Rifles:	1,179
S.M.L.E. Mk I Rifle:	1,447
S.M.L.E. Mk I* Rifle:	435
S.M.L.E. Cond. Mks I, II & II* Rifles:	105
S.M.L.E. Mk III Rifle:	1,128 +
S.M.L.E. Cond. Mk IV Rifle:	20 +

+ Incomplete figures, up to March 1914 only.
No figures are available for Ishapore and Lithgow.

Figures for 1914-18, per War Office contracts:-

S.M.L.E. Marks III & III*:	500 (Bonehill 1917)
Skeleton Action Mk III Cond. to Mk III*:	500 (Greener 1917)

During 1939-45, some No. 1 Mks III & III* were converted at R.S.A.F. Enfield and by H. Atkin.

Post war figures, No. 4 rifles:-	Enfield
1951-52:	300
1952-53:	2,020
1953-54:	259

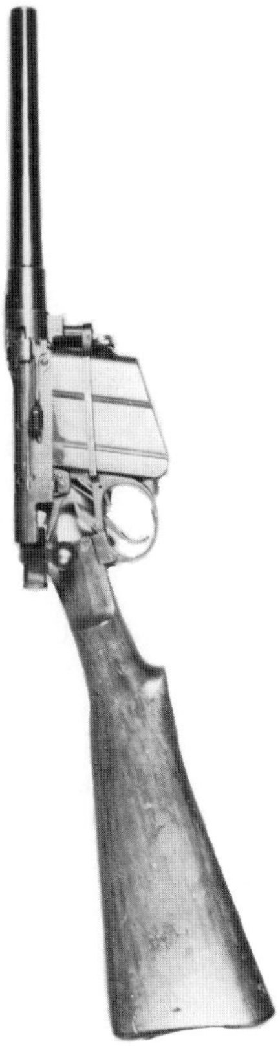

ACTION SKELETON:- M.L.M. & M.L.E. RIFLES

Barrel Length: *9.0 in. (228mm)*
Overall Length: *28.2 in. (716mm)*
Weight, approximate: *4 lb. 8 oz. (2.0 kg)*
No Designation Markings.

While the initial pattern of skeleton M.L.M. action introduced in 1890 was an armourers tool, such actions were later used for instructional purposes regarding the operation of the action. Only a short length of barrel was fitted, without any sights. No fore-end is required either, which exposes the action to view. Slots have been machined into selected parts of the action and components so as to permit viewing of the internal operations, such as engaging of the bolt lugs and working of the firing pin. The right side of the butt socket has been cut away, so there are no receiver markings left to assist in differentiation between the various models. This can only be done by examining the configuration of the action. The long range volley aperture sights were retained.

Technically incorrect butts may be found fitted, as many of the demonstration models were asssembled from reject component parts. Minor differences such as location of the viewing slots may be encountered, but these cuts are usually found in bolt, left side of the receiver, and magazine. Most of the lower mechanism including the trigger and sears is visible due to the absence of the fore-end. Butts have a small portion of wood removed on the right side near the socket, on the receiver. Bolts and barrels were numbered to the receivers. Some cavalry carbines, in addition to Lee-Metford Mark I* and II rifles and Lee-Enfield rifles, were thus modified.

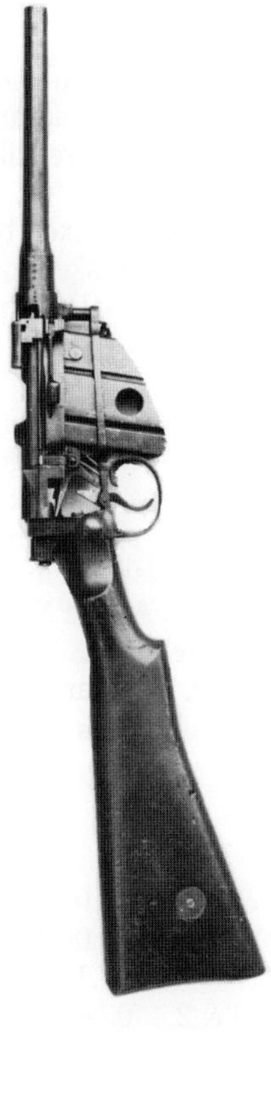

ACTION SKELETON:- S.M.L.E. Mk I & I*

Barrel Length: *8.9 in. (224mm)*
Overall Length: *28.4 in. (722mm)*
Weight, approximate: *4 lb. 10 oz. (2.1 kg)*
No Designation Markings.

Like some of the previous models, this skeleton action was originally assembled from reject or used components, to be used as a teaching aid. There are no markings on the receiver to indicate any particular designation, rifle mark number or manufacturer, as this was removed when the right side of the butt socket was machined away. The butt has also been cut away at this point, in line with the receiver body. No sights were retained, except for the long range aperture. The bolt and barrel were numbered to the receiver.

The Mk I S.M.L.E. series action may be identified by the sliding charger guide on the bolthead and the corresponding nib on the left side of the body. Viewing slots are cut into the left side of the receiver, bolt, bolthead, and sometimes the cocking piece. The left side of the magazine has three wide slots, exposing the magazine platform and spring. Some specimens have been noted with Metford barrels, from the Mk II or II* Lee-Metford rifle.

Some Mk I* S.M.L.E. skeleton actions are listed in manufacturing returns and differ from the Mk I model in having the later pattern butt with inletted swivel mount and brass buttplate with trap provision, sling swivel bracket on the trigger guard, magazine link loops deleted, coin slotted striker keeper screw, and the deeper No. 2 magazine case.

ACTION SKELETON:- S.M.L.E. Mk III & III*

Barrel Length:	*8.9 in. (224mm)*
Overall Length:	*28.0 in. (711mm)*
Weight, approximate:	*4 lb. 14 oz. (2.2 kg)*
No Designation Markings.	

Assembled from Mark III and III* S.M.L.E. parts for use as instructional aids, these shortened rifles were used to demonstrate the mechanism and the loading and firing sequences. Examples have been noted of British and Australian origin and it is most likely that similar examples were fabricated in India also. The receiver markings on the butt socket have been machined as this part is cut away to show the socket; therefore the differentiation between Mk III and III* patterns is decided by the provision for magazine cut-off and the long range aperture sight on the Mk III. Minor variations are likely to be noticed in the barrel lengths.

Serial numbers are usually marked on the receiver, bolt and barrel, although some have been noted without any serial numbers, only a rack number painted on the butt. Viewing slots were usually cut in the left side of the receiver, bolt, bolthead, cocking piece and magazine. Because of the absence of markings, the Mark III and III* may be differentiated from earlier models by the charger bridge as the M.L.M. and M.L.E. variants have no such facility and the Mk I S.M.L.E. has a different charger system. Some Mk III and III* models were assembled by the gun trade in Britain during the 2nd World War.

ACTION SKELETON:- RIFLE No. 4

Barrel Length:	*9.0 in. (228mm)*
Overall Length:	*28.1 in. (714mm)*
Weight, approximate:	*5 lb. 5 oz. (2.4 kg)*
Receiver Markings:	*Original stampings on L.H. side of Receiver.*

Like the previous models, these were also generally built up from condemned or downgraded rifles or components, but because the designation and manufacture markings are found on the left side of the receiver, these models can be readily identified. Examples have been noted of all the No. 4 variants, Marks I, I*, I/2 and I/3 and original factory models usually have the serial number prefixed by "SKN", also marked on the bolt and magazine. In recent years, these skeleton actions have been assembled by the commercial British trade, for disposal on the collectors market, these do not have the "SKN" serial prefix.

Viewing slots have been milled in the left side of the receiver, and in the bolt, bolthead, cocking piece and magazine; and as no fore-end is fitted, the trigger and sear mechanism is also exposed. The right side of the butt socket and portion of the butt have also been machined. Many of these actions were painted red where the metal was removed for viewing slots and a way to distinguish the more recent reproductions is by closely examining the age of such paint. As the No. 4 is fitted with aperture sights on the body, these have been retained. Some Stevens-Savage and Long Branch actions were sectionised like this, but apart from the possibility of some done at Long Branch or the Canadian ordnance system, most were modified at Enfield, England.

Dates of approval for the Skeleton Actions were:-

M.L.M. Mk I	24.11.1890	S.M.L.E. Mk III & IV	21.05.1908
M.L.M. Mk II	01.11.1893	S.M.L.E. Mk III*	17.06.1925
M.L.E.	05.02.1898	No. 4 Mk I & I*	11.12.1950
S.M.L.E. Mk I & II	27.02.1904	No. 4 Mk I/2, I/3 & 2	03.02.1955

FENCING MUSKETS

The terminology "musket" was generally carried over from its muzzle-loading predecessors and this variant was used for bayonet practice and exercise, mostly in the gymnasium. These muskets were constructed so as to resemble the balance and feel of the current service rifle with fixed bayonet, as closely as possible. On most such practice arms, the bayonet took the form of a spring-loaded rod with padded button on the end, so as to avoid injury to the opponent. Protective clothing was often worn in practice as well.

Only in a few instances did the fencing musket really resemble the appearance of the service arm, even though it may have duplicated its service counterpart rifle's handling qualities. Some obsolete rifles were converted for this role too, even though there was no resemblance whatsoever to the current service rifle; an excellent example being the Mark VII Fencing Musket of 1909 which was made up from the .450-in. Martini-Henry rifle. In other cases they were assembled from a representative stock and tubing with a spring-loaded rod, bearing scant likeness to the service rifle.

Some unofficial patterns were also widely used, especially during the latter part of the 19th century, as many volunteer and colonial forces had models locally assembled from available materials, usually outdated or unserviceable rifles and carbines. In the British armed forces, however, there were specific patterns; the table at the bottom of this page also provides some details of preceding service models as a background.

Those fencing muskets illustrated and described in this section have been chosen as representative models; all the original types and patterns are now not available and some such as those converted from Martini rifles do not really apply to this particular study. Conversions from Magazine Lee rifles, especially the early ones, have been especially selected. Quite a few fencing muskets were constructed and furnished by the trade, including the makers W.W. Greener, C.G. Bonehill, and Webley & Scott.

Approval	*Designation*	*L.o.C.*	*Brief Details*
1863	Musket, Fencing, Mk I	807	Fabricated like musket, triangular spring byt.
1867	Musket, Fencing, Mk II	1605	Fabricated like musket, triangular spring byt.
1891	Musket, Fencing, Mk III	6370	Fabricated, tubing & rod.
1894	Musket, Fencing, Mk IV	7406	Fabricated, tubing & rod.
1896	Musket, Fencing, Mk V	8369	Fabricated, tubing & rod.
1902	Musket, Fencing, Mk VI	11489	Fabricated, tubing & rod.
1905	Musket, Fencing, Short, Mk I	12761	Fabricated, tubing & rod.
1907	Musket, Fencing, Short, Mk II	13991	Converted from Martini-Henry rifle.
1909	Musket, Fencing, Mk VII	14757	Converted from Martini-Henry rifle.
1909	Musket, Fencing, Mk VII N.P.	14902	Converted from Martini-Henry rifle.
1912	Musket, Fencing, Mk VIII	16121	Converted from Martini-Enfield rifle.
1912	Musket, Fencing, Mk IX	16121	Converted from Mk I* Lee-Metford rifle.
1913	Musket, Fencing, Mk X	16547	Fabricated, tubing & rod.
1915	Musket, Fencing, Mk X, Nos. 1 & 2	17503	Fabricated, tubing & rod.

The nomenclature change which was applied to all small arms in 1926 resulted in new designations for the Fencing Muskets as well:

Musket, Fencing, Mk VI	became	Musket, Fencing, No. 1
Musket, Fencing, Mk X No. 1		Musket, Fencing, No. 2
Musket, Fencing, Mk X No. 2		Musket, Fencing, No. 3

CONVERTED M.L.M. RIFLE

V.R.
ENFIELD
1889
I

Overall Length: *61.4 in. (1560mm)*
Rifle Length: *48.5 in. (1232mm)*
Weight: *10 lb. 7 oz. (4.7 kg)*
Converted: *Royal Small Arms Factory.*

Receiver Markings, R.H. side of Butt Socket:

Though this example was not an approved specimen, it represents one of the first conversions of the Magazine Lee rifle for bayonet practice. It differs from the other patterns in that a mock blade with gunmetal button is used as the bayonet, whereas the approved patterns from this period onwards used steel rods rather than the representative bayonet form. The early fencing muskets had mock triangular bayonets somewhat similar to the service pattern bayonet, and operate on a rather similar principle to this musket. Obviously, the double-edged blade positioned below the line of bore was intended to be comparable to the Lee-Metford or Lee-Enfield rifle fitted with the Pattern 1888 sword bayonet.

On the illustrated sample, a Mark I Magazine Lee-Metford rifle was used as the basis for this bayonet exercise arm and the particular features of the Mark I .303 rifle which make it quite distinctive are the safety catch, Lewes sights and early type of handguard with squared front end. The action and all the external features to the rear of the lower band remain unchanged, although the action and catches have been rendered innocuous, which makes the bolt, cocking piece, magazine release catch and safety catch inoperable. A metal screw has been tapped through the safety catch lever into the receiver for this purpose.

The double-edged blade of the representative bayonet, with a rounded tab at the point to prevent injury in practice, slides in a groove cut below the fore-end and barrel. This groove has been reinforced with brass strips and securing screws either side for its 15-in. length, extending all the way to the lower band. The replacement nosecap at the front of the rifle is rounded in style and it also acts as a stop for the extreme end of travel of the bayonet.

Some similar conversion styles were applied by Colonial and volunteer forces to the Pattern 1853 series, Snider conversions and the Martini-Henry rifles and carbines, although a plain rod with rubber or padded button at the end was more often used, rather than a mock bayonet such as is illustrated here.

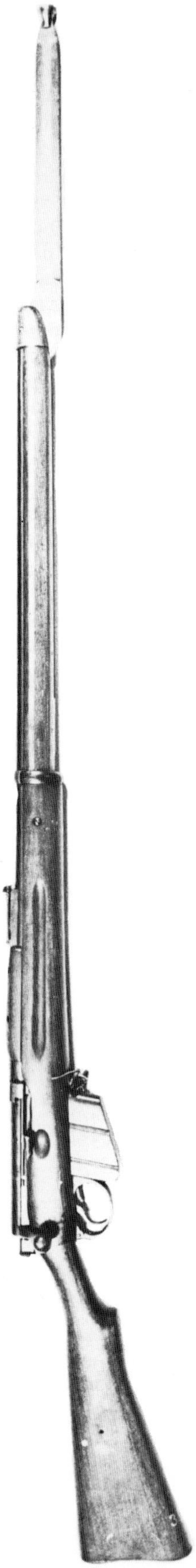

M.O.D. Pattern Room

MUSKET, FENCING, Mk IX

Overall Length:	*61.5 in. (1562mm)*
Rifle Length:	*47.25 in. (1200mm)*
Weight:	*10 lb. 1.5 oz. (4.6 kg)*
Converted:	*Royal Small Arms Factory, Ordnance Depot or Approved Contractor.*

Typical Markings, R.H. side of Butt Socket:

V. R.
ENFIELD
1891
I*

Of all the approved patterns of fencing muskets, this was the only one which was a conversion from the Magazine Lee rifle, and it was approved on 25th April 1912. Contemporary with the Mark III S.M.L.E. rifle, it was intended to duplicate the weight and balance of this rifle with bayonet fixed. Some balance weights were fitted into the butt to achieve this weight and correct balance, the point of balance from the butt was given in the official description as being 24 ins. from the butt end.

In this particular conversion, quite a few modifications were made to the original Mark I* Lee-Metford; the magazine was removed and the recess plugged with a wooden block, the bolt handle, cocking piece and striker were removed, magazine cut-off dispensed with and the trigger mechanism withdrawn. There are no sights at all; the front and rear sights were removed along with the long range volley dial sight plate, pointer and complementing folding rear aperture. The recess for the long range sights in the fore-end has been repaired and plugged. The handguard was also dispensed with, so the overall lines are quite smooth and there are no sharp corners likely to cause injury to clothing or either participant in the course of fencing exercises.

A special bulbous "nosecap" has been fitted, with a forward extension for the spring operated bayonet rod. The coil return spring for this plunger or rod is housed inside the bore, the barrel having been bored out to accommodate it. At the end of the plunger, a 1.25-in. diameter knob is fixed in order to avert injury to the opponent during the fencing practice.

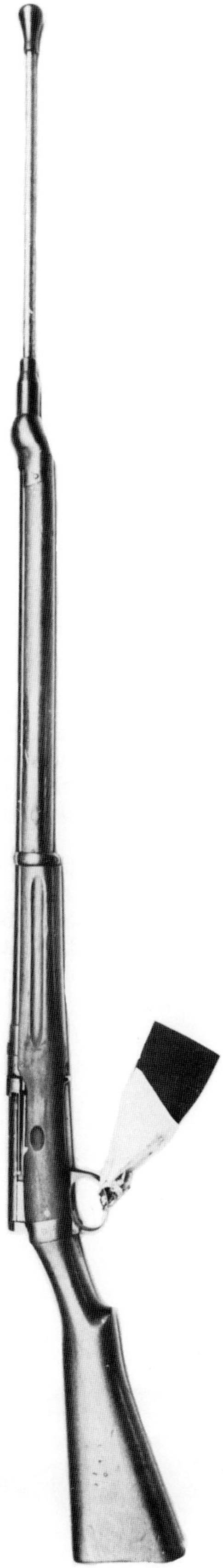

M.O.D. Pattern Room

MUSKET, FENCING, Mk X No. 1 (MUSKET, FENCING, No. 2)

Overall Length:	*63.3 in. (1608mm)*
Rifle Length:	*46.3 in. (1176mm)*
Weight:	*9 lb. 0 oz. (4.1 kg)*
Manufactured:	*Various Government Factories and Approved Contractors.*

No Receiver Markings, not converted from Rifle Action.

Unlike many of the previous service patterns of fencing muskets, the Short Mk II, Mk VII, Mk VIII and Mk IX, this model was not a rifle conversion, but fabricated from steel tubing, rod and wood. It was intended to resemble the balance and feel of the S.M.L.E. rifle with bayonet fixed. The Mk X Fencing Musket was approved on 28th August 1913 and upon the introduction and announcement of the Mk X No. 2 pattern in 1915, it became known as the "Mk X No. 1". In 1926, its nomenclature was further changed to "Musket, Fencing, No. 2".

Like the rifle which it was intended to represent, the butt and fore-end are wood, in separate sections, although their shape and mode of construction is different. The butt has a strong, cupped steel butt cap with tangs along both the top and bottom of the grip at the small of the butt for reinforcement; the top tang is chequered in a diamond design so as to provide a better grip. The butt and tang fit into a circular collar, about 2.75-in. in diameter. From this collar, a reinforcing strip of metal extends underneath the "fore-end" for its full length to the rounded nosecap.

The metal tubing has a spiral return spring inside for the rod plunger. At the front end of the tubing there is an internal thread which accepts a serrated collar, this can be unscrewed to release the rod plunger and spring. The diameter of the steel plunged rod is 5/8-in. and a 1 3/8-in. diameter steel button is fitted to its tip so as to avoid injury during the bayonet or fencing practice.

Apart from a few inspectors stamps, there are no markings on the metal, the only clue as to the particular designation of this fencing musket is found stamped on the butt. These types of bayonet practice muskets were fabricated by a number of contractors during the Great War; W.W. Greener, Webley & Scott, E.R. Watts & Son, G. Beaton & Sons, and C.G. Bonehill. A few were also assembled in the Government Ordnance establishments.

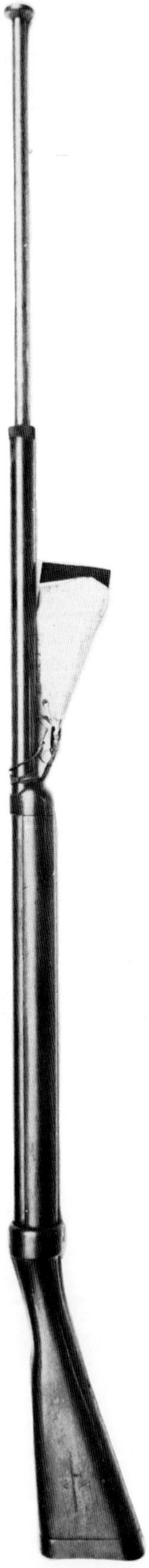

M.O.D. Pattern Room

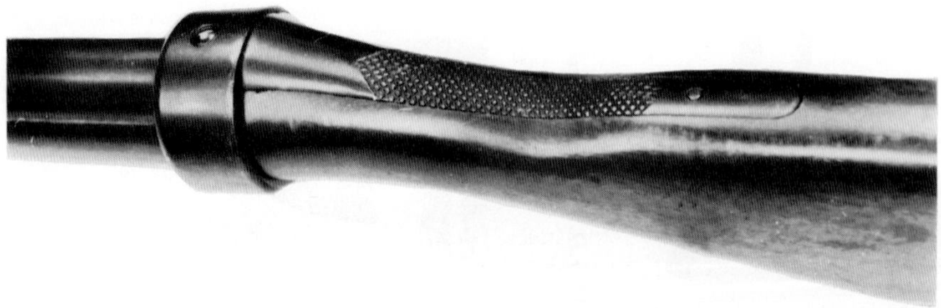

SILENCED COMMANDO CARBINES

These British Commando carbines were developed by William De Lisle in conjunction with Sir Malcolm Campbell of Combined Operations H.Q. from 1942 and officially approved on a secret list for limited production. The De Lisle carbine was essentially designed and built around the pistol cartridge because this is effectively the most important consideration for a silenced firearm, that the projectile should travel below the speed of sound (so as to avoid the distinctive "crack" heard when the bullet achieved supersonic velocities) yet still have adequate muzzle energy.

The U.S. military .45-in. Auto Colt Pistol cartridge was an obvious choice as it was also in British service at the time with the Thompson Sub-Machine gun; some M1911 Colt pistols were on British issue during the war also. Because this cartridge was chosen, it enabled the adaption of the Model 1911 Colt pistol magazine for use in the De Lisle carbine as these .45 magazines were a stock item with certain specialist units such as the Commandos.

Production of the De Lisle carbine was terminated before the airborne models were made and so only the prototype of this folding stock model was assembled. The planned production of 500 carbines being cut short, only about 130 of the standard model with wooden S.M.L.E. butt were manufactured, according to the official records. However, it appears that there may have been an additional contract or other service department outfitted as some serial numbers in excess of the 130 have been noted; nevertheless, total production of the .45 De Lisle Commando Carbines did not exceed a few hundred.

Some prototypes were fabricated at the Ford Dagenham factory for tests, the manufacture of the approved pattern was carried out by Sterling at Dagenham, East London. Two models were planned for production; a standard type using the S.M.L.E. rifle butt and an airborne version with folding metal stock; weighing 8 lb. 4 oz. and 7 lb. 2.5 oz. respectively. The silencer was no newcomer as it had been developed to quite an effective standard by the end of the 19th century by inventors such as Hiram Maxim, although De Lisle's design featured a number of refinements. A military application for the silencer or suppressor had been considered by Britain and trials were conducted in the United States by the turn of the century although the subject was not investigated further due to the doubtful requirement.

Although the .45 ACP round is only a pistol cartridge, the De Lisle bolt action carbine has a barrel length of 7.25 ins. compared with that of the Model 1911 Colt pistol's 5 ins., this noticeably improved the performance and effectiveness of the round. Along with the increased sighting radius, it also permitted shooting over longer ranges than was normally effective with the pistol. The .45-in. calibre bullet left the muzzle at just below the speed of sound so there was no super-sonic "crack" and with the muzzle blast also muffled, those only really noticeable sounds were the "thump" of the bullet upon striking the target as well as the manipulation of the bolt by the hand.

Production of the De Lisle was terminated before the airborne models were made and so only the prototype of this folding stock model was assembled. The planned production of 500 carbines being cut short, only about 130 of the standard model with wooden S.M.L.E. butt were manufactured, according to the official records. However, it appears that there may have been an additional contract or other service department supplied as some serial numbers in excess of the 130 have been noted; nevertheless, total production of the .45 commando carbines did not exceed a few hundred.

	De Lisle	*Ford Dagenham*	*Sterling*
.22RF De Lisle Prototype:	1	-	-
9mm De Lisle Prototype:	1	-	-
.45 ACP De Lisle Prototypes:	-	17	-
De Lisle Commando Carbine:	-	-	130+ ?
De Lisle Airborne Model:	-	-	prototype

DE LISLE PROTOTYPE

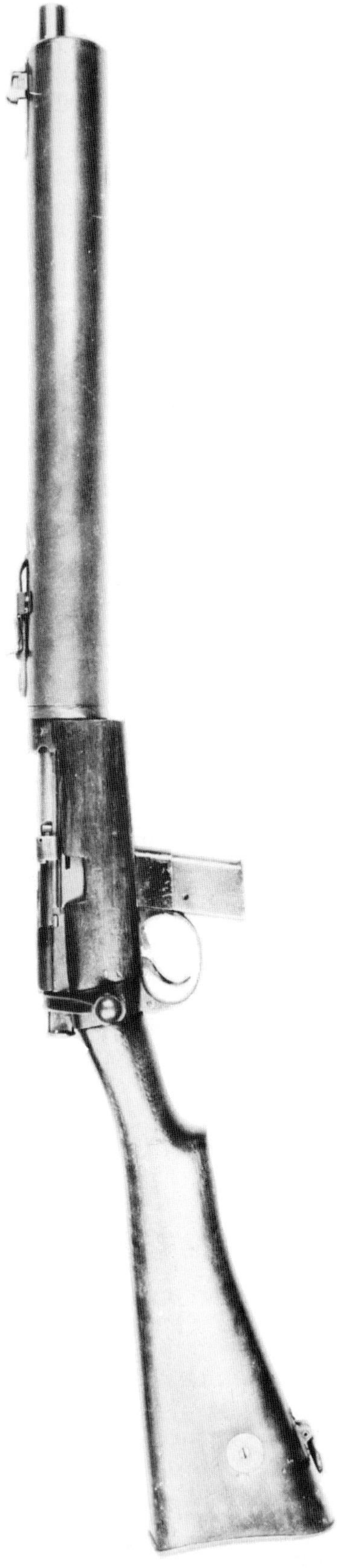

Calibre:	*.45 in.*
Rifling & Twist:	*6 groove, L.H.*
Barrel Length:	*7.25 in. (184mm)*
Overall Length:	*37.2 in. (945mm)*
Weight:	*8 lb. 5 oz. (3.8 kg)*
Magazine Capacity:	*7 rounds.*
Converted:	*Ford Motor Co., Dagenham.*

Typical Markings, R.H. side of Butt Socket:

This prototype was one of a small test batch assembled in 1943, using a reject Mk III S.M.L.E.; it is marked "THE DE LISLE COMMANDO CARBINE" along the top of the silencer jacket. Developed by W.G. De Lisle in association with Sir Malcolm Campbell (father of Sir Donald Campbell of "Bluebird" fame), it consists of a modified S.M.L.E. action with shortened bolt, .45-in. calibre barrel, and silencer body which encases the barrel and extends for a considerable distance beyond the true muzzle, so as to muffle the blast. This is done by means of a chamber and baffles inside the tube, positioned by longitudinal rods which align the bullet passage holes.

A steel casing was used for the silencer body, with a phosphate finish which gives a bronze appearance; this makes it somewhat heavier than the subsequent production models which utilised aluminium alloy. The backsight is a simple V-notch with sliding ramp to adjust elevation, similar to many of the Winchester sporting rifles. The foresight is a S.M.L.E. blade set in a ramp, with projecting wings.

Receiver modifications involved removal of the charger bridge and magazine cut-off and the fitting of a long ejector on the left side of the receiver, secured by a screw. The tip of this ejector protrudes through a hole in the wall of the receiver, into the boltway. The new .45 cal. barrel extends for more than 2 inches into the boltway, necessitating a shortened bolt body about 3.5-in. long. The face of the bolthead has been recessed and a special extractor fitted for the new rimless cartridge.

The magazine was modified from that of a U.S. Model 1911 Colt auto pistol, by the addition of a rib similar to that on the S.M.L.E. magazine; it is secured in the rifle in a similar manner. The new magazine housing extends for about 1-in. below the sheet metal plate attached to the bottom of the trigger guard. The rear portion of the S.M.L.E. fore-end has been used, trimmed level with the front of the receiver; there is no fore-end. Seventeen of these prototypes were fabricated, serial numbered from 1001 to 1017.

M.O.D. Pattern Room

DE LISLE COMMANDO CARBINE

Calibre:	*.45 in.*
Rifling & Twist:	*6 groove, L.H.*
Barrel Length:	*7.25 in. (184mm)*
Overall Length:	*35.75 in. (908mm)*
Weight:	*8 lb. 4 oz. (3.7 kg)*
Magazine Capacity:	*7 rounds.*
Converted:	*Sterling Armament.*

Typical Markings, R.H. side of Butt Socket:

G. R.
ENFIELD
1918
S^HT^ L.E.
III*
F.T.R.

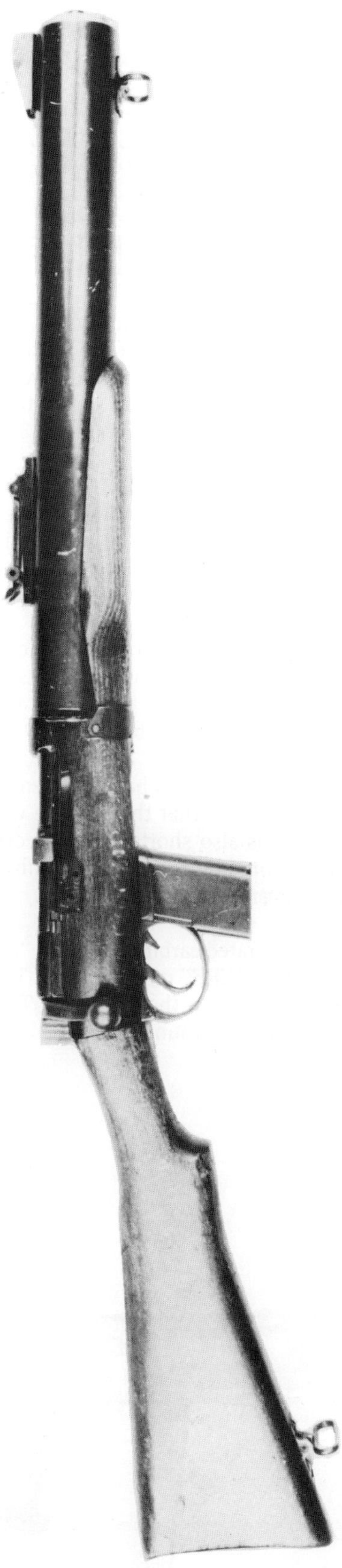

Used or condemned rifles were used as the basis for these conversions, possibly the only unmodified component would be the butt assembly. The receiver has had the charger guide removed and an ejector let into the left side of the body; the trigger guard has a tubular casing for the Colt Model 1911 magazine and the remainder of the .303 magazine slot has been filled in. The bolt has been shortened to 3.6 in. and the original bolthead face recessed and fitted with a modified extractor. A Bakelite insert has been dovetailed into the underside of the bolt handle so as to reduce the sound made when closing the bolt.

The 1911 Colt pistol magazine has been modified with an S.M.L.E. type rib down the back, the lips and platform have also been altered slightly. Extending nearly 2.5 in. into the boltway, the barrel has an effective length of 7.25 in. Beyond the barrel muzzle is an expansion chamber and a series of baffles, contained within a 2-in. diameter alloy tube. The thirteen baffles are positioned on two locating rods so as to align their bullet passage holes; most of the baffles are cut and bent above this hole to form a type of Archimedes screw effect. An alloy end cap for the silencer is secured by two bolts with Allen key recesses in their heads, and some small bleed holes from the silencer assembly are drilled at the bottom of this cap.

A conventional leaf backsight is graduated at 50, 100, 150 and 200 yards and the bed is riveted onto the silencer casing; the blade foresight has two protecting ears. An upper sling swivel is attached to the underside of the silencer body. The original S.M.L.E. fore-end was cut off level with the front of the receiver and an 8-inch long fore-wood is attached beneath the silencer casing.

A two-piece cleaning rod is housed in the butt trap, totalling 15-in. in length; the rear portion is tubular and accepts the smaller diameter front section for storage in the butt.

M.O.D. Pattern Room

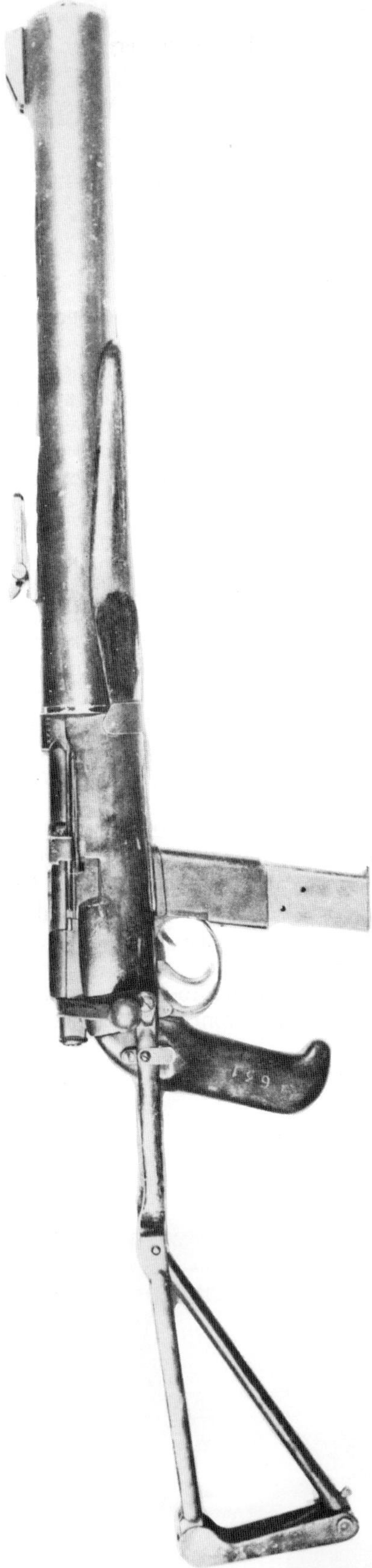

DE LISLE AIRBORNE COMMANDO CARBINE

Calibre:	*.45 in.*
Rifling & Twist:	*6 groove, L.H.*
Barrel Length:	*7.25 in. (184mm)*
Overall Length:	*25.7 in. (653mm) folded.*
	35.3 in. (897mm) extended.
Weight, approx.:	*7 lb. 8 oz. (3.4 kg)*
Magazine Capacity:	*11 rounds (optional).*
Converted:	*Sterling Armament.*

Markings, on R.H. side of Butt Socket:

G.R.I.
ISHAPORE
1931
S$^{\underline{HT}}$ L.E.
III

This was a special model of the Silenced Carbine designed for paratroops and airborne forces; the folding butt bears some resemblance to those of the Patchett and Sterling series machine carbines, some of which are also silenced. Orders were placed for 50 of these carbines but the contract was terminated before they were manufactured, so only this prototype appears to have been produced. The specimen illustrated was made from an Ishapore S.M.L.E.

A pistol grip is attached to the receiver and the folding metal butt is hinged though the receiver butt socket. A spring clip on each side of the pistol grip secures the butt assembly in the extended position and there is a slot in the fore-wood to secure the butt assembly in the folded down position.

In other respects, it is the same as the fixed stock pattern, with a 2-in. diameter alloy tube containing the silencer assembly and its thirteen duralium baffles. A 200 yard backsight is riveted onto this casing. Long high-tensile rods locate the baffles inside the silencer and the end cap is secured by Allen key bolts.

The barrel extends rearwards into the boltway and the bolt has been shortened. Modifications were made to the .303 S.M.L.E. bolthead in that the boltface was recessed and the extractor altered, the firing pin was also shortened. A special ejector is fitted into the left side of the receiver, secured by a screw, and a nib at the forward end projects through a hole in the receiver wall into the boltway.

A large capacity magazine was supplied with the illustrated carbine, stamped ".45 ACP NOT MORE THAN 11 ROUNDS" and "FILL TO HERE" on both sides. The holes in the casing enable the ammunition to be viewed in the magazine. This magazine is interchangeable with the more conventional 7-round model.

Infantry Museum, Warminster

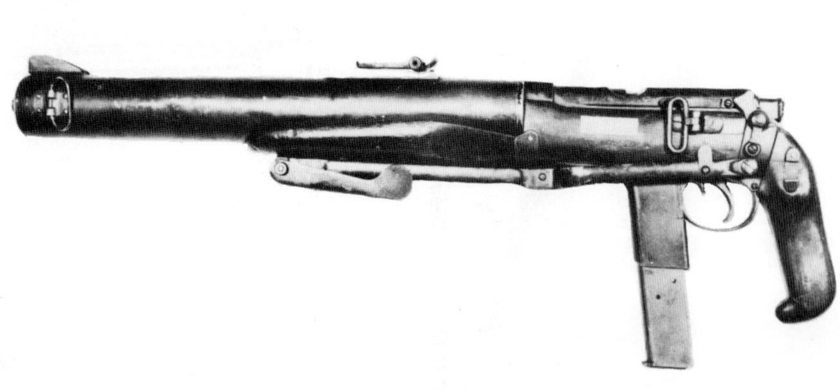

AUTOMATIC MODELS

Whilst the gas-operated conversions of the Lee-Enfield cannot generally be considered to have been very successful, they are nonetheless interesting developments which played a role in the Lee rifle history. The Charlton conversion was even considered as an emergency alternative to the light machine gun in New Zealand when the latter was in short supply during World War 2.

First patents for an "automatic S.M.L.E." were taken out in 1918 by Dawson and Buckingham; in this model the gas tube was mounted along the left side of the barrel, although a subsequent alternative arrangement specifies fitting of the gas tube underneath the barrel. Because Dawson and Buckingham took out numerous patents in the class of firearms and ammunition during the war and shortly afterwards, it is not likely that any prototype was assembled.

The Howell was another British model, basically similar to the Charlton, but less refined in a number of features and fittings. This model appears to date from the 1st World War although the example shown here more likely originates from the 1920's. The Howell was put up before the Ordnance Board for Home Guard use during the 2nd World War, but rejected. The South Africans also developed the Reider gas-operated conversion and notified the Board in 1941; samples were forwarded, but again no action was taken by the British.

A slightly different concept in S.M.L.E. autoloaders was put before the Ordnance Board for consideration late in 1943; the 7.63mm Howard Francis carbine. This took the form of a light automatic carbine which used the 7.63mm pistol cartridge, but it was also rejected and it is most likely that only the prototype was fabricated.

The Charlton however had a more noteworthy career in New Zealand where this conversion was issued. The light machine gun style conversion of outdated Lee-Metford and Lee-Enfield rifles was fabricated in New Zealand and the order for 1,500 units was filled, some serial numbers have been noted in excess of 1,500 which could indicate that even more were produced. Unfortunately, nearly all the guns were destroyed in a fire at Palmerston after the war.

In Australia, another Charlton design was produced by Electrolux, presumably on contract for New Zealand, although this maintained a more conventional rifle configuration as it was converted from the S.M.L.E. or No. 1 rifle. Only a few of these samples or prototypes appear to have been fabricated in Australia, so the Charlton, regarding its design and conversion, can be classified as unique to New Zealand.

However, the Lee-Enfield bolt and its rear locking lug configuration did not lend itself to these self-loading and automatic conversions, especially with the full-bore .303-in. cartridge. They were stop-gap measures at best, and thus cannot be considered to have been very successful. Except for the 7.63mm Howard Francis, all these designs are similar, with a gas piston exerting rearward force onto the bolt or bolthandle in order to unlock it and move it rearward. So the designs of World War 2 were only marginally more practical than the original Dawson and Buckingham 1918 design and the Howell. Only the Magazine Lee and Short Magazine Lee rifles were so adapted, the No. 4 action was not as smooth, nor able to be manipulated as rapidly.

Conversion & Country of Origin	*Number & Year*
.303 Dawson & Buckingham (U.K.):	Patent models only; 1918.
.303 Howell (U.K.):	Experimentals, prototypes; 1918, 1941.
.303 Reider (South Africa):	18; 1940-1941.
7.63mm Howard Francis: (U.K.)	Experimentals, prototypes only; 1943.
.303 Charlton (New Zealand):	Over 1,500; 1941-1943.
.303 Charlton (Australia):	Prototypes only; 1944.
.303 Ekins (Australia):	Prototypes only; 1944.

HOWELL AUTOMATIC RIFLE

Calibre:	*.303 in.*
Rifling & Twist:	*5 groove, Enfield, L.H.*
Barrel Length:	*25.2 in. (640mm)*
Overall Length:	*44.5 in. (1130mm)*
Weight, approx.:	*11 lb. 12 oz. (5.3 kg)*
Magazine Capacity:	*10 or 20 rounds.*
Converted:	*Base Workshops.*

Typical Markings, R.H. side of Butt Socket:

G. R.
B.S.A. C°
1918
S^HT L.E.
III*

Howell's conversion system features a gas tube and plunger along the right side of the rifle, it was developed during the Great War. The concept was again brought to the notice of the Ordnance Board early in 2nd World War, as a ready means of conversion to automatic fire, for issue to the Home Guard and application against German aircraft.

The gas take-off is positioned about 4-ins. from the muzzle, on a relatively cumbersome 2-stage tube which extends back to the charger bridge. Nearly 25-ins. in length, the forward 7-in. section has a diameter of 1-in., while the rear portion is larger, at about 1.25-in. A bulky rear mount for the gas tube is positioned through the fore-end, about 2.5-in. behind the backsight protector. The gas piston is 24-ins. long, with a 13-in. coil spring around the rear portion.

A tubular metal pistol grip is attached to the small of the buttstock and a sheet metal guard encloses the firer's hand around the pistol grip and it extends forward to the trigger. Another perpendicular guard on the top of the butt is mounted at the extreme point of rearward bolt travel. The action of the S.M.L.E. mechanism is quite simple; the bolt handle knob has been cut off and the bolt handle stem operates in an angular slot at the rear of the gas piston, so that the backward and forward movement of the gas piston unlocks, opens and then closes the bolt. Otherwise, the bolt action is unmodified. A new trigger has been fitted with a spring behind it and this incorporates a safety feature to prevent operation of the trigger until the action has been recycled.

This particular rifle retains the original backsight, although it was not intended to be used. The new offset sights comprise of a folding flap positioned in cuts in the rearsight protector wings, while the foresight folds out from a recess in the left side of the nosecap slots. The nosecap has been reduced in size, with the bayonet and sling mounts removed. The butt marking disc on the illustrated rifle is engraved "HOWELL AUTOMATIC".

M.O.D. Pattern Room

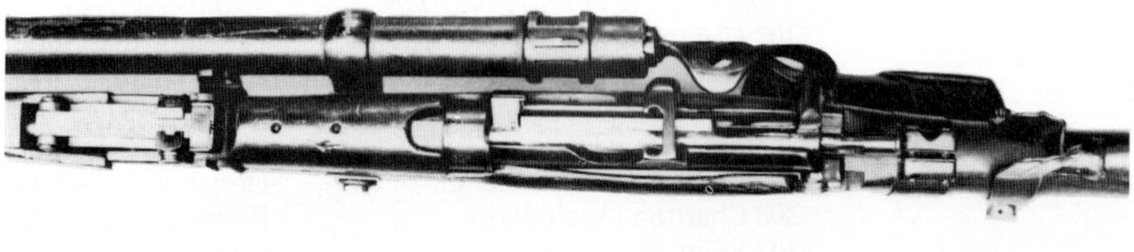

REIDER AUTOMATIC RIFLE

Calibre:	*.303 in.*
Rifling & Twist:	*5 groove, Enfield, L.H.*
Barrel Length:	*25.2 in. (640mm)*
Overall Length:	*44.5 in. (1130mm)*
Weight:	*11 lb. 8 oz. (5.2 kg)*
Magazine Capacity:	*10, 25 or 30 rounds.*
Converted:	*The Castle, Capetown.*

Typical Markings, R.H. side of Butt Socket:

G. R.
ENFIELD
1909
S^{HT} L.E.
III

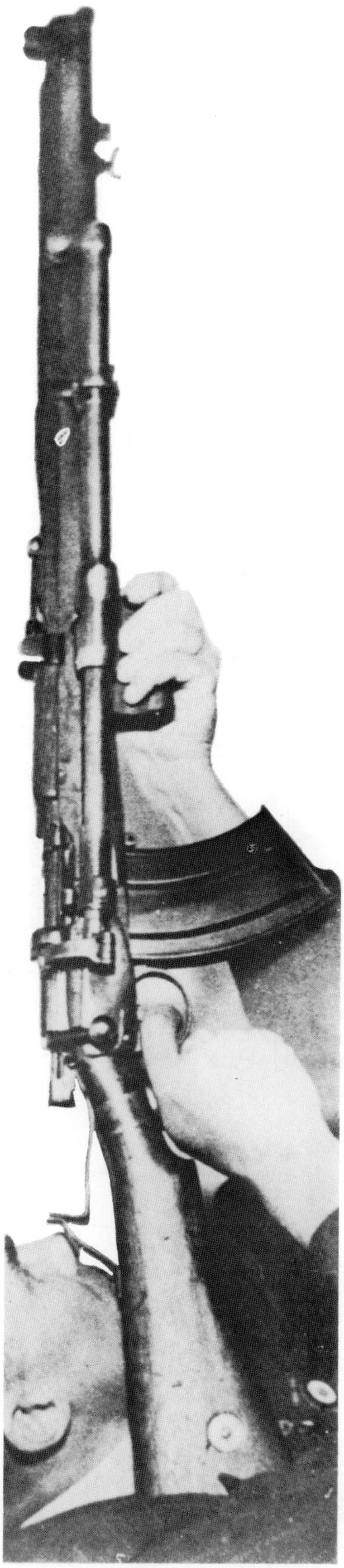

This South African automatic rifle attachment was produced by Henry J.R. Reider in 1940-41 and some eighteen rifles were converted for tests and trials, of which three were also sent to England for evaluation. Reider's gas system and the mechanical operation and rifle fitting are somewhat similar to the British Howell, described on the previous page.

The gas bleed port is positioned about 8-ins. from the muzzle and the tube is mounted along the lower right hand side of the rifle. This gas system is more finely developed than that of the Howell and the whole attachment can be quite easily removed; the bolt handle remains unmodified. There are three coil springs inside the gas tube, each about 12-ins. long, compressed by a relatively short gas piston at the front. The Reider automatic assembly attachment weighs about 2.5 lb. and it is relatively simple in construction, also permitting "single shot fire with automatic loading", better known as self-loading or repetition fire. It was found to overheat after firing about one hundred rounds.

A wood pistol grip was fitted behind the trigger guard and a forward grip screwed on underneath the fore-end. The rear portion of the gas tube is mounted onto a plate which fixes over the butt socket and rear portion of the fore-end, obscuring the original rifle designation and manufacturer's markings. Some 25 and 30-round modified Bren magazines were also applied to increase the firepower. A bipod attachment, too, was tried, comprising of the Pattern 1907 bayonet handle and hilt with some folding legs, this permitted ready attachment to the S.M.L.E. rifle nosecap.

Reider improved his system of conversion slightly in August 1941 and the last conversions in mid-1941 were made to this new plan. However, the project was shelved after no particular service requirement was stated and no interest was indicated from England.

Historical Arms Society of South Africa

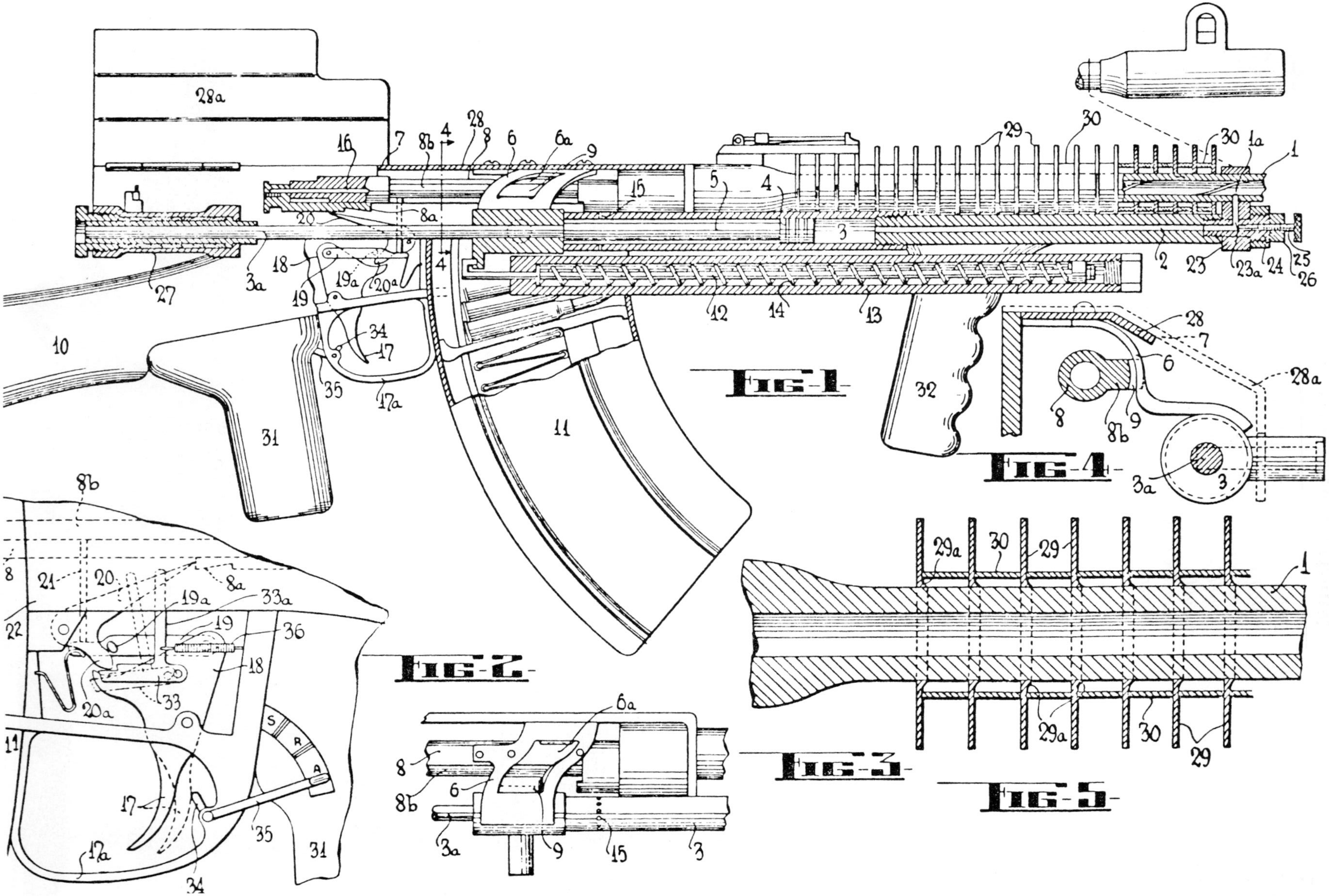

Charlton Conversion Patent Drawings
This form of auto conversion is generally that applied to all variants.

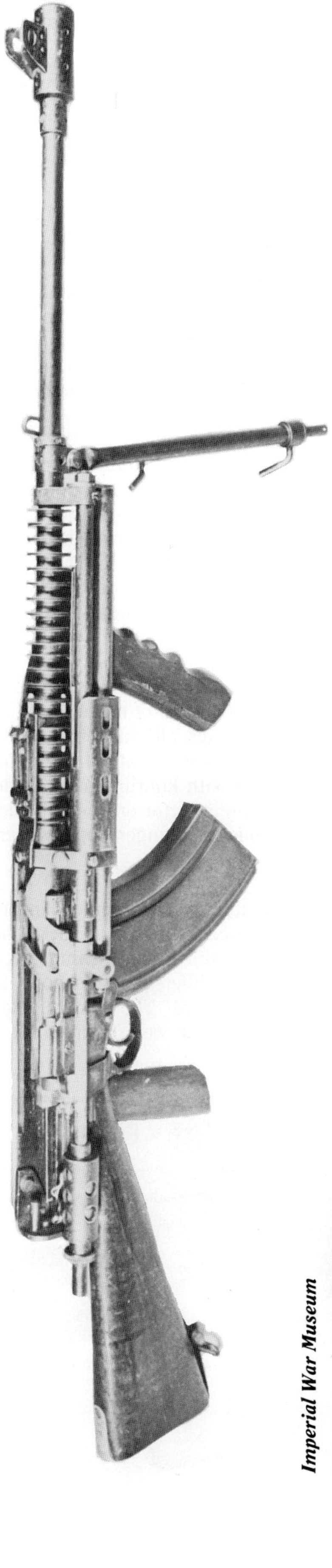

CHARLTON AUTOMATIC RIFLE

Calibre:	*.303 in.*
Rifling & Twist:	*5 groove, Enfield, L.H.*
Barrel Length:	*26.0 in. (660mm) (including compensator)*
Overall Length:	*45.5 in. (1156mm)*
Weight, approx.:	*16 lb. 0 oz. (7.3 kg)*
Magazine Capacity:	*10 or 30 rounds.*
Converted:	*Charlton's Motor Workshop.*

Typical Markings, R.H. side of Butt Socket:

V. R.
ENFIELD
1896
L. E.
I

Indigenous to New Zealand, this conversion was designed by Philip Charlton as a stop-gap light machine gun which utilised obsolete Lee-Metford rifles, although some Lee-Enfield models were similarly applied when no more of the Lee-Metfords were available. In principle, the Charlton is similar to the other British and South African gas-operated conversions of the Lee-Enfield rifle, although it is configured as a light machine gun which results in a different external form. The illustrated example was originally a B.S.A. manufactured Lee-Enfield.

Steel plates have been fitted along the left side of the receiver and around the receiver ring, so as to support the bolt cam and guide as well as the piston buffer assembly. Another lug has been welded onto the bolt rib so as to work in the angled cam slot of the operating slide; this causes the backward and forward movement of the slide to unlock and open the bolt, then close and lock it up again. The other gas-operated conversions utilise the original bolt handle or stem for this operation, instead of a lug on the bolt rib.

Of the two tubes along the right side of the gun, the upper one is the gas tube and the lower one houses the return spring and guide rod; a buffer assembly is mounted onto the rear of these tubes. A fire selector switch is positioned on the left of the receiver with "S", "R" and "A" positions for safe, repetition and automatic fire. Many of the rifle's original components were used, some in new roles, such as the long range aperture sight which was modified and used for the fire selector switch on some guns.

The original fore-end was removed and a forward pistol grip fitted while a rear pistol grip has been screwed onto the small of the butt. Previous New Zealand issue marks are usually evident on the butt or buttplate tang, as these parts were retained. Either the 10-round Lee Enfield or a modified Bren LMG magazine could be fitted.

A new backsight was mounted onto a strap over the barrel cooling fins which are of pressed sheet metal and extend to the bipod mount. A muzzle compensator assembly also incorporates the foresight and sight protector. New bipod legs are adjustable for length and secured by screws at the required length. A sling swivel is fitted onto the left side of the bipod mount.

Imperial War Museum

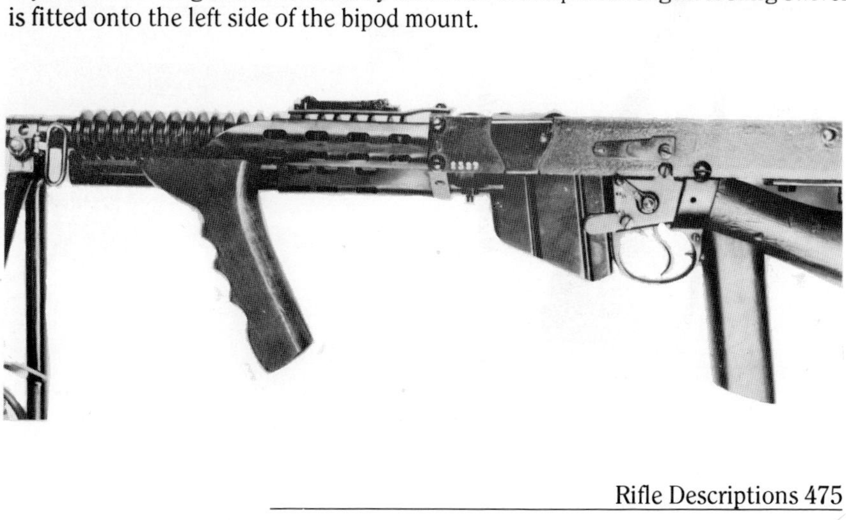

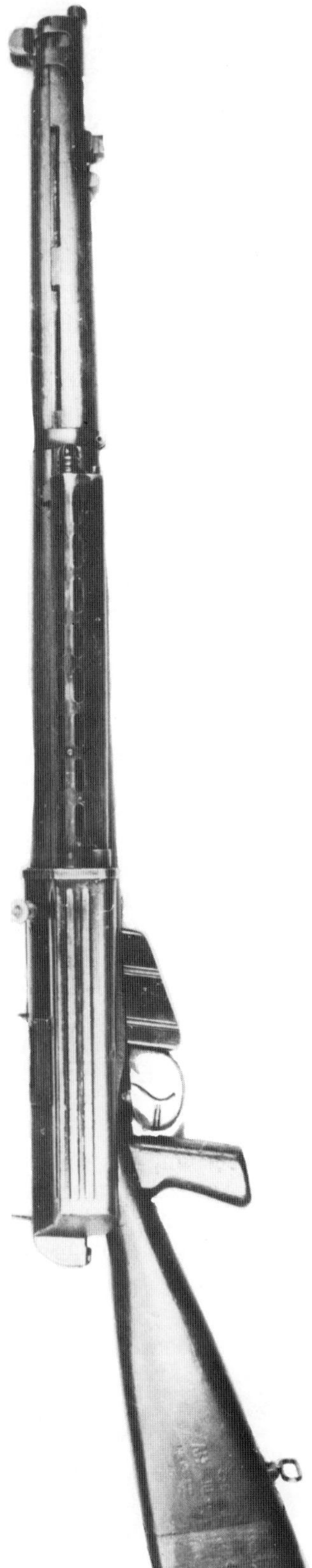

CHARLTON AUTOMATIC RIFLE, S.M.L.E. MODEL

Calibre:	*.303 in.*
Rifling & Twist:	*5 groove, Enfield, L.H.*
Barrel Length:	*25.2 in. (640mm)*
Overall Length:	*45.6 in. (1158mm)*
Weight:	*11 lb. 8 oz. (5.2 kg)*
Magazine Capacity:	*10 rounds.*
Converted:	*Electrolux, Australia.*

Markings on L.H. Sideplate of Receiver:

NZ
CHARLTON AUTOMATIC

Although this particular model is quite different in its external appearance to the previous-listed Charlton, the system and layout are very similar to the New Zealand-made version. This variant was made in Australia and utilised rifles of more recent vintage, which eliminated some of the problems encountered in New Zealand. Although, this appears to be a simplified conversion, the removal of its steel dustcovers reveals a very similar system.

Two machined plates have been attached to the receiver at the front, around the receiver ring and along the left side of the action body. On these are mounted the fire selector switch and much of the new mechanism, which is quite intricate and surprisingly well made. A felt oiling pad lubricates the bolt and a small dustcover mounted onto the bolt body reciprocates when the action is worked. The whole of the action is protected by a soldered sheet metal dustcover and another slotted cover encloses the gas piston assembly and return spring. The original charger guide bridge has been removed.

An 8-in. long, right side section of the fore-end has been squared off for clearance for the gas piston, guide and return spring plunger. Removal of metal from the butt socket has erased the rifle's original action markings, although this whole section is now enclosed by the dustcover. A cocking knob is attached to the forward portion of the bolt and it has an alloy knob with knurling around the circumference. Three positions are marked for the fire selector on the left side, "F", "L" and "S", and this operates on the cocking piece. A plunger stop has been inserted behind the trigger.

The original rifle furniture has been modified in that slots have been cut in the fore-end and a special wooden rear handguard fabricated as there is no backsight on the barrel. A wooden pistol grip has been fitted and a lengthening piece inserted into the butt. An aperture rearsight with 3 different range holes is mounted at the rear, although no ranges are indicated. The two samples examined are marked "N.Z.", so it is assumed that these were samples for a proposed New Zealand contract.

Infantry Centre Museum, Singleton

HOWARD FRANCIS SELF-LOADING CARBINE

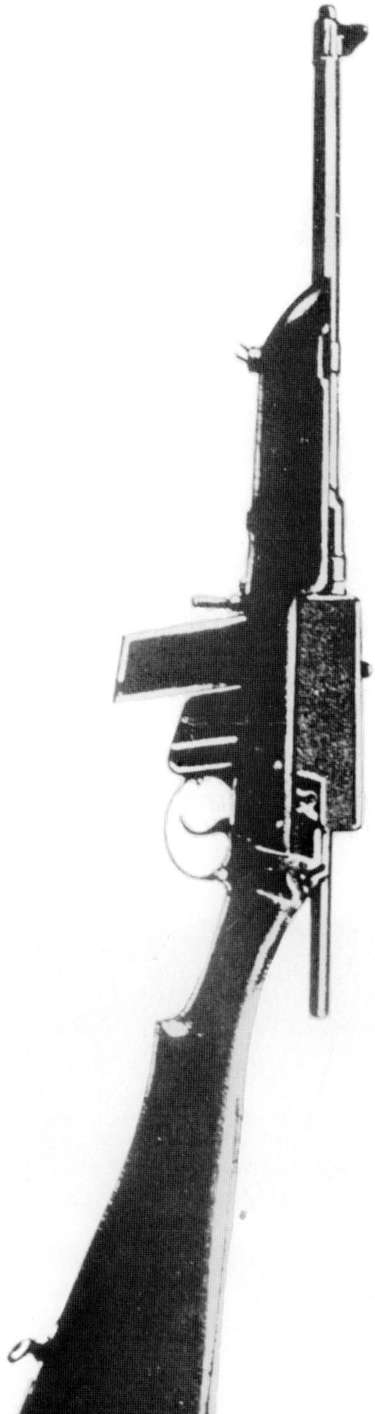

Calibre:	*7.63mm.*
Rifling & Twist:	*5 groove, Enfield, L.H.*
Barrel Length:	*12.75 in. (324mm)*
Overall Length:	*32.0 in. (812mm)*
Weight:	*8 lb. 1.5 oz. (3.7 kg)*
Magazine Capacity:	*12 rounds.*
Converted:	*Prototype only.*

Markings, R.H. Side of Butt Socket:

G. R.
B.S.A. Co
1918
SHT L.E.
III*

A British development in 1943, the Howard Francis carbine was submitted to the Ordnance Board for consideration later that same year. Generally an adaption of the No. 1 Mk III* rifle, the original barrel was plugged then re-chambered for the 7.63mm Mauser pistol cartridge, but it retained the original bore and rifling. The barrel was shortened and a "V"- backsight mounted onto the action cover; the new sighting radius was recorded as being 13.9 ins.

The magazine housing was modified so as to accept a long, rectangular box magazine while the trigger mechanism was replaced by a re-designed assembly in which the trigger disengages the sear from a lug on the firing pin, permitting forward travel. The simple blow-back action incorporated a heavy breech block with moving striker and coil striker spring and it required the trigger to be tripped each time to release the firing pin; thus it was a self-loader rather than an automatic or machine carbine.

Difficulties were experienced with the functioning of the carbine due to improper feed from the magazine and some extraction problems, however the shots were found to group consistently and reasonably close, being tested at ranges of 30, 50, 100 and 200 yards. The ejector was located in the left side of the body.

A standard rifle butt and modified fore-end were used and the rear portion of the magazine was retained so as to position the new, thinner 7.63mm magazine. Minimal modifications were required to the original No. 1 components used and the new breechblock, striker, springs, cover, magazine and catches were relatively simple designs.

However, no further action appears to have been taken on this conversion, even though its design concept and accuracy were quite satisfactory. The particular illustrations here have been reproduced from wartime Ordnance Board Proceedings reports and the originals are of poor quality.

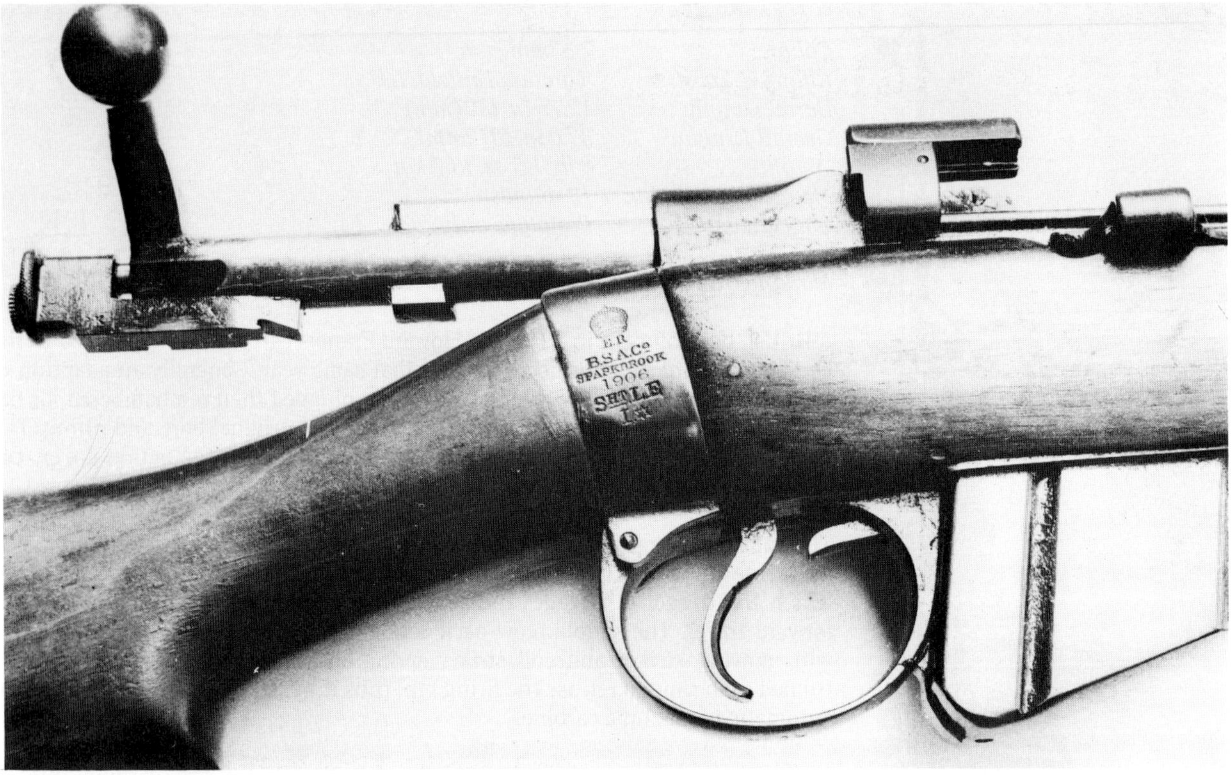

Typical receiver marks on service Lee-Metford, Lee-Enfield and Short Lee-Enfield rifles; Royal Cypher, factory, year of manufacture, and mark and model number.

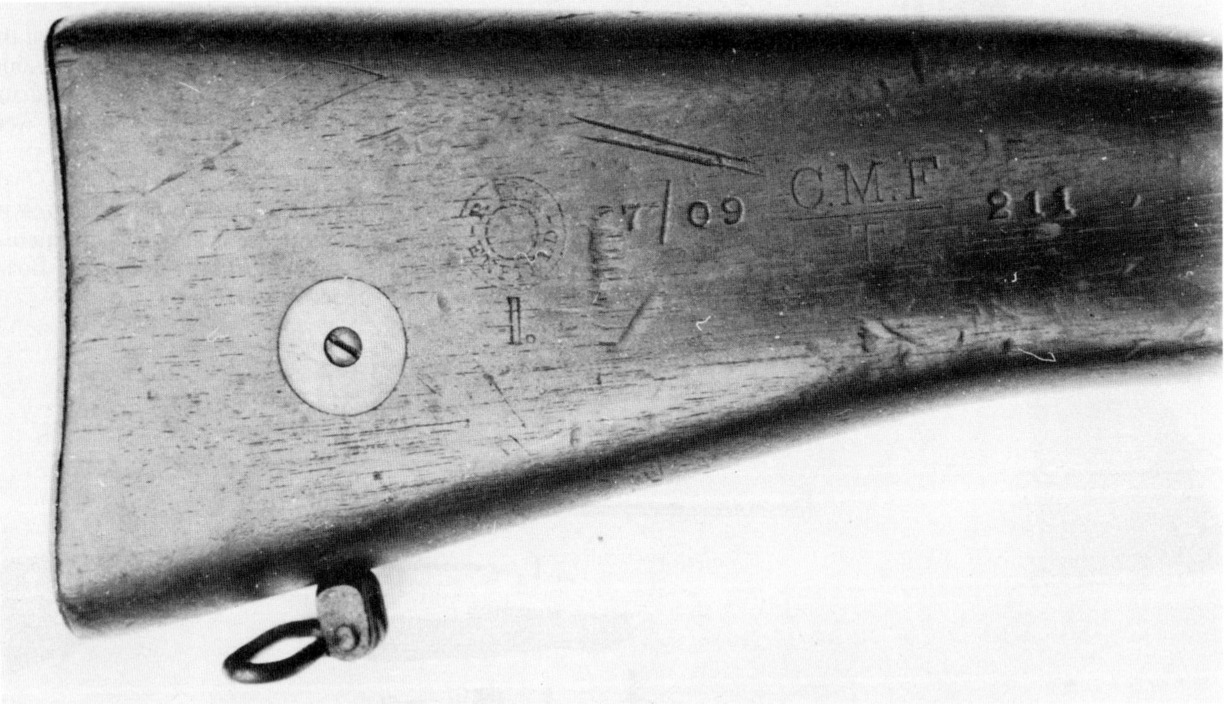

Typical butt markings on M.L.M., M.L.E. and S.M.L.E. models; the factory roundel and mark number underneath, issue date and unit (in this case, C.M.F. Tasmania) and unit rack number. Issue marks were often applied to brass butt marking disc.

MANUFACTURE, ARMOURER & ISSUE MARKINGS

CHAPTER 13

For convenience of reference, the principal assorted letter and numeric markings in this chapter have been generally tabulated in alphabetic sequence. The various proofs, makers' butt roundels and component manufacturers' codes and initials, however, have been listed in individual sections. Some are producer's markings, proof and inspection stamps, others are armourer and Ordnance Depot or workshop symbols and indicators, and yet more are issue markings. The last group can never be listed completely as the Lee rifles have been in service in many countries, for more than a century. We also note some proprietary markings as some ex-service and volunteer models have passed through surplus and commercial trades in British and Commonwealth nations as well as the United States and possibly some other countries.

Manufacture markings commonly consist of the factory name and year of production on the action body or receiver. The service Lee-Metford, Lee-Enfield and Short Lee-Enfield models incorporate the royal cypher above the factory name and year, comprising a royal crown and initials of the reigning monarch. On the first Lee rifles and carbines this was "V.R." (Victoria Regina) until 1901 when Edward VII ascended the throne and the initials were changed to "E.R.". Then in 1910 George V became King and it was modified to "G.R.", until his death in 1936. Then Edward VIII was to have become monarch in 1936 but he renounced the throne in favour of George VI who reigned until 1952. By the end of the 2nd World War and as applied to the .303 No. 4 and subsequent rifle production, the royal cypher was not marked and only the factory, year of manufacture and the rifle designation were stamped onto receivers.

Proof and Inspection marks are smaller and sometimes not as easily distinguished as they may have been mis-struck or the metal or wood surface refinished over the top of original marks, or even worn. From these marks, the country and factory of production or inspection can be ascertained, depending upon the particular letter or symbol incorporated. British and associated proof marks generally include crossed flags.

British factory proofs usually incorporate a small cypher in the upper quadrant of the crossed flags, Indian proofs incorporate an "I" suffix (e.g. GRI) until Independence, Australian proofs have an "L" for Lithgow or "O" for Orange, South Africa used a "U" over the crossed flags, and Canada used a "D C" combination or maple leaf sign. The letter "P" is nearly always included in the proof marks.

On the British inspection marks, "E" indicates Enfield, an italic "B" means B.S.A., the plain Roman "B" or sometimes "Sk" signifies Sparkbrook, and "x" denotes L.S.A. The inspection marks also incorporate the particular inspector's number.

Sample proof & inspection marks.

As all components were subject to inspection, a rifle or carbine may be checked to see if all the parts bear the same factory inspection code; in order to determine its originality and what parts may have been replaced. In service, as arms were repaired and serviced, it was usual practice to replace the parts as required and no attention was paid to retaining original factory parts as all these components are interchangeable.

Foreign inspection and production marks include "MA" for Lithgow, Australia, and "OA" for Orange, "BA" for Bathurst, "FA" for Forbes and "WA" for Wellington; the "A" suffix obviously indicates Australia and early Lithgow marks consist merely of the "A" in a seven-pointed star. "SLAZ" will be noted on Australian wood furniture, indicating "Slazengers" in Sydney. Canada's Long Branch factory usually used an entwined "LB" logo, although its post-war successor "CAL" will be noticed on later production components. On Indian production, "GRI" is often seen; "IS" or combinations thereof appear to have superseded this after India achieved Independence.

The 2nd World War codes found on the No. 1, No. 4 and No. 5 rifle components are comprehensively listed in the appropriate Chapters 5 and 6; only those marks commonly encountered have been tabulated here along with related proprietary trade name initials where appropriate.

Mark	Usual Location	Indication
A	No. 4 magazine.	Not interchangeable.
A.I.D.	Metal or wood.	Small Arms Inspection Department.
A.T.	Receiver, barrel, furniture.	Aiming Tube, usually .22-in. R.F.
A.T. N.I.	Receiver, barrel, furniture.	Aiming Tube, not interchangeable.
B	Heel of butt.	Bantam butt, ½-in. shorter than normal.
B	No. 4 magazine.	Acceptance mark, interchangeable.
B L R	Stencil paint on butt.	Beyond local repair.
[C]	Metal or wood components.	Stored for use on conversions.
Ⓒ	Barrel, Canada.	Competition grade. Usually 7.62mm conversions.
C L	Backsight, bottom RH corner, C.L.L.E. I	Sighted for Mk VII ball.
C L	Dial sight plate.	Plate for Charger Loading long rifles, converted to Mk VII ball.
CR xxx	Various components.	Parts numbers on late production rifles, e.g. No. 8, L8, L42, &c.
Ⓓ	Barrel.	Draw lapped barrel.
D	Barrel.	New standard of finish for bore.
D.P.	Barrel knox form, also receiver & furniture	Drill purpose.
E	Barrel, on knox form.	Enfield rifling.
E	Foresight & protector, C.L.L.E. rifle.	Set up for Enfield rifling.
E C	Bottom RH corner of backsight leaf on L.E. carbines.	Backsight graduated for Enfield carbine barrel.
EC/88	Bottom RH corner of backsight leaf on N.Z. M.L.E. carbines.	Lower backsight cap on N.Z. carbines designed for Patt. 88 bayonet.
E.I.*	Barrel, on knox form.	M.L.M. Mk I profile barrel with Metford rifling.
E.Y.	Barrel, knox form, receiver, furniture.	Only to fire ball in a case of emergency.
F.T.R.	Receiver.	Factory thorough repair.
F.R.	Receiver.	Factory repair: India.
H	Barrel, top of knox form.	Australian heavy barrel, No. 1 rifle only.
H	Top of butt, near socket.	Fitted with heavy barrel, Australian.
H.V.	Barrel, behind rearsight.	High Velocity; sighted for Mk VII ball.
JJ. CO.	Barrel, knox form.	Recent retailer's mark; John Jovino Co., New York.
L	Heel of butt.	Long butt, ½-in longer than normal.

RECEIVER MARKINGS:
Magazine Lee-Metford Rifle Mk II VR cypher, Sparkbrook, 1894.

Magazine Lee-Metford Rifle Mk II VR cypher, Enfield, 1895.

Magazine Lee-Enfield Rifle Mk I VR cypher. Sparkbrook, 1900 converted to .22 Patt. '14 No. 1.*

Short, Magazine Lee-Enfield Mk I ER cypher, Sparkbrook, 1906.*

*Short, Magazine Lee-Enfield Mk I*** ER cypher, Enfield, 1905; with the additional "*" on later conversion.*

Short, Magazine Lee-Enfield Mk III GR cypher, B.S.A. Co., 1915; later converted to .22 Patt '14 No. 2.

"Peddled Scheme" S.M.L.E. Mk III receiver, GR cypher and marks. Manufacturer; top, rear of action.*

S.S.A. (Standard Small Arms) mark on Peddled Scheme rifle. Note the adjacent Enfield inspection marks.

N.R.F. (National Rifle Factory) mark on Peddled Scheme rifle. Adjacent number is assembly No.

Lithgow 1925 S.M.L.E. Mk III with early style shield logo. Note row of receiver inspection marks, 1-10.

Australian Lithgow markings of later Mk III and III rifles, the most commonly encountered type.*

CONVERSION MARKS: left side of body. Charger Loading Lee-Enfield Mk I.*

S.M.L.E. Converted Mk II, converted at Enfield in 1905 from MLM or MLE.

.410 Musket, converted at the Rifle Factory Ishapore in 1949.

S.M.L.E. Converted Mk IV, converted at Enfield RSAF in 1911. Note later inspection marks over "Cond." mark.

Scarce Mk I M.L.E. rifle re-marked as Mk II M.L.M. after being fitted with Metford rifled barrel.*

.22RF AG Parker conversion marks on right side of receiver. The new rifle is a .22 Patt '14 No. 1.

Receiver designation markings (Royal cypher, year of manufacture, factory, mark number and designation abbreviation on the L.- E. model) for the Lee-Metford and Lee-Enfield carbines are stamped into the left side of the action, on the butt socket; this is the opposite side to the rifles. For C.L.L.M., C.L.L.E., and various S.M.L.E. conversions and some of the .22 RF training models, the conversion marks are also stamped on the left side of the receiver butt socket. In the case of these conversions, the original rifle designation markings will be found on the right side of the receiver butt socket. The date (year) stamps under the factory names can also be compared to confirm this.

INDIAN ISHAPORE RECEIVER DESIGNATIONS (below)

Edward cypher with full factory name. Edward did not become king.

Wartime No. 1 Mk III*, note G.R.I. cypher (George Rex Imperiatus)

Postwar Ishapore Mk III*, note use of "3" rather than "III".

Post Independence, cypher is replaced by Indian ashoka.

No. 4 rifle with Factory Repair mark. India did not make the No. 4.

7.62mm 2A rifle. Note ashoka and Rifle Factory Ishapore mark.

LES	Dial Sight Plate.	Plate for S.M.L.E. Marks I, I*, II & II*.
LES 2	Dial Sight Plate.	Plate for all Short Rifles converted to Mk VII ball.
LES III	Dial Sight Plate.	Plate for S.M.L.E. Mk III & Cond. IV.
L. T.	Usually stencilled on butt.	Line thrower, usually Canadian.
M	Bolthead, No. 1 rifle.	Difference in specification.
M	Foresight & protector, C.L.L.M. rifle.	Centrally mounted foresight for Metford rifling.
N	Receiver, left side (usually).	Naval Service.
N	Butt, underneath at wrist.	Normal length.
N.S.	Bolthead, No. 1 or No. 2 rifle.	Difference in specification.
O	Barrel, knox form.	Unlapped barrel.
P	Barrel, knox form.	Replacement barrel.
(P)	Barrel, L.H. side of knox form.	Parallel bore.
P	Small of butt, RH side, M.L.M. & M.L.E.	Compressed butt socket.
P 2	Small of butt, RH side, M.L.M. & M.L.E.	Socket treated with paraffin, compressed & fitted with keeper plate and squared stockbolt end.
P W	Various parts.	Part worn, used by Weedon on re-issued arms.
R	Barrel knox form.	Replacement or reserve barrel?
R	Barrel knox form.	Found rusty by armourer, to be checked by armourer.
R	Barrel knox form, receiver, butt & fore-end.	Unsafe barrel, usually drill purpose for Cadets.
S	Heel of butt.	Short butt, ½-in. shorter than normal.
S	Underneath fore-end, near nosecap.	No. 1 rifle fitted with barrel centering stud & spring under the nosecap.
S	Top or side of bolthead.	Spare component.
S	Underneath butt, near socket.	Indian service; fitted with spring washer on stockbolt.
S.C.	Barrel, behind rearsight.	Small cone; short lead from chamber to rifling for Mk VII ball.
S.K. A.	Various components.	Reject parts, suitable for use on skeleton actions.
S.L.	Receiver, left side of butt socket.	Indian service; Single loader.
S of M	Metal or wood.	School of Musketry.
→ S ←	Barrel, receiver or furniture.	Sale mark.
T	Receiver, No. 4.	Fitted for telescopic sight; No. 4 (T) sniper rifle.
U	Bolthead, No. 1 rifle.	Difference in specification.

No 4 series rifles are marked on the left side; but the early 1930's trials rifles retained the early cypher style.

British No 4 Mk I; "M47C" is BSA, "F" is Fazakerley, "M" is Maltby. "No 4 Mk I" on receiver, not clear.

Typical Canadian Long Branch markings; note serial number on butt socket with the letter "L" in the serial number.

Stevens Savage No 4. Note squared "S" logo, U.S. Ordnance flaming bomb mark and "No 4 Mk I". Serial number incorporates a "C" letter.*

No 4 Mk 2 made at Fazakerley in 1955. Serial number is A1471.

No 5 Mk I "jungle carbine" made at ROF Fazakerley, September 1945.

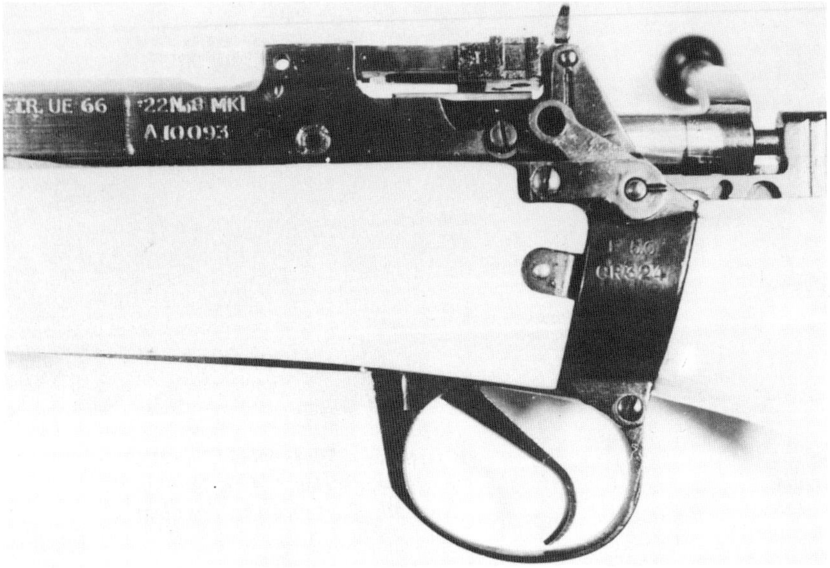

.22RF No 8 trainer. This sample has FTR (Factory Thorough Repair) mark, Enfield (UE) in 1966. On the receiver butt socket, is the original Fazakerley production mark and year (F 50) and service part catalogue number, CR324.

An original Canadian Long Branch 1945 No 4 Mk I rifle, upgraded (FTR) to No 4 Mk I/3 at Fazakerley in 1952. The original serial number (83L6587) on the butt socket was also engraved on the receiver wall during conversion.*

A series of conversions: from Long Branch 1943 No 4 Mk I, to Mk I/3 at Fazakerley in 1951, then to an L8A3 after the introduction of the 7.62mm round into service from 195*

U↑P	Receiver.	Unproofed.
$\overset{*}{\text{W}}$	Barrel knox form.	Cord worn, marked by A.O.D. rather than Inspection Dept.
Ⓦ	Barrel knox form.	Cord worn, Inspection Dept. mark.
W↑D	Barrel, receiver, components.	War Department ownership mark.
Z F	Main components.	Beyond local repair, except at factory.
Z - BLR	Main components.	Beyond repair.
Z - BER	Main components.	Beyond economic repair; reduced to/written off as scrap.
'18	Barrel.	Year of acceptance for service.
.22 PATT	Metal stamp.	.22 Rifles, Patt. 1914 of standard weight with .22 barrel.
2	Underneath S.M.L.E. backsight slide.	Backsight slide & bed converted to Mk VII ball.
2	S.M.L.E. bolthead charger guide.	No. 2 charger guide with stronger retaining screw.
2	Small of butt, RH side, M.L.M. & M.L.E.	Keeper plate fitted to fore-end, squared stockbolt end.
2	S.M.L.E. magazine case, back rib.	Magazine case No. 2.
3	S.M.L.E. magazine case, back rib.	Magazine case No. 3.
4	S.M.L.E. magazine case, back rib.	Magazine case No. 4.
I*2	Bottom R.H. corner of backsight, C.L.L.E. Mk I* rifle.	Rifle sighted for Mk VII ball.
II	Dial sight plate.	Plate for M.L.M. II & II*; M.L.E. I & I*; C.L. rifles sighted for Mk VII ball.
19T	Bolthead.	Reproofed for 7.62mm NATO cartridge.
⊗	Barrel.	Suitable for Mk VII ball.
*	Barrel knox form.	Slight rust or cut inside barrel.
*	Barrel, any other position.	Corrosion or rust near mark.
*	Bottom R.H. corner of backsight leaf, M.L.E. rifles.	Backsight V-notch offset .03" to left, to correct sighting of foresights set .05" to the left.
*	R.H. side of M.L.E. foresight.	Foresight offset .02" rather than .05" to compensate for bullet drift.
⊛	Barrel knox form.	Rust or cut mark, marked by A.O.D. rather than by Inspection Dept.
* *	Barrel knox form.	Moderate pitting of rifling or throat.
* * *	Barrel knox form.	Heavy pitting of rifling or throat.
↑ or ⋏	Various components.	Accepted for Government service, after inspection.
→←	Barrel, receiver or furniture.	Sale mark.
→ S ←	Barrel, receiver or furniture.	Sale mark.

→CC←	Barrel, receiver or furniture.	Sale mark to County Cadet corps.
→RC←	Action body & butt.	Sale mark to County Cadet corps.
	Barrel knox form.	Sale mark.
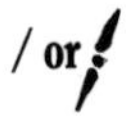	Receiver or wooden furniture.	Cancellation or separation mark.
o↑o	Metal or wood components.	Percentage stamp, where 10% of stores have been examined.
	Metal or wood components.	Stored for emergency use or repairs.
	Receiver.	U.S. Ordnance mark on Lend-Lease No. 4 rifles.

MANUFACTURERS' ROUNDELS

R.S.A.F. ENFIELD
(also with no centre)

R.S.A.F. ENFIELD

R.S.A.F. SPARKBROOK

R.S.A.R.F. SPARKBROOK

B.S.A. & M. Co.

B.S.A. Co.

L.S.A. Co.

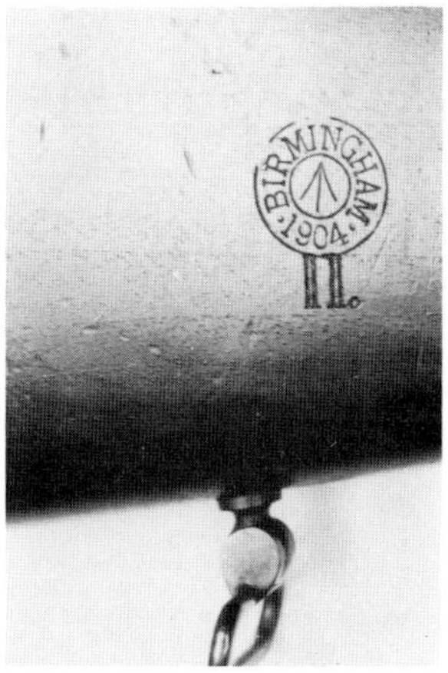

Birmingham RSARF

B.S.A. & M. Co.

Enfield

Sparkbrook

BUTT ROUNDELS, ARMOURY MARKS

Typical Manufacturer's Butt Roundel (right). This example (Royal Manufactory, Enfield) is representative of such markings on Magazine Lee-Metford, Magazine Lee-Enfield and Short Lee-Enfield butts. The factory roundel usually has the War Department broad arrow in the centre, with the rifle Mark number underneath. In some instances, a class mark is appended below; in this case it is a Mk I rifle, 1st class arm.*

Such class marks are commonly found on the Martini rifles and carbines, but much less often on the Lee series.

Wartime Lithgow Mk III S.M.L.E. stampings.*

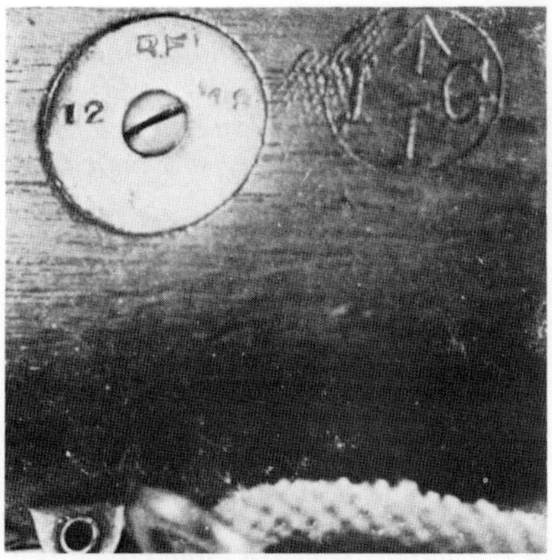

Early Ishapore marks on the marking disc; Gov't roundel on butt.

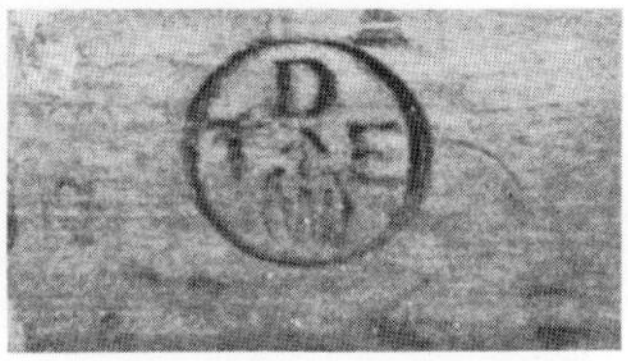

Above & right; further Indian ownership and arsenal butt stamps.

Above & below: Inspection and proof marks, matching serial nos. on receiver and barrel; the "D↑" mark indicates Australian ownership, Defence Department.

BARREL and RECEIVER MARKS.

British S.M.L.E. with "Z" series serial numbers, "DP" (Drill Purpose) and "P" (replacement barrel) marks. Inspection marks comprise a crown, number, and factory letter ("E" or "EFD" for Enfield, "X" for LSA, "B" for Sparkbrook, "β" for B.S.A.).

Early Australian S.M.L.E. proof and inspection marks. "A" star with Roman numerals underneath are inspector's stamps, the "ACP" enclosed by a shield is a proof mark. The "A" inside a star is a common Australian inspection mark. "P" in a heart indicates parallel bore rifling.

2nd World War Australian S.M.L.E. markings use some of the earlier style stamps, "MA" for Lithgow, "O" for Orange, and "L" over crossed flags and "P" for Lithgow proof. The "GR" and crown is a George Rex cypher incorporated in the proof mark.

MANUFACTURERS' MARKINGS

Factory	Markings	Rifle Models	Production +
R.S.A.F. ENFIELD Enfield Lock, Middlesex, U.K.	ENFIELD EFD Ð UE	M.L.M. All Marks Rifle & Carbine M.L.E. All Marks Rifle & Carbine S.M.L.E. All Marks .22 Sht. & Long Rifles, except for Patt. '14 & Patt. '18 No. 4 Mk 1 Trials & early (T) models L8 Series; L39, L42 & Envoy	1888 - 1941 1959 - 1970?
R.S.A.F. SPARKBROOK Birmingham, U.K.	SPARKBROOK Sk	M.L.M. All Marks of Rifle M.L.E. All Marks of Rifle S.M.L.E. Mark I	1889 - 1906
R.S.A.R.F. SPARKBROOK Birmingham, U.K.	[crown] BR	Repair & Conversion	1894 - 1906
R.S.A.R.F. BAGOT ST. Birmingham, U.K.	[crown] BR	Repairs only, until	1894
R.O.F. FAZAKERLEY Lancashire, U.K.	ROF(F) FY F UF	No. 4 Marks 1, 2, 1/2 , 1/3 No. 5 No. 8	1941 - 1960
R.O.F. MALTBY Yorkshire, U.K.	R.O.F.M. O.F.M. R.M. *or* RM M	No. 4 Mark 1	1941 - 1946
B.S.A., B.S.A.& M.Co. or B.S.A. Guns; Small Heath, Birmingham, U.K.	B.S.A. Co. B.S.A.& M.Co. M47A	M.L.M . Marks I, I*, II M.L.E. Marks I, I*, C.L.L.E. S.M.L.E. Marks I, I*, III, III*	1889 - 1943
B.S.A. SHIRLEY Birmingham, U.K.	B M47C 85B BS UB	No. 4 Mark I No. 5 .22 Nos. 7 & 8	1941 - 1952?
L.S.A. Co. Old Ford, London, U.K.	L.S.A. Co.	M.L.M. Marks I, I*, II M.L.E. Marks I, I*, C.L.L.E. S.M.L.E. Marks I, I*, III, III*	1889 - 1925
R.N.O.D. PORTSMOUTH, PLYMOUTH & CHATHAM, U.K.	N	S.M.L.E. Marks I**, II**, II*** #	1908 - 1911?
A.G. PARKER & Co. Ltd. Birmingham, U.K.	A.G. PARKER & Co. Ltd.	.22 Sht. Rifle Patt. '14 #	1915 - 1918
STANDARD SMALL ARMS Birmingham, U.K.	S.S.A.	S.M.L.E. Mark III*	1916 - 1918
NATIONAL RIFLE FACTORY No. 1, Birmingham, U.K.	N.R.F.	S.M.L.E. Mark III*	1918 - 1919

VICKERS Ltd. (Vickers Sons & Maxim)	V.S.M.	C.L.L.E. Marks I & I* #	1911 - 1912
W.W. GREENER Birmingham, U.K.	W.W.G.	.22 Sht. Rifle Patt '18 #	1918 - 1919
PARKER HALE LTD. Birmingham, U.K.	P.- H.	No. 9 #	1956 - 1960
S.A.L. / C.A.L. LONG BRANCH Ontario, Canada.	LONG BRANCH B (↑)	No. 4 Marks I, I*, I*(T) C No. 7	1941 - 1958?
SAVAGE ARMS Co. Chicopee Falls, Massachusetts, U.S.A.	S [S]	No. 4 Marks I, I*	1941- 1944
LITHGOW S.A.F. N.S.W., Australia.	LITHGOW MA	No. 1 Marks III, III* No. 2 Mark IV* No. 6	1913 - 1953
ORANGE RIFLE FACTORY No. 2, N.S.W., Australia.	OA MAO	Components & assembly, No. 1 Mark III*	1942 - 1945
ISHAPORE RIFLE FACTORY, Bengal, India	ISHAPORE RFI GRI IS ID	C.L.L.E. (I.P.) # S.M.L.E. Marks I*, I* I.P., III, III* No. 2 Mark IV*, .410 Musket, Single Loaders # 7.62mm 2A & conversions	1905 - post 1970
PAKISTAN ORDNANCE FACTORY, Wah, Pakistan.	POF	No. 4 &c. Factory repair programmes	c. 1957 - 1960's

\+ The Production Years listed are those during which the particular firm or factory was engaged in the manufacture or conversion of the rifles and carbines applicable to this study and it does not refer to the production period of other arms.

Conversions; not newly manufactured rifles.

NOTE:- This tabulation has been compiled from available documentation and specimen rifles and carbines; where some doubt exists, the symbol "?" has been applied.

South African production Lee-Enfield barrel. Note the "S A" logo; and "U", crossed flags and "P" proof mark.

S.M.L.E. COMPONENTS - CONTRACTORS

S.M.	Singer Manufacturing, Glasgow	B A	Bathurst Rifle Factory No. 2, Australia
S.S.A.	Standard Small Arms Co., Birmingham	W A	Wellington Feeder Factory, Australia
N.R.F.	National Rifle Factory (succeeded S.S.A.)	F A	Forbes Feeder Factory, Australia
LINLEY	Linley & Co., Birmingham	SLAZ	Slazengers Pty. Ltd., Sydney, Australia

Separate, comprehensive lists of the British Great War and 2nd World War contractors and parts are included in Chapter 5.

RIFLE Nos. 4 & 5 COMPONENTS - CONTRACTORS

A.W.	Austey & Wilson (M8)	S.M.	Singer Manufacturing Co. (N67)
C.E.W.	C.E. Welstead (S127)	S.M.C.	Singer Manufacturing Co. (N67)
C.C.M.	Canadian Cycle & Motor	V.N.S.	Viners Ltd. (N79)
C.F.M.	C.F. Moore & Son (S238)	W.L.	Wilkinson's Ltd. (N85)
D & M	Davis & Mawson Ltd.	WLM/260	Walls Ltd. (M260)
J.C.	John Curtis (N22)	W.W.G.	W.W. Greener (M94)
H.M.Co.	H. Morris Co. (N49)	W	
M.B.	Metal Box Co. (S76 & S77)	B	Wilkinson's Ltd., Bradford (N85)
R.C.	Raleigh Cycles	LTD	
S.L.	William Sykes Ltd. (N74)		

CODES: Prefix N = Northern Districts
M = Midlands
S = Southern District

N6	Backer Electrical Co.
N22	Curtis & Sons
N49	H. Morris
N56	Prince-Smith
N64	Sheffield Steel Products
N67	Singer Manufacturing
N74	Sykes Ltd.
N76	Tempered Springs
N77	T. Turner
N79	Viners Ltd.
N85	Wilkinson's Ltd.
M1	Accles & Pollock
M8	Austey & Wilson
M47A	B.S.A. Small Heath
M47B	B.S.A. Redditch
M47C	B.S.A. Shirley
M56	Cheney & Son
M78	Elkington & Co.
M158	Joseph Lucas
M216	Sandra Ltd.
M260	Walls Ltd.
M262	Ward & Son
M607	General Presswork & Stampings
M615	B.S.A. Leicester
M616	B.S.A. Mansfield
S3	Adams Bros. & Burnley
S12	L. Benn & Son
S21	British Salmson Aero
S51	Holland & Holland
S56	Imperial Engineering
S68	Lines Bros.
S76	Metal Box Co.
S77	Metal Box Co.
S88	National Cash Register
S98	Pruners Ltd.
S104	Sidcup Gauge Co.
S106	J.W. Spear
S113	Trafalgar Engineering Co.
S114	Trevor Stampings
S117	United Motor Co.
S123	H. Wall Ltd.
S124	Well Larsen Tool Co.
S126	C.E. Welstead
S128	Wembley Tool Co.
S131	Wright Bros.
S132	L.S. Mayer
S134	Addressograph Multigraph Ltd.
S223	E.S.S. Signs
S238	C.F. Moore
S252	Philco Radio & T.V.
S262	Roneo Ltd.
S323	Chambon Ltd.

A full listing of the 2nd World War contractors and the component parts is included in Chapter 6; also see Skennerton's "British Small Arms of World War 2 - The Complete Reference Guide to Weapons, Makers' Codes & 1936 - 1946 Contracts".

The "" forward of the knox form indicates rust in the barrel, "P" a replacement barrel in most cases, and "DP" a drill purpose downgrading.*

"W" and the "" indicate cord worn, and rust or cut in the barrel. Note the BSA inspection marks incorporating an italic "B".*

Dial sight assembly; note the BSA inspection mark on the pointer.

The ".22 No 2" on the bolthead indicate a rimfire conversion. Note longer extractor.

Rifle serial number underneath the backsight leaf on a Mk III S.M.L.E. rifle.

Fore-end markings on an early S.M.L.E. Note rifle serial number, Enfield inspection stamp, and "S" indicating a barrel centering stud and spring under the nosecap.

PROOF MARKS

	Typical British; on receiver and barrel. Cypher initials change with each reigning monarch (see pge 193).
	British Naval.
	Proof mark on bolthead.
	Lithgow, Australia.
	South Africa.
	Ishapore, India.
	Ishapore, India; post 1950.
	Long Branch, Canada.
	Long Branch, Canada; later.

INSPECTION MARKS

E 12	R.S.A.F. Enfield Inspection, England.
B 16	B.S.A. Inspection, Birmingham, England.
B 5	Sparkbrook Inspection, Birmingham, England.
X 8	L.S.A. Inspection, London, England.
GRI 12 / IS 7 B	Ishapore Inspection, India.
	Early Lithgow, Australia Inspection Mark.
	Early Lithgow, Australia Inspection.
	Lithgow Inspection, Australia.
L ↑ 3	Lithgow Inspection, Australia.
O ↑ 2	Orange Inspection, Australia.
5 S	Sydney Inspection, Australia.
	Long Branch Inspection, Canada.
	U.S. Inspection Mark.

OWNER/ISSUE MARKS

A ↑ F	Australian Forces.
A.L.H.	Australian Light Horse.
C ↑ G	Cape Gov't., South Africa.
	Canada.
D ↑ C	Dominion of Canada.
D ↑ D	Australia, post 1911.
	Australia, post 1911.
	Irish Free State.
N.S.W.	New South Wales, Australia.
N ↑ Z	New Zealand.
NZ ↑ D	New Zealand.
Q ↑ G	Queensland Gov't., Australia.
Q ↑ P	Queensland Police, Australia.
S.A.	South Australia.
S ↑ A I	India. post Independence.
TAS	Tasmania, Australia.
VIC.	Victoria, Australia.
U.S. PROPERTY	U.S.A. Lend Lease.
	South Africa.
W.A.	Western Australia.
	Iraq.
1MD – 7MD	1st – 7th Military Districts, Australia.

Canadian Mounted Police mark, "MP", on a No. 4 Mk I rifle.*

Above and below: Thai ownership marks on the receiver ring and butt of a No. 4 Mk I rifle.*

The "RCAF" mark indicates issue to the Royal Canadian Air Force.

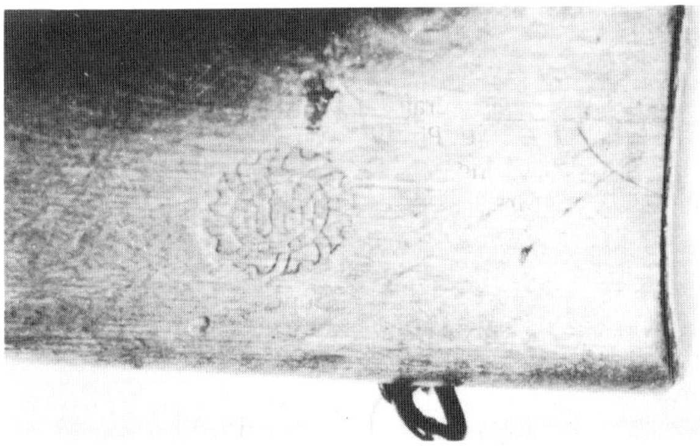

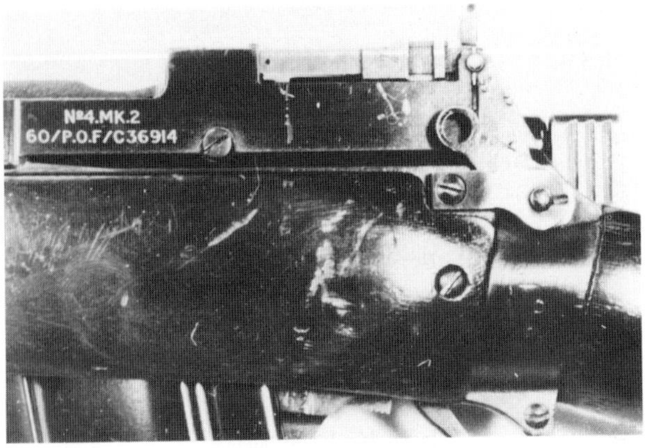

Pakistan Ordnance Factory rework only, not manufactured there. The "P" over "62" (below) indicates POF 1962.

Above and right: Indonesian refurbished Lee-Enfield rifles are found with a star logo.

New Zealand ownership marks on an M.L.E. carbine; rack no. 248 dated 1901.

ABBREVIATIONS

These abbreviations are often encountered with stops between the letters or at the end, e.g. "R.I.C." and "Cond.". Some letters will not only be encountered in the text of this reference, but also as markings on some arms, component parts and accoutrements.

AA	Anti-Aircraft.
AP	Armour Piercing.
AT	Aiming Tube.
BSA	Birmingham Small Arms.
C	Canada.
CAL	Canadian Arsenals Ltd.
CEAD	Chief Engineer & Superintendent of Armaments Design.
CISA	Chief Inspector of Arms.
CL	Charger Loading.
CLLE	Charger Loading Lee-Enfield.
CLLM	Charger Loading Lee-Metford.
Cond	Converted.
CSAD	Chief Superintendent Armament Design.
DP	Drill Purpose.
Efd; EFD	Enfield.
EY	Emergency Use.
FTR	Factory Thorough Repair.
H	Heavy (barrel).
HV	High Velocity.
ICI	Imperial Chemical Industries.
ILC	Indian List of Changes.
LE	Lee-Enfield.
LEC	Lee-Enfield Carbine.
LM	Lee-Metford.
LMG	Light Machine Gun.
Lng	Long.
LoC	List of Changes.
LSA	London Small Arms.
ME	Martini-Enfield.
Mk	Mark number.
MLE	Magazine Lee-Enfield.
MLEC	Magazine Lee-Enfield Carbine.
MLECLR	Magazine Lee-Enfield Charger Loading Rifle (CLLE).
MLER	Magazine Lee-Enfield Rifle (also see MLE).
MLERS	Magazine Lee-Enfield Rifle, Short (same as SMLE).
MLM	Magazine Lee-Metford.
MLMC	Magazine Lee-Metford Carbine.
MLMCLR	Magazine Lee-Metford Charger Loading Rifle (CLLM).
MLMR	Magazine Lee-Metford Rifle (also see MLM).
MM	Martini-Metford.
MPI	Mean Point of Impact.
MT	Morris Tube (sub-calibre).
MV	Muzzle Velocity.
M'	Model.
N	Naval.
NATO	North Atlantic Treaty Organisation.
NC	Nitro-cellulose.
NRA	National Rifle Association.
Pat.	Patent.
Patt.	Pattern.
P'	Pattern.
PH	Parker Hale.
REL	Research Enterprises Ltd. (Canada).
RF	Rimfire.
RFI	Rifle Factory Ishapore.
RFRL	Rimfire Rifle Long (.22 Long Rifle).
RFRS	Rimfire Rifle Short (.22 Short Rifle).
RIC	Royal Irish Constabulary.
RL	Royal Laboratory, Woolwich.
ROF	Royal Ordnance Factory.
RSAF	Royal Small Arms Factory.
RSARF	Royal Small Arms Repair Factory.
RSMLE	Rifle, Short, Magazine Lee-Enfield (same as SMLE).
SA	Small Arms.
SAA	Small Arms Ammunition.
SAC	Small Arms Committee.
SAL	Small Arms Ltd. (Canada).
SMLE	Short, Magazine Lee-Enfield.
SMRC	Society of Miniature Rifle Clubs.
WO	War Office.
XP	Experimental.
z	Suffix applied to ammunition loaded with a nitro-cellulose propellant.
?	Uncertain, or not known.
§	Paragraph number indicator, List of Changes.
§§	Paragraph numbers, plural of the above.

SELECT BIBLIOGRAPHY

OFFICIAL PUBLICATIONS & RECORDS:

Australian List of Changes in War Material (1942-1948)	Defence Department
Australian Munitions Digests, 1944, 1945 & 1946	Defence Department
British Rifles- Catalogue of the Enfield Pattern Room	H.M.S.O.
Canadian Army Local E.M.E. Instructions	[Various]
Contracts for Small Arms & Machine Guns (1915-1917)	H.M.S.O.
History of the Ministry of Munitions, Vol. XI	H.M.S.O.
Indian List of Changes in War Material (1890-1936)	India Office
Industries of Enfield	H.M.S.O.
List of Changes in British War Material 1920-1956	H.M.S.O.
Machine Gun Committee Minutes (1879-1880)	War Office Papers
Machine Guns, Small Arms, &c. from the Trade (1914-1915)	H.M.S.O.
Musketry Regulations, 1898	H.M.S.O.
Notes on Sniping Instruction - 1916	H.M.S.O.
Ordnance Board Proceedings & Minutes (1939-1945)	H.M.S.O.
Ordnance Factory Returns (1886-1914)	Parliamentary Papers
Pattern Book, R.S.A.F. Enfield	[Original Register]
Pattern Room (The)	M.O.D.
Rifle Committee Minutes (1879-1900)	War Office Papers
Papers & Record Books - Lend Lease Agreement	[Original Registers]
Patent Specifications & Abridgments 1887-1926	H.M.S.O.
Secret Production Reports (1915-1918)	H.M.S.O.
Small Arms Committee Minutes & Reports (1900-1939)	H.M.S.O.
Statistics of Military Effort of the British Empire	H.M.S.O.
Textbook of Military Small Arms & Ammunition - 1894	H.M.S.O.
Textbook of Small Arms - 1909	H.M.S.O.
Textbook of Small Arms - 1929	H.M.S.O.
Treatise on Military Small Arms & Ammunition, 1888	H.M.S.O.
Army, Navy and Air Ministry Handbooks, Pams, Armourers' Instructions, E.M.E.R.'s, &c.	[Various]

OTHER PUBLICATIONS:

A History of the Birmingham Small Arms Company	D. Nie
Arms & Explosives - assorted volumes 1892-1920	[Journals]
Australian Service Longarms	I. Skennerton
Australian Service Machineguns	I. Skennerton
Australian Military Rifles & Bayonets	I. Skennerton
Bayonet Book (The)	Watts & White
British Service Lee (The) 1st Ed.	I. Skennerton
British Small Arms of World War 2	I. Skennerton
British Sniper (The)	I. Skennerton
British Spike Bayonets	I. Skennerton
British War Production (1939-1945)	Postan, Ed.
British War Production (1939-1945)	"The Times"
British & Commonwealth Bayonets	Skennerton, Richardson
B.S.A. Catalogues (various)	B.S.A. Co Ltd.
B.S.A. Group News; 1861-1961 Centenary	B.S.A. Guns
De Lisle's Commando Carbine	I. Skennerton
EM2 - Concept & Design	T. Dugelby
Great War British Sniping Rifles (HMSA Journal)	R. Cleverly
History of the Dept. of Munitions & Supply	Kennedy
Identification Manuals on .303 British Service Cartridges	B. Temple
Introduction to British Grenades	I. Skennerton
Lee-Enfield Rifle (The)	Maj. E. Reynolds
Lee, James Paris (Hawick Archaeological Society, Nov. 1981)	W. Laughlan
List of Changes in British War Material Vol II (1886-1900)	I. Skennerton [ed.]
List of Changes in British War Material Vol III (1900-1910)	I. Skennerton [ed.]
List of Changes in British War Material Vol IV (1910-1920)	I. Skennerton [ed.]
Munitions of War (1914-1918) - B.S.A. & Daimler Co.	G. Frost [ed.]
Official History of the Indian Armed Forces in W.W.2	Prasad
Remington Firearms 1904-1905 Catalogue	Remington Arms
R.O.F. 1939-1948	I. May
Scouts & Sniping in Trench Warfare	Maj. F. Crum
Telescope Sighting No. 32 (The)	P. Laidler
The Other Battle - A History of B.S.A.	D. Ward
Treatise on the British Military Martini Vol. I	Temple & Skennerton
Treatise on the British Military Martini Vol. II	Temple & Skennerton
Treatise on the Snider	I. Skennerton
U.K. & Commonwealth F.A.L.'s	R. Blake Stevens
U.S. Enfield (The)	I. Skennerton

INDEX

Also by this Author:

".380 ENFIELD No. 2 REVOLVER" *Stamps & Skennerton*
Hard & soft covers, 9 1/2 x 6 in., 126 pages, 95 illustrations. 1992.
Its historical and technical evolution from the .38 Webley Mk IV revolver.

".303 No. 4 (T) SNIPER RIFLE - An Armourer's Perspective" *Laidler & Skennerton*
Hard & soft covers, 9 1/2 x 6 in., 126 pages, 80 illustrations. 1992.
A practical study of the sniping rifle through .303 and 7.62mm eras.

"BRITISH SMALL ARMS OF WORLD WAR TWO" *Ian Skennerton*
Hard cover, 9 1/2 x 6 ins., 110 pages, 36 illustrations. 1989.
The rifles, pistols, machine guns, &c., 1936-1946.
A full listing of wartime British codes and manufacturers' contracts.

"BRITISH & COMMONWEALTH BAYONETS" *Skennerton & Richardson*
Hard cover, 11 x 8 3/4 ins., 404 pages, approx. 1,300 illustrations. 1986.
The standard reference, from the 17th century to today.

"HANDBOOK OF BRITISH BAYONETS" *Ian Skennerton*
Soft, durable cover, 5 3/4 x 4 ins., 66 pages, not illustrated. 1986.
Pocket-sized companion to "British & Commonwealth Bayonets".

"THE BRITISH SNIPER" *Ian Skennerton*
Hard cover, 11 x 8 3/4 ins., 266 pages, 400 illustrations. 1984.
British, Canadian, Australian, Indian and South African sniping.

"AN INTRODUCTION TO BRITISH GRENADES" *Ian Skennerton*
Soft cover, 9 1/2 x 6 ins., 56 pages, 60 illustrations. 1988.
Descriptions of service grenades, especially the numbered series from 1908.

"AUSTRALIAN MILITARY RIFLES & BAYONETS" *Ian Skennerton*
Hard & soft covers, 9 1/2 x 6 ins., 124 pages, 207 illustrations. 1988.
200 years of rifles, carbines and bayonets; from Brown Bess to Steyr F88.

"AUSTRALIAN SERVICE MACHINEGUNS" *Ian Skennerton*
Hard & soft covers, 9 1/2 x 6 ins., 122 pages, 150 illustrations. 1989.
.450 Gatling to 5.56mm Minimi. Includes the Owen and Austen machine carbines.

"S.L.R., AUSTRALIA'S F.N. F.A.L." *Skennerton & Balmer*
Hard & soft covers, 9 1/2 x 6 ins., 122 pages, 100 illustrations. 1989.
Details of this military rifle, models, manufacture and serial number tables.

"TREATISE ON THE BRITISH MILITARY MARTINI" *Temple & Skennerton*
2 volumes, hard covers, 11 x 8 3/4 ins., c. 250 pages, 120 illust. 1983 & 1988.
Comprehensive studies of the Martini rifles and carbines.

"LIST OF CHANGES IN BRITISH WAR MATERIAL" *Official*
Hard covers, 8 1/2 x 5 1/2 ins., projected four volume series; edited Skennerton.
Vol. 1 (1860-1886) 170 pages, 100 illust. 1980.
Vol. 2 (1886-1900) 201 pages, 75 illust. 1977.
Vol. 3 (1900-1910) 216 pages, 80 illust. 1987.
War Office details of small arms and accoutrements. A collector's Bible.

"DE LISLE'S COMMANDO CARBINE" *Ian Skennerton*
Small Arms Series No. 1. 1981. Soft cover, 9 1/2 x 6 ins., 32 pages, 25 illust.
Britain's service .45 silenced carbine, development and construction details.

"THE BRITISH SPIKE BAYONET" *Ian Skennerton*
Small Arms Series No. 2. 1982. Soft cover, 9 1/2 x 6 ins., 32 pages, 18 illust.
A complete study of the No. 4 and Sten bayonets and scabbards of W.W.2.

OUT OF PRINT:
Australian Service Longarms
Australian Service Bayonets
A Treatise on the Snider
The U.S. Enfield
The British Service Lee